W9-CHQ-851

DIGITAL LINE-OF-SIGHT RADIO LINKS

A Handbook

DIGITAL LINE-OF-SIGHT RADIO LINKS
A Handbook

A. A. R. TOWNSEND

PRENTICE HALL
NEW YORK LONDON TORONTO SYDNEY TOKYO

First published 1988 by
Prentice Hall International (UK) Ltd,
66 Wood Lane End, Hemel Hempstead,
Hertfordshire, HP2 4RG
A division of
Simon & Schuster International Group

Printed and bound in Great Britain at the
University Press, Cambridge.

Library of Congress Cataloging-in-Publication Data

Townsend, A. A. R. 1944-
Digital line-of-sight radio links.

Includes bibliographical references and index.
1. Digital communications. 2. Line-of-sight radio links. I. Title.
TK5103.7.T69 1987 621.3841 87-17536
ISBN 0-13-212622-2

British Library Cataloguing in Publication Data

Townsend A. A. R.
Digital line-of-sight radio links: a handbook.
1. Mobile communication systems
2. Microwave communication systems
621.3841'65 TK6570.M6

ISBN 0-13-212622-2

1 2 3 4 5 92 91 90 89 88

ISBN 0-13-212622-2

To my wife Ngozi

CONTENTS

PREFACE

There has been since the late 1970s a trend in many countries towards the installation of digital transmission systems against the older and more established analog counterpart. The reasons are in part, ease in switching and routing, and cheaper processing equipment brought about by the reduced costs of large scale integration (LSI) technology for digital communications. Another advantage of using digital transmission, by radio, cable, or optical fiber, is the reduction possible in the noise on a long multi-hop transmission section. This is made possible by the regeneration circuits which may be located at repeater sites. In an analog system, the contrary occurs since noise is added at repeater stations by the drop and insert equipment, etc. A further reason for the trend towards digital transmission is the progress being made in office automation, producing an increasing demand for high-speed digital data transmission for inter-office communication and possibly for teleconferencing services. In addition, the domestic requirements of the near future will possibly create the demand for a digital video distribution system somewhat like an advanced cable television (CATV). This demand may be satisfied as video codec technology permits the reduction in video transmission bit rates. The domestic requirements may also be extended for digital data transmission services to cope with the expected digital traffic produced by microcomputers as they become more prevalent and acceptable for electronic mail, shopping, etc. As a consequence of the increase expected in digital traffic flowing into the switching centers, growth in digital traffic between switching centers is also expected. To cater for this growth, governments or authorities are installing digital broadband communications networks using such transmission mediums as optical fiber cables, digital radio links and satellite communication links, which have been developed to satisfy the requirements of large-capacity digital traffic. Also some governments have embarked upon a program which will provide the integration of separate overlay networks such as the private data network, package switching and telex into the digitalized public switched network. This integrated network is called the Integrated Services Digital Network (ISDN).

The immediate apparent disadvantage of digital transmission is the greater bandwidth necessary to transmit a voice channel, or an assembly of voice channels over a transmission medium, than that necessary for an analog transmission system. Even so, as radio systems at frequencies below 12 GHz evolve by increasing the number of modulation levels, better use is made of the frequency spectrum. This will permit a greater capacity of existing routes, and of those which are at present being installed. Radio systems at frequencies above 12 GHz, i.e. millimeter wave systems, are already widely used in the trunk network and will possibly be accepted for extensive use in the local network once the digital system develops. This will give small local area networks (LANs), and other businesses, access to the public network via a point-to-multipoint configuration. As advancing technology makes the higher frequency bands more readily available, the limitations on radio transmission imposed by available bandwidth will disappear. Optical fiber technology is rapidly developing – giving lower losses per kilometer (1 to 2 dB/km at present), and permitting the use of longer wavelengths. Its use in communications is becoming the main means of providing national networks with the high-capacity links which they may require at a competitive price. At the present time, bit rates of over 500 Mbit/s are in use, and the technology is still evolving. The inevitable outcome will be virtually unlimited transmission capacity. Thus the inefficient use of the spectrum by digital transmission, when compared with its analog counterpart, is not the disadvantage that it initially appeared to be. Digitial transmission is here to stay for some time, and is at the present time the major consideration for new communications networks or for the replacement of existing systems by national telecommunications authorities. The installation of equipment relating to digital transmission systems will continue to increase until it ultimately becomes the predominant means of world telecommunications

It is intended that this book will provide the engineer, technician, engineering manager or student, with a background to a complete digital system, with the eventual emphasis on the design of a digital radio-relay system. The digital radio-relay system, in this text, is the only medium considered for the transmission channel, after digital multiplex equipment has been discussed. To achieve these objectives, a model of a digital system is first established, and then each block within the model is discussed. The discussion of each of these blocks includes a description of the equipment used, its function and operation, any performance objectives which have to be met, impairments which may occur, the various items of equipment available, and where applicable, any testing which may be carried out to verify the correct operation of the equipment, ensuring that the objectives are met. The later chapters deal exclusively with the design of digital radio-relay systems and cover such subjects as the derivation of performance and availability objectives and the design of the radio link or system, so that these objectives can be met at a reasonable cost. In addition to these subjects, the last chapter presents the means of verifying by testing that some of these objectives have been achieved in practice. Exercises based on material in each chapter are given at the end of most chapters with suggested answers. Interwoven into the development of the system and equipment parameters are mathematical topics considered necessary for a better understanding of these parameters. The text is organized to begin at the analog voice channel, and then to develop the source encoder/decoder before considering the multiplex and demultiplex equipment, and digital modulator/demodulator. Information on the problems encountered by the radio channel due to the transmission path and its associated impairments are then provided. Frequency planning and interference considerations are set out to assist the digital radio system designer, and to present to the practicing engineer mechanisms which may cause problems on a new or on an existing radio system. Radio and multiplex equipment block diagrams are extensively used for different manufacturers' equipment to demonstrate the techniques developed, and to compare different types of modulation. The book aims also to provide a practical bridge between the theory of digital radio systems and systems encountered in the field. It is written primarily for practicing engineers, who require a practical understanding of the operation of the equipment, the reasons behind the tests recommended by the manufacturer, and the limitations of these tests, as well as information on the different types of modulation schemes used on commercially available equipment, etc. This book should also provide engineering managers, field technicians and students of communications engineering with a handy reference on the more practical points of digital radios and systems, and allow an understanding of the terms used in technical papers on this subject. Emphasis has been placed on American and European equipment and on CCITT and CCIR recommendations. Extensive references are used to permit further and more detailed reading.

Although the book is the second of two books on radio systems, the first being on analog systems, it has been designed to stand alone, without unnecessarily repeating information previously covered. The subject of interference and frequency planning has been briefly introduced in Chapter 10, and is to be used only as an introduction to this extensive field.

A.A.R.T.

ACKNOWLEDGEMENTS

To Mercury Communications Limited for giving me the opportunity to work for them and with their staff during most of the period in which this book was written, specifically: Messrs T. King, P. E. Morris and B. Goodyear of the frequency management section; Messrs J. J. Spearing, D. Chaloner and E. S. J. Knoop of the Trunk National Operations; and Mr. I. D. Clough. Special mention also must go to Messrs Russ Covey and David Heppel.

To the following equipment manufacturers: Telettra Telefonia Elettronica e Radio SpA, Ayadin Microwave Division; L. M. Ericson; GEC Telecommunications Ltd; GTE Telecommunicaziona SpA; Loral-Terrocom; NERA; Rockwell Collins; SIAE Microelettronica SpA; Telenokia; Thorn-EMI; Plessey Radio Systems; NEC Corporation; Hewlett-Packard; and Farinon Electric. Special mention also must go to Bruce Fennie, George Perzel, Dan Throop, Steve Johns and Bob Braggins of Harris Corporation, RF Communications Group.

1 INTRODUCTION TO DIGITAL SYSTEMS

1.1 THE ELEMENTS OF A DIGITAL COMUNICATIONS SYSTEM

A digital communications system comprises a series of signal processing units. These units may be broadly categorized under the following headings for telephony applications:

Conversion of an analog signal into a digital signal
Assembling the digital signals from various sources into a digital baseband signal
Processing the baseband signal for transmission over a communications channel
Transmitting the digital baseband signal over a communications channel
Receiving the digital baseband signal from a communications channel
Processing the received baseband signal for disassembly into its various source counterparts
Dissassembly of the processed baseband signal into its various source counterparts
Conversion of the digital signal into its analog counterpart.

Figure 1.1 provides a simple model of these processes, and this diagram will be referred to throughout the book.

The conversion or the encoding of an analog signal into a digital signal may be done by one of the following; pulse code modulation (PCM); logarithmic pulse code modulation [log(PCM)]; differential pulse code modulation (DPCM); adaptive differential pulse code modulation (ADPCM); delta modulation (DM), adaptive delta modulation (ADM) or linear predictive coding (LPC). Similarly, the same process of decoding the digital signal into its analog counterpart using these techniques may be used.

The assembly and disassembly of the digital signals from and into a digital baseband signal may be achieved by multiplexing and demultiplexing. At the present time, there exist many frequency-division-multiplex (FDM) systems. With digital transmission systems, it is often necessary to interface to these systems the *group*, *supergroup*, master-group, or 16 supergroup FDM assemblies of voice channels. This is achieved by using special codecs. For a completely new system, the assembly and disassembly of digital signals is usually accomplished using time-division-multiplex (TDM) schemes.

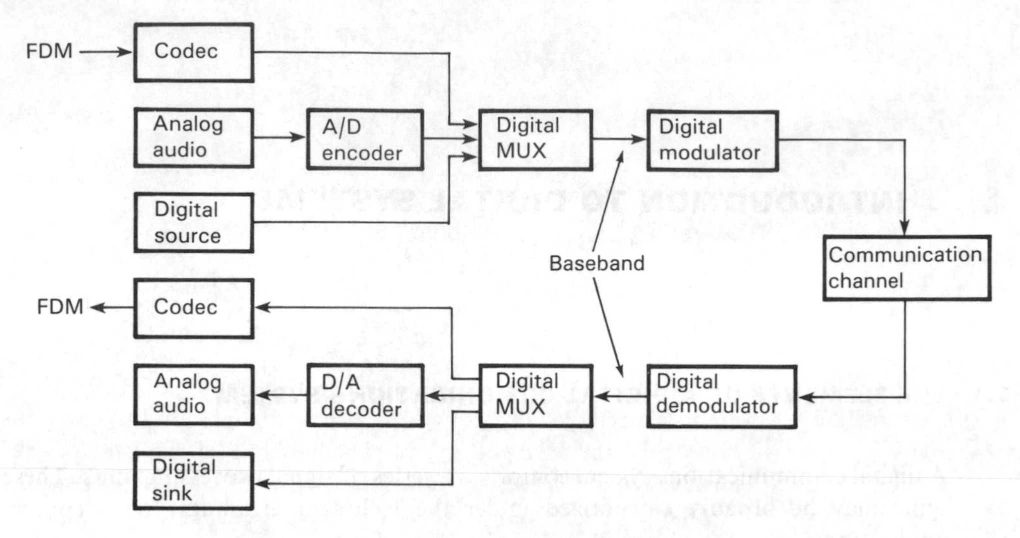

Figure 1.1 A model of a digital system

The processing of the digital baseband signal into a form which is suitable for transmission over the communications channel depends upon the transmission medium to be used. This is because particular types of communication channels have their own peculiar characteristics and limitations. To determine which modulation scheme to use of those available, the required signal-to-noise ratio for a given *bit error rate*, the spectral efficiency measured in bits/Hz, and the equipment complexity and cost must be considered.[1] Some of the modulation and demodulation methods presently in use, which are classed under the main modulation types and which are some of those considered in Chapter 5 and Chapter 6, are:

Amplitude modulation (AM)
- Pulse-amplitude modulation (PAM)
- *M*-level pulse-amplitude modulation (*M*-level PAM)
- On–off keying (OOK) – coherent detection
- On–off keying (OOK) – envelope detection
- Quadrature amplitude modulation (QAM)
- *M*-ary quadrature amplitude modulation (*M*-ary QAM)
- Quadrature partial response (QPR)

Frequency modulation (FM)
- Frequency shift keying – noncoherent detection (noncoherent FSK detection)
- Continuous phase – frequency shift keying – coherent detection (CP–FSK–CD)
- Continuous phase – frequency shift keying – noncoherent detection (CP–FSK–NCD)
- Minimum shift keying (MSK)
- Minimum shift keying – differential encoding (MSK-differential coding)

Phase modulation (PM)
- Binary phase shift keying (BPSK) – coherent detection
- Differentially encoded binary phase shift keying (DE–BPSK)
- Differential phase shift keying (DPSK)
- Quadrature phase shift keying (QPSK)
- Differential quadrature phase shift keying (DQPSK)
- *M*-ary phase shift keying (*M*-ary PSK) – coherent detection

Amplitude-modulation/phase-modulation (AM/PM)
- 16-ary amplitude phase keyed (16-ary APK)

This list is not exhaustive, and provides a representative sample of the modulation techniques presently available. The communications channel used to carry the modulated carrier may take the form of an optical fiber, line-of-sight microwave radio, troposcatter radio, coxial cable, satellite link, or audio transverse-screened pair cables. As the subject of this book concerns line-of-sight microwave radio, the other transmission mediums will be considered only briefly throughout the text.

1.2 DIGITAL SYSTEM PLANNING CONSIDERATIONS

System planning may be considered as the specification of the performance objectives of individual items of equipment or subsystems so that the overall circuit end to circuit end transmission performance objectives are met when they are integrated into a complete transmission system. This concept becomes more complex when a system already exists, and an overlay or replacement system is to be installed. The specification of the new equipment network configuration, the various noise contributions to the total transmission impairment of both the new and, if in existence, the old network, and the allocation of the noise source contributions of each of the separate items of equipment or subsystems, to the total noise budget are required to be determined.

The conversion of a network from one which is predominantly analog based to one which is to become ultimately digital may be based upon the rapid introduction of digital switching machines, or the increasing cost of analog-to-digital conversion equipment. Other factors are the requirement for an integrated digital network, which is based on the economics of keeping the link costs low, and the choice of using existing broadband transmission facilities or the new broadband transmission technologies which are presently available. Also important are rapid population growth and the subsequent overloading of analog junction and trunk cables as determined by population distribution surveys and traffic flows, together with demand forecasting of population growth and overloading of present juction and trunk networks. Those responsible must have confidence that the final network not only meets the sophistication required by the customer at a competitive cost, but also provides a continuing high performance in the event of equipment failure or damage. Finally, detailed planning of transmission networks for the short-, medium- and long-term operational needs also strongly influences the decision to convert to digital or not.

The decision whether to choose existing cables, or to install optical fiber or digital radio, after a decision to 'go digital', may depend on the availability of existing cables, both transverse screened pair and coaxial, and the relative cost of digitizing analog coaxial cable systems. One must consider the current usage of the existing analog coaxial cable and the comparative cost of optical fiber and digital radio systems, together with the mix of analog and digital systems required. For a digital radio-relay system, the availability of radio link infrastructures, such as towers and buildings, and relevant geographical considerations must be weighed against the alternative costs. The time for the conversion of an existing analog system to that of a digital sysem is important for the lifetime of existing equipment may render the cost of conversion prohibitive for some areas in the network or system. The problems of frequency coordination, of availability of frequency bands and of the possibility of meeting the overall performance and reliability objectives, together with the availability of equipment and the number of relevant manufacturers, may also complicate matters. Assuming these problems can be solved, additional factors affecting the decision are: the expected period before obsolescence of any equipment chosen; whether the equipment design and performance specifications are in accordance with CCIR recommendations or with some other recognized authority's recommendations; and whether these specifications agree with specifications which are reasonably likely to be formulated in future years. One should consider also the present and future link bit rate requirements, together with the need for the transmission of video information and television, and whether the information flow is unidirectional or bidirectional. Present and expected future digital regenerator spacings for optical fiber, the link usage for digital private circuits, television, satellite earth stations, submarine cable terminals and 'confravision' networks together with flexibility in the usage of the links may well be important.

The transmission planning of a new or hybrid network must cover not only the mix of medium-capacity long-range systems and very high-capacity shorter-range systems, on coaxial cable, digital radio or optical fiber, but also the continually changing economics of each type of system. To cope with all these complexities, transmission planning is being assisted by sophisticated computer systems, which permit the production at any time of a dynamic and cost-effective plan.

If the choice of the transmission system is digital radio, the factors to assist in the choice of the radio and equipment also appear to be nonexhaustive. Some of these are the frequency band to be used on proven propagation performance, the availability of that frequency band, equipment operating on that frequency at the required bit rate, which has been determined from network and traffic flow considerations. Choice of the modulation scheme is based on cost, availability, reliability, equipment complexity and robustness (resilience to interference or fading conditions), etc., for bandwidth efficiency. Choice of the equipment itself is based on size, power consumption, reliability, ruggedness, etc., and on the required protection, such as protection against fading, or other radio channel impairments. This means that there may be a requirement for diversity, adaptive equalization (IF or baseband or both), or both space diversity and adaptive baseband and IF equalizers. A major factor in built-up areas is the availability of buildings for antenna mounting and for equipment housing, the civil works which may be required and the erection of towers on 'green field' sites or the hiring of existing towers and the rents associated with such, as well as the requirements of other plant and services. The switching in a protected system which does not produce 'hits' or corruption of the bit stream must be taken into account in the system design amongst other things such as the quality or expected bit error rates (BER) and availability, both of which are based upon transmission performance objectives of the CCIR or other authorities. A quick mention of other items to be considered in addition to those already mentioned are; the amount of RF interference expected on the installation of a new radio link in an already congested radio node. This may be based on the results of a computer interference analysis. This analysis would, like the path calculations, or the system design, have to allow for the type of antenna systems to be used. The commissioning objectives for transmission quality based on end-to-end tests of bit error rate (BER) performance, and on long-term BER performance requirements also place a limit on the maximum practical hop length for a digital radio.

The approach used in the planning of analog systems, and more specifically, analog microwave radio systems remain valid. This permits concepts of the hypothetical reference circuit and hypothetical reference connection to be used, taking into consideration that the hypothetical reference circuit may be a mixture of analog and digital systems. That is not a problem as long as the impairments of digital systems which cause a particular value of bit error rate (BER), or jitter, are taken into account in addition to the accumulation of noise, distortion and loss peculiar to analog systems. The value of the signal-to-noise ratio S/N at the receiver is equally important to both analog and digital radio systems. In analog systems, carrier to noise ratio (C/N) and thus, S/N is one of the main performance-determining parameters. In digital systems, it has the same importance except that it is expressed in terms of the bit error rate to which it is directly related. In an analog system, noise introduced at an earlier stage in the system accumulates as repeater stations demodulate and modulate information according to the network requirements. This problem does not appear in a digital radio system to the same extent owing to redundant error-correcting codes and regenerators which virtually eliminate accumulated noise. The problems which do exist in a digital radio system or in fact any digital transmission system are:

1. Low-frequency jitter which cannot be attenuated by the digital regenerators, and which builds up along the transmission path. This process can be overcome as will be discussed later.
2. The BER performance of a digital transmission system is critically dependent upon the link with the worst BER performance. This is not so in an analog FDM system where the noisest link may only contribute a small percentage of the total system noise.

In an analog radio system one of the important design considerations of a transmitter is the effect which the baseband loading has on the transmitter deviation, and consequently on the the bandwidth and distortion due to transmitter overloading.

In a digital radio, once the analog signals have been converted into digital form, there is no direct relation between the characteristics of the digital bit stream and the baseband peak power level of the original analog signal. This permits the digital radio to be largely independent of signal loading and simplifies the design in this respect. As a consequence the concept of busy hour has no significance in digital radio systems.

1.3 DIGITAL NETWORK ARCHITECTURE

At the present time the architecture of any digital network will have evolved, or be evolving, from the analog network which has been developing in most parts of the work since the beginning of this century. In many countries, the rapid changes to the analog network to accommodate new digital network technologies indicate the conviction of the authorities concerned that advantages to the public exist. Over the past few years it has been seen that these advantages once shown to the public create a demand which in turn adds impetus to the change. An upward spiral of the supply of new services and the subsequent creation of demand by the users of these services has proved that the research, development and investment into the more innovative and sophisticated programmes has been worthwhile. In Sections 1.3.1 to 1.3.6 a brief description of the architecture required to support a digital network is given.

1.3.1 Integrated services digital network (ISDN)[2(p.249),3-6]

In 1972 the CCITT formulated two definitions to describe the evolution of a digital telephone from an analog network, and its further development into an ISDN:

> 'An integrated digital network (IDN) is a network in which connections established by digital switching are used for the transmission of digital signals.'

> 'An integrated services network (ISDN) is an integrated digital network (IDN) in which the same digital switches and digital paths are used to establish connections for different services.'

The second definition, of ISDN, was further modified in 1982 by the following definition;

> 'A network evolved from the telephone IDN that provides end-to-end digital connections to support a wide range of services, including voice and non-voice services, to which users have access by a limited set of standard multipurpose customer interfaces.'

Thus this network is being established to convey traffic in digitally coded form for the direct connection of digital transmission systems to digital switching systems and for the conveyance of a variety of services such as telephone, microcomputer, viewdata, facsimile and telex. The ISDN basic telephony service is 64 kbit/s, which is compatible with most modern digital switches. It has been recognized that implementation of a full ISDN network will require the gradual phasing out of the analog network. Three phases have been proposed. The first phase is the phasing in of the digital telephone network operating with the 64 kbit/s connectivity. The second phase is the interfacing of the nontelephony services to provide a multiservice network still with the 64 kbit/s connectivity. The third and final stage is where the bit rate is raised from 64 kbit/s for some service requirements such as sound programme transmission and moving pictures, and where the bit rate is dropped below 64 kbit/s for other services. Problems encountered at the present, and which can be easily dealt with by the switching machine software are;

The differentiation between a digital call and an analog call in a mixed network, and the identification, routing, etc., of each

The correct identification and routing of nontelephonic calls between bonafide users

Barring access to certain classes of call, and the suppression of the telephonic tones or announcements when inappropriate

The requirement of the local network is to extend to the customer digital transmission and a form of message-based signaling. This is commonly referred to as an integrated digital access (IDA). Two forms of IDA are being investigated. These are *basic* and *extended access*. The basic access uses single metallic-cable pairs and provides one or two 64 kbit/s channels plus a signaling channel. The information rate becoming accepted is 144 kbit/s. Extended access uses the present CCITT PCM multiplex standard to offer thirty 64 kbit/s channels together with a 64 kbit/s signaling channel. The advantage of this form of IDA is that it permits private digital PABXs to be connected by digital links to the digital public network.

The I-series recommendations of the CCITT[5] make it clear that a key element for an ISDN is the provision of a limited set of standard multipurpose user–network interfaces, and that the ISDN will be recognized by the service characteristics available through these interfaces.

1.3.2 The local network[2]

The local network may currently be defined as the network existing between the subscriber's local public exchange and the subscriber. This definition is being eroded as customers' private automatic branch exchanges (PABXs) increasingly offer services and operate more like public exchanges, especially since these PABXs may be tied to the public network by direct-inward-dial trunks using PCM links over the local network. In addition to the effects of PABX, the possible integration of a digital video distribution system or of cable television into the network may introduce television receiving stations as new nodal points in the cable network. The integration of the packet switching service (PSS), and the private circuit digital network (PCDN) into the public switched telephone network (PSTN) to give the ISDN will also produce radical changes to the future local network as various forms of network terminating equipment is introduced at the customers' premises to provide the local line termination, multiplex the different user and signaling information into the appropriate channel structure and transmit the data information back and forth between the customers' exchanges.

At the present time the bit-error-rate performance of the local loop at speeds of 64 kbits/s or higher[3,7,8] is being investigated; this is being done with the aim of providing an integrated services digital network (ISDN) directly to the subscriber and so reducing the possibility of developments leading to obsolescence of the existing metallic-pair local network. At 64 kbit/s, services such as telephony, telemetry, telegraphy and facsimile can be carried. The distribution of services such as videophone or television however are not possible using the existing local network. This is because the practical limit which is set by the attenuation characteristics of the metallic-pair cable is around 100 kHz, and that required by videophone and television is above 1 MHz.

Using optical fiber Leakey[9] has proposed that a 64-channel wavelength division multiplex scheme, using optical fiber, and with each channel carrying at least 140 Mbit/s digital streams, could be used for the local network distribution. This may be achieved using a four-fiber optical cable serving say 50 customers with full interactive services and still leaving a generous number of channels for wideband broadcast operation. This four-fiber optical cable would feed directly into the local exchange reducing considerably the present cabling problems. Another alternative to the future structure of the local network centers around the increasing use of cellular radio, and the cordless telephone. This could evolve to provide complete mobility of all telephones within a large building or residential area. The base station of each phone would be located in a street cabinet and so eliminate the final physical connection in the local distribution network.

Telecommunications authorities are currently considering (amongst other things) the digital local line transmission for the following areas of use:

Integrated services digital network (ISDN) access
Connection to private data networks
Carrier systems for rural networks

Transmission methods for duplex operation which may be used for the local metallic-pair network include the modulation of the forward and return signals onto different frequency shift keying (FSK) carrier pairs. These carrier pairs are then separated at the receivers by filtering prior to FSK demodulation. This method is usual in the design of voice-band modems. Some other methods which are used are:

Burst-mode or ping-pong or time compression multiplex system (Section 1.3.2.2)
Hybrid-separation or echo-canceling (Section 1.3.2.3)

An obvious way of providing duplex operation over metallic pairs is by four-wire operation, where one pair is used for one direction of transmission, and the other pair is used for the opposite direction. This of course is possible only if the subscriber has been allocated two pairs.

Sections 1.3.2.1 to 1.3.2.4 give a brief description of some of the more important uses and transmission methods of the digitalization of the local network, including a discussion of noise.

1.3.2.1 Carrier systems for rural networks[10]

In sparsley populated areas, the cost of providing each subscriber with an exchange line may be prohibitive. Low-capacity digital carrier systems have been developed to permit the connection of a limited number of subscribers. Reference 10 outlines such a system for up to ten subscribers, who may be connected to the local digital switching center either by symmetric-pair cables or a radio-relay system. The speech channels are multiplexed and demultiplexed at each end of the link.

Subscriber radio has been commercially available for some time. It distributes from a central location with digitally encoded and decoded voice channels to many points, and vice versa using time division multiplexing techniques[1] (TDM). (See Section 1.5.1 for more details.)

1.3.2.2 Time compression multiplex systems

In order to obtain duplex operation on a single metallic pair, data to be transmitted are collected together and sent as short bursts at regular intervals to the far end of the subscriber's loop. The far end replies with similar bursts. Provided that the received and transmitted bursts do not overlap in time, two-way digital transmission on one pair of wires can be achieved. Since a guard time is required between alternate transmissions to allow for propagation along the channel, burst of groups of bits rather than individual bits are sent in each direction. This mode of operation is thus alternatively referred to as *burst mode*, or *ping-pong* transmission. In the United Kingdom[8] a WAL2 line code is used in the integrated digital access (IDA) system, which transmits 4000 bursts/s, where each burst comprises two parcels of 11 bits each – a start bit, 8 bits from the 64 kbit/s, B traffic channel, 1 bit from the 8 kbit/s D signaling channel and 1 bit from the 8 kbit/s B traffic channel. The WAL2 line code is a modified Walsh function code[12] which has the advantage that it is free from DC components and contains a large number of transitions from which timing information can be recovered regardless of the transmitted data pattern. The bandwidth occupancy is almost twice that of the transmission rate.

1.3.2.3 Echo canceling[13–15]

Hybrids in analog circuits are used to convert a four-wire circuit into a two-wire circuit, and vice versa. With digital transmission it is imperative that the decoupling between the four-wire forward and return circuits is kept as large as possible. This decoupling is limited by the accuracy with which the input impedance of the two-wire circuit is matched by the balancing circuit. Rather than attempting to adjust constantly the balancing network for perfect match, an echo canceler may be

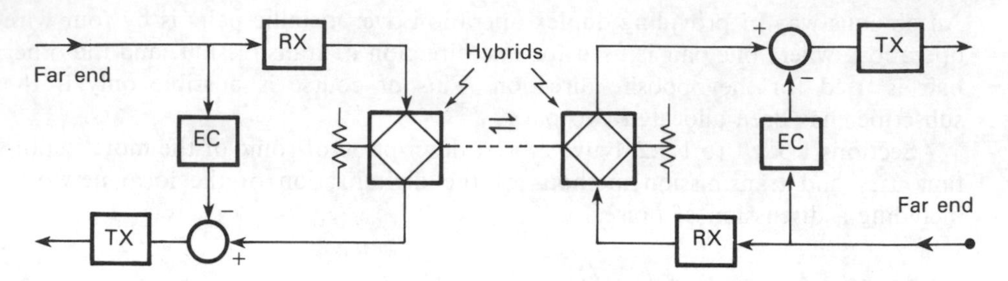

Figure 1.2 Echo canceler technique for two-wire digital transmission EC = echo canceler

inserted between the four-wire circuits to generate a copy of the echo signal produced when the four-wire received signal arriving at the hybrid from the far end leaks into the four-wire circuit transmitting to the far-end. The echo copy is then subtracted from the four-wire transmit signal which is the sum of the echo signal and of the signal going to the far end, so that the useful signal may be restored. Figure 1.2. shows the hybrid and echo canceler configuration. The echo canceler may be adaptive.

1.3.2.4 Types of noise found in the local network

The main types of noise which are experienced in the local network, and which have a detrimental effect, are:

Impulse noise
Crosstalk

IMPULSE NOISE

Impulse noise is characterized by having a large amplitude, and a short duration interspersed between long periods of no activity. It is introduced into the subscriber loop cable from the following sources:

Switching transients occurring on power lines
Electric traction systems
Lightning
Other cable pairs in the same cable which may be connected to an analog exchange, which produces noisy line signaling

Due to the sources from which this type of noise may originate, it is difficult to control. The effect on a speech circuit is to produce at the telephone earpiece 'clicks'. On a data connection it will produce an increase in the bit error rate. The development of the local network for digital transmission will be limited by the local line performance characteristics set by bodies like the Comité Consultatif International Télégraphique et Téléphonique (CCITT). The standards on bit-error-rate performance may in some cases inhibit the use of the local network for digital transmission due to this type of noise.

A disturbance analyzer system[16] (DAS) has been developed to determine the digital quality of a subscriber's line, and to estimate the noise margin between a satisfactory working pair and a degraded pair. This is achieved by measuring the bit-error-rate variation when the signal-to-noise ratio changes. The DAS has been designed to calculate the error distribution (percentage of error free seconds,..) in an attempt to identify the origin of the disturbing signals.

CROSSTALK

Crosstalk interference in voice-frequency metallic pairs is mainly caused by the electrostatic and magnetic coupling of adjacent pairs in the same cable which are carrying other information. There are two forms which are significant. These are near-end and far-end crosstalk. Near-end crosstalk results from the unwanted signal being received at the near end of a cable pair which is coupled from adjacent pairs that are transmitting information in the opposite direction to that of the wanted signal. It is usually expressed in terms of the near-end crosstalk attenuation (NEXTA):

$$\text{NEXTA} = 10 \log(\text{sending signal power from pair A})/(\text{received crosstalk power from pair B})\ \text{dB} \quad (1.1\text{a})$$

$$\text{NEXTA} \simeq 10 \log(K_N f^{-3/2}) = (10 \log K_N - 15 \log f)\ \text{dB} \quad (1.1\text{b})$$

where f is frequency in the same units as in K_N. Both powers are measured at the sending or near end. NEXTA tends to be flat over the voice-channel frequency spectrum except for deep narrow high-attenuation troughs interspersed at selected intervals. This form of crosstalk is that which provides the limitation on the maximum bit rate for a given line length before repeaters are required. For long transmission lines, of the twisted pair or coaxial types, and the higher carrier or baseband frequencies, equation (1.1b) shows a 15 dB/decade decrease in NEXTA.

Far-end crosstalk results from the unwanted signal appearing at the far end of the cable pair. It arises from the coupling of adjacent pairs that are transmitting their information in the same direction as the wanted signal pair. Far-end crosstalk (FXT) is usually defined in terms of the far-end-signal-to-crosstalk ratio (FESXTR):

$$\text{FESXTR} = 10 \log(\text{received signal power from pair A})/(\text{received crosstalk power from pair B})\ \text{dB} \quad (1.2\text{a})$$

$$\text{FESXTR} \simeq 10 \log K_f - 20 \log f - 10 \log L\ \text{dB} \quad (1.2\text{b})$$

where L is the cable length in the same units as in the constant K_f. Both powers are measured at the far end. FESXTR varies regularly over the voice band and above, before rapidly decreasing to zero around 2 MHz, as shown by equation 1.2b. The rate of decrease is 20 dB/decade of frequency and 10 dB/decade of length. The effect of FESXTR on the transmission of digital signals up to 2048 bits/s, with differing bit rates on adjacent pairs, has been found to be negligible[17] (see Section 1.4.8 for further details).

1.3.3 Digital switching machines[18–21]

These machines in the main use *time division switching*, which connects the signal coming into the exchange to the required outgoing path from the same exchange, only for a short period of time before it attends to other incoming and outgoing path requirements. Each input is in effect 'sampled' by the switch. The input signal into this portion of the exchange however is not an analog speech waveform, but is, due to line interface modules (LIMs), a digital bit stream representing the speech waveform. If the connection of an input to its required output is done repeatedly, and at a rate which is the same as the converted input signal's bit rate, no information is lost. Basically, the exchange operates by repeatedly sampling each bit of the input bit stream, and repeating this bit at the output of the exchange. As the sampling is of very short duration compared with that of the input bit, many inputs may be sampled one after the other before the necessary time has elapsed for the exchange to return to the first input and sample its next bit.

From the above discussion it is apparent that an input bit stream is connected to its required output only in one direction of propagation. This implies that speech can be made in one direction only. For speech to occur from the opposite end once the call has been set up, a return path must also be established which is processed in a similar manner. Therefore in a digital exchange using time division techniques a separate transmit and receive path must exist.

1.3.3.1 Line interface modules (LIMs)

The complete range of requirements of the line interface module to convert an analog line into a form which is compatible with the digital exchange is given by the acronym BORSCHT. This acronym is derived from the first letters of:

*B*attery feed
*O*verload protection
*R*inging
*S*upervision (signaling)
*C*odec and filter
*H*ybrid
*T*esting (access for line testing)

Figure 1.3 shows the interrelationship between these requirements in a line interface module.

As the line interface modules are expensive, and may account for more than half the capital cost of a digital telephone exchange, it is not unreasonable to assume that in the not too distant future the analog telephone will be replaced by the digital telephone. This will reduce the requirement for these line interface modules. Apart

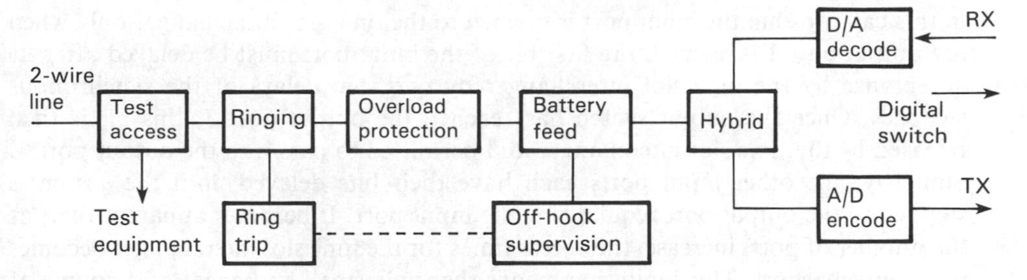

Figure 1.3 Interrelationships of a digital line interface module (Courtesy of Peter Peregrinus Ltd, Reference 2, p. 85)

from that, the advantages that the digital telephone offers over its analog counterpart are:

1. The transmission quality of speech is independent of local line length, since the local line carries digital information only.
2. As four wires are required to transmit and receive the digital information from the exchange, there is no requirement for a hybrid in the telephone.
3. The information signaling is digital, hence there is a simplification in the keypad circuitry. The line signaling is digital, so there is no need to send ring current from the exchange. Alerting of the called party and the various other tone requirements will be accomplished by tone producing circuits within the instrument itself.

1.3.3.2 Time-division switching

Figure 1.4 shows the principle of time-division switching. When analog signals enter the line interface module they are coded using pulse code modulation (PCM) techniques as described in Chapter 2. The resulting digital bit stream enters one of the digital input ports to the exchange, say port 1. Assume that the called subscriber is accessed via the output port 3. As the input and output switches are synchronized

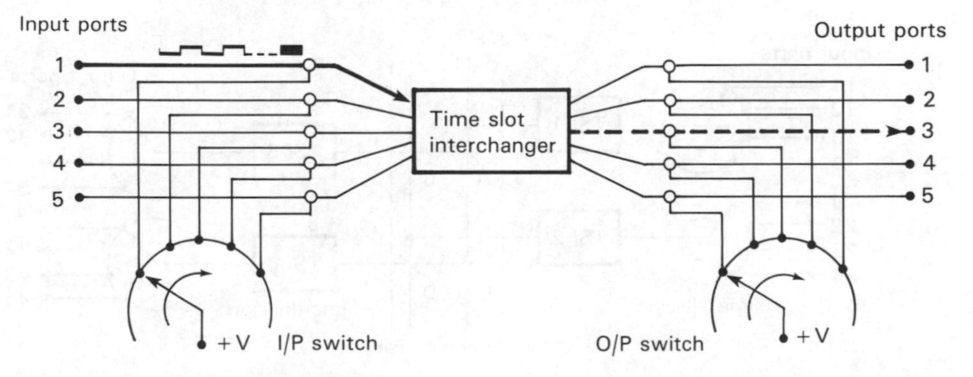

Figure 1.4 A basic time-division switch

in this case, so that the input port is opened to the time slot interchanger only when the output port 1 is opened, the first bit of the input port must be delayed after its acceptance by the time slot interchanger by two step delays of the synchronous switches. Once the output switch has reached the output port 3, this bit is then released by the time slot interchanger and permitted to pass into the output port 3. Similarly, the other input ports each have their bits delayed until the switch is stepped to the output port required by the input port. It becomes apparent that, as the number of ports increase, the access times for the time slot interchanger becomes prohibitively short. This limitation means that only small exchanges of a couple of hundred lines can use this method. For larger-capacity exchanges, the digital switch comprises a cascade of time-division and space-division switches. Space-division switches cause individual input ports to be physically switched by switching matrices to the output ports. These switches form the conventional method of switching analog signals.

A simplified time-space-time (TST) digital switch is shown in Figure 1.5. If a call is to be routed from port 11 through to the output port 44, the time slot interchangers will delay the bit to be transferred as in the simple time division switch described above. The space division crosspoint switch A is closed by the processor logic to permit the information transfer. The complication which arises in the transfer of this bit is when the space division crosspoint switch C on bus 1 is already closed due to there being a call already in progress. Then the attempted call is blocked. To overcome this difficulty, the space-division switch must have the facility for working at twice the time slot rate or twice the rate of the synchronized multiplexing switches. That in effect permits two bits, each of half their normal duration, to be transferred onto the same bus, but in different half time slots. The provision of a fixed time slot through the switch, which essentially relates to the maximum exchange capacity can be overcome by a method called *dynamic time slot assignment*. This method allocates time slots and paths through the exchange as required, rather than having these time slots and paths dedicated. By this method the exchange capacity may be increased to more than five times that of its nondynamic counterpart.

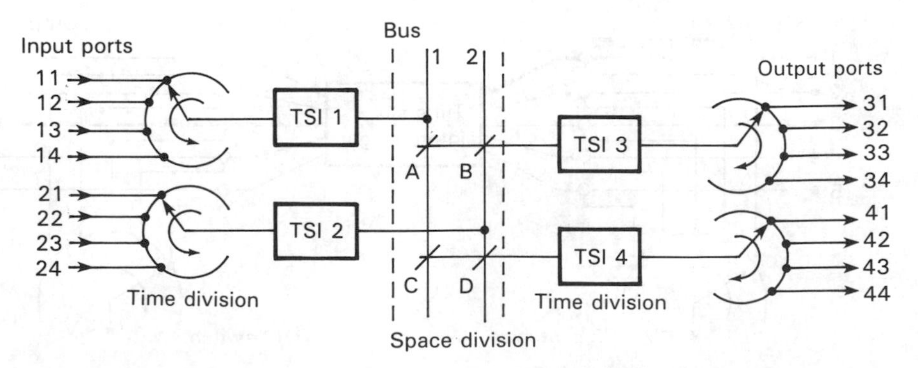

Figure 1.5 A simplified time-space-time (TST) digital switch

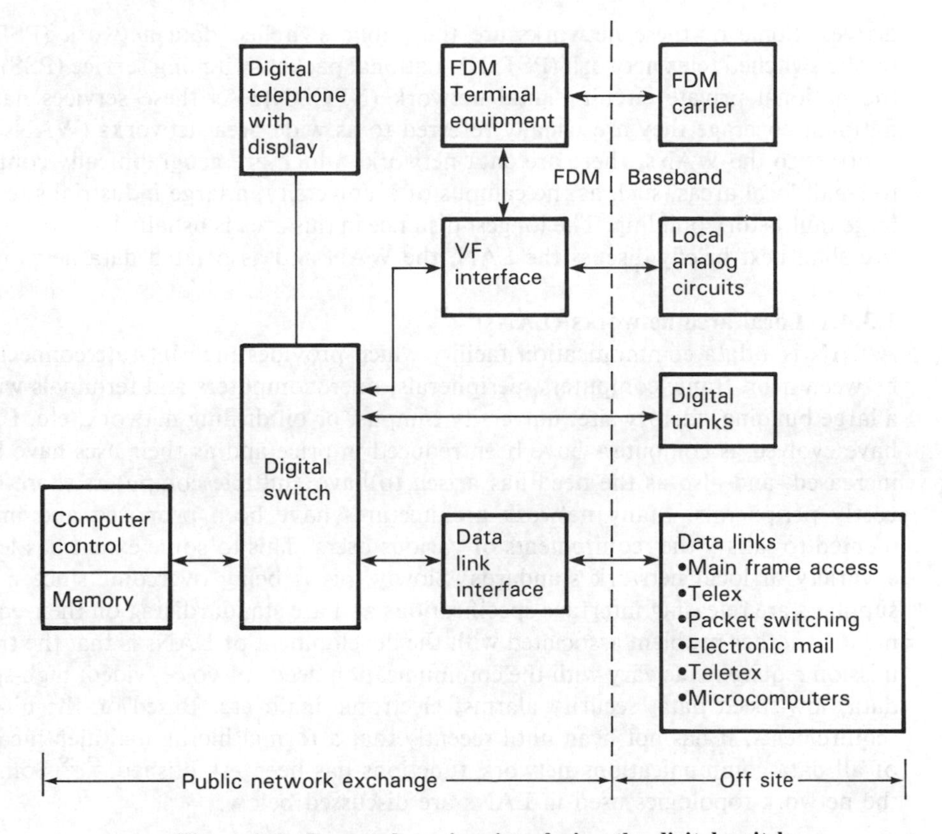

Figure 1.6 Expected services interfacing the digital switch

Other configurations, such as in the No.4 ESS (Electronic Switching System),[19,22] have a more complicated architecture. In this machine the architecture is T–SSSS–T, i.e. seven 105×128 time slot interchangers feeding into a 4 stage 1024×1024 space division switch which feeds into another seven 128×105 time slot interchangers. The resulting capacity is over 100 000 ports. Figure 1.6 shows the expected interrelationships between the present and future services and a local exchange (which may end up being an in-house PABX).

1.3.4 Data switching networks

Although many countries are embarking on the Integrated Services Digital Network (ISDN), it will be some time before it is fully implemented due to the heavy capital cost in replacing existing analog exchanges, solving technical difficulties associated with the local network, replacing existing broadband transmission equipment, having fully defined standards, etc. Meanwhile, until ISDN implementation, the increased demand for data services requires that existing overlay networks be expanded in a way that will enable easy integration into the ISDN when the time arrives. Some of these networks are the public switched data network (PSDN), public switched telex network (PSTxN), national packet switching service (PSS) and the national private circuit digital network (NPCDN). As these services have a national coverage they are usually referred to as wide area networks (WANs). In contrast to the WANs, there are data networks which are geographically confined to small local areas, such as the campus of a university, a large industrial site or a large multi-story building. The longest distance in this area is usually less than 5 km. We shall next briefly discuss the LAN, the WAN and associated data networks.

1.3.4.1 Local area networks (LANs)[23,24]

A LAN is a data communication facility which provides high bit-rate connections between main frame computers, peripherals, microcomputers and terminals within a large building, factory site, university campus, or oil drilling network, etc. LANs have evolved as computers have been reduced in price and as their uses have been increased, and also as the need has arisen to have multiple computers share their costly peripherals. Many network architectures have been proposed and implemented to satisfy the requirements of various users. This to some extent has led to a variety of local network standards. Slowly this is being overcome since major suppliers are releasing interface specifications and are standardizing on their equipment. Another problem associated with the development of LANs is that the transmission requirements vary with the communication needs of voice, video, high-speed data, low-speed data, security alarms, electronic mail, etc. Based on the bit-rate requirements, it has not been until recently that a formal hierarchial identification of all data communications network functions has been established.[23,29] Some of the network topologies used in LANs are discussed below.

ETHERNET* (CSMA/CD)[24,25]

The Ethernet architecture allows multiple distributed devices to communicate with each other over a coaxial cable using 10 Mbit/s baseband transmission within a local area spanning a kilometer or so. The Ethernet was designed and proven in the mid-seventies on the following assumptions:

1. Microprocessor-based work stations will be prevalent in the office and other environments in the late 1980s.
2. Distributed processors in the form of intelligent terminals will handle more tasks.
3. Intercommunications between processors and distributed information sources will increase the benefits to the user.
4. The next generation of user work stations will meet the challenges of pictorial display, manipulation and transmission, together with additional new functions such as facsimile, microfiche, teletext, etc.

The Ethernet operation is such that when one station is communicating with another, any other station wishing to transmit must wait until the common bus is free. Once it is free, a packet of data is transmitted. This data packet contains the

*Ethernet is a trademark of the Xerox Corporation

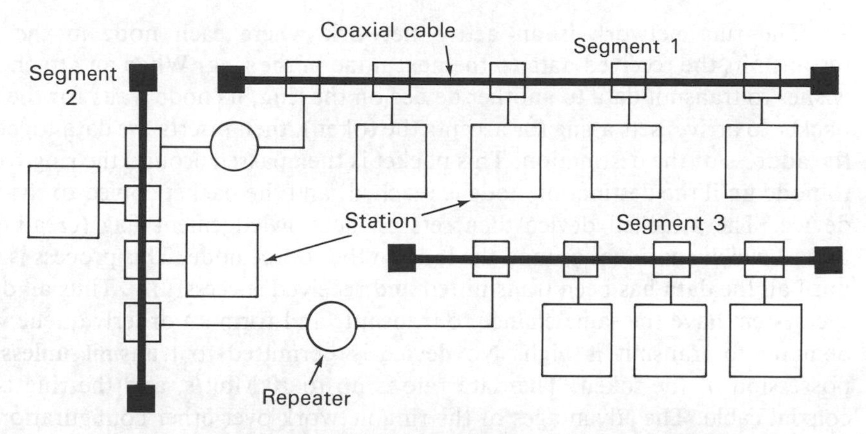

Figure 1.7 Ethernet bus architecture

destination address, source address, and parity bits for error detection. All stations continuously monitor the common bus for transmitted data, and accept only that packet which has the particular station's address and valid check sums. If the originating station does not get a response from any other station after a specified time has elapsed since transmitting, it will retransmit the information on the assumption that the original information has been corrupted by noise, or that another station was transmitting at an instant before it was. Although the data are transmitted at baseband, that is without the use of a carrier, the Manchester Line Code which is used to encode the data provides the equivalent of a carrier due to the continuous transitions which it provides. This line code also prevents the build-up of a DC component due to the use of a positive and negative voltage, which averages out to zero during transmission. The absence of a DC component assists the correct thresholds being detected by the zero-crossing detectors. The continuous transitions of the Manchester Line Code also permit easy clock recovery. Because the Ethernet is a baseband system, to connect it to the public switched telephone network (PSTN) would require an interface device called a *communication server*. The communication server not only matches the line signals, but permits matching of the protocols used. Another feature of the Ethernet is its ability to detect the existence of an already occurring transmission. This is referred to as *collision detection*. Should two or more stations decide to transmit simultaneously, a collision is detected and each device backs off, only attempting to transmit again after a randomly determined time interval. Because the time interval is random the probability of the same two terminals again attempting a simultaneous transmission is small. The three basic steps required for accessing an Ethernet are given in the acronym CSMA/CD which stands for carrier sense, multiple access with collision detection. Figure 1.7 shows a typical Ethernet structure.

TOKEN PASSING NETWORKS[26]

The advantages of a token passing local area network are: high capacity, flexibility, mixed applications, high reliability and low cost. Two types of token passing network are the ring and bus.

The ring network is an active network where each node in the network retransmits the received data to the next node in the ring. When an attached device wishes to transmit data to another device on the ring, its node waits for the first free packet to arrive, sets a flag (or accepts the token), then inserts the data together with the address of the destination. This packet is then passed around the ring from node to node until the destination node is reached, and the packet copied to the attached device. The attached device then sets an acknowledgement flag (creating a new token) which continues around the loop to the source node. This process is repeated until all the data has been transmitted and received successfully. Thus all devices in the system have the same chance to transmit, and form an orderly queue when the demand to transmit is high. No device is permitted to transmit unless it is in possession of the token. The data rate is up to 10 Mbit/s, and the ring is usually coaxial cable. The advantages of the ring network over other configurations is that it lends itself to duplication and can give a guaranteed response time to the attached devices. Its disadvantages are; a failure in a single node may result in the complete network failure on a non-duplicated system and secondly, all cables are required to be laid at the initial installation time – to cater for both the present and anticipated future requirements. The ring topology is, however, ideally suited to transmission using optical fibers.[27]

The token passing bus has more operational flexibility because the token passing order is defined by look-up tables programmed into each station. For example, if a station consisting only of a printer never originates traffic, it need not be in the polling sequence, or if a station has a high priority it can appear more than once in the polling sequence. Unlike the ring configuration, the bus does have a distance limitation due to the delay dependence in its operation, for the signal is not regenerated at each node. Other limitations are that it is difficult and expensive to duplicate and requires extensive cabling.

BROADBAND BUS

The topology for this LAN is usually likened to that of a tree. The data are transmitted in analog form using frequency division multiplexing (FDM) to establish multiple channels within the system. As on cable television, where multiple television channels are provided, the broadband bus can provide 50 to 60 separate 6 MHz wide channels when following the standards and technology developed by the cable television industry. For this extra data-handling capacity the nodes are more expensive than the baseband system counterparts, particularly if multi-frequency selective nodes are required so that devices may communicate on different frequencies. This technique has in its favor the ability to carry closed-circuit television simultaneously with data and voice. These networks are expensive and difficult to duplicate. They require extensive cabling and are subject to the different standards used by different equipment manufacturers.

DIGITAL PABX

The PABX may be used to switch digital information between interactive terminals or low data speed central processor units, or between modems, etc. Where the PABX is an analog exchange using space division techniques (crosspoint matrix), each of the devices is connected to the switch via dedicated twisted pair wires. Connection through the PABX is made by the conventional method of setting up a telephone call. That is, the off-hook signal and the dialling of digits is used to establish a connection. Once established, data transmission proceeds over the physical pair. Due to the use of twisted pairs, the capacity of the data channels are restricted to a range of 50 kbit/s to 1 Mbit/s, depending on the distance between the devices. Thus the PABX is not practical for high-speed processor-to-processor operations. As the digital PABX has now developed, voice circuits, data, word and image processing can all be done at one work station, providing an integrated voice and data services local network. The digitalization of speech at the line interfaces although being a major part of the cost of the exchange does permit a lower switch cost than its analog counterpart due to the reduction in the cost of the switching matrix. The digitalization of speech however, is moving away from the line interface equipment at the PABX towards the voice equipment itself. This permits a sophisticated multiline telephone/data terminal multiplexing typically 64 kbit/s channels to use one or two pairs of wires.

The integration of the LAN with the proposed ISDN may introduce problems due to the different philosophies of operation. The LAN is a packet switching network, whereas the ISDN is a circuit switching network. A high degree of standardization between the two networks is required before they will be able to interface satisfactorily.[28,29]

1.3.4.2 Private data networks[2(p.223),30]

These networks have evolved due to the needs of private organizations. Some of these needs are:

1. The connection of a large variety of terminals designed for different office requirements, with flexible access to existing and future public network services
2. The handling of the different kinds of office traffic, such as text-transmission, image processing, electronic mail and the dissemination of other information
3. Communication with separate parts of an organization which are located in different geographic locations

The more simple private networks consist of private lines which compete with the public network. Other networks may operate around a large centralized mainframe computer or they may be decentralized by the use of smaller computers. A definition of a private network could be: a set of nodes using for intercommunication a variety of transmission media which are private or public or mixed. This means that the network consists of switching nodes which can communicate with each other by dedicated transmission links, which in some cases are linked to the public telecommuncations network for reasons of traffic overflow, and security in the event of link failure, or are privately leased lines (for communication between distant nodes). The principal incentives for the development of these networks have been economic and improved quality of service. Private networks are able to offer the ISDN features in advance of the public network owing to lower cost of installation. Such an arrangement may be *feature transparency*, that is when the network does not inhibit the operation of any feature available at any node. To provide a full feature-transparent integrated voice/data network which also will access the public ISDN, is the ultimate aim of those developing private networks.[31] At present the linking of nodes via the public network is achieved when a digital link is cross-connected between two cross-connection sites containing multiplex and demultiplex (MULDEX) equipment. This MULDEX equipment converts thirty-one 64-kbit/s circuits, originating from the customer's network termination units (NTU), into a 2 Mbit/s bit stream for distribution, or demultiplexes a 2 Mbit/s bit stream down to 64 kbit/s for cross-connection or distribution. These private paths provide a high operational performance and high circuit reliability owing to the quality of the digital link and the extensive supervisory and error detection facilities built into the equipment. The present problems of private networks are due to the different protocols used by different computer manufacturers, and the number of different interfaces presently in use. In addition to these problems are those arising from the differences between packet switched networks and circuit switched networks.[32] The CCITT X–Series recommendations, which are in line with the developments of the open systems interconnection[29] (OSI), may be used to permit private organizations to construct private data networks which are independent of their present and future applications, and which through package switching techniques can interact with all the different transmission services and media, such as lease-line, the public switched telephone network, data networks (CCITT Recomendation X25 and X21), digital microwave radio, fiber optics, satellites, etc.

1.3.4.3 Wide area networks (WANs)

As the name implies, these data networks cover whole geographical regions, from large industrial concerns[33] to national and international operations, meeting the problems both of terminal interconnection protocols* and of long-distance transmission. Traditionally the customer's data terminal has been connected to a modem which has converted the data signals into an analog form for transmission over a network designed for voice services. Due to the limited bandwidth, the high error rates produced, and the expense of long-distance circuits, this method has not been entirely satisfactory. As mentioned in Section 1.3.1, to overcome the poor quality of service and the long fault restoration times, caused by the ad hoc manner in which these data services have been provided, the introduction of the ISDN is being adopted by most developed countries. Modems will continue to be used until the new network is fully implemented, and so we briefly consider them here in the context of wide area networks.

The limitations of the telephone network for the direct transmission of the data are: the restriction on bandwidth due to speech channel filters, permitting only a bandwidth in the range 0.3 to 3.4 kHz; the envelope delay distortion arising from long-distance circuits and associated filters; attenuation distortion, etc. Since the spectrum of a data signal may contain significant DC and low-frequency components, and be susceptible to the analog voice channel impairments, or impairments due to the frequency division multiplex (FDM) process, it is necessary to shift

*See Section 1.3.4.4.

the spectrum of the baseband digital signal using modulation. The modulator and demodulator necessary for this operation is referred to as a *modem* (MOdulator-DEModulator). The first modems operated by using two frequencies to represent the state of a binary bit (high or low). This operation is generally termed *frequency shift keying* (FSK). Two standard FSK modem systems are in general use. The first enables word processors and the like to communicate over the public switched network and private data circuits at signaling rates of up to 200 bits/s in a full duplex mode. The CCITT Recommendation V21[34] covers this requirement and provides for a possible extension of the rate up to 300 bits/s in each direction. The second standard was introduced to provide higher speed links, enabling the use of video display units to be served quickly by a host computer, and it permits keyboard access to the computer on a low-speed return channel. Signaling rates up to 1200 bits/s are possible. CCITT Recommendation V23[34] covers this requirement. For signaling rates higher than 1200 bits/s more efficient modulation schemes are necessary. Some of these modulation schemes are single-side-band (SSB) amplitude modulation, vestigial-side-band (VSB) amplitude modulation, and quadrature amplitude modulation (QAM), which is covered by CCITT Recommendations[34] V26, V26 *bis*, V27, V27 *bis*, V27 *ter* and V29. Baseband modems are used where modulation of the baseband is not performed, and the data are directly translated into the baseband after signal processing. These types of modems convert the input binary data into a suitably encoded line signal to insure that timing information is contained in the transmitted signal and that the signal energy is distributed correctly over the channel bandwidth.

Reference 2, Chapter 7, and Reference 24, Chapters 4 and 5, provide additional information on modems.

VESTIGIAL SIDEBAND (VSB) AMPLITUDE MODULATION

This form of modulation represents binary coding by the presence or absence of an audio tone or carrier. Its disadvantages are its susceptibility to sudden changes in the amplitude of the carrier due to changes in noise or gain, and its inefficiency in modulation. The modulation process filters out one of the two duplicated information-carrying sidebands present in amplitude modulation leaving only its 'vestige'. The carrier frequency is preserved to recover the DC component of the information envelope, together with the retained sideband. This method reduces the bandwidth to three-quarters of that required for a double-sideband modulated system. Data rates up to 2.4 kbits/s are possible using this form of modulation. Multilevel (*M*-ary) amplitude modulation techniques in which modems may achieve data rates up to 9.6 kbits/s will be discussed in Chapter 2.

1.3.4.4 Packet switching[2,18,24,35]

Packet switching networks are intended to offer, to private or public users, high-capacity and high-efficiency data transfer, with the ability to access very large data bases carrying mass information files or unusually specialized or complex software. The advent of real-time computer systems for airline reservations, banking systems and remote data processing in general has been based on the use of telecommunications networks to carry data between computer-type terminal devices and large main-frame real-time computers. Such applications are suited to a packet switched network. This system, prior to transmission, assembles the individual source digit streams into short protected blocks of a maximum specified number of digits called packets, and then transmits these packets in sequence through the network from one computer attached to one node to another, or to a remote terminal located at another node. Each packet has at its head a group of digits containing information on the packet, such as its destination, length and priority. By transferring data in packets, a time multiplexing of users may be achieved. This leads to the efficient use of each physical message-carrying circuit. The node computers do not store the data and the sending node forgets the message as soon as the receiving node acknowledges receipt of the information. This method is to be contrasted with the *circuit* switching technique which is at present prevalent in the public network. In this case the availability capacity of the circuit is allocated for the duration of the connection. Since data are not always being transferred on a circuit, the allocation of a full-time circuit is wasteful and expensive. Because the destination address information is an inherent part of the package, numerous messages may be sent to many different destinations quickly using the packet switching system, and, because of the intelligence built into each node by the computers, dynamic routing of data is possible throughout the system, allowing each packet to travel over the best available route at the time. This permits the network to maximize on information transfer and thus on efficiency; and to reduce congestion and down time caused by failures in part of the network. Because of the computers at the nodes in the network, services such as error detection and correction, message delivery verification, group addressing, reverse billing, message sequence checking and diagnostics are possible.

Whatever the topology of the network is, there must be flexibility in its hardware. This is achieved using a hypothetical node-to-node connection which is amenable to being divided into interaction levels. One common level (or layer) of interaction may be the transmission medium. A higher level may be the signaling format which again is something that is common between the nodes. A higher level still may be the recognition of the data block structuring, and so on. The highest level is possibly the communication between two users of the network. These level functions and interactions are referred to as *protocols*. More strictly defined, a protocol is a set of rules governing a time sequence of events that takes place at the same interaction layer or level in a network. The *architectural layers* are; the physical layer, or level-one protocol (mechanical, electrical, logical and functional relationships between the interfaces); the link layer (address, text, error checking fields in the same bit stream); and the network layer where the protocols may be used to route data from link to link through the network nodes. These protocol layers and the interaction between the terminal and the network are specified in the CCITT Recommendation X25.[23] This recommendation is a standard for packet switching systems and is one of the most significant networking architectures affecting data communications. The OSI[29] defines a complete architecture having seven layers. These are (starting from the lowest level); physical, link, network, transport, session, presentation and application.

GATEWAYS[30]

A gateway defines a facility which interconnects two networks so that users on one network many communicate with users on another network. Examples of these gateways are:

Telex-packet switching interworking. By means of the telex network adaptor (TNA), telex users may use their teleprinters to access national or international users with a similar adaptor via the package switching network.

*Packet network adaptor (PNA).*The package network adaptor permits access to and from the public switched network from and to the package switched network. This allows access to services which are available in either network.

Telex and teletex adaptor[36]. The telex-teletex conversion facility (TTCF) enables interworking between telex terminals and teletex terminals on both a public data switching network (PDSN) and the public switching telephone network (PSTN). By means of the TTCF it is possible to access the international telex and teletex services from a single terminal of either type.

1.3.4.5 Telex network

The telex network is a separate national *circuit* switched network which is overlayed onto the public switched telephone network (PSTN). It is a fully automated service which is internationally established for the interconnection of teleprinters. In the U.K. it is considered that the existing switching and transmission systems of telex are not suitable at the present time for adaptation to the integrated digital services network. The network has been upgraded with stored program control (SPC) exchanges and the low-voltage single channel voice frequency (SCVF) transmission system as per CCITT Recommendation R20[37] for customer access, and within network transmission conforming to CCITT Recommendation R111. The network in the UK contains exchanges which provide features such as:

1. Abbreviated dialling – a user is able to store in the exchange frequently used numbers and use a two-digit code to call any of these numbers.
2. Redirection of incoming calls – a user will be able to redirect calls to another specified number in the nation.
3. Recorded messages – a user may be able to store a prerecorded message in the exchange which will be returned to callers.
4. Advice on last call charging – upon request, advice on the number of chargeable call units used for the last call, or at any time, can be provided by the exchange.
5. Changed number interception – on the changing of a user's number, the old number will have associated with it the new number for returning to the caller.
6. Broadcast – a user transmits a message to the network which will pass it to a maximum of five other specified parties.
7. Data and time insertion – date and time may be automatically inserted at the beginning of all user outgoing calls.

8. Closed user groups – these groups will belong to private networks using the public telex network.
9. Answerback – this will be sent to the called user at the start of each calling user's call, if so desired.
10. Message separation – line feeds to separate messages will automatically be sent after each call.
11. PABX hunting – the private automatic branch exchange (PABX) is able to hunt for a free circuit for a calling user.
12. Store and forward – the exchange is able to store a message left by a user for delivery by the exchange as soon as possible or at a predetermined time. The message can be forwarded up to two hundred addresses. This avoids having repeatedly to key in address lists which are pre-stored in the exchange.
13. Low-speed data – the exchange has the capability of switching calls at 110, 200 and 300 bits/s.
14. Speed and code conversion – transmission speed and code conversion by the exchange permits interfacing between users with different terminals.

The exchange will be responsible for the keeping of call records, routing, billing and traffic analysis. It is likely, as the ISDN begins to offer the packet switched teleservices including videotex, electronic mail, facsimile, electronic funds transfer (and other forms of message handling such as teletex, which are offered by the circuit switched data network), that the telex network will become obsolete and cease to exist.

1.3.4.6 Public circuit switched data network

Leased and switched data connections in the telephone network have existed in many administrations since the 1960s. As demand for these circuits increased, data switching exchanges were installed. The data network resulting from the installation of these exchanges provided circuit-switched data transmission services according to CCITT Recommendation X1[39] (with user classes up to 9.6 kbits/s) and customer interfaces according to Recommendations X21, X21*bis*, and X22.[39] The services offered by this circuit-switched network include the teletex service. To couple into the network the customer requires modems as briefly described in Section 1.3.4.3. Sometimes the network is not separate from the voice or telex networks, but is an integral part of these networks. This is so in Italy.[40]

1.3.4.7 Teleservices[41,42]

The teleservices are those which would greatly benefit from the ISDN. They are: facsimile, teletex, videotex, message handling systems and telewriting, and are briefly described below;

FACSIMILE

Facsimile is the means of transmitting to a distant terminal a replica of a document. The earlier forms of facsimile could provide only a limited range of grey and black on a white background. This was due both to the printers used and to the rate of data transmission. The CCITT Recommendations F161 and T5[43,44] provide details on the international Group 4 facsimile service (FAX 4), which is relevant to the ISDN, as well as to the public data networks. It can be used also on the public switched telephone network (PSTN). The major advantage of Group 4 facsimile over the earlier Groups 1–3 is that the transmission is error checked and if necessary the message is repeated. This produces an error-free copy at the receiving terminal. Three classes of Group 4 facsimile are defined. They are: basic facsimile; teletex/mixed-mode reception capability; and full-fax/teletex/mixed-mode capability. As ISDN caters for both circuit- and package-switched network services, the network on which facsimile will best operate may depend on the comparative charges.

TELETEX[43]

The service that a basic teletex system provides is the transfer of a typewritten office document, using some form of word processing, to a distant terminal. Each document can comprise several pages, typically in ISO A4 form. CCITT Recommendation F200[43] provides the payout specifications to ensure that the text is compatible with a variety of equipment used in the service. During the transfer of the text, information is added for identification. This identification may be printed on the final copy, or used as information for electronic logging purposes. The teletex may be considered as a 'super-telex' service, and may end up by superseding it. Teletex has the attributes of transmission speed, error performance, character repertoire and document layout which are superior to those of telex. Mixed-mode teletex may contain facsimile images, and mixed-mode Group 4 facsimile may contain teletex messages. This subject is covered further in CCITT Recommendation T72.[42]

VIDEOTEX

Videotex differs from the facsimile and teletex services in that is is based upon a terminal-to-host and not a terminal-to-terminal configuration. Also unlike facsimile and teletex, it is usually confined to national boundaries. It is used for both domestic and business purposes, and may be provided with limited access where privacy is required such as in subscribed specialized information sources. The form of presentation used by this service is significantly related to the television broadcasting standards, since it is predominantly a video screen medium, placing emphasis on graphics and colour. The simple graphics used are based upon the definition of shapes known as *alphamosaic graphics*. Other images which may be formed such as those based upon enhanced graphics and photographs require more bits to create as defined in Recommendation F300.[43]

MESSAGE HANDLING SYSTEMS[44]

These systems operate on a store and forward basis and are used for both business and military purposes. The terminals used are usually teleprinters, but unlike the telex network they are not interconnected directly. Instead, when a terminal operator types a message destined for another terminal, the system stores the message temporarily and delivers it to the required terminal at a later time. The delay is caused by the queuing of the message with others wishing to use the line. This provides a much higher utilization of the link than that which would be offered by a *circuit-switched* network such as that of telex. The CCITT Recommendation[44] on message handling systems suggests a network structure in which both private office and public network message centers can cooperate in the 'intelligent' transfer of messages.

BIT PLANE TO BIT PLANE IMAGE TRANSFER
This has developed in different parts of the world under different names. The alternative names are 'telewriting' (U.S.A.), 'Vidiboard' (Holland), 'remote blackboard' (France) and 'Cyclops' (U.K.). The user writes with a stylus on a digitized pad and the line formed is reproduced on the user's video screen and at the distant end. The applications are for signature verification, distance teaching, etc.

1.3.5 Trunk switching network configurations

1.3.5.1 Noncentralized or distributed switching systems

Figure 1.8(*a*) shows a simple way to structure a distributed switched network. Every exchange directly links every other exchange by means of switches for incoming and

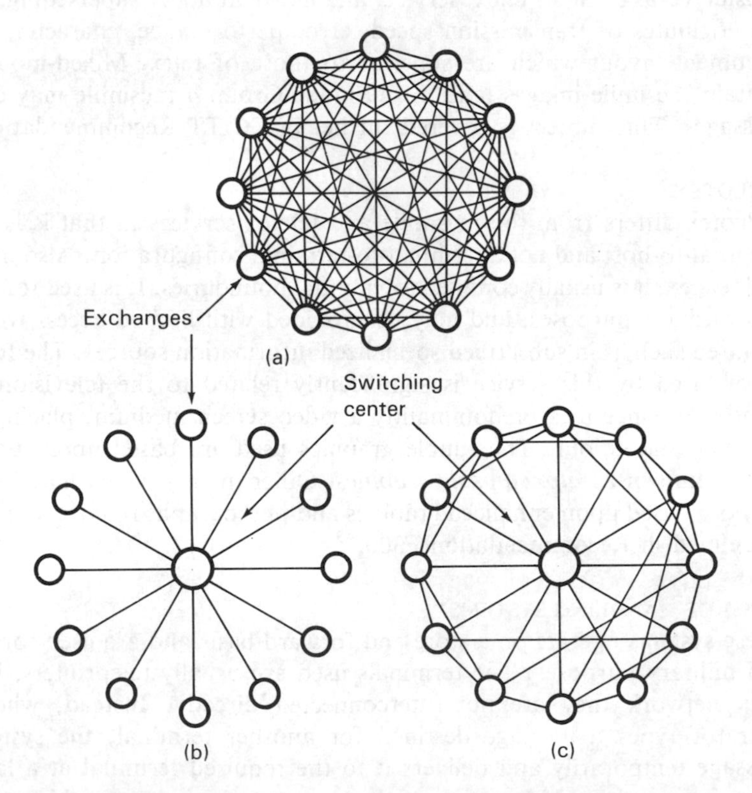

Figure 1.8 Basic network forms: (a) distributed network (mesh); (b) centralized network (star); (c) hybrid network with alternate routing (ARS)

outgoing trunk traffic. Thus for N exchanges where each of the exchanges requires $N-1$ trunks, a total of $N(N-1)$ trunks would be required if the outgoing and incoming trunks were separated. If the trunks are both-way trunks, the number of links required would be only half, that is $N(N-1)/2$. The interlinking of every exchange with every other exchange would be impracticable in a large network, although it would be satisfactory in a very small network of six or less exchanges. Impracticable because of the amount of equipment required to provide trunk groups to $N-1$ other switches, and also because of the cost of providing all of the $N(N-1)$ single-input and -output trunks. Usually a trunk group comprises more than one trunk, and the number of trunks in each group is determined by the traffic requirement; thus the number of trunks to serve all other $N-1$ exchanges in this configuration is far greater than the simplistic $N(N-1)$.

1.3.5.2 Centralized switching systems

Figure 1.8(*b*) shows a simple way to structure a centrally switched network, with each exchange directly linking a central exchange. Thus for $N-1$ satellite exchanges, $N-1$ input and $N-1$ output trunks are required, or $2(N-1)$ trunks. If both-way trunks are used then of course only $N-1$ trunks are required to provide each exchange with one trunk to the central exchange. Again the number of trunks required to each satellite exchange will depend upon the traffic requirements and will be far greater than $N-1$. A comparison with the distributed exchange configuration shows that the distributed system requires $N/2$ more single trunks than the centralized switching configuration, and the centralized switching system is thus less expensive in terms of the overall system switching equipment. This of course is offset by the complexity required of the central switch and the low system security. If the central switch fails, the system fails.

In practice a mixure of both distributed and centralized switching occurs, as is shown in Figure 1.8(*c*). In this case each switch is able to provide alternate routing to its adjacent satellite exchange. This means that if a call cannot pass from the central exchange directly to a satellite exchange because the trunks are all busy or have been damaged, the call can be routed to the satellite exchange via its adjacent exchange, which is then used as a 'tandem exchange'. Again depending on the traffic requirements and upon the size of the exchange, alternate routing paths in practice will differ from that shown in Figure 1.8(*c*).

1.3.5.3 Hierarchial systems

In the public switched telephone network (PSTN) all telephones in a particular area are served by a local exchange (see Table 1.1 for alternative names). All calls made to or from subscribers in the area are dealt with by this exchange. Local exchanges areas which are geographically close are regarded as a group, and each local exchange within this group is served by an exchange known as a primary center. This primary center is a trunk exchange which is used to connect a subscriber who wishes to place a call to a subscriber in a different local exchange area if the local-to-local exchange trunks are busy or do not exist for some reason. On the placing of the call, the local exchange will first try to place the call directly to the adjacent local exchange. If this does not work, the call will be attempted via the primary center.

Table 1.1

Function	International name	Australian name	North American name	U.K. name
Local switching center direct to subscriber	Local exchange	Terminal exchange	Class 5 end office	Local exchange
1st level trunk switching center	Primary center	Minor center or tandem	Class 4P toll point Class 4C toll center	Group switching center
2nd level trunk switching centre	Secondary center	Secondary center	Class 3 Primary center	District switching center
3rd level trunk switching center	Tertiary center	Primary center	Class 2 sectional center	Main switching center
4th level trunk switching center	Quaternary center	Main center	Class 1 regional center	Nonexistent
Local exchange connections only	Tandem exchange	Tandem exchange	Tandem office	Tandem exchange
International access switching center	Centre du transit 3 (CT3)			
International transit center	Centre du transit 2 (CT2)			
Fully interconnected international transit switching center	Centre du transit 1 (CT1)			

If that does not work, the call will be attempted via the secondary center. To reach the secondary center, which is a trunk exchange, the call has to pass through the primary center. This routing of the call upwards shows the difference between a trunk switching center such as the primary center and a 'tandem exchange', which

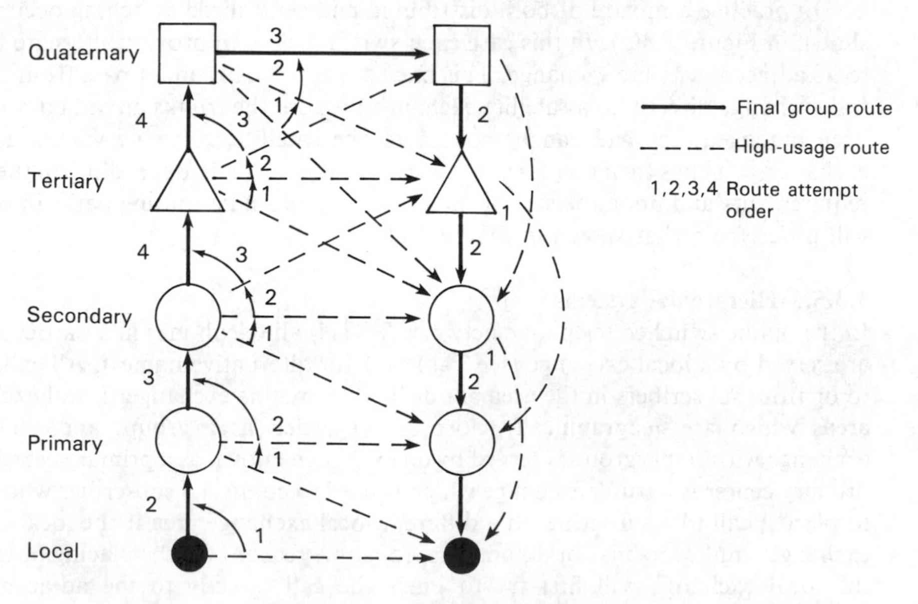

Figure 1.9 Hierarchical network of trunk switching centers

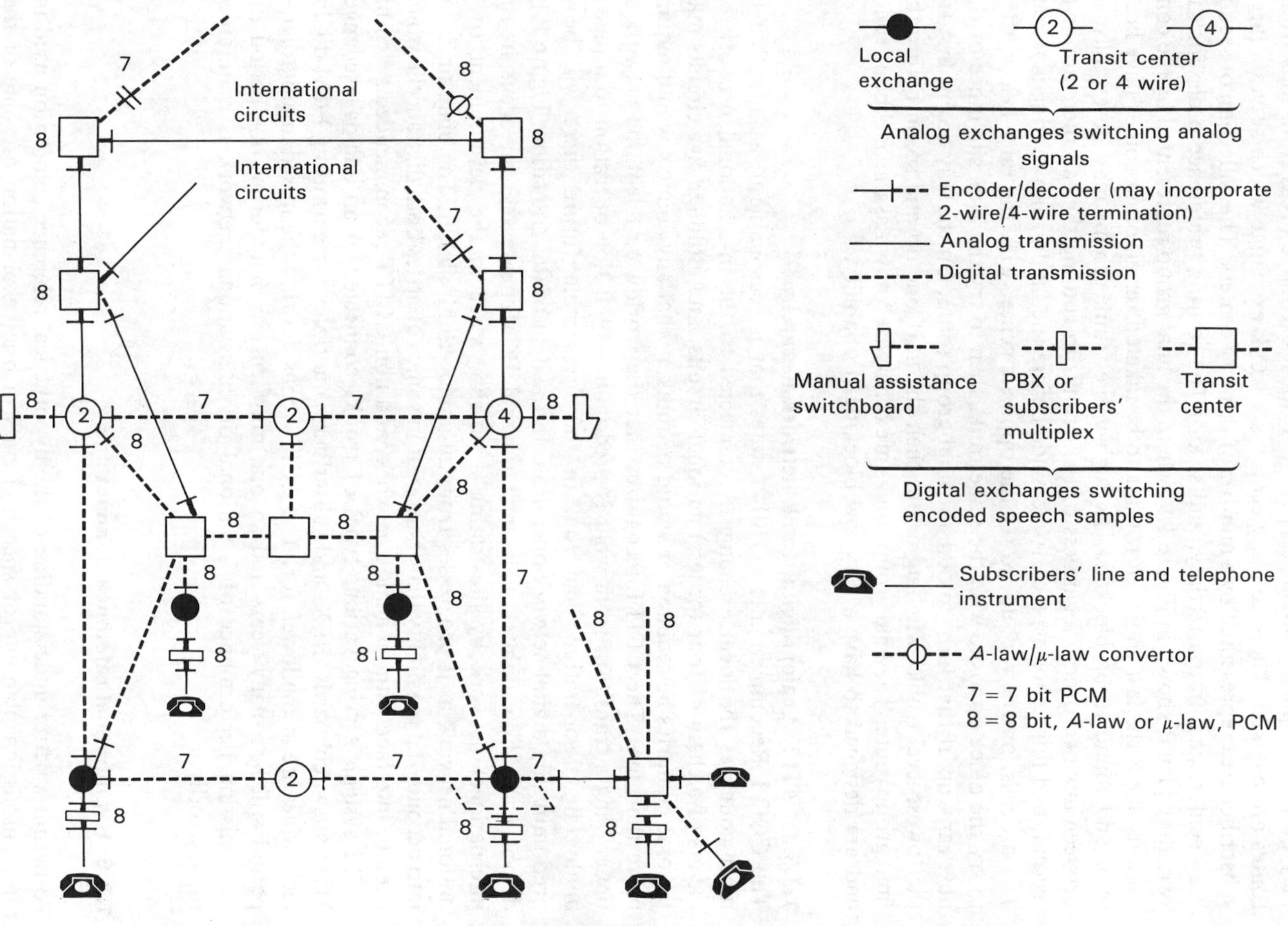

Figure 1.10 A possible intermediate stage of development in a national network (Courtesy of CCITT, Recommendation G101, Fig. 4, Reference 45)

is used only to provide connections between local centers and has no access to the trunk network. Figure 1.9 shows the upward and downward alternate routes which may be used in placing a call between two local exchange areas if various trunk routes are blocked. The secondary center is a trunk exchange which serves a large subscriber area represented by a number of primary centers. The collection of secondary centers pass into a tertiary, and so on. The routing preference order for the switching plan is shown in Figure 1.9, where the final group represented by the continuous line is the last choice of routing to be taken after all other routes have been tried and found to be blocked. The high-usage trunk groups may be established between any two offices regardless of their rank. In routing the call there are first, second, and third choices, etc., to be made before the final group routing is taken. The order is, on the way up, to try the opposite center with the rank below, then to try the office opposite with the same rank, then to try the opposite office of the next rank up; if that fails, to take the final group route. On the way down, first try the lowest-level trunk switching center then the next one up and so on before the final group route. The final group routes are otherwise known as the backbone route and are designed to have a very low blocking probability.

1.3.5.4 CCITT Analog/digital trunk network hierarchy

The CCITT Recommendation G101[45] states that is reasonable to assume that in most countries the local exchange is connected to the international network by means of a chain of four (or less) national circuits, and although five circuits may in some countries be required, it would be unlikely that any country would require more than five. The CCITT recognizes that the worldwide telephone network is undergoing a transition from what is predominantly analog operation to a mixed analog/digital operation, and that in the foreseeable future there will be a predominantly digital network operation. In the light of this transition, Figure 1.10 demonstrates how unintegrated analog/digital PCM processes can occur in the international network by illustrating a possible stage in the development of a national network as it progresses from all-analog to all-digital. The diagram takes into account the possibility that there exist in some countries both digital cells which need to interface with an analog network, and non-CCITT-recommended seven-bit PCM systems serving various types of trunks connected to an analog exchange. Manual switchboards, PBXs and subscribers' multiplex systems using PCM digital techniques are also allowed for. The CCITT consider that the mixed analog/digital period will last many years and so one must ensure that during this period the transmission performance of a national or international network is maintained satisfactorily.

1.3.6 Hypothetical reference connections

To permit a better understanding of the hypothetical reference connection used in the transmission planning of speech circuits, a brief explanation of some of the terms used is given below. For more details of these terms and their test set-ups, the reader is referred to CCITT P Recommendations.[46] The aim of a properly engineered telephone connection is to provide a means of exchanging information by speech which optimizes customer satisfaction within technical and economic constraints. One of the most important tools available to the transmission engineers for planning, design and assessment of telephone networks is the *reference equivalent*. The reference equivalent system has been used for determining the quality of speech based upon the loudness of speech emitted by a talker and as perceived by a listener. It provides a measure of the transmission loss, from mouth to ear, of a speech path. Reference equivalents as described below have currently been modified or 'corrected', to make them more useful to the transmission planning engineer than previously. In addition there is now an alternative approach. This is termed 'loudness ratings', and is also briefly described in Section 1.3.6.1.

1.3.6.1 The reference equivalent, corrected reference equivalent (CRE) and loudness rating (LR) NOSFER standard (CCITT Recommendation P42)[46]

When two or more different communication systems are interconnected, it is necessary to have some international standard which is adopted by all system authorities. This is to insure that the quality of speech or communications is maintained satisfactorily under the worst call routing conditions. These standards are set in the CCITT recommendations. A reference system for determining the quality of speech based upon loudness and having a nominally flat frequency response up to 4 kHz has been developed. This reference system is known as the 'Nouveau Système Fondamental pour la détermination des Equivalents de Réference', or NOSFER. Its sensitivity is such that it produces a signal of equal loudness to a 1 meter air path when its attenuator is set to 33 dB. Usually the attenuator is set somewhere between 15 and 24 dB, when subjectively comparing the loudness of a system under test, since this is the level at which most accurate speech comparisons can be made. The measurements of the loudness comparisons at different points in a system under test, result in the reference equivalents; sending, receiving, and overall. The tests and calculations are indicated in Figure 1.11, and are described below.

THE REFERENCE EQUIVALENT

The reference equivalent system is used to rate customer satisfaction according to the received level of a signal at the telephone handset. It is obtained from subjective listening tests of trained operators (CCITT Recommendation P72)[46] using a test language. Usually the system is divided to permit the evaluation of the sending reference equivalent (S) and the receiving reference equivalent (R), as well as the overall reference equivalent (OVE) and overall system reference equivalent (OSR). The test comprises an operator who speaks into the send circuit of the system under test, at a fixed distance (140 mm) from the microphone as determined by a guard ring, at such a level that the volume meter reading reaches a predetermined level. The three attenuators, of which one is in the NOSFER path and another in the path under test, are set to arbitrary values by the operator before start of the test. The listener at the far end listens alternately to the speech coming through the test channel and the reference channel (NOSFER) by means of the talker-controlled switch, and then adjusts the NOSFER circuit attenuator until the speech from both

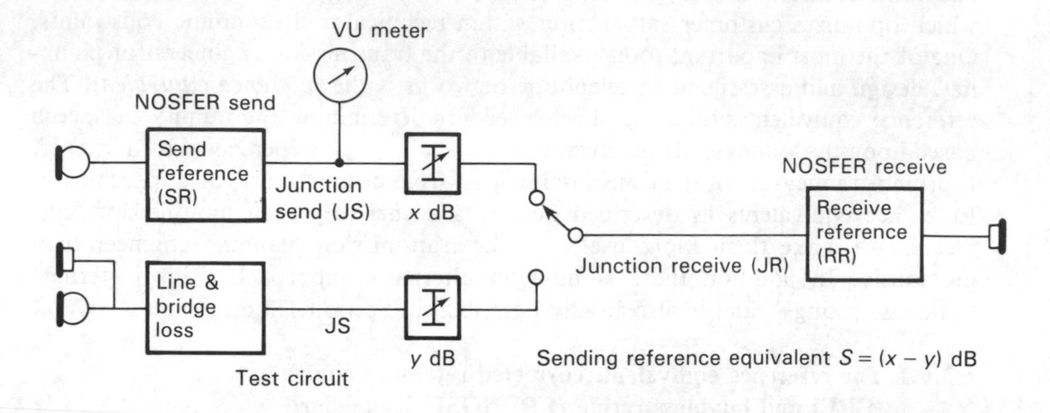

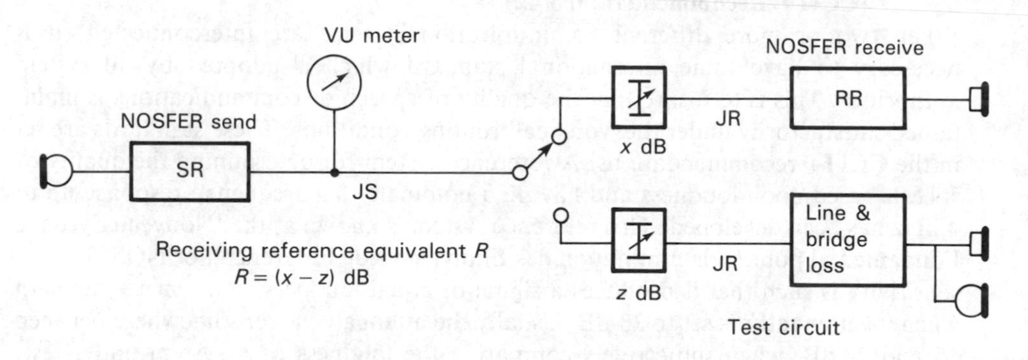

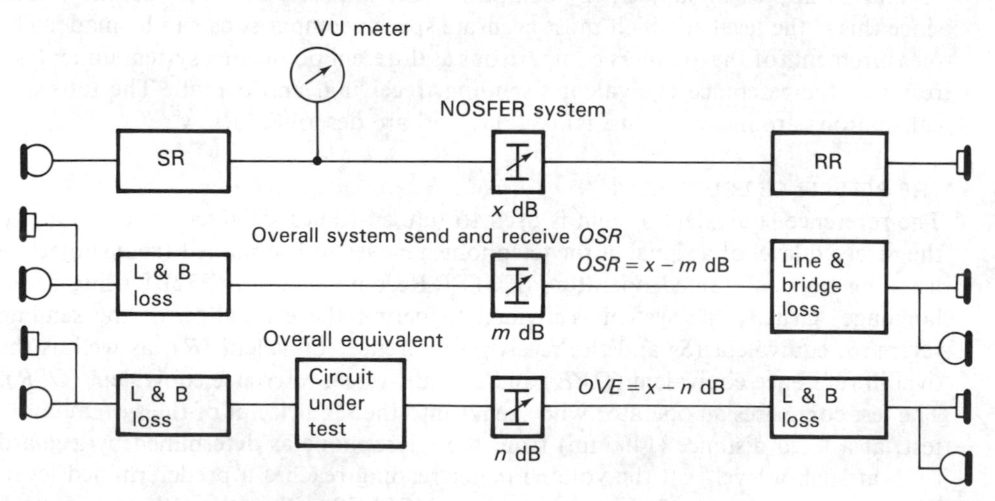

Notes:
1. Reference equivalents if x is varied for path balance
2. R25 equivalents if x is set at 25 dB

Figure 1.11 Connections and systems for the definition of reference equivalents

channels has equal loudness. The difference between the readings obtained from the NOSFER attenuator, and the test circuit is the measure of the sending reference equivalent $(S) = (x - A) - (y - A) = (x - y)$ dB.

The receiving reference equivalent is obtained similarly shown in Figure 1.11, where the balancing of each path is done by adjusting the NOSFER attenuator: $(R) = (x + A) - (z + A) = (x - z)$ dB

A common attenuator A (not shown in Figure 1.11) is set at twenty-five dB for the R25* equivalents. For the non-R25 equivalents it is located in the single leg of the sending or receiving circuit. The overall system (send + receive) reference equivalent (*OSE*), and the overall equivalent (send + receive + circuit) (*OVE*) may also be determined by the methods shown in Figure 1.11. The loudness insertion loss of a junction circuit (*LIL*) can be derived if both the *OSE* and *OVE* are known, for the relationship is given by

$$LIL = OVE - OSR = (x - n) - (x - m) = m - n \text{ dB} \tag{1.3}$$

The R25 equivalents are obtained when the NOSFER attenuator is fixed at 25 dB. The attenuator A (0–34 dB) is removed from the common-leg circuit, and is placed next to the attenuator in the circuit under test or the 'hidden loss attenuator' *S*, which also has an attenuation range of 0–34 dB. The 'hidden loss' attenuator is so called because it is preset and hidden from the listener who can only adjust attenuator A in the test system for the loudness balance.

THE CORRECTED REFERENCE EQUIVALENT (*CRE*)

Due to the differences in the planning values of the overall reference equivalent of a complete connection obtained by calculation and in practice, a corrected value has been derived. This corrected value, based on the fixed 25 dB NOSFER attenuator (R25), can be expressed in terms of the reference equivalents q, which are available for many types of local systems comprising the subscriber sets with their lines. The equivalence is given by:

$$CRE = 0.0082q^2 + 1.148q + 0.48 \text{ dB} = 0.0082(q + 70)^2 - 39.7 \text{ dB} \tag{1.4}$$

The terms send corrected reference equivalent (*SCRE*) and receive corrected reference equivalent (*RCRE*) are also used in the literature. Equation 1.4 is used to determine each from the send reference equivalent (*S*) or (*SRE*), and receive reference equivalent (*R*) or (*RRE*).

For an overall connection there is also need to determine an 'additivity' term *D* as described below on the planning methods based on *CRE*; in addition, CCITT Recommendation G111,[45] Annex B, provides details of the method used.

LOUDNESS RATING (*LR*) (CCITT RECOMMENDATIONS P48, P64, P65, P76, P78, P79[46]

The reasons for an alternative approach to that of the reference equivalents arise

*The R25 equivalent is the loudness loss determined as a reference equivalent, except that the listening level is constant, corresponding to a 25 dB fixed attenuator in NOSFER.

from problems with these equivalents. The problems are:

1. Reference equivalents cannot be added algebraically without discrepancies of ±3 dB occurring.
2. Changes in operators used to determine the reference equivalents produce different results. These differences can be as much as 5 dB.
3. Increments in real (distortionless) transmission loss do not produce equal increments of the reference equivalent. A 10 dB increase in loss results in an increase of only 8 dB in the reference equivalent.

Loudness ratings do not have these problems, and produce the same values no matter how they are determined – whether by subjective tests, by calculations based upon sensitivity/frequency characteristics, or by objective instrumentation. The loudness rating is defined by the amount of loss required to be inserted in a reference system to obtain the same perceived loudness as that obtained over the speech path being measured. Rather than being expressed directly in terms of perceived loudness, loudness ratings are expressed in terms of the frequency-independent transmission loss that must be introduced in an intermediate reference speech path and the unknown speech path to obtain the same loudness of received speech as that defined by a fixed setting of NOSFER. In practice the unknown speech path consists of a sending local telephone circuit coupled to a receiving local telephone circuit via a chain of circuits. The interfaces junction send (JS) and junction receive (JR) separate the three parts for the connection to which loudness ratings are assigned. These assignments are: sending loudness rating (*SLR*) from mouth to JS, receiving loudness rating (*RLR*) from junction JR to the ear reference point, and junction loudness rating (*JLR*) from JS to JR. The overall loudness rating (*OLR*) is assigned to the whole of the connection from the mouth to the ear reference points. The relationship between *OLR* and the other loudness ratings is given by:

$$OLR = SLR + RLR + JLR \tag{1.5}$$

The reference system used to define sending and receiving loudness ratings is not the wideband NOSFER used for the R25 equivalents (R25E), but a system rather similar to real telephone connections and termed the 'intermediate reference system (IRS)'. Details on the specifications for an IRS are given in CCITT Recommendation P48,[46] and has been chosen with the following in mind:

> It shall have a frequency response similar to that of existing and future national networks, and a nominal frequency range of 0.3–3.4 kHz.
>
> The absolute sensitivity has been chosen to reduce as much as possible differences between reference equivalents and loudness ratings.

Figure 1.12 shows details of the procedures used to determine the *LR* of a system.

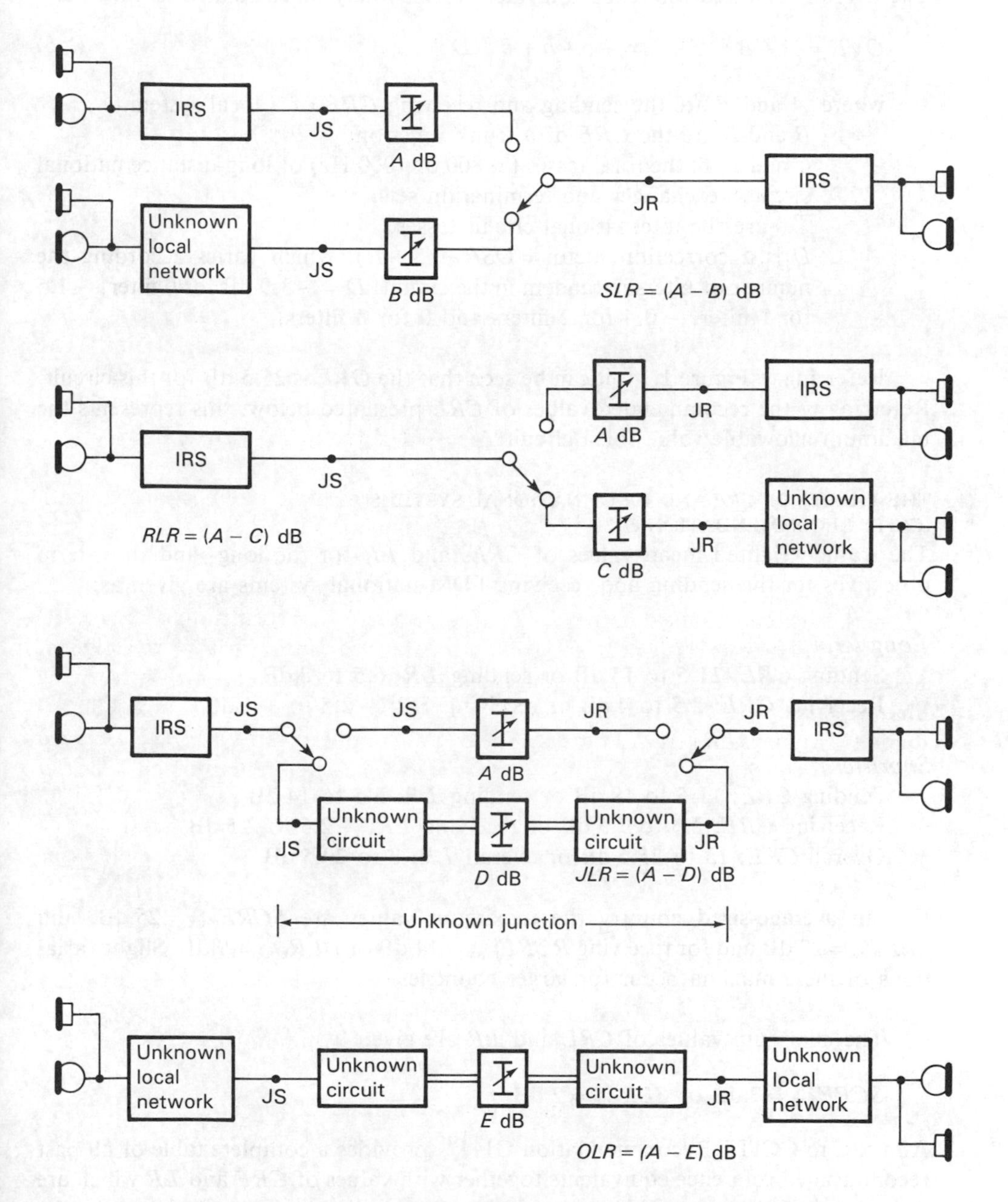

Note: Intermediate reference system (IRS) is used for defining loudness ratings (CCIR Recommendation P48, Reference 46).

Figure 1.12 Connections and systems considered for the definition of loudness ratings (Courtesy of CCITT, Recommendation G111, Reference 45)

PLANNING METHODS BASED ON *CRE*

The overall corrected reference equivalent (*OVE*) may be calculated as follows:

$$OVE = A + B + C + \Sigma x_i + c + b + d + D \tag{1.6}$$

where A and d are the sending and receiving *CRE* of a local system.
B and b are the *CRE* of a trunk junction.
C and c are the total losses (at 800 or 1000 Hz) of long-distance national circuits, exchanges and termination sets.
Σx_i are the international circuit losses.
D is a correction factor $= OSR - (S + R)$, which varies according the number of filters in tandem in the circuit. $D = -3.9$ dB for 0 filter, -1.5 for 1 filter, -0.4 for 2 filters and 0 for 3 filters.

Referring to Figure 1.14, it can be seen that the *OVE* is 25.5 dB for this circuit. Referring to the recommended values of *CRE* presented below, this represents the maximum allowable value for a circuit.

THE VALUES OF *CRE* AND *LR* OF NATIONAL SYSTEMS (CCITT RECOMMENDATION G121)[45]

The traffic-weighted mean values of *CRE*s and *LR* for the long- and short-term objectives for the sending and receiving FDM national systems are given as:

Long-term

Sending *CRE*: 11.5 to 13 dB or sending *LR*: 6.5 to 8 dB
Receiving *CRE*: 2.5 to 4 dB or receiving *LR*: -2.5 to -1 dB

Short-term

Sending *CRE*: 11.5 to 19 dB or sending *LR*: 6.5 to 14 dB
Receiving *CRE*: 2.5 to 7.5 dB or receiving *LR*: -2.5 to 2.5 dB
Overall *CRE*: 13 to 25.5 dB or overall *LR*: 8 to 20.5 dB

For an average-sized country the maximum values are $SCRE_{max} = 25$ dB, and $SLR_{max} = 20$ dB and for receiving $RCRE_{max} = 14$ dB or $RLR_{max} = 9$ dB. Slight variations of these maxima occur for larger countries.

The minimum values of *CRE* and *LR* are given by

$$SCRE_{min} = 7 \text{ dB or } SLR_{min} = 2 \text{ dB}$$

Annex C to CCITT Recommendation G111[45] provides a complete table of all past recommended reference equivalents together with values of *CRE* and *LR* which are the subject of the present recommendations.

1.3.6.2 Virtual analog switching points

The sending and receiving *CRE* or *LR* of the national systems are calculated to, and from, the virtual analog switching points of the international system, that is points

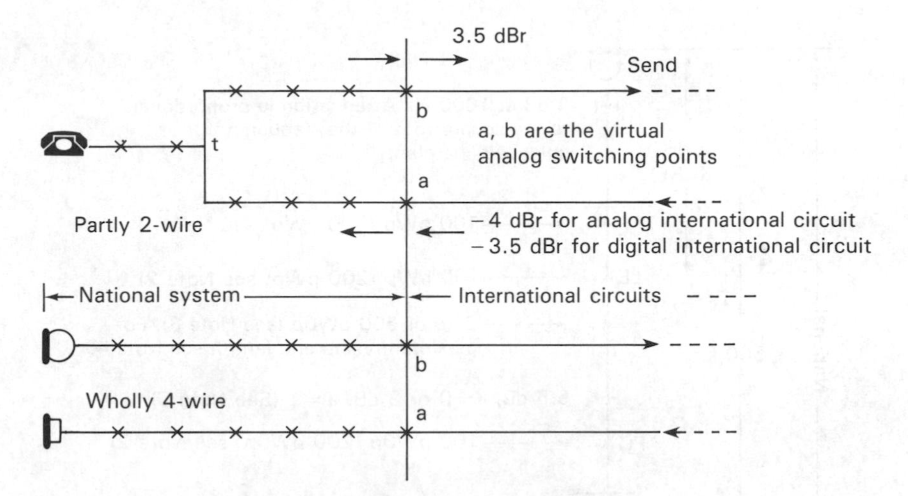

Figure 1.13 Virtual analog switching points

'b' and 'a' of Figure 1.13. The virtual analog switching points divide the international telephone connection into three parts. The first and third parts are the national circuit to and from the originating and terminating international switching centers (ISCs). These circuits may comprise one or more four-wire national trunk circuits with four-wire interconnection, as well as circuits with two-wire connection to the local exchanges and the subscriber sets with their subscribers lines. The second part is the international four-wire chain connecting two national chains. The virtual analog switching points are not necessarily the same points at which the international or national circuits terminate physically in the switching equipment.

1.3.6.3 Analog and mixed analog/digital hypothetical reference connection (HRX)

As stated in the CCITT Recommendation G103,[45] a hypothetical reference connection (HRX) is a model which describes the transmission impairments contributed by circuits and exchanges. The uses of the model are to examine the effect on transmission quality, brought about by varying the routing structure, the noise budget and the transmission losses in a national network, and also to test national planning rules to ensure compliance with any CCITT recommended criteria on statistical impairment. This connection however, is not intended to be used for the design of transmission systems. Figure 1.14 shows an HRX for an international connection of moderate length (say not longer than 2000 km) comprising the most frequent number of international and national circuits. This connection is given for an *analog* circuit, but is presented here because under certain conditions it is equally applicable to a *mixed analog/digital* system. One condition is that account must be taken of all possible unintegrated digital processes that might be present and which have an important effect on the overall transmission performance – in particular, quantizing distortion (Recommendation G113, Reference 45) as described in Section 1.4, and transmission delay. A second consideration is that in upgrading an analog

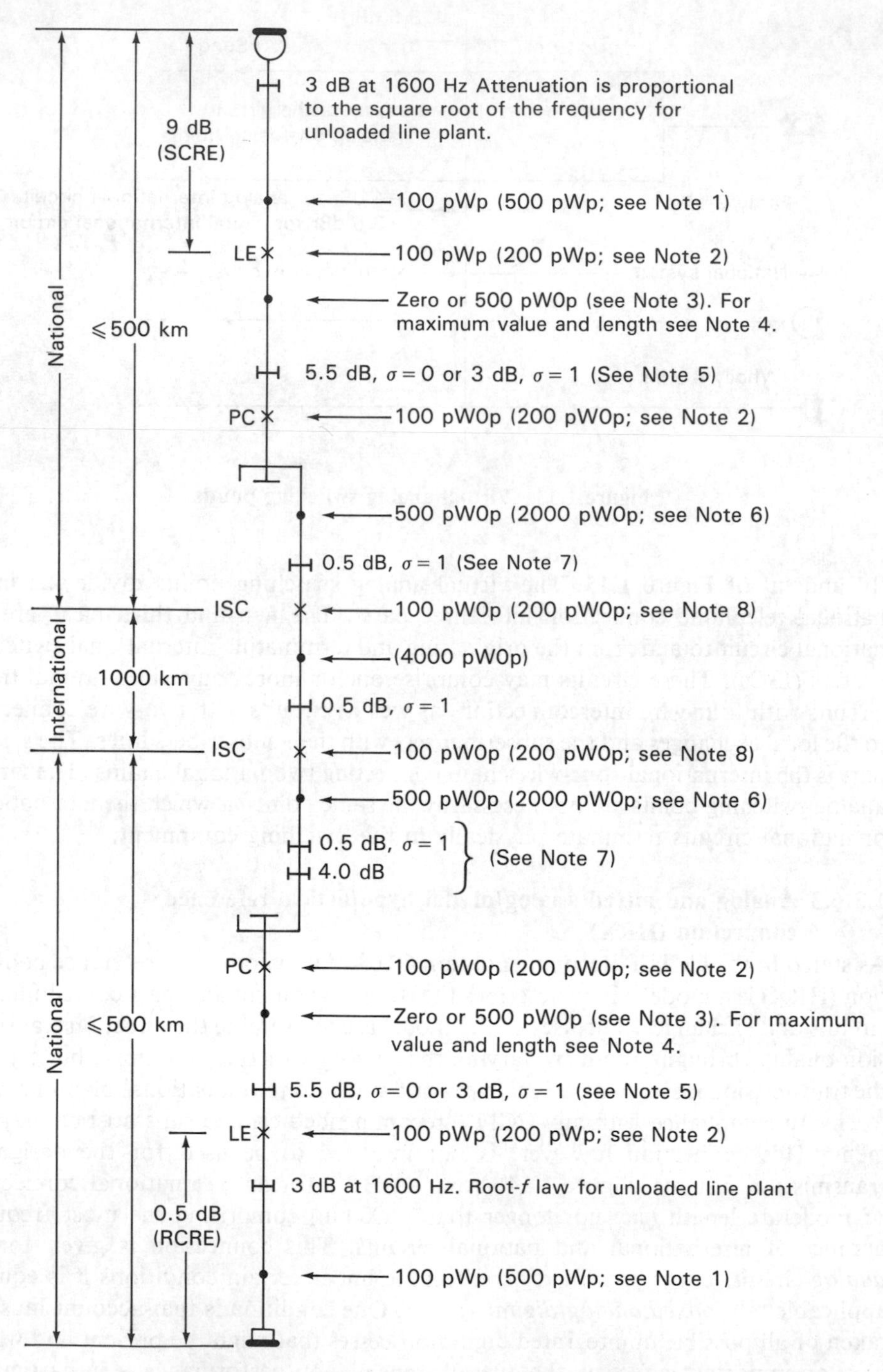

Figure 1.14 An example of an international connection of moderate length with only one international circuit (Courtesy of CCITT, Recommendation 103, Reference 45)

network so that there is a mixture of digital equipment and circuits, two approaches may be used. One is to maximize the number of digital processes, and the other to take more account of an evolving network. For the evolving network approach, islands of digital connectivity are established such that the number of independent digital processes in each connection is approximately one-half of the maximum, and all exchanges are assumed to be digital. Also the connections from the primary switching center to the international switching centers and back to the primary center are digital.

1.3.6.4 Digital hypothetical reference connection

Figure 1.15 provides the hypothetical reference connection for a digital network. Its purpose is the same as that for the analog case. This connection is primarily based on telephony applications and is that connection for moderate length. CCITT Recommendation G104[45] contains two connections beside that shown in Figure 1.15. These are: the longest reference connection, and a typical international connection within the international switching center (CT1) area, between subscribers situated near an international switching center (CT3). The CCITT recognizes that

---

NOTES TO FIGURE 1.14

Note 1. The average value of 100 pWp, for subscriber line noise is considered to be typical and is used by at least one administration as an objective for maximum noise at the receiver.
Note 2. The value of 200 pW0p as the design objective for the maximum noise power in a national four-wire automatic exchange is taken from Recommendation G123[45] §3. The same value, i.e. an absolute noise power of 200 pWp, has provisionally been assumed for national two-wire exchanges. No assumption has been made concerning the position of any national zero relative level point.
Note 3. The noise power level may be taken as negligible if the circuit is provided on physical line plant. A mean value of 500 pW0p is appropriate if the circuit is provided on a FDM or TDM short-distance carrier system.
Note 4. For FDM or TDM short-distance carrier circuits not exceeding about 250 km, the maximum value of noise power may be taken to be 1000 pW0p. See Recommendation G123.[45]
Note 5. For circuits on physical line plant the CRE may be taken to have a nominal maximum value of 6 dB with a standard deviation $\sigma = 0$. This value was arrived at in the following way: Recommendation G121[45] gives a 97 percent limit on 25 dB sending corrected reference equivalent (SCRE) referred to a point of −3.5 dBr on the international circuit at the international switching center of junction ab (see Figure 1.13). Referring this to a zero relative level point at the input to the chain of national and international circuits (i.e. to the primary center) gives 21.5 dB. The CCITT manual 'Transmission planning of switched telephone networks' (ITU, 1976) indicates that a 12 dB sending corrected reference equivalent, i.e. 15.5 dB *SCRE*, is typical for maximum local lines, thus leaving 6 dB for the circuit from the local exchange to the primary center switching losses being included.
Note 6. The maximum value of 2000 pW0p provides for a circuit length of about 500 km with some margin.
Note 7. Both countries are assumed to have the 2 + 0.5 + 0.5 + 0.5 dB type of plan. The nominal value of the 4 dB pad in the receiving direction at the switching center includes the loss of the terminating unit.
Note 8. The value of 200 pW0p as the design objective for the maximum noise power in an international exchange is that recommended by Recommendation Q45 (CCITT Recommendation 'Transmission characteristics of an international exchange', Vol. VI).

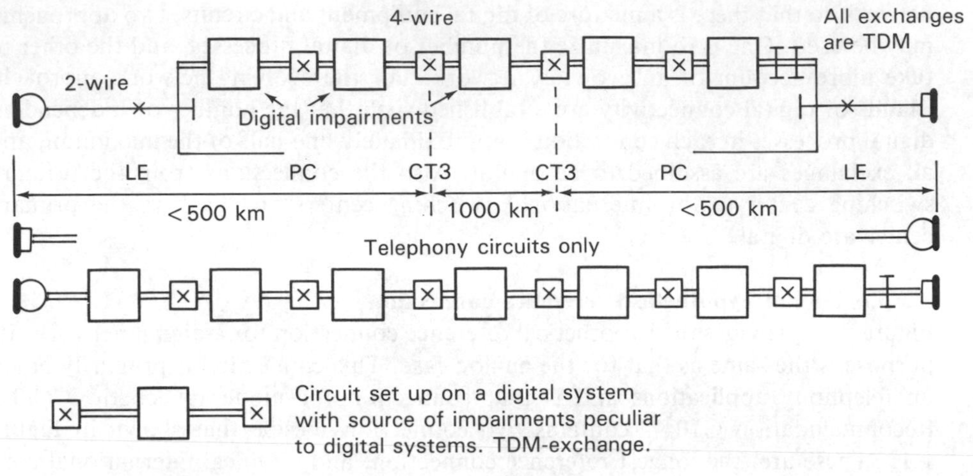

Figure 1.15 Typical international connection for telephony (64 kbit/s) digital transmission (Courtesy of CCITT, Recommendation G104, Reference 45)

the actual performance of a connection will only rarely be identical with the performance estimated by the calculations, due to the variations in the working conditions of some items of equipment and that it is unlikely that every item of equipment is experiencing the most adverse combination of specified conditions.

1.4 TRANSMISSION IMPAIRMENTS TO A DIGITAL NETWORK

The CCITT Recommendation G113[45] states that the network performance objective, for the one-minute mean values of signal-independent circuit noise power on a complete worldwide telephone connection that comprises for international circuits, should not exceed − 43 dBm0p at the output when referred to the input of the first circuit.

During the transition phase in which digital equipment and circuits are being integrated into an analog network, it is to be expected that there will be an appreciable accumulation of transmission impairments. Those which affect the digital network portion will be briefly discussed below.

1.4.1 Quantization distortion

This form of distortion will be dealt with in greater depth in Chapter 2, when the subject of pulse code modulation has been considered. Quantization distortion is pertinent only to digital systems and occurs when an analog signal has been sampled and encoded into one of a finite set of values, and then decoded at the receive end to obtain the recovered analog signal. During the encoding/decoding processes random errors in the encoding levels of the analog signal are produced by random noise present at the low-level signal points along the channel. The effect of this noise is to produce random jumps in the quantized signal causing the recovered analog signal to deviate significantly from the original analog input signal. The difference between the original analog signal and that which is recovered after quantizing is called quantization distortion or quantizing noise. To assist in network planning, weighting units of impairment quantization distortion units (qdu) have been assigned to any nonstandard analog/digital conversion process, transmultiplex pairs and processes introducing digital loss. The planning rule provisionally accepted is to allow 14 units of impairment for an overall international connection, with up to 5 units for each of the national extensions and 4 units for the international chain. Such a rule permits the tandeming of 14 unintegrated eight-bit processes, where 1 unit of impairment is assigned to an eight-bit A- or μ-law codec pair (see section 2.2.1.2) for the discussion on the A-law or μ-law digital encoding algorithms. A tentative value of 3 qdu has been assigned to the non-CCITT standard seven-bit system used for speech. These and other processes are listed below in Table 1.2 against the number of impairment units (qdu).

Table 1.2 Quantization distortion units (qdu) for different processes (Courtesy of CCITT, Recommendation G113, Reference 45)

Process	Number of impairment units, qdu
One eight-bit A-law or μ-law PCM codec-pair	1
One seven-bit A-law or μ-law PCM codec-pair	3
One 32 kbit/s adaptive predictor (ADPCM): PCM–ADPCM–PCM tandem	3.5
Transmultiplexer pair based on PCM A- or μ-law	1
One digital pad realized by manipulating 8-bit PCM code words	0.7
A/μ-law or μ/A-law conversion	0.5
$A/\mu/A$ law tandem conversion	0.5
$\mu/A/\mu$ law tandem conversion	0.25
PCM–ADPCM–PCM tandem conversion	2.5
8-7-8 bit transcoding (A- or μ-law)	3

1.4.2 Loudness loss

If the loudness in a network is increased from a preferred range, the listening effort is increased and customer satisfaction decreases. If it is too little, it becomes annoying to the user, and causes distraction. The method used to measure and express the loudness loss of a telephone connection has been the reference equivalent method. Table 1.3 gives values of the overall reference equivalent in a virtually noiseless environment shown against subjective opinions of how good the telephone connection was.

Table 1.3 Subjective results on the quality of a telephone conversation for different OREs (Courtesy of CCITT, Recommendation P11, Reference 46)

Planning value of the overall reference equivalent, dB	Representative opinion results	
	Percent 'good plus excellent'	Percent 'poor plus bad'
5 to 15	>90	<1
20	80	2
25	65	5
30	40	15

As discussed in Section 1.3.6.1, this method has been replaced by the corrected reference equivalent. Recommended values of loudness loss are presently under study by the CCITT, and in accordance with those figures already given in Section 1.3.6.1, are listed in Table 1.4, together with the previously recommended reference equivalents.

Note that equation 1.4 provides the value of *CRE* from the specific reference equivalent (q).

Table 1.4 Reference equivalents, corrected reference equivalent, and loudness loss (LR), recommended values (Courtesy of CCITT, Recommendation P11, Reference 46)

	RE(q) previously recommended	*CRE* currently recommended	Loudness rating or loudness loss (LR)
Preferred range for a connection (G111, #3.2)	Minimum: 6	4	2
	Preferred: 9	7 to 11	about 5
	Maximum: 18	16	11
Maximum value for the national system (G121, #2.1)	Send: 21	25	20
	Receive 12	14	9
Minimum value for the national sending system (G121, #3)	6	7	2
TRAFFIC WEIGHTED MEAN VALUES			
Long-term objectives			
Connection (G111, #3.2)	Minimum: 13	13	8
	Maximum 18	16	11
National system send (G121, #1)	Minimum: 10	11.5	6.5
	Maximum: 13	13	8
National system receive (G121, #1)	Minimum: 2.5	2.5	−2.5
	Maximum: 4.5	4	−1
Short-term objectives			
Connection (G111, #3.2)	Maximum: 23	25.5	20.5
National system send (G121, #1)	Maximum: 16	19	14
National system receive (G121, #1)	Maximum: 6.5	7.5	2.5

The G numbers in the table refer to CCITT recommendations, and the # numbers to sections in them.

1.4.3 Circuit noise

The forms of noise which affect an analog connection, but have a minimum effect on a digital connection are:

White noise
Intermodulation noise
Single-frequency interfering tones
Hum

The form of noise which may effect the bit error is impulse noise.

The CCITT Recommendation G143[45] stipulates circuit noise limits for the transmission of data at bit rates less than 1200 bits/s as:

Leased circuits without impulse noise interference = −40 dBm0p
Switched connections without compandors, except on international circuits = −36 dBm0p

Circuit noise is subject to CCITT and CCIR recommendations which provide the noise objectives for hypothetical reference circuits of varying length. These references are:

CCITT: G123, 152, 153[45] and 212, 222 and 226[48]
CCIR: 391 to 397[49]

The CCITT Recommendation G151[45], states that for single-frequency interfering tones in all modern international and national extension circuits, the interference level in a telephone circuit should not be higher than −73 dBm0p (provisional). Psophometric weighting should only be accounted for when the frequency of the interference is well defined.

INTERFERENCE AT HARMONICS FROM THE MAINS AND OTHER LOW FREQUENCIES

For all modern international and national extension circuits the hum or interference at harmonics from the mains and other low frequencies should not exceed a level of −45 dBm0, i.e. the maximum level of the unwanted side components should be 45 dB or more below the sine-wave test-tone whose level is at zero dBm0.

1.4.4 Attenuation distortion

Attenuation distortion is also referred to as 'frequency distortion'. It is the selective attenuation of signal levels at some frequencies or some bands of frequencies over the whole of the spectrum in use. If all the signal levels at frequencies in the pass-band being considered suffered the same attenuation or gain, this form of distortion would not be apparent. If frequency distortion appears within a voice channel and is excessive, the effect is noticeable. Where the low frequencies in the band are attenuated the resulting speech sounds 'tinny' and faint; whereas if the higher frequencies in the band are attenuated, the speech sounds 'drummy' and articulation of the speech is affected. Frequency distortion is characterized by the transmission loss or gain relative to the transmission loss of 800 Hz or 1000 Hz, at frequencies other than 800 Hz or 1000 Hz. Thus this form of distortion includes the low- and high-frequency roll-offs which determine the effective bandwidth of a telephone connection, as well as in-band variations in loss or gain as a function of frequency. The loudness loss and articulation (higher frequencies) are a function of the attenuation distortion. The network performance objectives for the variation in frequency of transmission loss in a terminal condition of a worldwide four-wire chain of 12 circuits comprising international and national extension circuits, each being routed over a single group link, are as shown in Figure 1.16. It is expected that the attenuation distortion in this case will not exceed 9 dB between 400 and 3000 Hz when the level of 800 Hz is taken as the reference. For a chain of six circuits, again international and national extension circuits in tandem, where each circuit is equipped with one pair of channel translating equipments, the attenuation distortion is expected to be usually less than 9 dB between 300 and 3400 Hz. For a single pair of channel terminal equipments in which the band is a little wider than 300 to 3400 Hz, the attenuation distortion at 300 and 3400 Hz should never exceed 3 dB and in a large number of equipments should not average more than 1.7 dB (CCITT Recommendation G151, #1.[45]

The effect of attenuation distortion on listening and conversation becomes increasingly more noticeable as the overall reference equivalent (ORE) of a connection decreases, and becomes even more marked as the circuit noise increases. CCITT Recommendation P11, Annex B[46] provides additional information on the effects of this form of distortion.

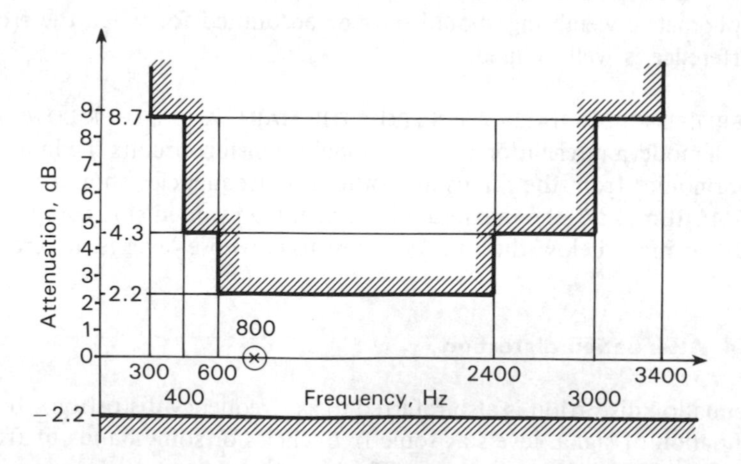

Figure 1.16 Permissible attenuation variation with respect to its value measured at 800 Hz (objective for worldwide 4-wire chain of 12 circuits in terminal service)
(Courtesy of CCITT, Recommendation G132, Reference 45)

1.4.5 Group-delay distortion

When a single frequency signal passes through a filter it does not pass through it at the speed of light, but is slowed down by the characteristics of the filter. Absolute delay may be considered as the delay that a signal at a reference frequency experiences in passing through a filter or channel. As the signal frequency is unchanged when passing through the filter, the slowing down may be represented by a phase delay. This delay changes however, as the signal frequency is changed from the center frequency of a filter to a frequency representing the edge of the filter. Usually the filter has a non-linear phase-to-frequency characteristic and is approximately parabolic in shape. If the delay versus frequency characteristic of the filter was linear a complex waveform representing the signal coming out of the filter would be a faithful replica of the waveform going in. In the non-linear phase-to-frequency case, the waveform coming out is distorted. Because a number of frequencies are involved when a complex waveform passes through the filter, the delay at any one of the frequencies in the band is known as the group delay at a particular frequency. *Group-delay distortion* is characterized by the group delay at other frequencies relative to the group delay at the frequency where the group delay is minimum. Another synonymous term used for group-delay distortion is *envelope-delay distortion* (EDD). This name is chosen because this type of distortion provides a measure of the time it takes to propagate a change in the modulation envelope of a signal. In a speech circuit, if the group-delay distortion is excessive at the upper and lower edges of the transmitted band, the effect is 'ringing' or 'blurred speech' respectively. When noise or attenuation distortion is not present, the effect of this distortion is conspicuous throughout the entire range of typical loudness loss values. The effect in a typical four-wire circuit chain is not serious due to the normally accompanying closely related attenuation distortion which masks the effect.

Group delay distortion significantly impairs digital transmission due to the limitation it imposes on the data bit rate because of the requirement that the duration of one information bit should be equal to or greater than the group-delay distortion over the band of interest. The network performance objectives of the group delay distortion as determined by the difference between the group delay at the lower and upper edges of the transmitted frequency band used and the minimum value of the group delay over this band is given by CCITT Recommendation G133[45] and is reproduced in Table 1.5.

Table 1.5 Group-delay distortion objectives

	Lower limit of frequency band, ms	Upper limit of frequency band, ms
International chain	30	15
Each of the national four-wire extensions	15	7.5
On the whole four-wire chain	60	30

1.4.6 Echo

Echo is the return of part of a transmitted signal after it has been transmitted. In a telephone system, talker echo is the return of the voice some time after it has spoken (typically more than about 30 ms), and which is of a high enough level to make it distinguishable from a normal sidetone. Whenever there is an impedance mismatch in a transmission system, some of the transmitted power will be reflected. The factors which may be considered to govern echo are:

The number of echo paths
The time taken by the echo signal to traverse these paths
The *CRE* of the echo path including the subscribers' lines
For a digital telephone: the accoustical coupling between the earpiece and the mouthpiece of the handset; and the electrical coupling in the flexible cord to the handset
The tolerance to echo exhibited by the subscribers

The curves shown in Figure 1.17 indicate the minimum value of nominal *CRE* in the echo path that must be introduced if no echo suppressor is to be fitted. The dashed curve is applicable to fully digital connections with analog subscriber lines as shown in Figure 1.18, and to fully digital conections with four-wire digital subscriber lines, as shown in the lower diagram of Figure 1.15. For the four-wire digital subscriber line, the echo path includes the acoustical path between the earpiece and the mouthpiece of the handset.

When a short and simple international circuit is used, the nominal transmission loss of this international circuit between the virtual analog switching points may be increased 0.5 dB for a route up to 500 km, 1.0 dB for lengths between 500 km and 1000 km, and 0.5 dB for every additional 500 km over 1000 km. This circuit may not form part of multicircuit connections unless the nominal transmission loss is restored to 0.5 dB. CCITT Recommendations G131, G122, and Supplement No. 2 all provide planning details on this subject. The examples given in CCITT Recommendation G131 permit the construction of analog, digital and mixed analog/digital connections.

As briefly described in Section 1.3.2.3, and more fully discussed in CCITT Recommendations G164 and G165, echo cancelers, which are voice-operated devices placed in the four-wire portion of a circuit, are used for reducing the echo by subtracting an estimated echo from the circuit echo. They may be classed according to whether they work into analog paths or into digital paths and according to how the canceling of the echo is done. They are generally used to reduce-near-end echo on the send path by subtracting an estimation of that echo from the near-end echo.

Another form which echo takes is called *listener echo*. It refers to a transmission condition where the main speech signal arrives at the listener, accompanied a little later by the same speech signal. The cause may be due to two-wire/four-wire hybrid inbalance which permits reflections at each of the four-wire ends of the circuit to occur. The main speech signal arriving at the near end is reflected back to the far end where it is again reflected, to return to the near end. The delay of the echo is

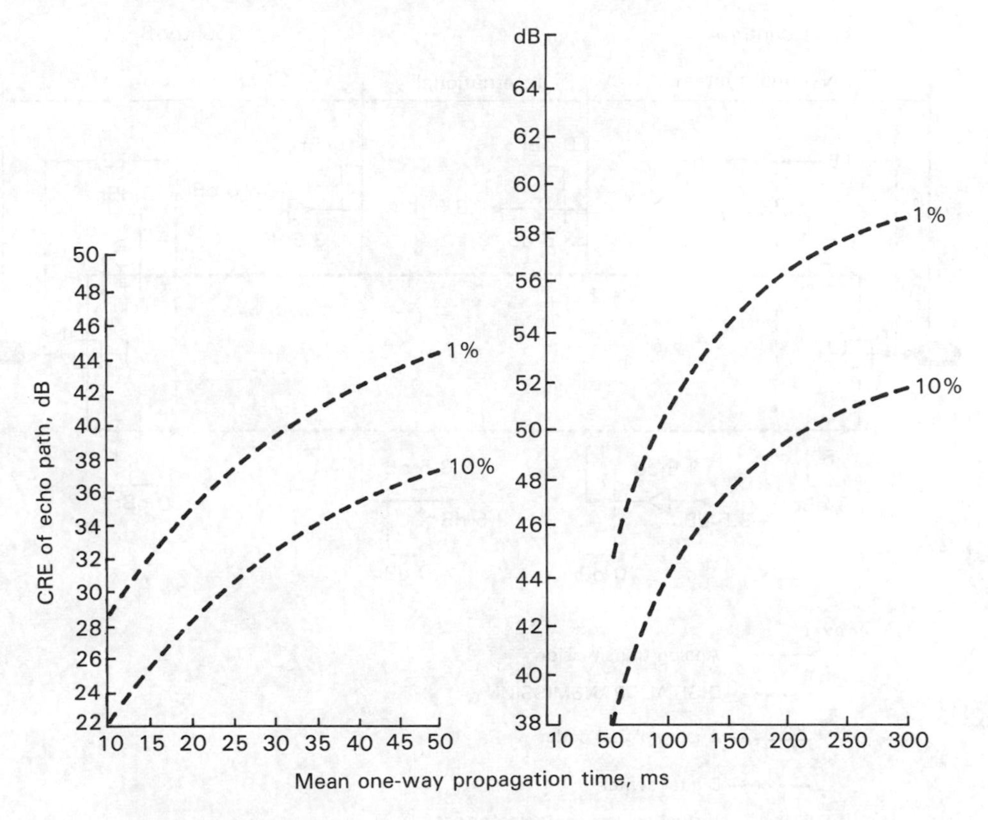

- - - Fully digital connection

Notes

1. The percentages refer to the probability of encountering objectionable echo.
2. The corrected reference equivalent (CRE) of the echo path is here defined as the sum of:

(a) The corrected reference equivalents in the two directions of transmission of the local telephone system of the talking subscriber (assumed to have minimum values of CRE);

(b) The corrected reference equivalents in the two directions of transmission of the chain of circuits between the two-wire end of the local telephone system of the talking subscriber and the two-wire terminals of the four-wire/two-wire terminating set (hybrid) at the listener's end; and

(c) The mean value of echo balance return loss at the listener's end.

Figure 1.17 Echo tolerance curves (Courtesy of CCITT, Recommendation G131, Reference 45)

determined by the time the signal takes to travel back and forth along the circuit, and the magnitude of the echo relative to the main speech signal depends on the two-way loss of the four-wire transmission path. For small delays there is a change in the spectral quality of the speech, but for longer delays it is more pronounced and causes considerable 'hollowness' in the speech. In addition to this it can impair the bit error rate (BER) of received voice-band data signals. As listener echo is characterized by the additional loss and delay in the listener echo path relative to

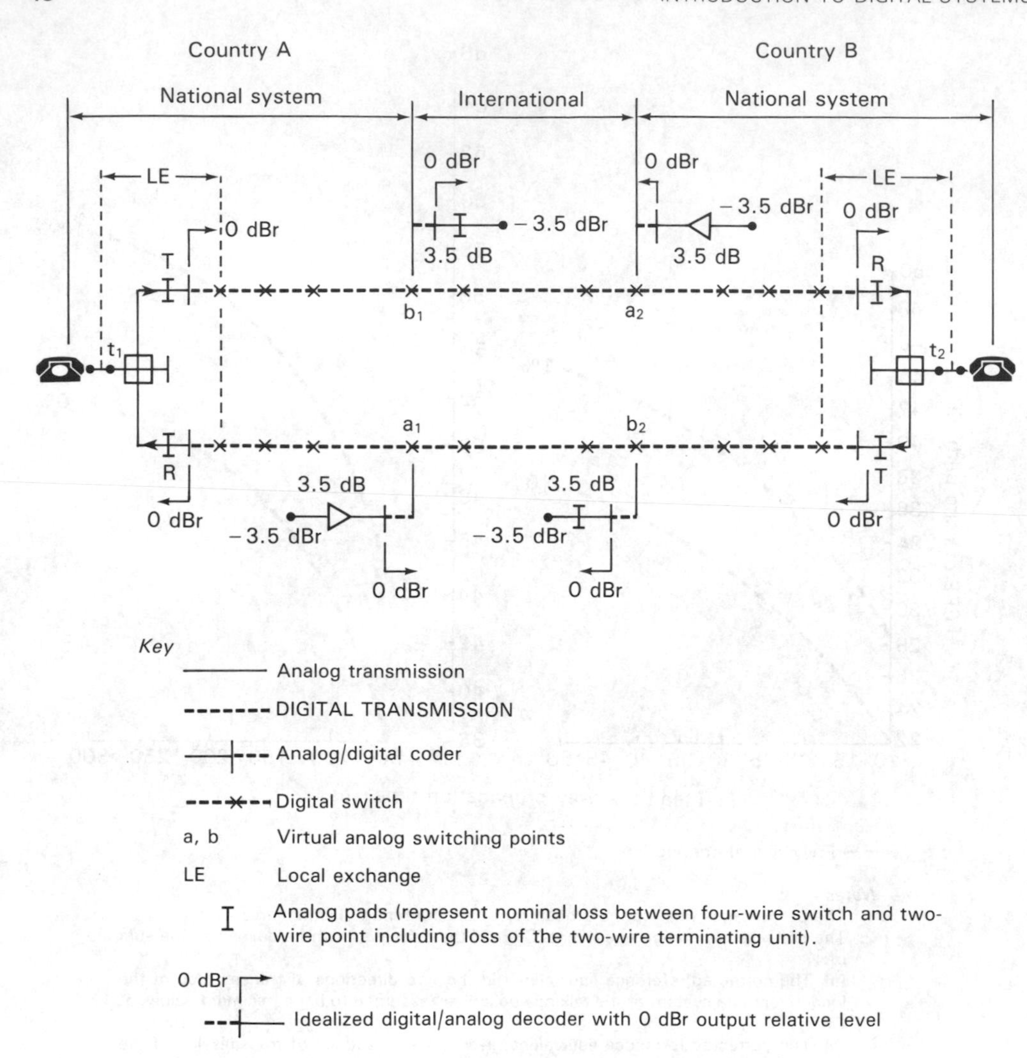

Figure 1.18 A fully digital connection with analog 2-wire subscribers' lines (Courtesy of CCITT, Recommendation G131, Reference 45)

that in the main speech path, the smaller the loss over the frequency band of interest, the greater the amplitude of the echo, and the closer the circuit approaches instability. As a result, this form of echo is sometimes referred to as *near singing distortion*. CCITT Recommendation G122[45], #5, considers this form of echo together with stability.

Balance return loss

The return loss of a hybrid or two-wire/four-wire bridge may be defined in terms of the impedance of the balancing network Z_b as seen from the four-wire side of the bridge itself, and the impedance of the two-wire line to the telephone Z_t, also as seen from the four-wire side of the bridge thus:

$$\text{Return loss} = 20 \log(Z_b + Z_t)/(Z_b - Z_t) \text{ dB} \tag{1.7}$$

Alternatively, return loss may be defined in terms of the reflection coefficient, which equals the reflected signal voltage divided by the incident signal voltage, thus:

$$\begin{aligned}\text{Return loss} &= 20 \log 1/(\text{reflection coefficient}) \\ &= 10 \log (\text{incident signal power}/\text{reflected signal power})\end{aligned} \tag{1.8}$$

The *balance return loss* with respect to stability is defined as the return loss as given by equation 1.7, in the frequency band 0–4 kHz. The *balance return loss* with respect to echo is defined as the return loss as given in equation 1.7 in the frequency band 300–3400 Hz.

1.4.7 Phase jitter[50–54]

Phase jitter on a digital signal may be defined as (CCITT Recommendation 823);[53] 'short term variations of the significant instants of a digital signal from their ideal position in time'. This may be interpreted as changes in the time positions of the zero crossings of the digital signal during the change of state of each of its bits. The zero crossing level is relative to some arbitrary zero reference level on a leading or trailing edge of a bit as its changes from a low to a high, or from a high to a low, state. The variations of the significant instants about the ideal time position implies that there is a peak-to-peak amplitude and a distribution of amplitudes together with a frequency or frequencies associated with these variations. This is partially true, for the process which causes the jitter is the phase or frequency modulation of the digital signal by some noise source during transmission. If the noise source is deterministic the rate of the variations will be deterministic; otherwise it will be random. The rate at which the variations occur is the modulation rate, and the peak-to-peak amplitude of the variations are a function of the deviation of the modulation, which often follows a Gaussian distribution. The jitter rates which are considered to be most significant occupy the frequency range from around 30–50 Hz to several kHz. The measure of phase jitter is the peak-to-peak amplitude expressed in degrees relative to one bit (180°) of the digital bit stream. The rate of the jitter is independent of the digital bit stream rate, and is quoted as a frequency if it is determinable; otherwise it is stated as nondeterministic fast, medium or slow jitter. For example, the jitter may have been measured as one-quarter of a bit, fast jitter, and so the result would be expressed as 45° peak-peak, at a fast rate. Alternatively, the jitter may be measured in terms of the proportion of a unit interval (UI), where a unit interval is the time duration of a bit. In the above example, the jitter would be expressed as 0.25 UI, at a fast rate. For a jittery data stream whose bit pattern cycle is small enough to be displayed on an oscilloscope the phase jitter may be determined as follows. The jittery signal is entered into a selected input channel of an oscilloscope, and a non-jittery square wave whose half-cycle time duration is the same as that of a single bit of the input data stream is used to trigger the oscilloscope. The display shows a stationary picture of the bit pattern with a variation on the leading and trailing edges of the pattern bits. The width of the variation can be directly read as a proportion of the bit duration, and the jitter determined, as can be the rate of amplitude variation, provided that the rate is not too fast for the eye to see. For the more normal case, where the bit pattern is not cyclic, is small and may be coded, the use of eye diagrams must be used to determine the jitter. This is discussed in Chapter 5, Section 5.2.5.

The peak-to-peak amplitude of the jitter, if not controlled, can cause the following problems in a digital network:

1. As jitter will cause the time that a zero crossing detection occurs in a digital regenerator, or other detection equipment, to vary, there will be an increase in the error rate, and consequently an increase in the bit error rate (BER).
2. The occurrence of uncontrolled slips (see below for an explanation) in the equipment at an interconnection interface due to its input tolerance to jitter not being compatible with jitter generated in preceding equipment.
3. The corrupting of data whose initial and final forms are analog. Such analog forms are speech, televised signals, sounds, etc. The end result of this degradation in the digitally encoded analog information is to produce distortion similar to that produced by quantization distortion. Encoded PCM speech is fairly tolerant to this type of effect, but digitally encoded television is much more sensitive.

CCITT Recommendation G823 contains information on the philosophy of jitter control adopted for digital networks based on the 2048 kbit/s hierarchy (see Figure 3.11, Chapter 3). Jitter limits have been derived for all the hierarchical base rates from 64 kbit/s up to 140 Mbit/s. This philosophy is to provide a maximum network limit which should never be exceeded anywhere in the network, national or international. To ensure compatibility of equipment and thus reduce uncontrolled slips, all digital equipments have to be able to accommodate a test condition where sinusoidal

Table 1.6 Maximum permissible jitter at a hierarchical interface

Digit rate, kbit/s	Network limit, unit intervals		Measurement filter bandwidth; band-pass filter having lower cut-off frequency f_1 or f_3 and upper cut-off f_4		
	Measured f_1–f_4	Measured f_3–f_4	f_1, Hz	f_3, kHz	f_4, kHz
64*	0.25	0.05	20	3	20
2048	1.5	0.2	20	18 (700 Hz)	100
8448	1.5	0.2	20	3 (80 KHz)	400
34368	1.5	0.15	100	10	800
139 264	1.5	0.075	200	10	3500

Notes
* Provisional limits for the codirectional interface only.
( ) Only allowed within national networks and used only by some administrations.

jitter is applied at the input port at a value which ensures compatibility when connected to any equipment output port. Table 1.6 contains the network jitter limits.

Slips

Slips may be classified into those which are controlled and those which are uncontrolled.

CONTROLLED SLIPS (CCITT RECOMMENDATION G822)[53]

Digital exchanges in a network require to be *plesiochronous*, that is the signals entering and leaving the exchange, and the exchange clock should all have the same frequency as closely as possible (1 part in 10^{11}). If part of the network loses its frequency synchronization with respect to the other equipments in the network, signals arriving at the faulty exchange are no longer arriving at the same rate as expected, although the exchange can, by means of the frame alignment signal associated with each digital stream, still identify which bits belong to which channel. Under these conditions, the exchange receives either too many or too few bits, and eventually implements a slip in a carefully controlled manner. For each 64 kbit/s channel, this controlled slip arranges for the repetition or deletion of one 8-bit byte, which in the case of PCM encoded speech in only one sample, and has a negligible effect. For voice-band data signals, especially those associated with high-bit-rate modems (e.g. 9.6 kbit/s), the slip can have a more significant effect.

UNCONTROLLED SLIPS

These slips are unforeseen events, where a bit or a number of bits are repeated or deleted in the interfacing equipment of the exchange. The slips can occur in the network as a result of excessive jitter or wander (see below for explanation), or by protection switching of equipment. The effect is often to lose frame alignment, which requires a long recovery time lasting several milliseconds whilst the demultiplexers re-align. No CCITT recommendations on uncontrolled slips exist at the present time.

WANDER (CCITT RECOMMENDATION 823)[53]

Wander is the very slow rate of variation of jitter. It may be caused as a result of temperature variation on the transmission media, producing propagation delay variations. Due to the long time-constants which would be required in the design of phase-locked-loops incorporated in digital regenerators, it is not practicable to eliminate it using digital regenerators. Digital exchanges incorporate additional buffer storage capacity at their inputs to absorb it.

SOURCES OF JITTER IN A DIGITAL NETWORK

We summarize below the main sources of jitter in a digital network:

Digital regenerators. Jitter from a digital regenerator shows a dependence on the pattern of the digital bit stream. The longer the bit stream continues without a change of state, the higher will be the jitter. This is because the timing information for the regenerator is derived from the input data stream. Incoming noise into the regenerator, and self-generated noise become predominant during the interval between the provision of timing information. The difference between the jitter on incoming data and that produced on the timing signal is jitter which must be restricted to an acceptable level. In addition to this, a second- or third-order phase-locked-loop, which is usually the heart of the regenerator, will amplify low-frequency jitter, or jitter which is of a frequency less than 1.414 times the natural frequency of the loop. The problem is the amplification or, at the best, the accumulation of this jitter as the digital signal passes from regenerator to regenerator.

Tank circuit mistuning. Due to the very sharp changes in the phase-to-frequency characteristic of high-Q tuned circuits, if the tuned circuit is slightly off frequency due to misalignment or ageing effects, the carrier will experience a phase shift. If the carrier is modulated, then, due to the difference in phase changes of the sidebands, phase modulation of the carrier results, and the outcome is jitter.

Justification operations. Justification operations normally form part of the multiplex process. Justification is basically that of the insertion of additional time slots into the incoming signal to a multiplex bank, in order to raise its net digit rate to the value employed in the synchronous part of the multiplex equipment. At the receiving terminal the added time slots are removed, with the result that the signal contains gaps which have a duration of one time slot. The retiming of this signal is achieved by passing it into a buffer store controlled by the operation of a local clock whose timing information is derived from the incoming signal. The result of this operation due to the imperfections in the processing is to produce jitter on the output signal.

Other sources. Jitter also can be produced by the threshold misalignment of zero crossing detectors, or the result of inter-symbol interference on the detector, as well as variations in the input signal's waveshape.

METHODS OF MINIMIZING JITTER

Either measures must be taken to prevent the generation of jitter or jitter already present must be reduced. To assist in preventing the generation of jitter in digital regenerators, digital scramblers can be used. These effectively randomize the signal, and reduce the sequence of long streams of marks or spaces in the digital bit stream (CCITT Recommendation V27, Appendix 1, Reference 34). To reduce phase jitter already present, a *dejitterizer* may be used. The operation of the dejitterizer has been briefly explained above in the description of retiming of an irregular signal under 'Justification operations'.

1.4.8 Crosstalk

Crosstalk has been briefly discussed in Section 1.3.2.4, and is considered further in this Section. Three types of crosstalk exist. These are *unintelligible crosstalk*, *babble*, and *intelligible crosstalk*.

Unintelligible crosstalk
This occurs when the voice band from the interfering circuit is translated in frequency, or appears in the disturbed circuits in an inverted form.

Babble
This is crosstalk from a number of different sources, which are either intelligible or unintelligible. The noise appears as if many people are talking due to the number of interfering sources. It is not considered as intelligible and usually occurs during busy traffic periods.

Intelligible crosstalk
This type of crosstalk occurs in the same frequency band as the disturbed circuit, but lower in amplitude than that of the wanted signal. The coupled signal is audible and intelligible to one or both of the participants on the second telephone connection. The major disturbance of this type of crosstalk is the loss in privacy which occurs. Design objectives for the various apparatus in telephone connections are such as to keep the probability of intelligible crosstalk occurrence below one percent in connections where interfering and interfered-with parties are unlikely to know each other and unlikely to suffer the same coupling again. A more stringent objective of 0.1 percent is typical for use in local equipment such as subscriber lines where the two parties may be neighbours. CCITT Recommendation P16[46] provides details of the subjective effects of direct crosstalk, which include the thresholds of audibility and intelligibility. The main factors that are considered to influence the intelligibility of vocal crosstalk are:

1. The quality of transmission of the telephone apparatus as decided by the sending and receiving corrected reference equivalents, and the reference equivalent of side tones when room noise is present.
2. Circuit noise on the connection of the disturbed call. If the level of line noise is assumed to be small, unrealistic demands may be placed on the required crosstalk attenuation; this difficulty is balanced by the assumption that the analog circuits and exchanges contribute 4 pW0p/km. If this is not the case the incidence of overhearing may be unacceptably high, particularly when the analog network is lightly loaded so that noise power levels are low. Usually a compromise between these two extremes is found, and supported by measurements of noise power levels in an established analog network during the light and busy hour periods. As mentioned previously busy hour and noise loading have no meaning in a purely digital network, and so the level of circuit noise in this case is unaffected.
3. Room noise which affects the hearing directly and indirectly by side tones. For the equipment design objective, this is disregarded, for it can be negligible in the most unfavourable case.
4. Conversation on the disturbed connection may always be in progress, and during pauses of speech, crosstalk may be heard.
5. Microphone noise may reduce the effect of crosstalk due to the side-tone noise produced. Usually for equipment design objectives, good quality microphones are assumed.

6. Crosstalk coupling is generally a function of frequency. The corrected reference equivalent of the crosstalk transmission path (*CRE*) can be divided into the sending *CRE* of the subscribers set causing the disturbance, the receiving CRE of the subscriber's set subject to disturbance, and the transmission loss of the crosstalk transmission path. The corrected reference equivalent of the crosstalk coupling may be taken to be the attenuation measured or calculated at a frequency of 1100 Hz (CCITT Recommendation G134, Reference 45)

To assist in the calculation of the probability of encountering intelligible crosstalk (CCITT Recommendation P16, Annex A, Reference 46), a hypothetical reference connection has been devised in CCITT Recommendation G105.[45]

CCITT Recommendation G151,[45] provides the circuit performance objective for the near-end or far-end crosstalk ratio for linear intelligible crosstalk only. This objective is that either of the two crosstalk ratios, measured at audio frequency at trunk exchanges between two complete circuits in terminal service position, should not be less than 65 dB, except when a minimum noise level of at least 4000 pW0p is always present in a system, such as in satellite systems. In this case the crosstalk ratio of 58 dB is acceptable. For cases involving the go and return channels of a four-wire circuit, the near-end crosstalk ratio for an ordinary telephone circuit should be at least 43 dB.

Methods for measuring crosstalk[55]

1. Single-frequency measurements at 800 or 1000 Hz.
2. At several frequencies (e.g. at 500, 1000 and 2000 Hz), the results being averaged on a current or voltage basis.
3. Measurements made using a uniform spectrum of random noise in which the disturbed circuit is terminated at its sending end, and the disturbing channel is loaded with the noise, the power applied to the channel being 0 dBm0, so as to avoid the influence of the channel limiter. Using a psophometer, the noise produced in the disturbed channed is then compared with the signal applied to the disturbing channel and the result is expressed as a crosstalk power ratio. The value obtained should be at least 60 dB
4. Voice/ear tests, in which speech is used as the distributing source and the crosstalk is subjectively measured by listening and comparing its level with a reference source

1.5 DIGITAL RADIO NETWORKS

The term *digital radio* is used to cover any of the radio modulation schemes used which alter, by discrete changes, the radio carrier characteristics. These discrete changes in the carrier are made in such a way that the modulated signal can assume only a finite number of unique changes in state.

Referring to Figure 1.1, the digital radio is one of the various types of transmission equipment which may be used in the blocks labeled 'Digital mod' and 'Digital demod'. These two blocks represent the radio transmitter and associated equipments, and the radio receiver and associated equipments respectively. The communications channel for digital radio communications is obviously the air. Other forms which communication channels may take are coaxial cable, metallic-pair cable, and optical fiber. Usually digital radio networks are those which connect together exchanges as a part of a national or private trunk network. Other uses are to connect 'spur' links, originating from various information gathering centres, to a main 'backbone' or broadband route, which carries the gathered information to one or more main processing centres. The 2 GHz band digital radios have found widespread use as feeders or 'spurs' for the more heavily loaded 6 and 11 GHz digital radios. Both 6 and 11 GHz radios are used for short-haul applications up to 1000 km or so. The 11 GHz digital radio was introduced because of the additional bandwidth it can provide compared with the 6 GHz radio, as well as the comparative lack of congestion in this band. However, the 6 GHz systems are better suited to longer hop lengths (50 km or so), thus reducing the number of repeaters required. Before discussing the broadband bearer system networks any further, it is worth while to consider another form that digital radio networks take, which is becoming more common.

1.5.1 Point-to-multipoint digital radio[56]

This type of digital radio network is becoming widespread in some suburban and rural areas. The network is configured as shown in Figure 1.19. The system comprises a central station which contains equipment for say 400 lines. The central station, which transmits on an omnidirectional antenna, serves a number of surrounding outstations that are all located within a single hop range from this central station. Where the distance of an outstation or outstations exceeds that required for a single hop, a repeater station is used, and this repeater station then distributes to the various outstations. The outstation radio equipment may be either built for pole mounting, etc., in an outside environment, or be housed in special huts built for the purpose. Each outstation may contain equipment for 15 or more extensions. For higher densities, equipment similar to that of each outstation may be added, and equipped with the channel equipment as required. The repeaters may be used in tandem to extend the range or the area to be served, or used as the first point in a 'spur' off of an existing digital trunk route. Equipment has been designed to operate in the 1.5 GHz, 1.8 GHz and 2.4 GHz bands using a single carrier for the complete system, with 64 kbit/s PCM voice trunks, and/or 2.4 kbit/s low-speed data trunks which are fully available to all extensions in the system. The ratio of trunks to extensions depend upon the grade of service required. The means of communicating is by using time-division-multiple-access techniques. The central station transmits to all outstations in a continuous time-division-multiplexing (TDM) mode. Each outstation, which is connected to the system, transmits to the central station one or more RF bursts, which are synchronized by the central station so that each station occupies an assigned nonoverlapping time slot in a time-division-multiple-access (TDMA) frame. The central station periodically interrogates each extension to

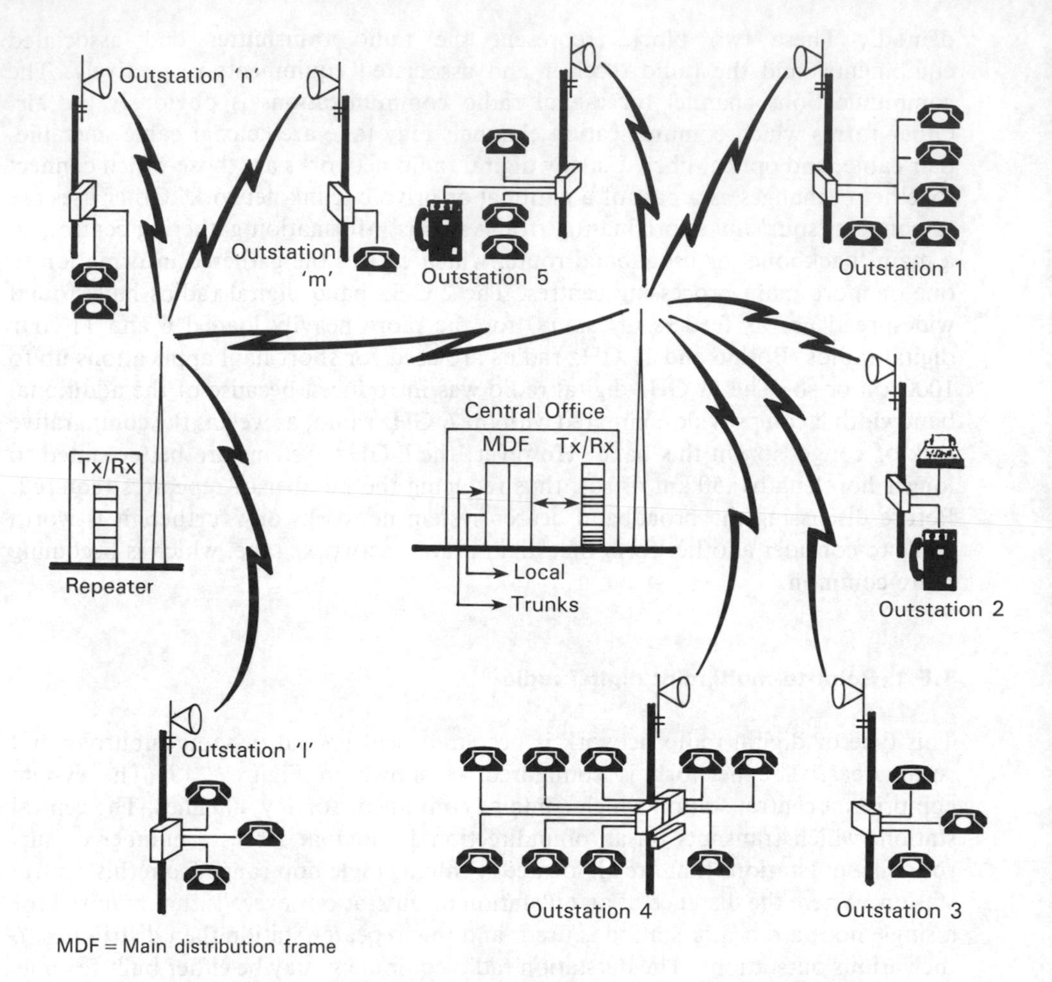

Figure 1.19 Point-to-multipoint or subscriber's radio

determine if any one extension requires a trunk, and if so, assigns a trunk to the demanding line.

At the present time 19 GHz point-to-multipoint systems are being manufactured and installed in the U.K.,[59] to provide 'kilobit/s' data rate services in the local network, over distances up to around 10 km, with an unavailability of 0.01 percent. The central station transmits an aggregate bit rate of about 8.2 Mbit/s, and addresses each station in turn using the TDMA technique. Each outstation can support several outlets operating at data rates from 12.8 up to 144 kbit/s.

1.5.2 Point-to-point digital radio

As administrations plan and start to implement their programs of conversion to digital systems as part of their network modernization, much effort will be given to

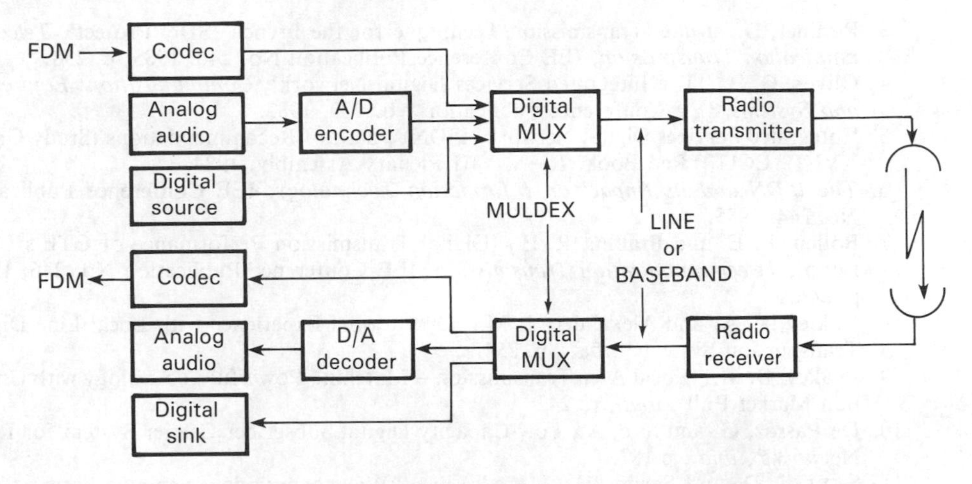

Figure 1.20 A model of a digital radio system

the replacement of the trunk network with optical fibers, and perhaps to a lesser extent with high-capacity digital radio. Low-capacity digital radio will, however, be an attractive solution for the serving of local exchanges in some cases. In addition to this as the demand increases for new data communication services within the local network, the use of point-to-multipoint, and point-to-point digital microwave radio becomes an attractive proposition against the replacement of existing unsuitable metallic-pair cable with transverse screen. With trunk networks there are many exchanges in a large national network which are at a distance 10 to 20 km from each other, and which may be linked more economically by digital radio. The costs of the digital radio are fixed up to 20 km for the 19 GHz band, against optical fibers which may be more expensive over 15 km. During the transition period between say metallic-pair cable replacement with optical fibers, digital radio may be used as a temporary measure to permit continuity of communications. The use of the 19 GHz (CCIR Recommendation 595)[49] and 29 GHz bands is growing, especially as the spectra of the 6 and 11 GHz bands are becoming more and more overcrowded. High-capacity digital radio systems have been designed with a spectral efficiency of 4.5 bits/s/Hz, which are capable of transmitting 140 Mbits/s in a 30 MHz channel within the 6 GHz band,[58] thus permitting more efficient use of the spectrum in this overcrowded band. Figure 1.20 shows the block diagram of the components of a digital radio system.

REFERENCES

1. Adel, A. A., 'On the Selection of an Optimum Digital Modulation', *Telecommunication Transmission*, IEE Conference Publication No. 246, 1985, p. 90.
2. Griffiths, J. M. (editor), *Local Telecommunications*, IEE Telecommunications Series 10, (Peter Peregrinus, 1983).

3. Berlinet, D. *et al.*, 'Transmission Technique for the French ISDN Project', *Telecommunication Transmission*, IEE Conference Publication No. 246, 1985, p. 258.
4. Oliver, G. P., 'The Integrated Services Digital Network', *Communications Equipment and Systems*, IEE Conference Publication No. 209, 1982.
5. 'Integrated Services Digital Network (ISDN)', Series I Recommendations (Study Group XVIII), CCITT Red Book, **III–5**, VIII Plenary Assembly, 1984.
6. *The ISDN and its Impact on Information Technology*, IEE Conference Publication No. 244, 1985.
7. Bollen, R. E. and Prabhu, R. P., 'Digital Transmission Performance of GTE's Local Loop', *Telecommunication Transmission*, IEE Conference Publication No. 246, 1985, p. 246.
8. Oakley, K. A. and Alexander, J. M., 'Operational Experience with Local Line Digital Transmission Systems', *ibid.*, p. 254.
9. Leakey, D. M., 'Local Area Transmission – Matching Powerful Technology with Uncertain Market Pull', *ibid.*, p. 242
10. De Passoz, G. and Sire, A., Low Capacity Digital Subscriber Carrier System for Rural Networks', *ibid.*, p 187.
11. Scott, J. D. and Sawinsky, G. K., 'A New Way to provide Telephone Services with Subscriber Radio', *Proceedings IEEE International Conference on Communications*, 1978, **1**, ICC–78.
12. Duc, N. Q. and Smith B. M., 'Line Coding for Digital Data Transmission', *Australian Telecom. Review*, 1977, **11**, pp. 14–27.
13. Milon, J., Maitre, X. and Rault, J. C., 'An Analogue Echo Canceller for Two-Wire Digital Transmission at 160 kbits/s', *Telecommunication Transmission*, IEE Conference Publication No. 246, 1985, p. 200.
14. Mueller, K. H., 'A New Digital Echo Canceller for Two-Wire Full-Duplex Data Transmission', *IEEE Transactions on Communications*, 1976, **24**, pp. 956–962.
15. Adams, P. F., Glen, P. J. and Woolhouse, S. P., 'Echo Cancellation applied to WAL 2 Digital Transmission in the Local Network', *IEE Telecommunications Transmission into the Digital Era* (London, 1981), pp. 201–204.
16. Deffin, P., Cousin, P. *et al.*, 'Definition of Engineering Rules for Digital Transmission Systems in the French Local Network', *Telecommunication Transmission*, IEE Conference Publication No. 246, 1985, p. 250.
17. de Bont, J. J. J., 'Effects of Crosstalk in Digital and Analogue Transmission Systems (ISDN, Hybrid ISDN) in VF Subscriber Cables', *ibid.*, p. 213.
18. Inose, H., *An Introduction to Digital Integrated Communications Systems* (Peter Peregrinus, 1981).
19. Briley, B. B., *Introduction to Telephone Switching* (Addison-Wesley, 1983).
20. Hills, M. T., *Telecommunications Switching Principles*, (Allen & Unwin, 1979).
21. Special Issue on 'Digital Switching', *IEEE Transactions on Communications*, 1979, **COM–27**, No. 7.
22. Vaughn, H. E., An Introduction to No. 4 ESS', *IEEE International Switching System Symposium Record*, 1972.
23. 'Interface between DTE and DCE from Terminals Operating in the Packet Mode on Public Data Networks', Recommendation X 25 (Study Group VII), CCITT Red Book, **VIII–3**, VIII Plenary Assembly, 1984
24. Friend, G. E., Fike, J. L. *et al.*, *Understanding Data Communications* (Texas Instruments, 1984).
25. Blake, N. B., 'Local Area Networks – Ethernet Objectives and Progress', *Communications Equipment and Systems*, IEE Conference Publication No. 209, 1982, p. 299.
26. Farrow, T. T., 'Local Area Communications – Which One to Back?' *IEE Telecommunications, Radio and Information Technology*, Conference Publication No. 235, 1984, p. 138.
27. Stevens, R. W. and Flatman, A. V., 'Implementation of and Experience with a High Per-

formance Optical LAN', *Telecommunications, Radio and Information Technology*, IEE Conference Publication No. 235, 1984, p. 73.
28. Davies, D. R. H., 'Local and Area Networks and ISDN', *The ISDN and its Impact on Information Technology*, IEE Conference Publication No. 244, 1985, p. 83.
29. 'Open Systems Interconnection (OSI)', Recommendations X 200–250 (Study Group VII), CCITT Red Book, **VIII–5**, VIII Plenary Assembly, 1984.
30. Thorning, G. P., 'Private Wide-Area Packet Switched Networks', *Telecommunications, Radio and Information Technology*, IEE Conference Publication No. 235, 1984, p. 58.
31. Davidson, T. S., 'The Role of the Small PABX in Digital Private Networks', *ibid.*, p. 64.
32. Hay. D;. 'Open All Hours – an Open Technology to Integrated Systems', *ibid.*, p. 131.
33. Baker, T. W. and Mostyn, E. R., 'Private Communications – the Evolution', *Communications and Systems*, IEE Conference Publication No. 209, 1982, p. 291.
34. 'Data Communications over the Telephone Network', Series V Recommendations (Study Group XVII), CCITT Red Book, **VIII–1**, VIII Plenary Assembly, 1984.
35. Coates, R. F. W., *Modern Communication Systems*, 2nd Ed. (Macmillan, 1983).
36. Bowker, R. A., 'A Gateway Facility to allow Intercommunication between Telex and Teletex Terminals', *Telecommunications, Radio and Information Technology*, IEE Conference Publication No. 235, 1984, p. 82.
37. 'Telegraph Transmission', Recommendations R Series (Study Group IX), CCITT Red Book, **VII–1**, VIII Plenary Assembly, 1984.
38. 'Data Communication Networks: Services and Facilities', Recommendations X1–X15 (Study Group VII), CCITT Red Book, **VIII–2**, VIII Plenary Assembly, 1984.
39. 'Data Communication Networks: Interfaces', Recommendations X20–X32 (Study Group VII), CCITT Red Book, **VIII–3**, VIII Plenary Assembly, 1984.
40. Cancer, E., de Julio U. and Romagnoli, M., 'The ISDN in Italy: Background, Experiments, Plans', *The ISDN and its Impact on Information Technology*, IEE Conference Publication No. 244, 1985, p. 41.
41. Bence, A. T., 'Teleservice Developments in CCITT – Migration to the ISDN', *ibid.*, p. 64.
42. 'Terminal Equipment and Protocols for Telematic Services', Series T Recommendations (Study Group VIII), CCITT Red Book, **VII–3**, VIII Plenary Assembly, 1984.
43. 'Telematic Services – Operations and Quality of Service', Recommendations F160–F350 (Study Group I), CCITT Red Book, **II–5**, VIII Plenary Assembly, 1984.
44. 'Data Communication Networks: Message Handling Systems', Recommendations X400–X430 (Study Group VII), CCITT Red Book, **VIII–7**, VIII Plenary Assembly, 1984.
45. 'General Characteristics of International Telephone Connections and Circuits', Recommendations G101–G181, (Study Groups XV, XVI, and CMBD), CCITT Red Book, **III–1**, VIII Plenary Assembly, 1984.
46. 'Telephone Transmission Quality', Series P Recommendations (Study Group XII), CCITT Red Book, **V**, VIII Plenary Assembly, 1984.
47. 'Noise in the Radio Portion of Circuits to be Established over Real Radio-Relay Links for FDM Telephony', CCIR Recommendation 395–2, **IX**, ITU, 1978.
48. 'International Analogue Carrier Systems – Transmission Media Characteristics', Recommendations G211–G652, (Study Groups XV and CMBD), CCITT Red Book, **III–2**, VIII Plenary Assembly, 1984.
49. CCIR Recommendations, **IX–1**, ITU, XVI Plenary Assembly, 1982.
50. Robins, W. P., *Phase Noise in Signal Sources*, IEE Telecommunication Series 9 (Peter Peregrinus, 1982).
51. Bennett, G. H., *Pulse Code Modulation and Digital Transmission*, 2nd Ed. (Marconi Instruments, 1983).
52. Bylanski, P. and Ingram, D. G. W., *Digital Transmission Systems*, IEE Telecommunications Series 4, Rev. Ed. (Peter Peregrinus, 1980).
53. '*Digital Networks – Transmission Systems and Multiplexing Equipments*', Recommen-

dations G700–G956 (Study Groups XV and XVIII), CCITT Red Book, **III–3**, VIII Plenary Assembly, 1984.

54. Cock, C. C., 'The Use of Pseudo-Random Sequences for Jitter Assessment in Digital Communication Networks', *Telecommunication Transmission*, IEE Conference Publication No. 246, 1985, p. 61.
55. 'Measurement of Crosstalk', CCITT Green Book, **VI–2**, Supplement No. 2.4, ITU, 1973.
56. Le-Ngoc, T., 'SR500 – A Point to Multipoint Microwave Radio System with a Random-Request Demand-Assignment Time-Division-Multiple-Access (RR–DA–TDMA) Scheme', *Telecommunication Transmission*, IEE Conference Publication No. 246, 1985 p. 93.
57. Hart, G. 'Local and Junction Network Applications for Microwave Radio', *ibid.*, p. 101.
58. Bates, C. P., Robinson, W. G. and Skinner, M. A., 'Very High Capacity Digital Radio Systems using 64-State Quadrature Amplitude Modulation', *ibid.*, p. 48.

2 SOURCE ENCODER/DECODER USING PULSE CODE MODULATION (PCM) AND PCM RELATED TECHNIQUES

The source encoder processes an analog signal taken into its input, and provides a digital signal at its output which represents that analog input signal. Figure 1.20 shows a block diagram of a digital radio communications system in which the source encoder may be located in the system configuration. The source decoder provides the reverse function, by taking into its input a digital signal, and decoding it into its analog representation. Typical analog signals entering the encoder and emerging from the decoder are electrical signals representing speech, video, facsimile, and instrumentation parameters such as pressure, temperature, etc. In analog systems these analog signals are directly modulated using amplitude, frequency or phase modulation before being transmitted over the radio channel. In a digital system, however, the analog signals are first converted into a digital form consisting of a sequence of binary digits, prior to further processing and modulation. The process of efficiently converting an analog signal into a sequence of binary digits is called *source encoding*. Several source encoding techniques will be described in this chapter, pulse code modulation (PCM) being the most important and predominant. As can be seen from Figure 1.20, there are information sources which are already digital in nature, and the need for a source encoder does not exist. Examples of these are computer terminals, teletex, microcomputers, etc., and the output of various codecs which may be used for converting television analog signals into digital signals, etc. The encoding and decoding schemes used in codecs (encoders and decoders), which are usually PCM-related will also be described in this chapter.

2.1 STATISTICAL METHODS USED IN DIGITAL COMMUNICATION SYSTEMS[1-6]

The incoming signal to a digital radio receiver contains the required signal together with noise, both of which are *random*, and are therefore unpredictable. Their waveforms are examples of realizations of *random processes*. In addition, these random processes are considered as *ergodic*, i.e. that the statistics of one sample function are representative over the whole random process, with a probability of approximately one that the sample function will be the same as the whole random process. Whilst random processes are not predictable in advance, as are *deterministic* processes, such as signals in which an explicit time function can describe their behavior, a prediction of the future performance of a random process can be made with a specific *probability* of it being correct.

2.1.1 Introduction to random variables and probability functions

In the study of statistical methods used in communications, random processes which have numerical outcomes, are of interest. If all the *possible* outcomes in an experiment form points in a sample space or a set of numbers $x_1, x_2, \ldots, x_n, \ldots$, the *observed* values, which may be represented by $X(x_n)$, may assume any of the possible values x_n. The variable $X(x_n)$ is the general symbol for the observed values of all of the possible outcomes of a trial or observation x_n, and is known as the *random variable*. Perhaps we may remember more easily what the random variable is, if we note that *the random variable is neither random nor variable*. The random variable may however, be discrete or continuous, and be used in probability functions which are not functions of the random variable. A simple example to illustrate its meaning is to take an observed value such as the number of people who attended a conference – that is the random variable X. If it were asked whether or not the number of people seen at this conference would be greater than x_p people, or less than or equal to x_q people at the next conference, the answer could be given in terms of a probability, whose value is, for this example, 0.8. The expression which is a function of x_q and x_p but not of X is given as

$$P(x_p < X \leqslant x_q) = 0.8$$

Cumulative probability distribution function

This function associated with a random variable is defined as the probability that one of the outcomes of an experiment will be $X \leqslant x$, where x is any given number. If $F(x)$ is used to represent the cumulative distribution function, then:

$$F(x) \triangleq P(X \leqslant x) \tag{2.1a}$$

$F(x)$ is simply the probability that an observed value will be less than or equal to the quantity x. The properties of the cumulative probability distribution function (CPDF) are:

$$0 \leqslant F(x) \leqslant 1 \tag{2.1b}$$

The CPDF is always a nondecreasing function of x, that is $\mathrm{d}F(x)/\mathrm{d}x \geqslant 0$

$$F(-\infty) = 0 \quad F(+\infty) = 1 \tag{2.1c}$$

$$F(x_1) \leqslant F(x_2) \text{ if } x_1 < x_2 \tag{2.1d}$$

$$P(x_1 < X \leqslant x_2) = F(x_2) - F(x_1) \tag{2.1e}$$

The CPDF applies to either discrete or continuous processes.

2.1.2 Discrete random variables and probability functions

This section will briefly review the subjects of discrete random variables and probability functions; the reader who is not broadly familiar with the material presented in this section may consider it worth while to consult reference 3 for more background information. A discrete random variable is one that can take on only individual values. Examples are the number of binary bits of information arriving at a receiver in a fixed interval of time, and the voltage quantization levels in a PCM system.

2.1.2.1 Probability frequency or mass function

The set of discrete or allowed values associated with a discrete random variable may be given by $x_1, x_2, \ldots, x_n$. The probability that the random variable equals one of these values is denoted by $P(X = x_i)$, and is called the *probability frequency* or *probability mass function*. It is assumed that of all the possible outcomes of an experiment one will equal an observed outcome. That is the probability mass function is greater than zero or $P(X = x_i) > 0$ for $i = 1, 2, \ldots, n$. Two important properties exist for this function, which are for n possible outcomes:

$$\sum_{i=1}^{n} P(X = x_i) = 1 \tag{2.2a}$$

$$F(x) = P(X \leqslant x) = \sum_{\text{all } x_i \leqslant x} P(X = x_i) \tag{2.2b}$$

From equation 2.2b, if x_{q-1} and x_q are two adjacent possible outcomes, then:

$$F(x_q) - F(x_{q-1}) = \sum_{i=1}^{q} P(X = x_i) - \sum_{i=1}^{q-1} P(X = x_i) = P(X = x_q)$$

This result is quite informative, since it means that the height between the steps in the discrete cumulative probability function at the values $x = x_i$ is the value $P(X = x_i)$.

Frequency or mass functions which are often used in digital communication systems are the uniform and binomial.

IMPULSES IN PROBABILITY MASS FUNCTIONS

In multilevel binary signals such as M-ary PSK and the like, which will be treated in Chapter 5, M levels or phase states may occur with equal probability of occurrence. This means that the random variable X has discrete possible outcomes,

$x_1, x_2, \ldots, x_M$, with the probability mass function associated with each outcome being equal, i.e. $P(X = x_1) = P(X = x_2) = \cdots = P(X = x_M)$. As the phase states are discrete, their representation as a *probability density function* (see below) must include the Dirac delta function, or impulse function $\delta(x - x_i)$. This means that each allowable discrete state is represented by a probability amplitude equal to $P(X = x_i)\,\delta(x - x_i) = 1/M$, for $i = 1$ to M, and in which all probability amplitudes are equally spaced over the range of occurring states, at intervals equal to $360°/M$, for M phase states.

The probability density function pdf may be represented by:

$$\text{pdf} = \sum_{i=1}^{M} P(X = x_i)\,\delta(x - x_i) \tag{2.3}$$

The cumulative probability distribution function (*CPDF*) can be constructed by progressively summing the equal probability amplitudes, at each of the discrete occurrence states ($X = x_1, X = x_2, \ldots, X = x_M$), with the step increase in probability occurring at each of the discrete occurrence states.

Example

For an 8-level PSK signal draw the probability density function *pdf*, and the cumulative probability distribution function CPDF

As $P(X = x_1) = P(X = x_2) = \cdots = P(X = x_M)$ and, from equation 2.2a, $\sum_{i=1}^{M} P(X = x_i) = 1$

we find:

$$P(X = x_i) = 1/M = \tfrac{1}{8}$$

For 360°, and 8 levels, each phase state has an allocation of 45°. If the phase states are centered around 0°, they may be represented by ±22.5°, ±67.5°, ±112.5° and ±157.5°. These states are the impulse instants $\delta(x - x_i)$, where $i = \pm 4$. Figure 2.1 shows the pdf and the CPDF for this example.

UNIFORM PROBABILITY MASS FUNCTION

A random variable X which has a uniform probability mass function is given by the expression:

$$P(X = x_i) = 1/Q, \quad \text{for } i = 1, 2, \ldots, Q \tag{2.4}$$

This means that, of the possible outcomes, each is equally likely to be an observed outcome over the range $i = 1$ to Q. This function is used in Section 2.2.1.2 in deriving the signal to quantization distortion for the case of uniform quantization in PCM systems, and it shows that, for each of the possible quantization levels, a sampled random input signal is equally likely to be quantized into any of Q quantization levels.

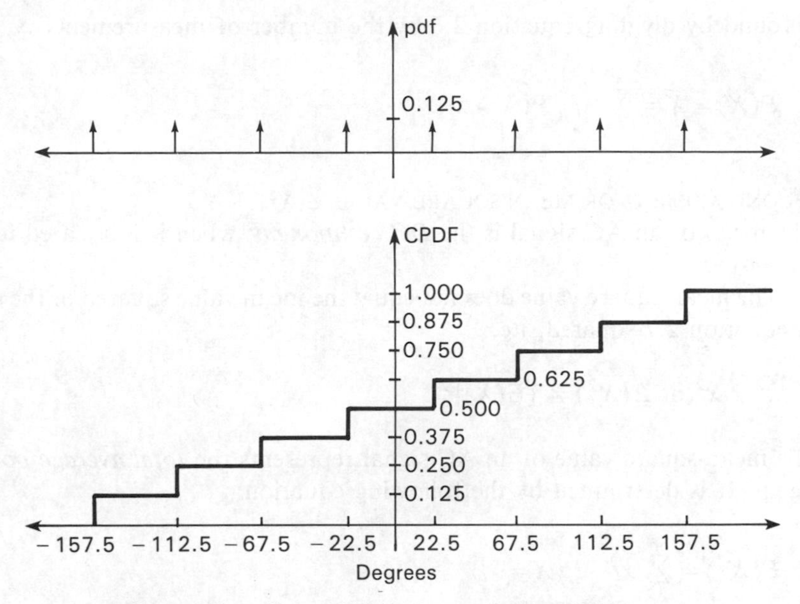

Figure 2.1 pdf and CPDF for an 8-phase PSK system

BINOMIAL PROBABILITY MASS FUNCTION

If an experiment or process is repeated n times, the probability that an event with a single probability of occurrence p occurs r times is given by:

$$P(X=r)=(n!)/r!(n-r)! \times p^r(1-p)^{n-r} \quad \text{for } r=0,1,2,\ldots,n \tag{2.5a}$$

the mean $\mu = np$ (2.5b)

and the variance $\sigma^2 = np(1-p)$ (2.5c)

2.1.2.2 Statistical moments of the random variable

FIRST MOMENT OR MEAN $E(X)$

Assume that the numerical values of the random variable X are $x_1, x_2, \ldots, x_n$, with probabilities of occurrence $P(X=x_1), P(X=x_2), \ldots, P(X=x_n)$. For a large number of measurements N, the possible outcomes $X=x_1, X=x_2, \ldots, X=x_n$, would each occur $N \cdot P(X=x_1), N \cdot P(X=x_2), \ldots, N \cdot P(X=x_n)$ times. The sum of the number of times each outcome actually occurred throughout the experiment is given by:

$$x_1 \cdot N \cdot P(X=x_1) + x_2 \cdot N \cdot P(X=x_2) + \cdots + x_n \cdot N \cdot P(X=x_n)$$
$$= N \sum_{i=1}^{n} x_i \cdot P(X=x_i) \tag{2.6}$$

The average or mean value $E(X)$ of the observed outcomes, or expected outcomes is found by dividing equation 2.6 by the number of measurements N, thus:

$$E(X) = \bar{X} = \sum_{i=1}^{n} x_i \cdot P(X = x_i) \tag{2.7}$$

SECOND MOMENT OR MEAN-SQUARE VALUE $E(X^2)$

The mean of an AC signal is the *DC component*; when it is squared it equals the *DC power*.

The mean-square value does not equal the mean value squared or the expressions in equation 2.7 squared, i.e.:

$$\bar{x}^2 \neq x^2 \text{ or } E(X^2) \neq [E(X)]^2 \tag{2.8}$$

The mean-square value of an AC signal represents the *total average power* in that signal. It is determined by the following equation:

$$E(X^2) = \sum_{i=1}^{n} x_i^2 \cdot P(X = x_i) \tag{2.9}$$

VARIANCE $E[(X - \mu)^2]$

A modified form of the second moment is the variance. It is defined as:

$$\sigma^2 = \sum_{i=1}^{n} [x_i - E(X)]^2 \cdot P(X = x_i) \tag{2.10}$$

The variance is the *AC power* in an AC signal, and may be alternatively expressed as:

$$\sigma^2 = E(X^2) - [E(X)]^2 \tag{2.11}$$

The square root of the variance is the *standard deviation* σ, which is a measure of the spread of observed values. It also represents the *RMS value* of an AC signal voltage or current.

EXAMPLE

Given that the number of uniform quantization levels in a PCM system quantizer is Q, and that its range is $-a$ to $+a$ volts, determine for an input sampled random signal, the following output parameters from the quantizer:

1. The DC component
2. The total average power
3. The RMS value of the signal

1. As the signal is random, and takes only discrete quantization levels, with each quantization level being equally probable, the probability mass function is uniform.

As given by equation 2.4, $P(X = x_i) = 1/Q$. From equation 2.7, the DC component $E(X)$ can be determined by:

$$E(X) = \sum_{i=1}^{Q} x_i \cdot P(X = x_i) = (1/Q) \sum_{i=1}^{Q} x_i$$

for x_i are the voltage values of the quantization levels; it may be expressed as levels above a reference level $(-a)$. If the voltage difference between levels is given by $\Delta = 2a/Q$, then:

$$x_i = -a + i\,\Delta - \Delta/2 = i\,\Delta - (\Delta/2)(Q+1)$$

Hence as $\sum_{i=1}^{Q} i = (Q/2) \cdot (Q+1)$

$$E(X) = 0$$

Thus the DC component is zero.

2. The total average power can be found using equation 2.9; hence

$$E(X^2) = \sum_{i=1}^{Q} x_i^2 P(X = x_i) = (1/Q) \cdot (\Delta^2/4) \left[Q(Q+1)^2 - 4(Q+1) \sum_{i=1}^{Q} i + 4 \sum_{i=1}^{Q} i^2 \right]$$

$$= (Q^2 - 1) \cdot \Delta^2/12$$

To derive this result, use is made of the expansion:

$$\sum_{i=1}^{Q} i^2 = Q(2Q^2 + 3Q + 1)/6 \text{ for } i = 1, 2, \ldots, Q$$

3. The RMS value of the signal can be found using equation 2.11. As the mean value is zero, the RMS value is the square root of the total average power:

$$\text{RMS value} = 0.29\,\Delta \cdot (Q^2 - 1)^{1/2}$$

2.1.3 Continuous random variables and probability density functions

A *continuous random variable* is one which can take on an infinite number of values over a specified range. Examples are the peak magnitudes of an analog waveform, or the measured values of 10-ohm resistors taken from a store. Experience will tell that in a bag of one hundred 10-ohm resistors, there will not be many whose resistance will read 10 ohms on a locally bought multimeter, and that if the international resistance standard is used as the reference, there would be possibly no resistor in the bag whose resistance equalled exactly ten ohms. Hence the concept of $P(X = x_i)$ used in the continuous random variable context has little use, since it is usually zero except in very special cases. As the 10-ohm resistors must have some value about ten ohms, it is convenient to represent the observed value, or random variable, as lying within a range of values, or by being less or equal to an upper value. That is:

$$x_1 < X \leqslant x_2, \text{ if } x_1 < x_2 \text{ or } X \leqslant x_2$$

from equations 2.1e and 2.2b we have:

$$P(x_1 < X \leqslant x_2) = F(x_2) - F(x_1) = P(X \leqslant x_2) - P(X \leqslant x_1)$$

Hence the concept of the cumulative distribution function is still applicable to continuous random variables.

2.1.3.1 Probability density function (pdf)

Because of the continuous functions which are involved, integration becomes the main mathematical tool for processing them. Consequently the derivative of the cumulative distribution function serves a more useful purpose. This derivative is called the *probability density function*, and is defined as:

$$f(x) \triangleq \mathrm{d}F(x)/\mathrm{d}x \tag{2.12}$$

If the probability density function is known, the cumulative probability function can be derived using:

$$F(x) = \int_{-\infty}^{x_1} f(x)\,\mathrm{d}x \qquad \text{as } F(-\infty) = 0 \tag{2.13}$$

This means that the area under the probability density function pdf from $-\infty$ to x represents the cumulative probability function, and that it is the probability that the observed value will be less or equal to x_1, i.e. $P(X \leqslant x_1)$. The probability density function has the following properties:

1. $f(x) \geqslant 0$ for $-\infty < x < \infty$ (2.14a)

2. $\int_{-\infty}^{\infty} f(x)\,\mathrm{d}x = 1$ (i.e. the total area under the $f(x)$ curve is unity) (2.14b)

3. $P(x_1 < X \leqslant x_2) = F(x_2) - F(x_1) = \int_{x_1}^{x_2} f(x)\,\mathrm{d}x$, for $x_1 \leqslant x_2$ (2.14c)

EXAMPLES OF PROBABILITY DENSITY FUNCTIONS

Uniform probability density function. A random variable X is said to have a uniform pdf if

$$f(x) = \begin{cases} 1/(x_2 - x_1), & \text{for } x_1 < X \leqslant x_2 \\ 0, & \text{elsewhere} \end{cases} \tag{2.15}$$

Using equation 2.14c, the cumulative probability distribution is unity for $X > x_2$, zero for $X \leqslant x_1$ and varies linearly from zero to unity for $x_1 < X \leqslant x_2$.

Example. Assume that the levels of a signal from a sampler are random and uniformly distributed, and that the dynamic range of the signal extends from $-a/Q$ to a/Q. Plot the pdf and CPDF.

Using equation 2.15, the value of the pdf within the range $-a/Q$ to a/Q is $Q/2a$. From equations 2.14b and 2.14c, we have:

$$\begin{aligned} F(x) &= 0.5\ (Qx/a + 1), && \text{for } -a/Q < X \leqslant a/Q \\ &= 0, && \text{for } X \leqslant -a/Q \\ &= 1 && \text{for } X > +a/Q \end{aligned}$$

Figure 2.2 shows the pdf and the CPDF for this example.

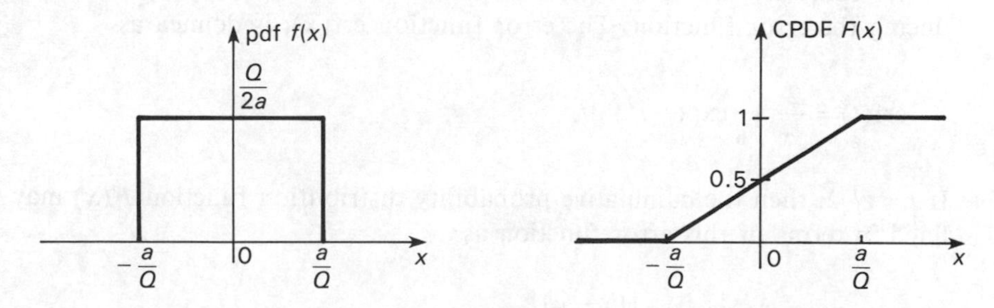

Figure 2.2 A uniform probability density function and its CPDF

The Gaussian or normal probability density function. This pdf is one of the more frequently used in the description of radio systems because it provides a good characterization of thermal noise, caused by the random motion of electrons. The pdf is given by:

$$f(x) = \frac{1}{\sqrt{2\pi\sigma^2}} \exp\left\{-\frac{(x-\mu)^2}{2\sigma^2}\right\}, \quad -\infty < x < \infty \tag{2.16}$$

where μ is the mean and σ^2 is the variance of the continuous random variable $f(x)$ or pdf. The value of the cumulative probability distribution function is found by using equation 2.13. That is:

$$F(x) = P(X \leqslant x) = \frac{1}{\sigma\sqrt{2\pi}} \int_{-\infty}^{x} \exp\left[-\frac{(x-\mu)^2}{2\sigma^2}\right] \mathrm{d}x \tag{2.17}$$

If the mean is zero, because there is no DC component to the signal, and because the standard deviation $\sigma = 1$ volt RMS, then, letting $x = z$, both the expressions for the pdf or CPDF are normalized to produce the standardized normal distribution. Alternatively, considering $f_x(x)\,dx = f_z(z)\,dz$, from the fact that the area under each curve is equal to unity, the value for z may be taken as:

$$z = (x - \mu)/\sigma \tag{2.18}$$

By either method equations 2.16 and 2.17 reduce to:

$$\sigma \cdot f(z) = \frac{1}{\sqrt{2\pi}} \exp(-z^2/2) \tag{2.19}$$

and

$$F(z) = \frac{1}{\sqrt{2\pi}} \int_{-\infty}^{z} \exp(-z^2/2)\,dz \tag{2.20}$$

Figure 2.3 shows the diagrams of the pdf and CPDF of the standardized normal distribution.

To complete this brief discussion of the Gaussian pdf, it is worth while introducing the error function. The error function erf(x), is defined as

$$\text{erf}(t) = \frac{2}{\sqrt{\pi}} \int_{0}^{t} \exp(-t^2)\,dt \tag{2.21}$$

If $t = z/\sqrt{2}$, then the cumulative probability distribution function $F(x)$ may be defined in terms of this error function as

$$F(x) = 1/2 + (1/2)\text{erf}\left[\frac{(x-\mu)}{\sqrt{2}\sigma}\right] \tag{2.22}$$

The complementary error function is given by the expression

$$\text{erfc}(t) = \frac{2}{\sqrt{\pi}} \int_{t}^{\infty} \exp(-t^2)\,dt = 1 - \text{erf}(t) \tag{2.23}$$

Thus

$$F(x) = 1 - (1/2)\text{erfc}[(x-\mu)/\sqrt{2}\sigma] \tag{2.24}$$

Values for both the probability density function and CPDF are given in Table 2.1.

Table 2.1 applies to a normalized Gaussian random variable, where the pdf is given by equation 2.16. As $1/\sigma = z/(x - \mu)$, this table gives the scaled values for $\sigma \cdot f(z) = \sigma \cdot$ pdf. For negative values of z, pdf values are the same, whereas, CPDF (for negative z) = 1 − CPDF(Table). The CPDF values are for $P(Z \leqslant z)$.

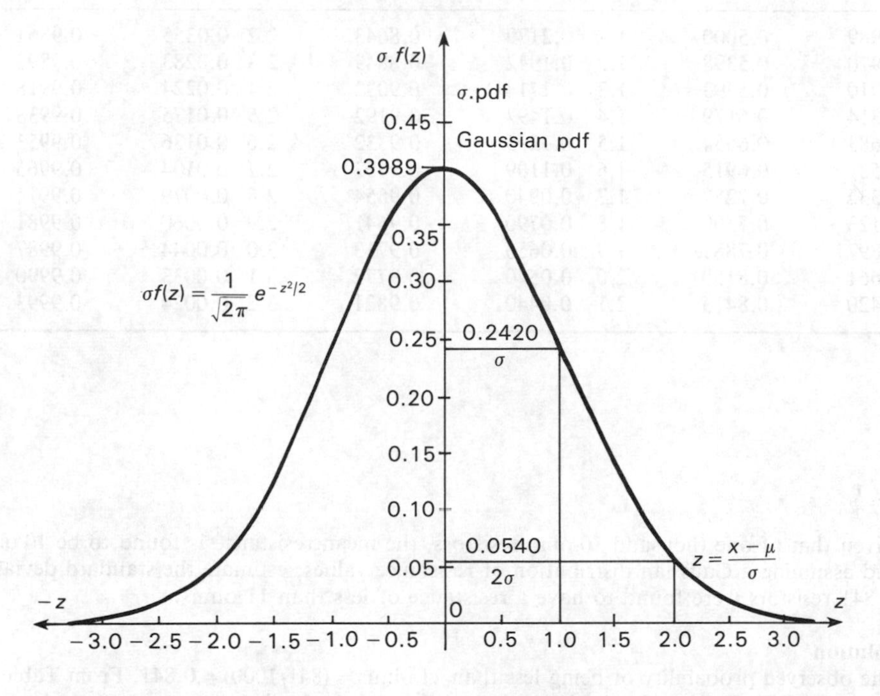

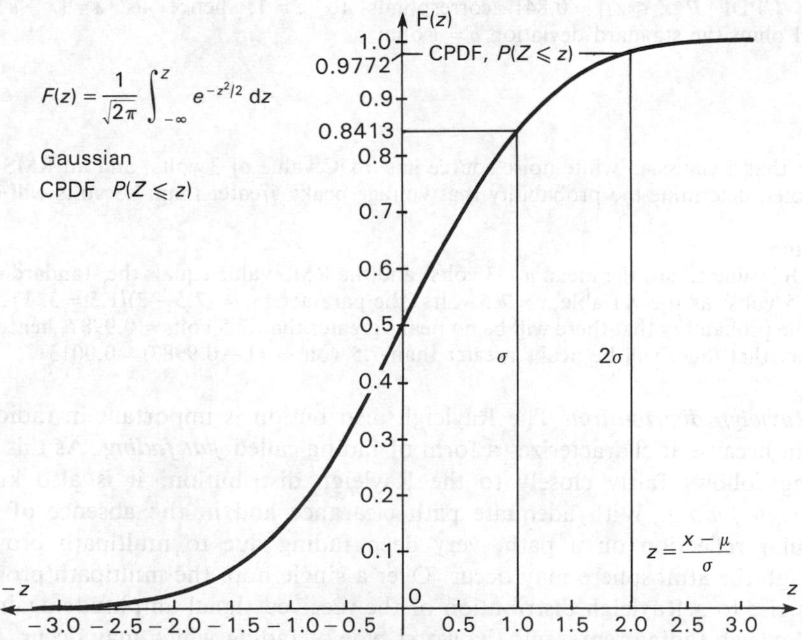

Figure 2.3 Standard normal pdf and CPDF

Table 2.1 Gaussian pdf and CPDF values

z	$\sigma \cdot$ pdf	CPDF[$P(Z \leqslant z)$]	z	$\sigma \cdot$ pdf	CPDF[$P(Z \leqslant z)$]	z	$\sigma \cdot$ pdf	CPDF[$P(Z \leqslant z)$]
0.0	0.3989	0.5000	1.1	0.2179	0.8643	2.2	0.0355	0.9861
0.1	0.3970	0.5398	1.2	0.1942	0.8849	2.3	0.0283	0.9893
0.2	0.3910	0.5793	1.3	0.1714	0.9032	2.4	0.0224	0.9918
0.3	0.3814	0.6179	1.4	0.1497	0.9192	2.5	0.0175	0.9938
0.4	0.3683	0.6554	1.5	0.1295	0.9332	2.6	0.0136	0.9953
0.5	0.3521	0.6915	1.6	0.1109	0.9452	2.7	0.0104	0.9965
0.6	0.3332	0.7257	1.7	0.0940	0.9554	2.8	0.0079	0.9974
0.7	0.3123	0.7500	1.8	0.0790	0.9641	2.9	0.0060	0.9981
0.8	0.2897	0.7881	1.9	0.0656	0.9713	3.0	0.0044	0.9987
0.9	0.2661	0.8159	2.0	0.0540	0.9772	3.1	0.0033	0.9990
1.0	0.2420	0.8413	2.1	0.0440	0.9821	3.2	0.0024	0.9993

Example 1

Given that of one thousand 10-ohm resistors, the mean resistance is found to be 10 ohms, and assuming a Gaussian distribution of resistance values, estimate the standard deviation, if 841 resistors were found to have a resistance of less than 11 ohms.

Solution
The observed probability of being less than 11 ohms = (841/1000) = 0.841. From Table 2.1, $F(z) = \text{CPDF}[P(Z \leqslant z)] = 0.841$ corresponds to $z = 1$; hence as $z = (x - \mu)/\sigma$, and $x = 11$ ohms the standard deviation $\sigma = 1$ ohm.

Example 2

Given that a Gaussian white noise source has a DC value of 3 volts, and an RMS value of 1.5 volts, determine the probability that voltage peaks greater than 7.5 volts will occur.

Solution
The DC value equals the mean $\mu = 3$ volts, and the RMS value equals the standard deviation $\sigma = 1.5$ volts; as the variable $x = 7.5$ volts, the parameter $z = (7.5 - 3)/1.5 = 3$. From Table 2.1, the probability that there will be no peaks greater than 7.5 volts = 0.9987, hence the probability that there will be peaks greater than 7.5 volts = (1 − 0.9987) = 0.0013.

Rayleigh distribution. The Rayleigh distribution is important in radio system design because it characterizes a form of fading called *flat fading*. As this form of fading follows fairly closely to the Rayleigh distribution, it is also known as *Rayleigh fading*. With adequate path clearance and in the absence of a single specular reflection on a path, very deep fading due to multipath propagation through the atmosphere may occur. Over a single hop, the multipath propagation gives rise to a Rayleigh distribution of the received signal amplitude against time. As Rayleigh fading represents the worst type of fading which may occur, it occurs only for a proportion of the worst month for radio propagation performance.

Actual observations of multipath fading indicate that for deep fades, the amplitude distributions have the same slope as the Rayleigh distribution but are displaced. These displaced distributions are the Nakagami–Rice distributions. The fading depth during periods other than those characterized by the Rayleigh distribution has been found to follow the log-normal distribution.[7,8] This distribution is discussed below. The mechanism of multipath fading is basically one of scattering the transmitted beam as it enters a highly stratified atmosphere, and the recombination of some of the scattered beams at the receiver site. The effect is to produce random additions and subtractions of the radio carrier arriving on the direct path, for the nondirect signals arrive with varying amplitudes and phases. The resulting received carrier is thus amplitude modulated, with a modulation envelope whose probability density function[2] is given by equation 2.25. This distribution is also important in the characterization of the channel dispersion of a troposcatter radio system

$$f(x) = \begin{cases} (x/\alpha^2)\exp - (x^2/2\alpha^2) & (0 \leqslant x \leqslant \infty) \\ 0 & (x < 0) \end{cases} \tag{2.25}$$

The CPDF is given by:

$$F(x) = P(X \leqslant x) = \begin{cases} 1 - \exp[-(x^2/2\alpha^2)] & (0 \leqslant x \leqslant \infty) \\ 0 & (x < 0) \end{cases} \tag{2.26}$$

If $z = x/\alpha$, then Table 2.2 shows values for $\alpha \cdot f(z)$, and $F(z)$.

Figure 2.4 shows the Rayleigh pdf and CPDF curves based on the values from Table 2.2.

Figure 2.3 shows plots of the Rayleigh pdf and CPDF for various values of z.

Log-normal distribution. As mentioned in the discussion of Rayleigh fading, the log-normal distribution also is used to describe multipath fading when it is not as severe as the worst-month case described by Rayleigh fading. The log-normal distribution is also used to describe approximately the statistical duration of fade

Table 2.2 Rayleigh pdf and CPDF values

z	$\alpha \cdot$ pdf	CPDF[$P(Z \leqslant z)$]	z	$\alpha \cdot$ pdf	CPDF[$P(Z \leqslant z)$]	z	$\alpha \cdot$ pdf	CPDF[$P(Z \leqslant z)$]
0.1	0.0995	0.0050	1.1	0.6007	0.4539	2.2	0.1956	0.9111
0.2	0.1960	0.0198	1.2	0.5841	0.5132	2.3	0.1633	0.9290
0.3	0.2868	0.0440	1.3	0.5584	0.5704	2.4	0.1347	0.9439
0.4	0.3692	0.0769	1.4	0.5254	0.6247	2.5	0.1098	0.9561
0.5	0.4412	0.1175	1.5	0.4870	0.6753	2.6	0.0885	0.9660
0.6	0.5012	0.1647	1.6	0.4449	0.7220	2.7	0.0705	0.9739
0.7	0.5479	0.2173	1.7	0.4008	0.7643	2.8	0.0556	0.9802
0.8	0.5809	0.2739	1.8	0.3562	0.8021	2.9	0.0433	0.9851
0.9	0.6003	0.3330	1.9	0.3125	0.8355	3.0	0.0333	0.9889
1.0	0.6065	0.3935	2.0	0.2707	0.8647	3.1	0.0254	0.9918

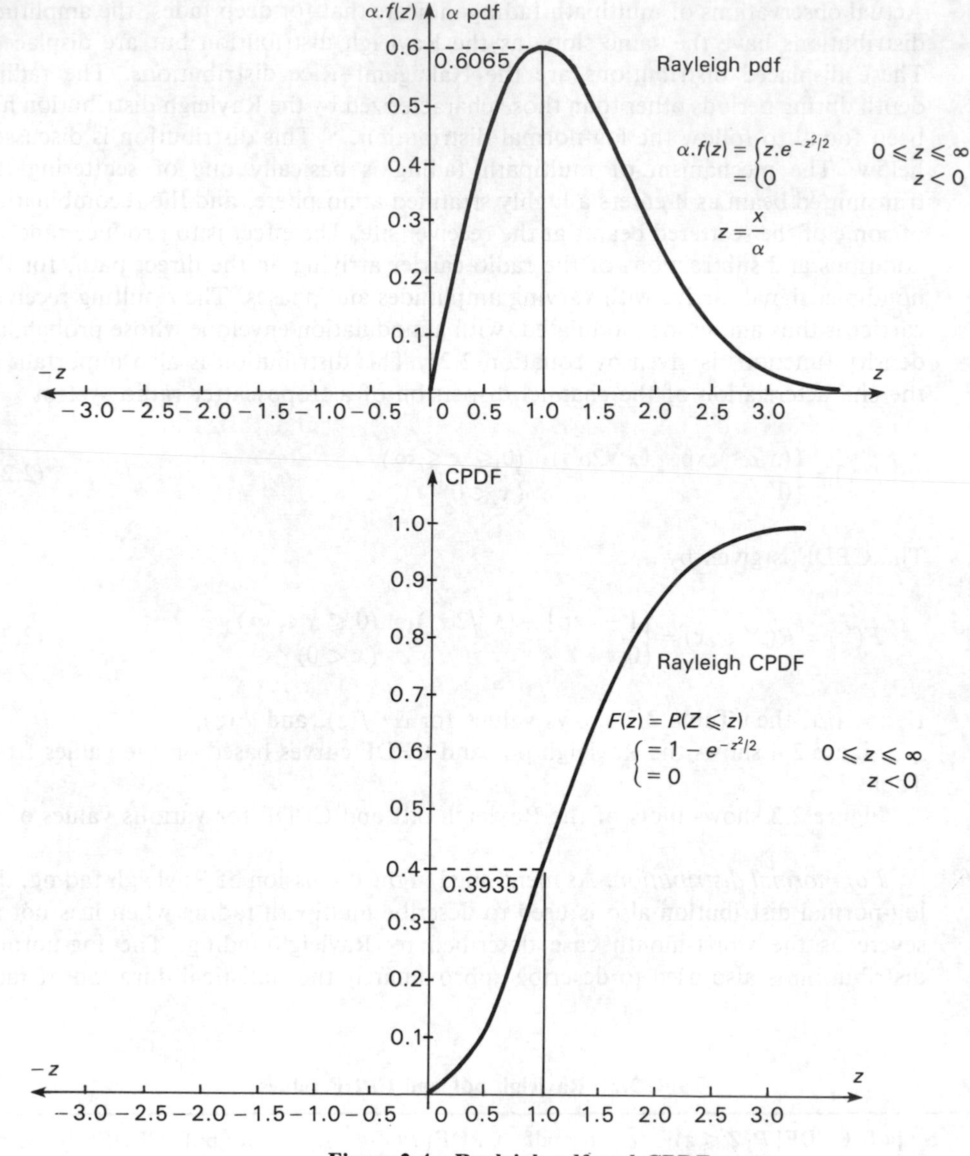

Figure 2.4 Rayleigh pdf and CPDF

durations. The median value of fade duration decreases when fading depth increases, decreases with increasing path length and increases with relative path clearance. The value of the standard deviation decreases when fading depth increases, and increases with frequency. Measurements in the U.S.A. also show that the median fade duration distribution is log-normal characterized by a standard deviation of 5.6 dB for nondiversity paths. Multipath median fade duration times,

$\langle t \rangle$ can be expressed for a nondiversity signal (after Lin 1971, and Vigants 1971, Reference 7) as:

$$\langle t \rangle = 56.6 \times 10^{-F/20} \times (d/f)^{1/2} \tag{2.27}$$

where d is the path length in km, f is the frequency in GHz, F is the fade depth (exceeding 20 dB) in dB, and $\langle t \rangle$ is the median fade duration at fade depth F, in seconds.

Deep multipath fading is selective due to its origin. To a first approximation it may be considered as the interference between the direct signal and a signal which has been refracted by the upper stratified layers of the atmosphere. The differential delay τ between the direct and refracted signal results in a 180° phase reversal for specific frequencies which are separated in the frequency spectrum by $\Delta f = 1/2\tau$. The effect on a digital signal is to produce inter-symbol distortion.

The log-normal distribution, like the Rayleigh distribution, belongs to the family of normal distributions, i.e. the distribution of $y = \exp(x)$ is log-normal when the distribution of x itself is normal.[8] If $y > 0$, then $y = \exp(x)$ has the single solution $x = \ln(y)$. Equation 2.16 shows the Gaussian pdf $= f(x)$. As $x = \ln(y)$, $dx = (1/y)\, dy$, and $f(x) \cdot dx = (1/y) \cdot f(\ln y) \cdot dy = f(y)\, dy$; hence:

$$\text{pdf} = f(y) = (1/y) \cdot (1/\sqrt{2\pi}) \cdot (1/\sigma) \cdot \exp[-(\ln y - \mu)^2/2\,\sigma^2] \tag{2.28}$$

$$\text{CPDF} = F(y) = \int_0^y f(y)\, dy = \int_{-\infty}^x f(x)\, dx$$

$$= F(x) = 1/2 + (1/2)\text{erf}[(x - \mu)/\sqrt{2}\sigma] \tag{2.29}$$

If

$$\mu = \mu_x = \ln \mu_y - \sigma^2/2 \tag{2.30}$$

then equations 2.28 and 2.29 may be given as:

$$\text{pdf} = f(y) = 1/(\sqrt{2\pi}) \cdot 1/(\sigma) \cdot (1/y) \cdot \exp\left[-[\ln y/\mu_y + \sigma^2/2]^2/2\sigma^2\right] \tag{2.31}$$

$$\text{CPDF} = P(Y \leqslant y) = 1/2 + (1/2)\text{erf}\left[[\ln(y/\mu_y) + \sigma^2/2]/\sqrt{2}\sigma\right] \tag{2.31a}$$

The value μ_y may be calculated as:

$$\ln \mu_y = \mu + \sigma^2/2 \tag{2.32}$$

and the median $\langle y \rangle$ as:

$$\langle y \rangle = (\mu_y)\exp[-\sigma^2/2] \tag{2.32a}$$

2.1.3.2 Statistical moments of the continuous random variable

FIRST MOMENT OR MEAN

This is given by expression:

$$E(X) = \bar{X} = \int_{-\infty}^{\infty} xf(x)\,dx \tag{2.33}$$

EXAMPLE

Determine the mean of the uniform distribution given in Section 2.1.3.

As $f(x) = 1/(x_2 - x_1)$, for $x_1 < X \leqslant x_2$, and $f(x) = 0$ elsewhere,

$$E(X) = \bar{X} = 1/(x_2 - x_1) \cdot \int_{x_1}^{x_2} x\,dx = (x_1 + x_2)/2$$

Thus for a uniform distribution

$$\bar{X} = (x_1 + x_2)/2 \tag{2.34}$$

EXERCISE 1

Prove that the mean of a Rayleigh distribution is given by $\alpha \cdot \sqrt{\pi/2})$

EXERCISE 2

Prove that the mean of a log-normal distribution μ_y is given by $\mu + \sigma^2/2$, where μ and σ are the Gaussian mean and standard deviation respectively.

SECOND MOMENT OR MEAN-SQUARE VALUE $E(X^2)$

This is given by the expression:

$$E(X^2) = \bar{X}^2 = \int_{-\infty}^{\infty} x^2 \cdot f(x)\,dx \tag{2.35}$$

EXAMPLE

Determine the mean square value of the uniform distribution where $f(x) = 1/(x_2 - x_1)$, for $x_1 < X \leqslant x_2$. We have

$$E(X^2) = \bar{X}^2 = 1/(x_2 - x_1) \cdot \int_{x_1}^{x_2} x^2\,dx = (x_1^2 + x_1 \cdot x_2 + x_2^2)/3$$

Hence the mean-square value is:

$$(x_1^2 + x_1 \cdot x_2 + x_2^2)/3 \tag{2.36}$$

EXERCISE 3

Show that the mean-square value of the Rayleigh distribution is given by:

$$E(X^2) = \bar{X}^2 = 2\alpha^2 \tag{2.37}$$

VARIANCE σ^2

The relationship between the mean and the mean-squared value of a distribution in order to obtain the variance is given by equation 2.11:

$$\sigma^2 = E(X^2) - [E(X)]^2$$

EXAMPLE

Determine the variance of the uniform distribution.

Knowing that the mean is given by $(x_2 + x_1)/2$, and that the mean-square is given by $(x_1^2 + x_1 \cdot x_2 + x_2^2)/3$, the variance becomes:

$$\sigma^2 = (x_2 - x_1)^2/12 \tag{2.38}$$

EXERCISE 4

Show that the variance of the Rayleigh distribution is given by

$$(2 - \pi/2) \cdot \alpha^2 \tag{2.39}$$

Also prove that the distribution reaches a maximum of $1/(\alpha\sqrt{e})$ at $z = 1$.

THE MEDIAN

The median of a random variable X is the value that x takes ($\langle x \rangle$) when $F(x) = P(X \leqslant x) = 0.5$. That is

$$\text{Median of } X = \langle x \rangle, \text{ when } P(X \leqslant \langle x \rangle) = 0.5 \tag{2.40}$$

EXAMPLE

Find the median of the log-normal distribution in terms of normal distribution standard deviation σ, and the log-normal mean μ_y.

From the definition of the median, and equation 2.32, the error function must equal zero; hence

$$\ln[\langle y \rangle/\mu_y\} + \sigma^2/2 = 0 \tag{2.41}$$

$$y = \mu_y \cdot \exp(-\sigma^2/2) \tag{2.42}$$

2.1.3.3 Central limit theorem

This theorem in effect states that, if a number of random variables are independent of each other, the probability density function of the sum of their pdfs usually tends to become Gaussian, even if the individual random variable pdfs are not Gaussian. In addition to this, the resultant probability density function has a mean which is equal to the sum of the means of the random variable pdfs, and a variance which is the sum of the variances of the independent random variable pdfs.

The resultant pdf may be considered as a convolution of the individual independent pdfs, that is

$$f(x) = f_1(x) * f_2(x) * \cdots * f_n(x) \tag{2.43}$$

where $f_1(x)$ is convolved with $f_2(x)$ and the resultant then convolved with $f_3(x)$ etc.

2.1.4 Autocorrelation and power spectral density function

When simultaneous measurements are made at time t_0 on a process, and these measurements are averaged, an *ensemble average* μ is obtained. But, if consecutive measurements are made on the same process, and the measurements averaged, a *statistical* or *time average* $\langle \mu \rangle$, is obtained. In general, statistical averages do not equal ensemble averages. A *stationary* random process is defined as one in which the statistical characteristics of the measurements made to a process do not change with time. An ergodic process is one in which the ensemble average equals the statistical average, and thereby infers a stationary process. Note however, that a stationary process does not infer ergodicity. The ensemble averages have previously been considered when dealing with the mean $E(X)$, and the mean-square $E(X^2)$.

A *stochastic* process is a continuous process in which an infinity of random variables exists, one for each value of t. It may be given by the expression $F(x, t) = P[X(t) \leqslant x]$.

Autocorrelation function $R(t_1, t_2)$

In a linear system, the output of the system $x_o(t)$ can be related to the input of the system $x_i(t)$ by the linear transfer function. If the input average power $E[x_i^2(t)]$ is known from a measurement taken at time t_1, the average output power $E[X_o^2(t)]$ cannot be found without a knowledge of the *autocorrelation function*, since there has been a delay before the measured input average power reaches the system output at time t_2. The autocorrelation function depends on the time difference and may be defined as either an ensemble or time average. As an ensemble average, it is given by:

$$R(t_1 \cdot t_2) = E[X(t_1)X(t_2)] = \int_{-\infty}^{\infty}\int_{-\infty}^{\infty} x_1 \cdot x_2 \cdot p(x_1, x_2) \cdot \mathrm{d}x_1 \cdot \mathrm{d}x_2 \tag{2.44}$$

If the realized process is time-averaged, the autocorrelation function can be defined as:

$$R(\tau) = E[X(t)X(t+\tau)] = \lim_{T \to \infty} 1/2T \int_{-T/2}^{T/2} X(t) \cdot X(t+\tau)\, \mathrm{d}t \tag{2.45}$$

For an ergodic process, equations 2.44 and 2.45 are equal. If the ensemble average is used then a knowledge of joint cumulative probability distribution functions must be gained. As the noise process is considered ergodic, much of the work requiring the use of the autocorrelation function will not require a knowledge of the *joint distribution function*, and so we shall not deal with it in this brief discussion, except to define the function $F(x, y)$ for two random variables X and Y. Thus

$$F(x, y) = P(X \leqslant x, Y \leqslant y) = \int_{-\infty}^{x}\int_{-\infty}^{y} f(u, v) \cdot \mathrm{d}u \cdot \mathrm{d}v \tag{2.46}$$

where u and v are dummy variables and $f(u, v)$ is the joint probability density function.[6]

Power spectral density function

The *Wiener-Kinchine* theorem transposes the autocorrelation function from the time domain into the frequency domain where it becomes known as the *power spectral density*. This theorem also permits the power spectral density to be transposed from the frequency domain into the time domain to form the autocorrelation function. The autocorrelation function and the power spectral density are known as a *Fourier transform pair*, and given by the relationships:

$$G(f) = \mathscr{F}[R(\tau)] = \int_{-\infty}^{\infty} R(\tau)[\exp(-j2\pi f\tau)]\,d\tau \quad (2.47)$$

$$R(\tau) = \mathscr{F}[G(f)] = \int_{-\infty}^{\infty} G(f)[\exp(j2\pi f\tau)]\,df \quad (2.48)$$

$$R(\tau) \Leftrightarrow G(f) = \text{Fourier transform pair} \quad (2.49)$$

The name power spectral density arises from the interpretation placed on the autocorrelation function when there is no delay τ present. In the case of a voltage v volts, across a 1 ohm resistor equation 2.48 gives:

$$R(0) = E[X^2(t)] = |v^2| = \int_{-\infty}^{\infty} G(f)\,df = \text{the average power dissipated by the random process across a 1 ohm resistor.}$$

2.2 MODULATION/DEMODULATION SCHEMES

Pulse modulation schemes which are of commercial importance are:

- Pulse amplitude modulation (PAM)
- Pulse duration modulation (PDM)
- Pulse position modulation (PPM)
- Pulse code modulation (PCM)
- Delta modulation (DM)

In this section only PAM, PCM and DM will be considered at all fully, but for the sake of completeness, a brief description of PDM and PPM are given below. Both of these modulation types are forms of pulse-timing modulation (PTM), which have advantages over the amplitude modulation types when exposed to noise. These advantages are that the pulse heights are of constant amplitude and do not contain information, and so any amplitude variations have no effect. Furthermore, since these pulses might be characterized by their rising and falling edges, if steep enough, they will be relatively unaffected by noise-induced amplitude variations.

Pulse duration modulation (PDM). The low-pass analog input waveform is sampled at twice the highest frequency of that which occurs in the input waveform. At each sample time, a pulse is created having a width proportional to the amplitude of the input signal at the sampling instant. The amplitude of the generated pulse is constant. Hence there has been a conversion of amplitude information into pulse-width or pulse-time information.

Alternative names given to PDM are pulse width modulation (PWM) and pulse length modulation (PLM).

Pulse position modulation (PPM). In this type of modulating technique, the pulses formed from the analog input signal are constant in width and height, but their position relative to a fixed sampling instant, or time reference, is adjusted in accordance with the amplitude of the sampled signal.

2.2.1 Pulse code modulation (PCM)[1,2,4,5,9,10]

PCM may be characterized by three processes, namely sampling, quantizing and coding. Sampling is the periodic scanning of the analog signal to obtain an instantaneous amplitude. The lower limit of the sampling rate is determined by the Nyquist sampling theorem. Quantizing is to assign the instantaneous amplitude obtained to one of a specified set of levels, and thereby convert a small continuous range of levels of the sample to its nearest fixed level. The combined operations of sampling and quantizing generate a quantized pulse amplitude modulated (PAM) waveform, whose amplitudes are restricted to a number of discrete magnitudes. Coding is to represent each of the fixed levels (or sum of the incremental levels) by a binary sequence usually called a *codeword*. All codewords consist of a certain number of binary digits which are transmitted in the interval between consecutive sampling instants. Typically seven-digit (non-CCITT) or eight-digit (CCITT recommended) codewords are sufficient to represent most analog waveforms with adequate accuracy. The combination of a quantizer and an encoder is often called an *analog to digital (A/D) convertor*.

Figure 2.5 shows the block diagram of the source encoder used for PCM systems.

2.2.1.1 Sampling

Figure 2.6 shows the sampling of an analog waveform. In the ideal case the width of the samples would be infinitesimal. In practice however, the widths are finite, but usually of a much smaller duration than the waveform being sampled. With sampling pulse widths which are short compared with the period of the sampled waveform, it would appear that to reconstruct the sampled waveform, the sampling pulses would have to be very closely spaced. This is not necessarily so. The Nyquist sampling rate, otherwise known as the sampling theorem, shows that it is possible for a low-pass band-limited signal $f(t)$ to be exactly specified by its sampled values provided that the time interval between the sampled values does not exceed half the period of the highest frequency, or, to express this alternatively, the lowest periodic sampling frequency can be no less than twice the highest frequency of the analog

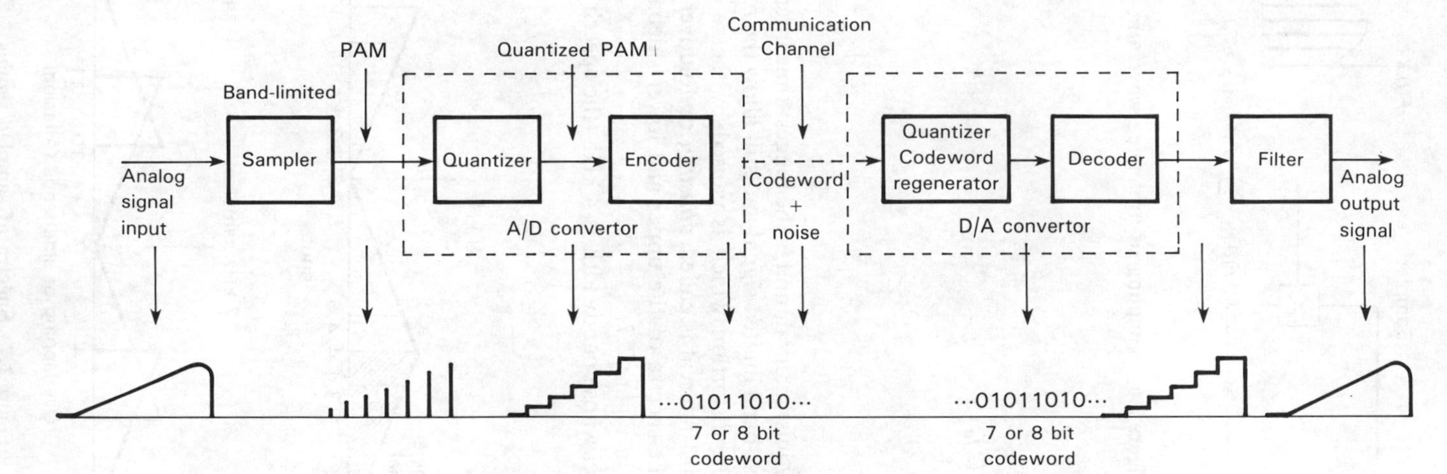

Figure 2.5 Block diagram of the source encoder/decoder in a PCM system

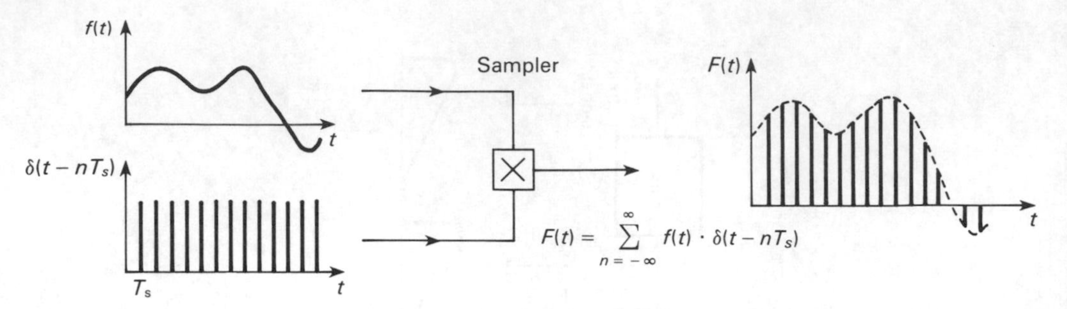

Figure 2.6 Sampling of an analog waveform

signal. That is

$$(T_s)_{\max} = T_a/2 = 1/(2f_a) \tag{2.50a}$$

or

$$(f_s)_{\min} = 2f_a \tag{2.50b}$$

$(T_s)_{\max}$ is called the Nyquist interval, and is the longest time interval that can be used for sampling a low-pass-band limited signal and still allow the recovery of the signal without distortion. The distortion which is caused by the sampling period being shorter than the Nyquist interval is called *aliasing*, or foldover distortion, and implies overlapping of the spectra of the upper and lower sidebands of the sampled signal. This is shown in Figure 2.7.

For a voice channel whose bandwidth is 4 kHz (the speech band ranging from

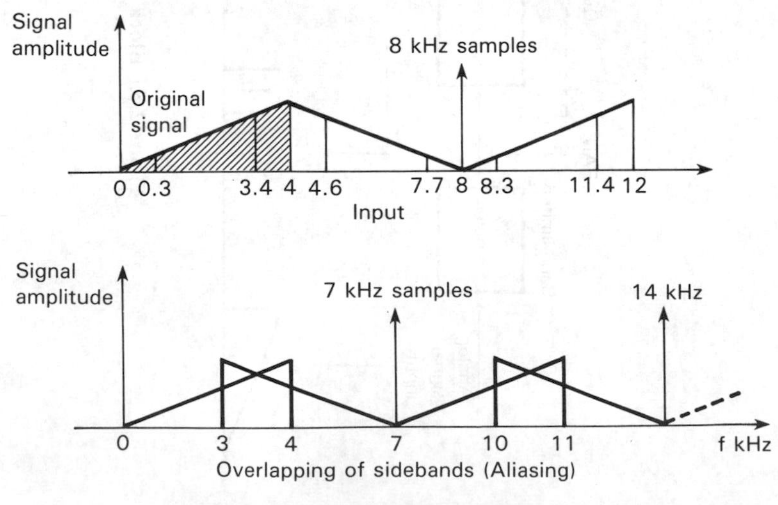

Figure 2.7 Spectra of sampled signals

0.3 to 3.4 kHz), the sampling rate is taken at 8000 samples/s or twice the highest band-limited frequency (which is 4 kHz). Hence the sampler will sample the input waveform every 125 μs, and the width of the sample must be much smaller than 125 μs (usually it is taken to be around 0.9 to 1 μs). The process of sampling is in effect a modulation of the analog waveform, and may be represented as the time multiplication of the input waveform $f(t)$ by a Dirac delta function $\delta(t - nT_s)$. The total resulting sampled waveform, being the sum of each of the multiplications at the sampling instants nT_s, where $n = 0, \ \pm -1, \ \pm -2, \ldots$, is given by

$$Y(t) = f(t) \cdot \sum_{n=-\infty}^{\infty} \delta[t - nT_s] \tag{2.51}$$

In Figure 2.7, the resulting frequency spectrum of the sampled input waveform shows upper and lower sidebands 4 kHz wide (or the highest analog waveform frequency), centered on 8 kHz and 16 kHz. Sidebands which are not shown in the figure would also center about all multiples of 8 kHz, up to $8n$ kHz, and would mathematically extend to the negative frequency region, for negative values of n.

The value of an 8 kHz sampling frequency, used in practice, is acceptable, for the speech band extends only to 3.4 kHz, hence providing a guard band between the sideband spectra as shown. This guard band prevents distortion occurring due to aliasing.

The circuits used to achieve the sampling process are usually sample and hold, in which a fast transistor used as a switch is turned on and off by a sampling pulse. The resulting sampled waveform level is then stored on a capacitor and sensed by a high-impedance buffer circuit before further processing. A high impedance is used to prevent the charge on the capacitor from leaking away during the interval between samples. The small amount of charge which in practice does leak away is called the 'droop'. Figure 2.8 shows the basic sample and hold circuit commonly used.

In contrast to the above discussion for the *low-pass* band of signals, the case for a *band-pass* band of signals sampling frequency is given below. For a bandwidth W, an an upper frequency f_h, the analog band-pass signal $x(t)$ can be represented by instantaneous values $x(nT_s)$ if the sampling rate is $2f_h/p$, where p is the largest integral value which is no greater than f_h/W. This is demonstrated in Table 2.3.

As $f_h/W \gg 1$, the minimum sampling rate approaches $2W$. This is an important result, because it shows that if an FM analog signal with a bandwidth of say 48 kHz,

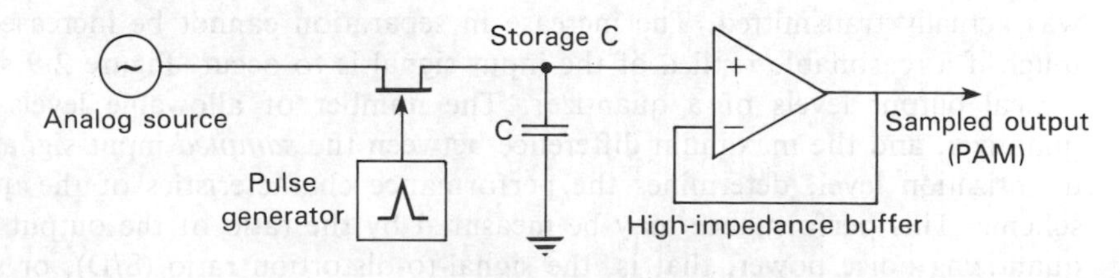

Figure 2.8 Sample and hold circuit basics

Table 2.3 Minimum sampling frequency for band-pass signals

Highest frequency in range f_h	Largest integer p $p < f_h/W$	Sampling frequency f_s $f_s = 2f_h/p$
W	1	$2W$
$1.5W$	1	$3W$
$1.8W$	1	$3.6W$
$2W$	1 or 2	$4W$ or $2W$
$2.5W$	2	$2.5W$
$2.75W$	2	$2.75W$
$3W$	2 or 3	$3W$ or $2W$
$3.6W$	3	$2.4W$
$4W$	3 or 4	$2.66\dot{6}W$ or $2W$
$5W$	4 or 5	$2.5W$ or $2W$
⋮	⋮	⋮
$10W$	9 or 10	$2.22W$ or $2W$
⋮	⋮	⋮
$25W$	24 or 25	$2.08W$ or $2W$

is in a frequency-division-multiplex system, with an upper frequency of 1052 kHz, then as $1052/48 = 21.9$, $p = 21$, and hence the minimum sampling frequency is $2f_h/p = 100.2$ kHz, and not twice the upper frequency (2104 kHz), which is true of the low-pass case. The sampling of an analog signal not only prepares it for quantization as described in Section 2.2.1.2, but permits that analog signal and others which have been sampled to be combined by interleaving, thus allowing the signals to be time division multiplexed. This subject is considered in Chapter 3.

2.2.1.2 Quantizing

Quantizing is the representation of a sampled analog signal by a finite set of amplitude levels, and is thus the conversion of a continuous time signal into a discrete amplitude signal. A signal which is to be quantized prior to transmission has usually been sampled. The advantage gained in quantizing the sampled signal lies in the reduction of the effect of noise on the system. Quantizing limits the number of permissible levels which can be taken by the sampled signal, and in effect prepares the original input signal for its transition from an analog to a digital form. If the separation between the quantizing levels is large compared with the perturbations caused by noise, the receiver may easily decide precisely which specific level was actually transmitted. The increase in separation cannot be increased by too much if a reasonable replica of the input signal is to occur. Figure 2.9 shows the typical output levels of a quantizer. The number of allowable levels Q, of a quantizer, and the maximum difference between the *sampled* input signal and the quantization level, determines the performance characteristics of the quantizing scheme. This performance may be measured by the ratio of the output signal to quantizing noise power, that is, the signal-to-distortion ratio (S/D), or signal-to-noise ratio (S/N). The difference between the original magnitude of the sample, and

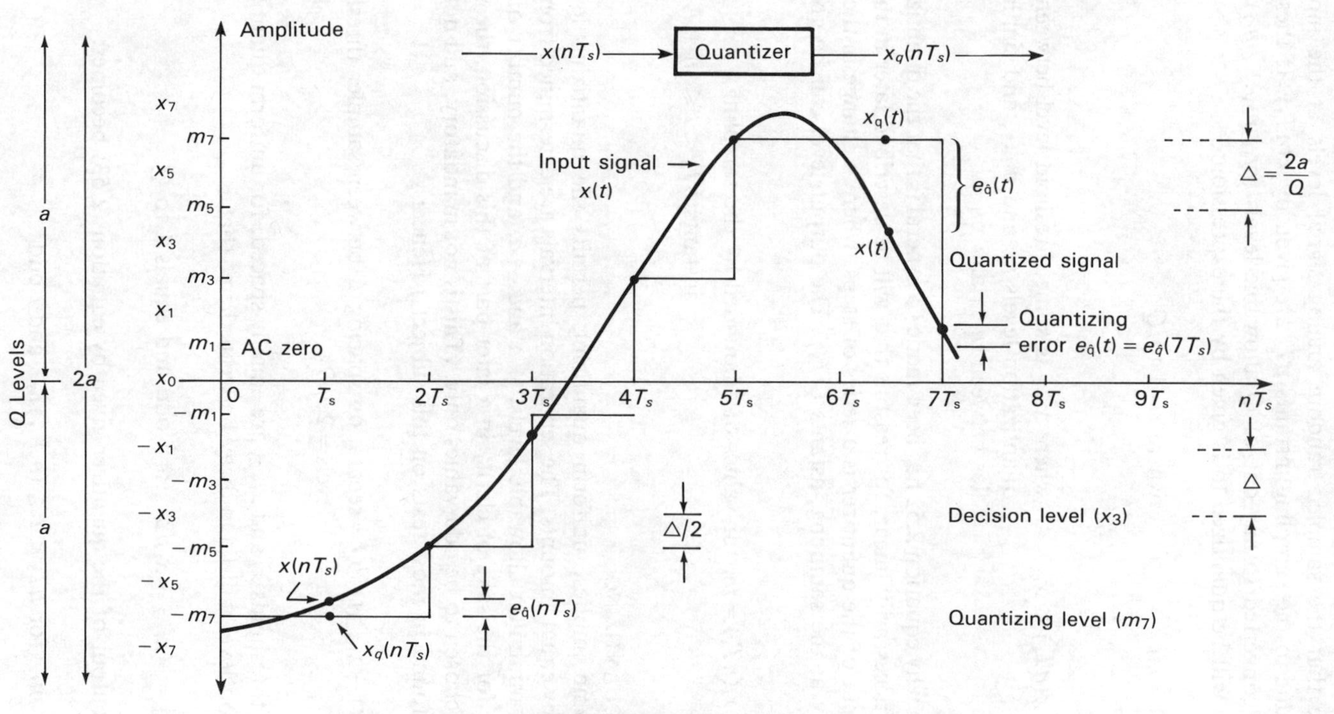

Figure 2.9 Quantization of sampled input waveform

the reconstructed value based on the nearest quantizing level, is the quantizing noise power, and is known as the *quantization distortion.*

Assume that the sampled random analog signal levels at the input to the quantizer due to the sampling instants nT_s are given by $x(nT_s)$. If these sampled values are converted to any one of Q allowable levels $(m_1, m_2, ..., m_Q)$, the level which they will be quantized to is given by the expression:

$$x_q(nT_s) = m_i \qquad \text{where } i = 1 \text{ to } Q \tag{2.52}$$

if

$x_{i-1} < x(nT_s) < x_i$ where x_{i-1} is some threshold level between the two quantization levels m_i and m_{i-1}, and similarly, x_i lies between m_{i+1} and m_i.

The reason why equation 2.52 has been made so general is that the quantization process is not necessarily linear, for reasons that will be described later in this section.

The output of the quantizer is a series of steps, where a change in the step level occurs only at the sampling instants (nT_s). The output level can be given by the expression

$$x_i(t) = x_i(nT_s) = m_i \text{ for values of } t \text{ bounded by the sampling instants } nT_s < t \leqslant (n + 1)T_s \tag{2.53}$$

UNIFORM QUANTIZING

As the name implies, uniform quantizing permits the quantizing levels to be separated by equal amounts. The separation intervals are determined from the maximum and minimum allowable levels, $+a$ and $-a$, and the number of intervals. Note that, for the sake of clarity in a later part of this discussion, the values of i have been chosen to be odd values only. (This is not mandatory, and not the usual approach found in most texts on this subject.) Hence:

$$[x_{\max}(t) - x_{\min}(t)]/Q = \text{separation spacing } \Delta \text{ between sampler threshold levels} = 2a/Q \tag{2.54}$$

and as the thresholds x_i and x_{i+2}, are equally spaced, for uniform quantization, the quantization level will lie midway between them, that is

$$|m_{i+2}| = |(x_{i+2} - x_i)/2| \text{ for } i \text{ odd and equals 1 to } Q \tag{2.55}$$

Thus the output of the quantizer given by equation (2.53) becomes:

$$x_q(t) = m_i \quad \text{for } nT_s < t \leqslant (n+1)T_s \quad \text{and } i \text{ odd}$$

On reconstruction of the quantized waveform at the receiver into an analog waveform, there will be a difference at any point in time t, between it and the original transmitted signal. This difference, as previously mentioned, is the quantization distortion or quantization error $e(t)$. Figure 2.9 shows this error $e(t)$, which may be expressed by

$$e_q(t) = x_q(t) - x(t) \tag{2.56}$$

This quantizing error appears as random noise, with a power $\overline{e^2}$ as calculated below.

Assume that the linear quantizer has a dynamic range equal to a value of $2a$ volts, from the values of $-a$ to $+a$. If Q equal levels are required by the quantizer designer, the spacing between the levels is given by

$$2a/Q = \Delta \tag{2.57}$$

and so

$$|e_q(t)| \leqslant a/Q \text{ or } |e_q(t)| \leqslant \Delta/2$$

This means that the maximum error that is permitted is made when the input signal waveform is at the decision threshold (x_i), and has an amplitude of $\Delta/2$ from the quantization level. If it is assumed that the input signal is random and has equal probability of occurrence over the range $(-a/Q, a/Q)$, its probability density function (pdf) is given by $f(e_q) = Q/2a$, and its probability mass function $P(X = m_i) = 1/Q$. Figure 2.2 shows the plots of the pdf and the CPDF. As shown in Section 2.1.3.2, the mean square value of a continuous function is given by:

$$\overline{X^2} = \int_{-\infty}^{\infty} x^2 f(x)\,dx$$

Hence

$$\overline{e^2} = \int_{-a/Q}^{a/Q} e_q^2 Q/2a\,de_q \tag{2.58}$$

$$= a^2/3Q^2$$

Since $\Delta = 2a/Q$

$$\text{Noise power } N = \overline{e^2} = \Delta^2/12 \tag{2.59}$$

This noise, although referred to the receiver by the reconstruction of the waveform, is a virtual noise appearing at the output of the quantizer. The signal power at the output can be calculated in a similar fashion, but not using the continuous variable analysis as used in deriving the quantization noise. This is because the signal power depends on the quantized level. Using the discrete random variable method, we have

$$\overline{X^2} = \sum_{i=1}^{n} x_i^2 P(X = x_i)$$

Hence

$$\bar{S} = \sum_{i=1}^{Q} m_i^2 \cdot 1/Q = \sum_{i=1}^{Q} x_q^2 \cdot 1/Q \tag{2.60}$$

As the quantization levels occur at only odd intervals of $\Delta/2$,

$$m_i = i \cdot \Delta/2 \tag{2.61}$$

In equations 2.60 and 2.61, $i = 1, 3, 5 \ldots, (Q-1)$ as shown in Figure 2.9. The number of quantization levels however, are equally divided about zero in practice, providing both positive and negative quantization levels; hence the average signal power given by equation 2.60 must be doubled to account for the doubling up of levels. Thus,

$$\bar{S} = 2/Q \sum_{i=1}^{Q} m_i^2 = 2/Q(\Delta^2/4) \sum_{i=1}^{Q} i^2 = 2/Q(\Delta^2/4) \cdot Q(Q^2-1)/6$$

$$= (Q^2-1) \cdot \Delta^2/12 \tag{2.62}$$

From equations 2.59 and 2.62 the signal-to-noise ratio at the output of the quantizer can be determined. That is

$$S/N = (Q^2-1) \cdot \Delta^2/12 \cdot 12/\Delta^2 = Q^2 - 1 \tag{2.63}$$

Hence for $Q \gg 1$, and as Q levels can be coded by a codeword of length b bits

$$S/N = 20 \log Q \text{ dB} = 20 \log 2^b = 6b \text{ dB} \tag{2.64}$$

Equation 2.64 shows that for every doubling of Q, or every additional bit added to a codeword, the value of S/N increases by 6 dB. Similarly, by expressing the noise power in dB using equation 2.59, with b equal to the number of bits, we have:

$$\text{Noise power } N = 10 \log \Delta^2/12 = 10 \log\left[\frac{(2a)^2 \cdot 2^{-2b}}{12}\right]$$

$$= 20 \log 2a - 10.8 - 6\text{b dB} \tag{2.65}$$

the noise being reduced by 6 dB for each additional bit added to a codeword.

Equation 2.64 implies that the larger the number of quantization levels, the higher the signal-to-noise ratio for a signal which has a uniform probability density function pdf. As the number of quantization levels increases, for a fixed output signal level, the lower the quantization distortion becomes because, as discussed above, an error of up to half a quantum may occur between the actual input sample and the quantized value. This sounds fine, except that as the quantization levels become more closely spaced, thermal noise, or other forms of noise on the input signal may cause erroneous quantization levels to be chosen. As always, an engineering compromise must be reached, because if the quantization levels are too widely spaced the reconstituted signal is not a near replica of the original input signal due to the excessive quantization distortion. In addition to the quantization distortion effect, another effect pertinent to the uniform quantization scheme occurs: for very low-level input signals occurring between the decision level representing AC zero, and the next level above or below, there will be no change in state of a binary number corresponding to the AC zero quantization level; hence the quantizing distortion for low-level signals may become intolerable, and yet for high-level signals be low enough to appear quite acceptable. If the spacing between the decision values is reduced for low-level signals, and increased for high-level signals, it is possible to reduce the effect that the quantization distortion has on these low-level signals, making the effect of the distortion quite acceptable over the required dynamic range of the input signal. There are three methods of achieving this improvement. These are:

1. Using a uniform quantizer preceded by sample compression, where nonlinear amplification of the signal at the quantizer input is chosen to force the Gaussian signal distribution to become a more uniform distribution. The complimentary nonlinear amplification to remove the imposed distortion on the input waveform occurs after transmission, decoding and signal reconstruction.
2. Using a nonlinear quantizer, with small spacings for low input levels.
3. Using a uniform quantizer in which the spacing between decision levels has been significantly reduced by increasing the quantization levels, and then manipulating the binary numbers to reduce the number of bits required.

COMPANDING

Companding is the first method as listed above, in obtaining a nonuniform quantization system. A logarithmic quantizing law is used in the COMPression and exPANDING process where the input variable X is transformed to another variable Y by the relationship $Y = \log X$ and the inverse of this relationship used when recovering the input variable at the system output when expanding. This relationship permits crowding of the levels at the lower levels, and wider spacing as the input level increases. The effect produced by amplitude compression of the speech waveform is to reduce its dynamic range and to produce a higher average signal-to-quantizing-noise-power ratio than that of the uniform quantizer for a Gaussian input signal. The difficulty encountered in the expanding process is to maintain exactly the complimentary compression characteristic, due to variation among diodes and the dependence of their characteristics on temperature.

The two commonly used logarithmic compression laws are: μ law and A law.

μ LAW

This law is given by equation 2.66:

$$y = \begin{cases} y_{\max} \dfrac{\ln[1 + \mu(x/x_{\max})]}{\ln(1 + \mu)}, & 0 \leqslant x \leqslant x_{\max} \\[2ex] -y_{\max} \dfrac{\ln[1 - \mu(x/x_{\max})]}{\ln(1 + \mu)}, & -x_{\max} \leqslant x \leqslant 0 \end{cases} \tag{2.66}$$

where x_{max} is the maximum input signal voltage, μ is a parameter which usually equals 100, but may be 255 for some systems (U.S. D1 system has $\mu = 100$, whereas the D2 system uses $\mu = 255$), and y is the compressed output signal which is to enter the uniform quantizer. It has a maximum value y_{max}. When $\mu = 0$, there is no companding, and the output is directly related to the input.

Table 2.4 provides values of y/y_{max}, for various values of x/x_{max}, for both $\mu = 100$ and $\mu = 255$.

A LAW

This law is given by equation 2.67:

$$y = \begin{cases} y_{max} \dfrac{A(x/x_{max})}{(1 + \ln A)}, & 0 \leqslant (x/x_{max}) \leqslant 1/A \\ y_{max} \dfrac{1 + \ln(Ax/x_{max})}{(1 + \ln A)}, & 1/A \leqslant (x/x_{max}) \leqslant 1 \end{cases} \tag{2.67}$$

A is usually chosen to be 87.6 (CCITT Recommendation for a 30-channel system plus one 8-bit slot for signaling, and one eight-bit slot for frame synchronization per frame).

The parameters μ and A both determine the severity of the compression

Table 2.4 Values of companded output signal for normalized input signal at $\mu = 100$ and 255

$\mu = 100$				$\mu = 255$			
x/x_{max}	y/y_{max}	x/x_{max}	y/y_{max}	x/x_{max}	y/y_{max}	x/x_{max}	y/y_{max}
0.01	0.1502	−0.01	−0.1502	0.01	0.2285	−0.01	−0.2285
0.02	0.2380	−0.02	−0.2380	0.02	0.3261	−0.02	−0.3261
0.03	0.3004	−0.03	−0.3004	0.03	0.3891	−0.03	−0.3891
0.04	0.3487	−0.04	−0.3487	0.04	0.4357	−0.04	−0.4357
0.05	0.3882	−0.05	−0.3882	0.05	0.4727	−0.05	−0.4727
0.06	0.4216	−0.06	−0.4216	0.06	0.5034	−0.06	−0.5034
0.07	0.4506	−0.07	−0.4506	0.07	0.5296	−0.07	−0.5296
0.08	0.4761	−0.08	−0.4761	0.08	0.5524	−0.08	−0.5524
0.09	0.4989	−0.09	−0.4989	0.09	0.5727	−0.09	−0.5727
0.10	0.5196	−0.10	−0.5196	0.10	0.5910	−0.10	−0.5910
0.20	0.6597	−0.20	−0.6597	0.20	0.7126	−0.20	−0.7126
0.30	0.7441	−0.30	−0.7441	0.30	0.7845	−0.30	−0.7845
0.40	0.8047	−0.40	−0.8047	0.40	0.8358	−0.40	−0.8358
0.50	0.8519	−0.50	−0.8519	0.50	0.8757	−0.50	−0.8757
0.60	0.8907	−0.60	−0.8907	0.60	0.9083	−0.60	−0.9083
0.70	0.9236	−0.70	−0.9236	0.70	0.9360	−0.70	−0.9360
0.80	0.9522	−0.80	−0.9522	0.80	0.9599	−0.80	−0.9599
0.90	0.9774	−0.90	−0.9774	0.90	0.9811	−0.90	−0.9811
1.00	1.0000	−1.00	−1.0000	1.00	1.0000	−1.00	−1.0000

characteristic. The greater their values, the greater will be the compression, and, as shown by Table 2.4, the greater the number of quantization levels appearing around the origin of the characteristic for quantizing lower amplitude levels of the input signal. As mentioned above, the use of diodes in generating the μ-law characteristic presents problems in the matching of the compression and expansion characteristics. To overcome this problem, it is generally accepted throughout the U.S.A. and Europe that digital compression techniques should be used, to produce a piece-wise linear approximation to the logarithmic laws. This will be further discussed later in the chapter. Before dealing with the two other forms of nonlinear quantization methods, we must first consider how the quantization levels are encoded into digital form. This is because quantization is an integral part of the encoding process when either a nonlinear quantizer, or a linear quantizer with manipulation of the binary codes, is used.

2.2.1.3 Encoding

Figure 2.5 shows the position of the encoder in a PCM system, and that for practical PCM systems the quantizer and encoder are effectively combined in what is known as an A/D convertor. The encoder, considered in isolation, is used to generate binary digits or codewords from the quantized values appearing at the output of the quantizer. As mentioned above in the discussion of the uniform quantizer, the number of quantization levels Q is equal to 2^b, where b equals the number of bits required to represent Q individual levels, each with a separate binary codeword comprising b bits.

For example, if 16 quantization levels were required, b would equal 4, and the code might run; 0000 = 0 V, 0001 = 1 V, 0010 = 2 V, ..., 1111 = 15 V. Naturally, if the maximum sampled level did not equal 15 V but some level such as $2a$, then each individual level would be spaced not by one volt, but by $Q/2a$, and the codeword would be allocated to each level in the same manner. The CCITT recommends that use be made of 8-bit PCM systems. This means that $b = 8$, and there are 256 levels which can be accommodated by 256 8-bit codewords. This would mean that if a uniform quantization scheme was to be used, for a maximum input signal level of ± 2.5 V, a change in the codeword would occur for approximately each 19.5 mV change in the input signal level.

The resulting series of codewords produced in this simple encoding process now represents digitally the analog signal. This digital signal may be either directly transmitted, which is unlikely except for very short distances, or may be further processed before direct transmission at baseband, or before it is used to modulate a carrier. The further processing of this signal, specifically for carrier modulation designed for the transmission of the digital information over a radio path, is the subject of subsequent chapters. However, before considering further this digital signal and the additional processes involved, we shall discuss the encoding of the information associated with the nonuniform quantization process.

A NONLINEAR QUANTIZER WITH SMALL SPACINGS FOR LOW INPUT LEVELS

This device, which is most commonly used throughout Europe and the U.S.A., uses digital compression techniques to produce a piece-wise linear approximation to

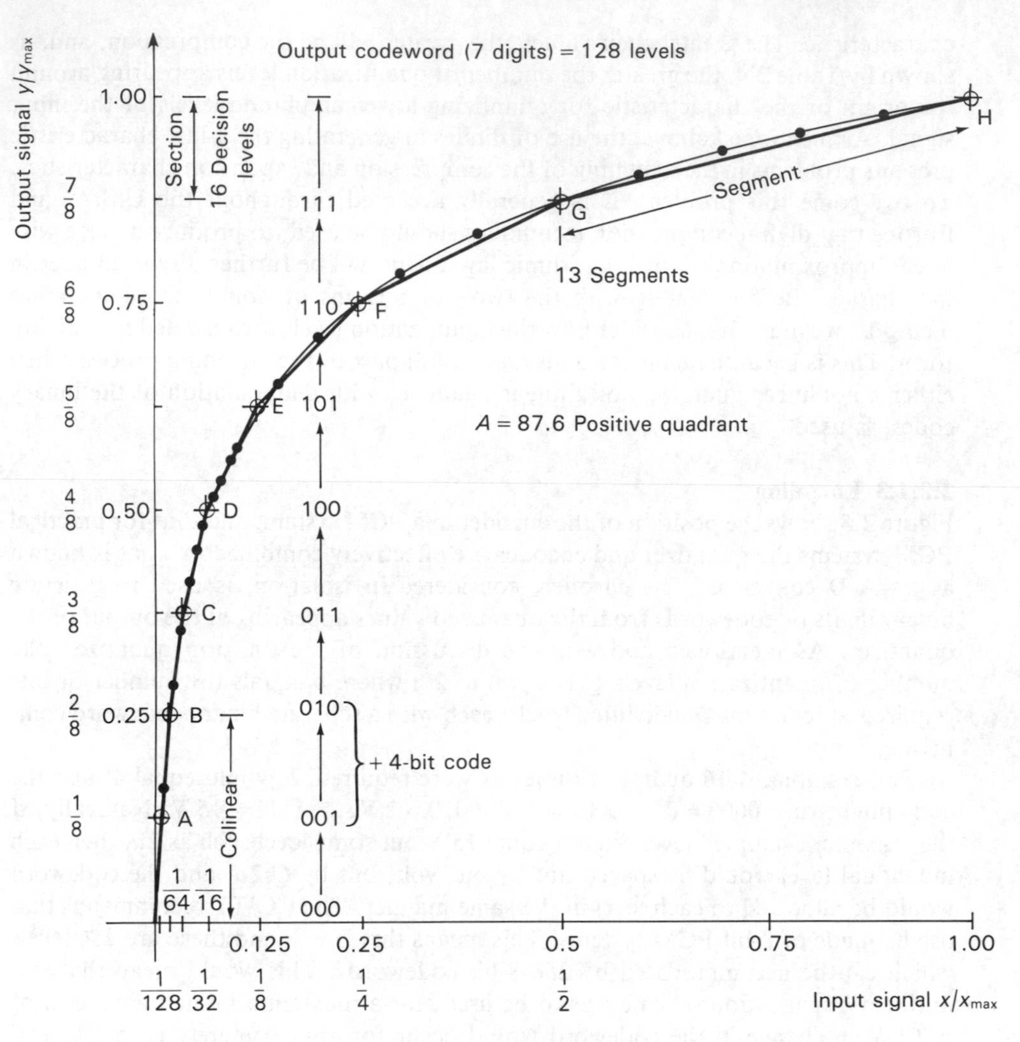

Figure 2.10 ***A*-law curve and piece-wise approximation (13 segments) for $A = 87.6$**

either the μ-law or A-law logarithmic companding characteristic. Figure 2.10 shows for demonstration the curve representing the A-law, for $A = 87.6$ according to values obtained in Table 2.5. In addition to the curve, piece-wise linear approximations to this curve can be seen. The approximated curve is used in the CEPT 30 + 2 PCM system (30 voice channels plus one 8-bit slot for signaling, and one 8-bit slot for frame synchronization per frame). The completely approximated curve comprises 13 segments, labelled A to H for the positive segment of the curve, and A′ to H′ for the negative. The region B to B′ is actually a straight line, and is not segmented. This is the reason for 13 segments only being used. For the sake of clarity Figure 2.10 does not show the negative region of the curve. This section of the curve lies in the negative x and negative y quadrant, and permits the total curve to be 'S' shaped. The y/y_{max} axis is divided into eight sections. This represents the region where each segment is enclosed. The fact that there are eight sections permits each section to be represented by a 3-bit code. Each section is divided into 16 decision levels, thus permitting a 4-bit code to be allocated to each of the levels. The total number of quantization levels in the positive region of the curve is thus 128. The positive and negative regions of the curve are allocated a 1-bit code (positive region = 1 and the negative region = 0). The total number of bits in the codeword is thus eight, and the total number of quantization levels is 256 ($2^8 = 256$). This means of course that the output digital representation of the input signal can be represented by 256 different 8-bit codewords.

The input amplitude is represented in Figure 2.10 by the x/x_{max} axis. This is logarithmically (base 2) divided into seven sections. As the input axis is normalized, with unity being the maximum value, the widest section represents 1/2, the next 1/4, followed by 1/8, 1/16, 1/32, 1/64 and 1/128. Each segment of the piece-wise curve (otherwise known as a *chord*) is contained in each of these sections, except for the segment B to zero, which as mentioned previously, is linear, and the 1/128 segment containment ignored. In the linear segment B to zero, there are 32 quantization levels. If a uniform quantizer was used which had the same resolution as the linear region B to zero in Figure 2.10, then, as the dynamic input signal range is the same for both nonlinear and linear processes, the number of quantization levels would be 32 for the region B to 0, doubling up to 64 for C to 0, again doubling up to 128 for D to 0, and continuing up to 2048 for the region H to 0. For the negative region, there are another 2048 equivalent linear quantization levels. This means that to achieve the same resolution and thus signal-to-noise ratio as a 256-level nonuniform quantizer, a uniform quantizer would require 4096 quantization levels or 12 codeword bits. Thus the nonuniform quantizer has the companding advantage of 20 log (4096/256) = 24.08 dB, which may be derived as the improvement in S/N for

Table 2.5 Values of companded output signal for normalized input signal at $A = 87.6$

$A = 87.6$							
x/x_{max}	y/y_{max}	x/x_{max}	y/y_{max}	x/x_{max}	y/y_{max}	x/x_{max}	y/y_{max}
0.01	0.1601	−0.01	−0.1601	0.02	0.2852	−0.02	−0.2852
0.03	0.3593	−0.03	−0.3593	0.04	0.4118	−0.04	−0.4118
0.05	0.4526	−0.05	−0.4526	0.06	0.4859	−0.06	−0.4859
0.07	0.5141	−0.07	−0.5141	0.08	0.5385	−0.08	−0.5385
0.09	0.5600	−0.09	−0.5600	0.10	0.5793	−0.10	−0.5793
0.12	0.6126	−0.12	−0.6126	0.15	0.6534	−0.15	−0.6534
0.20	0.7059	−0.20	−0.7059	0.25	0.7467	−0.25	−0.7467
0.30	0.7800	−0.30	−0.7800	0.40	0.8326	−0.40	−0.8326
0.50	0.8733	−0.50	−0.8733	0.60	0.9067	−0.60	−0.9067
0.70	0.9348	−0.70	−0.9348	0.80	0.9592	−0.80	−0.9592
0.90	0.9807	−0.90	−0.9807	1.00	1.0000	−1.00	−1.0000

the two quantization levels according to equation 2.64, i.e.

$$(S/N)_{\text{nonlinear}} - (S/N)_{\text{linear}} = \text{companding advantage} = 20 \log Q_{\text{nl}} - 20 \log Q_{\text{l}}$$

or

$$\text{Companding advantage} = 20 \log Q_{\text{nl}}/Q_{\text{l}} \tag{2.68}$$

In the nonuniform or nonlinear quantizer, the binary bits comprising the codeword are arranged as in Table 2.6.

Table 2.6 Summary of nonuniform *A*-law quantizer and encoder

Binary bits:	I_7	$I_6I_5I_4$	$I_3I_2I_1I_0$
Segment	Sign Bit no. 1	Segment selection Bit no. 2 to 4	Position within segment Bit no. 5 to 8
0–A(′)	1/(0)	0 0 0	16 codewords (0000 to 1111)
A–B(′)	1/(0)	0 0 1	16 codewords (0000 to 1111)
B–C(′)	1/(0)	0 1 0	16 codewords (0000 to 1111)
C–D(′)	1/(0)	0 1 1	16 codewords (0000 to 1111)
D–E(′)	1/(0)	1 0 0	16 codewords (0000 to 1111)
E–F(′)	1/(0)	1 0 1	16 codewords (0000 to 1111)
F–G(′)	1/(0)	1 1 0	16 codewords (0000 to 1111)
G–H(′)	1/(0)	1 1 1	16 codewords (0000 to 1111)
Bits used	1	3	4 Total = 8 bits

Totals: Regions = 2. Segments/region = 8. Codewords/region = 8 × 16 = 128
Thus: Total bits = 8. Total regions = 2. Total sections = 16. Total codewords = 256. Total segments = 13

Note
(′) indicates negative quadrant, that is segment 0–A′, A′–B′, etc.

Similar to the above scheme is the D2 (U.S.) system in which the μ law is used, with $\mu = 255$. The codeword structure is the same and there are 256 quantization levels. The difference in the piece-wise approximation to the μ-law curve for that used for the A-law curve is that 15 segments are used. This means that the region B–B′ shown in Figure 2.10 is not a straight line, but is segmented into the segments B–A, A–A′ and A′–B′. Another difference between the CEPT 32 system and the D2 system is that the sign bit I_0 is used for signaling in every sixth word of each frame.

The piece-wise linear approximation to the logarithmic compression law may be obtained by using a nonlinear digital-to-analog convertor in the feedback path of a feedback encoder. Another method is to use purely digital methods employing a 12-bit linear analog-to-digital convertor. Both methods are explained in Reference 10, Chapter 3, and in Reference 11, Chapter 6.

USING A UNIFORM QUANTIZER WITH REDUCED SPACINGS BETWEEN THE DECISION LEVELS

As mentioned above, if the nonuniform quantizer had been expanded to become a uniform quantizer, there would have been required for each positive and negative region 2048 equivalent uniform quantization levels. This represents for each region an 11-bit codeword. Due to the redundancies in groups of bits, within this 11-bit codeword, it may be recoded into 7 digits, to produce the same nonlinear effect as the nonuniform quantizer. Thus the binary numbers are manipulated to reduce the number of bits in the codeword. Reference 10, Figure 9, briefly shows the coding pattern for the 11-bit codeword.

REQUANTIZATION AND DECODING

Throughout this discussion only encoding has been dealt with, and emphasis has been placed on the fact that the quantization and encoding process in the actual analog-to-digital convertor itself cannot usually be separated. At the receiver end of the system, at the point where the digital bits have been produced by the receiver signal processing equipment, this coded information must be decoded. The first operation in the reception process of the PCM system is the separation of the received signal from the noise. This is usually performed by the operation of requantization, in which the sampler makes the decision as to whether the received digit is either a 1 or a 0 logic level. The reconstituted digital bit stream is then passed to the decoder, where a digital-to-analog (D/A) convertor performs the inverse operation of the encoder, by changing digital codewords to discrete quantization levels. The output of the decoder is thus a sequence of quantized multilevel sample pulses representing the quantized PAM signal. This signal is then filtered to reject the frequencies which lie outside the voice band. The resulting output is similar to that of the analog input signal at the transmitter end. To assist the receiver sampler in sampling at the correct instant, timing information is sent together with the codeword. This will be discussed in the next two sections.

2.2.1.4 Bandwidth requirements

The Nyquist theorems on minimum bandwidth transmission show that it is possible to transmit r_s independent symbols per second, with zero inter-symbol interference, through an ideal low-pass filter having a bandwidth $B = r_s/2$. Hence:

$$\text{Bandwidth } B \geqslant (\text{symbol rate } r_s)/2 \qquad (2.69)$$

In a PCM system, the symbol rate equals the number of bits in a codeword formed by the quantization of a sample, multiplied by the sample rate f_s. That is

$$\text{PCM symbol rate } r_s = \text{codeword bits } b \times \text{sample rate } f_s \qquad (2.70)$$

Since the number of codeword bits b depends upon the number of quantization levels Q, as shown in Section 2.2.1.3, we have:

$$\text{Number of binary bits } b \text{ in a PCM codeword} = \log_2(Q) \qquad (2.71)$$

Hence from equations 2.69, 2.70 and 2.71, the bandwidth of a PCM system using binary encoding is given by:

$$\text{BPCM} > (f_s/2) \cdot \log_2(Q) \text{ kHz}$$

where f_s is in kHz.

Thus for a sample rate of 8 kHz, and a system of 256 levels, a minimum bandwidth of 32 kHz would be required to transmit a voice channel. When compared to analog methods where only 3.1 kHz (0.3–3.4 kHz) is required to transmit the speech portion of the voice band (0–4 kHz), the bandwidth requirement is quite enormous and is approximately ten times more. To reduce this bandwidth, in preparation for transmission over the channel, use is made of M-ary schemes, where instead of a binary digit being transmitted, an M-level digit is formed and transmitted. This will be discussed in Chapter 5. However, for M-level transmission, the transmission bandwidth becomes:

$$B \geqslant (f_s/2) \cdot \frac{\log_2(Q)}{\log_2 M} = \frac{b \cdot (f_s/2)}{\log_2 M} \tag{2.72}$$

Example

Assume that a nonlinear A-law compander produces an output S/N of 72.25 dB. Determine the approximate bandwidth requirement of an ideal low-pass filter, if the PCM binary signal results in an 8-level signal prior to entering the filter.

Solution
From equation 2.68, S/N for a uniform quantizer is found to be 48.16 dB. From equation 2.64, this represents 265 quantization levels. From equation 2.72 this represents an approximate bandwidth of 10.667 kHz if $M = 8$.

2.2.2 Differential pulse code modulation[12] (DPCM)

One way of reducing the bandwidth by a factor of 2 is to reduce the number of bits b in a codeword by half. This can be done using DPCM, where only the digitally encoded difference between successive samples is transmitted. As the differences between successive samples are expected to be smaller than the values of the sampled amplitudes, fewer bits are required to represent the differences. This form of encoding can produce good results if the succesive sample variations are close in amplitude, or if there is a high degree of correlation between samples. This occurs especially in video signals in which background or tonal values do not appreciably change between sampling instants. Black-and-white television reception of acceptable quality can be achieved using ordinary PCM with 256 quantization levels and 8-bit codewords. Using DPCM, the same quality can be attained using 8 quantization levels and codewords of 3 bits. This means that the bandwidth requirement has been reduced to 3/8 that of ordinary PCM. For quantization levels greater or equal

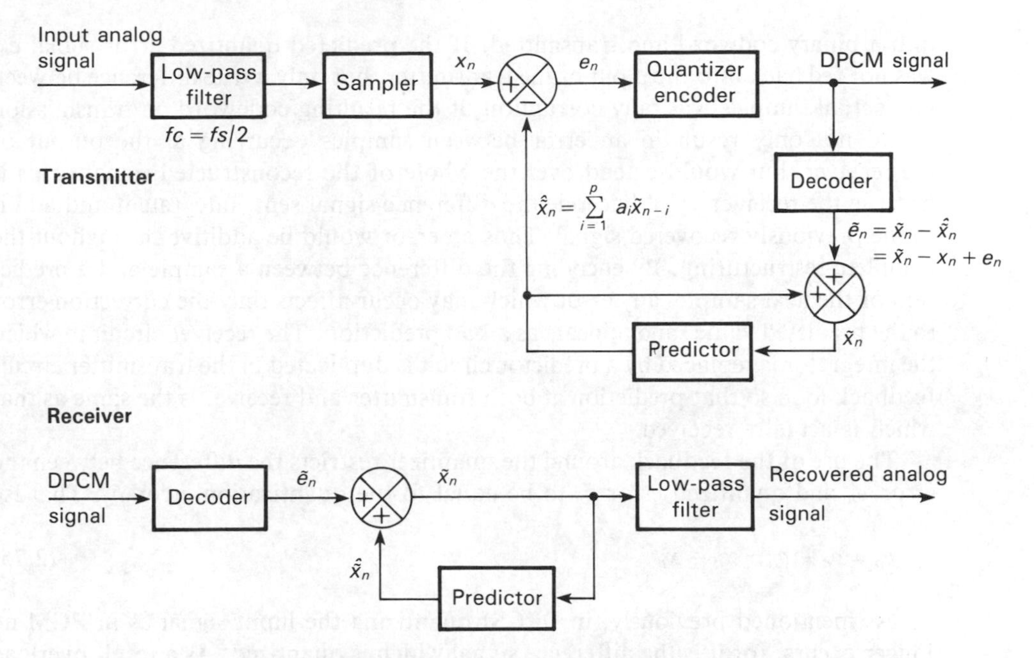

Figure 2.11 Block diagram of a DPCM transmitter and receiver

to 4, DPCM can result in a higher S/N than those offered by PCM using the same bit duration. However, analog signals that vary rapidly from sample to sample, are not highly correlated and cannot be accurately predicted, and so cannot be adapted to DPCM without large quantization errors occurring. Figure 2.11 shows the block diagram of the transmitter and receiver of a simple differential PCM system. After the low-pass filter limits the bandwidth of the input signal to one-half or less of the sampling rate f_s, the transmitter quantizes and codes the difference between an analog sampled signal x_n, and a predicted signal $\hat{\tilde{x}}_n$, which is fed back from the predictor output. The predicted value of the next sample is obtained by extrapolation from p previous sample values:

$$\hat{\tilde{x}}_n = \sum_{i=1}^{p} a_i \tilde{x}_{n-i} \tag{2.73}$$

where a_i are the predictor coefficients and are selected to minimize the error between the sampled value and the predicted value of the next sample. The input to the predictor is the value $\tilde{x}_n$, which represents the signal sample x_n modified by the quantization process. The difference between the sampled input signal and the quantized predicted output signal is the error e_n. This is given by:

$$e_n = x_n - \hat{\tilde{x}}_n \tag{2.74}$$

The output from the quantizer is the quantized value of this error. This is encoded into a binary codword and transmitted. If the predicted quantized error signal $\tilde{e}_n$, was not fed back into the input of the transmitter, but only a true difference between two actual samples was, any corruption of the resulting codeword on transmission would not only result in an error between samples occurring at the output of the receiver, but would extend over the whole of the reconstructed signal. This is because the receiver would decode the difference signal sent, integrate it and add it to the previously recovered signal. Thus an error would be additive throughout the complete restructuring. By encoding the difference between a sample and a prediction of the next sample, any error which may occur affects only the correction error to the predicted value, and appears as a bad prediction. The receiver circuit in which the integrator is replaced by a predictor circuit is duplicated in the transmitter circuit feedback loop so that prediction at both transmitter and receiver is the same as that which is actually received.

The use of the feedback around the quantizer restricts the difference between the error e_n and quantized error $\tilde{e}_n$ to be equal to the quantization error e_q. That is

$$e_q = \tilde{e}_n - e_n = \tilde{x}_n - x_n \tag{2.75}$$

As mentioned previously, in DPCM quantizing the input signal as in PCM no longer occurs, for it is the difference signal which is quantized. As a result overload errors which may be present in PCM are characterized, when they occur in DPCM, not as amplitude overload, but as *slope overload*. This is shown in Figure 2.12 which shows the waveforms of a delta modulation system, which is the simplest form of DPCM. Slope overload occurs whenever the largest step size of the quantizer is too small to keep up with the rate of change of slope of the input waveform $x(t)$. This results in a lag of the staircase function $\hat{\tilde{x}}_n(t)$ with respect to the fast changing input $x(t)$. The opposite situation to this, producing *granular noise*, occurs when the input variations are too small in comparison with the smallest step size of the quantizer. This is also shown in Figure 2.12. Granular noise may occur during the silent periods in speech, or of constant gray or nearly constant gray areas in images. A third kind of DPCM error occurs specifically with image coding. This occurs in the

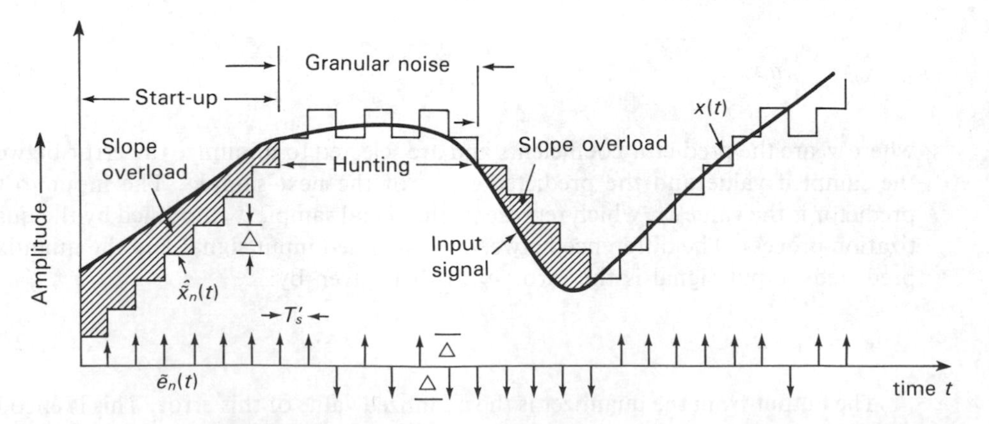

Figure 2.12 Waveforms of a delta modulation process

coding of low-constrast edges, where the quantizer output oscillates in a random fashion, taking on slightly different values from line-to-line or from frame-to-frame. This error is called *edge-business*. These three types of errors are different from those which may be produced by PCM, and may cause annoyance. Because they are not pertinent to PCM systems they are not included in any specified mean-square error for noise in a PCM system.

2.2.3 Adaptive differential pulse code modulation (ADPCM)

As shown by equation 2.70, the symbol rate or bit rate for an 8-bit codeword PCM signal in which the voice signal has been sampled at an 8 kHz rate is 64 kbit/s. If the number of codeword bits is reduced to 4, as for differential pulse code modulation (DPCM), the bit rate is halved and becomes 32 kbit/s. There is a trend towards internationally standardizing 32 kbit/s as the bit rate for voice-encoded signals using ADPCM. This is reflected in the CCITT Recommendation G721, which deals with 32 kbit/s Adaptive Differential Pulse Code Modulation,[13] and other papers.[14,15]

As mentioned in Section 2.1.4, a *stationary* random process is defined as one in which the statistical characteristics of the measurements made to a process do not change with time. Many real signal sources are not stationary, but quasi-stationary. This is indicated by their variance and autocorrelation function slowly varying in time. The encoders used in the PCM and DPCM systems are designed on the basis of stationary input signals and must be modified to handle quasi-stationary signal sources. If a uniform PCM quantizer is considered for the moment, as the mean of the quantization noise is zero, the variance or quantization noise power is given by equation 2.59, i.e. the variance $= \Delta^2/12$. If this variance is changing due to the changing quantization error caused by a quasi-stationary signal input, one way to counteract it would be to modify the quantization spacing Δ. This is one of the ways in which an adaptive quantizer operates. The adaptive quantizer varies its spacing in accordance with the variance of the past signal samples. Algorithms have been developed for differential pulse code modulation in the encoding of speech signals[12] using both an adaptive quantizer and an adaptive predictor, in which the coefficients are periodically changed to reflect the changing input signal statistics. Rather than transmit the predictor coefficients to the receiver, and thus increase the number of bits transmitted, and the bit rate, the receiver predictor computes its own coefficients.

There are two different types of adaptive systems. The first is the DPCM system with adaptive quantization (this is often termed DPCM–AQB), whilst the other system incorporates both an adaptive quantizer and an adaptive predictor. This is termed DPCM–APB–AQB. The 'B' in both APB and AQB stands for *backward* estimations and each will briefly be described below. For a fuller study of these techniques, see Reference 12.

'Forward' and 'backward' estimations in adaptive quantization–AQF, AQB

Figure 2.13 shows adaptive quantization with forward and backward estimation of input signal level. Forward estimates of step size are unaffected by quantization

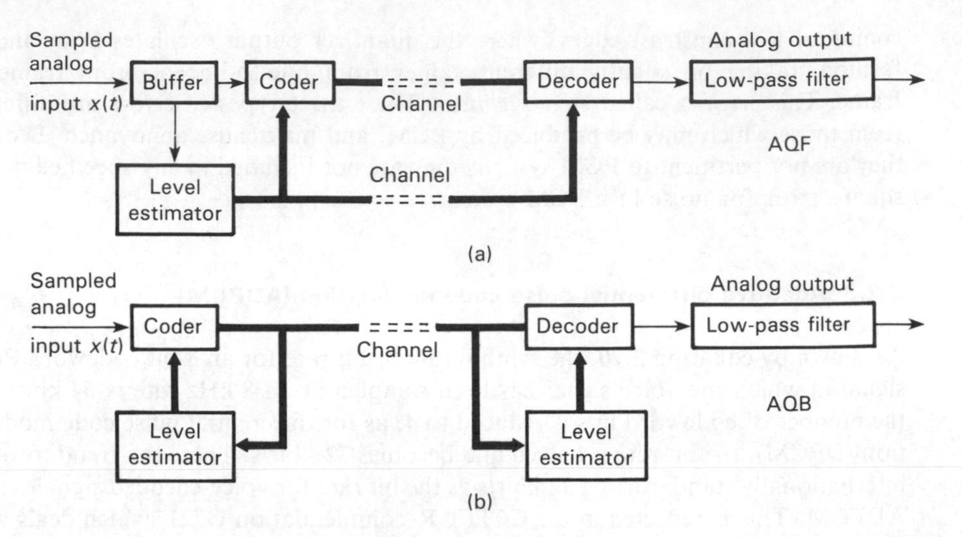

Figure 2.13 Adaptive quantization with: (a) 'forward', and (b) 'backward' estimation of input signal level

noise, and the estimates in AQF are more reliable than in AQB. The points of interest of each technique are listed briefly below:

AQF

- Signal level information transmitted to a remote decoder uses a sample of 5–6 bits per step size
- Permits special protection of step size information in transmission by added redundancy
- Estimation delay is produced in the encoding operation (up to 16 ms for speech)
- Requires buffering of 'unquantized' input samples
- Blockwise or periodically adaptive; i.e. its step size Δ gets updated once every block, and remains constant over the duration of a block of N samples
- Estimates are based on unquantized samples

AQB

- Step size information Δ is extracted from the recent history of the quantizer.
- No estimation delay
- Quantization noise degrades the level tracking performance, and degrades the performance further as the quantization step size is increased
- It is a nonlinear system with feedback and may have inherent stability problems

As AQF systems require expensive buffering, which is complex and also causes delay, low-complexity DPCM systems usually use AQB circuits, where the additional advantage of not needing extra bits for explicit transmission of step size information is provided. DPCM–AQB coders operating at 32 kbit/s are acceptable for certain speech transmission applications. The results are not as good as 7-bit PCM systems which use logarithmic quantization methods, but do compare favourably with a

6-bit log-PCM system. If a high-quality 32 kbit/s DPCM system is required, an adaptive predictor (APB) must be incorporated into the design.

'Forward' and 'backward' estimations of adaptive predictor coefficients

Figure 2.14 shows block diagrams of both the forward-adaptive prediction APF, and backward adaptive prediction APB of the adaptive predictor coefficients (a_i). DPCM coders which use adaptive predictive coding to determine a_i, as shown in equation 2.73, are more complex than adaptive quantizers, but are more bit-efficient. They will provide toll-quality speech at bit rates of 32 kbit/s. For image processing, inter-frame coders using low bit rates may be designed.

Forward adaptive prediction APF

As shown in Figure 2.14(a), p input values are processed and are used to determine a set of predictor coefficients. The salient points of this type of adaptive predictor are:

- A significant amount of encoding delay may occur.
- For speech the number of coefficients (a_i) which may need to be calculated is around 10.
- For speech the number of samples held at any time to calculate the coefficients is around 128.

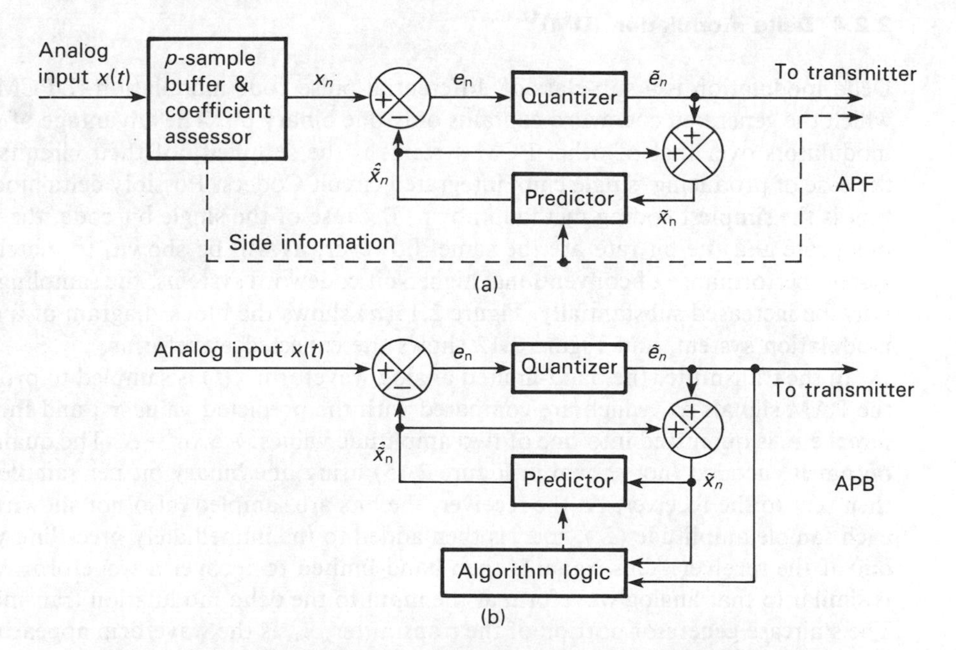

Figure 2.14 Block diagram of DPCM with: (a) forward (APF) and (b) backward (APB) adaptive prediction

Additional side information required to compute the predictor coefficients in the receiver increases the channel capacity.
Data buffering is required.
The technique lends itself to packet switching networks.

Backward adaptive prediction APB
The general block diagram is shown in Figure 2.14(*b*). The salient points are:

APB compares favourably with the APF in obtaining the same order of prediction without resorting to side information.
APB is subject to quantization effects at low bit rates, where APF is not.
Employs the use of AQB in all designs.
Number of predictor coefficients is around six.

CCITT standard for toll-quality ADPCM at 32 kbit/s
To construct the adaptive predictor circuit, an algorithm is required which permits the physical realization of a mathematical concept. Reference 12, pp. 308–311, shows two types of algorithms which may used in the CCITT standard for toll-quality ADPCM at 32 kbit/s. These algorithms are the AQB and APB, and the system designated as DPCM–APB–AQB for toll-quality speech at 32 kbit/s, or simply as ADPCM at 32 kbit/s. The resulting performance of this system is very signal-dependent. With speech, its S/N is around 25–30 dB.

2.2.4 Delta modulation (DM)[12,16,17]

Delta modulation is a sub-class of differential pulse code modulation (DPCM), in which the generated codeword contains only one binary bit. The advantage of delta modulators over that of other PCM systems is the simplicity of their circuits and the ease of producing 'single chip' integrated circuit Codecs. Possibly delta modulation is the simplest coding method known. Because of the single bit code, the sampling rate and the bit rate are the same; however, as will be shown, to match the system performance of conventional higher-bit codeword systems, the sampling rate must be increased substantially. Figure 2.15(*a*) shows the block diagram of a delta modulation system, and Figure 2.12 shows the expected waveforms.

In the transmitter the band-limited analog waveform $x(t)$ is sampled to produce the PAM signals x_n, which are compared with the predicted value $\hat{\hat{x}}_n$, and the difference e_n is quantized into one of two amplitude values $+\Delta$ or $-\Delta$. The quantizer output is encoded (not shown in Figure 2.15) using one binary bit per sample and then sent to the receiver. At the receiver, the bits are sampled (also not shown) and each sample amplitude ($\tilde{e}_n$), $\pm\Delta$, is then added to the immediately preceding value out of the receiver. This signal is then band-limited to recover a waveform which is similar to that analog waveform at the input to the delta modulation transmitter. The staircase generator portion of the transmitter, $\hat{\hat{x}}_n$, is the waveform appearing at the negative input to the differential amplifier. The delay T_s' circuit is otherwise known as a 'first-order predictor', 'accumulator', or 'integrator'. In Figure 2.15(*b*),

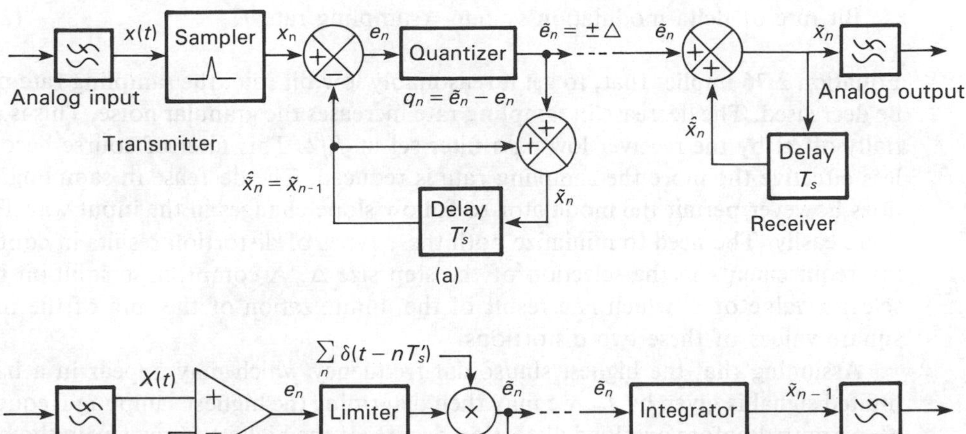

Figure 2.15 Delta modulation system: (a) discrete time model; (b) equipment realization

the equipment realization diagram shows this delay feedback circuit replaced by a simple integrator whose input is a train of impulses spaced at intervals of T_s' and due to $\tilde{e}_n$, having an amplitude of $\pm\Delta$. Figure 2.12 shows the input analog (unsampled) waveform $x(t)$. This waveform is compared against its stepwise approximation $\hat{\tilde{x}}_n$. The difference signal e_n is then quantized into two levels $\pm\Delta$ depending on the sign of the difference, and the output of the quantizer or two-state limiter. This quantized difference signal $\tilde{e}_n$, is next sampled to produce impulses of amplitude equal to $\pm\Delta$. The stepwise approximation $\hat{\tilde{x}}_n(t)$ is generated by passing the impulse waveform through an integrator which adds or subtracts a step of amplitude Δ to that level produced by the previous sample integration. The problems which may occur in the transmitter have already been discussed in Section 2.2.2. These problems are those of slope overload and granular noise or hunting. As the sampling rate in a delta modulation system is normally much higher than the Nyquist rate, the granular noise produced by hunting or idling around a constant amplitude $x(t)$ waveform is easily reduced by the receiver low-pass filter. However, the slope overloading is a problem not so easily overcome. Referrring to Figure 2.15(a), when $x(t)$ is changing, $\hat{\tilde{x}}_n(t)$ follows in a step-wise fashion as long as the successive samples of $x_n(t)$ do not differ by an amount greater than the step size Δ. Should this not be the case, $\hat{\tilde{x}}_n(t)$ cannot follow. As mentioned previously this type of overload is not determined by the amplitude of the input signal, but by the slope or the rate of change of the input signal amplitude with time. Equation 2.70 shows that the symbol rate of a PCM system equals the number of codeword bits multiplied by the sampling rate. As the number of codeword bits for delta modulation is only one, the bit rate equals the sampling rate:

Bit rate of delta modulation system = sampling rate f_s' (2.76)

Equation 2.76 implies that, to get a reasonably low bit rate, the sampling rate must be decreased. The decrease in sampling rate increases the granular noise. This is partially offset by the receiver low-pass filter set at $f_s'/2$. This filter of course becomes less effective the more the sampling rate is reduced. The decrease in sampling rate does however permit the modulator to follow slope changes in the input waveform more easily. The need to minimize both these types of distortion results in conflicting requirements in the selection of the step size Δ. A compromise solution is to select a value of Δ which is a result of the minimization of the sum of the mean square values of these two distortions.

Assuming that the highest sinusoidal frequency which may appear in a band-limited signal is given by f_m, we may then determine the highest sampling frequency f_s', permitted before overload distortion occurs. If the highest frequency in the band is given by f_m, the sine wave representing this frequency is given by:

$$x(t) = X \cdot \sin(2\pi f_m t)$$

of slope

$$\mathrm{d}x(t)/\mathrm{d}t = 2\pi \cdot f_m \cdot X \cdot \cos(2\pi f_m t)$$

This has a maximum value $2\pi \cdot f_m \cdot X$. To avoid slope overload, it is necessary that

$$\Delta / T_s' \geqslant 2\pi f_m \cdot X \tag{2.77}$$

Thus the condition for the prevention of slope overload is that the sampling frequency $f_s' = 1/T_s'$, is to be such that

$$f_s' \geqslant 2\pi \cdot f_m \cdot X/\Delta \tag{2.78}$$

Rearranging to determine the dynamic range X/Δ gives

$$X/\Delta < (1/2\pi) \cdot f_s'/f_m \tag{2.79}$$

where X = the peak amplitude of the input signal of frequency f_m kHz, in volts
Δ = the required step size, in volts.
f_s' = the sampling frequency, in kHz.
f_m = the input sinusoidal frequency, in kHz.

Example

In speech signals much of the power is concentrated in the lower frequencies, i.e. around 400 Hz. Considering the dynamic range in terms of these lower frequencies and equating the dynamic range of a delta modulation system with the dynamic range of an 8-bit uniform PCM system, determine the ratio of the sampling frequency to the maximum modulating frequency ratio for a DM system. How does this compare with the PCM system?

Solution
The maximum frequency f_{max}, of a speech signal is 3400 Hz. As the energy of this signal is mainly concentrated around 400 Hz (f_m), $f_m = f_{max}/8.5$. For a uniform PCM system of 8 bits, there are 256 quantization levels, which provides a dynamic range of 256 : 1, hence using equation 2.79, and $f_m = f_{max}/8.5$, we have

$$(f_s'/f_{max})_{DM} = (2\pi) \cdot (256) \cdot (1/8.5)$$
$$= 190$$

Compare this against PCM systems:

$$(f_s/f_{max})_{PCM} = 2$$

Thus the delta modulation system requires an exceptionally high sampling ratio to attain the same dynamic range as that provided by an ordinary PCM system.

To overcome the problem of having a high sampling rate to maximum signal frequency, the conversion of the delta modulated signal (one-digit coding) into a PCM signal (N-digit coding) is performed. For speech applications the PAM sampling rate (f_s) is chosen to be 8 kHz, and the delta modulation sampling rate (f_s') is chosen to insure an adequate signal-to-quantization noise ratio. The higher f_s', the better the quality of reproduced speech. The CCITT CEPT 32 standard for European PCM recommends, for 32 channel systems, $f_s'/f_s = 200$. This makes f_s' to be in the order of 50 megasamples/s for a 32 channel system.

The problem of slope overloading in delta modulation systems is reduced by filtering the signal to limit the maximum rate of change which may occur, or by increasing the step size as shown above in equation 2.78, or by reducing the sampling rate. However, filtering and increasing the step size only assist in reducing resolution. Increasing the sampling rate leads to additional bandwidth requirements, and, for a band-limited input signal, reduces the step size and consequently the granular noise.

In addition to these problems, the integrator at the receiver end permits an accumulation of error to occur if the system is disturbed. This particular problem may be reduced by a modification of delta modulation known as sigma-delta modulation, which is described in Section 2.2.6.

A solution to the problem of slope overload lies in the detection of the overload condition, and adaptively increasing the step size as a result. A system which provides a signal-dependent step size is called an *adaptive delta modulation system* (ADM).

2.2.5 Adaptive delta modulation (ADM)[18,19]

Alternatively named *continuously variable slope delta modulation* (CVSDM), ADM improves the relatively poor performance of the elementary delta modulator by

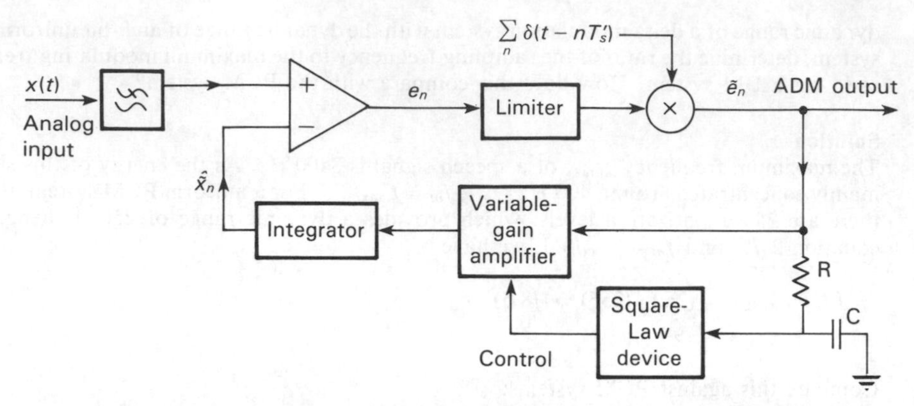

Figure 2.16 A basic adaptive delta modulation system (ADM)

dynamically varying the integrator gain in accordance with the average power level of the input signal. Figure 2.16 shows the block diagram of such a circuit. The step size is varied by controlling the gain of the integrator by an RC circuit and a square-law device. When the input signal is constant or varying slowly in time, the delta modulator will be hunting, producing at its output a sequence of alternate polarity impulses. The RC network averages these impulses, producing an average output of zero (theoretically). This means that the control signal controlling the variable gain amplifier is small, and the amplifier gain is small. This in turn produces a small step size. Where the input signal has a steep slope, the staircase function generated ($\hat{\hat{x}}_n(t)$), will try to keep up with the slope of the input signal. This will produce either a series of all-positive or all-negative impulses, which produces a large control voltage when averaged by the RC circuit. Consequently, the step size will be increased. Because of the squaring circuit, the amplifier control voltage will always be positive no matter what the impulse polarity is. By using this type of circuit, the adaptive delta modulator is able to provide a reduction in slope overload distortion and granular noise. The demodulator contains an adaptive gain control circuit similar to that used in the modulator. Reference 18 provides details on another method which may be used to vary adaptively the step size. Reference 12, pp. 400–417, deals in depth with adaptive delta modulation and the various configurations, types and algorithms used.

2.2.6 Sigma-delta modulation[12,20]

Voice-band data waveforms do not provide high correlation between successive samples. This lack of correlation is directly reflected in the corresponding values of quantization noise in the delta modulation coding of the data signals. To employ delta modulation to encode speech data, as well as preserving any DC level or LF content, the input of the modulator is integrated prior to DM encoding. In Figure 2.15(b) the integrator would be placed after the low-pass filter, at the positive input to the difference amplifier. This permits the low frequencies of the input signal to be pre-emphasized, which increases adjacent sample correlations. At the receiver the signal is de-emphasized by differentiating the signal, resulting in a decoder that is simply a low-pass filter. To simplify the hardware, the integrator in Figure 2.15(*b*) may be replaced by a short circuit, and moved to the output of the difference amplifier. This also eliminates the requirement of the integrator at the positive input to the amplifier.

2.2.7 Quantization noise in DM systems

We consider only the case where slope overload distortion does *not* occur. Referring to Figure 2.12, the quantization error e_n is the difference between the input signal $x(t)$ and the staircase signal $\hat{\hat{x}}_n(t)$. Thus

$$|x(t) - \hat{\hat{x}}_n(t)| = e_n \leqslant \Delta \tag{2.80}$$

The output of the demodulator is the sum of the reconstituted signal, assumed to be equal to the input signal $x(t)$, plus transmission errors $n_o(t)$, plus quantization noise $n_q(t)$, i.e.

$$\text{Demodulator output } y(t) = x(t) + n_o(t) + n_q(t) \tag{2.81}$$

where $n_q(t)$ is the output response of the demodulator low-pass filter to the quantization error $e_n(t)$. Assuming a uniform pdf for the quantization error ranging from $+\Delta$ to $-\Delta$, the pdf $= 1/(2\Delta)$; hence the mean-square error or total average quantization noise power is

$$E[e_n(t)^2] = 1/(2\Delta) \int_{-\Delta}^{\Delta} e_n(t)^2 \cdot \mathrm{d}e_n(t) = \Delta^2/3 \tag{2.82}$$

Thus the output power from the demodulator filter will be

$$N_q = E[n_q(t)^2] = (\Delta^2/3) \cdot (f_m/f_s') \tag{2.83}$$

where f_s' is the sampling frequency and f_m is the cut-off frequency of a brick-wall low-pass filter equal to the input sinusoidal frequency. From equation 2.78, it can be seen that, to avoid slope overload, the maximum peak amplitude of the highest frequency is X and this is given by

$$X \leqslant (1/2\pi) \cdot (f_s'/f_m) \cdot \Delta \tag{2.84}$$

As the average power of this signal is $X^2/2$, the signal-to-quantization distortion ratio can be determined from equations 2.83 and 2.84 as

$$\begin{aligned}(S/N_q) &= (3/8\pi^2)(f_s'/f_m)^3 \\ &= 30\log(f_s'/f_m) - 14.2\ \text{dB}\end{aligned} \tag{2.85}$$

In Section 2.3.2 it will be shown that S/N_q of a delta modulation system is not as good as that of a PCM system where equal bandwidths for both systems are assumed.

2.3 COMPARISONS BETWEEN DIFFERENT SYSTEMS

2.3.1 Thermal noise calculations for PCM and DM systems

2.3.1.1 PCM channel noise

When white noise enters the input of the receiver decoder circuits, it may cause an error in the detecting circuits which determine whether a 0 or a 1 logic state was transmitted. As discussed in Section 2.1.3.1, thermal noise is represented by a Gaussian distribution. Equation 2.16 is the probability density function of the Gaussian distribution, where σ^2 is the variance which can be equated to the average noise power $N_o(t)$ of a noise sample $n_o(t)$. If, during the bit interval, the input signal voltage to the receiver quantizer of Figure 2.5 is at a level of $-x$, then at the sample time T_s, the signal sample amplitude is $-x(T_s)$, whilst the noise amplitude is $n_o(T_s)$. If $n_o(T_s)$ is positive and larger than $x(T_s)$, the total sample voltage $x_o(T_s)$ will be positive. This positive sample voltage will result in an error as the receiver decoder mistakes this as an information bit of opposite polarity. The probability of this error is the probability that $n_o(T_s) > x(T_s)$, that is $P(n_o > x)$ or $1 - P(n_o \leqslant x)$; hence

$$P_e = 1 - P(n_o \leqslant x) = (1/2)\text{erfc}[x(T_s)/\sqrt{2} \cdot \sigma] \tag{2.86}$$

PCM systems operate typically with error probabilities small enough to assume that only a single bit error will occur within a codeword. Assume a codeword of b binary bits, with the smallest binary number corresponding to the most negative level, and the highest binary number corresponding to the most positive voltage within the uniform quantizer. An error which occurs in the least significant bit of the word corresponds to an error Δ, or the quantization level separation. The next significant bit corresponds to 2Δ. This continues up to the highest state of an error (2 to the power of $(b-1)$) times Δ. Assuming that an error may occur with equal likelihood in any bit of the word, the variance of the error can be determined as

$$E(e_q^2) = (1/b)\cdot[\Delta^2 + (2\Delta)^2 + (4\Delta)^2 + \cdots + (2^{b-1}\Delta)^2] = (1/b)\cdot\left[\sum_{i=1}^{b}[\Delta\cdot 2^{i-1}]^2\right]$$

$$= \frac{2^{2b}-1}{3b}\cdot\Delta^2 \simeq \frac{2^{2b}}{3b}\cdot\Delta^2 \tag{2.87}$$

for $b \geqslant 2$.

If P_e is the probability of a bit being in error, the mean separation between errors is given by $1/P_e$. Since there are b bits in a codeword, the mean separation between codewords in error is given by $1/bP_e$, and since the words are separated in time by the sampling interval T_s, the mean time between word errors, T_w, is given by

$$T_w = T_s/(b \cdot P_e) \tag{2.88}$$

The power spectral density of the thermal noise error impulse train is given by

$$G_{th}(f) = (1/T_w) \cdot E(e_q^2) = (P_e/T_s) \cdot (2^{2b}/3b) \cdot \Delta^2 \tag{2.89}$$

At the output of an ideal low-pass filter, the average noise power of the thermal noise impulse becomes

$$N_o = \int_{-f_m}^{f_m} G_{th}(f)\,\mathrm{d}f = [(2^{2b}\,\Delta^2 P_e)/(3T_s^2)] \tag{2.90}$$

since $2f_m = f_s$, hence $T_s = 1/2f_m$

2.3.1.2 DM channel noise

As with PCM channel noise, the integrator in the receiver circuits will occasionally make an error in determining the polarity of the transmitted signal. The effect of thermal noise again corrupts the transmitted signal by appearing as an impulse of polarity opposite to and greater than that of the transmitted bit. In delta modulation, the noise impulse however needs to be twice the opposite magnitude of the bit being corrupted, in order to reverse the bit magnitude. Reference 2, p. 406, derives the thermal noise power output from a delta modulation system as

$$N_o = \frac{(2\,\Delta^2 \cdot P_e \cdot f_s')}{\pi^2} \cdot [1/f_1 - 1/f_m] \tag{2.91}$$

This involves a band-pass filter, with an upper cut-off frequency f_m, and a lower cut-off frequency f_1. If a perfect low-pass filter were to be considered, the noise power would become infinite as $f_1 \to 0$, due to the power spectral density being infinite around the origin. In voice-band frequency applications, this is not a problem, because the voice band extends from 300 Hz to 3400 Hz, and the filter used is in effect a band-pass filter. For the case where $f_1 \ll f_m$, equation 2.91 reduces to:

$$N_o \simeq (2\,\Delta^2 \cdot P_e/\pi^2) \cdot (f_s'/f_1) \tag{2.92}$$

The important point to stress here is that the thermal noise power output of the delta modulation system depends on the *low-frequency* cut-off of the receiver filter, rather than the high-frequency cut-off as in PCM systems.

2.3.2 Calculations of output signal-to-noise ratio (S/N) for PCM and DM systems

To determine the total signal-to-noise ratio, the output signal power $E[X_o(t)^2]$ has to be found, together with the total noise power. The total noise power will be the sum of the quantization noise power N_q or $E[n_q(t)^2]$, and the channel noise power N_o or $E[n_o(t)^2]$. Thus:

$$(S/N)_0 = E[X_o(t)^2]/[E[n_o(t)^2] + E[n_q(t)^2]] = E[X_o(t)^2]/(N_q + N_o) \quad (2.93)$$

PCM SYSTEM OUTPUT S/N

From equation 2.62 the signal power S at the uniform quantizer output is equal to $(Q^2 - 1) \cdot \Delta^2/12$. From equation 2.59 the quantization noise power $E(e_q^2)$, also at the output of the quantizer, is given by $\Delta^2/12$. Both these equations are modified when they pass through an ideal low-pass filter with a cut-off frequency equal to twice the sampling frequency. Thus:

$$E[X_o(t)^2] = S/T_s^2 \qquad = (Q^2 - 1) \cdot (\Delta^2/12) \cdot (1/T_s^2) \quad (2.94)$$

$$E[n_q(t)^2] = E[e_q(t)^2]/T_s^2 \qquad = (\Delta^2/12) \cdot (1/T_s^2) \quad (2.95)$$

$$E[n_o(t)^2] \qquad = (Q^2 \cdot \Delta^2 \cdot P_e)/(3T_s^2) \quad (2.90)$$

Hence

$$(S/N)_o = (Q^2 - 1)/(1 + 4Q^2 P_e) \quad = (2^{2b} - 1)/(1 + 4P_e \cdot 2^{2b}) \quad (2.96)$$

where $Q = 2^b$ = the number of quantization levels in a uniform quantizer.
b = the number of binary bits in a PCM codeword.

DM OUTPUT SIGNAL-TO-NOISE RATIO

From equations 2.83, 2.84 and 2.92, the signal-to-noise ratio is given by:

$$(S/N)_o = [(3f_s'^3 \cdot f_1)/(8\pi^2 \cdot f_1 \cdot f_m^3 + 48P_e \cdot f_s'^2 \cdot f_m^2)] \quad (2.97)$$

where f_1, f_m = the lower and upper cut-off frequencies of the receiver filter, Hz.
f_s' = the sampling frequency of the delta modulation system (which is not the same as PCM).
P_e = the probability of bit error expected from the method of transmission of the data through the channel.

Note that the sampling frequency f_s', and hence the sampling period T_s', do not necessarily have the same values for PCM and delta modulation systems. More often they will differ as shown in the example given in Section 2.2.4.

Example 1

The probability of error P_e for an M-ary PSK system is defined to be approximately $\exp[-(C/N)\sin 180/M]$. Assuming an 8-PSK system is used, and the carrier-to-noise ratio at the input to the receiver is expected to be 20 dB, what would be the expected $(S/N)_o$ out of a speech channel if an 8-bit PCM system is used?

Solution
For $C/N = 20$ dB, the C/N factor = 100, hence P_e for an $M = 8$-PSK system equals 4×10^{-7}. For an 8-bit PCM system, $b = 8$, and from equation 2.96, $(S/N)_o = 47.7$ dB. The interesting result of this example is that it shows that, for small P_e, equation 2.96 reduces to approximately $S/N = 6b$ dB.

Example 2

If $(S/N)_o$ of a Delta modulation system voice channel is 18 dB, and (S/N_q) is found to be 22 dB, determine P_e for a sampling frequency of 80 kHz.

Solution
Rearranging equation 2.97 in terms of $(S/N)_o$ and (S/N_q), as given by equation 2.85, we find

$$(S/N_q)\text{ dB} - (S/N)_o\text{ dB} = 10 \log[1 + 6P_e \cdot f_s'^2/(\pi^2 \cdot f_1 \cdot f_m)]\text{ dB} \quad (2.98)$$

For a voice channel $f_1 = 300$ Hz, and $f_m = 3400$, and it is given that $f_s = 80000$; then P_e is determined as approximately 4×10^{-4}.

2.3.3 PCM versus analog

Listed below are the advantages and disadvantages that PCM systems have over analog systems. These are present advantages; however, there is a cyclical pattern to new techniques and new systems. Today's advantages become tomorrow's disadvantages.

Advantages of PCM
1. Because the analog signal has been converted into a binary codeword consisting of several logic bits, the recovery or regeneration of an information bit consisting of either a high or low logic state is not difficult. The ability to regenerate the complete signal, or time-division multiplex composite signal, has a distinct advantage for systems comprising many repeater stations. It is the prevention of noise accumulating along the long-distance telephone system that makes them so attractive. In a uniform quantizing scheme, the amplitude of the noise for complete regeneration must be less than half the separation between the quantization levels, as discussed in Section 2.3.1.1. Repeaters using analog modulation techniques, such as frequency division multiplex (FDM), may add considerable noise to the system, if the RF signal is brought down to voice.
2. A PCM system designed for analog message transmission may be adapted for carrying data signals. This provides flexibility in the use of the system, especially in the planning of unknown future needs.
3. In PCM baseband transmission systems, coding techniques may be used to reduce the effects of noise, and provide security.

4. PCM provides a means of storing voice or data information for providing non-real-time transmission, or for other processing circuits.
5. For low input signal-to-noise ratios, the output signal-to-noise ratios of PCM systems are better than those of analog systems.
6. For a given ratio of transmission bandwidth to message bandwidth (bandwidth expansion or compression ratio), at a P_e less than 1 in 10000, PCM is superior to all other forms of pulse modulation. This is an important factor if time division multiplex is to be considered.
7. As the transmission bandwidth is increased, a reduction in the receiver power requirements occurs. This exchange of bandwidth for power is easily accomplished in PCM systems. It should be noted, however, that the signal power cannot be increased indefinitely, since there is a threshold beyond which there is no improvement in output signal-to-noise ratio and little reduction in bandwidth.

Disadvantages of PCM

1. The PCM system circuitry is more complex than other forms of modulation systems, but does not vary as the number of channels increases. Thus this complexity is offset where large numbers of channels are required, because in other modulation techniques the complexity does increase as the number of channels increases.
2. The PCM system circuitry adds its own forms of impairments, such as jitter, etc.
3. The bandwidth requirements are greater than that of the analog systems (approximately 48 kHz against 4 kHz for an analog voice channel).

Consider Figure 2.17. Information entering into an ideal receiver has a bandwidth B_T, and a signal-to-noise ratio $(S/N)_i$. The channel capacity at the input is given by the Hartley-Shannon equation as

$$C_r = B_T \log_2 [1 + (S/N)_i] \tag{2.99}$$

At the output of the receiver, of bandwidth B, and signal-to-noise ratio $(S/N)_o$, the theoretical information capacity is given by

$$C_d = B \log_2 [1 + (S/N)_o] \tag{2.100}$$

At the receiver input the noise power will equal the input spectral density η watts/Hz multiplied by the bandwidth B_T. However, at the output of the receiver, the noise power will be the same single sided power spectral density η, multiplied by the receiver single sided bandwidth B, i.e.:

Noise power out of an ideal receiver $= N_o = \eta B$

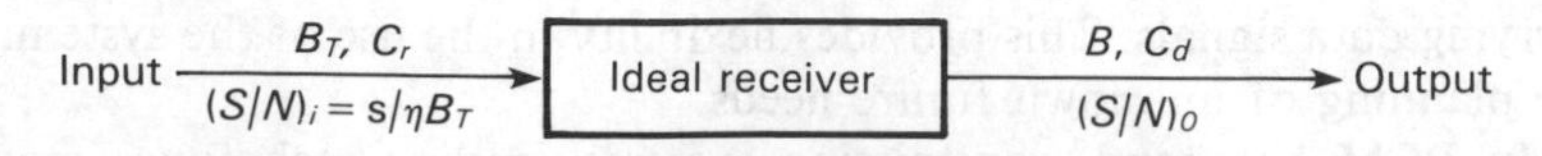

Figure 2.17 An ideal transmission system receiver

We define $\gamma = S_i/\eta\beta$, (2.101a)

$$\beta = \text{bandwidth expansion or compression ratio} = B_T/B \qquad (2.101b)$$

These two parameters permit a comparison to be made of all the different types of transmission systems, if each transmission system is defined in terms of its $(S/N)_o$. The effect of thermal noise or channel additive noise is not taken into account in the $(S/N)_o$ in these systems, since it is a comparison of the individual performance characteristics of different types of communication systems that is being considered. Table 2.7 lists the different types of communication systems and their thermal noise or external interference-free output signal-to-noise ratios, in terms of the bandwidth expansion ratio β, and the input system signal power to system output noise power ratio γ.

Table 2.7 Performance parameters of a communication system

System	$(S/N)_o$	Bandwidth expansion ratio β
Ideal	$(\gamma/\beta)^\beta$	β
AM	$\gamma/3$	2
SSB	γ	1
DSB	γ	2
Wideband FM	$3\gamma \cdot \beta^2/8$	$2(\Delta f/f_m + 1)$
PCM	$M^{2\beta}$	$\log_M Q$
PDM	$\gamma\beta/4$	>1
PPM	$\gamma\beta^2/16$	>1
DM	$3\beta^3/\pi^2$	b

Figure 2.18 shows the normalized power γ, plotted against the bandwidth expansion ratio β, for a high system output signal-to-noise ratio of 50 dB.

The expressions given in Table 2.7 are used to calculate $(S/N)_o$ for all systems except the PCM case. For PCM systems the value of γ needed to produce $(S/N)_o$ equal to 50 dB is calculated by defining the PCM threshold. In the listed advantages of PCM systems given above, Number 6 states, that for a P_e less than 1 part in 10000, PCM is superior to all other forms of pulse modulation. Taking P_e as less than 1 part in 10000 as the PCM threshold, and using the equation for an M-ary baseband PCM system, γ can be determined.

$$P_e \text{ for } M\text{-ary PCM system} = 2(M-1)/M \cdot F[\{(3\gamma)/[(M^2-1)\log_M Q]\}^{1/2}] \qquad (2.102)$$

where $F(\ )$ is the Gaussian cumulative probability distribution function as given by equation 2.20.

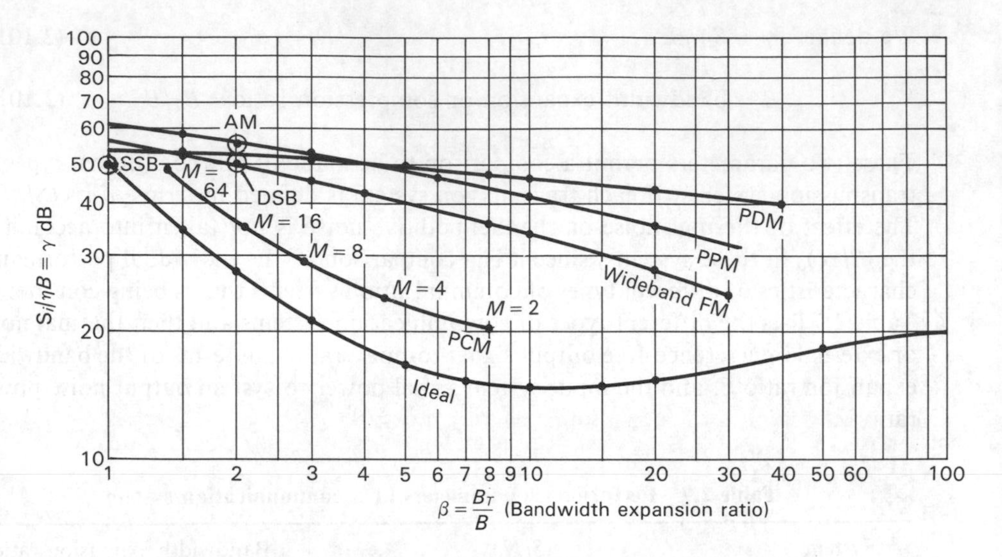

Figure 2.18 Normalized input power to bandwidth for various transmission systems

For $P_e < 10^{-4}$

$$F[\{(3\gamma)/[(M^2-1)\cdot \log_M Q]\}^{1/2}] < M\cdot 10^{-4}/2(M-1)$$

If z_1^2 is determined from $F(z_1) < M \cdot 10^{-4}/2(M-1)$ then

$$z_1^2 = (3\gamma)/[(M^2-1)\cdot \log_M Q]$$

Hence

$$\gamma \geqslant [(M^2-1)\cdot \log_M Q]\cdot z_1^2/3 \tag{2.103}$$

Table 2.8 shows values of z_1 and γ and β (as given in Table 2.7) for various values of M, for $P_e < 10^{-4}$, and for $(S/N)_o = 50$ dB.

From equation 2.63, $(S/N)_o = Q^2 - 1$. As this equals 10^5, Q may be determined as approximately 316.

Table 2.8 γ and β for different M values of a M-ary PCM system

M	$> F(z_1) \times 10^{-4}$	z_1	z_1^2	$(M^2-1)[\log_M Q]1/3$	γ	γ dB	$\beta = (\log_M Q)$
2	1	3.7	13.69	8.3038	113.67	20.56	8.3048
4	0.666	3.82	14.5924	20.7595	302.93	24.81	4.1519
8	0.5714	3.86	14.8996	58.1265	866.06	29.38	2.7679
16	0.5333	3.875	15.0156	176.4553	2649.6	34.23	2.0759
64	0.5079	3.89	15.1321	4349.837	65882	48.18	1.3840

2.3.4 Comparison of PCM and DM systems

To compare a PCM system with a DM system, the bandwidths of each are made equal. This is done by equating the pulse durations from each system. For the PCM system there are b bits produced to form the codeword for every sampling period T_s. This implies that a bit occupies a time interval equal to T_s/b. As the sampling frequency $f_s = 1/T_s$, and the filter upper cut-off frequency f_m must be $2f_s$, the bit duration equals:

$$\tau_{PCM} = 1/2f_m b, \text{ which equals } \tau_{DM} = 1/f_s' \tag{2.104}$$

where f_s' is the DM sampling rate, and τ_{DM} is the time interval that the DM pulse occupies. Equation 2.104 implies;

$$f_s'/f_m = 2b \tag{2.105}$$

Substituting this into equation 2.97 gives the output signal-to-noise ratio from a DM system, i.e.

$$(S/N)_{oDM} = 3b^3/[\pi^2 + 24b^2 \cdot P_e \cdot (f_m/f_1)] \tag{2.106}$$

Equation 2.106 can be compared with equation 2.96, which gives the output signal-to-noise ratio for a PCM system.

$$(S/N)_{oPCM} = 2^{2b}/[1 + P_e \cdot 2^{2b+2}] \tag{2.96}$$

for $2^{2b} \geqslant 1$.

As each system may be directly transmitted at baseband or transmitted using PSK modulation, P_e may be determined using equation 2.102, where $M = 2$ for binary transmission. Hence P_e in both equations 2.106 and 2.96 may be determined by:

$$P_e = F[(\gamma/b)^{1/2}] \tag{2.107}$$

where F implies that the Gaussian cumulative probability density function as given in equation 2.20 is to be used.

When the received signal level is raised, γ becomes larger and P_e becomes smaller. Hence for very small values of P_e, or for values of input signal level above the threshold of each system, equations 2.106 and 2.96 reduce to: $(S/N)_{oDM} = 3b^3/\pi^2$, and $(S/N)_{oPCM} = 2^{2b}$. Thus:

$$[(S/N)_{oPCM}/(S/N)_{oDM}] = 5.2 + 6b - 30 \log b \text{ dB} \tag{2.108}$$

Equation 2.108 shows the advantage of the PCM signal-to-noise ratio over that of DM in dB when both systems are above their receive signal level threshold. The thresholds may be determined by plotting equations 2.106 and 2.96 for $(S/N)_o$

versus γ in dB, to determine the values of γ for the knee of each curve before it levels out to a constant $(S/N)_o$.

The performance of DM systems can be improved using adaptive DM (CVSDM) (see Section 2.2.5). To compare the bandwidths required, consider the CCITT recommended 8-bit PCM system. The bit rate is thus 64 kbit/s. To obtain a quality similar to this using DM, the sample rate, and hence the bit rate, must equal 100 kbit/s. However, using CVSDM, a bit rate of 32 kbit/s will produce results similar to that of 8-bit PCM. Results like these do tend to make CVDM a strong contender for use of DM in preference to PCM, especially as the CVSDM circuits are easier to construct in single integrated circuit form.

2.4 CCITT SYSTEMS

This section will concentrate on the processing of analog waveforms in digital bit streams using the CCITT recommended methods and equipment. The reason for this is that it places some of the concepts already discussed in this chapter into a more realistic and practical framework, since the CCITT recommendations are more often than not incorporated into transmission system design by equipment manufacturers and system designers. Incorporated into this section will be for both two-wire and four-wire interfaces:

The recommended performance parameters
Descriptions of the tests to be carried out to verify that the system is performing to the recommended performance parameter values
Listings of the recommended levels and frequencies for correct system alignment
Recommendations on the test equipment to be used to verify, align and adjust the PCM system

2.4.1 Pulse code modulation (PCM) of voice frequencies

The relevant CCITT recommendations are G711–714,[13] O131–133, O152, Supplements to Series 0 Recommendations.[21] In Recommendation G711, the sampling rate f_s, is specified at 8000 samples/s with a tolerance of ± 50 parts per million. Eight binary digits per sample should be used with μ- or A-law encoding. Details of a 13-segment A Law and a 15-segment μ law are provided, together with the relationship between the encoding laws and audio signal level. The relationship is one in which a 1 kHz sine wave of a nominal level of 0 dBm0 should be present at any voice-frequency output of the PCM multiplex when the periodic sequence of codes relating to the A Law or μ Law is applied to the decoder input. The theoretical load capacity is defined as the level of an input sinusoidal signal whose positive and negative peaks coincide with the virtual decision values which define the working range of the quantizer. This theoretical load capacity $T_{\max}$ is 3.14 dBm0 for the A

law and 3.17 dBm0 for the μ law. In addition to the above, Recommendation G711 states that digital paths between countries which have adopted different encoding laws should carry signals in accordance with the A law, except where both countries have adopted the same law. In that case the adopted law should be used on digital paths between them. Any necessary conversion to the A law will be done by countries using the μ law; for the conversion the tables given in Recommendation G711 may be used.

2.4.2 Performance characteristics and testing of PCM channels

The performance characteristics and tests (where applicable) are listed under the following headings: attenuation/frequency distortion, group delay, impedance of audio ports, which includes:

Nominal impedance
Return loss
Longitudinal balance

The idle channel noise similarly can be divided into:

Weighted noise
Single frequency noise
Receive equipment noise

Other performance factors to be taken into consideration are: discrimination against out-of-band input signal; spurious out-of-band signals at the channel output; intermodulation; total distortion, including quantization distortion; spurious in-band signals at the channel output port; variation of gain with input level; inter-channel crosstalk; go-to-return crosstalk. The list continues to include: interference from signaling; relative levels at input and output; adjustment of relative levels; and short-term and long-term stability.

Four-wire interfaces at VF (Recommendation G712)
The performance limits given in Sections 2.4.2.1–2.4.2.16 should be met in all cases using two PCM multiplex terminal equipments connected back-to-back with input and output ports of the channels terminated, where not otherwise specified, in 600 ohms. The nominal test frequency used for attenuation/frequency distortion and relative level adjustments is 1000 Hz. In practice however, to permit faster measurements to be made and to avoid stroboscopic fluctuations, the actual test frequency should be chosen in the range 1004 to 1020 Hz.

Two-wire interfaces at VF (Recommendation G713)
In some situations PCM channels will be utilized on a two-wire basis in association

with two-wire/four-wire terminating units. Together with signaling functions, these four-wire/two-wire terminating units may sometimes form an integral part of the PCM multiplex terminal. The performance characteristics given below should be met when measured between the two-wire input and output ports of two PCM terminal equipments connected back-to-back and except, where required otherwise, with input and output ports terminated with their nominal impedances. In addition to the situation where the circuits are two-wire to two-wire, situations may exist where there is a mixture of two-wire to four-wire circuits, or four-wire to two-wire circuits. The limits presented below also apply in these mixed cases. The stated nominal frequency of 1 kHz should be in practice 1004 to 1020 kHz, to permit faster measurements and to avoid fluctuations due to the stroboscopic effect. It should be noted that this recommendation is not intended to cover PCM systems used in the subscriber's local network.

Sections 2.4.2.1 to 2.4.2.16 give the specifications for both the two-wire and four-wire PCM channel interfaces at voice frequencies.

2.4.2.1 Attenuation/frequency distortion

For both two-wire and four-wire circuits, the nominal reference frequency is 1000 Hz at a preferred input power level of either −10 dBm0 or 0 dBm0. Figure 2.19 shows the mask within which the variations of attenuation with frequency of any channel should remain.

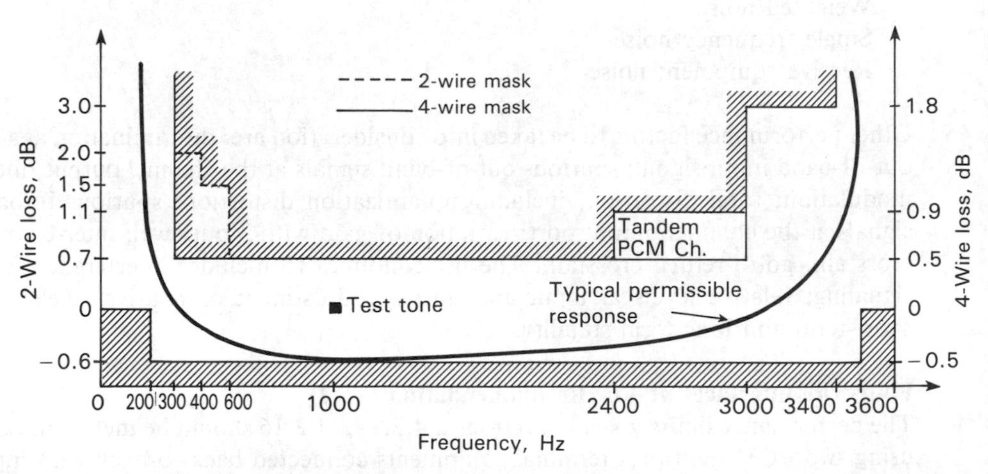

Figure 2.19 CCITT Attenuation/frequency distortion mask for 2-wire and 4-wire PCM channels (Courtesy of CCITT, Reference 13)

2.4.2.2 Group delay

The absolute group delay at the frequency of minimum group delay should not exceed 600 μs. The minimum value of group delay is taken as the reference for group delay distortion. Figure 2.20 shows the mask within which the group delay distortion

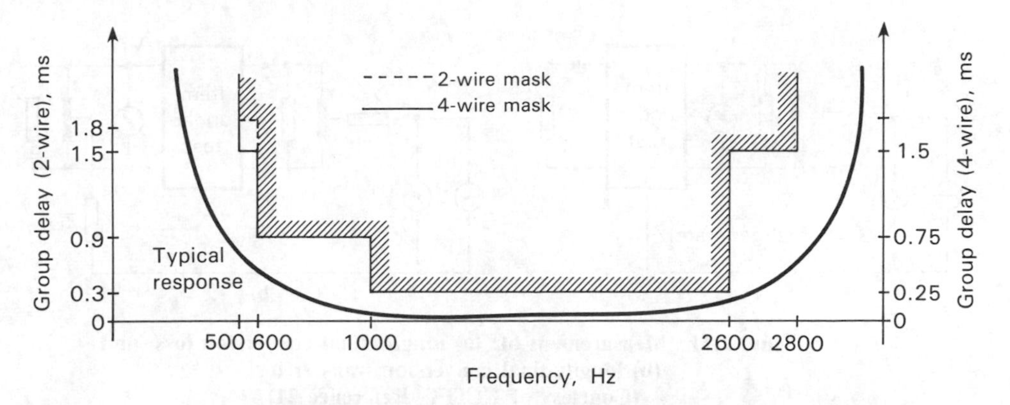

Figure 2.20 Group delay distortion with frequency for 2-wire and 4-wire PCM system interfaces (Courtesy of CCITT, Reference 13)

should remain. Both group delay and group delay distortion performance criteria should be met with an input power level equal to either −10 dBm0 or 0 dBm0.

2.4.2.3 Impedance of audio ports

For the four-wire circuits, the nominal impedance is 600 ohms. For the two-wire case no single value is recommended.

RETURN LOSS

The return loss either side of a junction point may be defined as:

$$\text{Return loss} = 20 \log [(z_1 + z_2)/(z_1 - z_2)] \quad (2.109)$$

where z_1 and z_2 are the complex impedances either side of the junction point. Alternative equations representing return loss are:

$$\begin{aligned}\text{Return loss} &= 10 \log [(\text{forward power})/(\text{reflected power})] \\ &= 20 \log [(\text{VSWR} + 1)/(\text{VSWR} - 1)]\end{aligned} \quad (2.110)$$

where VSWR is the voltage standing wave ratio.

The CCITT Recommendation is:

2-wire 300–600 Hz > 12 dB
600–3400 Hz > 15 dB

4-wire 300–3400 Hz > 20 dB

LONGITUDINAL BALANCE (RECOMMENDATION O121)[21]

Using an impedance of 750 ohms ±20 percent (Z_{L1}) in the driving circuit (Figure 2.21), and with the other port terminated in its nominal characteristic impedance,

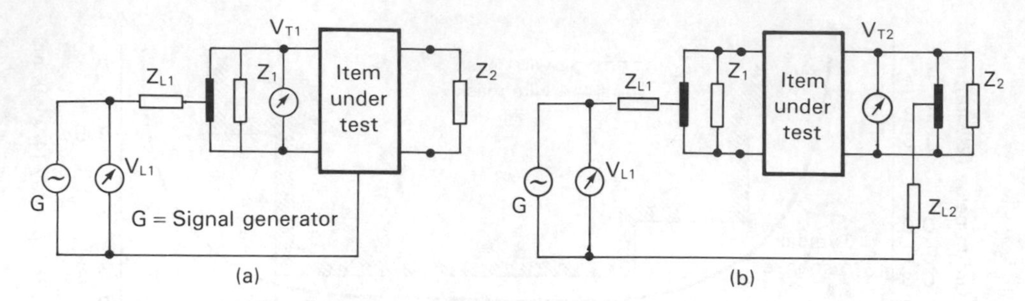

Figure 2.21 Measurement of: (a) longitudinal conversion loss, and (b) longitudinal conversion transfer loss (Courtesy of CCITT, Reference 21)

we have:

1. The longitudinal conversion loss as measured at the input port, output port or two-wire interface, should not be less than the limits shown in Figure 2.22. The test set up is shown in Figure 2.21(a).
2. The difference between the longitudinal conversion transfer loss at the specified frequencies and the insertion loss at the same frequencies should not be less than the limits shown in Figure 2.22. The test set-up is shown in Figure 2.21(b).

The longitudinal conversion loss of a one- or two-port network is a measure in dB of the degree of unwanted transverse signal produced at the terminals of the network due to the presence of a longitudinal signal on the connecting leads. It is measured as shown in Figure 2.21(a). This technique is applicable to either the input

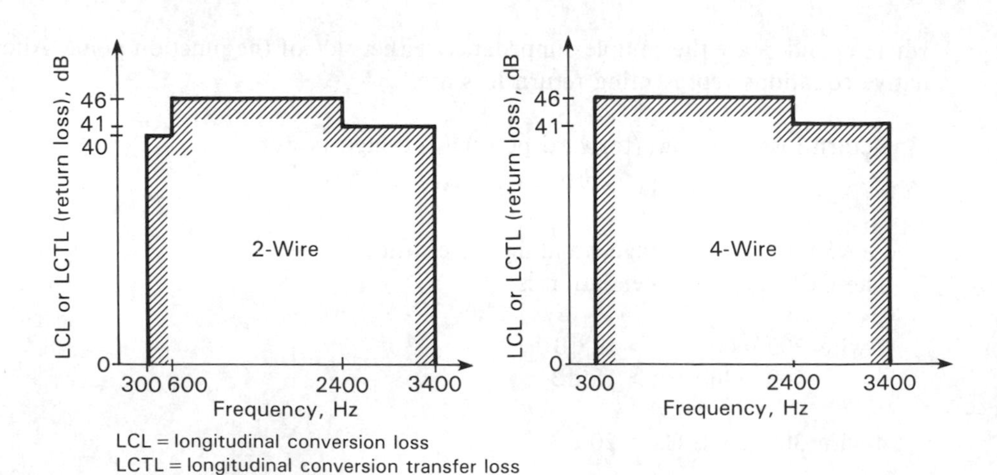

Figure 2.22 Longitudinal balance (Courtesy of CCITT, Reference 13)

or output terminals. Referring to Figure 2.21(*a*), we have:

$$\text{Longitudinal conversion loss (LCL)} = 20 \log | V_{L1}/V_{T1} | \text{ dB} \quad (2.111)$$

Referring to Figure 2.21(*b*), we have:

$$\text{Longitudinal conversion transfer loss (LCTL)} = 20 \log | V_{L1}/V_{T2} | \text{ dB} \quad (2.112)$$

2.4.2.4 Idle channel noise

For either a two-wire or a four-wire circuit, the following applies:

Weighted noise. With the input and output ports of the channel terminated in the nominal impedance, the idle channel noise should not exceed −65 dBm0p.
Single frequency noise. The level of any single frequency (in particular the sampling frequency and its multiples), measured selectively, should not exceed −50 dBm0.
Receiving equipment noise. Noise contributed by the receiving equipment alone should be less than −75 dBm0p when its input is driven by a PCM signal corresponding to the decoder output value number 0 for the μ law, or decoder output value number 1 for the *A* law.

2.4.2.5 Discrimination against out-of-band input signals

With any sine wave signal at a frequency above 4.6 kHz applied to the input port of the channel at a suitable level such as −25 dBm0, the level of any image frequency produced at the output port of the channel should, as a minimum requirement, be at least 25 dB below the level of the test signal (−50 dBm0). In additon to this, under the most adverse conditions encountered in a national network, the PCM channel should not contribute more than 100 pW0p of additional noise in the 0–4 kHz band at the channel output, as a result of the presence of out-of-band noise.

2.4.2.6 Spurious out-of-band signals at the channel output port

With any sine wave signal in the range 300–3400 Hz, at a level of 0 dBm0 applied to the input port of a channel, the level of spurious out-of-band image signals measured selectively at the output port should be lower than −25 dBm0. In addition to this, the spurious out-of-band signals should not give rise to unacceptable interference in equipment connected to the PCM channel. In particular, the intelligible or unintelligible crosstalk in a connected FDM channel should not exceed a level of −65 dBm0 as a consequence of the spurious out-of-band signals at the PCM channel output.

2.4.2.7 Intermodulation

Two sine wave signals of different frequencies f_1 and f_2 not harmonically related, in the range 300–3400 Hz and of equal levels in the range −4 to −21 dBm0, which are applied simultaneously to the input port of a four-wire channel should not produce any $2f_1 - f_2$ intermodulation product having a level greater than −35 dB relative to the level of one of the two input signals. Furthermore, a signal having a level of −9 dBm0 at any frequency in the range 300–3400 Hz and a signal of 50 Hz at a level of −23 dBm0 applied simultaneously to the input port should not produce any intermodulation product of a level exceeding −49 dBm0.

2.4.2.8 Total distortion, including quantizing distortion

Two methods of testing are described below, together with the test set-up and recommended performance values.

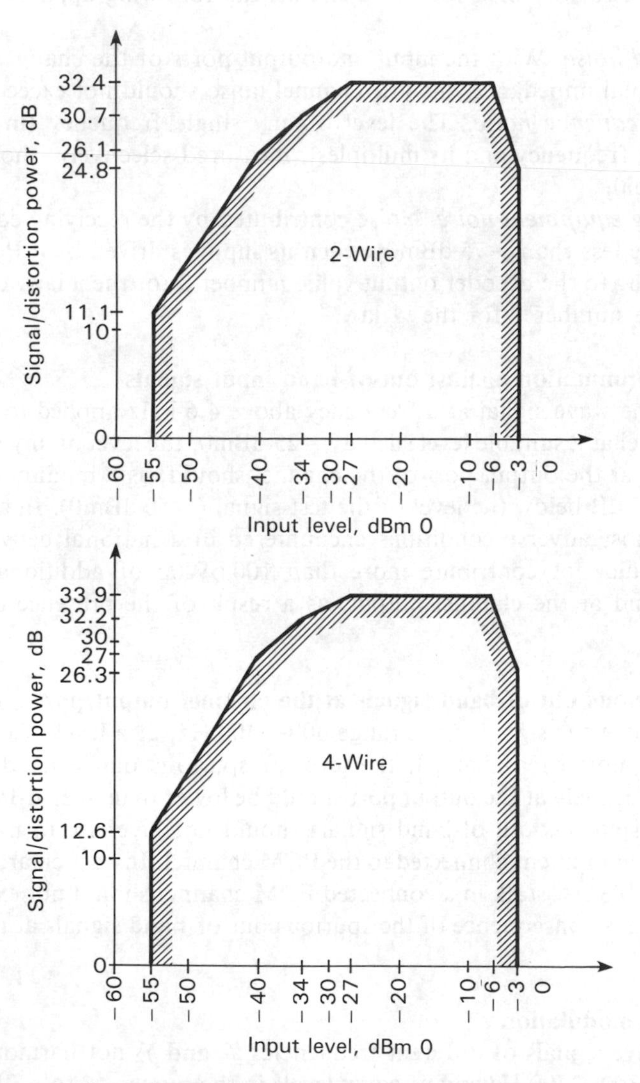

Figure 2.23 Signal-to-total distortion ratio as a function of input level – Method 1 (Courtesy of CCITT, Reference 13)

METHOD 1

A noise signal, as completely specified in Recommendation O131[21], (which briefly is a 100–200 Hz bandwidth Gaussian noise signal with a center frequency of 350–550 Hz, having a 10.5 peak-to-RMS ratio, and having a level ranging from 0 dBm0 to − 55 dBm0) is applied to the input port of a channel. The resulting ratio of signal-to-total-distortion-power should lie above the limits shown in Figure 2.23.

The test method used is shown in Figure 2.24. The signal-to-distortion power ratio, including quantizing distortion, is measured as the ratio of the power of received stimulus in the reference band to the noise power in the measured band. A correction is included to relate the measurement to the full PCM speech channel bandwidth. Recommendation O131 provides details of the transmit and receive reference path filter masks to be used when making these measurements.

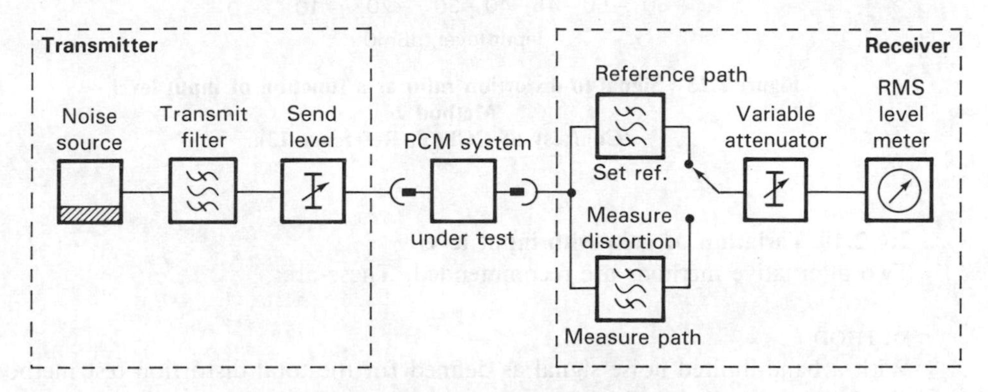

Figure 2.24 Principle of quantizing distortion measurement (Courtesy of CCITT, Reference 21)

METHOD 2

With a sine wave signal at a nominal frequency of 820 or 1020 Hz (see Recommendation O132),[21] or at a nominal frequency of 420 Hz (see Recommendation O131, Section 2,[21] applied to the input port of a PCM channel, the ratio of signal-to-total-distortion power measured with psophometric weighting or alternatively C-message weighting (which may require a calibration correction factor) should lie above the limits shown in Figure 2.25. The method also requires the use of a narrow-band rejection filter in the receiver equipment to block the sinusoidal test signal from the distortion measuring circuits so that the distortion power may be measured. The level of test signal is at least − 45 to + 5 dBm0 with a setting accuracy of ±0.2 dB.

2.4.2.9 Spurious in-band signals at the channel output port

With a sine wave signal in the frequency range 700–1100 Hz and at a level of 0 dBm0 applied to the input port of a channel, the output level at any frequency other than the frequency of the applied signal, measured selectively in the frequency band 300–3400 Hz, should be less than − 40 dBm0.

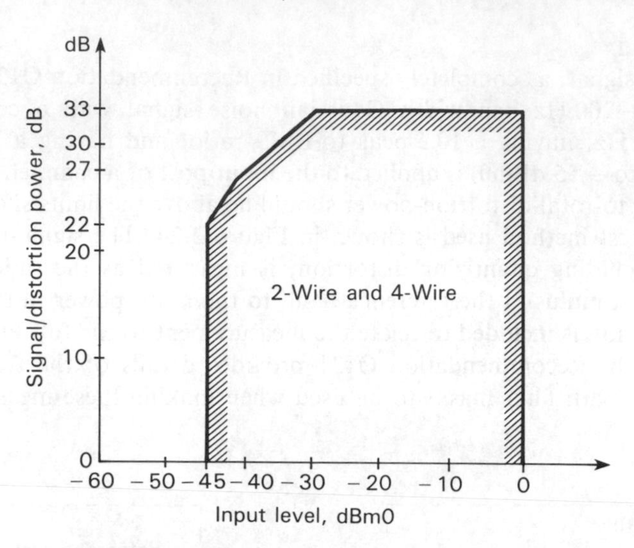

Figure 2.25 Signal-to-distortion ratio as a function of input level – Method 2 (Courtesy of CCITT, Reference 13)

2.4.2.10 Variation of gain with input level

Two alternative methods are recommended. These are:

METHOD 1

With a band-limited noise signal as defined for the total distortion test method 1 given above, applied to the input of any channel at a level between −55 dBm0 and −10 dBm0, the gain variation of that channel, relative to the gain at an input level of −10 dBm0, should lie within the limits of the mask given in Figure 2.26(*a*). The measurement should be limited to the frequency band 350–550 Hz. Furthermore, with a sine wave signal in the frequency range 700–1100 Hz applied to the input of any channel at a level between −10 dBm0 and +3 dBm0, the gain variation of that channel relative to the gain at an input level of −10 dBm0 should lie within the limits of the mask given in Figure 2.26(*b*). The measurement should be made selectively.

METHOD 2

With a sine wave signal in the frequency range 700–1100 Hz applied to the input port of any channel at a level between −55 dBm0 and +3 dBm0, the gain variation of that channel relative to the gain at an input level of −10 dBm0 should lie within the limits of the mask given in Figure 2.26(*c*). The measurements should be made selectively.

2.4.2.11 Inter-channel crosstalk

The crosstalk between individual channels of a multiplex equipment should be such that, with a sine wave signal in the frequency range 700–1100 Hz and at a level of

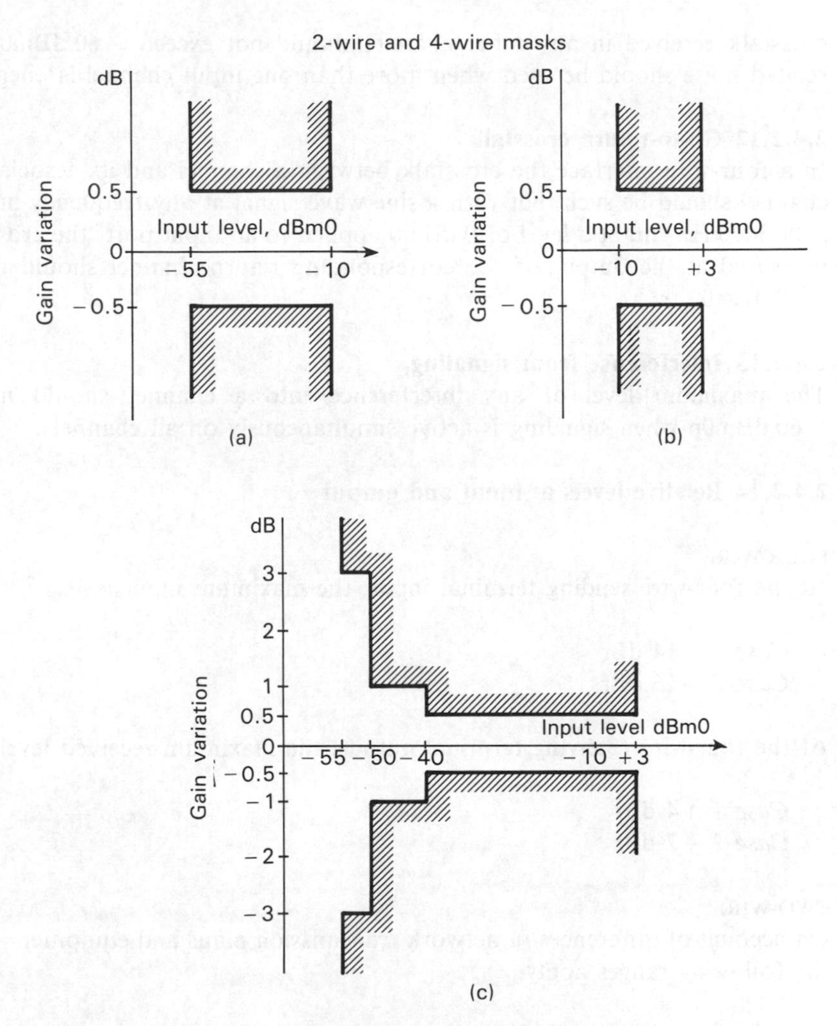

Figure 2.26 Variation of gain with input level: (a) Method 1: White noise test signal; (b) Method 1: Sinusoidal test signal; (c) Method 2: Sinusoidal test signal (Courtesy of CCITT, Reference 13)

0 dBm0 applied to an input port, the crosstalk level received in any other channel should not exceed −65 dBm0. In order to overcome fundamental gain enhancement effects, associated with PCM encoders, which can mask the true crosstalk, an activating signal may be injected into the disturbed channel when implementing crosstalk measurements with sine wave signals. Suitable activating signals are band-limited noise as used in the total distortion measurements given above, at a level in the range −50 to −60 dBm0, or a sine wave at a level in the range −33 to −40 dBm0.

When a white noise signal whose peak-to-RMS ratio does not exceed 3.5, at a level of 0 dBm0, is applied to the input port of up to four channels, the level of crosstalk received in any other channel should not exceed −60 dBm0p. Uncorrelated noise should be used when more than one input channel is energized.

2.4.2.12 Go-to-return crosstalk

In a four-wire interface the crosstalk between a channel and its associated return channel should be such that with a sine wave signal at any frequency in the range 300–3400 Hz and at a level of 0 dBm0 applied to an input port, the crosstalk level measured at the output of the corresponding return channel should not exceed −60 dBm0.

2.4.2.13 Interference from signaling

The maximum level of any interference into a channel should not exceed −60 dBm0p when signaling is active simultaneously on all channels.

2.4.2.14 Relative levels at input and output

FOUR-WIRE

At the four-wire sending terminal input, the maximum input level is:

Case 1 −14 dBr
Case 2 −16 dBr

At the four-wire receiving terminal output, the maximum received level is:

Case 1 +4 dBr
Case 2 +7 dBr

TWO-WIRE

On account of differences in network transmission plans and equipment utilization the following ranges apply:

At the two-wire input, the sending level ranges from 0 to −5 dBr in 0.5 dB steps.
At the two-wire output, the receiving level ranges from −2 to −7.5 dBr in 0.5 dB steps.

It is, however, recognized that a particular design of equipment need not be capable of operating over the entire range.

2.4.2.15 Adjustment of relative levels

Adjustment of the relative levels and hence the gain of the separate send and receive sides of PCM channels, should be made at typical values of environmental conditions. The adjustments should lead to a deviation of the actual values against the nominal values not exceeding ±0.3 dB for four-wire circuits and ±0.4 dB for two-wire circuits, and should be made as follows:

The receive side should be adjusted to provide a sine wave of 1 kHz at a nominal level of 0 dBm0 ± 0.3 dB for a four-wire and ±0.4 dB for a two-wire circuit, when a periodic sequence of character signals as defined in Recommendation G711 is applied to the decoder output.

The send side should be adjusted by connecting its output to the input of a receive side which has been adjusted to have precisely nominal gain, and by applying a sine wave signal at a nominal frequency of 1000 Hz at a level of 0 dBm0 to the voice-frequency input of the send side. The send side should then be adjusted so that the resulting sine wave signal at the voice-frequency output of the receive side is at a level of 0 dBm0. In practice the adjustment should be made with a tolerance of ±0.3 dB for a four-wire circuit, and ±0.4 dB for a two-wire circuit. Alternatively, a receive side with a known error, within the limits defined above, may be used, provided that account is taken of this known error in adjusting the send side.

The load capacity T_{max}, of the send side may be checked by applying a sine wave signal at a nominal frequency of 1000 Hz at its voice-frequency input. The level of this signal should be initially set well below T_{max} and should then be slowly increased. The input level should be measured at which the first occurrence is observed of the character signal corresponding to the extreme quantizing interval for both positive and negative values. T_{max} is taken as being 0.3 dB greater than the measured input level.

This method allows T_{max} to be checked for both positive and negative amplitudes and the values thus obtained should be within ±0.3 dB for a four-wire circuit and ±0.4 dB for a two-wire circuit of the theoretical load capacity (+3.14 dBm0 for *A* law or +3.17 dBm0 for μ law). As an alternative, the occurrence of the largest pulse amplitude at the decoder output may be used as a means of identifying T_{max}.

2.4.2.16 Short-term and long-term stability

When a sine wave signal at a nominal frequency of 1000 Hz and at a level of either −10 dBm0 or 0 dBm0 is applied to any voice-frequency input, the level measured at the corresponding voice-frequency output should not vary by more than ±0.2 dB during any 10 minute interval of typical operation, nor by more than ±0.5 dB during any one year under the permitted variations in the power supply voltage and temperature.

CCITT Recommendation G714 deals with the performance characteristics of international or non-point-to-point PCM systems between four-wire voice-frequency interfaces. This recommendation treats separately the performance characteristics of send and receive sides of PCM channels applicable to four-wire voice-frequency interfaces. The reader is referred to Recommendation G714 for detailed information. Briefly the differences between this system and the point-to-point four-wire circuits of Recommendation 712 outlined above are:

Attenuation/frequency distortion:	The values for loss in Figure 2.19 for four-wire should be halved.
Group delay	The values of group delay in Figure 2.20 for four-wire should be halved and taken to the nearest whole integer.
Longitudinal balance	Remains the same as that of Figure 2.22 for four-wire.

Idle channel noise	Should not exceed −66 dBm0p.
Total distortion, including quantizing distortion	Method 1, as shown in Figure 2.23, is required to be raised by 0.5 dB for the send side, and 1.5 dB for the receive side in the signal-to-total-distortion ratio shown for 4-wire. Method 2, as shown in Figure 2.25, requires to be raised 2 dB higher for both send and receive sides.
Variation of gain with input level	Shows significant departures from the point-to-point diagrams of Figure 2.26, and the reader is referred to CCITT Recommendation G714.
Interference from signaling	Not to exceed −63 dBm0p.
Crosstalk	

Inter-channel measured with analog test signal

Should not exceed −73 dBm0 for near-end-crosstalk (NEXT)

Should not exceed −70 dBm0 for far-end-crosstalk (FEXT)

Go-to-return measured with analog test signal

Should not exceed −66 dBm0

Inter-channel measured with digital signal inserted into the output of a PCM A/D convertor and simulating a sine-wave signal 700–1100 Hz at a level of 0 dBm0

Should not exceed −73 dBm0 for NEXT and −70 dBm0 for FEXT

Go-to-return measured with digital signal inserted into the output of a PCM A/D convertor and simulating a sine-wave signal 300–3400 Hz at a level of 0 dBm0.

Should not exceed −66 dBm0.

2.4.3 32 kbit/s adaptive differential pulse code modulation (ADPCM)[13]

CCITT Recommendation G721 provides the characteristics of an ADPCM coding technique to be used in converting a 64 kbit/s A-law or μ-law PCM channel to and from a 32 kbit/s channel.

2.4.3.1 ADPCM encoder

After the A-law or μ-law PCM input signal has been converted to a uniform PCM signal, a difference signal is obtained by subtracting an estimate of the input signal from the input signal itself as already described in Sections 2.2.2 and 2.2.3. An adaptive 16-level quantizer is used to assign four binary digits to the value of the difference signal for transmission to the decoder. An inverse quantizer produces a quantized difference signal from these same four binary digits. The signal estimate is added to this quantized difference signal to produce the reconstructed version of the input signal. Both the reconstructed signal and the quantized difference signal

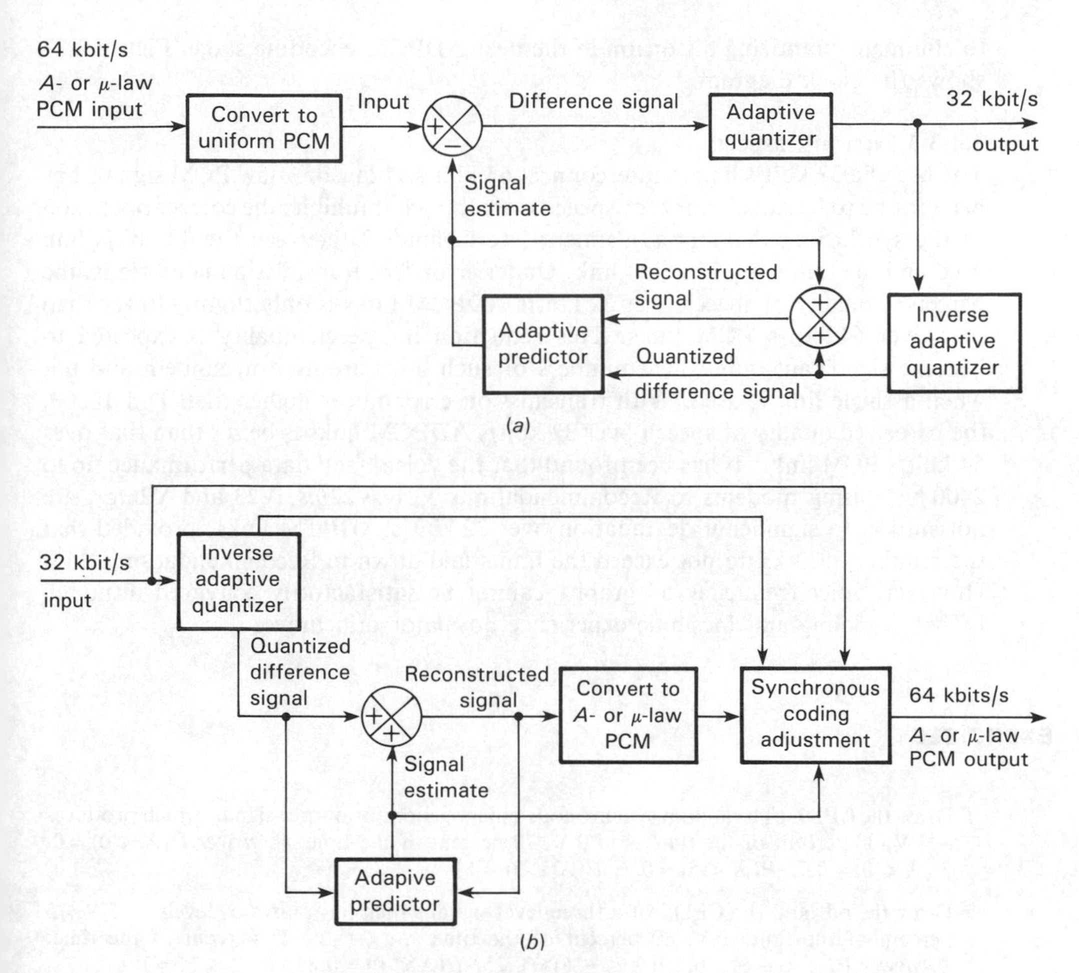

Figure 2.27 The ADPCM encoder and decoder: (a) Encoder; (b) Decoder (Courtesy of CCITT, Reference 13)

are operated upon by an adaptive predictor which produces the estimate of the input signal, thereby completing the feedback loop. Figure 2.27 shows a block diagram of the encoder.

2.4.3.2 ADPCM decoder

The decoder includes a structure identical to the feedback portion of the encoder, together with a uniform PCM to A-law or μ-law conversion and a synchronous coding adjustment. The synchronous coding adjustment prevents cumulative distortion occurring on synchronous tandem codings (ADPCM–PCM–ADPCM, etc., digital connections) when both ADPCM and PCM systems are error-free and are not disturbed by digital signal processing devices. The synchronous coding adjustment is achieved by adjusting the PCM output codes in a manner which attempts to eliminate quantizing distortion in the next ADPCM encoding stage. Figure 2.27 shows the block diagram.

2.4.3.3 Network aspects

The use of a 32 kbit/s link to interconnect 64 kbit/s A-law or μ-law PCM signals has been found to be satisfactory for speech even though it inhibits the correct operation of the synchronous coding adjustment (see Chapter 3) between the 32 kbit/s link used and the following 32 kbit/s link. Under error-free transmission conditions, the perceived quality of speech over 32 kbit/s ADPCM links is only slightly lower than that over 64 kbit/s PCM links. This reduction in speech quality is expected to become significant only when numbers of such links are used in tandem and not when a single link is used. With transmission error ratios higher than 1 in 10000, the perceived quality of speech over 32 kbit/s ADPCM links is better than that over 64 kbit/s PCM links. It has been found that the voice-band data performance up to 2400 bit/s using modems to Recommendations V21, V22*bis*, V23 and V26*ter*, are not subject to significant degradation over 32 kbit/s ADPCM links, provided that the number of links do not exceed the limits laid down in Recommendation G113. However, voice-frequency telegraphy cannot be satisfactorily conveyed although DTMF signaling and facsimile experience no major difficulties.

EXERCISES

1 Draw the CPDF of a random synchronous binary generator output signal, which produces +5 V, 30 percent of the time, and 0 V, 70 percent of the time. [*Answer* $P(X < 0) = 0$, $P(X \leqslant 0) = 0.7$, $P(X < 5) = 0.7$, $P(X \leqslant 5) = 1$]

2 Draw the pdf and the CPDF of a three-level signal which has pulses of levels: +5 V, 15 percent of the time, 0 V, 60 percent of the time and −5 V, 25 percent of the time. [*Answer* $P(X < -5) = 0$, $P(X \leqslant -5) = 0.25$, $P(X \leqslant 0) = 0.85$, $P(X \leqslant 5) = 1$]

3 Show that the mean and variance of a binomial random variable X, are $\mu = np$, and $\sigma^2 = np(1 - p)$.

4 If X is a Gaussian random variable with $\mu = 3.2$, and $\sigma^2 = 16$, find $P(-3.2 < X \leqslant 8)$. (*Answer* 0.8301)

5 When a Gaussian white noise source is passed through a filter, the output noise still remains Gaussian. Assume that the DC component has been removed by the signal passing through the filter, but that the RMS component remains at the original 3 V RMS. Determine the probability of a +6 V DC signal being reduced to 0 V or less by the noise. (*Answer* 0.0228)

6 Assuming that a Rayleigh distribution pdf is given by equation 2.25, derive the CPDF as given by equation 2.26, and show that the median $\langle x \rangle = 1.1774\alpha$.

7 For a uniform quantizer with 8 bits used in the codeword, determine the signal-to-noise ratio of the signal appearing at the output of the quantizer. (*Answer* 48.165 dB)

8 Using equation 2.66 plot:
(*a*) The compressor characteristic for $\mu = 100$
(*b*) The expander characteristic for $\mu = 100$

9 Assuming a uniform quantizer, what would be the number of quantization levels Q, if the quantization distortion was to be no greater than 5 mV for a maximum peak-peak sampled input signal swing of 5 V? (Answer $Q = 500$)

10 Assume that an ideal low-pass filter has a bandwidth of 28 kHz, determine the expected S/N from an A-law ($A = 87.6$) nonuniform quantizer. (*Answer* 66.22 dB)

11 Determine the bit rate of a PCM encoded voice signal in which the encoding process requires only four bits to represent the analog voice signal. (*Answer* 32 kbit/s)

12 Given that the thermal noise in a communication channel produces an error bit once in every 10000 bits transmitted (on the average), determine the number of uniform quantization levels required in a PCM system if the required $(S/N)_o$ out of a voice channel is to be no less than 30 dB. (*Answer* 41)

13 If eight bits are used in the PCM codeword of question 12 determine, for the same P_e, the new output $(S/N)_o$. (*Answer* 33.8 dB)

14 For an 8-bit PCM system plot the curves of $(S/N)_o$ versus γ for a PCM system and a DM system as given in equations 2.91, 2.101, and 2.102 (using if needs be Normal Distribution tables), and confirm that the threshold for DM occurs at $\gamma = 20.5$ dBm and for PCM occurs at 22.5 dB.

Assume that both systems pass through a voice frequency band-pass filter, whose $f_m = 3000$ Hz and $f_l = 300$ Hz.

15 Determine, for the same parameters as given in Question 14, the $(S/N)_o$ advantage of PCM over DM. (*Answer* 26.1 dB)

16 Show that for a sinusoidal input to a linear quantizer of peak-to-peak amplitude $2a$, the input signal to distortion (noise) ratio is given by $6b + 1.8$ dB, where b is the number of codeword bits.

REFERENCES

1. Shanmugam, K. S., *Digital and Analog Communication Systems* (Wiley, 1979).
2. Taub, H. and Schilling, D. L., *Principles of Communication Systems* (McGraw-Hill, 1971).
3. Papoulis, A., *Probability, Random Variables, and Stochastic Processes*, 2nd Ed. (McGraw-Hill, 1984).
4. Proakis, J. G., *Digital Communications* (McGraw-Hill, 1983).
5. Carlson, A. B., *Communication Systems – an Introduction to Signals and Noise in Electrical Communication,* 2nd Ed. (McGraw-Hill, 1975).
6. Poularikas, A. D. and Seely, S., *Signals and Systems* (PWS Publishers, 1985).
7. 'Propagation Data required for Line-of-Sight Radio-Relay Systems', CCIR Report 338–3, 1966, 1970, 1974, 1978.
8. Pearson, K. W., 'Method for the Prediction of the Fading Performance of a Multisection Microwave Link', *Proc. IEE*, 1965, **112**, No. 7, p. 1291.
9. Oliver, B. M., 'The Philosophy of PCM', *Proc. IRE*, 1948, **36**, No. 11, pp. 1324–1331.
10. Bennett, G. H., *Pulse Code Modulation and Digital Transmission*, 2nd Ed. (Marconi Instruments., 1983).
11. Coates, R. F. W., *Modern Communication Systems*, 2nd Ed. (Macmillan Press, 1983).
12. Jayant, N. S. and Noll, P., *Digital Coding of Waveforms – Principles and Applications to Speech and Video*, Signal Processing Series, Bell Telephone Laboratories (Prentice-Hall, 1984).

13. 'Digital Networks – Transmission Systems and Multiplexing Equipments', Recommendations G700–G956 (Study Groups XV and XVIII), CCITT Red Book, **III–3**, VIII Plenary Assembly, 1984.
14. Gruber, J. and Psimenatos, N., 'Network Performance and Transmission Planning for 32 kbit/s ADPCM', *Telecommunication Transmission*, IEE Conference Publication No. 246, 1985, p. 27.
15. Wright, T. C., 'Speech Coding Standards for Public Telecommunications Applications', *ibid.*, p. 266.
16. de Jager, F., 'Deltamodulation: a method of PCM Transmission using a 1-unit Code', *Philips Research Reports*, 1952, No. 7, pp. 442–446.
17. Steele, R., *Delta Modulation Systems* (Pentech Press, 1975).
18. Jayant, N. S., 'Adaptive Delta Modulation with a One-Bit Memory', *Bell System Technical Journal*, March 1970, pp. 321–342.
19. Greefkes, J. A., 'A Digitally Companded Delta Modulation Modem for Speech Transmission', *Proceedings IEEE International Conference on Communications*, June 1970, pp. 7.33–7.48.
20. van der Kam, J. J., 'A Telephony CODEC using Sigma-Delta Modulation and Digital Filtering', *Communications Equipment and Systems*, IEE Conference Publication No. 209, 1982, p. 49.
21. 'Specifications of Measuring Equipment', Recommendations O Series (Study Group IV), CCITT Red Book, **IV-4**, VIII Plenary Assembly, 1984.
22. Cattermole, K. W., *Pulse Code Modulation* (Iliffe, 1969).

3 DIGITAL MULTIPLEXING

In Chapter 2 we discussed how the analog voice signal was processed into a digital bit stream. This chapter deals with the combining of digital bit streams from various sources by means of the digital multiplexing process. Referring to Figure 1.1, in which a model of a digital system is shown, we can see that the digital multiplex equipment, labeled 'Digital MUX', collects data bit streams from sources as well as those arising from analog audio-PCM processes. Once all the digital bit streams from Codecs, digital sources and analog audio-PCM circuits are combined and processed in the digital multiplex equipment, the resulting baseband output is ready for further processing in order to prepare it for transmission over the communication channel via (in this case) the digital radio. After reception and demodulation by the radio receiver equipment, the resulting baseband signal is conveyed to the digital multiplex equipment, which analyzes it and passes it to the various sink equipments similar to those from which the transmitted signal originated. It is worth while mentioning at this point, that the multiplex equipment itself may contain in its circuits, PCM equipment. If so, voice channels in analog form may be presented, at the multiplex terminal, to the input of the multiplex equipment directly, for processing into a PAM signal and interleaving with other PAM-converted voice channels before conversion into a composite PCM bit stream. Where the digital source is a 64 kbit/s PCM data bit stream, which originated from voice, the multiplex equipment inputs accepting this data stream will treat it as a pure digital source and will multiplex the bit stream directly as in Figure 1.1. In addition to the transmission of many sources over a single channel, multiplexing provides the additional benefit of processing every source so that they have almost identical formats. This permits a single design type of transmission equipment which carries many different types of traffic, as will become more apparent in discussing the multiplex hierarchies.

We begin this chapter by discussion of the basics of time division multiplexing.

3.1 BASICS OF TIME DIVISION MULTIPLEXING

When more than one analog signal is to be transmitted over a communication channel, usually one of the two classical schemes of combining these separate signals is used. The first scheme is where the signals are processed to occupy separate segments in the frequency domain, but share a common time in which they are transmitted. In other words, all signals may be transmitted at the same time, but each signal is translated in frequency. This is known as frequency division multiplex. The second classical scheme is where all source signals share the same frequency, but occupy different segments of the time domain. In this case each band of frequencies, such as a voice band, remains the same, but is transmitted over the communication channel at different times. To do this, however, each analog signal is sampled at a different time and the resulting pulse containing the amplitude information of each separate signal is then placed on line. The result is a sequence of interleaved PAM pulses, where each amplitude-modulated pulse in the cycle originates from a different signal. This is made possible because the width of the sampling pulse used to sample signal 1, say, is of much shorter duration than the time which elapses before signal 1 is sampled again. Consider Figure 3.1(*a*), in which four analog signals enter on different lines into a fixed-speed rotating switch called a *commutator*. The fifth line is held at some voltage V, which is greater than the maximum voltage from any of the four signals. The resulting waveforms appear as in Figure 3.1(*b*). The largest pulse F is called the *framing pulse*, and it is used to define a frame or set of signals during one switch rotation. The group of five pulses in this example is called a *frame*. In practical systems usually many more pulses comprise a frame.

If each analog signal has a bandwidth B Hz as defined by a low-pass filter, the sampling rate must be at least $2B$ Hz, if aliasing is not to occur. This, as mentioned in Chapter 2, is the Nyquist bandwidth. For n time-division multiplex PAM signals, the pulse transmission rate is given as $2Bn$ pulses/second. This means that the minimum bandwidth of the proceeding ideal filter is nB. In practice, it will be larger than this, as it is not possible to construct such a filter, and also larger to prevent inter-symbol interference, or interference between adjacent pulses.

At the receiver, the samples from each individual channel are separated and distributed by another rotary switch called a *distributor*. The samples from each channel are then filtered to reproduce a version of the original analog signal. The rotary switches at both transmitter and receiver are carefully synchronized. This synchronization is the most critical of all the time-division-multiplexing (TDM) processes. Two forms of synchronizing in TDM exist, these being frame and sample (or bit) synchronization. Frame synchronization is necessary to determine correctly the start of a group of samples, and bit synchronization is necessary to separate correctly the samples within each frame. The synchronization process also may permit the synchronization of both the transmit and receive terminals since the timing signals within both terminals will occur at the same average rate. In the transmit terminal equipment, the clock which generates the required timing pulses to control the various functions and to control the digital signal bit rate for transmission purposes will not, in most cases, be running at the same rate as the clock built into the receive terminal. To overcome this difference in clock rates, the receive terminal usually derives its timing from the received digital bit stream. This insures that both terminals, in respect of the particular bit stream, are functioning at the same average rate, and may be considered to be synchronized. As such, the receive terminal is 'slaved' to the transmit terminal. The transmit clock in this slaved terminal may also

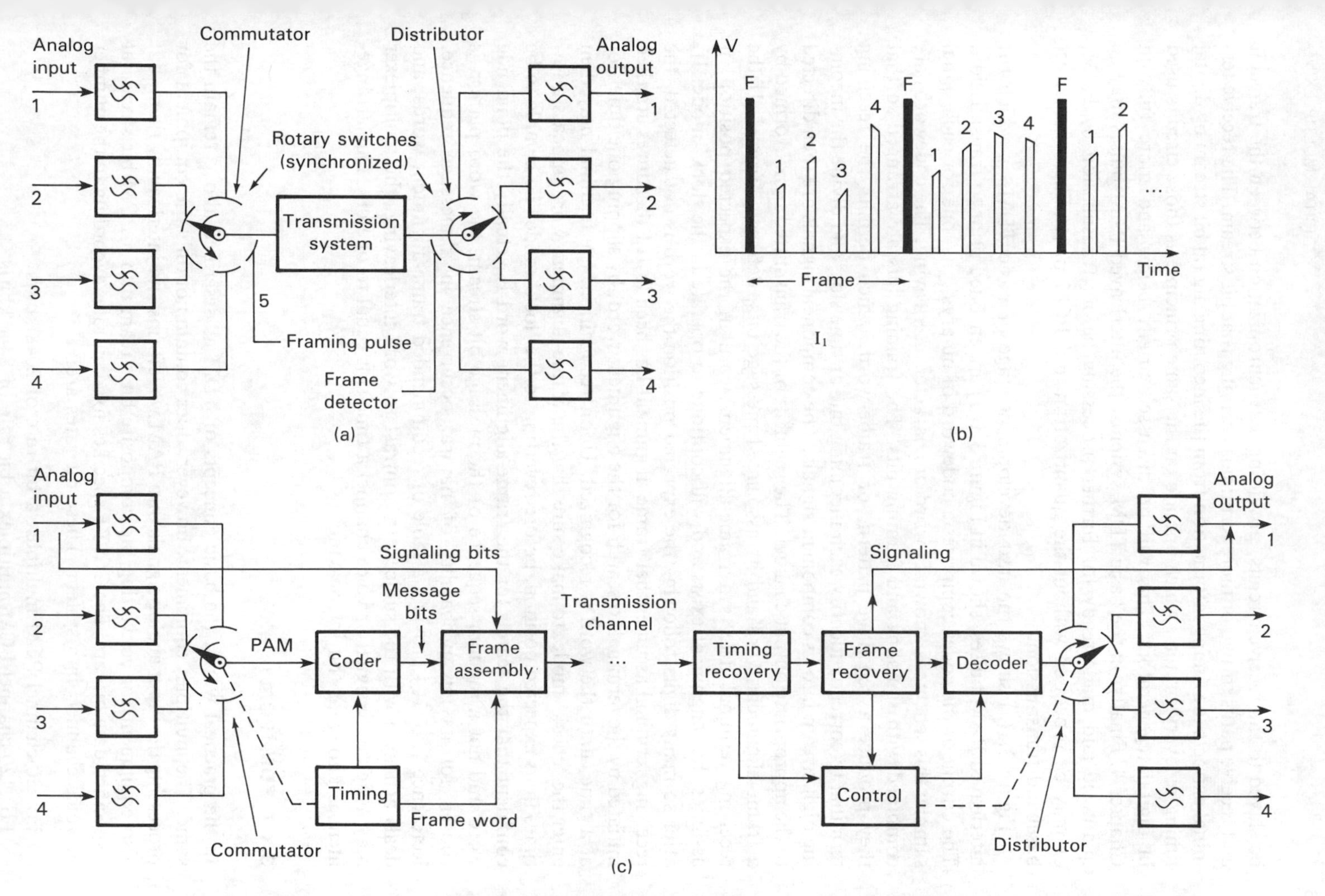

Figure 3.1 Time-division multiplexing of PAM signals: (a) four-channel TDM system; (b) TDM waveforms; (c) four-channel encoded TDM system.

be slaved to the receive circuits, providing a synchronization between the transmit and receive paths of two or more terminals carrying that bit stream. The receive terminal extracts the timing information from the incoming digital bit stream by digital timing recovery circuits, which operate on the same principle as those circuits used in digital regenerators. This will be discussed briefly below and more fully in Chapter 4. Finally, in the basic TDM systems, the interleaved PAM pulses may be quantized and made ready for further processing before transmission by PCM, or may be processed without the quantization and PCM in readiness for direct baseband transmission.

Figure 3.1(*c*) shows the next development stage of basic TDM, in which the interleaved PAM pulses, shown in Figure 3.1(*b*) are encoded prior to transmission. The output of the coder comprises a codeword of binary digits, one for each signal sample; this scheme is known as word or character interleaving. The codewords are combined with signaling and framing bits. The framing bits are arranged so that they produce a repetitive pattern, or frame-word, which permits the receiving terminal to correctly identify each incoming bit, or time slot, and divide the incoming digit stream into its component signals. The component signals are then directed to the appropriate output channel. The segment of the transmitted signal formed by a frame-alignment word and its associated message bits, defines a 'frame'. If the receiving terminal detects a frame-alignment word in the expected position, it assumes that alignment exists and will continue to operate in the 'lock' mode. If, after so many digits following the expected position the word is not detected, the receiving terminal assumes that frame-alignment has been lost. Procedures are then initiated by the terminal to search for the alignment word, by slipping one time slot at a time, until the word is recognized. Upon recognition, the terminal may then enter the 'check' mode, to make sure that the frame-alignment word appears a few times in its expected position, before reverting to the 'lock' mode. To be able to be confident that the search for the frame-alignment word is actually the bona fide word and not a random sequence of the message bit stream, the word pattern is chosen for its low probability of natural occurrence in a message sequence. Reference 1, p. 104 provides a table of recommended frame-alignment words, and deals at length with the subject. In contrast to word interleaving, the multiplexer may assign each signal or voice channel a time slot equal to one bit. This arrangement is known as 'bit interleaving'.

3.1.1 TDM–PCM

In the practical world, the basic concepts of TDM as described above remain the same. The multiplex equipment, however, does contain not only the equipment for processing the input analog signals into PAM waveforms and interleaving them, but also equipment for converting the composite bit stream into a PCM bit stream at a specified data bit rate. The CCITT specifies fixed bit rates according to the number of voice channels in a system. These bit rates are:

For 30-channel PCM multiplex a bit rate of 2048 kbit/s
For 120-channel PCM multiplex a bit rate of 8448 kbit/s

or

For 24-channel PCM multiplex a bit rate of 1544 kbit/s
For 96-channel PCM multiplex a bit rate of 6312 kbit/s

The 1544 kbit/s system is the same as the North American and Japanese hierarchy of 24 64-kbit/s PCM voice channels, which, when multiplexed, result in the DS–1 level, and similarly with 6312 kbit/s resulting in the DS–2 level. Both 24- and 30-channel structures are described below, as both are equally important in today's practice.

The multiplexing of a number of voice channels into a digital bit stream may be considered by referring to Figure 3.2, which shows a typical PCM multiplex terminal block diagram. A two-wire connection enters the hybrid or 'term set' of a telephone exchange, or PCM terminal. This hybrid separates the voice into a transmit circuit and a receive circuit. The transmit circuit is first band-limited to 3400 Hz by a low-pass filter whose output is connected to a sampling circuit. Groups of 24, 30, 96 or 120 voice channels (depending on the system in use), all similarly converted into a pulse amplitude modulated (PAM) waveform, are in turn added onto the PAM highway or bus, under control of the transmit clock which provides the sampling pulse for the PAM signal and controls the information onto the PAM bus. The encoder input is presented at any time with only one PAM signal from the PAM

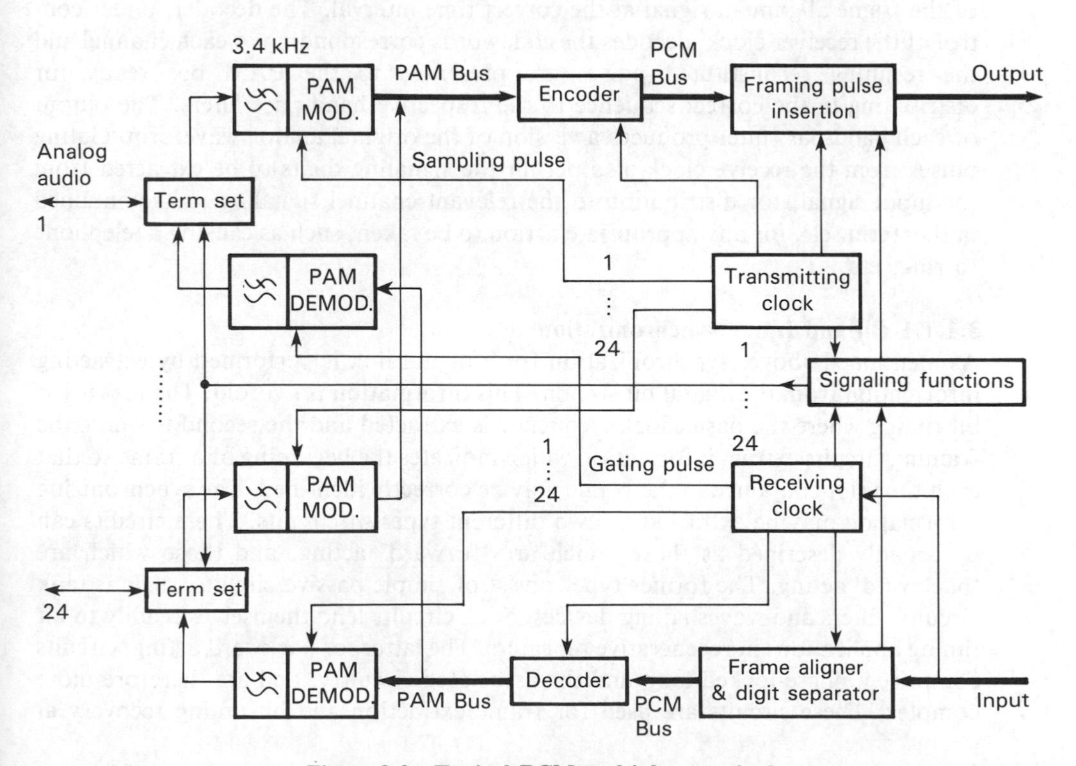

Figure 3.2 Typical PCM multiplex terminal

modulator and under the control of the system or transmit clock, quantizes this pulse and generates the pertinent codeword. The output of the encoder is a digital bit stream which is at the system bit rate. The frame alignment digit is inserted in the appropriate time slots as is also the signaling information, and the total information packet is presented at the output. If a 24-channel system is considered for the moment, each of the 24 speech signals is sampled every 125 μs (8 kHz sampling rate). The separation between successive samples from each of the 24 speech signals is 5.18 μs, and thus the 24 PAM pulses obtained by sampling are aligned every 5.18 μs at the common bus. The encoder with the 15-segment μ-law converts each of the PAM pulses into an 8-bit PCM codeword in a 5.18 μs time slot. At the end of time slot 24, the timing circuit inserts another pulse which is used for framing. Thus, $8 \times 24 + 1$ frame bit = 193 pulses, which are aligned in a frame of 125 μs. In this system, twelve successive frames are combined to form a multi-frame* of 1.5 ms. The state of the framing pulses is such as to permit the identification of the frames and the multi-frames. This will be dealt with further when considering each of the 24- and 30-channel systems.

The receiving side of the PCM multiplex accepts the incoming data, regenerates this signal and extracts the timing information in the process. The receive clock uses this timing information for the operation of the demultiplexing circuits, assured that its rate corresponds to the average rate of the remote transmit terminal. The frame aligner and digit separator examines the received signal and determines the presence of the frame alignment signal at the correct time interval. The decoder, under control of the receiver clock, decodes the codewords corresponding to each channel and the resulting reconstituted pulses are presented to the PAM bus ready for distributing in the correct sequence to the respective band-pass filters. The output of each band-pass filter produces a version of the original analog waveform. Gating pulses from the receive clock also permit the signaling digits to be extracted from the input signal, for distribution to the relevant channel signaling units contained in the 'term set', for any appropriate action to be taken, such as causing a telephone to ring, etc.

3.1.1.1 Bit and frame synchronization

As mentioned above, synchronization for a digital link is performed by extracting information from the digital bit stream. This information is twofold. The first is for bit timing where the basic clock frequency is extracted and the second is where the framing circuits extract information which indicates the beginning of a frame so that each time slot contained in the frame may be correctly identified. The synchronizing information may be extracted by two different types of circuits. These circuits can be broadly described as those which are 'forward' acting, and those which are 'backward' acting. The former type consist of simple passive circuits, such as tank circuits, filters and waveshaping devices. Such circuits lend themselves readily to bit timing applications in regenerative repeaters. The latter, or backward acting, circuits comprise a phase-locked loop and the associated circuitry, and are therefore more complex. These circuits are used for frame extraction and bit-timing recovery at

* See Section 3.2.1.1

multiplex terminals. Forward-acting framing is employed when a pattern sequence is preceded by a predetermined framing pattern sequence permitting easy recognition by logic circuits. The timing circuit is started by the arrival of the framing pattern which then initiates the timing circuit in providing timing pulses for the identification of the information sequence. Backward-acting circuits consist of a pattern-matching circuit and a frame counter. The frame counter steps up one for each frame received and adjusts the pattern-matching circuit accordingly to the next input pattern expected. The output from this type of circuit provides all the timing information for the receiver.

In order that the receiving terminal may correctly segregate the incoming bit stream into channels, correct identification of each incoming time slot must be possible. No problem would exist if there were no transmission impairments to corrupt the bit stream. However, the bit stream is from time to time corrupted, with the result that the frame alignment is lost. Frame alignment is considered to be lost when three or four frame alignment signals have been received with an error. The restoration of the frame alignment by automatic means is an essential part of the circuit or system design which ensures that frame alignment is recovered after a valid frame alignment signal is available at the receiving terminal equipment. Another real-world problem, which arises and which is taken account of in the circuit design or in the system design, is the occurrence of a series of eight zeros in the digital bit stream. This long stream of zero logic states causes a loss in the timing of the timing recovery circuits. To overcome this problem some networks eliminate the need for this codeword, whilst others arrange for the coder to generate a binary output in which alternate digits are inverted (ADI). Thus all zeros become 10101010 and binary 106, which is 10101010, becomes all zeros. The advantage of binary 106 becoming all zeros, and vice versa, is that binary 106 rarely occurs, whereas all unoccupied or unused channels, which may occur quite frequently, will transmit streams of zeros. Reference 1, p. 88, deals at length with frame-alignment systems.

3.2 PRACTICAL PRIMARY OR LEVEL 1 MULTIPLEX

We now discuss practical multiplex systems to the level 1 bit streams, which are 1544 kbit/s, and 2048 kbit/s. The 1544 kbit/s bit stream is the DS–1 level in the North American multiplex hierarchy, and level 1 in the Japanese systems. The CCITT now recommends 1544 kbit/s as hierarchical level 1 in CCITT Recommendation G702.[2] The 2048 kbit/s bit stream is also designated as level 1 in the same recommendation (G702), although it is a completely different hierarchial structure from that of the hierarchy based on a first-level bit rate of 1544 kbit/s. Figure 3.3 shows the various sources which are multiplexed up to these bit stream levels. Note that, in the blocks representing each of the interfacing or processing equipment, is the relevant CCITT recommendation to be consulted for that particular interface or process. The overall hierarchy for both 1544 kbit/s and 2048 kbit/s systems will be given in Section 3.3. Figure 3.3 may be compared with Figure 1.1, to obtain a more detailed concept of what is being considered in this chapter.

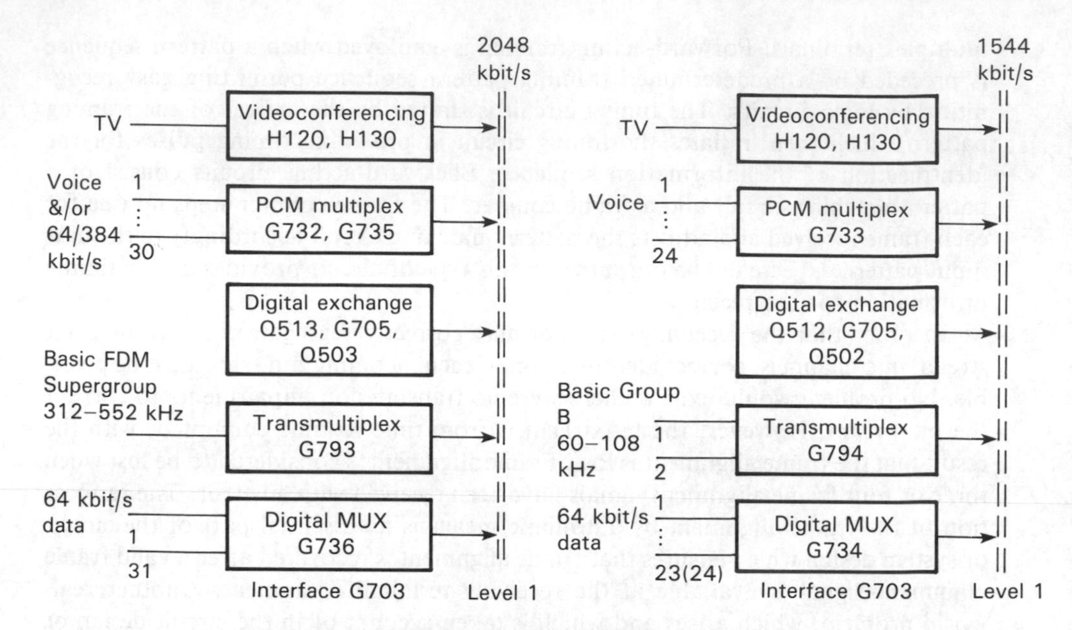

Figure 3.3 CCITT first level multiplex hierarchies and recommendations[2]

The remainder of this Section will deal briefly with each of the interfaces and processes as presented in Figure 3.3, before considering level 2 and higher levels of the hierarchies, and their associated equipment and processes.

3.2.1 24-channel PCM multiplex

The 24-channel TDM multiplexer is used as the basic T1 carrier system by the American Telephone and Telegraph Company and by the Bell Telephone Company. This system is designed primarily for short distance and heavy usage in metropolitan areas. The maximum length of the T1 system is now limited to 90–160 km with a repeater spacing of 1.5 km or thereabouts. The transmission system used is metallic-pair cable, although optical fibers have and are being used. The original Bell T1 system uses 7 digits for each codeword with an encoding law based on $\mu = 100$. The CCITT 24-channel system (Recommendation 733),[2] which has been established to provide compatibility of international networks, supersedes the Bell T1 system by recommending that an 8-digit codeword and $\mu = 255$ be used. The bit rate of this 24-channel system is 1544 kbit/s ± 50 parts per million (ppm), and may be used as the input bit stream to higher-level multiplex bit streams; or as for frequency-division multiplex (FDM), placed directly onto a carrier system for transmission by radio; or transmitted directly without a carrier as in the T1 metallic-pair system. For comparison, note that in Figure 3.3, the transmultiplex equipment takes two groups, or 24 VF channels, where the basic group is a building block comprising 12 VF channels in the FDM hierarchy. Similarly, 24-channel PCM multiplex may be considered as a building block in the TDM hierarchy. The timing signal for the CCITT multiplex equipment should be derived internally from the received signal and also from an external source. Under fault conditions, the multiplex equipment should be so constructed as to detect a failure in the power supply, the loss of incoming signals at 1544 kbit/s, the loss of frame alignment and an alarm indication received from the remote PCM multiplex equipment. This alarm indication is provided by the remote end sending a digit stream in which bit 2 in every channel time slot is forced to 0 or by changing the *S* bit to 1 for the 12-frame multi-frame or by sending a frame alignment alarm sequence 111111100000000 in the *m* bits of the 24-frame multi-frame system. Sections 3.2.1.1 and 3.2.1.2 describe the 12- and 24-frame multi-frame arrangements for the 24-channel CCITT PCM primary multiplex.

3.2.1.1 12-frame 24-channel PCM multiplex (CCITT Recommendations G733, G704)[2]

In Figure 3.4, the sampling rate for a VF channel, as previously discussed, is 8000 samples/s. This means that the interval between samples of that same voice channel is 125 μs. As 24 separate channels are to be accommodated in that interval, and each channel upon sampling, quantization and encoding, provides an 8-bit codeword, there will be in the 125 μs frame, $24 \times 8 = 192$ information bits. To provide framing alignment, one bit in addition to the 192 bits is also required. The frame alignment bit is *always* bit 1 in the frame. Bits 2–193 are the remaining octet-interleaved information bits. If we consider that, due to a transmission error, the frame alignment has suddenly been lost, two steps must be taken by the detection circuit. These are, first, to detect the misalignment, and secondly, to realign the system. In the detection of the misalignment, the frame alignment *word* is checked,

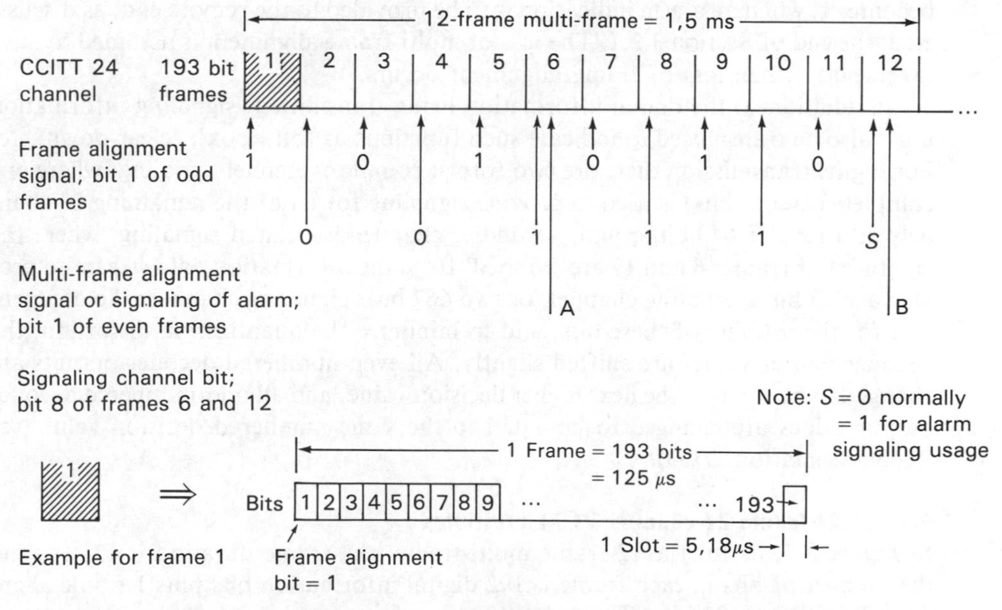

Figure 3.4 12-frame multi-frame CCITT 24-channel PCM primary multiplex system

and, if after x successive tests the equipment fails to confirm the presence of the word, the circuit changes from the 'lock' mode into the 'search' mode. After unsuccessful attempts to try and find the correct sequence of the codeword, the circuits initiate an 'out of frame alignment' alarm. Since the risk of simulating the correct frame alignment pattern by random incoming data can be high during the realignment process, it is normal practice to provide a codeword whose probability of being duplicated by the random signal is very small. The longer the frame alignment codeword, the less the chance of random data duplicating it, but the longer the word, the longer the realignment takes. However, frame alignment codewords have been found which exhibit a high tolerance to transmission errors, and which are not over long.[3] One of these patterns is 1110010. In the light of this very brief discussion on a complicated subject, it is apparent that the first bit of a frame is not sufficient to detect loss of alignment or to detect realignment in the presence of the random digital data which comprise the information data stream. To overcome this problem, a frame alignment codeword is constructed from the first bit of each frame in a fixed number of frames. This fixed number of frames is called a *multi-frame*. The creation of a multi-frame structure, however, produces its own problems when frame misalignment occurs, because multi-frame misalignment also occurs. To overcome this problem, a multi-frame alignment codeword is also generated. To accommodate the frame alignment codeword and the multi-frame alignment codeword and yet still retain the condition that the first bit only of any frame is to be used for alignment, means that the two alignment codeword bits alternate between successive frames. This is shown in Figure 3.4, where the frame alignment codeword is 101010 on all odd frames, and the multi-frame alignment codeword is 00111S on all even numbered frames within the multi-frame structure; S is normally 0, and becomes 1, when an alarm indication is to be provided to the remote end, as discussed at the end of Section 3.2.1. The loss of multi-frame alignment is assumed to have taken place when loss of frame alignment occurs.

In addition to the digital information being transmitted, signaling information must also be transmitted to indicate such functions as 'off-hook', 'clear-down, etc. For digital transmission there are two forms; common-channel signaling, where one complete octet (8 bits) is used to provide signaling for all of the remaining 23 channels at a rate of 64 kbit/s; and, secondly, channel-associated signaling, where the eighth bit of frames 6 and 12 are 'robbed' from the information data stream to provide a 1333 bit/s signaling channel, or two 667 bit/s signaling channels. To compensate for the robbing of these bits, and to minimize the quantization distortion, the decoder output values are shifted slightly. All even-numbered decoder outputs are changed to be equal to the next higher decision value, and all odd-numbered decoder output values are changed to be equal to the same numbered decision value (see Recommendation G733).[2]

3.2.1.2 24-frame 24-channel PCM multiplex

In Figure 3.5, as for the 12-frame multi-frame, the frame duration is 125 μs, and the number of bits in each frame is 192 digital information bits plus 1 frame alignment bit, giving 193 bits. The only difference between this system and that of the 12-frame multi-frame is the frame structure itself. As can be seen from Figure 3.5,

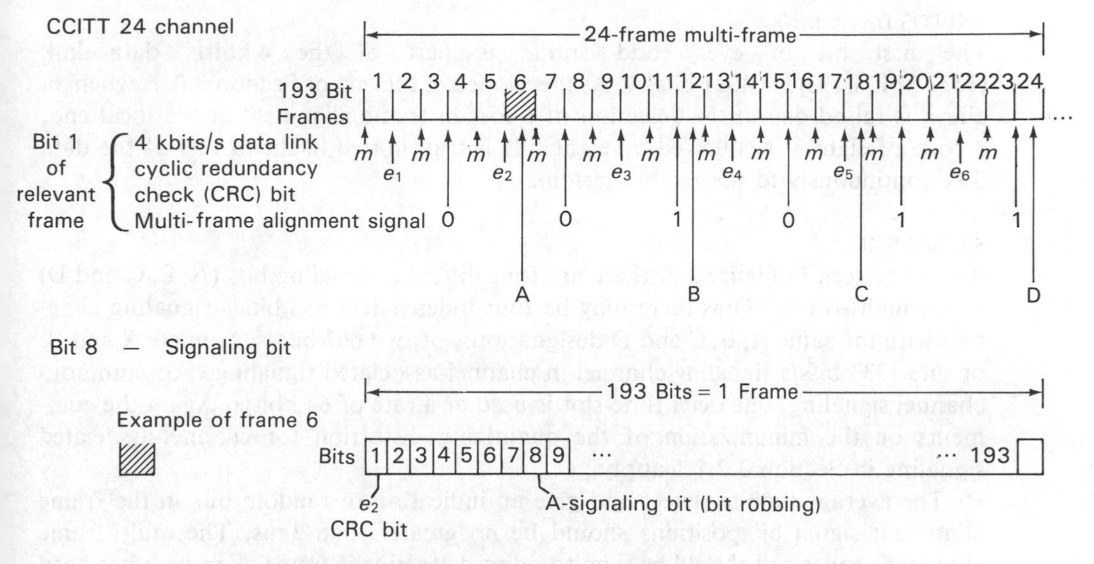

Figure 3.5 24-frame multi-frame CCITT 24-channel PCM primary multiplex system

the multiframe comprises 24 frames with a duration of $24 \times 125\ \mu s = 3$ ms. Bit 1 of each frame is used to provide three purposes in this case. These are:

To provide a multi-frame alignment codeword
To provide a cyclic redundancy check code
To provide a 4 kbit/s data link

Bit 8 of frames 6, 12, 18 and 24 are again used for signaling by robbing data information bits.

MULTI-FRAME ALIGNMENT CODEWORD

The framing bit (*F*-bit), bit 1, of every fourth frame forms the pattern 001011... ...001011. This multi-frame alignment signal is used to identify where each particular frame is located within the multi-frame in order to extract the cyclic redundancy check code (CRC), and the data link information as well as to identify those frames which contain signaling.

2 KBIT/S CYCLIC REDUNDANCY CHECK

The CRC (CRC–6) is a method of providing additional protection against simulation of the frame alignment signal, and/or where there is a need for an enhanced error monitoring capability. The CRC–6 message block checks bits e_1, e_2, ..., e_6 are contained in the first frame bit of frames 2, 6, 10, 14, 18 and 22 of every multi-frame (see CCITT Recommendation G704[2] for further details).

4 KBIT/S DATA LINK

The first bit of every odd frame is part of the 4 kbit/s data link (1552 kbit/s/193/2 = 4021 bits/s). As mentioned at the end of Section 3.2.1, when an alarm is raised due to the detection of a loss in frame alignment at the local end, a code of eight '1's followed by eight '0's is transmitted in the *m* bits of the data link continuously to the remote terminal.

SIGNALING

As can be seen in Figure 3.5, there are four different signaling bits (A, B, C and D) in the multi-frame. Thus there may be four independent 333 bits/s signaling channels with the same A, B, C and D designations, or two 667 bits/s channels A and B, or one 1333 bits/s signaling channel in channel-associated signaling. For common-channel signaling, one octet time slot is used at a rate of 64 kbit/s. Again the comments on the minimization of the quantizing distortion for channel-associated signaling in Section 3.2.1.1 apply.

The average time to detect and give an indication of random bits in the frame alignment signal bit positions should be no greater than 3 ms. The multi-frame alignment codeword should be monitored to determine if frame alignment has been lost, and recovered after a valid frame alignment signal is available at the receiving terminal equipment.

3.2.2 30-channel PCM primary multiplex (Recommendations G732, G704)[2]

The CCITT 30-channel primary PCM multiplex equipment operates at a bit rate of 2048 kbit/s. The encoding law used is the 13-segment *A* law (Recommendation G711), where *A* is given as 87.6 (see equation 2.67), and the number of quantization levels is 256. Referring to Figure 3.6, each multi-frame comprises sixteen frames of 2 ms total duration, consecutively numbered 0 to 15. Each frame 125 μs long is made up of a set of 32-channel time slots numbered 0 to 31 and each of the channel time slots 3.9 μs long comprises an eight-bit codeword, or an eight-digit time slot 488 ns long. The gross digit rate generated by the system = sampling rate of 8000 samples/s $\times$ 32 time slots $\times$ 8 bits = 2048 kbit/s $\pm$ 50 parts per million (ppm). The transmitting timing signal is derived from either an internal or external source or from the incoming digital signal. The frame structure is given in Figure 3.6.

Multi-frame alignment

With channel-associated signaling, the multi-frame alignment codeword is 0000 and occupies digit time slots 1 to 4 of channel time slot 16 in frame 0. This of course means that, every 16 frames, the codeword appears in a bunched form, and is not distributed throughout different frames as with the 24-channel system. The remaining four-digit time slots 5 to 8 of channel time slot 16 in frame 0 comprise spare bits (5, 7 and 8), which are set to 1 if they are not being used, and bit 6, which is used to indicate 1, when there is a loss of multi-frame alignment. Multi-frame alignment is assumed to have been lost when two consecutive multi-frame alignment signals

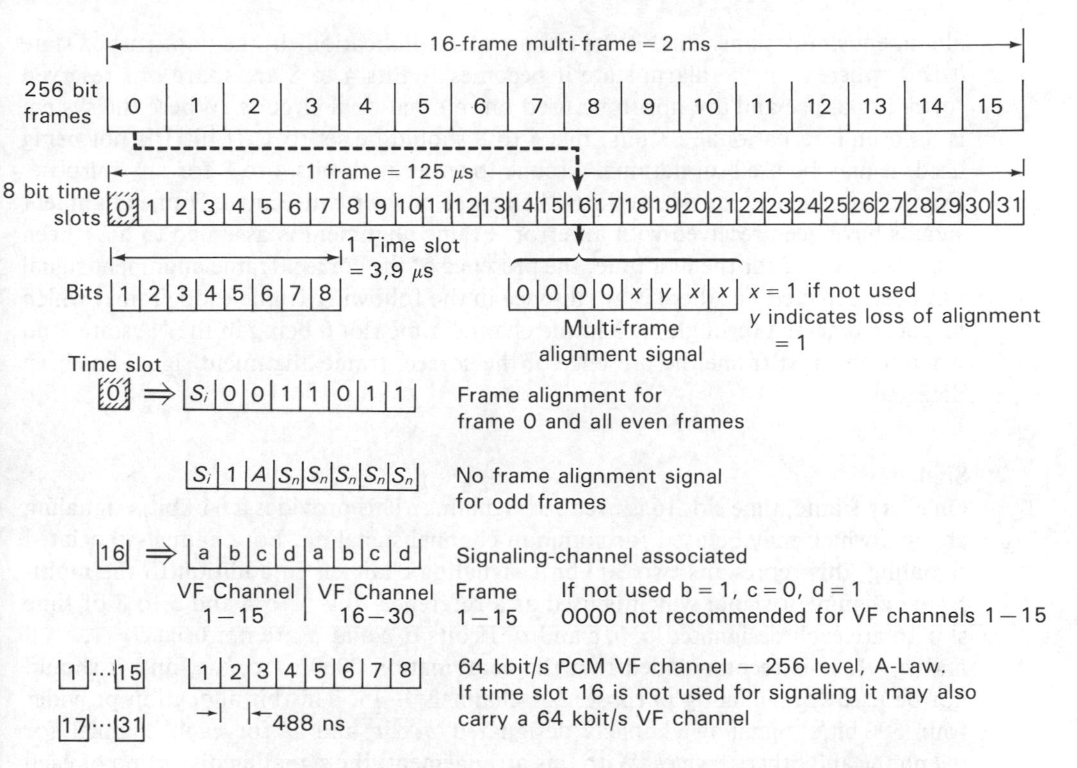

Figure 3.6 CCITT 30-channel PCM primary multiplex system

have been received with an error, and when for a period of 1 or 2 multi-frames, all of the bits in time slot 16 are in the 0 state. This second condition avoids the spurious multi-frame alignment condition. Multi-frame alignment is assumed to have been recovered as soon as the first correct multi-frame alignment signal is detected, and when at least one bit in time slot 16 is at logic level 1 preceding the multi-frame alignment signal being first detected.

Frame alignment

The frame alignment signal occurs in time slot 0 of every even-numbered frame, including frame 0, i.e. frames 0, 2, 4, ..., 30. In time slot 0, bits 2 and 8 contain the frame alignment codeword 0011011. Bit 1, however, is not used as part of this frame alignment signal and is reserved for international use. One application is in the cyclic redundancy check which may be implemented in the system should there be a need to provide additional protection against simulation of the frame alignment signal, and/or where there is a need for an enhanced error-monitoring capability. Details of the procedure are given in CCITT Recommendation G704.[2] For odd-numbered frames, no frame alignment signal is contained in the bit positions 2 to 8. Bit 1 is used as part of the cyclic redundancy check if so required, or is reserved for international use. If no use is made of bit 1, of time slot 0 in either the odd or even frames, it should be fixed at 1. Bit 2 is always fixed at 1 to assist in the avoidance of frame alignment simulations. Bit 3 is the remote alarm indication. In the undisturbed state it is 0, whereas in the alarm state it becomes 1. Bits 4 to 8 are spare bits reserved for national use and are not to be used on international circuits. Where the system is used on international circuits, bits 4 to 8 should be set to 1. If bit 1 is not being used, it may be used on national circuits together with bits 4 to 8 for any purpose. Frame alignment is considered lost when three or four consecutive frame alignment signals have been received with an error. Frame alignment is assumed to have been recovered when, for the first time, the presence of the correct frame alignment signal has been detected but there is an absence in the following frame (odd frame) which has been detected through bit 2 in the channel time slot 0 being in the '1' state, and when in the next frame the presence of the correct frame alignment signal has been detected.

Signaling

On every frame, time slot 16 is used for signaling. This provides a 64 kbit/s signaling channel which may be used for common-channel signaling. For channel-associated signaling, this represents two 30 kbit/s signaling channels in addition to the multi-frame alignment signal which is used as a reference. Bits 1 to 4 and 5 to 8 of time slot 16 are each designated a, b, c and d. If bits b, c and d are not used $b = 1, c = 0$ and $d = 1$. It is recommended that the combination 0000 of a, b, c, and d should not be used for signaling purposes for channels 1–15. This bit allocation provides four 500 bit/s signaling channels designated a, b, c and d for each channel for telephone and other services. With this arrangement, the signaling distortion of each signaling channel introduced by the PCM transmission system will not exceed ± 2 ms.

Fault conditions and subsequent actions

The PCM multiplex equipment should detect the following fault conditions:

1. Failure of the power supply
2. Failure of a codec, except when single-channel codecs are used
3. Loss of the 64 kbit/s incoming signal, which is not mandatory for channel-associated signaling, or when the signaling multiplex is located within a few meters of the PCM multiplex equipment, or when contradirectional interfaces are used
4. Loss of the incoming signal at 2048 kbit/s when a loss of frame alignment is not indicated
5. Loss of frame alignment
6. A bit error rate less than 1 in 10000
7. Alarm indication received from the remote PCM multiplex equipment

The appropriate actions to be taken are as follows:

Service alarm to be raised for fault conditions 1, 2, 4–7
Prompt maintenance alarm indication generated for fault conditions 1–6

Transmission of alarm indication to the remote end for fault conditions 1, 2, 4, 5 and 6
Transmission suppressed at analog outputs for fault conditions 1, 2, 4, 5 and 6
Alarm indication signal applied to 64 kbit/s output, time slot 16 for fault conditions 1, 4, 5 and 6
Alarm indication signal applied to time slot 16 of 2048 kbit/s output for fault conditions 1 and 3

Jitter
Where the transmitting timing signal is derived from an internal oscillator, the peak-to-peak jitter at the 2048 kbit/s output should not exceed 0.05 UI (unit interval, equal to time duration of a bit; see Section 1.4.7), when it is measured within the frequency range 20 Hz–100 kHz.

3.2.3 64 kbit/s digital multiplex (CCITT Recommendations G734, G735, G736)[2]

3.2.3.1 Synchronous digital multiplex equipment operating at 1544 kbit/s

This multiplex equipment accepts 23 separate 64 kbit/s digital signals, called *tributaries*, and combines them into a 1544 kbit/s bit stream. For applications within the integrated services digital network (ISDN), it is expected that 24-channel multiplex equipment with a frame structure conforming to Figure 3.5 will be used. The timing signals for this multiplexer are derived from a composite clock signal of a centralized clock which provides timing signals to both the transmitting and receiving equipment sides. The centralized clock itself may be derived from certain incoming line signals to it, or derived from a frequency standard such as a caesium beam or rubidium atomic clock. The codirectional interface or centralized clock interface should be used for synchronized networks and for plesiochronous* networks to ensure an adequate interval between the occurrence of slips.† One way of distributing the timing information of a centralized clock is to use a *two tone* system, which operates at the frequencies of 1.7 kHz and 2.7 kHz, both sinusoids being derived from an atomic clock. Both of these signals are then distributed over the FDM carrier network, and, at the remote side, the equipment subtracts the 1.7 kHz from the 2.7 kHz frequency. The remaining 1 kHz frequency is reasonably free of the effects of transmission due to the interferences being roughly the same on both frequencies. By using equipment with a high common-mode rejection ratio, the 1 kHz frequency can be considered as a standard to be used as the input to a synthesizer for the generation of all required frequencies. Experimental results on this system show that over a three-minute interval the frequency stability compared with that of the atomic clock used was within 1 part in 10^{10}, and over 3 days, five parts in 10^8. This system was developed by Telecom Australia in the 1970s and used on an experimental basis

* See Section 3.2.7.
† See Section 3.2.8.

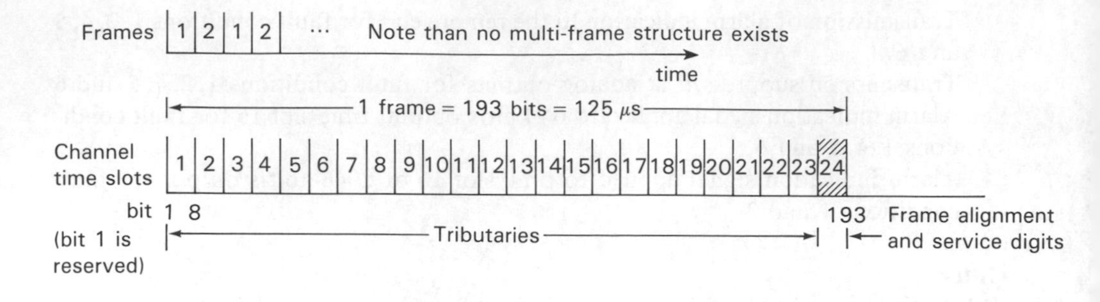

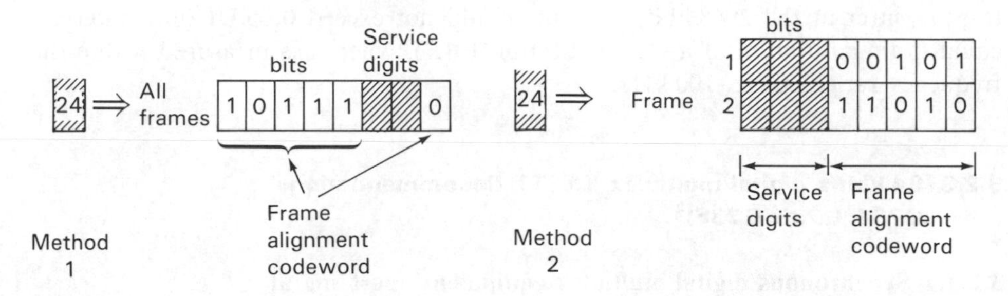

Figure 3.7 1544 kbit/s synchronous digital multiplex

to provide a plesiochronous network. In addition to the above, the multiplexer timing signals may be derived from the 1544 kbit/s incoming digital bit stream. The multiplexing method used in this system is to interleave cyclically each 8-bit byte in tributary number order, to produce the frame structure as given in Figure 3.5, or as that given in Figure 3.7.

The frame structure shown in Figure 3.7 has 8 bits per channel time slot, numbered 1 to 8, and 24 time slots per frame, numbered 1 to 24. Successive bits for bytes 1 to 24 are consecutively numbered from 2 to 193 and the first bit is reserved for optional use. The frame repetition rate is 8000 Hz. Channel time slots 1 to 23 are assigned to tributaries, and channel time slot 24 is assigned to frame alignment and service digits. As can be seen from Figure 3.7, there are two methods of allocating digits to time slot 24.

Method 1 uses bits 1, 2, 3, 4, 5 and 8 to carry the frame alignment signal 101110, and uses bits 6 and 7 for service digits. The loss of frame alignment is assumed when more than three of twelve successive frames have an error in the frame alignment signal and/or in bit 1 of the 193-bit frame. It is assumed that frame realignment has occurred when four consecutive correct frame alignment signals have been received.

Method 2 uses alternate frames, where bits 1, 2, and 3 of time slot 24 are used for service digits in any frame and bits 4–8 provide the code 00101 for frame 1, and 11010 for frame 2. The loss of frame alignment is assumed when seven consecutive pairs of the frame alignment signal (00101, 11010) have been incorrectly received in their predicted positions. Frame alignment restoration is assumed to have occurred when two consecutive correct pairs of frame alignment signals have been received.

3.2.3.2 Synchronous digital multiplex equipment operating at 2048 kbit/s

This multiplex system combines up to 31 tributaries, or 64 kbit/s bit streams to produce at its output a digital stream at a bit rate of 2048 kbit/s. As with the 1544 kbit/s system, a data stream of 2048 kbit/s is also demodulated into its component 64 kbit/s tributaries by this equipment. The term 'multiplex equipment' is used throughout this chapter to include the multiplexing and demultiplexing processes, unless otherwise stated. The timing source for this system is derived from any of the following sources:

The received 2048 kbit/s signal
An external source at 2048 kbit/s derived from a centralized clock
An internal source

The provision of a timing signal output for the purpose of synchronizing other equipments may be required in some national synchronizing arrangements. The frame structure for this system has already been described in Section 3.2.2 and is shown in Figure 3.6. If this equipment is interconnected with multiplex equipment which uses time slot 16 for internal purposes, this time slot cannot be used as for a 64 kbit/s tributary. The loss and recovery of the frame alignment, the fault conditions and consequent actions and the jitter at the 2048 kbit/s output, are all the same as that discussed in Section 3.2.2 for the 30-channel PCM primary multiplex. The digital interfaces at 64 kbit/s should be of either the codirectional or contradirectional type. The codirectional type is used to describe an interface where the information and its associated timing signal are transmitted in the same direction, i.e. the timing signal is transmitted alongside the transmitted information signal. Likewise, alongside the received signal, is received the timing signal from the far end. At 64 kbit/s, three signals are carried across the interface in both the transmit and receive directions. They are:

The 64 kbit/s information signal
The 64 kbit/s timing signal
The 8 kbit/s timing signal

The 8 kbit/s timing signal must be generated at the transmit side, but it is not mandatory for the receive side to use it. The detection of an upstream fault can be transmitted across a 64 kbit/s interface either by transmitting an *alarm indication signal* (AIS) and/or by interruption of the 8 kbit/s timing signal received. The contradirectional interface is used to describe a situation where the timing signals associated with both directions of transmission are in one direction only. This means that both the receive signal side and the transmitting side have a timing signal being transmitted. CCITT Recommendation G703 provides details on these types of interfaces.[2]

Some PCM multiplex equipment operating at 2048 kbit/s provides internal digital access options (CCITT Recommendation G735).[2] These options include bidirection 64 kbit/s data channels and unidirectional synchronous or asynchronous

Table 3.1 National time slot allocations for 384 kbit/s channels

Country	Type of insertion	384 kbit/s channels (Note 1)				
		A	B	C	D	E
Norway West Germany Denmark Switzerland	Synchronous	1–2–3 17–18–19	4–5–6 20–21–22	7–8–9 23–24–25	10–11–12 26–27–28	13–14–15 29–30–31
France United Kingdom	Asynchronous	1–7–11 17–23–27	3–9–15 19–25–31	4–8–12 20–24–28	5–10–13 21–26–29	2–6–14 18–22–30

Note 1. The five possible 384 kbit/s channels in a 2048/s bit stream are numbered A to E. Preferably the channel pairs A–B and C–D should be used for stereophonic transmission.

384 kbit/s channels for the setting up of sound-programme circuits. The 384 kbit/s channel is based upon the allocation of six 64 bit/s time slots. The frame structure is the same as that provided in Figure 3.6, and the allocation of the time slots are in accordance with that in Table 3.1, for various government administrations.

3.2.4 Transmultiplexers (CCITT Recommendations G791, G792, G793, G794)[2]

The CCITT definition 4020 (Recommendation G701)[2] of a transmultiplexer is 'An equipment that transforms a frequency-division multiplexed signal (such as a group or supergroup) into a corresponding time-division multiplexed signal that has the same structure as if it had been derived from PCM multiplex equipment, and that also carries out the complementary function in the opposite direction of transmission'. The signaling and supervisory protocols are different in a digital network from those found in an analog network as is the nature of the composite baseband signal. One method of overcoming these differences is to demultiplex the composite baseband down to voice frequencies using FDM channeling and primary PCM multiplex. This, however, is very expensive especially since signaling conversion must take place at channel level. The transmultiplexer method removes the need to perform this demultiplexing and converts from one format to the other easily and inexpensively. The direct conversion from one format to the other is done using digital signal processing, and the conversion between signaling systems is performed by a single unit serving all channels. The basic transmultiplexer[4] is shown in Figure 3.8.

The PCM interface provides a conversion between the level-1 digital bit stream used in the digital network and a linear '2's complement representation which is frame synchronous with the digital signal processing (DSP) unit. This interface also provides the generation of the alarm indication signal (AIS) and the detection of such, the monitoring of errors and loss of input, and access to housekeeping bits

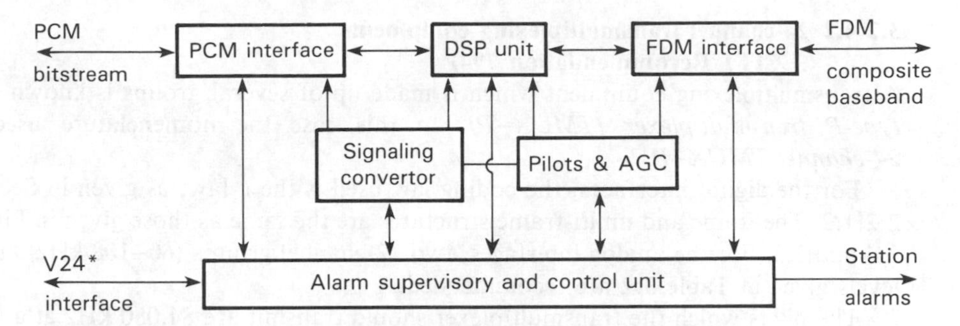

Figure 3.8 Block diagram of a transmultiplexer
(Courtesy of IEE, Reference 4)

in the frame structure. The frame structure will be that given in Figure 3.6 for 2048 kbit/s, or Figure 3.4 or 3.5 for 1544 kbit/s. The FDM interface provides conversion between the digital representation of the supergroup or groups required by the digital signal processing (DSP) unit and the analog form required in the FDM network. The DSP unit translates the 60- or 24-speech channels sampled at 8 kHz into a single FDM signal. Digital signal processing techniques are used to perform the translation and a digital-to-analog convertor reconstructs the analog FDM signal. The FDM to PCM conversion is also performed by this unit using similar hardware.

3.2.4.1 60-channel or supergroup transmultiplexing equipments (CCITT Recommendation G793)[2,4–6]

A transmultiplexing equipment in which the analog interface is made up of one or more supergroups is known as *type-S transmultiplexer (TMUX–S)*. In this case where only one supergroup is being considered, the nomenclature used is *60-channel TMUX–S*.

For the digital interface, the coding law used is the *A* law, which was dealt with in Section 2.2.1.2. The frame and multi-frame structure is that given in Figure 3.6. For the analog interface, the supergroup (312–552 kHz) levels at the supergroup distribution frame are recommended to be:

For sending	– 36 dBr
For receiving	– 30 dBr

at impedances of 75 ohms (unbalanced). The pilots which the transmultiplexer should transmit are a supergroup pilot with a frequency 411 920 Hz at a level of – 20 dBm0, and one pilot per group also at a level of – 20 dBm0 with frequencies: Group 1, 335 920 Hz; Group 2, 383 920 Hz; Group 3, 431 920 Hz; Group 4, 479 920 Hz; and Group 5, 527 920 Hz.

3.2.4.2 24-channel transmultiplexing equipments (CCITT Recommendation 794)[2,7–9]

A transmultiplexing equipment which is made up of several groups is known as a *type-P transmultiplexer (TMUX–P)*. In this case the nomenclature used is *24-channel TMUX–P*.

For the digital interfaces, the coding law used is the μ law, as given in Section 2.2.1.2. The frame and multi-frame structures are the same as those given in Figure 3.4 and 3.5. For the analog interfaces, two 12-channel groups (60–108 kHz) at the levels given in Table 3.2 are recommended.

The pilots which the transmultiplexer should transmit are 84.080 kHz at a level of −20 dBm0; 84.140 kHz at a level of −2 dBm0; and 104.080 kHz at a level of −20 dBm0.

3.2.5 Codecs

The term *Codec* (COder/DECoder) has been adopted to describe integrated circuits which are primarily used for voice-telephony applications and which are constructed using one or two chip processors. However, codecs using integrated circuits are now being developed, and these conform to the digital standards for the component coding of television signals as provided in the CCIR Recommendation 601[10–12]. The conventional approach to PCM encoding of telephony speech signals has already been discussed in Section 2.2.1, where an anti-aliasing filter is followed by sampling at 8 kHz, quantizing and A-law or μ-law companding. This conventional approach, if replaced by one in which the analog filtering is replaced by digital filtering after quantizing and the sampling rate is increased, lends itself to easier integrated circuit fabrication of more sophisticated circuits to satisfy present and future requirements. The preference for digital filters lies in their inherent long-term stability, temperature independence and confined tolerances in the manufacturing process. Also digital filters may be designed which are software controllable and which contain adaptive signal-processing functions.

A host of manufacturers[16] at present fabricate codecs, which fall into three main categories. These are:

Continuously variable slope delta modulation (CVSDM)
Sigma-delta-based codecs using interpolation or digital filtering[13,14]
PCM using successive approximation logic[15]

Of these three main types, the last two are perhaps the most popular, since the CVSDM coder provides neither *A*-law nor μ-law companding and therefore would not be a likely candidate for inclusion into the major public telephone networks.

The CCITT states in its Recommendation 795, that codecs capable of encoding/decoding FDM assemblies will be a useful element in the transmission networks of some administrations during the period of transition from analog to digital working, and that these codecs will have a limited life and application, not forgetting that already they are available in a number of different forms. An annex

Table 3.2 Relative power levels at the basic group and supergroup distribution frames†

Country	Relative power level at group distribution frame (GDF)		Basic group distribution frame	Impedance at group distribution frame	Relative power level at supergroup distribution frame (SGDF)		Impedance at supergroup distribution frame
	Transmit dBr	Receive dBr		Ohms, balanced	Transmit dBr	Receive dBr	Ohms, unbalanced
West Germany	– 36	– 30	B	150	– 35	– 30	75
Australia System 1	– 36.5	– 30.5	B	150	– 35	– 30.5	75
Australia/Denmark System 2	– 42	– 5	B	135	– 35	– 30	75
Austria	– 37 – 36	– 8 – 30	B	75 unbalanced 150	– 35	– 30	75
Belgium	– 37	– 8	B	150	– 35	– 30	75
Bulgaria	– 36	– 23	B	150	– 36	– 23	75
Spain, Ireland, N.Z., Norway, U.K.	– 37	– 8	B	75 unbalanced	– 35	– 30	75
U.S.A. (ATT)	– 42	– 5	B	135	– 25	– 28	75
France	– 33	– 15	B	150	– 45	– 35	75
Hungary, Italy, Netherlands	– 37	– 30	B	150	– 35	– 30	75
India	– 36.5	– 30.4	B	150	– 34.8	– 30.4	75
Japan (NTT)	– 36	– 18	B	75	– 29	– 29	75
Mexico (T de M)	– 47	– 10	B	150	– 47	– 24	75
Poland	– 36	– 23	B	150	– 36	– 23	75
German D.R.	– 36	– 23	B	150	– 36	– 23	75
Sweden					– 35	– 30	75
Switzerland	– 36.5	– 30.5	A or B	75 unbalanced	– 35	– 26	75
U.S.S.R	– 36	– 23	B	150	– 36	– 23	75

† This table was taken from CCITT Recommendation G233.[2]

Table 3.3 FDM codecs[2]
(Courtesy of CCITT)

Administration	Analog interface	Digital interface	Noise performance
British Telecom	Supergroup (312–552 kHz) 15 SG assembly (312–4025 kHz)	8448 kbit/s 68736 kbit/s	140 pW0p <700 pW0p
China	Master-group (812–2044 kHz) (60–1300 kHz)	34368 kbit/s	<783 pW0p
Japan NTT	Group (60–108 kHz)	1544 kbit/s	<340 pW0p

to this recommendation is provided in Table 3.3; this gives details supplied by some administrations of a number of FDM codec realizations.

The CCITT recommends that the noise produced by a codec should be no more than 800 pW0p. This magnitude is expected to occur only on codecs for the higher-order FDM assemblies and that significantly lower values will be achieved with smaller FDM assemblies. The noise allowance of 800 pW0p takes into account the analog processing before the coder and following the decoder, quantization, and errors and jitter on the received digital signal. The phase jitter on a signal caused by a coder/ decoder pair should not exceed 1° peak-to-peak when measured on the output of the equipment, on each side of the signal in the frequency band 20–300 Hz (see CCITT Recommendations G230 and O91).

3.2.6 Digital exchange interfacing (CCITT Recommendations G705, Q502, Q503, Q512, Q513)[2,17]

The digital exchange terminal is a piece of synchronous equipment which has a frame aligner circuit. Its timing signal is derived from the incoming signal at the receive end if the remote end is PCM multiplex equipment. If the network is not synchronized, the exchange's transmitted signal is derived from an office clock. The frame structure has already been discussed, and is given in Figures 3.4–3.6, for the 1544 kbit/s and 2048 kbit/s level 1 data streams. For the 2048 kbit/s data stream, additional time slots may be utilized for common-channel signaling, if there is a need for more signaling capacity. On routes between exchanges comprising more than one 2048 kbit/s digital path, it may be possible to provide an adequate signaling capacity without using time slot 16 of all systems on the route. In these circumstances time slot 16 in those systems not carrying signaling can be allocated to speech or other services. This does not, however, apply to time slot 0, which is reserved for frame alignment, alarms and network synchronization information.

Figure 3.9 shows a network structure in which digital links provide the interconnection of digital exchanges and higher-level multiplex systems. The digital exchanges which themselves are served and serve PCM multiplex equipment are also shown. The length of the network is not dependent upon the accumulation of noise, as is a similar network using FDM and analog techniques. Because only two PCM

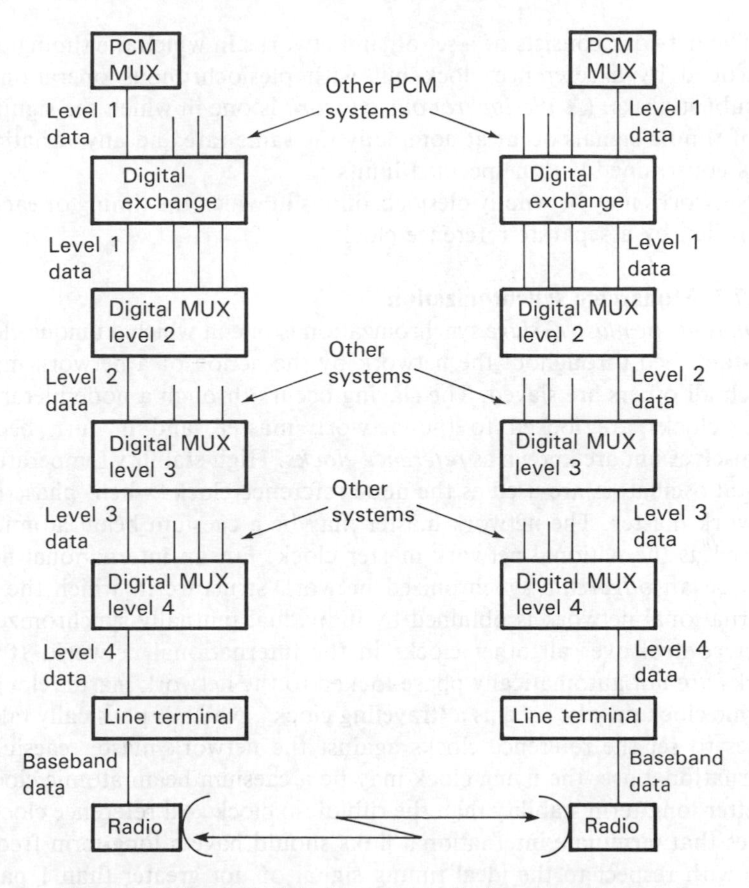

Figure 3.9 Typical digital network utilizing digital exchanges

terminals are required for an analog connection (such as voice-to-voice), the overall attenuation in the circuit end-to-end will be stable, as will the frequency response.

3.2.7 Network synchronization

Section 3.1.1.1 gave a method of providing synchronization for a digital link by extracting the timing information from the incoming digital bit stream. Rather than depend on incoming bit streams to provide the timing between terminals in a network, and never knowing if these bits streams arriving from different sources are derived from the same clock, it is preferable to establish a unique clock frequency throughout the system. It is also preferable to compensate for transmission delays between the switching centers or terminals and variations due to temperature changes, etc., to an integral multiple of a frame period so that the frame phase of each of the time-division buses coincide. Networks which are synchronized may be such that:

1. Timing is controlled by a single reference clock, or master clock.

2. The network consists of a set of subnetworks in which the timing of each is controlled by a reference clock but with plesiochronous operation between the subnetworks. (A *plesiochronous network* is one in which the significant instants of timing signals occur at nominally the same rate and any variation in this rate is constrained within specified limits.)
3. Networks may be wholly plesiochronous in which the timing of each node is controlled by a separate reference clock.

3.2.7.1 Monarchic synchronization

Monarchic or *master–slave* synchronization is one in which a unique clock frequency is established throughout the network by the action of a network master clock to which all others are slaved. The slaving occurs through a node hierarchy, in which node clocks are locked to the network master, and in turn become masters themselves but are known as *reference clocks*. High-stability temperature-controlled crystal oscillators are used as the node reference clocks when 'phase-locked' to the network master. The network master may be a caesium beam atomic clock which is used as the national network master clock. For an international network, there may be an *oligarchic* synchronized network structure in which the synchronized international network is obtained by individual mutually synchronized clocks that exert control over all other clocks in the international network. If the reference clocks are not automatically phase-locked to the network master clock, a rubidium atomic clock may be used as a 'traveling clock', which is physically taken to various nodes to set the reference clocks against the network master caesium beam. For international use, the flying clock may be a caesium beam atomic clock, which has a better long-term stability than the rubidium clock. All reference clocks at network nodes that terminate international links should have a long-term frequency departure with respect to the ideal timing signal of not greater than 1 part in 10^{11}, as specified in CCITT Recommendation G811 on the timing requirements at the outputs of reference clocks and network nodes suitable for plesiochronous operation of international digital links. Typical long-term frequency departures from the ideal timing signal are provided in the table below for various frequency standards used in an international or national network:

Type of clock	Frequency departures ($\Delta f/f$ = Time interval error/Measurement time)		
	Short term	Medium term	Long term
	1 second	1 day	1 year
Caesium beam atomic primary standard	5×10^{-12}	5×10^{-14}	1×10^{-13}
Rubidium vapor secondary standard	1×10^{-11}	1×10^{-12}	1×10^{-11}
Temperature-controlled precision crystal oscillator	1×10^{-10}	5×10^{-10}	5×10^{-10}
Temperature-controlled glass-encapsulated crystal oscillator	5×10^{-10}	1×10^{-9}	1×10^{-9}
Temperature-controlled mass-produced-metal-case encapsulated crystal oscillator	1×10^{-9}	1×10^{-8}	5×10^{-8}

The method of measuring these frequency departures is by use of a *vector voltmeter* (Hewlett-Packard), in which a primary or secondary standard signal is placed into the reference input port, and the source to be compared is placed into the signal input port. Both the standard and compared frequencies should nominally be the same. The output from the vector voltmeter is fed into a chart recorder. The width of the chart is set for $\pm 180^\circ$, and left to run at a predetermined speed, such as 5 cm/h. As the phase of the source changes, from the starting phase, across the chart, and back to the starting phase, one cycle difference between reference and that source being compared has occurred. The time taken for this to occur is shown on the chart recorder. Thus 1 cycle (time to complete one cycle) is the measurement of Δf. If the frequency being compared is f, the frequency departure is given by $\Delta f/f$.

3.2.7.2 Independent synchronization

As shown above, the availability of highly accurate and stable clock sources suggests that it is possible to provide each main switching node in a national network with a highly stable reference clock without providing any automatic overall control. The disadvantage of this system is that slips will occur due to the differences in these reference clocks; also pulse stuffing techniques may be required. These effects will be discussed in Sections 3.2.8 and 3.5.1, respectively. Plesiochronous operation between each of these nodes is a means of overcoming the problem of the accumulation of frequency errors which may accrue between the reference nodes, and keeping these differences within tolerable limits as determined by the slip rate.

3.2.7.3 Mutual synchronization[18,19]

This synchronization method is to establish a unique clock frequency throughout the national network by the mutual interaction of mutiple-input phase-locked high-stability temperature controlled crystal oscillators, which are provided in each of the network nodes and which are fed from each of the other switching nodes connected to that node. A slight variation in the frequency of one clock in the network causes a control signal to be interchanged between the other centers at the same hierarchial level in the network, so that all the clocks are brought back together into phase. The changes in the control signal are quantized and the multiple-input phase-locked oscillators are weighted to provide discrete steps for the control of the other clocks. The control arrangements to higher and lower hierarchical levels are designed such that the information on relative clock frequency (or phase) differences is always controlled in the downward direction of the hierarchy.

3.2.8 Slips

Consider the synchronization case of independent oscillators, as discussed briefly in Section 3.2.7.2. If the frequency of the node clock is slightly higher than that of the timing information contained in the digital bit stream, pulses which are read into a single frame buffer at the node by the node clock, may occasionally be read out

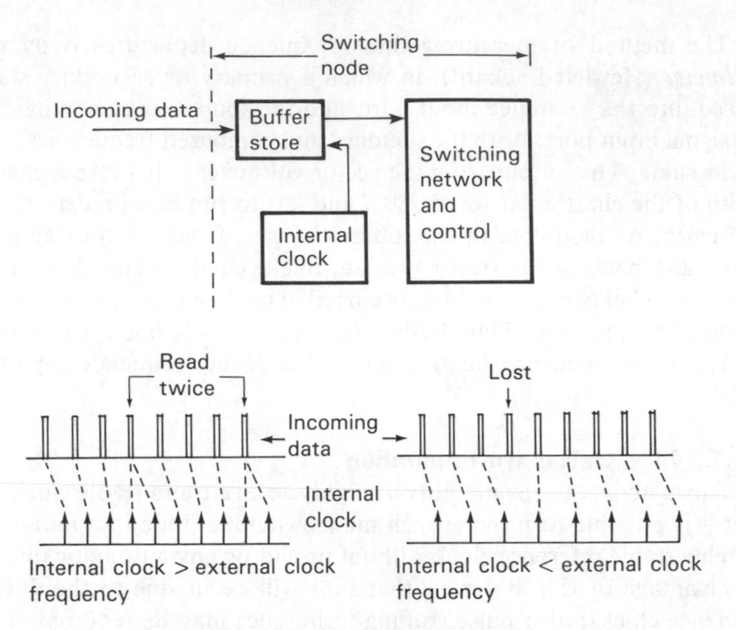

Figure 3.10 Slip mechanisms

twice as shown in Figure 3.10. If the node clock, however, is lower in frequency than the timing information contained in the bit stream, some of the pulses may be lost. This loss or gain of pulses is referred to as a *slip*. More specifically, in practice there are two forms of slip. The first is a *controlled* slip, which is the irretrievable loss or gain of a set of consecutive digit positions in a digital signal, in which both the magnitude and instant of that loss or gain are controlled, to enable the signal to accord with a rate different from its own. These slips may be an octet, or a frame slip, where a specified set of consecutive digits are gained or lost. The second form of slip is *uncontrolled*, where the loss or gain of a digit position or a set of consecutive digit positions in a digital signal is a result of an aberration of the timing process associated with transmission or switching of a digital signal; the magnitude or the instant of that loss or gain is then not controlled. The occurrence of a slip depends upon the frequency difference between the node clock and the external clock controlling the timing information contained in the digital bit stream. If the frequency difference is kept to one part in 10^9, and the sampling frequency of speech is 8 kHz, a slip may occur once every 34 hours.[20] With an increase in the buffer memory capacity an uncontrolled slip may be eliminated by shifting it to a pause in a conversation, or to an interval between data blocks being transmitted. The CCITT Recommendation G822[2] provides the controlled slip objectives for an octet of digits for 64 kbit/s international digital connections, recognizing that, even though a synchronous digital network may have been formed, the defined transmission characteristics can be exceeded under operating conditions and cause a limited number of slips to occur. Also under temporary loss of timing control within a synchronized network, additional slips may be incurred. In a plesiochronous network, the number of slips on the international links is dependent on the sizes of the buffer stores and the accuracies and stabilities of the interconnecting international clocks. For international plesiochronous inter-exchange links, one slip in 70 days is the recommended maximum theoretical slip rate. For an international hypothetical reference circuit comprising 13 nodes in a plesiochronous mode, the nominal octet slip rate is 1 in every 5.8 days (70/12). The acceptable controlled mean slip rate on a 64 kbit/s international connection or bearer channel is less than or equal to 5 slips in 24 hours for greater than 98.9 percent of time which is greater than one year; greater than five slips in 24 hours and less than or equal to 30 slips in 1 hour for less than 1 percent of time greater than one year; and greater than 30 slips in 1 hour for less than 0.1 percent of time greater than one year.

3.3 DIGITAL HIERARCHIES

The CCITT in Recommendation G701, definition 4003,[2] defines a digital multiplex hierarchy as 'a series of digital multiplexers (otherwise known as MULDEXes) graded according to capability so that multiplexing at one level combines a defined number of digital signals, each having the digit rate prescribed for a lower order, into a digital signal having a prescribed digit rate which is then available for further combination with other digital signals of the same rate in a digital multiplexer of the next higher order'.

The digital hierarchy is arranged as is the frequency-division multiplex (FDM) hierarchy in order that some international agreement can be followed when setting up a digital system.

3.3.1 European system

With reference to Figure 3.11, two systems are recommended in CCITT Recommendation G702, based on differing first-level bit rates. Figure 3.11(*a*) is for a first-level bit rate of 1544 kbit/s, whereas Figure 3.11(*b*) is for a first-level bit rate of 2048 kbit/s. The internationally agreed maximum level is level 4 for international interconnections. Levels higher than this are not mentioned in the recommendation. Annex B to Recommendation G954 does, however, give the digital multiplexing strategy for a $4 \times 139\,264$ kbit/s or 564 992 kbit/s system. As can be seen from the diagram, if PCM voice channels alone are considered, 1920 would be able to be carried by level 4 for the 2048 kbit/s primary level. For the 1544 kbit/s primary level, level 3 may be split into 32064 kbit/s, or 44736 kbit/s. The 44736 kbit/s is the highest level (level 3) for that stream, and is able to carry 672 PCM voice channels. The level 3 bit rate of 32064 kbit/s may be further multiplexed to level 4 with the resulting data bit rate of 97728 kbit/s. This can carry up to 1440 PCM voice channels. A 564992 kbit/s can carry up to 7680 PCM voice channels for level 5 of the 2048 kbit/s primary level.

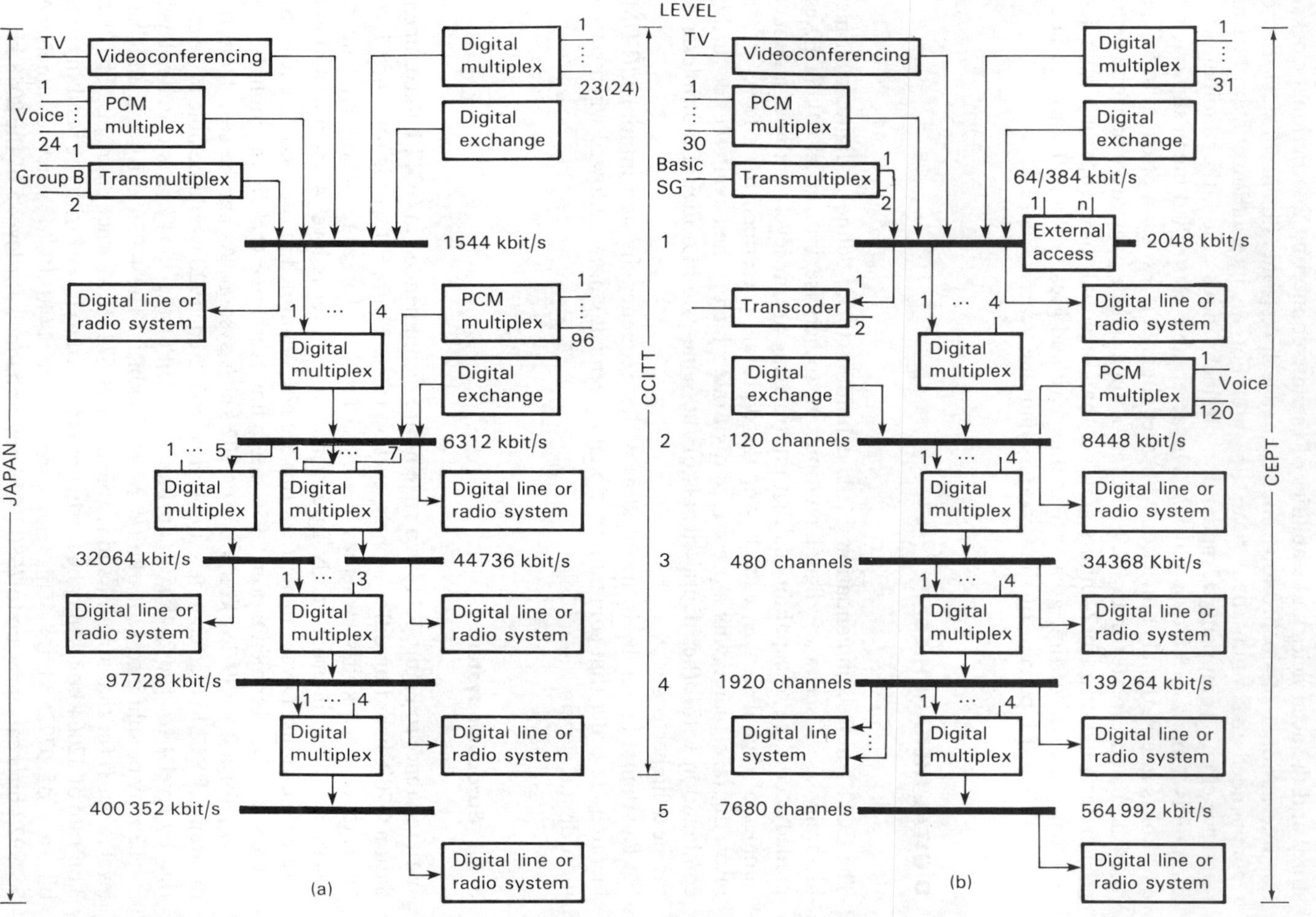

Figure 3.11 CCITT hierarchical bit rates for networks with the digital hierarchy based on primary-level bit rate: (a) 1544 kbit/s; (b) 2048 kbit/s

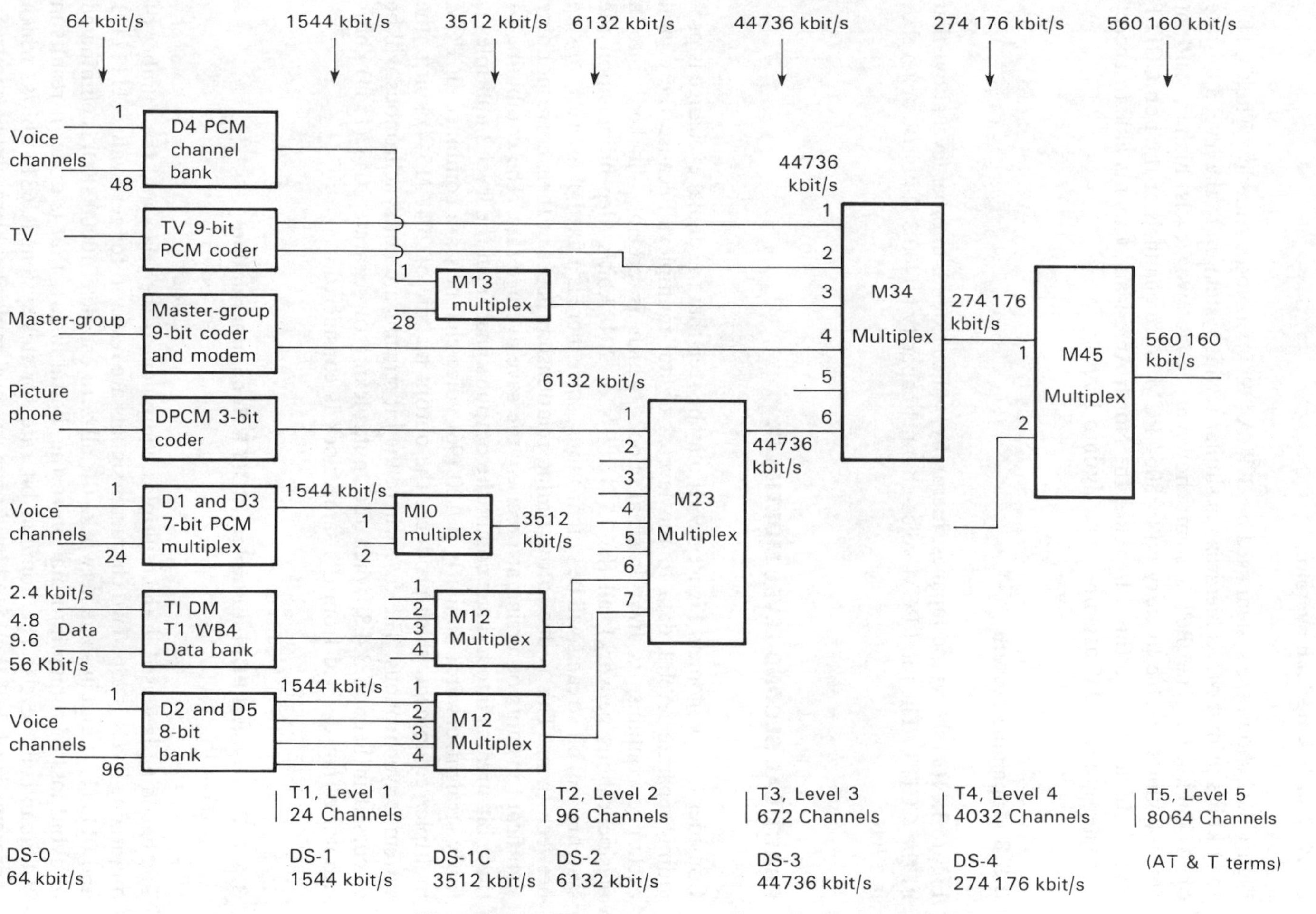

Figure 3.12 North American digital hierarchy

3.3.2 North American system

Figure 3.12 shows the system used by Bell (AT & T) in North America. The CCITT 1544 kbit/s primary-level hierarchy is similar to this system up to the level 3 bit rate of 44736 kbit/s. The Bell system, however, goes two levels higher, and at 560 160 kbit/s, is able to carry up to 8064 PCM voice channels at the non-CCITT level 5. In addition to this difference, the North American system is able to accept a master-group FDM assembly directly into level 3.

3.3.3 Japanese system

Figure 3.11(*a*) shows the Japanese hierarchy, which extends another level above that of the CCITT. The total PCM voice-channel capacity at 400 352 kbit/s, or level 5, is 5760.

3.4 PRACTICAL SECOND-LEVEL MULTIPLEX

Consider for the moment Figure 3.9. It may be seen that the digital exchange is providing separate level 1 data to the level 2 digital multiplex. Because the digital exchange contains its own internal clock (which is possibly 'locked' into a plesiochronous network), all the 1544 kbit/s or 2048 kbit/s (depending upon the system used) bit streams will be synchronous, since there is a fixed phase relationship between them. This is due to the timing relationship between their corresponding significant instants occurring at precisely the save average rate. The combining of these bit streams into a single composite output signal from the level 2 multiplex is fairly straightforward. Similarly if 120 (96) voice channels are combined in PCM multiplex equipment which produces the output bit rate of 2048 (1544) kbit/s, the system is synchronous. The case where the bit streams are not synchronous will be discussed in Section 3.4.3, when the multiplexing of separate 2048 (1544) kbit/s tributaries not derived from the same clock is considered.

3.4.1 96-channel PCM multiplex (CCITT Recommendation G746)[2]

The encoding law used in the multiplex equipment is the μ law and the number of quantizing levels is 256. Two character signals are reserved for zero value (11111111 and 01111111) and in some networks the all-zero character (00000000) is eliminated to avoid loss of timing information to the digital line. If that is done, it results in 254 quantizing levels. The nominal bit rate for this system is 6312 kbit/s, which represents level 2 of the 1544 kbit/s primary level multiplex hierarchy as shown in Figure 3.11(*a*). The tolerance on this bit rate is ± 30 parts per million (ppm). The timing signal is derived from an internal source, from the incoming digital signal or

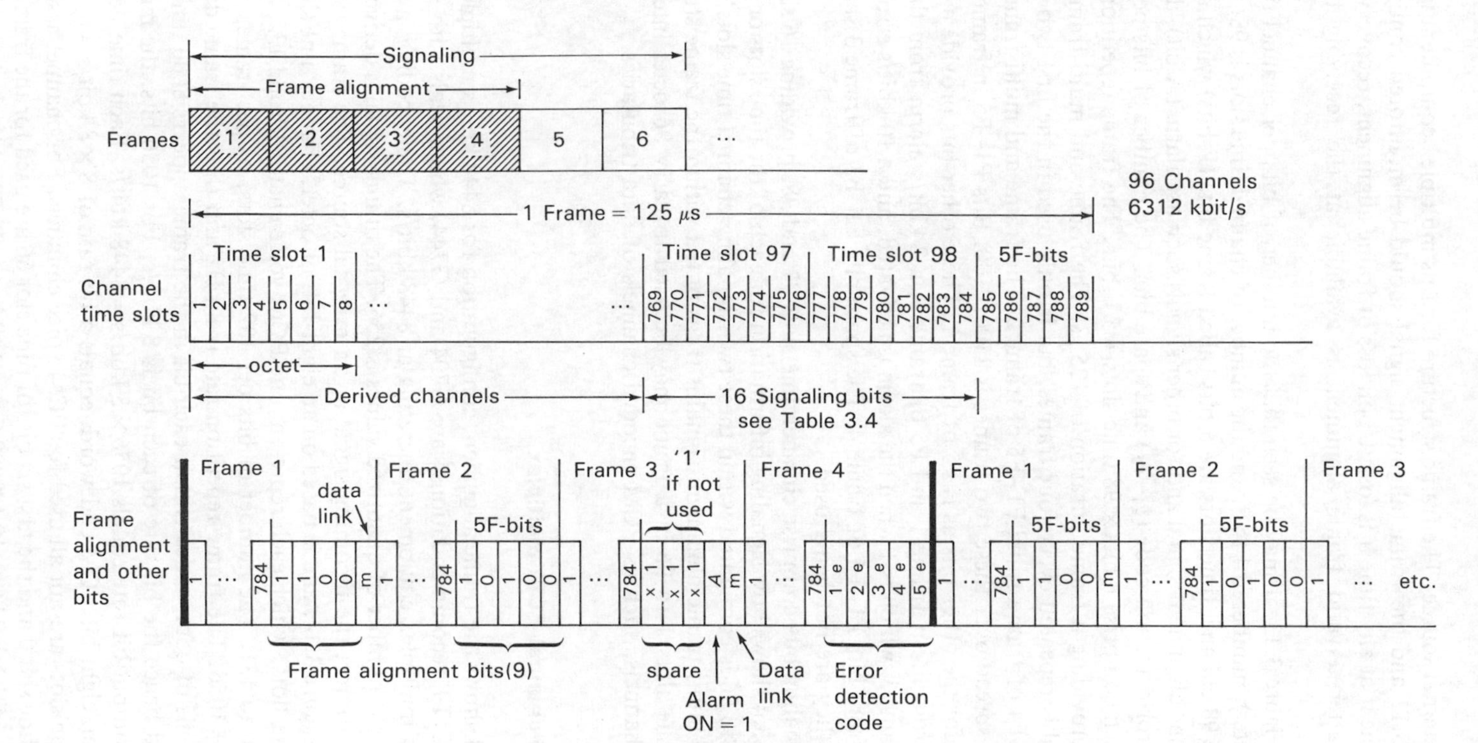

Figure 3.13 **CCITT 96-channel PCM secondary-level multiplex system**

from an external source. The frame structure for this multiplex equipment is shown in Figure 3.13 and the frame alignment signal should be monitored continuously to determine if at any time it is lost. Upon loss of frame alignment, recovery should be made after a valid frame alignment is available at the receiving terminal equipment.

From Figure 3.13, it may be seen that the number of bits per channel time slot is eight (octet), numbered 1 to 8, and the number of channel time slots is 98. In addition to the 98 channel time slots are 5 bits called frame bits (F-bits) which are used for the frame alignment signal and for other signals such as: alarm (A-bit), data link (m-bit), error detection code (e_1-e_5) and spare bits. The number of bits per frame is 789, which comprise 8 bits × 98 time slots + 5 F-bits. The frame repetition rate is 8000 Hz, providing a frame duration of 125 μs. The frame and multi-frame alignment signal is repeated every four frames, but is only used in the first two frames, in bits F1–F4 of frame 1 and F1–F5 of frame 2. The frame and multi-frame alignment code spread over these two frames is 110010100. Bits F1–F5 in frames 3 and 4 are used for the other bits as is F5 of frame 1. These other bits provide a 4 kbit/s data link from F5 of frame 1 and F5 of frame 3, a 2 kbit/s alarm from the A-bit of F4 of frame 3, which is 1 when the alarm is activated, and a 10 kbit/s error detection code from bits F1–F5 of frame 4. The spare bits F1–F3 of frame 3 should be set at 1 if they are not being used.

The signaling bits comprise channel time slots 97 and 98, providing 16 signaling bits. Table 3.4 shows how signaling information for each of the time slots or derived channels 1 to 96 is configured within these two signaling channel time slots, as well as how the multi frame alignment signal is transmitted within the frame structure. Channel time slots 1 to 96 in a frame may be used to carry 96 octet interleaved 64 kbit/s channels, such as PCM encoded channels of data tributaries.

3.4.2 120-channel PCM multiplex

Figure 3.14 shows the frame alignment configuration for 8448 kbit/s multiplex. The relevant CCITT recommendations[2] are G704 and G744, which deals with second-order PCM multiplex equipment operating at 8448 kbit/s. The encoding law is the *A* law, and the number of quantized values is 256. The timing signal is derived from an internal source, the incoming signal or an external source. The frame structure is given in Figure 3.14, and is based on an eight-bit, or octet, channel time slot. The channel time slots, which each represent one PCM voice channel, total 132, and are numbered 0 to 131. The number of bits per 132 time slots, or per frame, is equal to $132 \times 8 = 1056$. The frame repetition rate is 8 kHz and thus the frame duration is 125 μs. This of course means that each bit in the frame appears 8000 times each second, and hence the bit rate of each bit is 8 kbit/s. For 1056 bits, the multiplex composite output bit rate equals 1056×8 kbit/s = 8448 kbit/s. Each time slot, as it represents an eight bit PCM codeword, equals a bit rate of 8×8 kbit/s = 64 kbit/s. The 132 time slots are not all used for PCM voice channels. For channel-associated signaling, time slot 0 and the first six bits of time slot 66 are used for the frame alignment signal, which is 11100110 100000 (i.e. a 14-bit word). The bit rate is determined

Table 3.4 Signaling bit allocation for the 6312 kbit/s multi-frame structure

	Time slot 97								Time slot 98								Time slot occupied
	769	770	771	772	773	774	775	776	777	778	779	780	781	782	783	784	Bit no. occupied
	ST1	ST2	ST3	ST4	ST5	ST6	ST7	ST8	ST9	S10	S11	S12	S13	S14	S15	S16	Signaling bits
Frame number n	0	1	0	1	0	1	0	1	0	1	0	1	0	1	0	1	Multi-frame alignment
$n+1$	1	2	3	4	5	6	7	8	9	10	11	12	13	14	15	16	Signaling
$n+2$	17	18	19	20	21	22	23	24	25	26	27	28	29	30	31	32	information
$n+3$	33	34	35	36	37	38	39	40	41	42	43	44	45	46	47	48	for
$n+4$	49	50	51	52	53	54	55	56	57	58	59	60	61	62	63	64	individually
$n+5$	65	66	67	68	69	70	71	72	73	74	75	76	77	78	79	80	numbered time
$n+6$	81	82	83	84	85	86	87	88	89	90	91	92	93	94	95	96	1–96
$n+7$	Reserved for optional use																

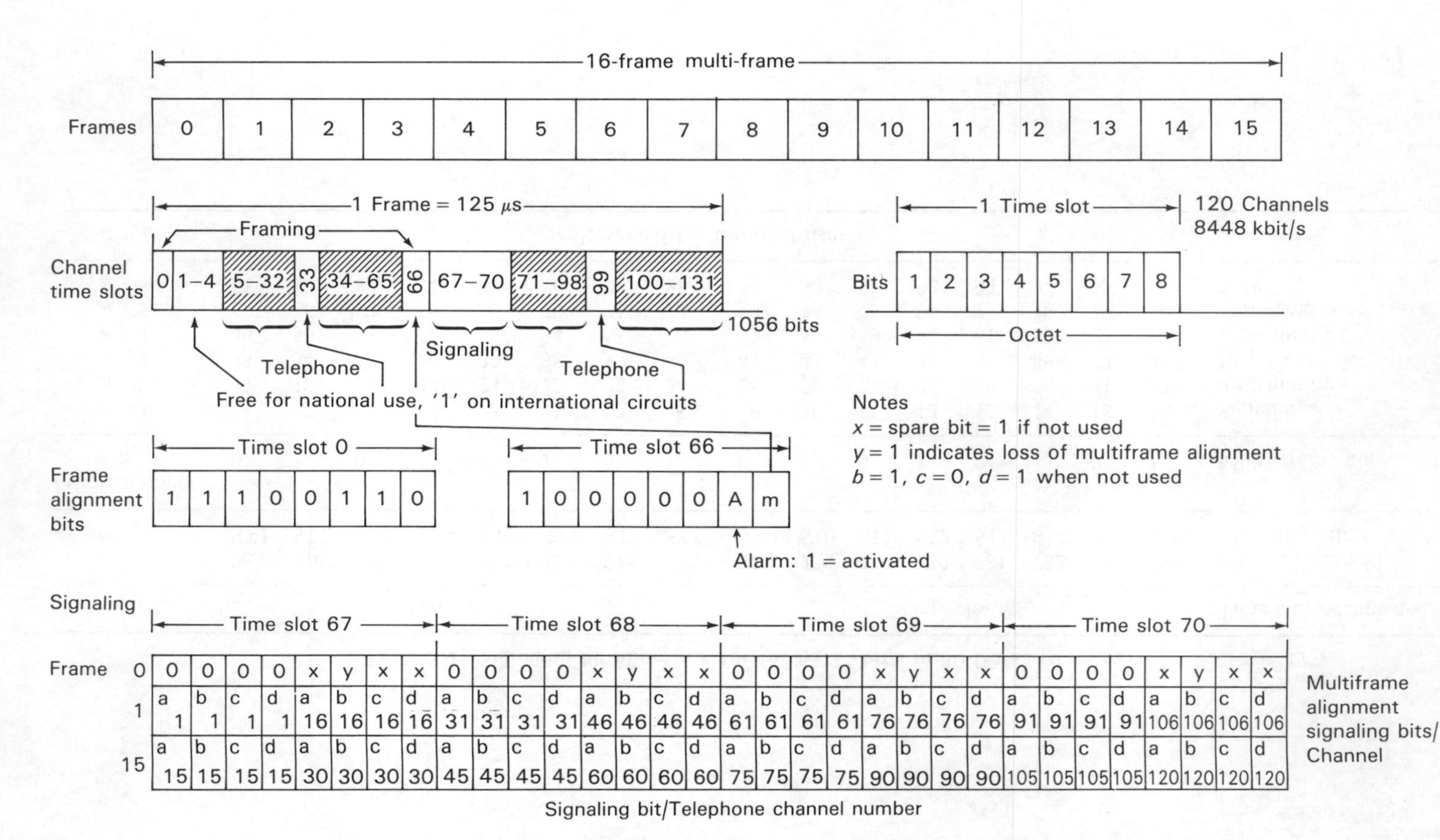

Frame	Time slot 67								Time slot 68								Time slot 69								Time slot 70							
0	0	0	0	0	x	y	x	x	0	0	0	0	x	y	x	x	0	0	0	0	x	y	x	x	0	0	0	0	x	y	x	x
1	a 1	b 1	c 1	d 1	a 16	b 16	c 16	d 16	a 31	b 31	c 31	d 31	a 46	b 46	c 46	d 46	a 61	b 61	c 61	d 61	a 76	b 76	c 76	d 76	a 91	b 91	c 91	d 91	a 106	b 106	c 106	d 106
15	a 15	b 15	c 15	d 15	a 30	b 30	c 30	d 30	a 45	b 45	c 45	d 45	a 60	b 60	c 60	d 60	a 75	b 75	c 75	d 75	a 90	b 90	c 90	d 90	a 105	b 105	c 105	d 105	a 120	b 120	c 120	d 120

Figure 3.14 CCITT 120-channel PCM multiplex system

from 14×8 kbit/s = 112 kbit/s. The seventh bit in time slot 66 is used for the out-of-frame alignment alarm. This bit becomes 1, when an alarm is activated. The eighth bit of time slot 66 is used as for national use together with time slots 1 to 4, 33, and 99. These bits are set to 1 if the system is used to cross international borders. Time slots 5 to 32, 34 to 65, 71 to 98 and 100 to 131 are assigned to 120 telephone channels numbered 1 to 120. Note that the telephone channels are not the same as the channel time slot numbers. Channel time slots 67 to 70 are used for signaling. The signaling information is transmitted over a 16-frame multi-frame numbered 0 to 15. Frame 0 is used exclusively for the multi-frame alignment signal. It is 0000 for the first four bits of time slots 67, 68, 69 and 70. Bit 6 of these four time slots is used to convey a loss of multi-frame alignment and becomes '1' when loss of multi-frame alignment occurs. Bits 5, 7 and 8 are spare bits for signaling information, and are set at 1 if not used. Frames 1 to 15 each contain time slots 67 to 70. This makes a total of $15 \times 4 = 60$ eight-bit time slot availability. If only four bits (a, b, c and d) are required to convey the signaling for a telephone channel, the availability is $60 \times 2 = 120$ four-bit time slots. Thus, as can be seen from Figure 3.14, each telephone channel number is allocated four bits in each time slot 67 to 70 in frames 1 to 15.

With common-channel signaling the additional time slots 2 to 4, 67 to 70 (total 7), are used for PCM voice channels, providing 127 telephone channels, against the 120 in channel-associated signaling. Time slot 1 however may be used if required either for a telephone channel, or for a service channel, or may be left free for service purposes within a digital exchange. Channels 67 to 70 are in descending order of priority available for common-channel signaling. Time slots 33 and 99 are left free for national use.

3.4.3 Digital multiplexing

In the evolvement of digital networks, the need for synchronous digital multiplex equipment arises, especially as digital switching comes more into the foreground in network design. Figure 3.14 provides the frame structure of such equipment as per CCITT Recommendation G741,[2] at a bit rate of 8448 kbit/s ± 30 ppm. The position of this equipment in the multiplex hierarchy is shown in Figure 3.11(b), which indicates that this equipment has four tributary inputs only, each tributary being at 2048 kbit/s. The multiplexing method used is cyclic time-slot interleaving in the tributary numbering order and synchronous multiplexing. The tributary frame structure is that given in Figure 3.6. Time slots used for the frame alignment signals of the tributaries should be identified at the multiplexer input and multiplexed into the pre-assigned time slots, positions 1 to 4 of the 8448 kbit/s frame. As with the 120-channel PCM multiplex, time slot 66, bit number 7 is used to transmit an alarm indication to the remote multiplex equipment when specific fault conditions are detected in the multiplex equipment, and time slots 33 and 99 are left for national use, or set to 1 if crossing an international border. It is assumed that loss of frame alignment has occurred when four alignment signals have been incorrectly received from their predicted positions. Recovery is assumed when the presence of three consecutive frame alignment signals have been detected. The changes to make in Figure 3.14 to adapt it to this multiplex are to time slots 5 to 32, 34 to 65, 71 to 98 and 100 to 131. These time slots are no longer for telephony, but are reclassified to the 2048 kbit/s tributaries.

3.5 NONSYNCHRONOUS MULTIPLEXING

Second-level digital multiplex and higher-order multiplex equipments usually work in a non-synchronous mode even though the tributary signals may be constrained to be within the same timing tolerance limits and operate at the same nominal bit rate in a plesiochronous network. The multiplexing of these tributaries involves a more complex process known as *justification*, in which different tributary bit rates are permitted to be properly related to the multiplex equipment clock. Consider the case, as shown in Figure 3.15, of one input tributary signal entering its corresponding channel in the multiplexer. The multiplexer clock is assumed to be running slightly faster than the incoming signal, and there will be periodically a surplus time slot in the transmitted signal that will contain either a repetition of one of the incoming digits, or a random digit. Either way, there will be a digital error at the system output. The idea behind *justification*, or *pulse stuffing*, systems is to identify the time slots containing these errors, transmit to the receiving terminals information on them and arrange for their deletion from the received signal. Justification can be considered as the process of changing the digit rate of a digital signal in a controlled manner so that it can accord with a digit rate different from its own inherent rate, usually without loss of information (CCITT Definition 4022, in Recommendation G701).[2]

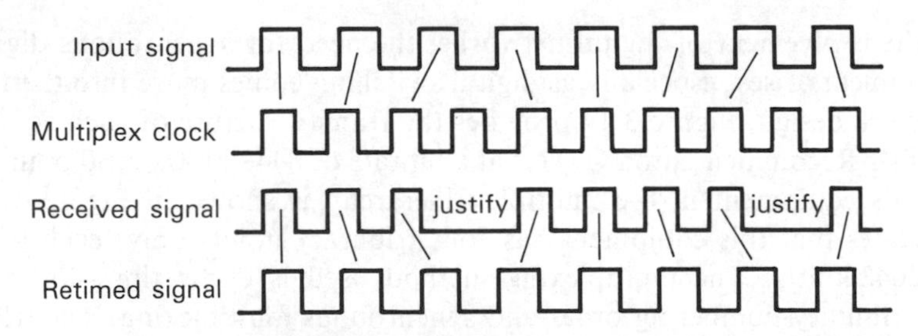

Figure 3.15 Positive pulse stuffing or positive justification (Courtesy of IEE, Reference 1, p. 109)

3.5.1 Positive justification (positive pulse stuffing)

Let the bit rate at the output of the multiplexer be F kbit/s, and the input data stream be f kbit/s. In a positive justification system the multiplexer output bit rate

F is made higher than the sum of the maximum bit rates of the input tributaries, i.e.:

$$F > 4f \tag{3.1}$$

CCITT systems and most practical multiplex systems restrict justification operations to pre-assigned time slots in order to simplify the system design. With this pre-assignment the allowable range of digit rates for the tributary inputs can be made according to equation 3.2:

$$4f_{\max} - 4f_{\min} = 4J \tag{3.2}$$

where J is the number of justifiable time slots per second per tributary.

In a positive justification system such as that shown in Figure 3.15, time slots in the outgoing multiplex signal will become available at a rate exceeding that of the total incoming data bit rates as given by equation 3.1. As this represents a bit rate difference, or a frequency difference between output (or multiplex clock frequency) and input signals, it may be represented as a phase change per unit time of one of the signals with respect to the other. If the reference signal is considered to be the multiplex clock signal, the input signal will be continuously shifting against this reference. The shift will continue until such time as the system decides that it has gone far enough and justification is required. At this point a message will be sent to the receiving terminal by means of the justification service digits, informing it that the next time slot will be justified. On receipt of this information and the justified signal, the receive terminal deletes the next justifiable time slot from the signal and the original input tributary signal is restored.

To overcome the problem of transmission channel errors affecting the transmitted justification service digit, and causing a false deletion or a spurious insertion of information in a time slot at the receive end, which in turn would cause an error and possibly loss of frame alignment, the justification service digits are transmitted in triplicate. The code 111 is for positive justification and 000 for no justification. On corruption of one of the three digits, the majority of the bits in the code produce the decision. That is, if 110 is received, it will be interpreted as 111, but if 100 is received it will be interpreted as 000 and no justification will result. Reference 1, equation 5.37, shows that if the system error rate is P_t, the error rate caused by incorrect decoding of the justification service digits, is given by:

$$P_j = P_t^3 + 3P_t^2 \tag{3.3}$$

In the same reference it is stated that: It also follows that since each justified digit and its control signal occupy four time slots, and allowing a framing content α, the total digit rate F required for a multiplex of c channels will be

$$F = c(f_{\min} + 4J)/(1 - \alpha) \tag{3.4}$$

Example

Consider four tributaries ($c = 4$), and minimum bit rate of the input tributary ($f_{\min}$) of 2048 kbit/s – 50 ppm = 2 047 897.4 bit/s, with a maximum justification rate J of 10 kbit/s, and a framing content α, which is the number of bits in the frame alignment word divided by the number of bits in the frame = 10/848 = 0.01179245. Determine the total digit rate required for a second-level multiplex system.

From equation 3.4 $$F = 4(2047897.4 + 40\,000)/(1 - 10/848) = 8451.25 \text{ kbit/s}$$

If the justification rate per tributary is assumed to be 9799.235 bit/s, corresponding to a justification ratio of 9799.235/10000 = 0.9799, the bit rate at the multiplex output becomes 8448 kbit/s as expected. In the justification process, the bit stream arriving at the receiver is irregular. This irregularity produces timing jitter comprising two components, namely 'justification jitter', introduced into the receiver circuits by the justification, and 'waiting-time jitter', introduced into the receiver because justification does not take place upon demand, but is delayed until one of the pre-assigned timing slots appears. Figure 3.16 shows a conceptual block diagram of a digital multiplex system of four tributaries.

Referring to Figure 3.16, we consider only one of four tributaries which may enter this multiplex equipment and, as shown, let this be input 1. The level 1 bit stream (2048 or 1544 kbit/s) entering the input is detected by the phase-locked loop which extracts the timing information f from the bit stream and permits the bit stream to be entered into the *buffer store*, or *elastic store*. The stored bits are read out of the elastic store with the timing of the internal clock whose frequency is the multiplex clock frequency (F) divided by 4. This frequency ($F/4$) is slightly higher than the external clock frequency f, since it has to accommodate, in the multiplex output frame structure, the justification control bits, justification bit and frame alignment signal, etc., i.e. all the overhead bits. As a result of this the reading of the pulses would appear to occur at a faster rate than the rate at which the pulses are stored, resulting in a gradual reduction of stored bits in the elastic store. This is not quite so, for the read pulse is effectively muted by the positions where the overhead bits would occur and thus another input pulse during this muting period is read into the store. The threshold detector supervises this, and when the number of stored bits is reduced to a predetermined value by the remaining difference in frequency of the two clocks after the read muting has been taken into account, the threshold detector informs the stuffing control to inhibit reading; as a result, bits with the zero state are stuffed into designated time slots known as 'justifiable digit time slots' contained in the digit stream, to make up the differences in the clock frequency. If the store is filling or is full when a fixed justifiable time slot occurs, the time slot is used. If the store is near empty, the time slot is ignored. Information concerning the status of the bits (that is justification or no justification) is also inserted into predetermined time slots in the digital bit stream together with other service bits and transmitted to the remote end. At the receiving end, another elastic store is provided. Writing into this store is inhibited by the destuffing control at the justified positions in order to remove the null bits inserted during stuffing or

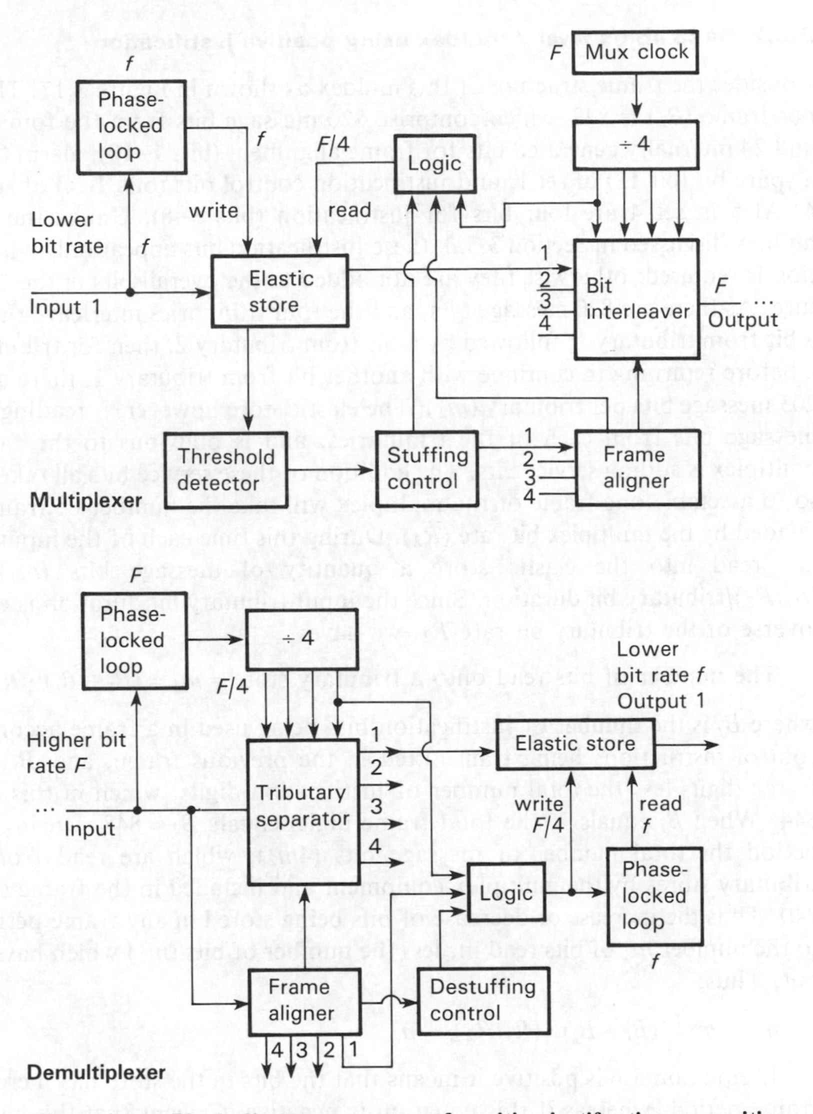

Figure 3.16 Block diagram concepts of positive justification or positive pulse stuffing in a level 2 mutiplex system

justification. As the removal of the justifications cause an abrupt change in the receiving clock, a second phase-locked loop is provided for smoothing and reading out the pulses in the elastic store. This phase-locked loop also has been designed to permit proper operation over the required range of variations in digit rates. CCITT Recommendations G742 and G743 provide details of second-order digital multiplex equipment operating at 8448 kbit/s and 6312 kbit/s respectively using positive justification. These are briefly discussed below and diagrams of their frame structure are presented.

3.5.2 8448 kbit/s level 2 muldex using positive justification

Consider the frame structure of this muldex as shown in Figure 3.17. The total bits per frame (B_2) is 848, which comprise 820 message bits from the four tributaries, and 24 internally generated bits for frame alignment (bits 1–10), alarm (bit 11), and a spare bit (bit 12) of set 1, and justification control bits (bits 1–4) of sets 2, 3 and 4. Also in set 4 are four bits for justification (bits 5–8). Unlike the 6312 kbit/s muldex discussed in Section 3.5.3, these justification bits appear only when justification is required; otherwise they are not added to the overall bits in the frame structure. As there are 820 message bits, and the four tributaries interleave their bits, i.e. a bit from tributary 1, followed by a bit from tributary 2, then for tributaries 3 and 4 before returning to continue with another bit from tributary 1, there are 820/4 or 205 message bits per tributary (m_2). The elastic store however, is reading in only the message bits from each of the tributaries, and is oblivious to the fact that the multiplex is adding service bits. The addition of these service bits all takes time, and so to assemble one frame of the multiplex will take the number of frame bits (B_2) divided by the multiplex bit rate (R_2). During this time each of the input tributaries has read into the elastic store a quantity of message bits (m_1) equal to (B_2/R_2)/tributary bit duration. Since the input tributary bit duration is equal to the inverse of the tributary bit rate R_1, we have:

The number of bits read onto a tributary store $= m_1 = (B_t + B_j) \cdot R_1/R_2$ (3.5)

where B_j is the number of justification bits being used in a frame according to the control instructions being transmitted in the previous frame, and B_t is the total frame digits less the total number of justification digits, which in this case equals 844. When B_j equals 4, the total frame digits equals $B_2 = 848$. During any frame period the total number of message bits ($4m_2$), which are read from the four tributary stores by the multiplex equipment and included in the frame structure, is 840. Thus the increase or decrease of bits being stored in any frame period is equal to the number m_1 of bits read in, less the number of bits (m_2) which have been read out. Thus:

$$m_1 - m_2 = (B_t + B_j) \cdot (R_1/R_2) - m_2 \qquad (3.6)$$

If equation 3.6 is positive it means that the bits in the store has increased in the frame period, whereas if this equation is negative it means that the bit number is decreasing for the frame period. A value of zero implies that what goes out is being replaced by that coming in. Consider the five cases where the number of justification bits B_j used in the frame is 0, 1, 2, 3 and 4. Using equation 3.6 with $B_t = 844$, $R_1 = 2048$ kbit/s, $R_2 = 8448$ kbit/s, and $m_2 = 205$, the number of bits increased or decreased in the tributary store is:

Number of justification bits used in frame	*Store change in 1 frame*
0	−0.3939 bits
1	−0.1515 bits
2	+0.0909 bits
3	+0.3333 bits
4	+0.5758 bits

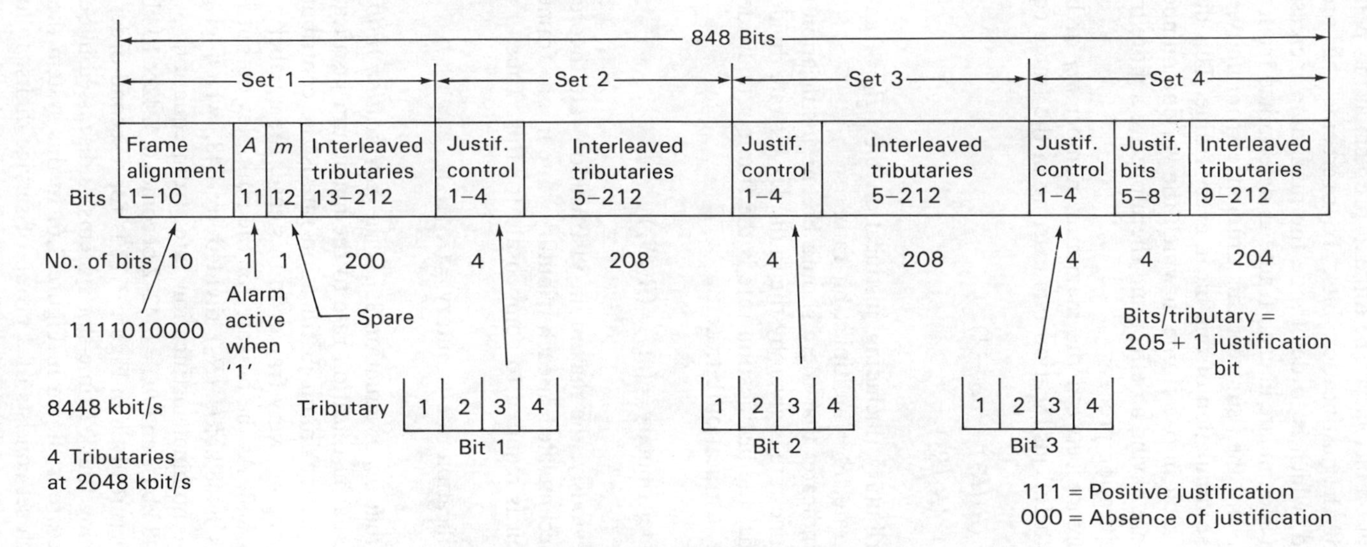

Figure 3.17 8448 kbit/s level 2 multiplex system using positive justification, frame structure

This result shows that the store is quite elastic, stretching and contracting as the number of justification bits vary. The difficult problem at this point is to determine the distribution of the justification bits. To overcome this problem, an easier method is adopted, which is to determine the maximum increase in bits that the tributary store can accumulate in each frame as determined from equation 3.6 with B_j at a maximum, and then subtract this difference from unity, which is the case where there is an accumulation of one bit in every frame. This difference indicates the proportion of a bit in one frame in which the varying number of justification bits in the system are having an effect and therefore are a more true representation of the system justification process.

Thus the difference between the maximum number m_d of bits read into the tributary store with full justification bits existing in the frame equals:

$$m_d = 1 - (B_2) \cdot (R_1/R_2) + (m_2' - 1)$$
$$= m_2' - (B_2) \cdot (R_1/R_2) \qquad (3.7)$$

where m_2 = bits/tributary (including justification bit); B_2 = frame length in bits; R_1 = tributary bit rate; R_2 = multiplex bit rate.

Equation 3.7 indicates that there should be one justification bit in every $1/m_d$ frames. As the frame rate is given by the multiplex bit rate divided by the number of bits in the frame, the justification rate is the frame rate divided by the number of frames for one justification bit, i.e.:

$$\text{Justification-rate/tributary} = (m_d) \cdot (R_2/B_2) \qquad (3.8)$$

As mentioned previously, the maximum justification rate theoretically possible is where one bit is accumulated in each tributary store in one frame period and thus one justification bit is required to compensate for this. Thus:

$$\text{Maximum justification-rate/tributary} = (R_2/B_2) \qquad (3.9)$$

This indicates that m_d of equation 3.7 is the *justification ratio* and is defined as the ratio of the actual justification rate to the maximum justification rate as given by equation 3.9. The maximum justification rate is the contribution made by the tributary store of one bit every frame. In this case, the number of bits transmitted by the frame B_2 is 848. As the bit rate for the multiplex output is 8448 kbit/s, the rate of one bit is 8448 kbit/s (R_2) divided by 848, which gives approximately 10 kbit/s. Thus the nominal justification rate, from equations 3.7 and 3.8, where $m_2 = 206$, $R_1 = 2048$ kbit/s and $m_d = 0.4242$ becomes 0.4242×10 kbit/s = 4.24 kbit/s.

The frame structure given in Figure 3.17 is for a second-order multiplex system formed by combining four 32-time-slot systems as discussed in Section 3.2.3.2 and whose frame structure is shown in Figure 3.6. In this system the frame is divided into four sets, each set consists of a group of service digits, followed by tributary interleaved message channels. Set 1 contains a bunched (against distributed) frame-alignment codeword 1111010000 of ten bits. Following this in bit position 11 is an alarm bit which is transmitted to the remote multiplex equipment when specific fault conditions are detected in the multiplex equipment. Upon alarm activation this digit is '1'. On detecting the alarm, the receiving terminal will disconnect the tributary outputs and replace the data stream with some predetermined pattern, minimizing disturbance to the rest of the network. Bit 12 is reserved for national use, and is set to 1 when crossing the border on an international path. In sets 2 to 4, the first four bits numbered 1–4 in each case are used to provide the positive justification control for each tributary. Bit 1 of each tributary is found in set 2, bit 2 of each tributary in set 3, and the third control bit for each found in set 4. The signal '111' is used to signify justification, as is '110' or any group of bits where the '1's outweigh the '0's. In contrast to this, '000' signifies that no justification is required, as is so when the '0's outweigh the number of '1's.

Justification is carried out in set 4, bits 5–8. Loss of frame alignment is assumed when four consecutive frame-alignment signals have been incorrectly received in their predicted positions. In this case the frame alignment equipment considers that frame alignment has effectively been recovered only when it detects the presence of three consecutive frame alignment signals. Once a single correct frame alignment signal has been detected the equipment begins a new search if the signal fails to appear in any one of the following two frames. The multiplex timing signal should be derived from an external source as well as from an internal source if it is economically feasible.

3.5.3 6312 kbit/s level 2 muldex using positive justification

As mentioned in Section 3.5.2, in this multiplex equipment the frame structure is such that the time slot containing the justification bit for each tributary is distributed over four frames, and thus is always transmitted whether justification is required or not. The reason why this time slot is always transmitted is because when it is not being robbed for use by the justification bit, it is used to transmit part of the message signal. In a multi-frame of four frames as shown in Figure 3.18, the input store reads data from the tributaries according to equation 3.5, where $(B_t + B_j) = 1176$ bits. Thus the elastic store fills to $1176 \times (1544/6312) = 287.6654$ bits over the four-frame multi-frame period. As the bits per tributary per multi-frame, read from the store into the bit interleaver is 288 without justification, there will be a decrease in the contents of the store equal to $287.6654 - 288 = -0.3346$. For the case where justification is required, the number of bits read from the store is 287 in the same period of four frames. This produces an increase in the number of bits entering the store equal to $287.6654 - 287 = 0.6654$ bits. This case is quite simple in comparison with that of Section 3.5.2, in that there is only one fixed rate of increase and one of decrease. For every multi-frame transmitted which contains a single tributary justification bit, the tributary store fills with a maximum of 0.6654 bits. This means that to compensate for this storage increase, the next two multi-frames (actually on the average 1.9886 multi-frames) should not contain a justification bit for that tributary. Thus the total multi-frames on the long-term average to reach this situation is 2.9886. One frame for the increase in bits stored and 1.9886, for the compensating decrease. This implies that (1/2.9886) or 0.3346 justification bits are required

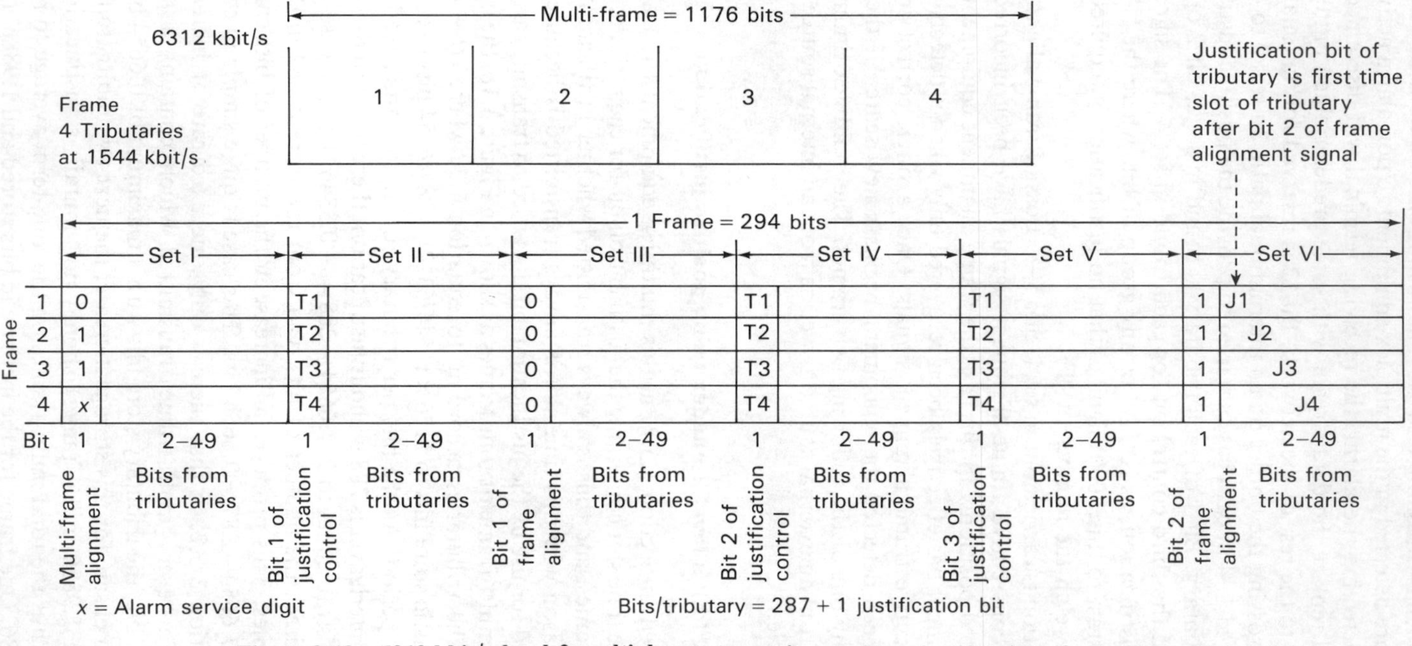

Figure 3.18 6312 kbit/s level 2 multiplex system using positive justification, frame structure

per multi-frame for a single tributary. Again equations 3.7 and 3.8 apply with B_2 being equal to the number of bits in the multi-frame. Thus the *nominal justification ratio* is equal to 0.3346, or 0.334 to confirm to the CCITT Recommendation G743.[2] The maximum bit rate per bit is equal to the multiplex output bit rate corresponding to the number of multi-frame bits. Thus the rate for one bit equals 6312/1176 = 5367 bit/s. As the maximum justification bit transmitted from each tributary is one bit per multi-frame, the maximum justification bit rate is 5367 bit/s. The nominal justification rate may be determined from the equation 3.8:

$$\begin{aligned}\text{Nominal justification rate} &= \text{Nominal justification ratio}\\ &\quad \times \text{Maximum justification rate/tributary}\\ &= 0.334 \times 5367 \text{ bit/s} = \textit{1792 bit/s}\end{aligned}$$

This second-order digital multiplex equipment is intended for use on digital paths between countries using 1544 kbit/s primary multiplex equipment. Figure 3.7 shows the frame structure for the 1544 kbit/s primary multiplex and Figures 3.11(*a*) and 3.12 show the CCITT and North American hierarchies which include this equipment. The tolerance on the 6312 kbit/s bit rate is ±30 ppm. The frame alignment recovery time is limited to 16 ms, and once frame alignment is established, multi-frame alignment should be recovered in less than 420 μs. The frame structure is repeated in a four-frame multi-frame arrangement which comprises 1176 bits. The reason for the multi-frame is to distribute the multi-frame alignment signal, justification control signal, frame alignment and justification bits. Bit 1 of set I of frames 1–3 in the four-frame multi-frame is used to transmit the multi-frame alignment code 011. The fourth bit (x) may if required be used as an alarm service digit. Bit 1 of sets II, IV and V is used to transmit the justification control bits. Frame 1 is for tributary 1, frame 2 for tributary 2, etc. This code is 111 for justification and 000 for no justification. With error, the majority of bits in the three-bit code determine the decision. Bit 1 of sets III and VI comprise the frame alignment code. For each tributary this is 010101010 etc. In sets I–VI, bits 2–49 (48 bits) are used for the interleaved tributaries. The number of bits per tributary is 288, and one of these bits is 'robbed' if justification is required. The timing signal if economically feasible may be derived from an external source as well as that of the internal source.

3.5.4 Negative justification (negative pulse stuffing)

So far we have considered the case where a low data bit rate has been multiplexed up to a higher bit rate in the multiplexor. If the time slots in the multiplexed signal appear at a slightly lower rate than those of the tributaries, it is necessary to insert or 'stuff' an extra bit or time slot into the bit stream before it is processed by the relevant tributary remote-end equipment. As with positive justification, this is achieved by transmitting both the control signals signifying that a bit has or has not been inserted together with the added time slot, to the remote end of the tributary. In practical systems, however, usually a pulse is deleted from the incoming bit stream as it enters the multiplexor. This permits the lower clock rate of the multiplex equipment to accept data without losing bits and reduces the complexity in the system design. This leads to the CCITT definition of negative justification as: a method of justification in which the digit time-slots used to convey a digital signal have a digit rate which is always lower than that of the original digital signal. The digits which have been deleted from this bit stream are transmitted by separate means (that is in a separate set of time slots in the transmitted signal frame structure) to the remote end for insertion, whereas information which facilitates the recovery of the deleted digits is conveyed by means of the justification service digits.

3.5.5 8448 kbit/s muldex using positive/zero/negative justification

If the nominal bit rate of the channels in the multiplex equipment are made ideally equal to those of the tributaries, there will exist in practice, due to the different independent clock sources, a situation where the digit time slots in the multiplex equipment used to convey the digital signals have a digit rate that may be higher than, the same as, or lower than the digit rate of the original signal. This means that a combination of both positive and negative justification methods are required. For positive justification, justifiable time slots are provided in the resultant signal frame structure, which transmit either information from the original signal, or no information, according to the relative digit rates of the resultant signal and the original signal. For negative justification, a separate means of transmitting the deleted digits is arranged, and the justification service digits are used to provide information for facilitating the recovery of the deleted digit. The advantage of this form of multiplexing system is that it can easily handle synchronous inputs or mixed systems, in which only some of the input tributaries are synchronized to the multiplex digit rate, the other input tributaries being asynchronous. Figure 3.19 provides the frame structure for this system.

CCITT Recommendation G745 provides the details of this system which is intended to be used on digital paths between countries using 2048 kbit/s primary multiplex equipments. The nominal bit rate for this system is 8448 kbit/s ± 30 ppm. A loss in frame alignment is considered to have occurred when five consecutive frame alignment signals have been incorrectly received in their predicted positions. Recovery of this frame alignment signal is considered to have taken place when at least two consecutive frame signals have been detected without errors in their predicted positions. As soon as frame alignment has been lost and until it has been recovered, a signal with a definite pattern is transmitted to all tributaries. This pattern is called the 'alarm indication signal' (AIS) and is used to minimize the disturbance to the rest of the network. The equivalent binary content of this pattern at 2048 kbit/s is a continuous stream of '1's. The multiplexing method is cyclic bit interleaving of the tributaries in their numbering order. The justification control signal as per Figure 3.19 is distributed and positive justification is indicated by the signal 111, transmitted in two consecutive frames. Negative justification is indicated by the signal 000 transmitted in two consecutive frames, and no justification

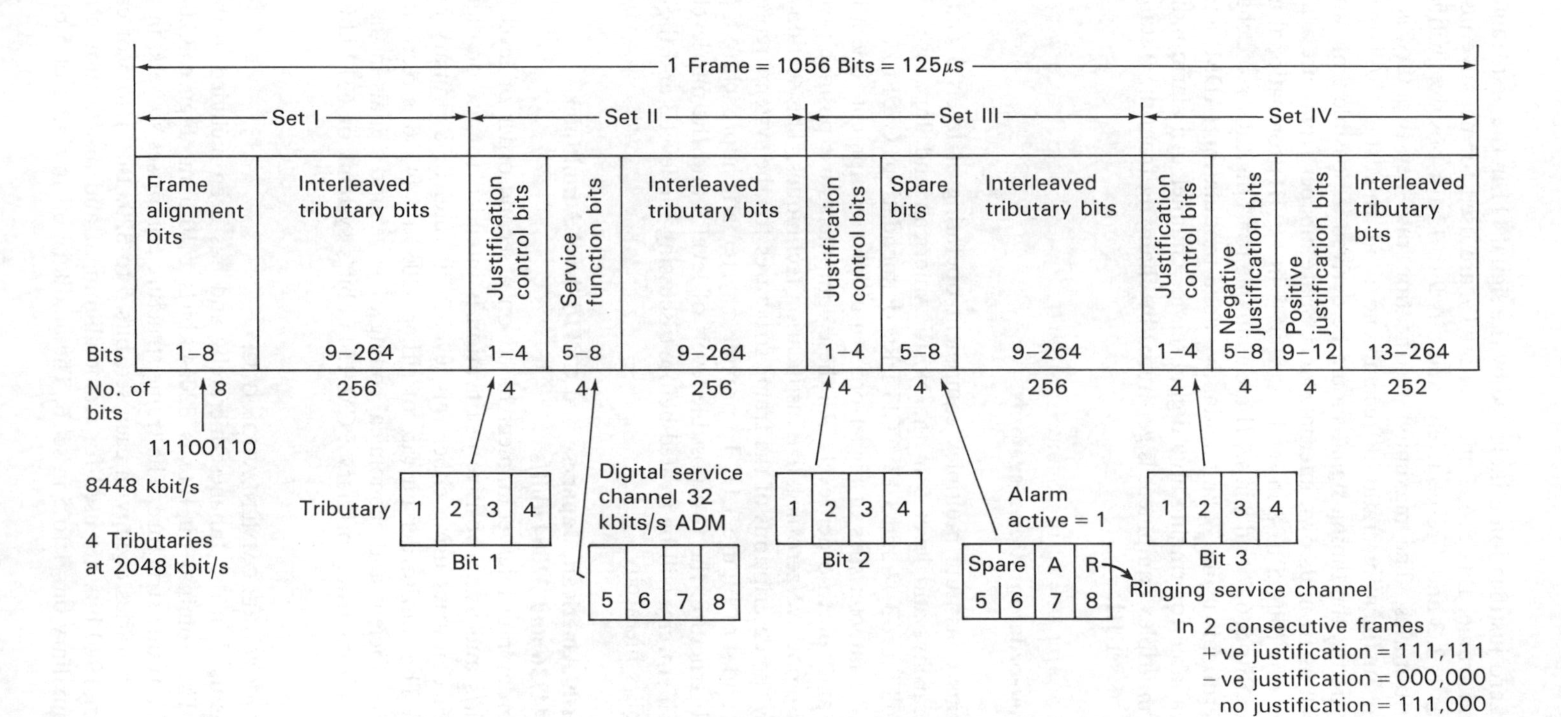

Figure 3.19 8448 kbit/s level 2 multiplex system using positive/zero/negative justification, frame structure

required or zero justification is indicated by the signal 111 in the first frame and 000 in the second frame. Bits 5, 6, 7 and 8 in set IV are used for negative justification of tributaries 1, 2, 3 and 4 respectively, and bits 9 to 12 for positive justification of the same tributaries. The maximum justification rate per tributary is given as 8 kbit/s. No justification ratio is recommended at this stage. As in previous multiplex systems, the timing signal should be derived if possible from an external source as well as from its own internal clock. Spare bits per frame are available for service functions (bits 5 to 8 in set II and bit 8 in set III) for national and international use. Bits 5, 6, 7 and 8 in set II are available for a digital service channel between two terminals using 32 kbit/s adaptive delta modulation (ADM) and bit 8 in set III is available for ringing up a digital service channel. An alarm indication to the remote multiplex equipment is generated by changing from the 0 state to the 1 state, bit 7 of set III.

3.5.6 Higher-order multiplex systems

These systems constitute multiplex equipment operating at levels 3 and 4 in the CCITT hierarchy, and level 5 in the North American and Japanese hierarchies according to Figure 3.11 and 3.12. CCITT Recommendations G751 to G754 provide the principal characteristics of higher-order multiplex equipments operating at the third-order and fourth-order levels. All these equipments use positive justification, and some use positive/zero/negative justification techniques. This section will comprise mainly the descriptions of the frame structures for the systems represented in the levels 3 and 4 of Figure 3.11. To provide a systematic approach to a particular hierarchical structure, the two possible levels of level 3 and the one level 4 of the 1544 kbit/s hierarchy will be treated before discussing the level 3 and the level 4 of the 2048 kbit/s hierarchy.

3.5.6.1 North American, Japanese and CCITT level 3 multiplex at 44736 and 32064 kbit/s

The third-order multiplex equipment based on a second-order or level 2 bit rate of 6312 kbit/s and using positive justification is intended for use on digital paths and between countries using either this level 2 bit rate, or a primary bit rate of 1544 kbit/s. The third-order bit rate of either 32064 or 44736 is recommended to allow for the efficient and economical coding of wideband signals in the networks of administrations using primary PCM 1544 kbit/s systems or 6312 kbit/s level 2 systems.

32064 KBIT/S (CCITT RECOMMENDATION G752)[2]

This level is used in the Japanese hierarchy and has been included in the CCITT hierarchy. The nominal bit rate is 32064 kbit/s ± 10 ppm. Figure 3.20 provides details of the frame structure. The frame structure comprises six sets, in which the interleaved tributaries occupy in each set, bits 6 to 320. Bits 1 to 5 of each set in the frame of 1920 bits are used for frame alignment and justification control bits as well as auxiliary bits 1 to 5 in set 6. The auxiliary bit in time slot 5 of set VI is

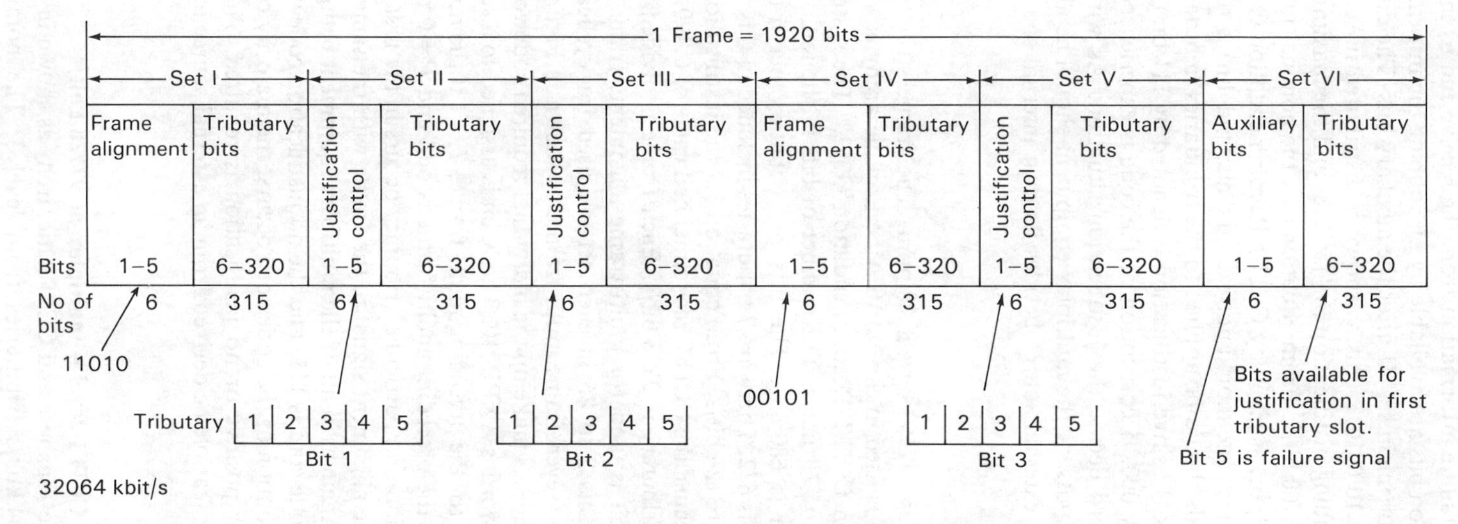

Figure 3.20 32064 kbit/s level 3 multiplex system using positive justification, frame structure

used for transmitting failure information from the receive end to the transmit end. The positive justification bits are available by bit robbing the information bit in the first tributary slot occurring in set VI after the auxiliary bits. The distributed frame alignment signal is 11010 00101. This as shown is distributed in set I and IV. The bits per tributary including justification (m_2') is 378, and the maximum justification rate per tributary is 16700 bit/s. From equation 3.7, the nominal justification ratio (m_d) is given as 0.0359, or, as given by CCITT Recommendation G752, 0.036. The frame alignment recovery time should not exceed 8 ms, and during the out-of-frame alignment period signals should be applied to the tributaries in order not to disturb the rest of the network. Positive justification is denoted by 111 on the control bits, and no justification by 000. If corruption of the code is experienced, the majority of bits in the codeword decide the justification situation. As with all multiplex equipments, wherever possible the multiplex equipment should be able to derive its timing signal from an external source besides its own internal source.

44736 KBIT/S

The nominal bit rate is 44736 kbit/s ± 20 ppm. The frame structure is shown in Figure 3.21. This level is used in the North American hierarchy as shown in Figure 3.12 and is included in CCITT Recommendation G752.[2] The system comprises a multi-frame structure of 7 frames. The number of bits in the multi-frame is 4760, and the frame length is 680 bits. The bits per tributary per multi-frame (m_2') including justification is 672. There are 7 tributaries included in this frame structure each with a bit rate of 6312 kbit/s. From equation 3.7 the justification ratio is determined to be 0.3906, against the CCITT recommended value of 0.390. The maximum justification rate per tributary is 9398 bit/s. Each frame is divided in to eight sets of 86 bits. The first bit is used for multi-frame alignment, frame alignment and justification control. Bits 2 to 86 in each set are used for cyclically interleaved tributary bits. Over the seven frame multi-frame, as shown in Figure 3.21, each of the bit 1 positions of the sets make up the individual required codewords. The multi-frame alignment codeword is XXPP010. The X bit is assigned to a service function, and P is the parity bit for the preceding multi-frame. $P = 1$ if the number of marks in all tributary bits in the preceding multi-frame is odd and $P = 0$ if even. The bit available for justification of a particular tributary occurs in the first slot of set VIII after bit 4 of the frame alignment signal in the frame with the same designation as that of the tributary. Again as in all of the multiplex equipment being discussed, the positive justification codeword is 111, and the no justification codeword in the control bits is 000. If corruption of the codeword occurs, the majority of bits make the decision as to whether positive or no justification is required. The timing signal should if economically feasible be derived from an external source in addition to its own internal clock.

3.5.6.2 Japanese and CCITT level 4 multiplex at 97728 kbit/s

This level is used in the Japanese and CCITT hierarchy as shown in Figure 3.11(a). It accepts three 32064 kbit/s tributaries from level 3. The nominal bit rate is

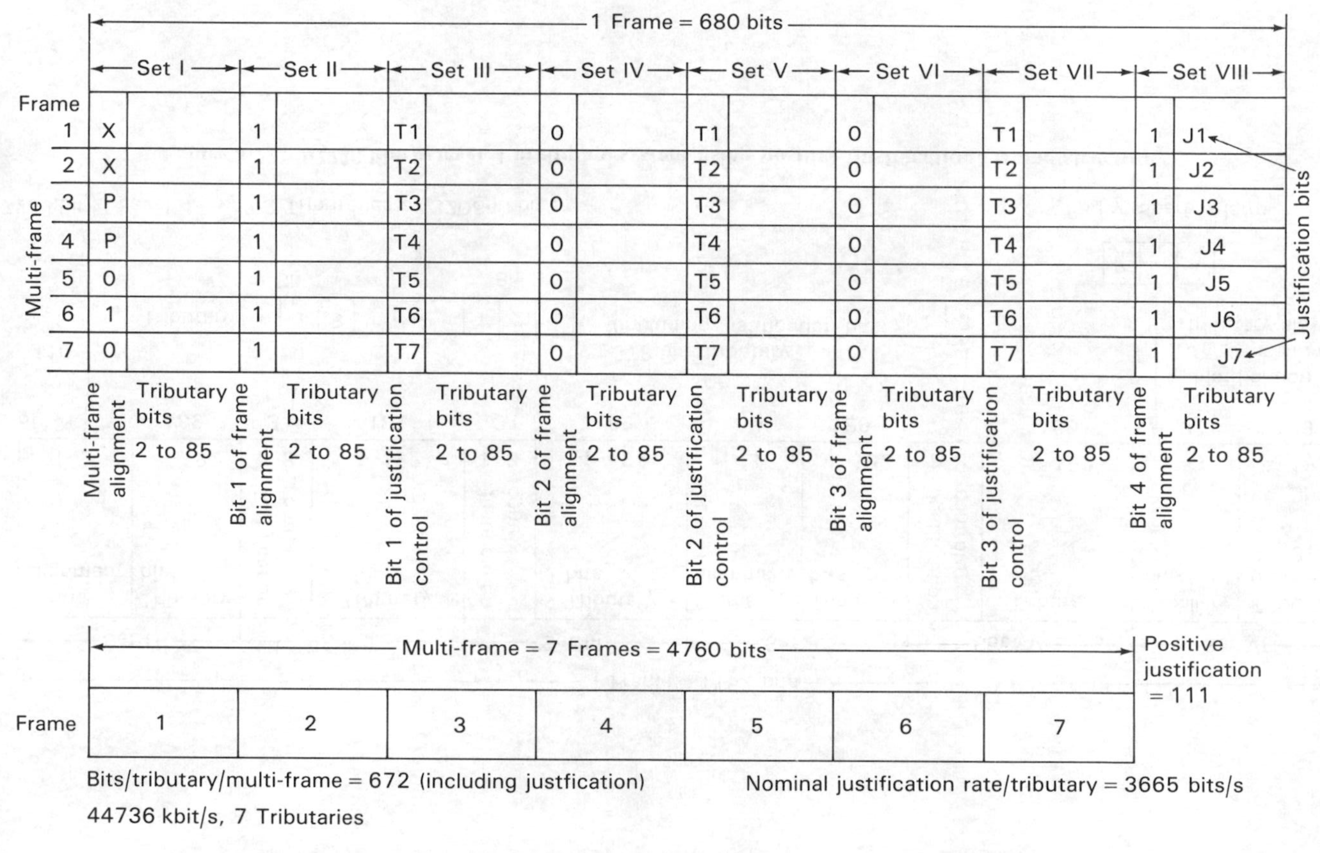

Figure 3.21 44736 kbit/s level 3 multiplex system using positive justification, frame structure

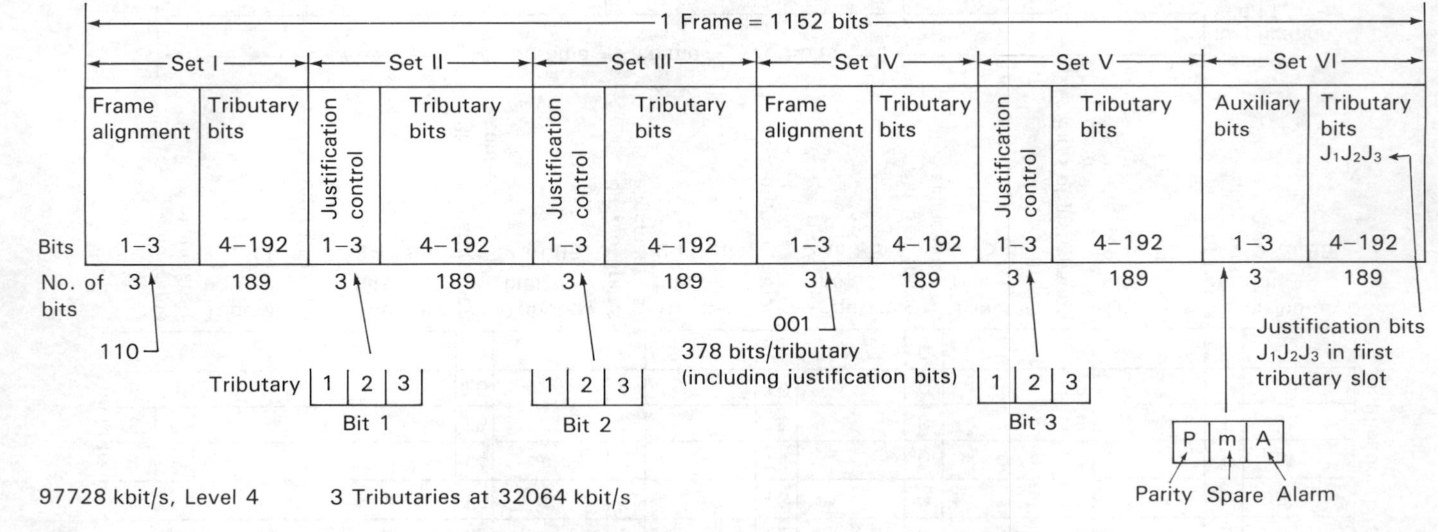

Figure 3.22 97728 kbit/s level 4 multiplex system using positive justification, frame structure

97728 kbit/s ± 10 ppm. The frame structure is shown in Figure 3.22. The frame comprises six sets of 192 bits per set. In each set bits 4 to 192 are used for interleaved tributary channels which including justification provide 378 bits per tributary (m_2'). The number of bits in each frame is 1152, and the maximum justification ratio per tributary from equation 3.9 equals 97728/1152 = 84833 bit/s, and from equation 3.7 the nominal justification ratio equals 0.0354, or, as given by CCITT Recommendation G752, 0.035. The frame alignment codeword is 110 001 and is distributed over the frame. The first three bits 110, occurring in bit positions 1 to 3 in set I and the bits 001 in bit positions 1 to 3 in set IV. Likewise, the justification control bits are distributed over the frame and appear in bit positions 1 to 3 in sets II, III and V, as bit positions 1, 2 and 3. Positive justification is indicated as 111, whilst no justification is indicated by 000. When an error in transmission occurs, the majority decision is made. The auxiliary bits occurring in bit position 1 to 3 in set VI are used for parity check and alarm. Bit position 1 is the parity bit for the preceding frame, whilst bit 3 is a bit used to transmit the failure information from the receiving end to the transmitting end. Bit 2 is a spare bit used for national use. The justification bit for each tributary is the first appearing tributary bit in set VI after the auxiliary bits.

3.5.6.3 CCITT and European level 3 multiplex at 34368 kbit/s

34368 KBIT/S MULTIPLEX USING POSITIVE JUSTIFICATION

This equipment multiplexes four digital signals at 8448 kbit/s. The nominal bit rate for the equipment is 34368 kbit/s ± 20 ppm. The frame structure is that given in Figure 3.23. This frame comprises four sets totaling 1536 bits. Bits 1 to 10 in set I are used to convey the bunched frame alignment codeword 1111010000. Bit 11 in set 1 is used for the alarm indication to the remote digital multiplex, and bit 12 is reserved for national use but set to 1 on crossing international borders. The remaining 372 bits in set I is used for interleaved tributary bits of four tributaries. Bits 1 to 4 in sets II, III and IV are the justification control or service bits. Again when justification is to take place in the following frame, the code 111 is transmitted for the relevant tributary. If no justification is required, 000 is transmitted. The justification bits occur in bit positions 5 to 8 in set IV, in the order of the tributary numbering. In sets II and III, bit positions 5 to 384 are interleaved tributary bits, whereas, in set IV, bits 9 to 384 are interleaved tributary bits. The frame comprises 1536 bits (B_2), and the bits per tributary including the justification bits (m_2') are 378 bits. The maximum justification rate according to equation 3.9 is 22375 bit/s, and the nominal justification ratio according to equation 3.7 is 0.4358. The CCITT Recommendation G751 gives the nominal justification ratio as 0.436. The loss of frame alignment should be assumed to have taken place when four consecutive frame alignment signals have been incorrectly received from their predicted positions. Recovery is assumed when three consecutive frame alignments have been detected correctly. When a single frame alignment signal has been detected, a search begins once again if an incorrect alignment signal is detected in either of the following two frames.

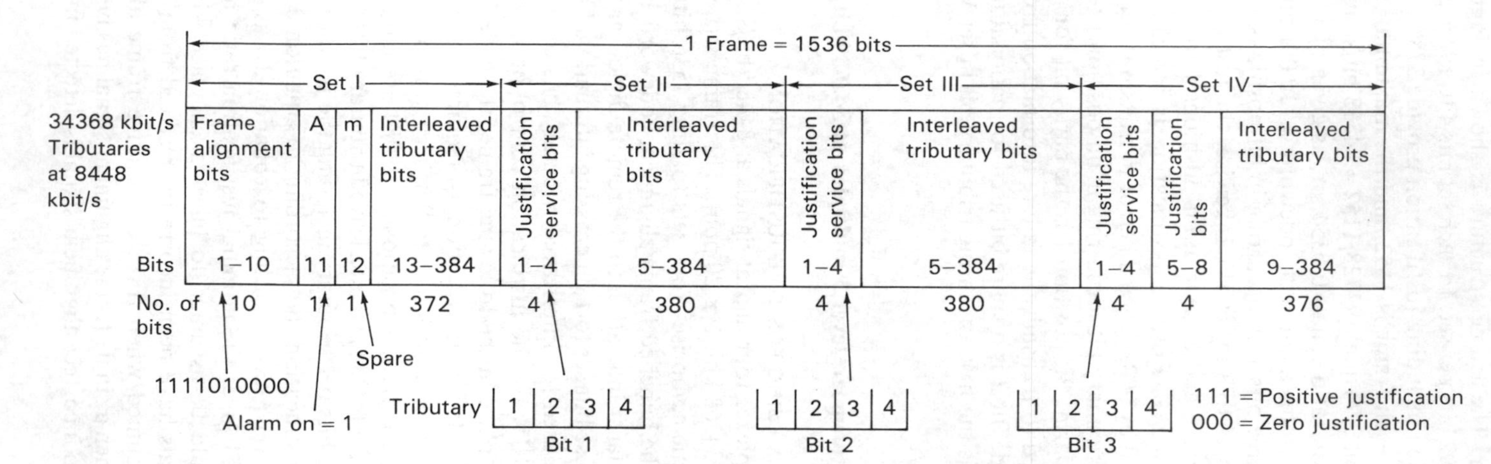

Figure 3.23 34368 kbit/s level 3 multiplex system using positive justification, frame structure

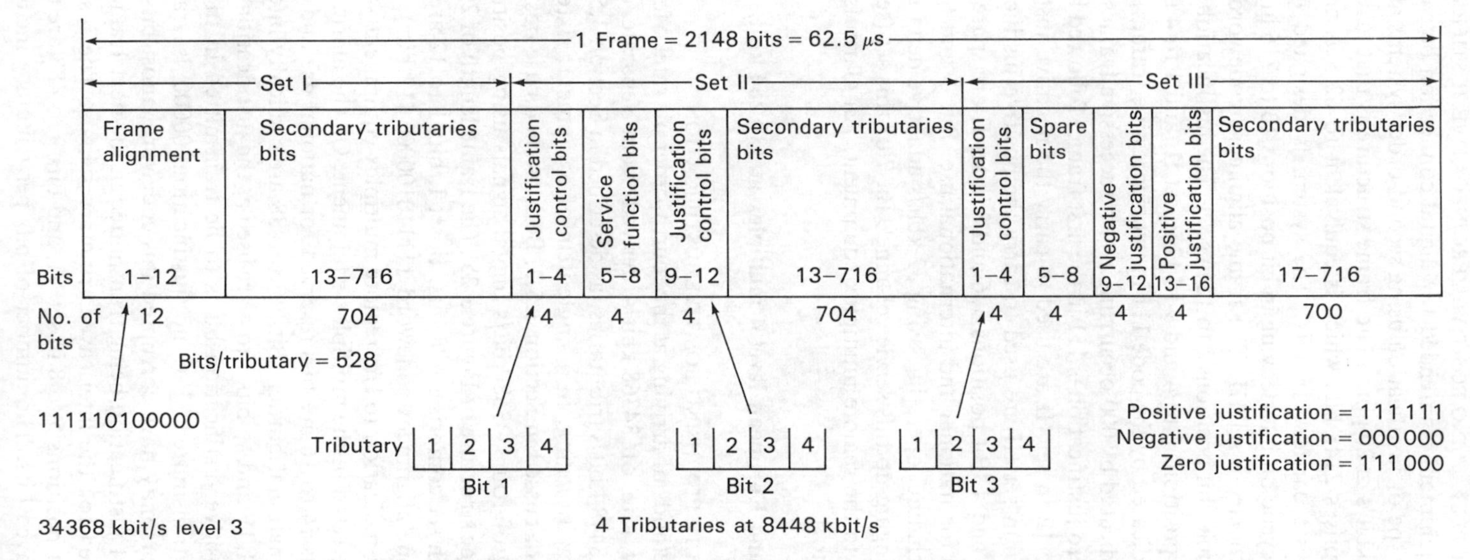

Figure 3.24 34368 kbit/s level 3 multiplex system using positive/zero/negative justification, frame structure

34368 KBIT/S MULTIPLEX USING POSITIVE/ZERO/NEGATIVE JUSTIFICATION

This multiplex equipment is intended for digital connection between two countries having the same type of justification using second-order systems at 8448 kbit/s. The bit rate is 34368 kbit/s ± 20 ppm. The frame structure is that given in Figure 3.24. Each frame comprises 2148 bits which is equivalent to 62.5 μs and is divided into three sets. Each set carries 716 bits. The system is designed to multiplex four tributaries using cyclic bit interleaving in bit positions 13 to 716 in sets I and II, and bit positions 17 to 716 in set III. The frame alignment codeword is 111110100000 (12 bits) and occupies bit positions 1 to 12 in set I. The justification control or service digits occupy bit positions 1 to 4 and 9 to 12 in set II, and 1 to 4 in set III. Positive justification is indicated by the code 111 in two successive frames, whereas negative justification is indicated by 000 occurring in two successive frames. The code for no justification or zero justification is 111 in the first frame, followed by 000 in the next. Bit positions 9 to 12 in set III are used to carry the negative justification bits. Bit positions 13 to 16 in set III are used to carry the positive justification bits for the tributaries 1, 2, 3 and 4. Bit positions 5, 6 and 8 in set II are spare bits available for service functions for national and international use. Bits 5 and 6 are available for a digital service channel operating using 32 kbit/s adaptive delta modulation. Bit 8 is available for ringing up this service channel. Bit 7 of this set (set II) is used as an alarm indication to the remote multiplex equipment. An alarm is raised when this bit goes from 0 to 1.

3.5.6.4 CCITT and European level 4 multiplex at 139 264 kbit/s

139 264 KBIT/S MULTIPLEX USING POSITIVE JUSTIFICATION

There are two methods of arriving at the fourth-order bit rate. Method 1 is by using a third-order bit rate of 34368 kbit/s derived from separate equipment which multiplexes four 8448 kbit/s tributaries, as described in Section 3.5.6.3, and method 2 is by directly multiplexing, in a single item of equipment, sixteen digital signals at 8448 kbit/s as discussed in Section 3.5.2. Both methods use positive justification. The multiplexing of four 34368 kbit/s digital signals using positive justification is shown in the frame structure of Figure 3.25. The frame length of 2928 bits is divided into six sets. Each set comprises 488 bits. In set I, bits 1 to 12 are used to transmit the bunched frame alignment codeword 111110100000 (12 bits), whereas bit 13 is used as an alarm indication to the remote multiplex equipment when specific fault conditions are detected in the multiplex equipment. On the alarm being raised, this bit goes to 1. Bits 14 to 16 are bits reserved for national use and are set to 1 when crossing an international border. Bits 17 to 488 are cyclically interleaved tributary bits. In sets II, III, IV and V, bits 1 to 4 are used as the justification control or service bits. When a positive justification bit is to be transmitted in the next frame, the single 11111 is transmitted. For no justification 00000 is transmitted. Where transmission errors may have corrupted this codeword, most bits in the codeword decide the type of justification being transmitted in the next frame. Bits 5 to 488 in these sets are used for the interleaved tributary signals. The justification bits are transmitted in bit positions 5 to 8 in set VI, and bits 9 to 488 are used for tributary cyclically interleaved bits. The number of bits per tributary including justification

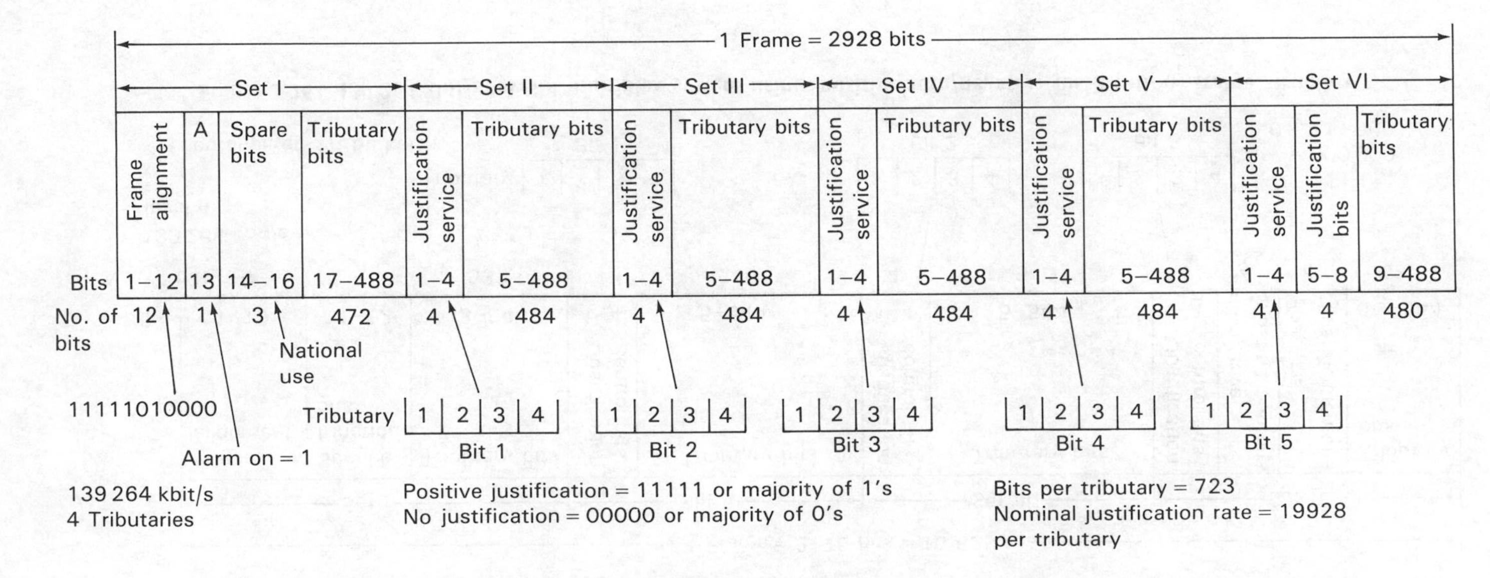

Figure 3.25 139 264 kbit/s level 4 multiplex system using positive justification, frame structure

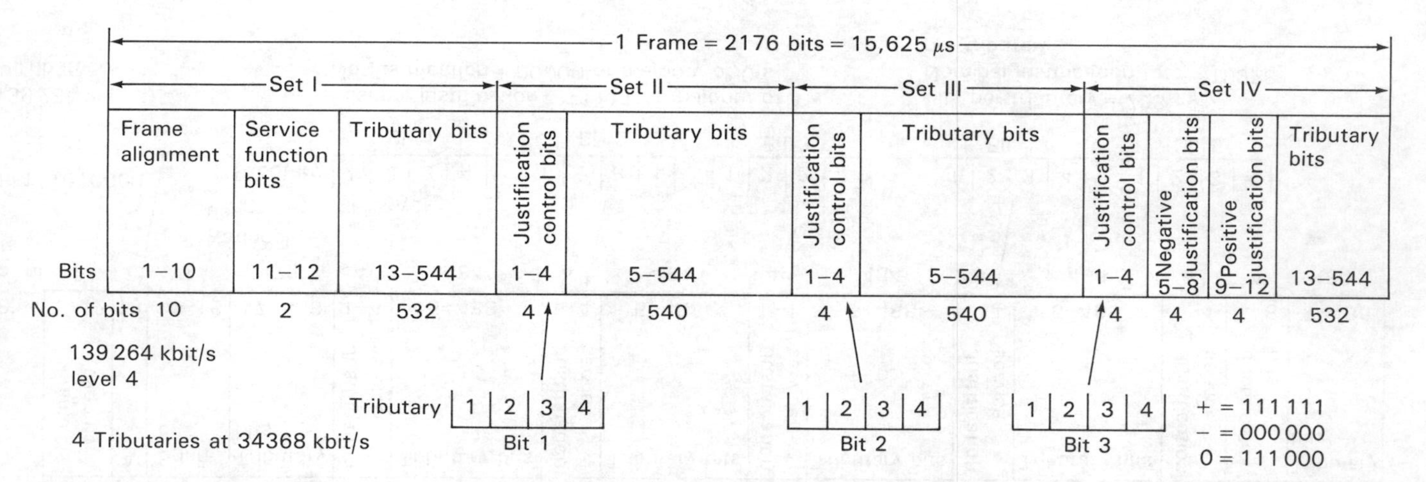

Figure 3.26 139 264 kbit/s level 4 multiplex system using positive/zero/negative justification, frame structure

($m_{\frac{1}{2}}$) is 723, and the maximum justification rate per tributary is 47560. The nominal justification ratio as given by equation 3.7 is found to be 0.4191, and, as in CCITT Recommendation G751,[2] it is 0.419.

139 264 KBIT/S MULTIPLEX USING POSITIVE/ZERO/NEGATIVE JUSTIFICATION

This level 4 multiplex system is intended for a digital connection between countries having the same type of justification system using any third-order digital system at 34368 kbit/s. The nominal bit rate is 139 264 kbit/s ± 15 ppm. Figure 3.26 shows the frame structure. The frame length is 2176 bits and has a duration of 15.625 μs. The four sets which comprise the frame each have 544 bits allocated to them. Bits 1 to 10 in set I are used to convey the bunched frame alignment signal. The codeword for this signal is not mentioned in CCITT Recommendation G754. Bits 11 and 12 are available for service functions for national and international use. Bit 11 is available for a 32 kbit/s adaptive delta modulation digital service channel, and bit 12 may be used for ringing up a digital service channel. Bits 13 to 544 are cyclically interleaved tributary bits of the four 34368 kbit/s tributaries accepted by this system. In sets II–IV, bits 1 to 4 are used as the justification control bits. Positive justification should be indicated by the signal 111, transmitted in two consecutive frames, whilst negative justification is indicated by 000 transmitted in two consecutive frames. If 111 is transmitted in the first frame and 000 in the second, it should be interpreted as a zero justification condition. Digit time slots 5, 6, 7 and 8 in set IV are used for information carrying bits or for negative justification. Digit time slots 9 to 12 in set IV are used only for positive justification bits for the tributaries 1, 2, 3 and 4. When information from tributaries is not being transmitted, bits 5 to 8 in set IV are available for transmitting information concerning the type of justification (positive or negative) in frames containing commands of positive justification control and the intermediate amount of jitter in frames containing commands of negative justification. The maximum justification rate per tributary is 64 kbit/s, and the number of bits per tributary is 537 (including negative justification bits 5 to 8 in set IV). The timing information for the system should be able to be derived from an external source. An alarm indication to the remote multiplex equipment is generated by changing from the state 0 to the state 1, bit 12 of set I at the output of the multiplexer.

3.5.6.5 Level 5 multiplex at 565 Mbit/s[28]

At the time of writing (1986), the maximum bit rate which is able to be carried by commercially available digital radio is 140 Mbit/s (level 4). For use over monomode optical fibers, 564.992 Mbit/s ± 15 ppm systems are presently being used to transmit up to 7680 voice channels (or equivalent) over trunk or high-capacity networks. This system combines in the optical line terminating equipment, four 139.264 Mbit/s ± 15 pm CMI coded data channels (see Section 5.1.5), which then produces the justified multiplexed output signal of 564.992 Mbit/s. This signal is converted into the 5B6B code (see Section 5.1.6.4) and combined with overhead bits to produce a line rate of somewhere around 680 Mbit/s depending on the system used. The overhead bits comprise the service data channel, protection switching control, supervisory data and remote service data signals. Repeater spacings are typically

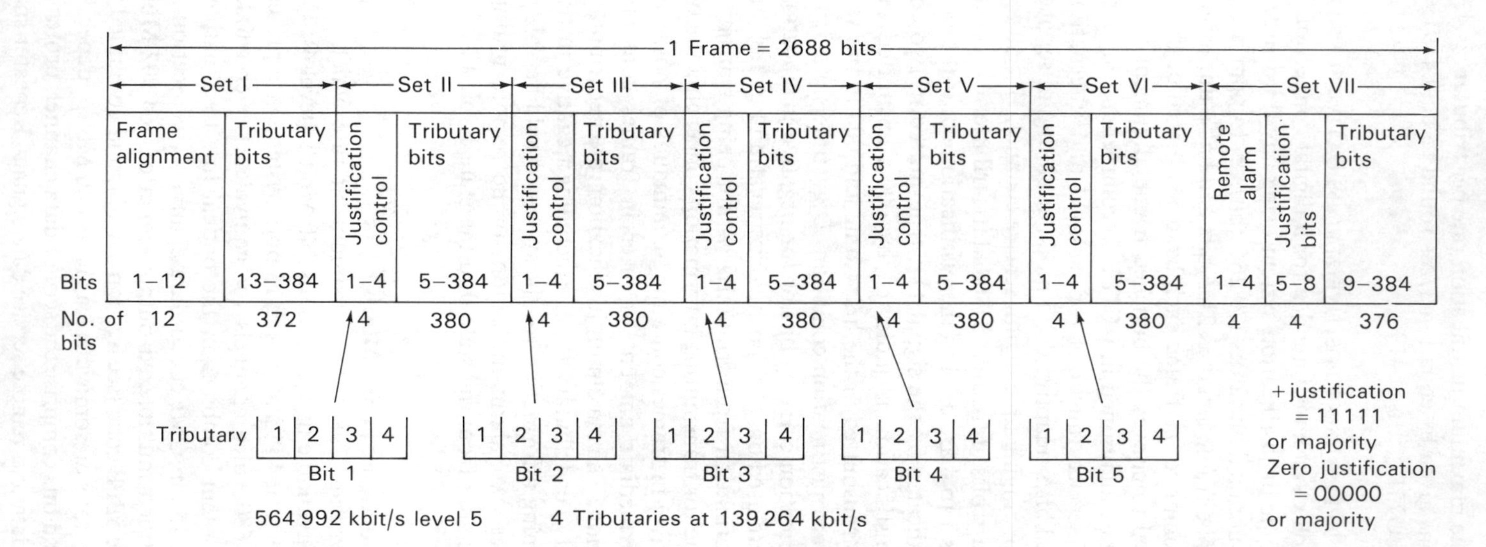

Figure 3.27 564 992 kbit/s level 5 multiplex system using positive justification, frame structure

of the order of 30 km assuming practical design values of 0.6 dB/km optical fiber cable loss, and an operating laser wavelength of 1310 ± 20 nm. The remote service data may include information on the bit error rate, laser diode bias current, alarm information on the loss of the optical transmit and receive signals. The service data channel which may be an optional feature can provide up to twelve 64 kbit/s data channels. In the receiving equipment the overhead bits are dropped and the 5B6B codes are decoded from the received signal. The final bit rate of the main data signal is thus converted back to 564.992 Mbit/s before being demodulated into the four separate 140 Mbit/s (139.264 Mbit/s) tributaries. Typical power transmitted into the optical fiber at the appropriate bit rate and 5B6B line code is −3 dBm and the typical receiver sensitivity is −35 dBm. Figure 3.27 shows the CCITT Recommendation G954 frame structure.[2]

Other multiplex frame structures at 564 992 kbit/s which still retain the same per tributary frame structure as implied in Figure 3.27 are possible. These alternative frame structures are based on the cyclic interleaving of groups of bits from tributaries and, with line codes such as 6B4T, may have implementation advantages. The multiplexing method used in Figure 3.27 is cyclic bit interleaving in the tributary numbering order and positive justification is recommended. The justification control signal should be distributed and the C_{jn} bits used. The C_{jn} bit is the nth justification service bit of the jth tributary. Positive justification should be indicated by the signal 11111 (five ones) and no justification by the signal 00000 (five zeros). As in the previous structures, if the control bits are corrupted during transmission, the majority of received bits determines the decision as to whether positive or zero justification is required. The first four bits in set VII of the pulse frame are available for service functions. The first of these bits is used to indicate a prompt alarm condition. In this system, bit sequence independence or the removal of pattern-dependent jitter (see Section 4.1.4) can be achieved by the use of a line code in combination with a scrambler (see Section 4.5.1) which is reset by the frame alignment word of the multiplexor. The CCITT Recommendation G954 Appendix I to annex B[2] provides a possible 'reset scrambler' proposal in which there would be no error multiplication and no necessity to provide additional measures to avoid periodic output signal sequences, especially where an alarm indication signal (AIS) is providing an all '1's signal at the tributary inputs. The scrambler operates at approximately 141 Mbit/s and after resetting generates the frame alignment signal. The circuit of this scrambler is provided in the same appendix as mentioned above.

3.6 SUPERVISION AND ALARMS

Alarms are usually raised when there is a loss in power to some part of the system; the input bit stream to the multiplex equipment is disrupted; there has been a loss in frame alignment; the tributary bit stream is disrupted; or there exists an excessive error rate condition, with for example the bit error rate greater than one bit error in one thousand. In a multiplex center there may exist a system whereby alarms are classed according to their importance. A power failure may be considered as an urgent or prompt alarm, whereas an error rate of one bit in 100 000 raises a nonurgent or deferred alarm which is provided as an 'early warning' that attention is required before conditions worsen to a bit error rate of 1 error in 10000 to raise an action alarm. In some systems there may be no personnel at the multiplex site and all the alarms are conveyed back to a control center which monitors the conditions of many sites throughout the network. Once an alarm is registered at this center, personnel then travel out and rectify the fault, knowing exactly what and where the fault is. As systems become more heavily loaded with traffic and more complex there is a trend towards such network management centers.[22–25]

3.6.1 Fault conditions

The fault conditions which initiate alarms (where applicable) are:

Failure of power supply
Loss or degradation of incoming signal from tributary or high-bit-rate signal
Loss of frame or multi-frame alignment.
Alarm indication from the remote signaling multiplex equipment
Receipt of the service alarm indication from the PCM multiplex
Failure of a codec
Excessive error ratio as detected in the frame alignment signal
Failure of standby equipment

The importance of these alarms is indicated by the consequent action taken as discussed in Section 3.6.2.

3.6.1.1 Error ratio

Error ratio can be measured either on the digital transmission system interconnecting two multiplex terminals so that errors due to the communication channel and the radio equipment can be determined, or at a multiplex terminal itself, in order that the performance of the overall transmission path can be determined. In addition to these two tests, multiplex equipment may be connected on a back-to-back basis (local loop-back mode), and the error ratio determined so that the effect of the multiplex equipment on the system can be determined. The determination of the error ratio in the multiplex equipment is achieved by examining the frame alignment signal, which is shown in Figures 3.4–3.6, 3.13–3.14 and 3.17–3.26 for the various multiplex levels in the two CCITT hierarchies. As the bit pattern is fixed, logic circuits in the measuring equipment can determine, by checking the incoming bits, if any error has occurred. Any error is added to a counter which indicates the ratio of the number of errors that have occurred to the total number of bits received, over a fixed predetermined time interval. For the activation of prompt maintenance alarms an excessive bit error ratio, as detected in the frame alignment signal, is used;

The CCIR Report 613-2[26] defines the error ratio as:

$$\text{Error ratio} = N_e/N_t = N_e/(Bt_o) \tag{3.10}$$

where N_e = number of bit errors on time interval t_o; N_t = total number of transmitted bits in the time interval t_o; B = bit rate of binary signal at the point where the measurement is performed; t_o = measuring time interval (error counting time).

When the error generation process is random and stationary and the errors are counted in a sufficiently long interval t_o, equation 3.10 can give an estimate of the error probability. The accuracy of this estimation increases as N_e increases, but practical requirements on the measuring time interval usually limit the values of N_e. The minimum acceptable value of N_e seems to be about 10 and then the true error probability is contained within a range equal to ± 50 percent around N_e/N_t with a confidence factor of 90 percent. The data required in equation 3.10 may be obtained using different procedures. If the procedure is used where the number of errors detected is limited to a fixed time interval t_o, the measured data can be directly expressed in figures of availability. If the method where the time interval required to detect exactly the error N_e (or more conveniently, the number of transmitted bits N_t, in such an interval), a nearly constant measurement accuracy for any value of the error ratio in a minimum of time is obtained. As discussed below, it is in some cases more convenient to measure the bit error ratio over a sufficiently large number of consecutive one-second time intervals. Test instruments which are commercially available to measure the error ratio are, among the many, the Hewlett-Packard 3764A digital transmission analyzer, and the Wandel and Goltermann PF-2.

CCITT EXCESSIVE ERROR RATIO VALUES IN THE 2048 KBIT/S HIERARCHY TO LEVEL 2

The criteria for activating an alarm are:

1. Error ratio $\leqslant 10^{-4}$
   The probability of activating an alarm in 4–5 s should be less than 10^{-6}.
2. Error ratio $\geqslant 10^{-3}$
   The probability of activating an alarm in 4–5 s should be higher than 0.95.

The criteria for deactivating the indication of the fault condition are:

3. Error ratio $\geqslant 10^{-3}$
   The probability for deactivating the alarm in 4–5 s should be almost zero.
4. Error ratio $\leqslant 10^{-4}$
   The probability of deactivating the alarm in 4–5 s should be higher than 0.95.

Above level 2 in the 2048 kbit/s primary hierarchy and for all of the 1544 kbit/s primary hierarchy, no recommendations to the error ratio are given. Other terms used in error analysis are briefly defined in this section and discussed again in later chapters.

ERROR SECONDS (ES) AND ERROR-FREE SECONDS (EFS)

These are defined as the ratio of the number of one-second intervals during which *no* bits are received without error (ES) or with error (EFS) to the total number of one-second intervals in the measurement time period. They are measured using two different methods. The first method is a synchronous method, where the start of a measurement period is coincident with an error, and the second is an asynchronous method, where intervals are independent of the receipt of errors. The second method is normally used because it gives a more realistic indication of a system's quality when bursts of errors are present. Both ES and EFS, plus bit-error-rate (BER), percentage of time (T) that the BER does not exceed a given threshold, and error-free blocks of data (EFB), are measures of the quality of a digital transmission system.

PERCENTAGE AVAILABILITY

This measurement gives an indication of a system's outage time. The period of unavailable time begins when the bit error ratio in each second is worse than 10^{-3} for a period of ten consecutive seconds, which are considered to be unavailable time. The period of unavailable time terminates when the bit error ratio in each second is less than 10^{-3} for a period of ten consecutive seconds, this period being considered as available time. Percentage availability is one of two ways of expressing the system's performance objectives. The other way is by means of the system's quality. The two areas affecting availability are; propagation reliability and equipment reliability.

3.6.1.2 Loss of frame or multi-frame alignment

Loss of frame alignment usually only occurs under conditions of system failure or excessive error ratio values. The CCITT requires that the frame alignment will be assumed to have been lost when so many consecutive frame or multi-frame alignment signals have been received with an error. Table 3.5 lists the conditions for assumed loss and recovery of frame and multi-frame alignment for different bit rates.

As can be deduced from Table 3.5, each time a frame alignment signal is incorrectly received a store is updated by one digit until the maximum permissible count is reached. This store is reset to zero on receipt of a valid frame alignment signal, insuring that on the next loss of the frame alignment signal (fas), the system waits until the maximum is reached before taking action. When the store reaches the maximum permitted, a search for the frame alignment signal is commenced. This is achieved with the 2048 kbit/s primary PCM multiplex, by stopping the timing signal in the receive PCM multiplex for a 1-digit interval, thus changing the relationship between the transmit and receive terminals. This action continues until a valid fas is acquired. As pointed out in the notes to Table 3.5, safeguards against accepting a false fas are built into the equipment.

3.6.2 Consequent actions

3.6.2.1 Service alarm

This alarm indication is generated to signify that the service provided by the PCM multiplex is no longer available, and the indication should be forwarded at least to the switching and/or signaling multiplex equipment depending on the arrangements

Table 3.5 Conditions for loss and recovery of frame and multi-frame alignment

Bit rate (kbit/s)	Number of frame alignments in loss of alignment	Frame alignment recovery sequence (see footnotes)	Number of multi-frame alignments in loss of alignment	Multi-frame recovery sequence (see footnote 3)
2048	3 or 4	(A)	2	one bit '1' in time slot 16 before multi-frame alignment signal
1544	1	after 1 valid fas	–	–
8448 (S, P)	4	3 consecutive fas (B)	–	–
8448 (N)	5	2 consecutive fas (C)	–	–
6312	1 (<16 ms)	1 (<420 μs)	–	–
34368 (P)	4	3 consecutive fas (B)	–	–
139 264 (P)	4	3 consecutive fas (B)	–	–
32064 (P)	1	1 (<8 ms)	–	–
44736 (P)	1 (<2.5 ms)	1 (<250 μs)	–	–
97728 (P)	1	1 (<1 ms)	–	–
34368 (N)	3 or more (D)	>2	–	–
139 264 (N)	3 or more (D)	>2	–	–
564 992 (P)	4	3 (B)		

Notes

1. P = positive justification. S = synchronous. N = positive/zero/negative justification. fas = frame alignment signal.
2. (A) Frame alignment will be assumed to have been recovered when the following sequence is detected:

   For the first time, the presence of the correct frame alignment signal in the even frame and the absence of the fas in the following (odd) frame detected by verifying that bit 2 in channel time-slot 0 is a 1.
   For the second time, the presence of the correct frame alignment signal in the next frame.

   For further details, see Section 3.2.2.
   To overcome fas imitation, when fas is detected in frame n, fas does not exist in frame $n+1$, but exists in $n+2$. If either of these conditions is not met, begin new search.
   (B) The frame alignment device, having detected the appearance of a single correct frame alignment signal, should begin a new search for the frame alignment signal when it detects the absence of the frame alignment signal in one of the two following frames.
   (C) As soon as frame alignment has been lost and until it has been recovered, a definite pattern should be sent to all tributaries from the output of the demultiplexer. The equivalent binary content of this pattern, called the alarm indication signal (AIS), at 2048 kbit/s is a continuous stream of binary '1's.
   (D) The frame alignment system should be adaptive to the error ratio in the line link. Until frame alignment is restored the frame alignment system should retain its position. A new search for the fas should be undertaken when three or more consecutive fas have been incorrectly received in their positions.
3. Loss of multi-frame alignment is assumed to have taken place when loss of frame alignment has occurred. To avoid a condition of spurious multi-frame alignment the following procedure may be used for the primary PCM 2048 kbit/s primary multiplex equipment, in addition to that given in the table.
   Multi-frame alignment should be assumed to have been lost when, for a period of one or two multi-frames, all the bits in time slot 16 are in state 0.
   Multi-frame alignment should be assumed to have been recovered only when at least one bit in state 1 is present in the time slot 16 preceding the multi-frame alignment signal is first detected.

provided. The indication should be given as soon as possible and not later than 2 ms after detection of the relevant fault condition, assuming that the average time to detect a loss of frame alignment and give the relevant indication takes no longer than 3 ms.

3.6.2.2 Alarm indication signal (AIS)

This signal is one that replaces the normal traffic signal when a maintenance alarm indication has been activated. The binary content of this signal applied to the input tributaries from the demultiplexer and to the multiplexer output is usually a continuous stream of '1's. When the demultiplexer equipment detects this signal it causes the prompt maintenance alarm associated with the loss of frame alignment to be inhibited.

3.6.2.3 Prompt maintenance alarm

A prompt maintenance alarm indication is generated to signify that performance is below acceptable standards and that maintenance attention is required locally. When the alarm indication signal (AIS) is detected, the prompt maintenance alarm indication associated with loss of frame alignment and excessive error ratio should be inhibited.

3.6.2.4 Alarm indication to the remote end

This is transmitted by changing bit 3 of channel time-slot 0 from the state 0 to the state 1 in those frames not containing the frame alignment signal for the 2048 kbit/s primary multiplex and by changing the service bit of time slot 24 from the state 1 to the state 0 for the 1544 kbit/s primary multiplex. In the frame structures shown in Sections 3.4 and 3.5 this alarm has been described. This alarm indication should be effected as soon as possible.

3.6.2.5 Transmission suppressed at the analog outputs

The situation applies only in the case of PCM multiplex.

3.7 SIGNALING[27]

Signaling in a network is the transfer of the information required to control the various functional elements within that network. The methods of signaling are many. This is partly due to the the development processes of an automatic network and the nonstandardization of equipment between administrations. Present trends are however towards standardization, as communications throughout the world are becoming more integrated and the tendency is towards a single complex world network. In a telephone network the signaling requirements may be divided into three well-defined divisions:

1. The transfer of signaling information from and to the subscriber from the local exchange
2. The processing of signaling information within exchanges
3. The transfer of signaling information between exchanges within a network or between networks

Subscriber-to-exchange signaling

The transfer of signaling information from subscriber to exchange takes two forms. The first is *line* signaling, where an on-hook or off-hook condition provides the local exchange with the necessary information on the state of the call. The second is *information* or *address* signaling, in which information in the form of dialled digits is passed to the local exchange.

Exchange processing of signaling information

The local or other type of exchange, after receiving information at its input ports, decides whether this information is to be processed into a form which is compatible with the outgoing port signaling requirements, and what signaling conditions (if any) are to be expected back from the outgoing port, etc.

Transfer of signaling information between exchanges

This information is usually revertive and in a compelled sequence, which means basically that when a sending exchange sends information on the routing of the call there is expected an answer from the distant exchange before the next piece of information is forwarded. In some arrangements it is not practicable to wait for an answer due to the time involved (as in satellite circuits), hence a different format is used.

In this section two telephone signaling techniques are referred to. These are conventional or *channel-associated signaling*, where the signaling information is conveyed on the same channel as that of the voice information and *common-channel signaling*, where the signaling is not sent over the same channel as the speech path, but over a channel which is shared with the signaling information of other voice channels.

In PCM the DC signals from the subscriber or from an analog switching machine are converted into bit-coded form and transmitted over the transmission link after multiplexing, only to be demultiplexed and reconverted into the original DC at the far end. The signaling is implicitly four-wire and continuous as the signaling information is given by the bit-code pattern of the signaling bits. As mentioned above, the PCM signaling information may be channel-associated or common-channel signaling according to the system in operation. Sections 3.2, 3.4 and 3.5 deal briefly with the placement of bits within the various frame structures for common-channel and channel-associated signaling, as per CCITT Recommendation G704.[2]

3.8 IN-STATION TESTING OF MULTIPLEX EQUIPMENT

In-station testing is usually done during the commissioning phase of new equipment, but may occur after replacement of faulty equipment or completed in part, as regular routine maintenance dictates. The philosophy behind these tests is:

1. To insure that all equipment is provided and that the strappings links or any relevant cards are correctly selected from the strapping options available and set.
2. To confirm that the equipment is operating to specifications and within the recommendations laid down by such bodies as the CCITT.
3. To verify that all alarms are operational and that the threshold levels are correctly set.
4. To provide initial commissioning values to parameters which may be used by the preventive maintenance personnel in determining at a future time if any equipment is degrading and requires special attention or replacement. To assist with this, each card is correctly labeled and marked and its position is recorded in a register with all relevant details.
5. To assist in the isolation of faults which may appear on end-to-end tests.

Usually on commissioning of newly installed equipment a *burn-in* period of 24 hours is required before any testing or commissioning. This period, although not entirely adequate, does permit early faults to appear. After commissioning, there is usually a stipulation that the equipment is to operate continually for three months to a year before final acceptance. This is to insure, for reliability, that any faults which appear after this stipulated period, are not burn-in faults.

During pre-commissioning tests, or if necessary during commissioning, should any card fail, the tests associated with that card are to be performed in their entirety and a record of the failure and details are to be kept as part of the test results. This also assists in quality control, in the case where a recurrent fault appears.

3.8.1 Test equipment required

For testing multiplex equipment up to bit rates of 140 Mbit/s, the following list of test equipment is provided as a guide.

Pattern-generator/error-detector WG–PF4 or digital transmission analyzer HP 3764A
Frequency counter
300 MHz oscilloscope
Digital multi-meter
Necessary patchcords
Necessary card extenders

3.8.2 Preliminary inspection

This inspection covers such items as properly terminated connections, proper placement of cards, correct labeling of the multiplex cards and racks, and proper recording of card serial numbers and types entered up into the records. It also covers the confirmation that the strappable links are correctly set.

3.8.3 Voltage and frequency stability

Using a digital multi-meter, confirm that all power supply voltages into the various racks of equipment are correct, and that on each card in the rack the regulated voltage supply is correct. All readings are to be entered into the appropriate test records and compared against the permitted tolerances for the measured values. Using a frequency counter check at points, for particular frequencies as recommended by the manufacturer, that the multiplex clock frequencies at the output (e.g. 34368 kHz) and input (e.g. 8448 kHz) are correct and within the tolerance as specified in Table 3.6. For the 140 Mbit multiplex equipment, the clock frequency may be a submultiple such as 34816 kHz. In this case the tolerance of ± 15 ppm will still apply to give 34816 ± 0.522 kHz.

Using the oscilloscope with a lead as short as possible and terminated in impedances as given in Table 5.5, so as not to unduly cause distortion of the waveforms, check the waveforms at the multiplex output and the tributary output. The waveforms should conform to those provided in CCITT Recommendation G703.

Table 3.6 Multiplex clock frequencies tolerance (CCITT Recommendation G703)[2]

Multiplex clock rate kHz	CCITT tolerance ± ppm	Actual frequency tolerance ± Hz
64	100	6.4
1544	50	77.2
6312	30	189.36
32064	10	320.64
44736	20	894.72
2048	50	102.4
8448	30	253.44
34368	20	687.36
139 264	15	2088.96
564 992	15	8474.88

3.8.4 Alarm tests

These tests usually are the bulk of the in-station testing, for there are many alarms as discussed in Section 3.6. Usually the alarm circuits require strapping for the alarm categories. These categories are 'Prompt' for an alarm which requires immediate attention, 'In-station' for an alarm which requires attention but is not in the urgent category and 'Deferred' for an alarm which provides an early warning that if the defect is not attended to it may become a problem later.

3.8.4.1 Prompt alarms
These alarms are recommended for:

Failure of power supply
Loss of incoming multiplex signal
Loss of incoming tributary signal
Loss of frame alignment
Error ratio for the alignment signal of 10^{-3}
Failure of codec (if installed)

3.8.4.2 In-station alarms
These alarms are recommended for:

Loss of incoming multiplex signal at the remote end
Loss of incoming tributary signal
Alarm indication signal (AIS) received from the remote multiplex equipment
Error ratio for the alignment signal of 10^{-4}

3.8.4.3 Deferred alarms
These are suitable for:

Loss of incoming tributary signal

The loss of the incoming tributary signal may be regarded as a prompt alarm by some administrations and not by others (it is regarded as a prompt alarm by the CCITT).

The initial tests are to test any lamps. It is common these days to have a lamp testing circuit built into the rack alarm system. Once all lamps are shown to be operable and all alarm category strapping has been completed, each individual alarm is required to be tested. Simulation of the system may be done by arranging for the multiplex equipments to be connected back-to-back at the highest level and manufacturer's procedures followed for the testing of each of the above simulated alarm conditions. By interrupting the highest multiplex level back-to-back connection, a system with an alarm indication signal (AIS) capability will have this alarm activated. If a pattern generator output with the appropriate pattern length is connected to the highest-level multiplex input port a loss of frame alignment alarm should occur. If a second multiplex system is available and is connected to the system under test it is possible to test for the distant alarm activation by removing the multiplex output link from the multiplex under test. As it is the output of the multiplex which is removed and not the input, the alarm generated may not necessarily be a prompt alarm, since the fault does not affect the station's incoming data.

3.8.5 Mechanical tests

One of the most frequent causes of failure is due to improper cable terminations. A simple way to test the system is to connect the highest level of the multiplex equipment back-to-back and to daisy-chain the input tributaries, i.e. to connect the output of tributary 2 to the input of tributary 3 and the output of tributary 3 to the input of tributary 4. The input of tributary 1 is then fed from a pattern generator with the appropriate pattern length as given in Table 4.10. The output of tributary 4 is then connected to the error detector and the error ratio recorded. Percussion tests on all cables and connectors are then carried out one-by-one and the error detector observed for any increase in the error ratio. If the error ratio increases, the cable or connector is investigated for the fault. Once the fault is found and cleared it must be confirmed by further percussion tests. This process is continued if there are lower levels of multiplex feeding into the previously tested tributaries.

3.8.6 Error checks

With the equipment daisy-chain as described in Section 3.8.5, the pseudo-random test pattern is applied with an error rate of 1 part of 10^6. The error detector at the output of the daisy-chained tributaries is then observed for this error rate. On confirmation, the injected error rate is removed and the system is monitored for one hour or more for error-free performance.

EXERCISES

1 In Section 3.2.2 it was explained that six bits (bits 1, 4–8) in odd-numbered frames may be used in national circuits for any purpose. Determine the bit rate if these bits are used to carry some form of digital information. (*Answer* 24 kbit/s)

2 In Section 3.2.2 determine the bit rate of the multi-frame alignment signal. (*Answer* 2 kbit/s)

3 In a 30-channel PCM system, how long is required for each voice-channel codeword to pass through a path in a digital exchange? (*Answer* 3.9 μs)

4 If 30-channel PCM systems enter a digital exchange, would the exchange switching matrix need to be larger or smaller than the same capacity analog exchange? (*Answer* Smaller – why?)

5 For the 96-channel PCM multiplex, what is the bit rate of the error detection code? (*Answer* 10 kbit/s)

6 For the 96-channel PCM multiplex, what is the bit rate of the signaling channel for the PCM channel 3? (*Answer* 1.666 kbit/s)

7 For the 120-channel PCM multiplex there are 49 spare bits for national use if the system is not used to cross international borders, plus 3 spare bits for the x bits in time slot 67–70 in frame 0. Determine the bit rate for a subsystem which makes use of all of these spare bits. (*Answer* 398 kbit/s)

8 What is the bit rate for the loss of frame alignment alarm in the 120-channel PCM multiplex system? (*Answer* 8 kbit/s)

9 Assuming four input tributaries have the same data rate of 2048 kbit/s and are justified at a rate of 5.736 kbit/s in a second-level multiplex equipment requiring positive justification, which has a framing content (a) equal to the number of frame alignment bits (10) divided by the total number of frame bits 848. Determine the theoretical output bit rate of the multiplexer. (*Answer* 8382.628 kbit/s)

10 A fourth-order multiplex system has an output bit rate of 97728 kbit/s, and an input tributary bit rate of 32064 kbit/s. The frame length is 1152 bits and the bits per tributary per frame including justification is 378 bits. Determine the maximum justification rate per tributary and the nominal justification ratio. (*Answer* 84833 bit/s, 0.0354)

11 Determine the nominal justification ratio of a 139 264 kbit/s multiplex equipment with a tributary bit rate of 34368 kbit/s and having a multiplexing frame length of 2928 bits and 723 bits per tributary (including justification bits). (*Answer* 0.4191)

12 Determine the nominal justification ratio of a 139 264 kbit/s multiplex equipment with a tributary bit rate of 34368 kbit/s and having a multiplexing frame length of 2176 bits and 537 bits per tributary. Why is your answer so? (*Answer* 0, see Section 3.5.5 and 3.5.6.4)

REFERENCES

1. Bylanski, P. and Ingram, D. G. W., *Digital Transmission Systems*, IEE Telecommunications Series 4, Rev. Ed. (Peter Peregrinus, 1980).
2. *Digital Networks – Transmission Systems and Multiplexing Equipments*, Recommendations G700–G956, (Study Groups XV and XVIII), CCITT Red Book, **III-3**, VIII Plenary Assembly, 1984.
3. Barker, R. H., 'Group Synchronizing of Binary Digital Systems', in *Communication Theory*, Edited by Jackson, W. (Academic Press, 1953), pp. 273–287.
4. Sexton, M. J. and Drudy, J. J., 'A 60 Channel FDM/PCM Transmultiplexer', *Telecommunication Transmission*, IEE Conference Publication No. 246, 1985, p. 270.
5. Narasimha, M. J., Yang, P. P. N., Byeong, G. L. and Abell, M. L., 'The TM7800–M1: A 60 Channel CCITT Transmultiplexer', *Telecommunications, Radio and Information Technology*, IEE Conference Publication No. 235, 1984, front pages.
6. Bonnerot, G., Coudreuse, M. and Bellanger, M. G., 'Digital Processing Techniques in the 60-Channel Transmultiplexer', *IEEE Transactions on Communications*, 1978, **COM–26**, pp. 698–706.
7. Special Issue on 'Transmultiplexers', *IEEE Transactions on Communications*, July 1982, **COM–30**.
8. Rossiter, T. J. M., Chitsaz, S., Gingell, M. J. and Humphrey, L. D., 'A Modular Transmultiplex System using Custom LSI Devices', *ibid.*, 1982, **COM–30**, No. 7
9. Narasimha, M. J. and Peterson, A. M., 'Design of a 24-channel Transmultiplexer', *IEEE Transactions on Acoustics, Speech, Signal Processing*, 1979, **ASSP–27**, December pp. 752–762.
10. Gerrard, G. A., 'A Review of the Applications of Different Types of Television Codecs', *Telecommunication Transmission*, IEE Conference Publication No. 246, 1985, p. 282.
11. Redstall, M. W. and White, T. A., 'Subjective Quality of a 70 Mbit/s Digital Codec for Colour Television', *Proc. IEE*, 1983, **130(F)**, pp. 477–483.
12. Devereux, V. G. and Jones, A. H., 'Pulse-Code Modulation of Video Signals: Codes Specifically Designed for PAL', *Proc. IEE*, 1978, **125**, pp. 591–598.

13. van der Kam, J. J., 'A Telephony Codec using Sigma-Delta Modulation and Digital Filtering', *Communications Equipment and Systems*, IEE Conference Publication No. 209, 1982, p. 49.
14. Misawa, T., Iwersen, J. E., Loporcaro, L. J. and Ruch, J. G., 'Single-Chip per Channel Codec with Filters utilizing Delta-Sigma Modulation', *IEEE Journal of Solid-State Circuits*, 1981, **SC–16**, No. 4, pp. 333–341.
15. Brewer, R. J., Janssen, D. J. G. and van der Meeberg, L., 'A Single Chip PCM Codec with Digital Filters', *Proceedings of the 9th International Congress on Microelectronics*, Munich, November 1980.
16. Coates, R. F. W., *Modern Communication Systems*, 2nd Ed (Macmillan Press, 1983), p. 233.
17. 'Digital Transit Exchanges in Integrated Digital Networks and Mixed Analogue-Digital Networks. Digital Local and Combined Exchanges', Recommendations Q501–Q517 Study Group XI, CCITT Red Book, **VI-5**, VIII Plenary Assembly, 1984.
18. Inose, H., *An Introduction to Digital Integrated Communication Systems*, Rev. Ed. (University of Tokyo Press © 1979, and Peter Peregrinus, 1981).
19. Inose, H. *et al.*, 'Theory of Mutually Synchronized Systems', *Electronics and Communications in Japan*, 1966, **49**, No. 4, pp. 263–272.
20. Osatake, T. *et al.*, 'Repeater for Pulse Code Modulation Systems in Independent Synchronization', *ibid.*, 1964, **47**, No. 2, pp. 71–82.
21. Duttweiler, D. L., 'Waiting Time Jitter', *Bell System Technical Journal*, 1965, **44**, No. 9, pp. 1843–1886.
22. Portejoie, J. F. and Le Moan, S., 'A New Approach of Performance Monitoring in a Centralized Digital Transmission Network Maintenance', *Telecommunication Transmission*, IEE Conference Publication No. 246, 1985, pp. 70–73.
23. Chahley, J. R. and Reid, J. T., 'Surveillance and Maintenance of the DRS–8 Long Haul Digital Microwave Radio System', *ibid.*, pp. 36–39.
24. Robinson, III, W. G., 'Maintenance of Digital Radio Using Remoted Performance Monitoring', *Links for the Future*, IEEE International Conference on Communications, May 1984, pp. 456–461.
25. Guenther, R. P., 'Planning for Centralized Maintenance Control of Digital Facility Networks', *ibid.*, pp. 462–465.
26. 'Bit Error Performance Measurements for Digital Radio-Relay Systems', *Fixed Service using Radio-Relay Systems*, CCIR Recommendations and Reports (Green Book), 1982, **IX**, Part 1, pp. 276–279.
27. Welch, S., *Signalling in Telecommunications Networks*, IEE Telecommunications Series 6, Rev. Ed. (Peter Peregrinus, 1981).
28. Gier, J. and Stummer, B., 'A 565 Mbit/s Line Equipment for Single-Mode Optical Waveguides', *Links for the Future*, IEEE International Conference on Communications 1984, pp. 783–786.

4 JITTER AND WANDER IN DIGITAL NETWORKS

Jitter and wander are defined respectively as the short-term and long-term variations of the significant instants of a digital signal from their ideal positions in time. The significant instants may be taken as the midpoint or any fixed arbitrary point which is clearly identifiable on each of the pulses. The variations in the significant instants affect a digital signal in a way which is similar to the case where the original timing signal used to generate the bit stream is phase-modulated with a modulating signal which is that of the jitter waveform. The jitter waveform represents jitter as a continuous time function with properties independent of the digital signal which it affects. This is shown in Figure 4.1 as the results of the effects of jitter on a simple timing signal at a receiver. Jitter signals considered to be most significant, occupy the frequency range from a few tens of Hz to several kHz. The units in which the amplitude of the jitter is specified depend upon the environment in which it is being measured and may be specified as units of time or phase, or as unit intervals (UI).

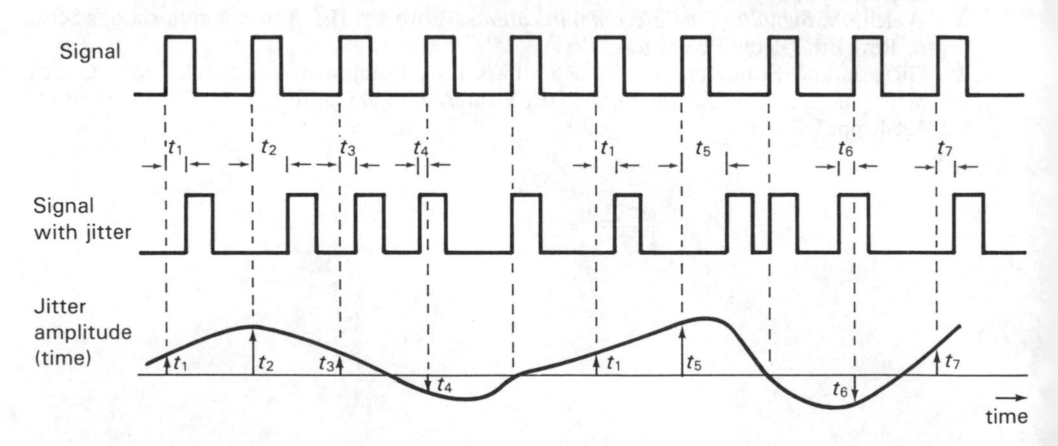

Figure 4.1 Pictorial representation of jitter and its effect on a digital signal

Throughout this text, the UI will be used as it is that adopted by the CCITT in its recommendations. The definition provided by the CCITT in Recommendation G701 (Definition 2018),[1] for the unit interval, is; 'the nominal difference in time between consecutive significant instants of an isochronous signal'. Paraphrased, this means that the width of a digit period is equal to the unit interval. Suppose, for example, the instantaneous jitter amplitude was measured to be one microsecond in a 100 kHz square wave; since the period of a frequency of 100 kHz is 10 μs, for a timing signal to differentiate between what is the mark and what is the space, the significant instants would have to be separated by 5 μs. Thus the unit interval would be 5 μs and the jitter amplitude would be $1\ \mu s/5\ \mu s = 0.2$ UI.

4.1 THE SOURCES OF JITTER

Jitter in many cases is indeterminable and can only be evaluated by using statistical tools such as the amplitude probability density function (pdf). This permits the probability of a jitter amplitude being less or greater than a certain value to be determined when used with the variance and mean. When the jitter waveform is determinable, the rate at which the jitter on a digital signal changes is an important parameter. Jitter can arise from a variety of sources in a digital transmission system as listed in the Sections 4.1.1 to 4.1.5. The majority of these sources however, fall into one of the following categories:

1. Very low-frequency jitter or wander caused by variations in the propagation delays of cables, etc., resulting from slowly changing temperature variations
2. Low-frequency jitter resulting from the inherent instabilities of clock sources
3. Noise-induced jitter arising from phase noise in crystal-controlled oscillator circuits used in clocks throughout the system as well as noise in logic circuits causing jitter on digit and clock transitions as a result of restricted rise times
4. Multiplex-induced jitter arising from the insertion and removal of justification bits and framing digits
5. Intersymbol interference produces a distortion in the pulse shape of each digital bit. This produces a variation in the detection level of each pulse, giving rise to jitter on the regenerated bit stream
6. Regenerator jitter arising from imperfect timing recovery in digital regenerative repeaters.

With the exception of jitter arising from regenerators, the sources as listed above provide jitter which is at a low level and is uncorrelated. In these cases the jitter adds on a power basis as it accumulates along a digital transmission system. Regenerator jitter is however, pattern-induced and hence correlated. As the same pattern or changes in pattern is applied to each regenerator, jitter in this case will be cumulative on an amplitude basis.

4.1.1 Jitter produced by timing elements

This jitter is relatively low in amplitude and frequency and arises from the phase noise inherent in the oscillator circuitry of the clock and in the variations in rise time associated with the logic circuits. The jitter produced is in most cases negligible. The factors which may produce the long-term form of jitter or wander are slow variations in the power supply and changes in temperature, for these in turn affect the triggering threshold levels in logic circuits.

4.1.2 Wander produced by temperature variations

This slowly time-varying jitter often referred to as *wander* arises from temperature variations of the transmission equipment and channel. This in turn produces a change in the time a signal takes to pass through them. Because of the low frequency of wander, phase-locked loops, unless designed with very long time-constants, tend to amplify it and so it is not practicable to use phase-locked loops to remove or reduce wander. Normally buffer stores of sufficient capacity are used instead, since they are able to absorb these longer-term jitter variations. Table 4.1[3] provides some estimated values of wander for various systems used by British Telecom. Wander amplitudes, like jitter, can be expressed in unit intervals but it is more common to use units of time.

Wander is usually accommodated at the input ports of digital equipments by the use of elastic stores in order to minimize controlled or uncontrolled slips.

Table 4.1 Estimated values for wander

System	Typical wander	
	Monthly variation	Annual variation
2048 kbit/s symmetrical pair	30–42 ns/km	62–77 ns/km
8448 kbit/s symmetrical pair	24–33 ns/km	48–61 ns/km
139264 kbit/s coaxial cable	0.5–0.7 ns/km	1.2–1.3 ns/km
139264 kbit/s fiber optic	0.3–0.4 ns/km	0.6–0.8 ns/km
139264 kbit/s radio	Very small	Very small
Digital Mux/demux pair	Negligible	Negligible
Digital transit exchange	Negligible	Negligible
Satellite geostationary orbit	Daily variation 0.15–1.5 ms	

4.1.3 Jitter produced by justification

As described in Section 3.5.1, justification is a process in which bits are regularly 'stuffed' into an incoming bit stream in order to raise the bit rate to that required by the multiplexing equipment. At the remote end and with the aid of justification control signals, the bits are removed in order to restore the information signal to its nominal input value. The removal of a justification bit at the receive end means that the digital signal arrives in bursts, separated by gaps of one time-slot duration. This irregular signal, besides entering the elastic store, is used to provide the input to a phase-locked loop circuit which derives the timing signal from the average rate of this irregular signal. The resultant timing signal is employed to read out the buffered signal from the elastic store (which has absorbed the irregularities), as shown in the demultiplexer block diagram of Figure 3.16. The retiming operation is not perfect, and as a result a residual jitter component remains superimposed on the output signal. The RMS value of this jitter may be shown to be (pp. 265, Reference 2) equal to $0.2887/(f_{mux})$, where f_{mux} is the multiplex output bit rate. In addition to the justification jitter, there is an added jitter component known as 'waiting time' jitter.[4,5] This arises as a result of the delay between the time that justification should occur and when a specific time slot allocated to justification operations becomes available. This form of jitter is the type which accumulates significantly in muldex equipment and which may have components at frequencies within the passband of the demultiplexer's phase-locked loop, that is, it is a low frequency jitter. During the waiting time the jitter amplitude increases linearly, providing the characteristic saw-tooth pattern. The expectations are that the accumulation of waiting time jitter will be at a rate between $(N)^{1/4}$ and $(N)^{1/2}$, where N is the number of cascaded multiplexer/demultiplexer pairs. A model which has been reported to fit measurements taken on a digital section that includes the associated multiplex equipment[6] shows that repeater jitter adds 'powerwise' to the waiting time jitter from the demultiplexer, and that in x digital sections for a line system operating at 140 Mbit/s and for a radio system operating at 34 Mbit/s, the total jitter J_x equals this power from a single section, multiplied by the cube root of the number of sections for through connections. Also it was found that waiting time jitter amplitude is greater in value than that of the pattern-dependent jitter.

4.1.4 Jitter resulting from regenerator operation[7-11]

Most digital regenerators used at present are self-timed, i.e. the output signal is retimed under the control of a timing signal which has been derived from the incoming signal. The most significant form of jitter arises from imperfections in the circuitry, which cause jitter that is dependent on the sequence of pulses in the digital signal being transmitted. This is termed pattern-dependent jitter.[7-9] In digital bit streams containing data, the timing information cannot be extracted directly. This is because the bit stream does not contain pulses in every time slot. To extract the timing information, further processing is required and it is during the processing needed to generate a regular train of clock pulses that timing jitter is introduced. One method used to extract the timing information is to use the input bit stream to set oscillating a high-Q tuned circuit. If the Q is high enough, by the time the oscillations start to subside, another pulse has arrived and started the circuit oscillating afresh. Because the Q of the circuit in practice is finite its output will consist of the extracted spectral line plus a jitter component due to the amplitude/phase conversion of random noise on the input signal. This noise-induced jitter will be purely random and bear no correlation with the input signal. Because of this, the jitter independently produced by further repeaters down the line will tend to accumulate even though each regenerator will suppress to some extent the phase modulation on the incoming signal. Any jitter or phase noise which falls within the clock recovery circuit bandwidth will, however, appear on the retimed signal in addition to the jitter produced by the clock recovery itself. The amount of phase modulation suppressed is thus dependent on the Q of the tuned circuit and the amount the circuit is off nominal frequency. From the viewpoint of jitter, a regenerative repeater acts as a low-pass filter to the jitter present on the input signal, but as mentioned above, it also generates jitter. This can be represented by an additional jitter source at the input. If this jitter were truly random, as distinct from pattern-dependent or correlated jitter, the total RMS jitter J_N, present on the digital signal after N regenerators, would be given by the approximate expression:

$$J_N \simeq J \cdot N^{1/4} \tag{4.1}$$

where J is the RMS jitter from a single regenerator due to uncorrelated jitter sources. This equation assumes that the jitter added at each regenerator is uncorrelated. The jitter produced by a random pattern is itself random in nature and the amplitude probability distribution function (pdf) is considered to be close to Gaussian. Hence, for a given standard deviation or RMS jitter amplitude, the probability of exceeding any chosen peak-to-peak amplitude can be calculated. For specification purposes a peak-to-peak to RMS ratio of 12–15 may be assumed, for this has a very low probability of being exceeded.

The uncorrelated contributions from regenerators are not regarded as being the only contributions made towards the total system jitter. This is because the message signals which were assumed to be random are not necessarily random but change from one repetitive pattern to another. In contrast then to uncorrelated jitter, pattern-dependent jitter *is* correlated. Since pattern-dependent jitter from regenerated sections is the dominant type of jitter in a network, the manner in which it accumulates is of importance. The changing of repetitive patterns in a regenerator chain causes a relative change in clock timing, which is magnified by the number of regenerators in the chain. Progressing down the chain there is a build-up of jitter as the relative change in clock timing appears to be increased. Consequently, regenerators at the distant end operate with an apparent mistuning of their circuits for an appreciable period after the pattern change. In addition there is the effect of an amplitude transient occurring at the output of each regenerator due to the associated change in pulse density arising from the pattern change. The combined result is to produce a long-duration transient disturbance in both phase and amplitude. The amplitude changes are dealt with by the regenerator limiter circuits, but the phase shift has the effect of degrading the operating margins of the regenerator for the period of the transient disturbance. It is accepted that most jitter added is pattern-dependent and, since the pattern is the same at each regenerator, it can be assumed that the same jitter is added at each regenerator in a chain of similar regenerators. For this case, it can be shown that the low-frequency jitter components add linearly, whereas the higher-frequency components are increasingly attenuated by the low-pass filtering effect of successive regenerators. If a random signal is being transmitted, the RMS jitter J_N, present on the signal after N regenerators would be given by the expression:

$$J_N = J_1 \cdot (2N)^{1/2} \text{ for large values of } N \tag{4.2}$$

where J_1 is the RMS jitter from a single regenerator due to pattern-dependent mechanisms.[10] The CCITT Recommendation G823[1] states that, based on operational experience to date, values of J_1 in the range 0.4 to 1.5 percent of a unit interval are achievable using cost-effective designs.

The comparison of equations 4.1 and 4.2 shows that pattern-dependent jitter accumulates more rapidly than non-pattern-dependent jitter, as the number of regenerators is increased, and that with such an increase the amplitude of jitter produced by a chain of regenerators increases without limit. If a phase-locked loop is used in timing recovery, the rate of accumulation is marginally greater. The expression for the RMS jitter J_N becomes in this case:

$$J_N = J_1 \cdot (2NA)^{1/2} \tag{4.3}$$

where A is a factor dependent upon both the number of regenerators and the damping factor of the loop. The damping factor is generally chosen in this application to produce a value of A which is marginally greater than unity.

The use of a transversal surface acoustic wave (SAW) filter in timing recovery[12] produces a rate of accumulation approaching that obtained for uncorrelated jitter sources. This is because of the large delay between the time the signal enters the filter, and the time that the recovered timing signal emerges at the output. This delay effectively randomizes the systematic pattern-dependent jitter allowing the jitter to accumulate in a similar manner to that arising from uncorrelated jitter sources. The only noticeable side effect is a marginal degradation in the alignment jitter. This type of jitter is described in Section 4.1.5. Another method of reducing pattern-dependent jitter to that of uncorrelated jitter involves the use of repeater circuitry which produces pattern transformations. An example of this is a pattern transformation based on the modulo 2 addition and its delayed version which permits the RMS jitter to accumulate as that given by equation 4.1.[11]

4.1.5 Alignment jitter[10]

Alignment jitter is the instantaneous difference between jitter on the timing signal and jitter on the message signal. It is a relative jitter as against an *absolute* jitter. It may also be considered as the difference between the input and output jitter of any one regenerator in a long chain of regenerative repeaters, and will have a constant value for all repeaters, the value being the maximum value occurring in the chain.

4.2 EFFECTS OF JITTER (CCITT Recommendation G823)[1]

At an interconnection interface, the input tolerance to jitter of the subsequent equipment should be designed to accommodate the jitter generated by the preceding equipment; otherwise the accumulation of jitter can, if not properly controlled, give rise to the following impairments:

1. An increase in the probability of introducing digital errors into digital signals at points of signal regeneration caused by the timing signals being displaced from their optimum positions in time.
2. The introduction of uncontrolled slips into digital signals as a result of the digital store capacity designed to cater for other effects being used up, thus causing store spillage and, for the opposite effect, store depletion. Spillage and depletion occur in certain types of terminal equipment incorporating buffer stores and phase comparators, e.g. jitter reducers and certain digital multiplex equipment.
3. A degradation of digitally encoded analog information as a result of phase modulation of the reconstructed samples in the digital to analog conversion device at the end of the connection. The timing jitter in this case which affects the regularity of the spacing between samples of the reconstructed PAM signals is sometimes known as *absolute* jitter. Encoded PCM speech is fairly tolerant to this type of effect, but digitally encoded television is much more sensitive.

4.3 JITTER AND WANDER LIMITS

In a digital communication network, the accumulation of jitter and wander must be controlled. The reasons for this control is to insure that the digital error and slip objectives as specified in the CCITT Recommendations G821 and G822 are not exceeded and that the quality of encoded analog information from reconstructed samples in the digital-to-analog conversion process is not significantly impaired.

The control of jitter to contain the digital error rate to within that recommended is achieved by limiting the alignment error in every retiming process. This may be achieved by designing the clock-recovery circuitry to have a jitter bandwidth comparable to that of the incoming jitter bandwidth.[13,14] The absolute jitter magnitude is usually not significant in the occurrence of jitter-induced errors.

The control of slips to within that recommended can only be achieved if the jitter generation and accumulation in all equipments throughout the network are considered. This means that there is the requirement for overall jitter control and jitter performance specification.

As the control of jitter means 'the control of jitter to within specified limits', there is a need to specify these limits in such a way that overall control can be realized. The specification of the limits are thus made for:

Different interconnecting elements such as digital radio links or individual pieces

of equipment such as multiplex equipment, digital exchanges, etc.
The network in its totality under different network configurations.
International connecting networks

Also this jitter control philosophy should ensure that the error and slip objectives, etc. will be met for any network element experiencing jitter on its input, irrespective of its location in the network or from where in the network the signal originates. Furthermore, these objectives must continue to be met as the network expands and its configuration changes.

The CCITT Recommendation G823[1] is concerned with the control of jitter and wander within digital networks which are based on the 2048 kbit/s hierarchy and is dealt with before considering the 1544 kbit/s hierarchy.

4.3.1 Limits of jitter and wander for the 2048 kbit/s digital hierarchy (CCITT Recommendation G823[1])

The basic philosophy in this case is to specify the maximum network limit that should not be exceeded at any hierarchical interface and to provide a consistent framework for the specification of individual digital equipments. In addition to this, guidelines and information, but not limits, are provided so that jitter accumulation studies and measurements can be made.

4.3.1.1 Network limits for the maximum output jitter at any hierarchical interface

Table 4.2 gives the CCITT Recommendation G823[1] limits which represent the maximum permissible levels of jitter at hierarchical interfaces within a digital network.

Table 4.2 Maximum permissible jitter at hierarchical interface based on the 2048 kbit/s hierarchy (tolerable input jitter)

Parameter value		Network limit of pk–pk jitter		Measurement filter bandwidth		
Digit rate of 1 UI		B_1 UI measured f_1–f_4	B_2 UI measured f_3–f_4	Band-pass filter having a lower cut-off frequency f_1 or f_3 and an upper cut-off frequency f_4		
kbit/s	ns			f_1, Hz	f_3, kHz	f_4, kHz
64 (Note 1)	15600	0.25	0.05	20	3	20
2048	488	1.5	0.2	20	18 (700 Hz)	100
8448	118	1.5	0.2	20	3 (80 kHz)	400
34368	29.1	1.5	0.15	100	10	800
139 264	7.18	1.5	0.075	200	10	3500

Notes
1. For the codirectional interface only.
2. The frequency values shown in parentheses apply only to certain national interfaces.

These limits should be met for all operating conditions and regardless of the amount of equipment preceding the interface.

In operational networks signals at an interface can contain jitter up to the maximum permissible network limit. This is important in the design of equipments incorporating jitter reducers where this jitter, together with any additional jitter generated in the system prior to the jitter reducer, needs to be accommodated.

4.3.1.2 Jitter limits appropriate to digital equipments

For individual digital equipments, the jitter performance is specified in three ways:

Jitter and wander tolerance of digital input ports (tolerable input jitter).
Maximum output jitter in the absence of input jitter (intrinsic jitter).
Jitter transfer characteristic.

JITTER AND WANDER TOLERANCE OF DIGITAL INPUT PORTS

Table 4.2 defines the network jitter limits for different bit rates. These limits are used to insure that, under the worst conditions, the input ports of equipments are capable of accommodating these levels of jitter when connected to any recommended hierarchical interface within the network.

MAXIMUM OUTPUT JITTER IN THE ABSENCE OF INPUT JITTER

As it is necessary to restrict the amount of jitter which an individual item of equipment can generate, CCITT Recommendations defining the maximum permissible levels of jitter generated in the absence of any input jitter are provided when dealing with specific systems. These limits never exceed the maximum permitted network limit.

JITTER TRANSFER CHARACTERISTIC

Jitter present at the input port of an equipment may in many cases be partially transmitted to the output port. In passing through the equipment the higher jitter frequencies are usually attenuated, whereas the lower-frequency jitter may not be. The jitter transfer characteristic provided in the CCITT recommendations take the form of a mask which indicates the permissible jitter gain versus frequency for particular items of equipment or systems.

4.3.1.3 Jitter limits for digital sections

To provide the jitter limits for digital sections it is necessary to introduce a hypothetical reference digital section (HRDS) whose lengths have been chosen to be 50 km or 280 km. These are representative of digital sections likely to be encountered in real operational networks, and are sufficiently long to permit a realistic performance specification for digital radio systems. The model is homogeneous in that it does not include other digital equipments such as muldexes. Figure 4.2 shows this HRDS, and the jitter specifications in Table 4.3 relate to it. Table 4.2 gives the lower limit of tolerable input jitter. The jitter transfer function recommended (CCITT Recommendation G921) limits the maximum gain to a value of 1 dB with the lower frequency limit of 5 Hz being acceptable. For line sections at 2048 kbit/s complying with the alternative national interface option as indicated by the figures

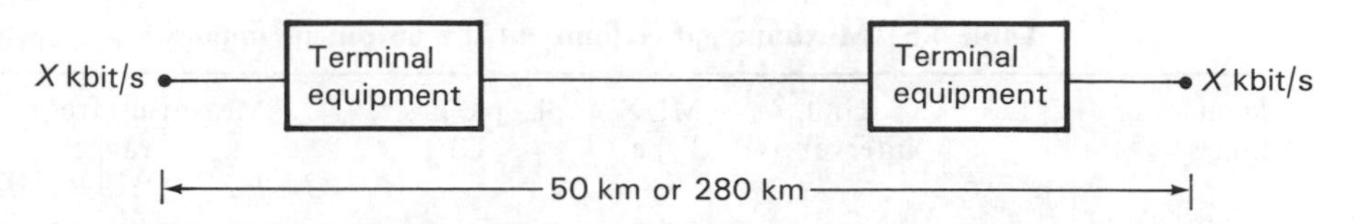

Figure 4.2 Hypothetical reference digital section (HRDS)

Table 4.3 The maximum output jitter in the absence of input jitter for a digital section up to the length of a HRDS (intrinsic jitter)

Bit rate, kbits/s	HRDS length, km	Maximum output pk–pk jitter for digital section up to the length of a HRDS: Low-frequency limit (f_1–f_4) UI	High-frequency limit (f_3–f_4) UI	Measurement filter bandwidth: Band-pass filter having a lower cutoff frequency f_1 or f_3 and an upper cut-off frequency f_4: f_1 Hz	f_3 kHz	f_4 kHz
2048	50	0.75	0.2	20	18 (700 Hz)	100
8448	50	0.75	0.2	20	3 (80 kHz)	400
34368	50	0.75	0.15	100	10	800
34368	280	0.75	0.15	100	10	800
139 264	280	0.75	0.075	200	10	3500

Note. For interfaces within national networks the values for f_3 shown in parentheses may be used.

in parentheses in Table 4.2, a jitter gain of 3 dB is acceptable. The output jitter in the absence of input jitter for any valid signal condition should not exceed the limit provided in Table 4.3.

4.3.1.4 Jitter limits for muldexes

The jitter transfer characteristic is given in Table 4.4 for the various hierarchical bit rates. The equivalent binary content of the test signal should be 1000. For digital muldexes the jitter limits are summarized in Table 4.5.

Table 4.4 Muldex jitter transfer characteristics

Digit rate, kbit/s	Jitter gain, dB	Frequency: Lowest	Frequency: Cut-off f_2	Roll-off after cut-off, dB/decade
2048 is under study				
8448 (P)	0.5	As low as possible	40 Hz	20
34368 (P)	0.5	As low as possible	100 Hz	20
139 264 (4 × 34368, P)	0.5	As low as possible	300 Hz	20
139 264 (16 × 8448, P)	0.5	As low as possible	100 Hz	20

P = Positive justification

Table 4.5 Maximum jitter limits at the output of muldexes

Digit rate, kbit/s	Unit interval *ns*	MUX Type*	pk–pk jitter, UI	Measuring frequency range f_1, Hz	to	f_4, kHz
2048 ($)	488	*S*	0.05	20		100
64 or 384 tributary is under study						
8448 ($)	118	*S*,*P*	0.05	20		400
2048 tributary (#)			0.05	18 kHz (700 Hz)		100
2048 tributary, no input jitter			0.25	up to		100
64 is under study						
34368 ($)	29.1	*P*	0.05	100		800
8448 tributary (#)			0.05	3 kHz (80 kHz)		400
8448 tributary, no input jitter			0.25	up to		400
139 264 ($)	7.8	*P*	0.05	200		3500
34368 tributary (#)			0.05	10 kHz		800
34368 tributary, no input jitter			0.3	up to		800

Notes
$ Transmitting timing signal is derived from an internal clock
* *S* = synchronous; *P* = positive justification
\# Use is made of a band-pass filter with a roll-off of 20 dB/decade. The jitter should not exceed the value tabulated with a probability of 99.9 percent during a measurement period of 10 s. The frequencies in parentheses indicate the national low-*Q* option as per CCITT G703

4.3.2 Limits of jitter and wander for the 1544 kbit/s digital hierarchy (CCITT Recommendation G824)[1]

The jitter control philosophy for this hierarchy is based on the need to have a maximum network limit for jitter which would meet the service quality on a 1544 kbit/s based network and to provide the jitter specification of individual digital equipments.

4.3.2.1 Network limits for the maximum output jitter at any hierarchical interface

Table 4.6 provides provisionally recommended values and is given for guidance.

4.3.2.2 Jitter and wander limits for digital equipments

It is assumed that digital networks are operating in the plesiochronous mode for international links and the plesiochronous or synchronous mode (which may be master–slave or mutual) for national networks.

Table 4.6 Maximum permissible jitter at hierarchial interfaces

Parameter value	Network limit of pk–pk jitter		Measurement filter bandwidth		
Digit rate of 1 UI, kbit/s	B_1 UI measured f_1–f_4	B_2 UI measured f_3–f_4	Band-pass filter having a lower cut-off frequency f_1 or f_3 and an upper cut-off frequency f_4. (B_1; f_1 & f_4, B_2; f_3 & f_4)		
			f_1, Hz	f_3, kHz	f_4, kHz
1544	2	0.05	10	8	40
*6312 (Note 1)	1	0.05	60	32	160
*6312 (Note 2)	2	0.05	10	24	120

*The intention is to arrive eventually at a single set of values for 6312 kbit/s interfaces.
Note 1. Represents values appropriate for networks using the 32064 kbit/s third hierarchical level.
Note 2. Represents values appropriate for networks using the 44736 kbit/s third hierarchical level.

DIGITAL EQUIPMENTS TERMINATING LINKS WITHIN A SYNCHRONIZED DIGITAL NETWORK

It is assumed that digital equipment provided at nodes will accommodate short-term clock variations together with jitter and wander in transmission plant such that, under normal synchronized conditions, slip will not occur. In the event of performance degradations, failure conditions, etc., the time-interval-error (TIE) between the incoming signal and the internal clock of the equipment may exceed the wander and the jitter tolerance of the equipment; then controlled slip with switches (telephone exchanges) and uncontrolled slip with an asynchronous multiplexer, will result. The level of jitter and wander which must be accommodated at the input to equipment without causing slips is given in Figure 4.3. Values for 6312 kbit/s and higher are not included in CCITT Recommendation G824. The value of 18 μs for A_0 represents

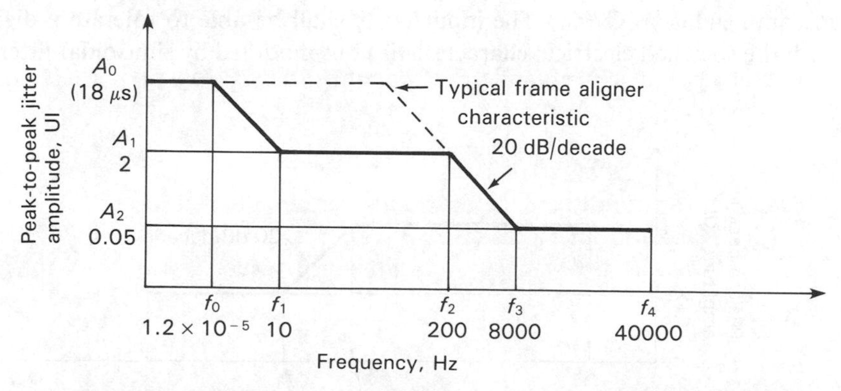

Figure 4.3 Mask of peak-to-peak jitter and wander which must be accommodated at the input of a node in a digital network (the digit rate in this case is 1544 kbit/s)

a relative phase deviation between the incoming signal and the internal timing local signal derived from the reference clock. This value for A_0 corresponds to an absolute value of 21 μs at the equipment input port (node) and assumes a maximum wander of the transmission link between the two nodes of 11 μs. The difference of 3 μs corresponds to the 3 μs allowed for long-term phase deviation in the national reference clock.

DIGITAL EQUIPMENT TERMINATING LINKS INTERCONNECTING INDEPENDENTLY SYNCHRONIZED NETWORKS

At nodes terminating links which are interconnecting independently synchronized networks, or where plesiochronous operation is used in national networks, the time-interval-error between the incoming signal and the internal clock of the equipment will eventually exceed the wander and the jitter tolerance of the equipment, and slip will thus occur. The maximum permissible long-term mean controlled slip rate is one slip in 70 days in this case.

4.3.2.3 Jitter limits appropriate to digital line sections

The CCITT Recommendations G911–G915 state that at the present time the following subjects for the 1544 kbit/s hierarchy are under study:

Lower limit of maximum tolerable jitter at the input
Limit of output jitter
Maximum output jitter in the absence of jitter at the input
Jitter transfer functions

4.3.2.4 Jitter limits for muldexes

SPECIFICATIONS AT THE INPUT PORTS

For all hierarchical levels up to level 4 (97728 kbit/s), the digital signal present at the input ports shall be in part as given in Table 5.5, modified by the transmission characteristic of the interconnecting cable. For a fuller specification refer to CCITT Recommendation G703.[1] The input ports shall be able to tolerate a digital signal with the specified electrical characteristics but modified by sinusoidal jitter up to the

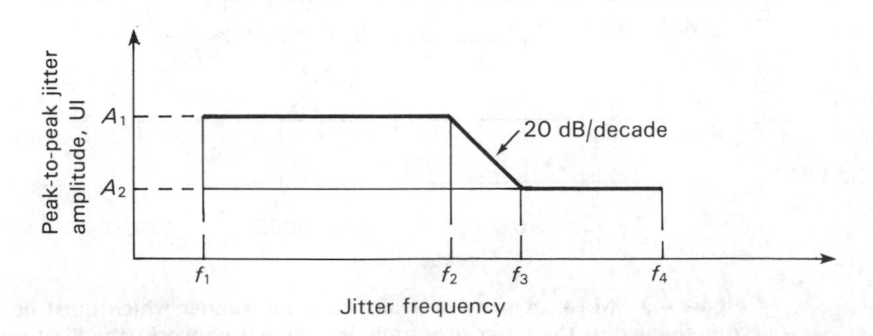

Figure 4.4 Mask for the jitter limits at the input ports of muldexes

Table 4.7 Lower limit of maximum tolerable input jitter

Input, kbit/s	A_1, UI	A_2, UI	f_1, Hz	f_2, Hz	f_3, kHz	f_4, kHz
1544	2	0.05	10	200	8	40
6312 (provisional)	8	0.05	10	200	32	160
6312	1	0.05	60	1600	32	160
32064 (provisional)	5	0.05	60	1600	160	800
6312	2	0.05	10	600	24	120
44736	14	0.05	10	3200	900	4500
32064 (provisional)	2	0.2	10	800	8	400
97728 (provisional)	6	0.1	10	5500	325	1000

limits specified by the amplitude-frequency relationship as shown in Figure 4.4. The equivalent binary content of the signal, with jitter modulation, applied to the inputs shall be a pseudo-random bit sequence of length $2^{15} - 1$. The signal with the jitter modulation applied to the demultiplexer input shall contain the bits necessary for framing and justification in addition to the information bits.

Table 4.7 shows the values of the lower limit of maximum tolerable jitter of various muldexes in the 1544 kbit/s hierarchy. The table should be used in conjunction with Figure 4.4, in order that the peak-to-peak jitter in unit intervals and various break-point frequencies can be interpreted.

MULTIPLEX SIGNAL OUTPUT JITTER

Table 4.8 provides the jitter limits at the output of the multiplexer. Also provided in this table is the demultiplexer output jitter with no multiplexer or demultiplexer input jitter.

DEMULTIPLEXER JITTER TRANSFER CHARACTERISTIC

The tributary signal modulated by sinusoidal jitter should be subject to a demulti-

Table 4.8 Multiplex output jitter and demultiplexer output jitter for no input jitter

Digit rate, kbit/s	RMS output jitter, UI	pk–pk tributary output jitter for no input jitter, UI
6312	0.01	
1544		0.333
32064	0.01	
6312		0.2
43736	0.01	
6312		0.2
97728	0.01	
32064		0.25

Table 4.9 Demultiplexer transfer characteristic

Tributary digit rate (output with jitter) kbit/s	Jitter gain at f_0, dB	First break frequency f_1, Hz	Second break frequency f_2, Hz	Roll-off $>f_1$, (dB/decade)	Roll-off $>f_2$, (dB/decade)
1544 (6312 input)	0.5	350	2500	40	20
6312 (32064 input)	0.5	65	–	20	–
6312 (44736 input)	0.1	500	2500	40	20
32064 (97728 input)	0.5	320	–	20	–

plexer jitter transfer characteristic within the gain limits given in Table 4.9. The frequency f_0, which is not specified, should be as low as possible taking into account the limitations of the measuring equipment.

4.4 JITTER MEASUREMENT METHODS[15,16]

In order that equipment and systems can be assessed for their jitter performance and compared against the limits provided in Section 4.3, a means of measuring jitter must be found. Before describing the test set-ups and the manner of testing, some equipment which may be used to make the actual measurements is briefly described in Section 4.4.1. Section 4.4.2 describes the testing methods commonly in use and in accordance with the CCITT Recommendations G823[1] and O151.[18]

4.4.1 Jitter measuring equipment

Each item of equipment discussed requires the use of a reference clock and a repetitive bit stream. This is not at variance with the measurement set-ups which use pseudo-random bit streams to induce some semblance of pattern-induced jitter, as the clock recovery circuits extract from this data bit stream the timing signal which contains on it the same jitter.

4.4.1.1 Oscilloscope

This is perhaps the simplest of all methods but can be used only to determine the peak-to-peak value of the phase jitter for repetitive sources such as the timing signal. It cannot be used for nonrepetitive input data streams due to the inability of the oscillosocope to trigger. The method involves the use of a reference clock signal which is at a submultiple of the input frequency, or is at the same frequency. This reference is used as the oscilloscope trigger input. The repetitive timing signal is entered on the oscilloscope input port and the oscilloscope adjusted for a steady display. The fuzziness on the edge of the displayed signal represents the peak-to-peak jitter, which can be determined by reference to the oscilloscope timebase settings. This method may be used to determine 'wander' by noting the movement of the timing signal leading or trailing edge over a long period of time, or by noting peak-to-peak time amplitude for slow variations. One disadvantage of this method is in its inability to determine the frequency components of any jitter measured. The frequency beneath which wander effects are considered to be predominant and taken into account in the design stage of equipment is f_1, as shown in Figure 4.3.

4.4.1.2 Calibrated phase detector[15]

One of the most useful instruments of this type which is commercially available is the Hewlett-Packard 'vector voltmeter'. Like the oscilloscope, it can be used only for clock timing signals or for repetitive bit streams, where the reference signal is either at the same rate or is a submultiple of the input bit stream. Basically, this type of detector compares the phase of the reference clock and the signal clock. Each input is processed prior to phase comparison in order to remove any amplitude variations and provide the proper waveshape. According to the design, the phase comparator may be an 'exclusive or' type of circuit or a sample-and-hold type of circuit, as used in the vector voltmeter. The output of the phase comparator is further processed and then fed to a meter, spectrum analyzer or chart recorder, each of which has been calibrated together with the instrument, to give a direct reading of the peak-to-peak jitter.

4.4.1.3 Digital processing oscilloscope[15–17]

This technique provides a time-and-amplitude-quantized histogram estimator for the probability density function of the transition times of a digital signal, which unlike many commercial jitter measuring techniques permits measurements at all hierarchial level and intermediate data rates. In addition this technique also provides a detailed description of the jitter phenomena which permits measurements to be made on either a single transition in a periodic data sequence, or on an ensemble of all transitions. This method also provides the means for separate identification and study of uncorrelated and pattern-dependent jitter contributions to be made. The jittered signal is displayed on the digital processing oscilloscope (DPO) using a jitter-free reference signal for the trigger input. By placing the maximum slope of the signal at the center of the screen, the signal is made to pass through a voltage window in the minimum time. Using high-speed plug-ins the time axis can be extended until the jitter on the signal just fills the whole display. Once this stage has been reached, the voltage sensitivity is decreased, thus creating a voltage window which takes up only part instead of the whole display as previously. The purpose of this voltage window is to provide amplitude quantization and thus determine the time interval in which points on the waveform fall within this window. If a weighting is given where any transition which occurs within the window equals 1 and outside the window equals 0 an estimate of the jitter pdf can be attained. This pdf is constructed by plotting its value after every m transitions. The discrete pdf values are obtained by accessing, after every m transitions, an n-word store which accumulates at each signal transition the number of transitions that occur within the window $N(t)$ and by dividing this value by the time taken since the experiment started, which is the total number of transitions multiplied by the time between transitions. In addition to the pdf the standard deviation or RMS jitter can also be determined. For a higher confidence factor, the pdfs for different parts of the input signal can be constructed as well as their RMS jitter values and the appropriate statistics applied. Additional experiments may be found to be necessary as the results are dependent on operator-setting for triggering and dot response optimization. This subjective element leads to difficulties in repeating results obtained.

4.4.1.4 Some commercially available test equipment

HEWLETT-PACKARD 3764A, DIGITAL TRANSMISSION ANALYZER
This equipment permits error measurements to 139.264 Mbit/s to CCITT Recommendation O151, and jitter generation and measurement to CCITT Recommendation O171 at 139 Mbit/s. Three main versions are, at the time of writing, available. In addition to other features, the following jitter measurements may be performed using this equipment:

- Maximum peak-to-peak output jitter
- Tolerance to input jitter
- Jitter transfer function
- Totalized jitter hit count
- Totalized jitter hit seconds count
- Totalized jitter hit-free seconds count

All jitter measurements are made simultaneously and the desired result is selected for display.

WANDEL & GOLTERMANN PF–4 BIT ERROR MEASURING SET AND PFJ–4 JITTER MODULATOR
Both these instruments when used together permit jitter investigations to be made on digital communications systems up to the 140 Mbit/s level.

In addition to the above two test equipment manufacturers, manufacturers which no doubt provide test equipment that would be suitable are Siemens, Marconi, etc.

4.4.2 Jitter measurement set-ups

Under normal operating conditions the undefined traffic signal may be considered as *quasi-random* and have some dependency on traffic loading. To identify abnormalities in the system under these conditions may take considerable time, especially if the system is new. The network limits given in Tables 4.2 and 4.6 are so derived that the probability of exceeding such jitter levels is very small; however, the practical observation time required to observe such levels with a high degree of confidence requires an unacceptable measurement interval. For laboratory, factory and commissioning purposes, measurements are made on the system using specific test sequences. The results of tests using these pseudo-random test sequences need to relate to an operational situation in which the information content is likely to be more random. Annex A of CCITT Recommendation G823[1] provides information on applying a suitable correction factor if the spectral content of a pseudo-random binary sequence (PRBS) within the bandwidth of the system is insufficient for a valid measurement of jitter to be made. In addition to the problems which may be encountered in using a pseudo-random binary sequence, the inclusion of additional signal processing devices integral to a transmission system often influences the observed jitter performance. The factors which have been found to be of influence are: the feedback connections on both the PRBS test signal generator and the transmission system's scrambler (see Section 4.5), the number of stages on the PRBS test signal generator and the scrambler and also the presence of a code convertor in the transmission system (see Section 5.1). Consequently, the following points should be considered when choosing a signal for equipment validation purposes: the PRBS test signal generator should not have a cycle length that has common factors with the transmission system scrambler; and the equal configuration of the PRBS test signal generator and the transmission system's scrambler should be avoided if a random signal is required.

The remainder of this section will consider the use of a pseudo-random bit sequence in the testing of a digital system. The aim of the measurement set-ups is to permit a set of tests to be performed which are in alignment with the jitter limits discussed in Section 4.3. Most tests involve simulating jitter. This is done by using a single frequency sinusoid which modulates either a clock generator that in turn drives a pattern generator, or the pattern generator directly. The modulating sinuosoid frequency can then be altered to determine the response of equipment to this jitter at different frequencies. Thus a range of sinusoidal modulating frequencies may be used to simulate a complex frequency jitter signal. The pattern generator is used to simulate as closely as practicable a random bit stream. In practice, however, the bit stream is not random but pseudo-random. The main reason for this is to induce pattern-dependent jitter since this is the greatest jitter contributor in a network. Once the data bit stream containing the pseudo-random pattern is received, clock extraction circuits recover the timing signal. This timing signal contains on it the same (if alignment jitter is negligible) absolute jitter amplitude as the data bit stream, and by means of the jitter detector and appropriate band-pass filters this jitter amplitude can be measured for a specific jitter frequency.

4.4.2.1 CCITT recommended test set-ups[1,18]

Figure 4.5(*a*) shows the test set-up for measuring timing jitter, which is derived[18] from Recommendations O171, and Figure 4.5(*b*) shows the test set-up for measuring output jitter from a hierarchial interface or an equipment output port (G823).[1]

TEST SIGNAL SOURCE

Figure 4.5(*b*) is effectively the jitter measuring circuit shown in the test set-up of Figure 4.5(*a*). The tests made on digital equipment may be made with either a jittered or non-jittered digital signal. To transmit this signal a pattern generator, clock generator and modulation source as shown in Figure 4.5(*a*) are required. The modulation source may be provided within the clock generator and/or pattern generator or it may be provided separately as shown. The clock generator is able

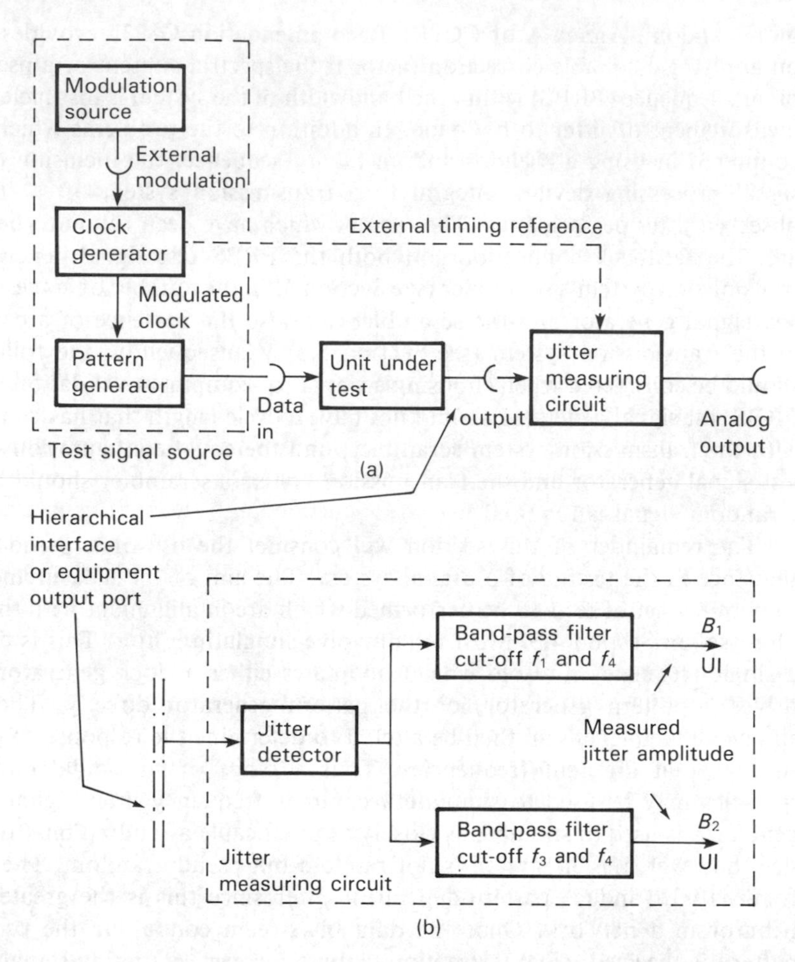

Figure 4.5 CCITT measurement arrangements for jitter

to be phase modulated from the modulation source and the peak-to-peak deviation of the modulated signal indicated. The clock generator outputs comprising the modulated clock signal and a timing reference signal are required to be not less than 1 volt peak-to-peak into 75 ohms. The modulated clock signal is used to drive a pattern generator which usually is able to provide a frame alignment signal and justification control bits if the pattern test signal is to enter the input of a digital demultiplexer. For use at digit rates of 64 kbit/s, the pattern generator is to provide a pseudo-random pattern of $(2^{11} - 1)$ bit length in accordance with CCITT Recommendation O152,[18] and encoded as discussed in Section 5.1.7. For levels 1,2 and 3 of both hierarchies, as shown in Figure 3.11, except for 34368 kbit/s, a pseudo-random pattern of $(2^{15} - 1)$ bit length is recommended. For 34368 kbit/s and 139 264 kbit/s, a pseudo-random pattern of $(2^{23} - 1)$ bit length is recommended. Both of these pseudo-random patterns are generated in accordance with Recommendation O151. In addition to these patterns, a 1000 1000 repetitive pattern, two freely programmable 8-bit patterns capable of being alternated at a low rate such as between 10 Hz and 100 Hz, or a freely programmable 16-bit pattern, is available from the pattern generator.

It is normally assumed that live traffic on transmission systems consists of essentially random digit sequences and, for this reason, test patterns normally consist of pseudo-random binary sequences (PRBS), suitably encoded and of a length adequate to give a reasonable distribution of power within the bandwidths of any timing recovery filters in the system. Preliminary studies on the relationships between test-pattern jitter and live-traffic jitter have revealed that jitter arising from the use of pseudo-random test sequences is dependent not only upon the sequence length, but also upon which of the set of allowable generator configurations (see Section 4.5.1) is chosen for a given sequence length. It is therefore possible that, for a given sequence length, some generator configurations will be better than others for live traffic simulations. The test pattern sequences most often used are pseudo-random binary maximal length [$(2^{15}-1)$bits], encoded into the appropriate interface code (see Section 5.1.7). Every sequence of $(2^n - 1)$ bits contains $(0.5)\times(2^n) - 1$ zeros and $(0.5)\times(2^n)$ ones. Every complete sequence will also contain all possible n-bit combinations of ones and zeros except the n-zeros combination and the $(n-1)$-ones combination. Thus, for moderate values of n, a sequence generated will for practical purposes approximate to a random binary sequence. The 2^{15} sequence frequently used for testing 2 and 8 Mbit/s systems does on some sequences[19] exhibit a tendency for bunching of the longer runs of zeros and even after HDB3 coding the effects of this bunching can be detected on a regenerative repeater. On live systems, traffic jitter usually varies from day to day, with changes in the traffic carried.

JITTER MEASURING CIRCUIT

In addition to measuring the peak-to-peak jitter, the measuring circuit is able to count the number of occasions and the period of time for which a given selectable threshold of jitter is exceeded either by using an internal or external counter. As an option on the equipment, the RMS jitter may be measured up to 3.0 UI at jitter frequencies up to f_2, as shown in Figure 4.4, and 0.15 UI from f_3 to f_4. In cases where this option is not available, the analog output can be used to make RMS measurements with an external meter. As shown in Figure 4.5(*b*) the basic jitter measuring circuit contains filters to limit the band of the jitter frequencies. The band of frequencies vary according to the bit rate and are given in Table 4.3. Recommendation O171 provides additional details on the high-pass filter cut-off frequencies and jitter measurement bandwidths. The reference timing signal for the phase detector is derived in the jitter measuring circuits from any input pattern for the case of end-to-end tests, but for back-to-back or loop measurements it may be derived from a suitable clock source.

As discussed in Section 4.4.1.4, equipment which will perform tests according to the arrangements given in Figure 4.5 is commercially available. One such test set is the Hewlett Packard 3764A digital transmission analyzer. When fitted with the appropriate options, this test set will generate jitter on bit rates of 2, 8 and 34 Mbit/s (Option 001), and 140 Mbits/s (Option 002), with the appropriate line interface code as recommended by the CCITT and provided in Table 5.5 and with the recommended CCITT pseudo-random binary sequence according to the bit rate, as given in Table 4.10. Used as a receiver, this instrument can determine the maximum peak-to-peak jitter, tolerance of equipment to input jitter, jitter transfer function and totalized jitter hit measurements. In testing a radio link and the associated multiplex equipment, to determine the measured value of maximum tolerable input jitter, two such sets of equipment are required: one at the sending end to transmit jittered data into the system, and the other at the receiving end which is used to measure the value of the jittered data received.

4.4.3 Jitter measurements[20-22]

System components are often designed to tolerate specific levels of jitter over designed frequency ranges; if the jitter on a digital signal exceeds these limits, transmission errors will occur. Recording the jitter amplitude and marking the instances when jitter exceeds a specific threshold and then correlating these instances against system performance can provide an insight into the effects of jitter. In some applications it is useful to consider a frequency-weighted jitter measurement, since systems typically tolerate more jitter amplitude when the frequency components of the jitter are at low frequencies than when they are at high frequencies. These tolerance characteristics and the jitter transfer functions of system components are also important parameters that can assist in the development of an error-free performance. In longer digital systems that are interconnected through repeaters the tolerance of a repeater input to jitter amplitude excursions is a function of the jitter frequency. If the system performance is to be correlated against this jitter frequency, it may not be possible to obtain an assessment of this performance due to the lack of information on the jitter frequencies. The frequency-weighted unit of jitter measurement (*wjits*) may be used;[22] it represents a single numerical representation of the jitter magnitude that is proportional to the detrimental effect on the system caused by the jitter, regardless of the frequency content of the jitter.

Three main measurements have to be performed with jitter test instruments on a digital system or digital equipment which comprise the system, to provide an effective means of controlling jitter. These measurements are:

The determination of the 'intrinsic jitter' which is the output jitter in the case of zero input jitter
The determination of the 'jitter transfer function'
The measurement of the 'maximum tolerable input jitter'

The basic test equipment as mentioned above consists of a pattern generator with a jitter modulation facility provided, a jitter meter and a bit error ratio detector. The device under test can be a system component such as a piece of multiplex equipment, a regenerator or a complete transmission system. Intrinsic jitter is measured with the jitter modulation on the test equipment turned off. To determine the jitter transfer function, the generator is sinusoidally phase modulated with different frequencies and with a tolerable jitter amplitude. The maximum tolerable input jitter is determined with the aid of the bit error ratio detector. The input jitter is increased until bit errors occur. This is repeated for different jitter frequencies. Normally a pseudo-random binary sequence is used as specified in Table 4.10 for all three measurements.

4.4.3.1 The testing of the jitter limits appropriate to digital equipments for the 2048 kbit/s digital hierarchy

As discussed in Section 4.3.1.2, Table 4.2 defines the network jitter limits for the different bit rates at the input ports of equipments; for convenience of testing, sinusoidal jitter is used to modulate a test pattern. The amount of jitter which should be used and which should not cause any degradation in the operation of the equipment, when superimposed on a signal with a line code, impedance, level, etc., as defined in Table 5.5, is given in terms of its amplitude and frequency in Table 4.10 and Figure 4.3. This test condition is not, in itself, intended to be representative of the type of jitter to be found in practice in a network, since the pseudo-random test pattern may in no way be representative of the different jitter-producing patterns which may occur. The test does, however, insure that the Q-factor associated with the timing signal recovery of the equipment's input circuitry is not excessive and, where necessary, an adequate amount of buffer storage has been provided.

Table 4.10 Sinusoidal test jitter amplitude/frequency characteristics at digital input

Digit rate, kbit/s	Pk–pk amplitude			Frequency					Pseudo-random test sequence
	A_0 UI	A_1 UI	A_2 UI	f_0 Hz	f_1 Hz	f_2 Hz	f_3 kHz	f_4 kHz	
				$\times 10^{-5}$					
64 (Note 1)	1.15(18 μs)	0.25	0.05	1.2	20	600	3	20	$2^{11}-1$
2048	36.9 (18 μs)	1.5	0.2	1.2	20	2400(93)	18(0.7)	100	$2^{15}-1$
8448	152 (18 μs)	1.5	0.2	1.2	20	400(10700)	3(80)	400	$2^{15}-1$
34368	–	1.5	0.15	–	100	1000	10	800	$2^{23}-1$
139 264	–	1.5	0.075	–	200	500	10	3500	$2^{23}-1$

Note 1
For codirectional interface only.
Note 2
For interfaces within national networks the values of f_2 and f_3 shown in parentheses may be used.
Note 3
(18 μs) represents relative phase deviation between the incoming signal and the internal timing local signal derived from the reference clock. This value for A_0 corresponds to an absolute value of 21 μs at the equipment input port and assumes a maximum wander of the transmission link between two nodes (equipment input ports) of 11 μs. The difference of 3 μs corresponds to the 3 μs allowed for long-term phase deviation in the national reference clock.
Note 4
Table 4.10 is to be used in conjunction with the designated jitter amplitudes A_0–A_2, and frequencies f_0–f_4 of Figure 4.3.

4.4.3.2 The level of output jitter in the absence of input jitter

Tables 4.3, 4.5 and 4.8 provide details of the recommended masks at various bit rates for the jitter limits at the output of a multiplexer with no jitter entering the input. The test method is to record selectively the value of the jitter at the output of the jitter receiver with no modulation on the jitter transmitter. The configuration of the test set up is shown in Figure 4.6. In systems involving regenerators or repeaters, the results will be dependent on the type of traffic being transmitted, since, if the pattern has a repetition rate falling within the passband of the clock extraction circuit, the jitter measured will contain a systematic component. For shorter patterns the jitter will be comprised entirely of random sources.

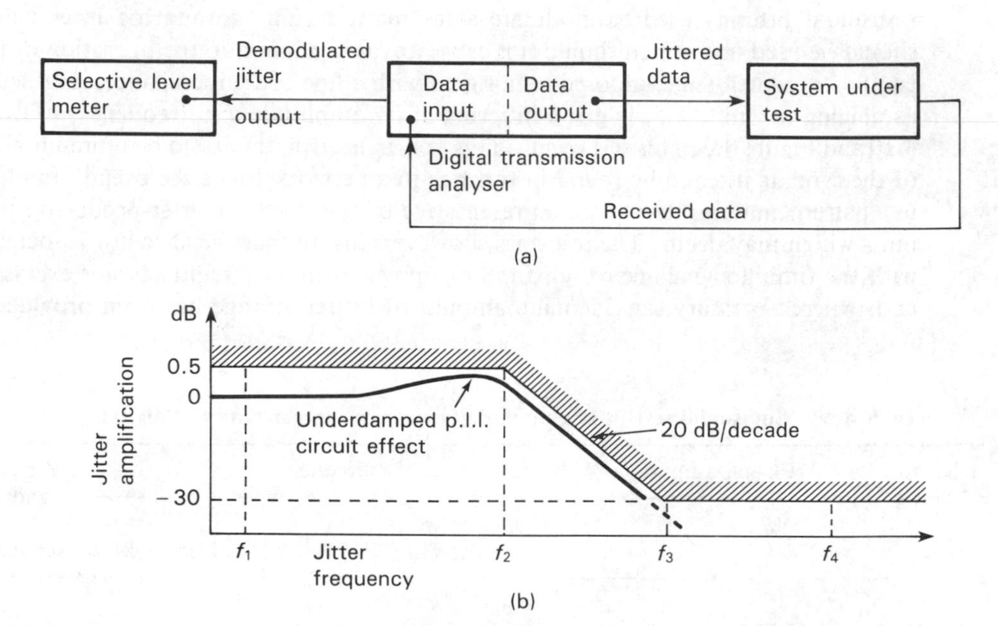

Figure 4.6 Test set-up: (a) for measuring jitter gain, and (b) expected results

4.4.3.3 Jitter transfer characteristic

The jitter transfer function is a measure of the gain or loss in jitter amplitude over a range of jitter frequencies. This is a very important specification for the control of jitter in cascaded digital equipments and systems within an overall network. Jitter gain is also an indication of the accumulated jitter introduced when systems are cascaded, or when the jitter contribution is from a single piece of equipment. This gain must be controlled and kept to a minimum, even at the lower jitter frequencies, and the jitter attenuated as much as possible if data errors due to timing jitter are to be avoided. The jitter transfer function is usually expressed as:

$$\text{Jitter gain} = 20 \log [(\text{Jitter measured at a system's output port}) / (\text{Jitter measured at a system's input port})] \quad (4.4)$$

The jitter measured at a system's output may be either wideband or selective. If selective, the demodulated output of the jitter receiver is to be measured at selected frequencies using a selective level meter. The results of the jitter gain for various multiplex equipments operating at specific bit rates are recommended to be no greater than that provided by the masks given in tabular form in Table 4.4 or Table 4.9. A spectrum analyzer may also be used to quantify the demodulated jitter output by providing the jitter spectral density. Figure 4.6 shows the test set-up for obtaining the jitter transfer function.

4.4.3.4 Measurement of jitter tolerance

The maximum tolerable input jitter is the level of timing jitter at which a system under test begins to introduce errors. This really means that the maximum jitter tolerance of an equipment input port is a measure of the ability of a digital equipment to accept jitter from a preceding equipment within a digital network without producing errors. As provided in Table 4.2 and Table 4.6 the lower limit of maximum tolerable input jitter is specified. This limit is a measure of the minimum jitter tolerance that all equipment must exhibit without introducing errors. The measured value of maximum tolerable input jitter is the level of input jitter at which a system under test begins to introduce errors. The tolerance to input jitter which is defined in terms of a 'lower limit of maximum tolerable jitter' which must be met, infers that the 'upper limit' is the actual tolerance of a particular digital equipment which is design-dependent. The test set-up is such that data with jitter on it are transmitted

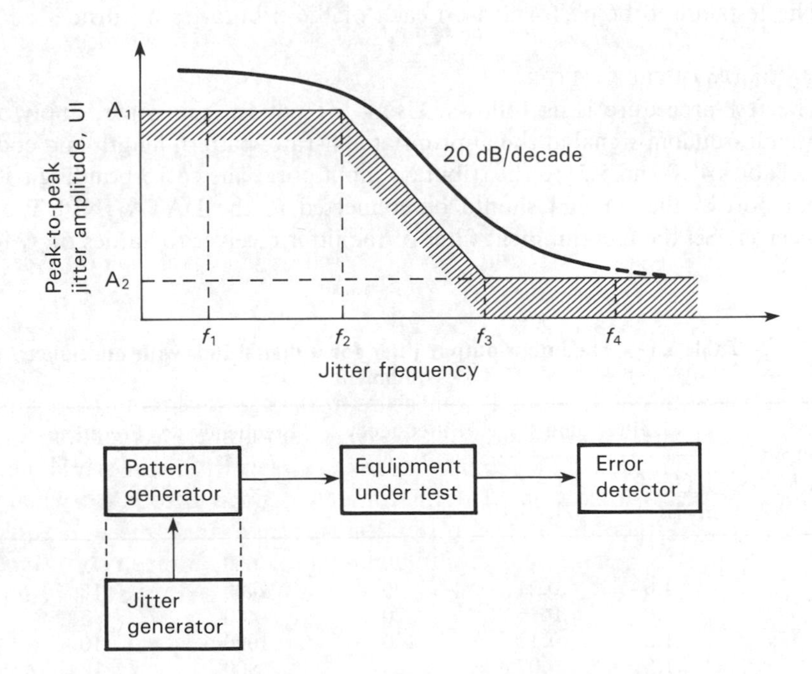

Figure 4.7 Test set-up and expected results for input jitter tolerance measurement

into the input of the system under test. The level of this jitter is then increased up to the level for the network limit for the mask frequencies as given in Table 4.10 and Figure 4.3. At the system output, readings are taken until errors begin to appear. The level of input jitter at which this occurs is then noted and should be below or equal to the network limit as defined by the mask values given in Table 4.2 and 4.6. It should be noted that the system mentioned includes digital sections operating at the specified bit rates besides equipment, except for the 1544 kbit/s hierarchy (Table 4.6) which is still under study. Figure 4.7 shows the test set-up and the expected results when compared against the relevant mask.

4.4.3.5 Multiplex system end-to-end jitter tests

In a multiplex end-to-end test, the majority of tests performed to demonstrate that the system performance conforms to expected parameters are jitter tests. The tests are usually performed at the digital distribution frame (DDF). The test equipment required at each end of the link or system is:

An oscilloscope terminated in 75 ohms to test that the waveform at the high bit rate input to the multiplex (MUX IN) conforms to CCITT Recommendation 703 waveform masks.[1]
A pattern generator which will operate at the multiplex bit rate.
An error detector.
A jitter generator and jitter detector.

The tests are to be performed on each of the tributaries in turn.

MAXIMUM OUTPUT JITTER

The test procedure is as follows. Using the pattern generator, apply a jitter-free pseudo-random signal at the appropriate bit-rate, pattern length and code (as given in Tables 4.10 and 5.5) to the tributary input port. The corresponding tributary output port at the far end should be connected to the DATA INPUT of the jitter receiver. Set the measurement filter of the jitter receiver to values of f_1 for the high

Table 4.11 Maximum output jitter for a digital link with multiplex equipment

Tributary input/output bit rate, kbit/s	Jitter gain		Frequency f_1, Hz	Frequency f_2, Hz	Frequency f_3, kHz	Frequency f_4, kHz
	(f_1-f_2) A_1, UI	(f_3-f_4) A_2, UI				
64	–	–	20	600	3	20
2048	1.0	0.21	20	2400	18	100
8448	1.0	0.16	20	400	3	400
34368	1.2	0.12	100	1000	10	800
139 264	1.5	0.075	200	500	10	3500

Note. Refer to Figure 4.4 for general mask to which Table 4.11 relates.

pass, and f_3 for the low pass and measure the peak-to-peak jitter. Repeat for the f_3 high-pass filter and the f_4 low-pass filter and record the peak-to-peak jitter. Values of f_1, f_3 and f_4 are shown in Table 4.10 and Figure 4.4 against the bit rates of the tributaries. As the expected values which comprise the mask are derived from that jitter arising from a digital section and from the multiplex equipment, Table 4.11 shows the expected values. These values are only approximate in the absence of any formal recommendations, but should provide an indication of what is expected.

LOWER LIMIT OF MAXIMUM TOLERABLE INPUT JITTER: TEST PROCEDURE

To the external clock input of the pattern generator, at the tributary bit rate, apply a jittered clock signal from the jitter generator. Connect the data output of the pattern generator to the TRIB IN point. Set the data output pattern length to that indicated in Table 4.10 and the output code as indicated by Table 5.5. Using Table 4.10 or 4.7, together with Figure 4.4, check that no errors are received on the error detector test equipment at the receiving end, for values of jitter just at or above the mask. See Figure 4.7 for the test set-up and the expected results. Repeat the test for the other direction of transmission.

JITTER TRANSFER CHARACTERISTIC: TEST PROCEDURE

Apply a repetitive pattern of 1000, jittered by a manually selected frequency. Insert this signal into the TRIB IN point in the multiplex equipment. At the distant end, measure the jitter received at the spot frequencies. Compare the results using equation 4.4. The results should lie within the mask as determined by Table 4.4 and 4.6, and the frequencies used in the mask and the measurement are the same as those used in determining the lower limit of tolerable input jitter. Above the frequency f_3, it is expected that the jitter gain levels off to -30 dB.

ERROR RATE MEASUREMENT: TEST PROCEDURE

Using the pattern generator set-up as for the lower limit of maximum tolerable input jitter measurement, connect, at the receiving end, the error rate measuring set and confirm that no errors are received at the TRIB OUT point within 20 minutes. Repeat for each tributary and for each direction of transmission. Looping back at one end and undertaking the test from the other is acceptable.

4.5 METHODS OF MINIMIZING JITTER

Two basic methods are commonly used. The first is to take steps to prevent the generation and systematic accumulation of the jitter by the use of scramblers. These devices effectively cause the signal to become random and so reduce the effect of any pattern-dependent jitter-causing mechanisms. The second method is to reduce the magnitude of the jitter already present by using a retiming circuit which has a bandwidth which is less than that of the signal bandwidth. This circuit is known as a 'jitter reducer'. As mentioned above, the jitter frequencies which are below the jitter reducer's cut-off frequency are not reduced and may cause, in some circuits, an uncontrolled slip, due to the jitter amplitude at these frequencies accumulating and becoming large enough to affect digital equipments which are not transparent to jitter. The choice of buffer stores with sufficient capacity at the input of digital equipments in the practical situation does, however, much alleviate this problem.

4.5.1 Scramblers[23]

Before discussing scramblers, the advantages of using them as an integral part of a digital transmission system are listed. These advantages are that they:

Insure that the accumulation of jitter is not correlated with the signal.
Permit equation 4.1 to apply in the interconnection of digital sections.
Reduce the effects of low frequency jitter accumulation.
Reduce the effects of crosstalk levels on symmetric-pair cable produced by synchronous systems, by suppressing discrete spectral components of periodic bit patterns.

One disadvantage of using them is that they cause error extension effects. That is, any transmission error which under nonscrambled conditions would cause only one error in a system, will when the system is in the scrambled condition cause further errors. This is due to the way the scrambler operates and to the use that it makes of a feedback path. The feedback path, as discussed in Section 4.5.1.1, feeds any error appearing on it back into the scrambling circuit's input, thus compounding the error.

4.5.1.1 The basic scrambler

The basic form of a data scrambler and descrambler is shown in Figure 4.8. The incoming binary signal is added to a second binary signal in a modulo-2 adder which takes the form of an exclusive-OR gate, having the Boolean expression $(\bar{X}Y + X\bar{Y})$. This second binary signal is also derived from a modulo-2 operation on the delayed version of the gate output. In the analysis of these devices an operator $1/x$ is introduced, which denotes the effect of delaying a bit sequence by one bit. If D_i is the input sequence, then $D_i \cdot (1/x) \cdot (1/x)$ means that the input sequence has been delayed by two bits, as $(1/x) \cdot (1/x) = (1/x^2) = x^{-2}$.

Analyzing the descrambler of Figure 4.8, $D_i' = D_s' \oplus (1/x^3)D_s' \oplus (1/x^5)D_s' = D_s'(1 \oplus x^{-3} \oplus x^{-5})$. In the scrambler D_s is operated on by $x^{-3} \oplus x^{-5}$ and added to D_i. That is $D_s = D_i \oplus D_s(x^{-3} \oplus x^{-5})$ or $D_i = D_s(1 \oplus x^{-3} \oplus x^{-5})$ and so,

$$D_s = D_i/(1 \oplus x^{-3} \oplus x^{-5}) \tag{4.5}$$

where $\oplus$ denotes the 'exclusive OR' operation and where division (indicated by /) stands for the inverse operation. In the absence of errors the transmitted D_s equals the received D_s' and hence the unscrambled output D_i' becomes:

$$D_i' = D_s'(1 \oplus x^{-3} \oplus x^{-5}) = D_i(1 \oplus x^{-3} \oplus x^{-5})/(1 \oplus x^{-3} \oplus x^{-5}) = D_i \tag{4.6}$$

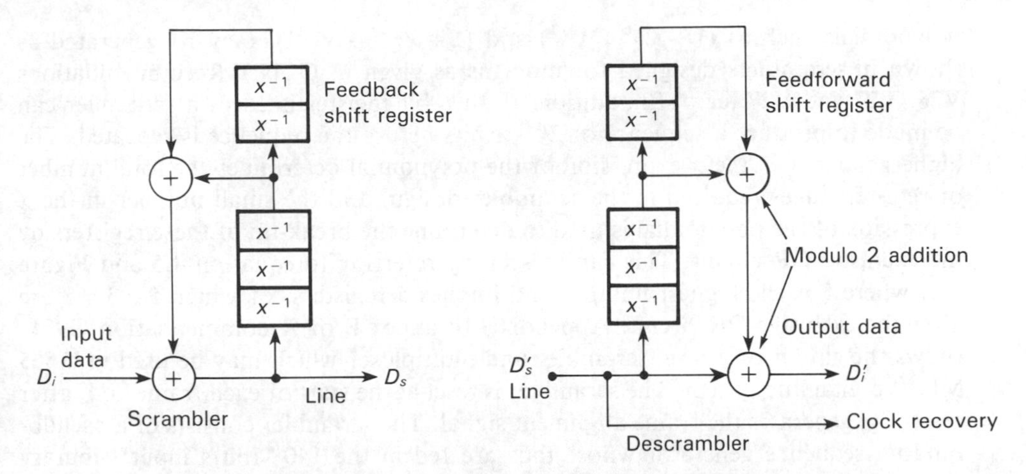

Figure 4.8 The basic scrambler and descrambler

Thus the input sequence is exactly duplicated at the descrambler output in the absence of transmission channel errors. The polynomial in x is known as the *generating polynomial* and is important in describing pseudo-random generators as well as scramblers. To demonstrate how the scrambler affects period sequences and long strings of ones or zeros, consider the input sequence as shown in Table 4.12, and assume that the initial content of the register is '1's (as for an AIS signal). Note the periodic sequences in the input corresponding to 10101010 and the effect of the scrambler in breaking these sequences up before transmission to line. As can be seen from Table 4.12, even for this simple scrambler long sequences of '1's and '0's are broken up and thus any pattern removed: as for 1010..., or seven '0's followed by ten '1's, followed by seven '0's, followed by ten '1's.... The scrambler thus produces a pseudo-random sequence. With an appropriate choice of taps, various

Table 4.12 Example of scrambler with polynomial $1 \oplus x^{-3} \oplus x^{-5}$

Initially all '1's	
Input D_i to scrambler	1 0 1 0 1 0 1 0 1 0 1 0 0 0 0 0 0 0 1 1 1 1 1 1 1 1 1 1 0 0 0 0 0 0 0 1 1 1 1 1 1 1 1 1 1
$D_s \cdot x^{-3}$	1 1 1 1 0 1 0 0 0 1 1 1 1 0 0 0 1 1 0 1 1 0 1 0 0 1 1 0 0 0 0 1 0 0 1 0 1 1 1 1 0 1 1 0 1
$D_s \cdot x^{-5}$	1 1 1 1 1 1 0 1 0 0 0 1 1 1 1 0 0 0 1 1 0 1 1 0 1 0 0 1 1 0 0 0 0 1 0 0 1 0 1 1 1 1 0 1 1
Output D_s to line and input to descrambler D_s'	1 0 1 0 0 0 1 1 1 1 0 0 0 1 1 0 1 1 0 1 0 0 1 1 0 0 0 0 1 0 0 1 0 1 1 1 1 0 1 1 0 1 0 0 1
Initially all '1's	
$D_s \cdot x^{-3}$	1 1 1 1 0 1 0 0 0 1 1 1 1 0 0 0 1 1 0 1 1 0 1 0 0 1 1 0 0 0 0 1 0 0 1 0 1 1 1 1 0 1 1 0 1
$D_s \cdot x^{-5}$	1 1 1 1 1 1 0 1 0 0 0 1 1 1 1 0 0 0 1 1 0 1 1 0 1 0 0 1 1 0 0 0 0 1 0 0 1 0 1 1 1 1 0 1 1
Output D_s' from descrambler	1 0 1 0 1 0 1 0 1 0 1 0 0 0 0 0 0 0 1 1 1 1 1 1 1 1 1 1 0 0 0 0 0 0 0 1 1 1 1 1 1 1 1 1 1

polynomials such as $(1 + x^{-6} + x^{-7})$ and $(1 + x^{-18} + x^{-23})$, may be generated as shown in scramblers designed for modems as given in CCITT Recommendations V26, V27 and V27 ter.[24] In addition, if an n-bit register is used, a scrambler can be made to produce a sequence of $2^n - 1$ bits before the sequence is repeated. The highest number in the x expression of the polynomial determines the total number of register stages required in the scrambler design, and the small number in the x expression of the polynomial is used to determine the break-up of these registers by the exclusive-OR circuit. This can be seen by referring to equation 4.5 and Figure 4.8, where 5 is the highest number and implies 5 registers, of which $5 - 3 = 2$ are after the exclusive-OR circuit. Appendix I to annex B of Recommendation G954[1] shows the circuit of a reset scrambler and multiplexer which may be used in a 565 Mbit/s digital line system. The scrambler is reset at the start of each frame and, after resetting, generates the frame alignment signal. The scrambler consists of a pseudo-random sequence generator whose taps are fed in the 140 Mbit/s input tributary circuits prior to multiplexing. The advantages of such a scrambler[25] over a free-running or self-synchronized type are that there is no error multiplication or necessity to provide additional measures to avoid periodic output signals.

REFERENCES

1. 'Digital Networks – Transmission Systems and Multiplexing Equipments', Recommendations G700–G956 (Study Groups XV and XVIII), CCITT Red Book, **III–3**, VIII Plenary Assembly, 1984.
2. Bylanski, P. and Ingram, D. G. W., *Digital Transmission Systems*, IEE Telecommunications Series 4, Rev. Ed. (Peter Peregrinus, 1980).
3. McLintock, R. W., 'Overview of International Standards for Transmission Impairments affecting Digital Telecommunications Networks', *Telecommunication Transmission*, IEE Conference Publication No. 246, 1985, pp. 19–22.
4. Duttweiler, D. L., 'Waiting Time Jitter', *Bell System Technical Journal*, 1965, **44**, No. 9, pp. 1843–1886.
5. Duttweiler, D. L., 'The Generation and Accumulation of Timing Noise in PCM Systems', *ibid.*, 1972, **51**, No. 1, pp. 165–207.
6. Frangou, G. J. and Patel, N., 'Jitter Accumulation in Digital Networks', *Measurements for Telecommunication Transmission Systems*, IEE Conference Publication No. 256, 1985, pp. 43–47.
7. Bennett, W. R., 'Statistics of Regenerative Digital Transmission', *Bell System Technical Journal*, 1958, **37**, pp. 1501–1542.
8. Rowe, H. E., 'Timing in a Long Chain of Regenerative Binary Repeaters', *ibid.*, 1958, **37**, pp. 1543–1598.
9. Manley, J. M., 'The Generation and Accumulation of Timing Noise in PCM Systems – An Experimental and Theoretical Study', *ibid.*, 1969, **48**, pp. 541–613.
10. Byrne, C. J., Karafin B. J. and Robinson, D. B., 'Systematic Jitter in a Chain of Digital Regenerators', *ibid.*, 1963, **42**, pp. 2679–2714.
11. Zeger, L. E., 'The Reduction of Systematic Jitter in a Transmission Chain with Digital Regenerators', *IEEE Transactions on Communications Technology*, 1967, **COM–14**, No. 4.
12. Hirosakai, B., 'Jitter Accumulation Property in a Regenerative Repeater System with ASWF as a Timing Extracting Filter', *NEC Research and Development*, 1976, No. 43.

13. Cock, C. C., 'Jitter Tolerances in Digital Equipment', *IEE Colloquium Digest*, 1977, No. 1977/45 ('Jitter in Digital Communication Systems').
14. Cock, C. C., Edwards, A. K. and Jessop, A., 'Timing Jitter in Digital Line Systems', *IEE Conference Proceedings*, 1975, No. 131.
15. Hill, S. E. and Butler, D. J., 'Jitter Measurements on a 324 MBaud Optical System', *Measurements for Telecommunication Transmission Systems*, IEE Conference Publication No. 256, 1985, pp. 38–42.
16. Cochrane, P., Barley, I. W. and O'Reilly, J. J., 'Direct Jitter Measurement Technique', *IEE Electronic Letters* **15**, No. 24, pp. 774–775, 1979.
17. Rogerson, S. P. and Sendell, A. A., 'Jitter Measurement using a Digital Processing Oscilloscope', *Measurements for Telecommunication Transmission Systems*, IEE Conference Publication No. 256, 1985, pp. 33–37.
18. 'Specifications of Measuring Equipment', Recommendations O series (Study Group IV), CCITT Red Book, **IV–4**, VIII Plenary Assembly, 1984.
19. Cock, C. C., 'The Use of Pseudo-Random Sequences for Jitter Assessment in Digital Communication Networks' *Telecommunication Transmission*, IEE Conference Publication No. 246, 1985, pp. 61–64.
20. Hoyer, W., 'Jitter Measurements at Bit Rates up to 167 kbit/s', *Measurements for Telecommunication Transmission Systems*, IEE Conference Publication No. 256, 1985, pp. 166–170.
21. Thomson, R. D., Huckett, E. P. and Hillery, P. J., 'Design and Application of Jitter Measuring Instrumentation', *ibid.*, pp. 171–175.
22. Rollins, W. W. and Adams, J. G., 'Jitter and Digital Transmission Systems', *ibid.*, pp. 181–186.
23. Savage, J. E., 'Some Simple Self Synchronizing Digital Data Scramblers', *Bell System Technical Journal*, 1967, **46**, pp. 449–497.
24. 'Data Communication over the Telephone Network', Recommendations V series (Study Group XVII), CCITT Red Book, **VIII–1**, VIII Plenary Assembly, 1984.
25. Muller, H., 'Bit Sequence Independence through Scramblers in Digital Communication Systems', *Nachr. Techn. Z.*, 1974, **27**, pp. 475–479.

5 BASEBAND SIGNAL PROCESSING

Although baseband signal processing is mainly required if the communication channel is comprised of either a coaxial cable or a symmetric-pair cable, it is also required for carrier-borne transmission methods where the signal is brought down to baseband at repeater sites. In a digital radio system, at a repeater site, there is need for filtering, equalization and regeneration. Line codes, however, are required to convert the binary signals from the multiplex equipment, for transmission to the radio equipment in order that the transmission channel is matched to reduce the errors which would otherwise be incurred.

CCIR Report 938 deals with the baseband interconnection of digital radio-relay systems.[13] Referring to Figure 5.1, the point TT′ or interconnection of digital radio sections at baseband frequency should be chosen in conformity with this scheme, and have signal characteristics which are those given in Table 5.5. Due to the addition of overhead bits, and to such operations as code conversion and insertion of justification bits, parity bits and service bits, the bit rate within a digital radio section may be different from the CCITT recommended hierarchial level, as provided in Figure 3.11. The CCIR has taken these factors into account and the occurrence of a nonhierarchial bit rate does not provide any conflict. For systems based on the 2048 kbit/s hierarchy, the signals and signal levels are defined at the input and output ports of the corresponding equipment in accordance with CCITT Recommendation G703, as outlined in Table 5.5. The losses in the station cables required to connect the input and the output ports of two successive equipments must be taken into account in engineering each installation. The digital distribution frame (DDF) may be located at any point along the station connection cables. Because of this the digital signal levels are not defined by the CCITT at the digital distribution frame. For systems based on the 1544 kbit/s hierarchy, the signal and levels are usually defined at the DDF. Therefore, the interface conditions as defined in CCITT Recommendation G703[1] and which are summarized in Table 5.5 coincide with the termination of a digital line section or a digital radio section.

When the digital bit rate carried over the radio system is an integral multiple of a hierarchial bit rate, multiplexing is usually necessary. This function may be carried out in one of the pair of boxes given in Figure 5.1. That is: (2, 3), (1, 4), (5, 6) or

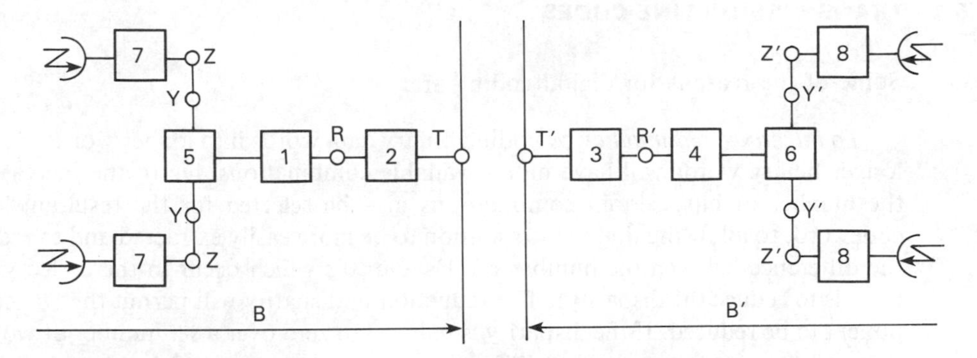

B,B digital radio section
R,R possible interconnection points between radiofrelay systems within a digital radio section
T,T output and input ports or in-jacks and out-jacks (as defined by CCITT Recommendation G703)
Y input of the receiving end of the switching equipment
Y′ output of the transmitting end of the switching equipment
Z output of the terminal receiver
Z′ input of the terminal transmitter
1,4 possible signal processing units
2,3 possible signal processors and interface units
5,6 possible receive and transmit ends of protection switching equipment
7 receiver (including demodulator) and possible signal processing units
8 transmitter (including modulator) and possible signal processing units

Figure 5.1 Baseband interconnection of digital radio systems (Courtesy of CCIR, Report 938, Reference 13)

(7, 8). Where multiplexing is required, bit insertion for BER monitoring may be performed in the same units in which the multiplexing processing is carried out. The sequence of digital pulses at the input of the radio modulating circuit must be such that, after modulation, a reliable timing signal can be derived from the radiated signal and the radiated signal spectrum does not contain intense spectral components which might cause harmful interference to other systems, particularly to co-channel or adjacent-channel analog systems. Most modulation methods provide a reliable timing signal if it is insured that sufficient transitions occur in the signal at the input to the modulating circuit. Similarly, intense spectral components are avoided by removing undesirable periodic sequences from that point. The alarm indication signal (AIS) when activated produces, as discussed in Section 3.6.2.2, a continuous stream of '1's at the output of the multiplexer. If this signal appears at the point T′ of Figure 5.1 and is applied unchanged to the radio modulator, undesirable spectral characteristics are produced. Before being modulated, the signal may therefore be required to be processed so that proper operation of the digital radio relay system is not dependent in any way upon the incoming bit sequence at point T′. This can be accomplished by using a scrambling device which may operate in any of the various boxes shown in Figure 5.1. Scramblers have been briefly discussed in Section 4.5.1, where the emphasis was placed on the reduction of jitter, rather than on spectral spreading.

5.1 TRANSMISSION LINE CODES

Some of the reasons for digital coding are:

To introduce redundancy by coding binary data words into longer words. As the longer binary words will have more available combinations due to the increase in the number of bits, certain combinations may be selected for the resultant legal codeword, to allow the timing information to be more easily extracted and to reduce the difference between the number of '1's and '0's which occur in the coded word (that is to reduce the disparity). The reduction in disparity will permit the DC component to be reduced. If the disparity is reduced to zero over a set number of words, that will permit a zero resultant DC component to occur over the same number of bits. Since the DC component of a digital signal cannot be transmitted this zero resultant is necessary. The additional usable words arising from a redundant code may also be used for transmitting nonmainstream digital data, such as parity bit transmission for error-control coding and auxiliary channels' transmission, as discussed under the 5B6B code in Section 5.1.6.4. The increase in the word length of a binary word does, however, result in an increase in the bit rate and hence the bandwidth. The bit rate is increased by the ratio of the output codeword length to the input codeword length. In the 5B6B code this is an increase in binary bit rate of 6/5 at the output.

To code a binary signal into a multilevel signal where a reduction in the bandwidth is required. This is discussed in Section 5.2.3. This type of encoding is important in the transmission of data at high bit rates over metallic-pair wires which are strictly band-limited. The cost of the reduction in required channel bandwidth, or the increase in bit rate for a given bandwidth, is offset by the increase in signal-to-noise ratio for a given probability of error (Section 5.2.4).

To encrypt a message where additional security is required, especially over data lines of data-processing computer systems.

To shape the signal spectrum for particular applications such as synchronization, reduction of the zero-frequency amplitude component to zero, or reduction of the high- and low-frequency components prior to filtering. Spectral nulls may be introduced as for interleaved high-order bipolar coded and interleaved digit streams.

In the PCM encoding process discussed in Chapter 2 and in Chapter 3, all the information bits that have been used have been implicitly assumed to be unipolar binary. This assumption is valid, as long as the information bits are located within a single item of processing equipment and do not have to travel over distances of more than a couple of meters. Once distances greater than this must be traversed using, say, screened twisted-pair cable, or coaxial cable, transmission of the bit stream in binary form is not recommended. Some reasons why distances greater than two meters might be involved are:

Primary PCM multiplex equipment is located in a different place from that of

multiplex equipment of level 2 or above.
Primary PCM or multiplex equipment of level 2 or above may have to be connected to the transmission equipment for further processing of the signal before it is transmitted over the communication channel.
Primary PCM information or multiplexed information of level 2 or above may have to be directly transmitted over a screened-pair or coaxial cable to the remote end.

Whatever the reasons may be, once the digital signal leaves a single item of processing equipment it requires the correct interfacing to match it to the transmission channel, even if this transmission channel is only a connecting link between equipments used in the signal processing before final transmission over the communication channel. The reasons why a unipolar binary signal, i.e. a signal which has only two states, of which one is at zero potential, is not suitable are:

1. It contains a DC component.
2. It has high-level frequency components at low frequencies.
3. It has a lack of signal transitions when long strings of '0's are transmitted.

Transformers are recommended in the use of interface equipment in order to prevent DC, which may be used to operate regenerative repeaters, from entering the input and output ports of the processing equipment. The unipolar binary signal has a DC component, and is thus unsuitable for transmission through transformers. In addition to this, as there is in a unipolar binary signal a high energy content in the lower portion of the frequency spectrum, it can only be expected that transmission through a transfomer of practical size will cause distortion to the pulses and thus introduce transmission errors. As the timing information at the receiver is required to be extracted from the bit stream, the occurrence of strings of '0's in the unipolar binary signal will disturb the timing extraction process and lead to loss of synchronization.

Two forms of unipolar binary signals occur in processing equipment. These are the 'return to zero' (RZ) and the 'non-return to zero' (NRZ). Both forms experience the same difficulties as just outlined if used in direct transmission. Figure 5.2 shows the waveforms of these two different unipolar binary signals. Notice that the RZ signal level, representing the bit value 1, is high only for the first half of the bit interval, after which the signal returns to zero for the remainder of the bit interval. The advantage in RZ signals lies in the increased transitions compared to NRZ. To overcome the problems associated with the unipolar binary signal, a series of *line codes* has been developed. The line codes do not have a significant DC component nor do they have their spectral energy concentrated at the lower frequencies. Sometimes the line codes may have the added advantage of not being susceptible to long strings of '0's or '1's, and are thus able to be used confidently with timing extraction circuits.

The requirements which a line code has to fulfil, in order that the binary signals are converted into a format more suitable for transmission over a communication

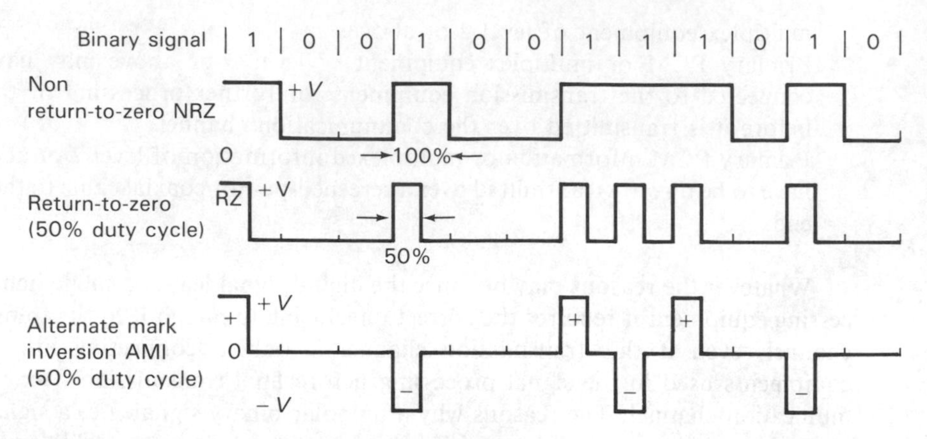

Figure 5.2 NRZ and RZ binary representation and AMI binary code conversion

channel, are:

- To match the signal spectrum to the properties of the channel (such as noise, bandwidth, etc.)
- To insure that bit sequences are independent, therefore reducing the amount of pattern-dependent jitter generated by repetitive patterns
- To facilitate clock extraction and signal regeneration
- To provide sufficient redundancy for error monitoring of the transmission path and for locating faulty equipment
- To keep redundancy as low as possible in order to minimize bit rates and therefore circuit complexity as well as to minimize the transmission bandwidth requirements
- To reduce the DC content of the signal to zero
- To reduce the low-frequency components in order to reduce crosstalk and component sizes
- To be a standard which is used by different manufacturers and is recommended by an appropriate authority, such as the CCITT.

It should be noted that two types of situation exist in digital systems which call for different codes. The first situation is that of an 'interface code', in which the output of digital multiplex equipment becomes an input tributary to the next higher-level multiplex equipment. The line length in this case is no longer than 100 meters. In this situation the line codes are those provided in Table 5.5. The second situation is where the output of the multiplex or other digital equipment is being sent directly to line. In this case the line codes do not necessarily follow those provided by Table 5.5, but depend on the equipment used to transmit the information signal over the transmission channel. This is shown in Table 5.1.

If the output of the multiplex equipment is being fed into the input of a digital radio, the above system line codes do not apply, and Table 5.5 is used, since the

Table 5.1 Examples of system line codes

Bit rate, Mbit/s	System type	Transmission medium	Line code
2.048	line	Symmetric pairs	MS43
		Optical line	1B2B
8.448	line	Optical line	2B3B
		Optical line (2nd Generation)	1B2B
34.368	line	Optical line	5B6B
139.264	line	Coaxial line	6B4T
		Optical line	7B8B
564.992	line	Optical line	5B6B

radio input is considered the same as another higher-level multiplex input tributary; the distance to be traversed is not far and after being patched at the digital distribution frame (DDF) is possibly the same as that for multiplex output to higher level multiplex tributary input.

Refer to Figure 1.1, in which the model of a digital system was shown. The baseband shown may be considered as the output of the multiplex equipment in the above discussion, and the 'Digital MOD', with transverse screened cable or coaxial cable, as the *line terminating equipment* (LTE) and, with optical line systems, the *optical line terminating equipment* (OLTE). It is common practice for the proposed input tributaries of level 2 or second-order multiplex equipment and above to be conveyed to a digital distribution frame (DDF), for manual patching to the relevant multiplex input tributary coaxial cable terminations, which are also located on this DDF. This permits flexibility and also permits restoration of important circuits via other equipments should a failure occur on the normally allocated equipment. For the input tributaries to the first-level multiplex [primary digital multiplex (PDM)] usually screened metallic-pair cable is used. The distribution circuits at 64 kbit/s and the 64 kbit/s tributary inputs to the multiplex equipment do not go to the DDF, but to a separate frame for patching due to the different types of cable used. On a DDF, it is usual to insure that each vertical used is always maintained to the same bit rate and that the top part of the vertical is the multiplex outputs from the preceding stages and the lower part of the vertical is for the input tributaries of the next higher level multiplex equipment.

5.1.1 Alternate mark inversion (AMI) code

By coding the unipolar binary signal into several pre-transmission levels, it is possible to eliminate the DC component and reduce the low-frequency content of the coded signal, thus keeping to a reasonable size any components used in the design of regenerator equalizers. This is done without extending the required transmission bandwidth, and in principle it is possible, in the 'binary-*n*ary conversion, even to reduce this. Ternary codes operating with three levels, in which the middle level of the signal is at zero voltage, have found wide application as line codes. Because the zero-voltage level is not a true logic level, the code is termed a 'pseudo-ternary'. Coders with balanced outputs $+A$ (by convention +) and $-A$ (by convention −) and with the zero-voltage level corresponding to the system ground level can easily be constructed. Of these ternary codes, the best known is the alternate mark inversion (AMI) code, otherwise known as 'bipolar'. The encoded sequence is obtained by coding the '0's as absence of pulsing and the '1's by alternate positive and negative pulses with the alternation taking place at every occurrence of a '1' in the binary signal. The alternation occurs irrespective of the number of zeros in between. The conversion of binary (RZ which has a 50 percent duty cycle) to AMI is shown in Figure 5.2. The AMI signal, which may also be of the NRZ type, with a 100 percent duty cycle, has an average value of zero, meaning that there is no DC component and that AC coupling in the transmission path has little effect on the transmitted digits. A maximum spectral density at half the bit rate and a very small spectral density at low frequencies is also a feature of this type of coding. Unfortunately, the conversion does nothing to reduce *disparity*, or the difference between the number of '1's and the number of '0's in a codeword, or reduce the timing extraction problems associated with such. Usually disparity is important since it indicates the propensity of a code to produce a DC level. If the disparity is greater or less than zero, it means that there is at any time more ones than zeros, or vice versa. As the zeros in the pseudo-ternary code may be at a negative voltage, there will be either a residual positive or a residual negative voltage. For the AMI code, this does not apply if the binary zero is set at ground potential. The importance of disparity becomes apparent if a line code is to be designed in which the number of ones equals the number of zeros in a sequence of bits. If an error is introduced in the line transmission system due to impulsive noise or crosstalk, it will cause a pulse either to be omitted or to be falsely added. In either case, there will be two pulses of the same polarity, permitting detection of the error by appropriately designed test equipment. This condition is known as *bipolar violation* and is a pertinent feature of alternate mark inversion (AMI).

5.1.2 Paired selected ternary (PST)[1] code

The conversion from RZ unipolar binary to paired selected ternary is made by pairing two neighbouring bits and by applying the conversion rules as specified in Table 5.2. For both 'positive mode' and 'negative mode' outputs, the ternary combinations resulting from binary pairs 00 and 11 have no DC level, although binary pairs 01 and 10 do produce a DC level equivalent to $+A/2$ and $-A/2$ depending on the mode of the output. The coder operates to provide positive-mode output until either of the states 01 or 10 occurs at the input. This then produces a $+A/2$ bias at the output which causes a change to the negative mode. The next occurrence of either 10 or 01 produces a bias of $-A/2$, removing the $+A/2$ bias and changing the mode back to positive. In addition to the balancing out of the DC component, the timing extraction is made easier at the regenerative repeaters and at the receiving end because long sequences of '0's or '1's do not occur.

Table 5.2 Conversion rules for paired selected ternary code

Binary input code adjacent pairs	Output code	
	Positive mode	Negative mode
00	− +	− +
01	0 +	0 −
10	+ 0	− 0
11	+ −	+ −

5.1.2.1 4B3T Code

This code makes more efficient use of ternary levels by compressing four input data bits into three output signal data bits. Consequently the baud rate is 0.75 times the input signal rate. Table 5.3 shows the conversion algorithm. The spectral energy for this code is fairly evenly spread across the spectrum, but has a rapid drop in energy for spectrum frequencies less than 0.1 times the bit rate. Attempts to overcome the large low-frequency content of this code have been made by the introduction of modified 4B3T codes such as the MS43,[2-4] discussed in Section 5.1.2.2 and given in Table 5.3B, and the VL43 code.[2] The 4B in the name 4B3T refers to the blocks of four binary digits of the input signal. The 3T term is the block of three ternary digits which is to represent the output code. This code belongs to a group of codes normally described as *nNm*T codes, where $3^m > 2^n$. Here *m* is the number of ternary digits, and *n* is the number of binary digits, in the block. From Table 5.3A

Table 5.3A Conversion rules for 4B3T code

Binary input code	Ternary output code		
	Positive mode	Negative mode	Digital sum variation (disparity)
0000	+ 0 −	+ 0 −	0
0001	− + 0	− + 0	0
0010	0 − +	0 − +	0
0011	+ − 0	+ − 0	0
0100	+ + 0	− − 0	+/−2
0101	0 + +	0 − −	+/−2
0110	+ 0 +	− 0 −	+/−2
0111	+ + +	− − −	+/−3
1000	+ + −	− − +	+/−1
1001	− + +	+ − −	+/−1
1010	+ − +	− + −	+/−1
1011	+ 0 0	− 0 0	+/−1
1100	0 + 0	0 − 0	+/−1
1101	0 0 +	0 0 −	+/−1
1110	0 + −	0 + −	0
1111	− 0 +	− 0 +	0

Table 5.3B MS43 code

Binary words	Ternary alphabets R1	R2	R3
0000	+ + +	− + −	− + −
0001	+ + 0	0 0 −	0 0 −
0010	+ 0 +	0 − 0	0 − 0
0100	0 + +	− 0 0	− 0 0
1000	+ − +	+ − +	− − −
0011	0 − +	0 − +	0 − +
0101	− 0 +	− 0 +	− 0 +
1001	0 0 +	0 0 +	− − 0
1010	0 + 0	0 + 0	− 0 −
1100	+ 0 0	+ 0 0	0 − −
0110	− + 0	− + 0	− + 0
1110	+ − 0	+ − 0	+ − 0
1101	+ 0 −	+ 0 −	+ 0 −
1011	0 + −	0 + −	0 + −
0111	− + +	− + +	− − +
1111	+ + −	+ − −	+ − −

the maximum disparity or digital sum variation is ±3 for an input code of 0111. As with the 2B2T code discussed in Section 5.1.2 there are two modes. If the average output signal is positive, the mode is switched to negative, and so on. The mode alphabet is organized to bias the output sequence towards a positive or negative accumulation of the digital sum variation, and to switch the modes in order to maintain the digital sum variation as close to zero as possible, thus minimizing the low-frequency spectral energy. This code may find applications in the Ethernet systems, the ISDN and in digital data-transfer systems.

5.1.2.2 MS43 code

Another 4B3T code is the MS43,[2] which has found its use in transverse screened cable digital transmission systems. This code employs three alphabets, R1, R2 and R3, which are selected according to the disparity of the digital sum variation of the code. At any time the digital sum can vary over 6 values, from 0 to 5, with the extreme values occurring during either the first or second elements of a ternary word. This restricts the terminal states to 1 to 4. If the terminal state is 1, alphabet R1 is selected for the next code word. If it is 2 or 3, alphabet R2 is selected and if it is 4, alphabet R3. Table 5.3B shows the conversion rules for this code. The alphabets are arranged so that each group of three ternary symbols is uniquely associated with one group of four binary bits and hence it is not necessary for the decoding equipment to identify the mode or alphabet employed by the transmitter. Four successive zeros in the ternary sequence, or up to five successive positive or negative marks, are permitted. The ternary word consisting of three zeros is not used and so any sequence of three or four zeros must span the boundary between two adjacent words. This feature may be used to achieve frame alignment. In addition to this, because each word contains one or more '1's, the timing extraction is simplified.

5.1.2.3 6B4T code

Table 5.1 shows that the 6B4T code is used for coaxial line systems at 140 Mbit/s. As the name implies, the code converts 6 binary digits into 4 ternary digits. This code will have 2^6 or 64 six-bit binary words and 3^4 or, with the exclusion of 0000, 80 ternary words.

The group of codes described as *nNm*T are a class of non-linear ternary alphabetical codes, and as shown above the simplest of these codes is when $n = m = 2$ as with the paired selected ternary (PST).

5.1.3 BNZS codes

Codes have been developed which are *B*ipolar and have in a string of *N Z*eros a particular pattern, containing bipolar violations, *S*ubstituted in order to reduce the disparity. These codes are a class of nonlinear ternary nonalphabetical codes and may be considered as modified AMI or bipolar codes. The three most important codes in this class are the B3ZS, B6ZS and the B8ZS. All these codes are recommended by the CCITT in Recommendation G703,[10] to be used in certain digital interfaces as discussed below.

5.1.3.1 B3ZS code

As stated in the CCITT Recommendation G703,[10] the B3ZS bipolar with three-zero substitution code is a modified bipolar or AMI pulse format. Logical 1 bits have a

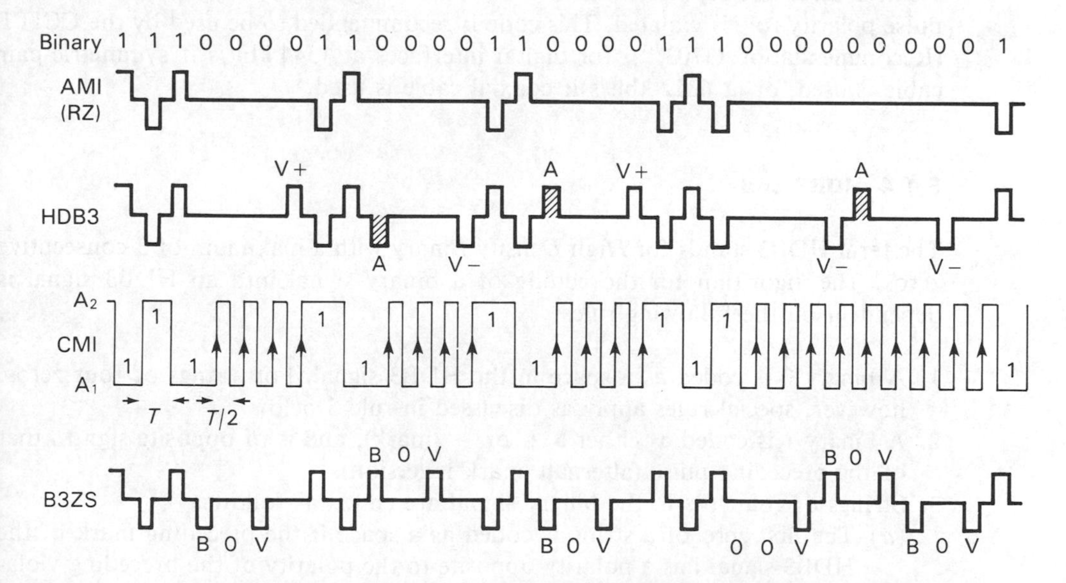

Figure 5.3 HDB3, CMI and B3ZS coded binary signals

50 percent duty cycle and are generally alternately positive and negative with respect to the logical 0 level. Exceptions are cases where three logical '0's appear together in the bit stream. In the B3ZS format, each block of three consecutive zeros is removed and replaced by B0V or 00V, where B represents a pulse conforming with the bipolar or AMI alternating mark polarity rule and V represents a pulse violating this rule. The choice of B0V or 00V is made so that the number of B pulses between consecutive V pulses is odd. Where this code is used in the interface equipment of a 44736 kbit/s level 3 multiplex as described in Section 3.5.6.1, the frame alignment bits as shown in Figure 3.21 are included in the coding. Figure 5.3 shows an example of this code. The spectral energy of this code peaks at 0.5 the bit rate, and most of the energy lies between 0.2 and 0.8 times the bit rate.

5.1.3.2 B6ZS code[4]

In this code a group of six consecutive zeros in the input signal is replaced with 0+ −0− + if the preceding pulse was +. If the preceding pulse was −, six consecutive zeros are replaced by the code 0− +0+ −. Thus long runs of zeros which may occur in conventional AMI coding are broken up by the substitution of pulses or groups of pulses that violate the AMI alternating mark pulse polarity rule. The spectral energy of this code peaks at 0.5 the bit rate, and most of the energy lies between 0.2 and 0.8 times the bit rate. The code is recommended to be used in the 6312 kbit/s digital interface if one symmetric pair of wires is used in each direction of transmission.

5.1.3.3 B8ZS code

In this code a group of eight consecutive zeros in the input signal is replaced with 000+ −0− + if the preceding pulse was +. If the preceding pulse was −, eight consecutive zeros are replaced with 000− +0+ −. Again the AMI alternating mark pulse polarity rule is violated. This code is recommended to be used by the CCITT (Recommendation G703[10]), for digital interfaces at 1544 kbit/s if symmetric-pair cable is used, or at 6312 kbit/s if coaxial cable is used.

5.1.4 HDB3 code

The term HDB3 stands for *H*igh *D*ensity *B*inary with a maximum of 3 consecutive zeros. The algorithm for the coding of a binary signal into an HDB3 signal is dependent on the following rules:

1. A binary 0 is coded as a space in the HDB3 signal. For strings of four zeros, however, special rules apply as discussed in rule 3 below.
2. A binary 1 is coded as either a + or − (mark), and is of opposite sign to that of the preceding pulse (alternate mark inversion).
3. Strings of four '0's in the binary signal are coded as follows:
   (*a*) The first zero of a string is coded as a space if the preceding mark of the HDB3 signal has a polarity opposite to the polarity of the preceding violation and is not a violation by itself.

(*b*) The first zero of a string is coded as a mark (A), that is not a violation (+ or −), if the preceding mark of the HBD3 signal has the same polarity as that of the preceding violation or is by itself a violation.

Rules 3(*a*) and 3(*b*) ensure that successive violations are of alternate polarity so that no DC component can accumulate.

(*c*) The second and third zeros of a string of four binary zeros are always coded as spaces.
(*d*) The fourth zero in a string of four binary zeros is always coded as a mark, the polarity of which is such that it violates the rule of alternate mark inversion. Such violations are denoted as V− or V+ according to their polarity.

The spectral energy distribution of a random input signal coded to HDB3 code is similar to that of AMI, in that maximum spectral energy lies about half the bit rate. Its shape is like that of a volcano, with a dip at 0.5 times the bit rate, and two small peaks at approximately 0.45 and 0.55 times the bit rate. This code is used mainly for 2048, 8448 and 34368 kbit/s multiplex interfaces as recommended by the CCITT (Recommendation G703).[10] Its uses are in the Ethernet local area network configurations and for the transmission of data. Figure 5.3 shows an example of HDB3 coding from a binary signal.

5.1.5 CMI code

CMI stands for *C*oded *M*ark *I*nversion. It is a 2-level non-return-to-zero (NRZ) code in which binary 0 is coded so that both amplitude levels A_1 and A_2 are attained consecutively, each for half a unit time interval $T/2$. Binary 1 is coded by either of the amplitude levels A_1 or A_2, for one full time unit interval T, in such a way that the level alternates for successive binary '1's. Figure 5.3 shows an example of this coding. It should be noted that, for binary 0, there is always a positive tradition at the midpoint of the binary unit time interval, and for binary 1, there is a positive transition at the start of the binary unit time interval if the preceding level was A_1, and a negative transition at the start of the unit time interval if the last binary 1 was encoded by level A_2. It may be noted that a binary zero is mapped to 01 and a binary one is mapped to 11 and 00, over a time slot period.

5.1.6 Other codes

The codes to be discussed are all binary codes, which have been found to be useful for the transmission of data in local area networks[5] and in the storage of data on magnetic tape or disk. They have been included because of the possibility that they, amongst others, may be used in the Integrated Services Digital Network (ISDN) for the transmission of digital signals in subscribers' loops or the local area network.

5.1.6.1 Wal1 and Wal2 codes[6,7]

These are modified binary codes in which the coded signal is free from a DC component and contains a large number of transitions from which timing information may be derived, irrespective of the input binary signal pattern. The bandwidth occupance of the Wal2 code is greater than that of Wal1 and extends to approximately twice the bit transmission rate, where Wal1 extends to approximately 1.5 times the bit rate. The Wal2 spectrum has no significant spectral energy content below 0.2 times the bit rate, and hence may be used for such applications as data-over-voice. There are at least two transitions for each data bit, providing a good signal for synchronization; however the baud rate is four times the data rate for Wal2, and three times the data rate for Wal1.

5.1.6.2 Manchester code

This code is otherwise known as *twinned binary coding*, *split phase* or *diphase* coding and is another form of the 1B2B code. It is commonly used for coding binary PAM data transmission and is often used for storing data on magnetic tape or disk. It has also been adopted for the Ethernet local area network. In this method, each bit is represented by two successive pulses of opposite polarity, the two binary values having the representation + − and − +. A transition from a negative to a positive represents a binary 1, whereas a transition from a positive to a negative represents a binary 0. The bit rate is half that of the unrestricted binary pulses and hence requires twice the bandwidth. As there is at least one transition with every data bit, clock recovery is easily obtained. The binary logic state 0 is mapped to 01, and 1 is mapped to 10. The states 00 and 11 are not used, in order to insure that long strings of zeros and ones do not occur which would introduce difficulties in timing recovery. The power spectral density peaks at 0.8 times the bit rate and falls to zero at twice the bit rate and has zero DC content.

5.1.6.3 S code[8]

This *S*equential code is a nonlinear code which has excellent synchronization properties. Fourteen output signal bits are used to code four data bits. The baud rate is 7/2 times that of the data rate. The start of each 14-bit sequence is 10110 and is followed by a sequence of 9 bits in which there is never more than two consecutive '1's or '0's occurring. This permits both bit and byte synchronization.

A brief description of these codes and their spectral density functions may be found in Reference 9 and on p. 414 of Reference 11.

5.1.6.4 5B6B code[2]

This code has found favor with 34 Mbit/s and 565 Mbit/s optical line systems, partly because:

A two-level signal is best suited as the transmission signal on optical fibers.

The low-frequency content in the signal spectrum is very small.

The density of signal transitions is high and distributed uniformly and the number of consecutive symbols of the same polarity is restricted to five. This provides an easily extracted timing signal with a low jitter amplitude.

The code permits a continuous monitoring of a bit error ratio (BER) of dependent repeaters by detecting error-induced overflows. The relationship between detected code violations and actual errors is very close. This correlation extends from low to high bit error ratios and is largely independent of signal and error statistics.

The encoder and decoder are easily fabricated using existing technology.

Table 5.4 provides the details of a typical mapping table for this code.

It may be noted that as the output code word is six bits and the input codeword is five bits at a bit rate of 565 Mbit/s, the output bit rate will increase by 6/5 times the input bit rate, to give 678 Mbit/s. Because there are 65 different words which can be accommodated using six bits, there are 32 additional words which can be used for the output codewords over the 32 input words. This redundancy permits codewords which have a disparity of more than 1 to be eliminated from use. Even when this has been done there are still superfluous codewords remaining. Use can be made of these additional codewords in providing an auxiliary data channel. The way that this may be done is as follows. Consider the six additional codewords in alphabet 1 and 2 for 17B, 19B and 28B. If the binary codeword 10001 can be obtained at the receiver output for any one of the three codewords 100011, 100001 or 011110, data representing decimal codeword 17 pass across the transmission channel in any of the three six-bit word forms. If alphabet 1 codeword 100011 with a disparity of zero is used to code a logic zero for the auxiliary channel and 100001 and 011110 as the auxiliary channel 1 (these two codes alternating to maintain a disparity of zero), additional data can be sent by an auxiliary channel together with the data

Table 5.4 5B6B code

Input word		Output word		Input word		Output word	
Decimal	ABCDE	Alphabet 1 LLLLLL	Alphabet 2 LLLLLL	Decimal	ABCDE	Alphabet 1 LLLLLL	Alphabet 2 LLLLLL
0	0 0 0 0 0	1 0 1 0 0 0		1	0 0 0 0 1	0 1 1 0 0 0	1 0 0 1 1 1
2	0 0 0 1 0	1 0 0 1 0 0		3	0 0 0 1 1	0 0 0 1 1 1	
4	0 0 1 0 0	0 1 0 1 0 0	1 0 1 0 1 1	5	0 0 1 0 1	0 0 1 0 1 1	
6	0 0 1 1 0	0 0 1 1 0 1		7	0 0 1 1 1	0 0 1 1 1 0	
8	0 1 0 0 0	0 0 1 1 0 0	1 1 0 0 1 1	9	0 1 0 0 1	0 1 0 0 1 1	
10	0 1 0 1 0	0 1 0 1 0 1		11	0 1 0 1 1	0 1 0 1 1 0	
12	0 1 1 0 0	0 1 1 0 0 1		13	0 1 1 0 1	0 1 1 0 1 0	
14	0 1 1 1 0	0 1 1 1 0 0		15	0 1 1 1 1	0 1 0 0 1 0	1 0 1 1 0 1
16	1 0 0 0 0	1 0 0 0 1 0	0 1 1 1 0 1	17A	1 0 0 0 1	1 0 0 0 1 1	
				17B		1 0 0 0 0 1	0 1 1 1 1 0
18	1 0 0 1 0	1 0 0 1 0 1		19A	1 0 0 1 1	1 0 0 1 1 0	
				19B		1 1 0 0 0 0	0 0 1 1 1 1
20	1 0 1 0 0	1 0 1 0 0 1		21	1 0 1 0 1	1 0 1 0 1 0	
22	1 0 1 1 0	1 0 1 1 0 0		23	1 0 1 1 1	0 0 1 0 1 0	1 1 0 1 0 1
24	1 1 0 0 0	1 1 0 0 0 1		25	1 1 0 0 1	1 1 0 0 1 0	
26	1 1 0 1 0	1 1 0 1 0 0		27	1 1 0 1 1	0 0 0 1 1 0	1 1 1 0 0 1
28A	1 1 1 0 0	1 1 1 0 0 0					
28B		0 0 0 0 1 1	1 1 1 1 0 0	29	1 1 1 0 1	0 1 0 0 0 1	1 0 1 1 1 0
30	1 1 1 1 0	0 0 1 0 0 1	1 1 0 1 1 0	31	1 1 1 1 1	0 0 0 1 0 1	1 1 1 0 1 0

information. For example, if the auxiliary channel input to the 5B6B coder is a 1, when data are being sent which represent 5B code 10001, the 6B code will be 100001. If the next auxiliary channel bit is a zero, when 5B code data 10001 are transmitted, the 6B code will be 100011. If the next auxiliary code is to be again a 1, then, on arrival of the 5B code 10001, it is coded as 011110 to maintain a zero disparity over the two representations of auxiliary bit 1. Up to three separate auxiliary data channels may be used, by employing binary codewords for 17, 19 or 28. These of course may be combined to treble the auxiliary channel capacity.

5.1.7 CCITT interfaces (Recommendation G703)[10]

Table 5.5 details the codes, impedances, type of transmission pairs, signal levels and pulse shape at the digital distribution frame for various bit rates.

Table 5.5 Digital interfaces at various bit rates

Bit rate, kbit/s	64	1544	6312	44736	2048	8448	34368	139264
Cable type*	S	S	S/C	C	S/C	C	C	C
Test load impedance, ohms	120	100	110/75	75	120/75	75	75	75
Line code recommended	†	AMI or B8ZS	B6ZS / B8ZS	B3ZS	HDB3	HDB3	HDB3	CMI
Nominal pulse shape	Rect.	Rect.	Rect.	Complex	Rect.	Rect.	Rect.	Rect.
Nominal pulse width	3.9 μs	#	#	#	244 ns	59 ns	14.55 ns	<2 ns
**Level in dBm at given frequency	–	12 to 19	0.2 to 7.3 / 6.2 to 13.3	−1.8 to +5.7	–	–	–	–
Frequency, kHz		772	3156	22368				
**Level in dBm at given frequency	–	−13 to −6	<−20 / <−14	−21.8 to −14.3				
Frequency, kHz		1544	6312	44736				

S = Symmetric pair cable
C = Coaxial cable
Rect. = rectangular

Notes

see CCITT Recommendation G703 for mask details.

†The 64 kbit/s codirectional interface (that is the interface across which the information and its associated timing signal are transmitted in the same direction) uses one balanced pair in each direction of transmission with transformer coupling at each terminal. A 64 kHz and an 8 kHz timing signal are transmitted with the information signal by means of the way the information signal is coded. The conversion rules are:

Step 1. A 64 kbit/s bit period is divided into four unit intervals
Step 2. A binary one is coded as a block of the following four bits: 1100
Step 3. A binary zero is coded as a block of the following four bits: 1010
Step 4. The binary signal is converted into a three-level signal by alternating the polarity of consecutive blocks
Step 5. The alternation in polarity of the blocks is violated every eighth block. The violation block marks the last bit in the octet.

These conversion rules are illustrated in Figure 5.4.

* For coaxial cable there is one cable for each direction of transmission, similarly, for symmetric pair cable there is one pair for each direction of transmission

** The signal level is the power level measured in a 3 kHz bandwidth at the in-jack for an all '1's pattern transmitted.

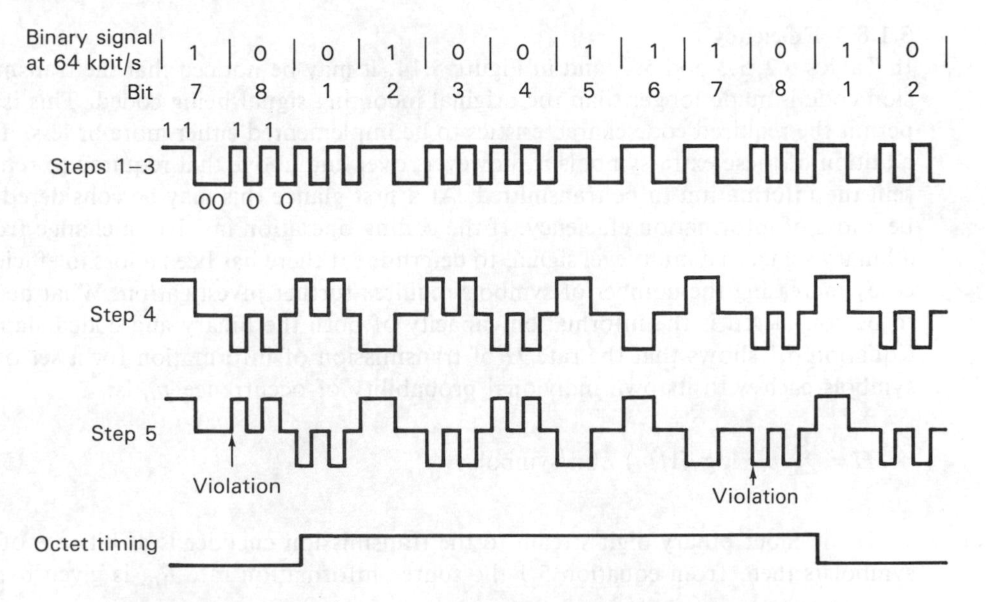

Figure 5.4 64 kbit/s line signaling coding for a codirectional interface

5.1.8 Transmission codes[11]

Section 5.1 listed the requirements which a line code has to fulfil in order that the binary signals processed by the various equipments are converted into a format more suitable for transmission over a communication channel. The line code however, is involved mainly to match the digital bit stream to the transmission channel. Line codes may be considered to belong to a class of *transmission codes*, whose reason for existence is not only to match the transmission channel, but to overcome restrictions which may be placed on binary digits by associated transmission equipment, such as that discussed in Section 5.2 and in later chapters. The characteristics which a transmission code should exhibit in order to transfer digital information at the highest bit rate, with the lowest probability of error and with the greatest information transfer (all of which are conflicting requirements), are given below in Sections 5.1.8.1 to 5.1.8.7.

5.1.8.1 Transparency

This implies that no restrictions are imposed on the information content of the transmitted signal. Thus a sequence of binary digits presented to the encoder are faithfully decoded at the receiving end, regardless of the length of the sequence, or the digital make-up of the sequence.

5.1.8.2 Unique coding

An input binary digit or block of binary digits must be uniquely coded such that there can be no ambiguity during the decoding process. Examples of uniquely coded binary digits are given in Tables 5.2, 5.3 and 5.4, and in Figure 5.14.

5.1.8.3 Efficiency

In Tables 5.2, 5.3 and 5.4, and in Figure 5.14, it may be noticed that the transmission code is made longer than the original incoming signal being coded. This is to permit the required code characteristics to be implemented either more or less. The addition of these extra symbols is, however, over and above that required to represent the information to be transmitted. At a first glance this may be considered to be a loss of information efficiency. If the coding operation involves a change from a binary signal to a multilevel signal, to determine if there has been a loss in efficiency by increasing the number of symbols requires further investigation. What needs to be considered is the information capacity of both the binary and coded signal. Equation 5.1 shows that the rate H of transmission of information for a set of γ symbols each with its own individual probability of occurrence p_i, is:

$$H = \sum_{i=1}^{\gamma} p_i \cdot \log_2(1/p_i) \quad \text{bit/symbol} \tag{5.1}$$

If the input binary digit stream to the transmission encoder is at bit rate of r_b symbols/s then, from equation 5.1 the source information rate R_b, is given by:

$$R_b = r_b \sum_{i=1}^{\gamma} p_i \cdot \log_2(1/p_i) \quad \text{bit/s} \tag{5.2}$$

If the output code of the encoder comprises r_s symbols/s, where each symbol is one of k characters, each with its own individual probability of occurrence q_i, then the rate of transmission of information is given by:

$$R_s = r_s \sum_{i=1}^{k} q_i \cdot \log_2(1/q_i) \quad \text{bit/s} \tag{5.3}$$

The efficiency is given by:

$$\varepsilon = (\text{Source information rate})/(\text{encoder output information rate}) = R_b/R_s \tag{5.4}$$

Example

If the input signal is a binary signal at 34 Mbit/s (r_b), with a probability of a logic 1 equal to 0.2 (p_1), and is converted into a quaternary (4-level) output signal at a bit rate of 30 Mbit/s with all four symbols having probabilities equal to 0.25, then $R_b = (34\ \text{Mbit/s}) \cdot (0.2 \log_2 5 + 0.8 \log_2 1/0.8) = (34\ \text{Mbit/s}) \cdot (0.7219) = 24.5456\ \text{Mbit/s}$, and $R_s = (30\ \text{Mbit/s}) \cdot 2 = 60\ \text{Mbit/s}$. Thus, the efficiency $\varepsilon = 24.5456/60 \times 100 = 40.9$ percent.

5.1.8.4 Redundancy

The redundancy is the unused transmission capacity of the coded signal. This may be given by the expression:

$$\text{Redundancy} = (R_s - R_b)/R_b = (1 - \varepsilon)/\varepsilon \tag{5.5}$$

For the above example, the redundancy or unused data information capacity which does not include overhead bit information is given as (60 – 24.5456)/24.5456 = 144.4 percent.

5.1.8.5 Error-control coding

This type of coding is distinct from line coding, source encoding or coding for transmission. As the name implies, it is specifically for the control of errors and thus a reduction in the error rate. In Section 5.3 we discuss this subject briefly, for it is a specialized subject on which there are many excellent texts.[30–36]

5.1.8.6 Spectrum shaping

The PAM signal spectrum in many applications is shaped to match the channel characteristics so that it is possible to design practical filters in both transmitter and receiver. If the signal was not shaped before filtering, the majority of the signal energy in its spectrum would not be transmitted, due to the restrictions in bandwidth produced by the action of the filters. This bandwidth restriction would tend to produce an unacceptable amount of distortion. The shaping of the spectrum ensures that the loss in energy caused by the action of the filter is small. In addition to the above, shaping the signal spectrum reduces the interference between different channels, such as crosstalk, which tends to increase with increasing frequency. The average energy distribution in a signal's spectrum may be estimated by autocorrelation techniques, such as those briefly discussed in Section 2.1.4. Again, as there are many excellent texts,[11,16,33] covering this aspect of transmission, it will not be dealt with in this book, except to make reference to the results for particular codes which are discussed.

5.1.8.7 Timing content

The timing content of digital bit streams is discussed in Sections 3.1, 3.5 and Chapter 4, as well as in Section 5.1. The importance of using a code which has sufficient transitions, so that timing information can be easily extracted by clock recovery circuits, has already been considered.

5.2 BASEBAND DATA TRANSMISSION

5.2.1 Channel capacity[14–18]

The maximum data rate or 'capacity' C, of a transmission channel having a bandwidth B and additive Gaussian band-limited white noise introduced into it, is given by:

$$C = B \log_2 [1 + (S/N)_o] \quad \text{bits/s} \tag{2.100}$$

Where S and N are the average signal power and noise power, respectively, at the output of the channel. The noise power can be expressed as:

$$N = \eta \cdot B = \eta/2 \cdot 2B = \eta/2 \cdot W \tag{5.6}$$

if the two-sided power spectral density of the noise is $\eta/2$ W/Hz, and the double-sided bandwidth is W.

This fundamental theorem of information is referred to as the Shannon-Hartley theorem and provides two important considerations for communication system design:

1. The upper limit that may be attained for the data transmission rate over a Gaussian channel
2. The exchange of signal-to-noise ratio for bandwidth

The second consideration is apparent from equation 2.100 for a fixed channel capacity. As the signal-to-noise ratio is reduced, the bandwidth must increase.

Another useful outcome of this theorem is that an analog signal occupying a band from DC (zero frequency), up to a cut-off frequency f_m, can be transmitted, after suitable coding, through a channel with a bandwidth less than f_m. For example, assume that the analog signal is quantized into Q quantization levels after being sampled at x times the Nyquist sampling frequency $(2f_m)$ in samples/s. Then, the number of coded binary bits is $\log_2 Q$ and the bit rate is; $2xf_m \log_2 Q$ bits/s, which is the required channel capacity. From equation 2.100, the theoretical channel capacity C_n can be made to be greater than that required for a reduced analog signal bandwidth of, for example, $f_m/2$, by increasing the signal-to-noise ratio, or just by increasing the signal power level.

Example

Assume that a 4 kHz bandwidth analog signal, extending effectively from 0 to 4 kHz, is required to be transmitted through a channel of 1 kHz. Determine how this can be done.

Possible solution

Consider that the analog signal is quantized into 256 levels, and is sampled at 8 kHz. This means that the channel capacity C bit rate is 64 kbit/s. From equation 2.100 with B equal to 1 kHz, we find that $1 + (S/N)_o = 6$, or $(S/N)_o = 5$ (or 16 dB) for an error-free channel. In reality, this figure of $(S/N)_o$ would be closer to 50 dB for a probability of error of 1 in 10 000, as discussed in Section 2.3.3.

Thus, crude as this example may be, it does demonstrate that it is possible to transmit a signal over a communication channel using less bandwidth than that of the original signal.

Another aspect of equation 2.100 is that it appears that if the bandwidth approaches infinity the channel capacity will also do so. This is not so, because, as the bandwidth increases, so does the noise power. The channel capacity reaches a finite upper limit with increasing bandwidth if the signal power is fixed. This limit may be calculated as follows. With $N = \eta \cdot B$ from equation 5.6, where $\eta/2$ is the noise power spectral density, equation 2.100 becomes:

$$C = B \log_2(1 + S/\eta B) = (S/\eta) \cdot (\eta B/S) \cdot \log_2(1 + S/\eta B)$$
$$= (S/\eta) \log_2(1 + S/\eta B)^{\eta B/S}$$

Since $\lim_{x \to 0} (1 + x)^{1/x} = e$

Letting $x = S/\eta B$

we have $\lim_{W \to \infty} C = (S/\eta) \cdot \log_2 e = 1.44(S/\eta) = 1.44B \cdot (S/N)$ (5.7)

Thus the upper limit of the channel capacity is $1.44S/\eta$.

Without delving too deeply into the subject of information theory,[14-18] it suffices to say that, if a source has an alphabet of M symbols, and each symbol has the same probability of occurring, the information content H per symbol is given by $\log_2 M$ bits/symbol, and if the symbols are emitted from the source at a fixed rate of r_s symbols/s, the source information rate R is:

$$R = r_s \cdot H \tag{5.8}$$

$$= r_s \cdot (1/M) \sum^{M} \cdot \log_2 M = r_s \cdot \log_2 M \text{ bits-of-information/second} \tag{5.9}$$

If the symbol rate is not uniform, but each of a set of γ symbols has its own individual probability of occurrence p_i, then

$$H = \sum_{i=1}^{\gamma} p_i \cdot \log_2(1/p_i) \text{ bits/symbol} \tag{5.10}$$

In a practical digital transmission system the data rate and the error probability define a channel which is transmitting discrete bits of information. This practical transmission system is also one whose capacity C will be less than that given by equation 2.100. Through this digital channel, bit rates approaching C are aimed for, together with a probability of error approaching zero. The zero-error probability limit, however, is never reached in practice, but may be made arbitrarily small using digital error-control coding techniques. This is a direct consequence of Shannon's theorem which says that it is possible, in principle, for M equally likely messages, with $M \gg 1$, to devise a communication system that will transmit information with an arbitrary small probability of error, provided that the information rate R (the number of messages per second r_s, multiplied by the information content per message H, as given by equation 5.8), is less than or equal to the channel capacity C. That is:

$$R \leqslant C \tag{5.11}$$

The important feature of this theorem is that it indicates that transmission may theoretically be accomplished without error in the presence of noise.

The inverse of Shannon's theorem is that the error probability will increase towards unity as the number of equally probable messages M (bits or words) increases, for $M \gg 1$, if the information rate is greater than the channel capacity. That is

$$R > C \tag{5.12}$$

This means that increasing the complexity of the coding results in an increase in the probability of error.

The concepts of data rate and error probability are to a digital transmission system what bandwidth and signal-to-noise ratio are to an analog system. In addition to this, there is a close relationship between data rate and bandwidth as shown by equation 2.100, and error probability and signal-to-noise ratio, as will be discussed in Chapter 6.

Example

Suppose an analog signal is quantized into Q quantization levels. Determine the channel capacity C required to directly transmit these levels over a transmission path, with negligible error.

Solution

From equation 2.62, the average signal power of the quantized signal is:

$$S = (Q^2 - 1) \cdot \Delta^2/12 \tag{2.62}$$

where S = the quantized analog signal average power.
Q = the number of quantization levels.
Δ = the difference in level between decision thresholds in the quantizer in the absence of noise.

If Gaussian noise is introduced, the difference in levels must by $\sigma\Delta$, where σ is the RMS voltage of the noise, and the noise power $N = \sigma^2$.

Hence in the presence of Gaussian noise, the signal power S as given by equation 2.62 is modified to become

$$S = (Q^2 - 1) \cdot \sigma^2 \cdot \Delta^2/12 \tag{5.13}$$

Rearranging, we have

$$Q = [(1 + (12/\Delta^2) \cdot (S/N)]^{1/2} \tag{5.14}$$

For equations 5.8, 5.9 and 5.14, we find

$$H = \log_2 Q = (1/2) \cdot \log_2 [1 + (12/\Delta^2) \cdot (S/N)] \quad \text{bits/message} \tag{5.15}$$

To transmit a pulse through a low-pass filter requires, for a 10–90 percent rise-time t, that $t = 0.44/B$; hence if the time allocated to permit a message to pass through the filter of bandwidth B is $0.5/B$, the message rate r_s is given by

$$r_s = (1/t) = 2B \quad \text{messages/s} \tag{5.16}$$

since the transmission of any of the messages or levels is equally probable.

$H = \log_2 Q$ and the channel is transferring information at a rate $R = r_s \cdot H$. If the channel is just able to allow the transmission with acceptable probability of error, then $R \simeq C$. From equations 5.15 and 5.16:

$$C \simeq R = r_s \cdot H = B \cdot \log_2 [1 - (12/\Delta^2) \cdot (S/N)] \tag{5.17}$$

This result can be made equal to the Hartley-Shannon equation (equation 2.100) if $12/\Delta^2 = 1$.

In addition to the use made of quantizing an analog signal, this example may be extended to a multilevel signal which is to be transmitted over a channel.

5.2.2 Inter-symbol interference

In Section 5.1 various signaling formats were discussed, and in some cases a brief description of the spectral energy distribution was given for a random pulse sequence to provide an idea of its band occupancy. In each of the various codes described, the output signal after coding is assumed to be a rectangular pulse. If this pulse shape is modified, the manner in which the digital waveform utilizes the baseband channel may be altered. Usually modification of the pulse shape occurs with filtering, so that out-of-band signal components may be removed, thus reducing the potential high-frequency crosstalk components which may couple into other baseband systems. Pre-detection filtering, like pre-transmission filtering, also produces pulse shaping and reduces the noise components lying outside the essential signal bandwidth. Section 1.4 dealt with transmission impairments which can arise in a digital network as the signal is transmitted through a communication channel. One of the most important effects of the combination of the transmission impairments due to the transmission medium and the filtering of the signal at the baseband is the extent to which *inter-symbol interference* (ISI) is produced. If it is assumed that the digit duration is T then, from Section 2.2.1.1 or from the Nyquist bandwidth/signaling-rate relationship, the essential signal bandwidth is $1/2T$. Any random digital sigal, however, occupies an infinite bandwidth and the effect of band limiting can only introduce attenuation distortion and therefore echoes as a result. These echoes spread the energy from one pulse into another adjacent pulse, thus causing what is known as inter-symbol interference. The effect of pulse spreading is known as *dispersion*. The Nyquist bandwidth signaling-rate relationship is described as follows:

'Given an ideal low-pass channel of bandwidth B it is possible to send independent symbols at a rate $r_s \leqslant 2B$ symbols/s without inter-symbol interference. It is not possible to send independent symbols at a rate $r_s > 2B$'.

In addition to this, Section 2.2.1.1 shows that for the upper frequency of a band-pass signal which is far greater than the bandwidth B, the symbol rate approaches

$2B$. Thus, inserting a random digital signal into a Nyquist band-limited channel of bandwidth $1/2T$, i.e. the essential signaling bandwidth, will result in a corrupted output signal, unless there is an absence of channel noise and a synchronous transmitter and receiver pair, when the inter-symbol interference may be contained so that digit errors do not occur. In practice, however, the influence of noise, crosstalk, attenuation distortion, phase distortion, etc., causes ISI to be a dominant factor in the transmission efficiency of a digital system.

5.2.3 Baseband filtering

Figure 5.5(a) shows a pulse which has a width of $2T/b$, centered about an amplitude axis at $t = 0$. The Fourier transform of this pulse is shown also in Figure 5.5(a) as $(1/\pi f) \cdot \sin(fT/d)$, for $b = 2\pi d$, which is derived from the sinc function; $\text{sinc}(x) = \sin(\pi x)/\pi x$. The sinc function is also shown to have a zero crossing at the frequencies $\pm k \cdot b/2T$, where $k = 1, 2, \ldots, n$. This frequency spectrum is unbounded. Where the spectrum is limited by an ideal low-pass filter, the resulting inverse Fourier transform produces a pulse which also has a sinc function response. Referring to Figure 5.5(b) if the ideal low-pass filter cut-off frequency is $1/2T$, the resulting output pulse is spread out in time and does not settle back to the zero level until a time greater than T has elapsed. The original pulse is shown in the dotted portion as having a width of $T/2$ about the zero time axis. This spreading out of the band-limited pulse is the cause of inter-symbol interference for it may interfere with adjacent symbols. Figure 5.5(c) shows the case of an input signal which is a sinc function. The Fourier transform of this sinc pulse is the same as that of the ideal low-pass filter. If a filter with a frequency characteristic between 0 and $1/2T$ which is the inverse of that shown in Figure 5.5(b) is produced, the resulting frequency spectrum would be square topped, and as shown by Figure 5.5(c) the output pulse would be a sinc pulse, as also shown in Figure 5.5(b), and have zero crossings at kT, where $k = 1, 2, \ldots, n$. This output pulse would not display any inter-symbol interference at these crossings. The utilization of the waveform is still not entirely adequate, for the pulse tails decay slowly and stringent requirements are placed upon the design of the filter in terms of it being rectangular and accurately tuned for all time. That of course is impracticable. In practical systems where the bandwidth available for transmitting data at a rate of r_s bits/s is between $r_s/2$ and r_s, a realistic compromise using a 'raised-cosine' filter is commonly used. Figure 5.5(d) shows the spectrum and the dotted band-limited curve shows the band-limited sinc function spectrum. The Fourier transforms of these spectra are also shown in Figure 5.5(d), as a modified sinc function in the time domain, together with the equivalent ideal pulse resulting from a full sinc function spectrum. The raised-cosine spectrum consists of a flat amplitude portion and a roll-off portion that has a sinusoidal form (not shown in figure). The pulse frequency spectrum is given by equation 5.18 as:

$$P(f) = v(f) = \begin{cases} 1/r_s & |f| < (r_s/2) - \alpha \\ (1/r_s)[\cos^2(\pi/4\alpha)[|f| - r_s/2 + \alpha]], & (r_s/2) - \alpha < |f| < (r_s/2) + \alpha \\ 0 & |f| > (r_s/2) + \alpha \end{cases} \tag{5.18}$$

where α is the roll-off factor (see equation 5.22).

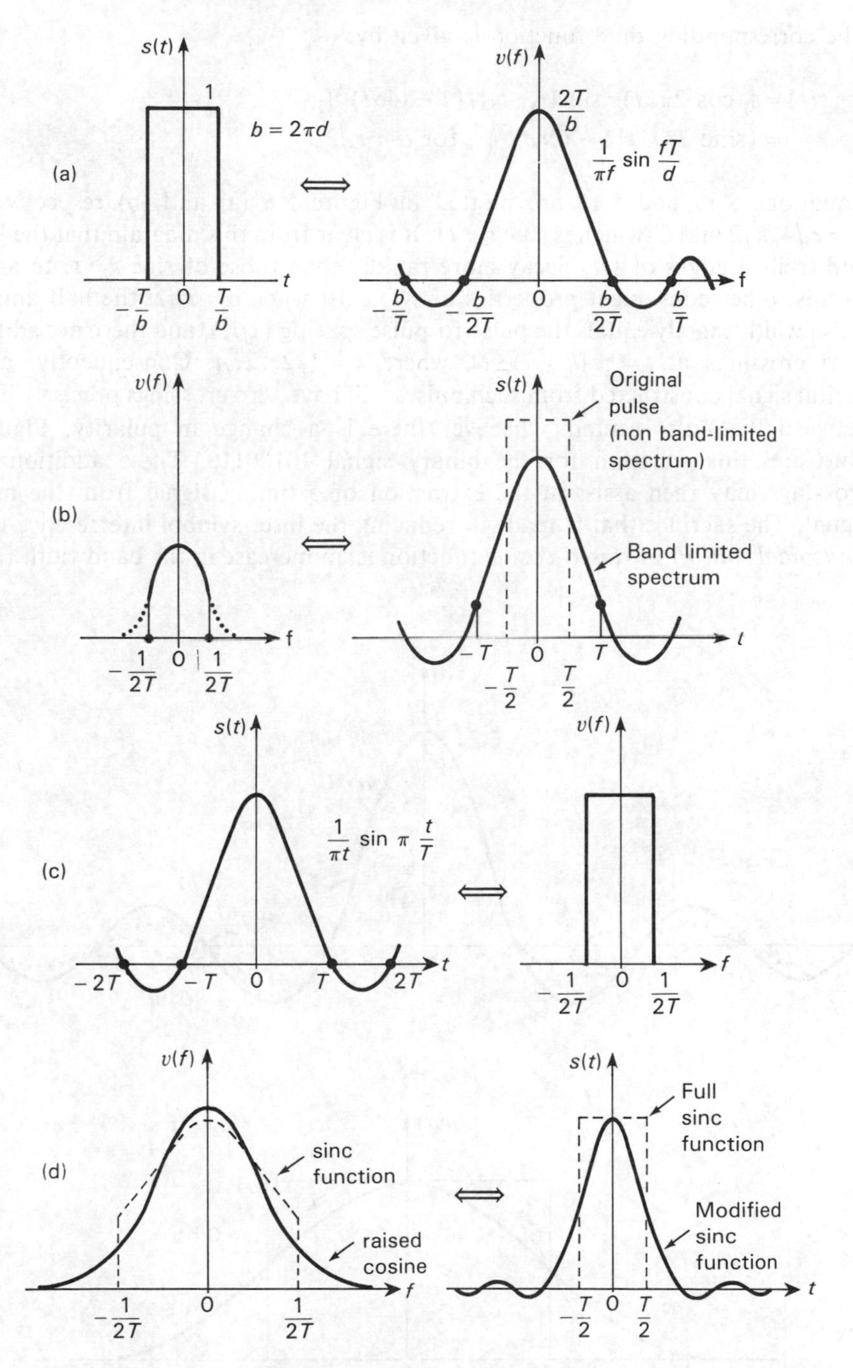

Figure 5.5 Fourier transform pairs of pulse and sinc function signals

The corresponding time function is given by:

$$s(t) = [(\cos 2\pi\alpha t) \cdot \text{sinc}(r_s \cdot t)]/[1-(4\alpha t)^2]$$
$$= [\text{sinc } 2r_s t]/[1-(2r_s t)^2] \quad \text{for } \alpha = r_s/2 \tag{5.19}$$

Equations 5.19 and 5.18 are plotted in Figure 5.6 (a) and (b) respectively for $\alpha = r_s/4$, $r_s/2$ and 0 (which is sinc $r_s \cdot t$). It is clear from this diagram that the leading and trailing edges of $s(t)$ decay more rapidly than those of sinc $r_s \cdot t$. In addition to this, other convenient properties of $s(t)$ exist when $\alpha = r_s/2$: the half-amplitude pulse width exactly equals the pulse-to-pulse spacing ($1/r_s$), and there are additional zero crossings at $t = \pm(k+1) \cdot r_s$, where $k = 1, 2, \ldots, n$. Consequently, a polar digital signal constructed from such pulses will have zero crossings precisely halfway between the pulse centers whenever there is a change in polarity. Figure 5.7 illustrates this situation for the binary signal 10100110. These additional zero crossings may then assist in the extraction of a timing signal from the message signal. The sacrifice that is made in reducing the inter-symbol interference using a *sinusoidal roll-off* or *raised-cosine* function is an increase in the bandwidth require-

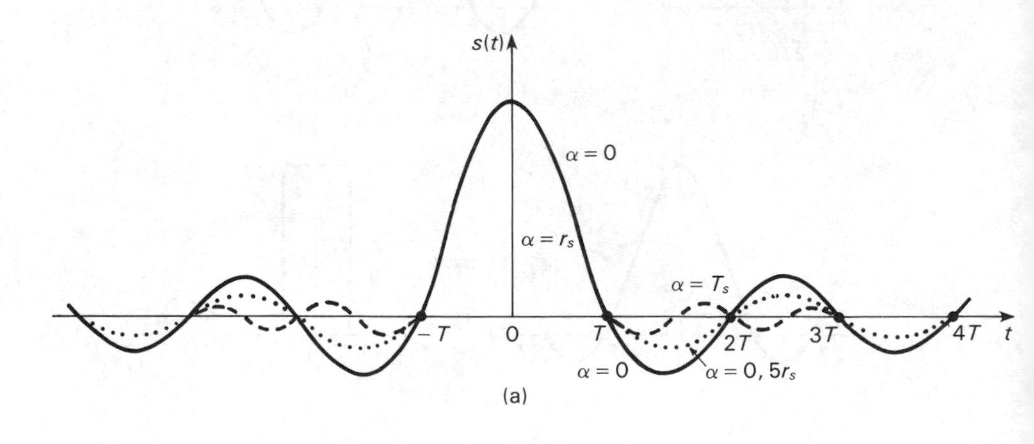

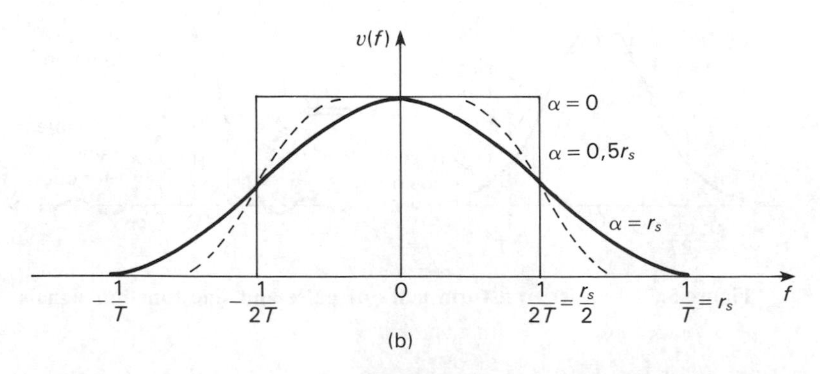

Figure 5.6 Raised-cosine (a) pulse shape and (b) frequency spectrum

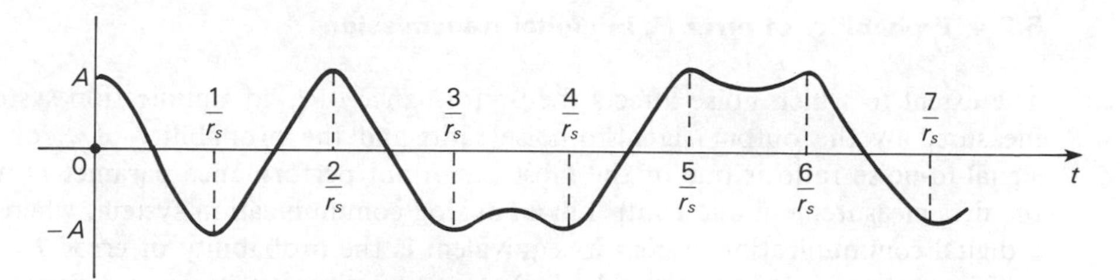

Figure 5.7 Raised-cosine output bit stream for $\alpha = r_s/2$

ment, or, for the Nyquist pulses with $\alpha = r_s/2$, a reduction in signaling speed, since $r_s = B$ rather than $2B$.

When r_s bits/s are transmitted by a particular type of equipment in a baseband channel, which has a bandwidth of B Hz, it is often necessary to know whether this transmission bit rate is better or worse than others produced by other types of equipment, using a different bandwidth. If r_s, the transmission bit rate, is normalized to a bandwidth B, of 1 Hz, the system efficiency can be determined in terms of the transmitted *bits-per-second-per-hertz* or bits/s/Hz. This is frequently employed in comparing different types of digital transmission systems which may utilize different modulation schemes.

In the Nyquist theorems on minimum bandwidth, discussed above, it has been shown that it is possible to transmit r_s independent symbols in a low-pass filter having a bandwidth B of:

$$B = r_s/2 \tag{5.16}$$

In binary transmission, one transmitted symbol contains only one information bit, and so the bit rate r_b equals the symbol rate r_s. Therefore

$$B = r_s/2 = r_b/2 \text{ for binary transmission} \tag{5.20}$$

In M-ary, or M-level, systems, each transmitted symbol contains m information bits, where $m = \log_2 M$. The symbol rate r_s, is given by r_b/m, and so the essential signal bandwidth becomes

$$B = r_b/2m = r_b/(2 \cdot \log_2 M) \text{ for } M\text{-ary transmission} \tag{5.21}$$

As simple as these three equations are, they are most important as the first step in determining the bandwidth necessary to transmit a data stream in coded or uncoded form and permit the spectrum efficiency to be determined in bits/s/Hz. Use of the raised-cosine filter increases the Nyquist bandwidth to B':

$$B' = B + \alpha = r_s/2 + \alpha = \text{practical bandwidth} \tag{5.22}$$

where α is often termed the *roll-off factor* and B is the Nyquist bandwidth.

5.2.4 Probability of error P_e in digital transmission

The extent to which noise affects the performance of a communication system is measured by the output signal-to-noise ratio and the probability of error. The signal-to-noise ratio is one of the most important performance parameters which requires measurement and control in an analog communication system, whereas in a digital communication system its equivalent is the probability of error P_e.

The probability of error may be understood by taking the case of a digital communication system which has, at its input, a sequence of symbols. The output of the system due to the influence of channel noise (which is assumed to be Gaussian) will be a different sequence of digits. In an ideal or noiseless system, both input and output sequences are the same, but in a practical system they will occasionally differ. The overall performance of a digital communication system is measured in terms of the probability of symbol errors P_e, which may be defined as the probability that the input sequence of symbols is *not* equal to the output sequence of symbols. In a practical digital communication system, the values of P_e range from 10^{-4} to 10^{-7}. Throughout the remainder of this book, the P_e will be used to compare the expected theoretical performance of the various modulation schemes for different baseband and radio modulation systems. There are alternative expressions for the probability of error. These alternatives, which are all the same quantity, are:

Average bit-error-rate (BER) = Average bit-error-ratio (BER) =
Probability of error P_e = Error probability P_e

The P_e performance specification is a measure of the *average* performance of a system, but is not indicative of the frequency of error occurrence. It is usually a theoretically derived quantity which does not provide any indication of how many errors occur in one second, or the measurement interval in which measurements of errors occurring in each second is taken. In practice, the bit error ratio (BER) is used together with time intervals to provide performance objectives for digital systems. The CCITT Recommendation G821[10] states that: 'The performance objective is stated in terms of error performance parameters each of which is defined as the percentage of averaging periods each of time interval T_0 during which the BER exceeds a threshold value. The percentage is assessed over a much longer time interval T_L. The time interval T_L, over which the percentages are to be assessed, has not been specified, since the period may depend upon the application. A period of the order of any one month is suggested as a reference.'

The total time T_L, is split into two parts, namely, time for which the connection is deemed available and that time when it is unavailable. The following BERs and intervals are used in the statement of the quality performance objectives:

1. A BER of less than $1:10^{-6}$ for $T_0 = 1$ minute
2. A BER of less than $1:10^{-3}$ for $T_0 = 1$ second
3. Zero errors for $T_0 = 1$ second. (This is equivalent to the concept of error-free seconds (EFS).)

The performance objective for degraded minutes is that fewer than 10 percent of one-minute intervals are to have a BER worse than $1:10^{-6}$. This is based on an averaging period of one minute. The averaging period and the exclusion of errors occurring within severely errored seconds which occur during this one-minute interval period may allow connections with frequent burst errors to meet this particular part of the overall objective, but such events will be controlled to a certain extent by the severely errored seconds objective. The one-minute intervals are derived by removing unavailable time and severely errored seconds from the total time and then consecutively grouping the remaining seconds into blocks of 60. The basic one-second intervals are derived from a fixed time pattern. The performance objective for severely errored seconds is that fewer than 0.2 percent of one-second intervals are to have a BER worse than $1:10^{-3}$. The performance objective for errored seconds is that fewer than 8 percent of one-second intervals are to have any errors. This is equivalent to 92 percent error-free seconds.

As these objectives relate to a very long connection (the Hypothetical Reference Connexion (HRX) which is 27500 km in length) most practical systems which are shorter even on international connections, are expected to offer a better performance than the limiting values given above. The CCITT Recommendation G821 defines the beginning of a period of unavailable time, when the BER in each second is worse than $1:10^{-3}$ for a period of ten consecutive seconds. These ten seconds are considered to be unavailable time at this point, the performance objectives for circuit quality change, to become the performance objectives for circuit availability. The period of unavailable time terminates when the BER in each second is better than $1:10^{-3}$ for a period of ten consecutive seconds, which are considered to be available time. Section 3.6.1.1 discussed the error ratio as defined in the CCIR Report 613–3.[13] For determining the average BER, the theoretical derivation which includes the signal-to-noise ratio out of the receiver will be discussed in this section for PAM baseband systems. Consider a received polar binary signal in which the positive amplitude is treated as a logical one and the negative amplitude, the logical zero. Suppose Gaussian noise is added to this waveform, which has an average amplitude of zero; then, at the optimum sampling instants, which occur when the inter-symbol interference is zero, the effect of this noise can either produce an amplitude close to the intended values of 1 or 0, or an error in which the unintended value of 1 or 0 is produced. If a 1 was intended but the converse, i.e. a 0, was received, the conditional probability of error may be given as:

$$P_{e1} = P(\text{error 1 sent}) = P(0 \mid 1) \tag{5.23a}$$

and similarly

$$P_{e2} = P(\text{error 0 sent}) = P(1 \mid 0) \tag{5.23b}$$

Hence the error probability in a binary transmission system can be written:

$$P_e = P(1) \cdot P(0 \mid 1) + P(0) \cdot P(1 \mid 0) \tag{5.24}$$

where $P(1)$ and $P(0)$ are the digit probabilities at the source and are the occurrence probabilities of 1 and 0 respectively. These are usually equal, but not necessarily so. As either a 1 or a 0 must be transmitted $P(1) + P(0) = 1$. The most basic disturbance which affects the true condition of the 1 or 0 state is that of random thermal noise. As previously mentioned in Chapter 2, this thermal noise is closely modeled by a process having Gaussian statistics. If the mean of the noise is set midway between the bipolar levels representing the logic conditions, then $P(0\ 1) = P(1\ 0)$. Figure 5.8 shows the probability density function for the occurrence of a particular level occurring about the voltage level x_1 which represents a 1, and the voltage level $-x_1$, which represents a 0. From equation 2.16, the probability density function of a zero-mean ($\mu = 0$) Gaussian process is given by:

$$f(x) = \frac{1}{\sigma\sqrt{2\pi}} \cdot \exp(-x^2/2\sigma^2) \quad -\infty < x < \infty \tag{2.16}$$

We develop the theory more generally: if the difference in the two logic levels, x_1 and $-x_1$, is the voltage level difference 2Δ and m_0 is the decision threshold, for the case of Gaussian noise and of a linear signal processing before the decision element, we have:

$$P(1\ 0) = P[-x_1 + \sigma_0 > m_0] = P[\sigma_0 > x_1 + m_0] = Q[(x_1 + m_0)/\sigma_0] \tag{5.25a}$$

$$P(0\ 1) = P[x_1 - \sigma_1 < m_0] = P[\sigma_1 > x_1 - m_0] = Q[(x_1 - m_0)/\sigma_1] \tag{5.25b}$$

where x_1 and $-x_1$ are the amplitudes of the signal corresponding to the logic levels 1 and 0 respectively, σ_1 and σ_0 are the noise standard deviations corresponding to the states 1 and 0 respectively and $Q(X)$ is the *complementary* Gaussian probability

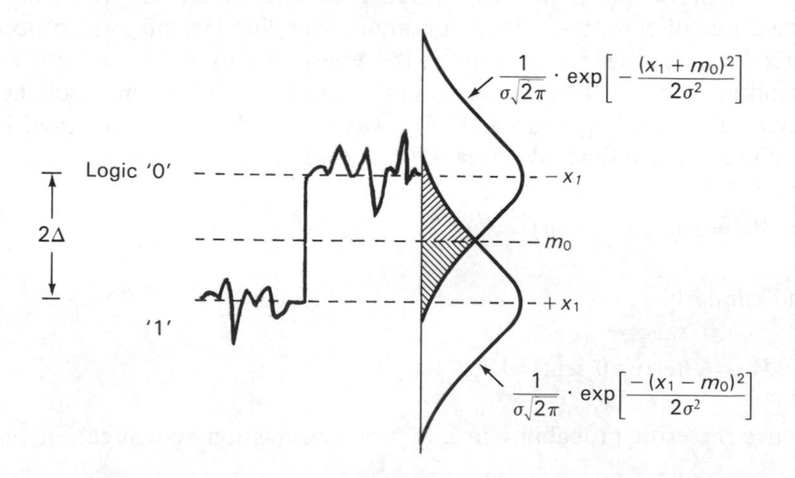

Figure 5.8 Thermal noise distributions about the logic levels x_1 and $-x_1$

density function, given by:

$$Q(X) = Q(x > X) = \int_{x}^{\infty} f(x)\, \mathrm{d}x = \frac{1}{\sqrt{2\pi}} \int_{x/\sigma}^{\infty} e^{-x^2/2}\, \mathrm{d}x \qquad (5.26a)$$

$$= (1/2) \cdot \mathrm{erfc}\,[X/(\sqrt{2}\sigma)] \qquad (5.26b)$$

where X/σ may take on the values $(m_0 + x_1)/\sigma_0$, or $(x_1 - m_0)/\sigma_1$, to produce distributions each having a nonzero mean. From equation 5.24, the probability of error is given as:

$$P_e = P(0) \cdot Q[(x_1 + m_0)/\sigma_0] + P(1) \cdot Q[(x_1 - m_0)/\sigma_1] \qquad (5.27)$$

There is an optimum threshold level $m_{0_{\mathrm{opt}}}$ for which the error probability P_e is minimal. For the case where the binary digits 0 and 1 are assumed to be equally probable, $P(0) = P(1) = 1/2$.

For the special case where *noise is independent of the signal*, $\sigma_0 = \sigma_1 = \sigma$ (as in cable systems). The optimum threshold $m_{0_{\mathrm{opt}}}$, is given when $(x_1 + m_0)/\sigma = (x_1 - m_0)/\sigma$, which means that $m_{0_{\mathrm{opt}}} = 0$ and $x_1 = \Delta$; hence, from $P(1) + P(0) = 1$, equations 5.26b and 5.27 and $X = \Delta/\sigma$, we find:

$$P_e = (1/2)\ \mathrm{erfc}\,[\Delta/(\sqrt{2}\sigma)] = (1/2) \cdot \mathrm{erfc}\,[X/\sqrt{2}\sigma)] \qquad (5.28)$$

Equation 5.28 relates the probability of error P_e to the peak signal/RMS-noise ratio for a binary baseband signal subject to additive Gaussian noise, and thus P_e may be expressed in the form

$$P_e = (1/2)\ \mathrm{erfc}\{[(S/N)/2]^{1/2}\} = (\tfrac{1}{2})\ \mathrm{erfc}\{[C/N]^{1/2}\} \qquad (5.29)$$

where

$$\Delta^2/\sigma^2 = S/N, \text{ and } \left[\frac{\Delta}{\sigma\sqrt{2}}\right]^2 = C/N \qquad (5.30)$$

and C = average signal or carrier power.

When *noise is dependent on the signal*, $\sigma_0 < \sigma_1$ as with optical fiber systems. In this case, $m_{0_{\mathrm{opt}}}$ is therefore lower than zero and the probability of error given in equation 5.27. If the threshold m_0 is chosen to be in the region of $-x_1$, the second term in equation 5.27 becomes negligible, to give:

$$P_e = P(0) \cdot Q[(x_1 + m_0)/\sigma_0] \quad \text{for } m_0 \ll 0 \qquad (5.31)$$

Similarly, for m_0 chosen to be in the region of $+x_1$, the first term in equation 5.27 becomes negligible, to give:

$$P_e = P(1) \cdot Q[(x_1 - m_0)/\sigma_1)] \text{ for } m_0 \gg 0 \qquad (5.32)$$

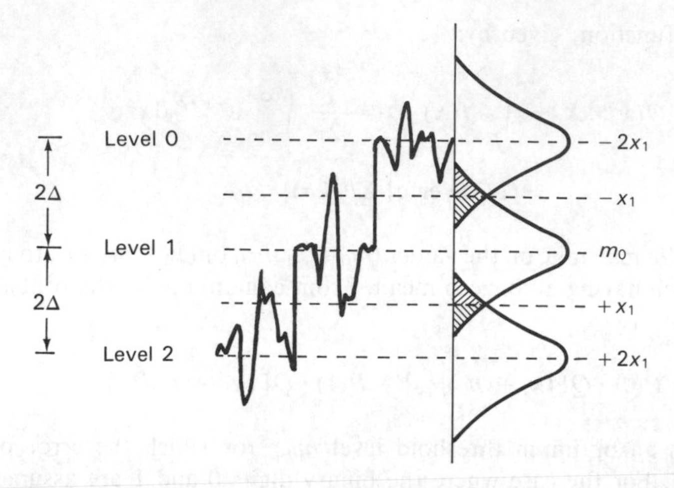

Figure 5.9 Thermal noise distributions about a three-level signal

These two expressions (equations 5.31 and 5.32) infer that a general expression for the probability of error P_e can be established. Equation 5.33 gives this general expression, where the coefficients A and X are the two characteristic constants of the system, and S/N is the signal-to-noise ratio of the receiver output.

$$P_e = A \cdot Q[X \cdot (S/N)] \tag{5.33}$$

If the binary signal is extended to a three-level signal with outputs $2x_1$, 0 and $-2x_1$ representing the digits 2, 1 and 0 respectively, there are three Gaussian probability density functions and two threshold levels as shown in Figure 5.9. If each of the digits is equiprobable, the optimum thresholds are found to be $\pm x_1$, for which:

$$P_{e_2} = P_{e_0} = (1/2)\,\mathrm{erfc}[x_1/\sqrt{2}\sigma] \tag{5.34}$$

whereas

$$P_{e_1} = \mathrm{erfc}[x_1/\sqrt{2}\sigma] \tag{5.35}$$

and

$$P_e = P_2 \cdot P_{e_2} + P_1 \cdot P_{e_1} + P_0 \cdot P_{e_0} = (4/3) \cdot (1/2) \cdot \mathrm{erfc}[x_1/\sqrt{2}\sigma] \tag{5.36}$$

5.2.4.1 P_e for M-ary baseband systems

In M-ary baseband PAM systems, the output of a pulse generator produces any one of M possible levels, where $M > 2$. Each level corresponds to a separate input symbol of which there are M. From equation 5.9 the source information rate for equiprobable and independent symbols is $r_s \cdot \log_2 M$ bits/s and requires a minimum bandwidth of $r_s/2$ Hz. In comparison, a binary scheme transmitting data at the same rate, that is $r_s \cdot \log_2 M$, will require a bandwidth of $(r_s \cdot \log_2 M)/2$ Hz, which is in accordance with equation 5.20. The advantage of bandwidth reduction in M-ary signaling schemes is offset by the increase in the power requirements and complexity of equipment.

Extending equation 5.36 to an arbitrary M levels with equal digit probabilities gives:

$$P_e = [1 - (1/M)] \cdot \text{erfc}[x_1/\sqrt{2}\sigma] \tag{5.37}$$

where the spacing between adjacent output pulse amplitudes is $2x_1$. This means that $2(M-1)x_1$ is the peak-to-peak range and there are $M-1$ threshold levels centered between the pulse amplitudes as discussed in the PCM quantizing arrangement of Section 2.2.1.2.

In Section 2.3.3 the Gaussian noise power N_0 out of an ideal receiver was defined as

$$N_0 = \sigma^2 = \eta B \tag{5.38}$$

where η is the power spectral density and is measured in W/Hz, and σ is the RMS value of the noise or the Gaussian distribution standard deviation. If the nominal received levels for a separation of 2Δ between adjacent levels are equally likely to assume any of the M-possible levels given by

$$\Delta_i \begin{cases} \pm\Delta, \pm 3\Delta, \pm 5\Delta, \ldots, \pm(M-1)\cdot\Delta & M \text{ even} \\ 0, \pm 2\Delta, \pm 4\Delta, \ldots, \pm(M-1)\cdot\Delta & M \text{ odd} \end{cases} \tag{5.39}$$

then the signal power for these level spacings of 2Δ is found, with the aid of equations 2.9, 2.13 and 2.62 (where the spacings are given as Δ) to be:

$$S = E(X^2) = (M^2 - 1) \cdot (\Delta^2/3) \tag{5.40}$$

Hence, from equations 5.38 and 5.40, the peak signal-to-noise ratio at the output of the receiver is

$$S/N = S_0/\eta B = (M^2 - 1) \cdot (1/3) \cdot (\Delta^2/\sigma^2) \tag{5.41}$$

As $x_1 = \Delta$ in equation 5.37, the P_e for a M-ary baseband system becomes, by substitution from equation 5.41

$$P_{e_{M\text{-ary}}} = (M-1) \cdot (1/M) \cdot \text{erfc}\left[\frac{1.5\,(S/N)}{(M^2-1)}\right]^{1/2} = \frac{M-1}{M}\,\text{erfc}\left[\frac{3(C/N)}{M^2-1}\right]^{1/2} \tag{5.42}$$

where C/N is the average carrier power to noise ratio.

Note that when $M = 2$, i.e. the polar binary case, equation 5.42 reduces to that of equation 5.29, as expected. This equation is important and is shown plotted in Figure 5.10 for various values of M. To ease calculations, a table of values for the complementary error function erfc(t), is provided in Table A1 of Appendix A.

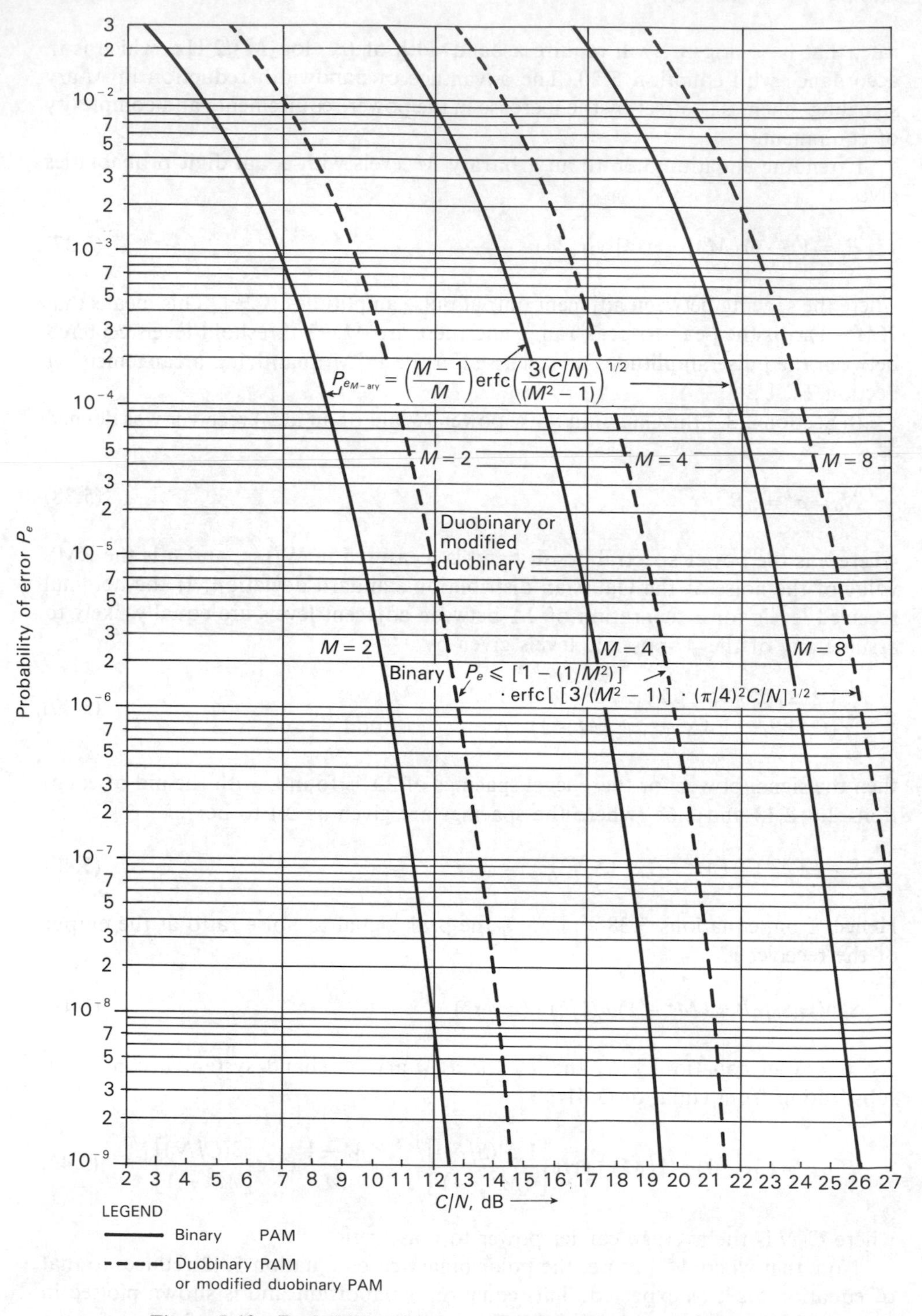

Figure 5.10 Probability of error curves for M-ary PAM systems versus C/N dB

Example 1

Determine the value of S/N required at the output of a quaternary system ($M = 4$), if the probability of error is to be no less than 3.51 parts in 1000.

Solution

From equation 5.42, with $M = 4$, $\text{erfc}(t) = 0.00468$. From Table A1 of Appendix A, when $\text{erfc}(t) = 0.00468$, $t = 2$. Hence $S/N > 40 = 16$ dB.

Example 2

With a signal-to-noise ratio of 28 dB, determine the expected P_e, for $M = 8$.

Solution

$S/N = 631$, $M = 8$, and $\text{erfc}(3.8759249) = 3.93 \times 10^{-8}$. Hence, from equation 5.32, $P_e = 3.44 \times 10^{-8}$.

5.2.4.2 P_e for partial response baseband systems[19–24]

Partial response or correlative techniques introduce a controlled amount of inter-symbol interference over one or more bits to produce a spectral reshaping of the binary or multilevel pulse trains. Consequently, for a given bandwidth and power input to the channel, the bandwidth efficiency is raised higher than that of the Nyquist system, to give a bandwidth equal to the digit rate ($1/T$), for a specified error probability. The Nyquist limit was devised in terms of binary transmission and thus the reduction in bandwidth using a multilevel signal does not contravene this limit. Partial response encoding also provides a method of introducing a DC spectral null in the spectral energy of the resulting bit stream. This technique[19] involves replacing each 1 in the incoming binary signal by a sequence of multilevel symbols in a number of time slots, such as replacing each 1 by the sequence 1, 0, − 1 as in the 'class 4 partial response encoding'.

In the Nyquist theorems outlined in Sections 5.2.1, 5.2.2, and 5.2.3, the inter-symbol interface at the sampling instants and the transition points between successive digits can be eliminated. Thus the digits are considered independent and uncorrelated and may be recovered without depending on the past history of the signal. This type of system is known as a *zero-memory system*. The partial response concept involves correlation between groups of digits by the introduction of a controlled amount of inter-symbol interference. Due to this correlation between digits, the system has a nonzero or finite memory. One of the specific techniques of partial response techniques is that of the 'duobinary' baseband PAM system, in which the 'duo' means that the bit rate is double that of a binary baseband system. Other classes of signaling schemes are modified 'duobinary' and 'polybinary' schemes. As discussed in Section 5.2.3, the baseband binary PAM data transmission system requires a bandwidth of at least $r_b/2$ Hz in order to theoretically transmit data at a rate of r_b bits/s, with zero inter-symbol interference. If the bandwidth available is exactly $r_b/2$, then the only way in which binary PAM data could be transmitted at a rate of r_b bits/s without inter-symbol interference would be by using an ideal rectangular low-pass filter, with a cut-off frequency $r_b/2$, at the transmitter and receiver. This would insure that there would be no inter-symbol interference at the sampling instants $1/r_b$ seconds apart. Such a filter is physically unrealizable. But if it could be made, its practical use would be precluded because of the excessive inter-symbol interference that would occur at the transition points, owing to the over-shoots of the pulse tails, which slowly decay as shown in Figure 5.6. Any such system would be extremely sensitive to perturbations in bit rate, timing or channel characteristics. By removing the constraint of having a bit rate less than r_b bits/s, physically realizable pulses can be made to exist which satisfy the constraint of no inter-symbol interference at the sampling instants. If the condition is made that the bit rate is to be at r_b bits/s, but the constraint that there should be no inter-symbol interference at the sampling instants is removed, a class of physically realizable pulses called *partial-response signals* are obtained, which have the capability of being transmitted in the same bandwidth as the Nyquist bandwidth, but at the expense of a higher carrier power into the receiver. One special case of these partial response signals is the duobinary signal pulse. It is specified in the frequency domain as:

$$S(f) = \begin{cases} (1/r_b)(1 + e^{-j2\pi f/r_b}) = (2/r_b) \cdot e^{-j2\pi f/r_b} \cos(\pi f r_b) & |f| \leq r_b/2 \\ 0 & |f| > r_b/2 \end{cases} \tag{5.43}$$

and

$$|S(f)| = (2/r_b) \cdot \cos(\pi f/r_b) \tag{5.44}$$

The duobinary signal pulse $v(t)$ and the frequency characteristic of the transmitted pulse are shown in Figure 5.11. The duobinary scheme introduces controlled

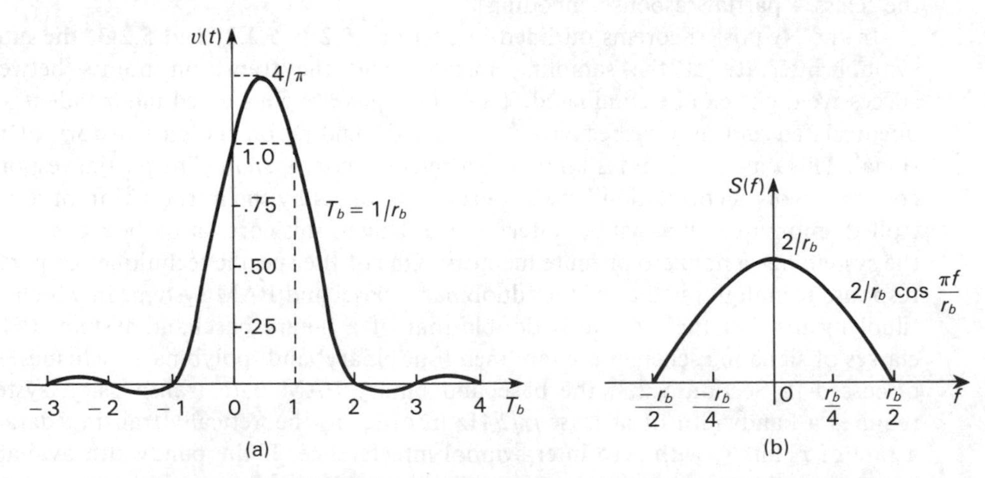

Figure 5.11 (a) Time and (b) frequency domain characteristics of a duobinary signal

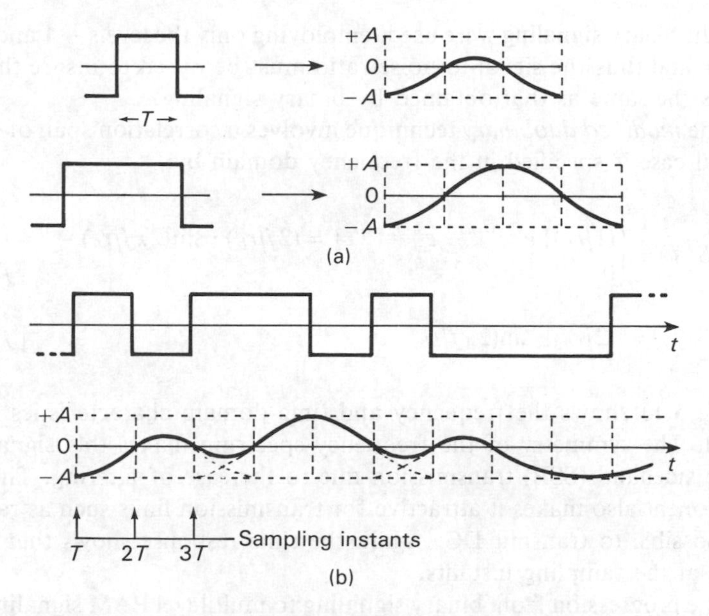

Figure 5.12 (a) Basic response waveforms for partial-response signaling; (b) the partial-response technique applied to a random sequence of binary digits
(Courtesy of Macmillan Press, Reference 25)

amounts of inter-symbol interference for transmitted data at a rate of r_b bits/s over a channel with a bandwidth of $r_b/2$ Hz. The filters for the duobinary system are not difficult to realize as shown by the absence of sharp discontinuities in the frequency characteristic. However the scheme does require more transmitter power than the binary scheme. Figure 5.12 shows how, for a random sequence of non-return-to-zero (NRZ) pulses, these pulses may be recovered when there is a controlled amount of inter-symbol interference. Another point about the duobinary system is that there is inter-symbol interference between digits over a correlation span of one.

The process of recovery lies in the response of the overall transfer function of the system being that given in equation 5.43. This may, for example, comprise a transmit filter which provides the raised cosine characteristic followed by a sharp cut-off low-pass filter with a transfer function equal to $2/r_b$. The multiplication of both transfer functions provides the system response of equation 5.44. A single digit of the NRZ signal will just intersect the zero voltage level. The direction of intersection will depend on the sign of the previous digit. Hence the term 'partial response'. The response of the entire transmission system is shown in Figure 5.12(*b*). To recover the original waveform from the partial response waveform, the latter is sampled at equal intervals of time T as indicated in the diagram. The reconstructed waveform is -1 if the partial response waveform is at a negative peak; it is $+1$ for a positive peak and is of the opposite sign to the previous value if the partial response waveform is at zero level. As the separation between the three levels of decision $+1$, 0 and -1 is only one-half of that which would occur if straight binary signaling were used, employing only the levels +1 and −1, the signal power and thus the signal-to-noise ratio must be raised to insure that the bit error rate is the same as that obtained by binary signaling.

The *modified duobinary* technique involves a correlation span of two digits. This special case is specified in the frequency domain by:

$$S(f) = \begin{cases} (1/r_b)[e^{j2\pi f/r_b} - e^{-j2\pi f/r_b}] = (2j/r_b) \cdot \sin(2\pi f/r_b) & |f| \leqslant r_b/2 \\ 0 & |f| > r_b/2 \end{cases} \quad (5.45)$$

$$|S(f)| = (2/r_b) \cdot \sin(2\pi f/r_b) \qquad |f| \leqslant r_b/2 \quad (5.46)$$

Figure 5.13 shows the frequency and time domain characteristics of this class of signal. The symmetry of the frequency spectrum makes this signal attractive for single sideband (SSB) transmission due to the ease of filtering. The lack of a DC component also makes it attractive for transmission links such as radio, where it is not possible to transmit DC. Again the time response shows that there are three levels at the sampling instants.

The progression from binary signaling to multilevel PAM signaling using partial-response pulses of the duobinary and modified duobinary signals, is relatively simple. When either of these two pulses is used for signaling, the transmission of one pulse of M equally spaced levels operating at the Nyquist rate gives rise at each sampling instant, in the absence of noise, to one of $2M - 1$ equally spaced signal levels. At the receive end, the $2M - 1$ levels are mapped back into an M-level signal pulse. A description of the precoding before transmission, to avoid error propagation and decoding at the receiver, is given on pp. 344–346 of Reference 26, and on pp. 155–168 of Reference 24.

The probability of error for a duobinary and a modified duobinary M-level signal is given by

$$P_e \leqslant [1 - (1/M^2)] \cdot \text{erfc}[[3/(M^2 - 1)] \cdot (\pi/4)^2 \cdot C/N]^{1/2} \quad (5.47)$$

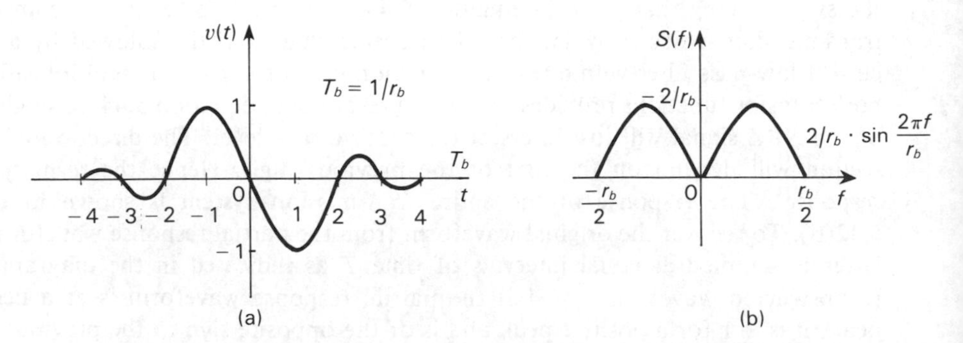

Figure 5.13 (a) Time and (b) frequency domain characteristics of a modified duobinary signal

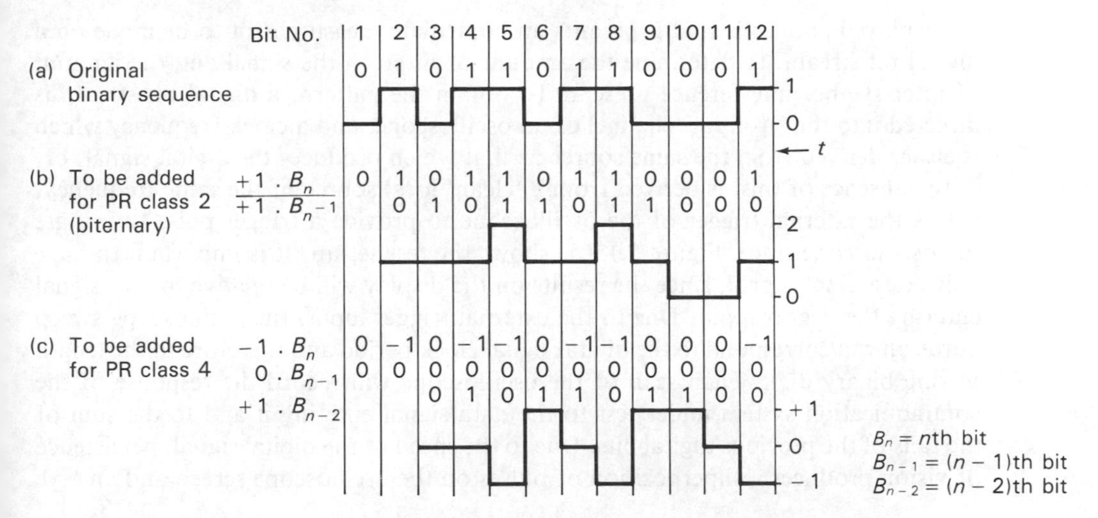

Figure 5.14 Two examples of partial response (PR) codes

Any group of k binary digits can be uniquely mapped into one of the M numbers; $0, 1, 2, \ldots, (M-1)$.

In comparison with the curves in Figure 5.10 for the M-ary and binary PAM systems, the duobinary and modified duobinary signals result in an increase of $20 \log(\pi/4)$ or 2.1 dB in signal-to-noise ratio for large values of M, if the same P_e is to be attained. Figure 5.10 shows the plots of P_e for both duobinary and modified duobinary for different average carrier-to-noise ratios (C/N). These curves were derived from equation 5.47.

In some radio systems, two bit rates r_b and $2r_b$ can be transmitted with each of the system filters: the pre-modulation filter, the IF filter and the post-modulation filter. The signal within the transmitter or receiver is transmitted and received at the bit rate r_b in binary form and at the bit rate $2r_b$ in '*biternary*' form. Figure 5.14 shows two examples of partial response (PR) codes. These codes are PR class 2 and PR class 4. The partial response code may be a line code in which each signal element is generated from m consecutive elements of another code, in such a way that these m elements, each provided with a weighting factor, are linearly added together. Partial response codes are used for shaping the signal spectrum for transmission as desired.

5.2.5 Eye diagrams[11,27]

When a new digital transmission system is planned, amongst other things the system degradations caused by inter-symbol interference (ISI) must be specified. This is usually done by a computer, which produces a printout of the predicted eye diagram. In real-life measurement, an eye diagram is a pattern which is shaped like an eye displayed on the screen of an oscilloscope. The importance of this diagram or displayed pattern is that it permits an immediate measurement to be made on a digital bit stream, to determine the amount of jitter on the signal and the amount of inter-symbol interference present. To obtain the pattern, a digital bit stream is directed into the 'A-input' channel of an oscilloscope, and a clock frequency which is either derived from the same source as that which produces the digital signal, or, in the absence of this, is derived from a 'clean' local source of the same frequency, enters the external trigger of the oscilloscope to provide a trigger pulse to initiate the oscilloscope trace. Figure 5.15(*a*) shows the test set-up. It is important to have a jitterless clock signal, since the results on the display will be relative to that signal entering the trigger input. Due to the external trigger input, the oscilloscope sweep duration will correspond to the digital signal clock period and therefore to the length of one binary digit. Each scan of the oscilloscope will record the response of the communication system under test to the data signal entering it and to the sum of the tails of the previous digital bits. Due to the speed of the digital signal, persistence of vision produces a superposition of pulses on the oscilloscope screen and an 'eye

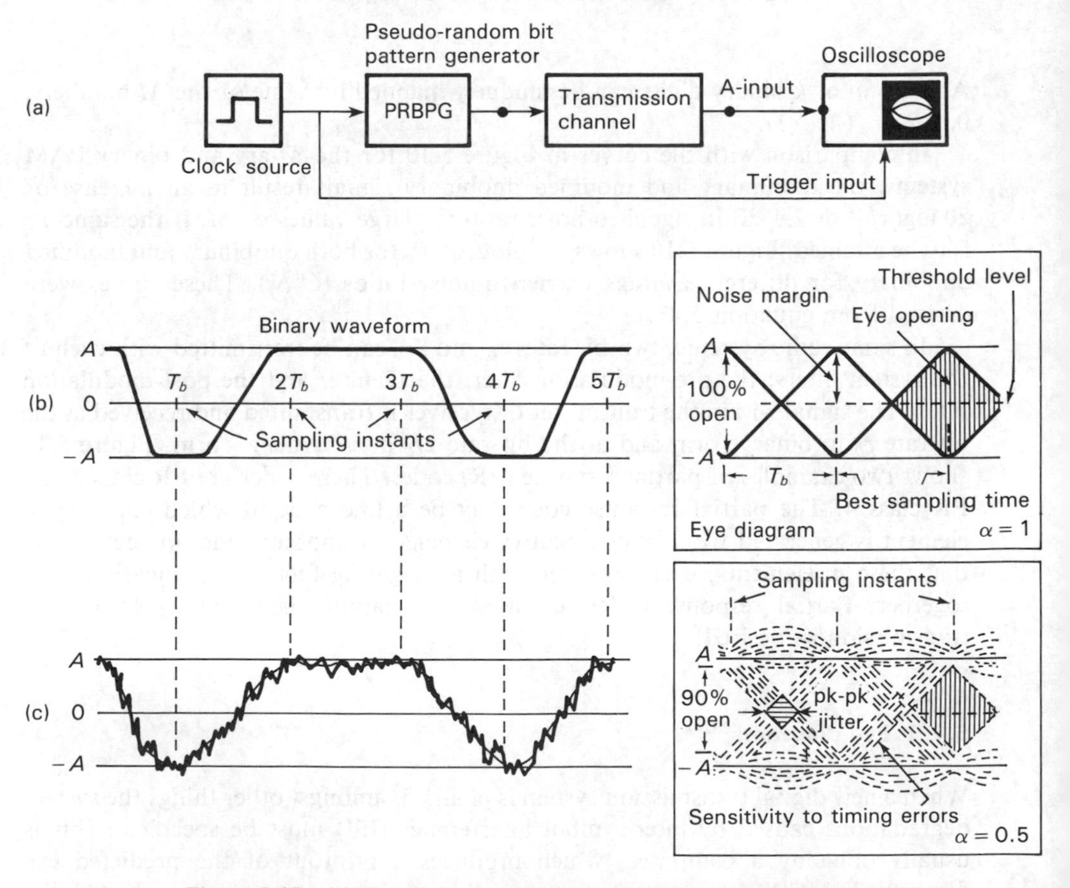

Figure 5.15 (a) Eye diagram test set-up, and (b) and (c) binary PAM system eye diagrams

pattern' is produced. If each digital bit displayed is identical to the last in amplitude and no overlapping of bits occurs, a clean pattern is produced, resulting in an eye which is 'wide' open, i.e. 100 percent open. The superposition of waveforms so obtained is such that the eye opening defines a boundary within which no waveform trajectories can exist under any condition of code pattern. If the transmission system produces a response to a digital bit which overlaps with the previous and the next bit (inter-symbol interference), the effect is to produce a 'closing' of the eye. The more the inter-symbol interference, the more the eye closes. This effect can be produced in practice by decreasing the bandwidth of a filter, or by adding white noise to the system. The amount of inter-symbol interference can be measured in terms of the *inter-symbol interference (ISI) degradation* and is given by:

$$\text{ISI degradation} = -20 \log[\text{percentage of eye open}/100] \quad \text{dB} \tag{5.48}$$

For example, it the eye is only 10 percent open, the inter-symbol interference degradation is 20 dB, which is so bad that the system would be unusable. A 90 percent eye opening, or a 0.9 dB ISI degradation, is tolerable. The inter-symbol interference degradation has the same effect as increasing the error rate when it occurs with additive white noise. Consequently it also has the effect of shifting the curves of Figure 5.10 to the right, for a given probability of error. In other words, the signal-to-noise ratio must be increased by approximately the inter-symbol interference degradation if the probability of error is to be the same as that without ISI degradation. In addition to the opening of the eye, the eye diagram shows the amount of jitter contained in the digital signal. If the digital signal contains no jitter relative to the oscilloscope trigger input signal, the corners of the eye come to a point and are sharp. This is because the zero-crossings or receiver detector threshold occurs at the same point on the slope of the digital bit pulse response. With jitter on the digital signal, at each of the zero crossings, the time position of the digital waveform differs. This gives rise to a fuzziness at the edges of the eye. The width of the fuzziness is a measure of the peak-to-peak jitter.

Figure 5.15(*b*) shows the eye diagram for a binary PAM system with a noiseless and jitterless digital signal, free from inter-symbol interference and Figure 5.15(*c*) shows it for a jittery signal with noise and inter-symbol interference. By referring to Figure 5.15(*b*) and (*c*) the following information can be gained from an eye diagram:

The optimum sampling instant of the received waveform is that instant when the eye opening is largest.

The optimum sampling instant is midway between the zero crossings. If the clock signal is derived from zero crossings the width of the eye pattern at these zero crossings determines the amount of peak-to-peak jitter occurring on the digital signal.

The nominal decision point for the receiver detector is the crosspoint between the zero crossing or threshold level and the optimum sampling instant.

The inter-symbol interference degradation is at a maximum, and is measured, at the optimum sampling instant.

The immunity to noise or the *noise margin* is proportional to the width of the eye opening and is:

20 log [(peak-to-peak amplitude of the eye opening)/2]

The sensitivity of the transmission system to timing errors is directly proportional to the rate of eye closure with the variation of sampling time.

The eye diagram should be symmetrical horizontally and vertically and each pattern identical. If this is not so, the transmission system is nonlinear and has a fault.

Practical data channels always contain jitter noise, which shifts the sampling instants from their ideal positions, or from their positions where inter-system interference is minimum. This in turn increases the inter-symbol interference. A symbol is insensitive to timing offset if the rate of change of the eye closure near the peak eye opening is small. Due to the interrelationship between jitter and inter-symbol interference, for comparing eye diagrams with different peak openings and jitter factors, a combined distortion factor[28] F_m, is defined, in which the sensitivity to timing offset J, otherwise known as the jitter factor, and the peak distortion $D(t_o) = (1 - \text{peak eye closure})$ are related by the expression:

$$F_m = D(t_o) + WJ \tag{5.49}$$

where W is a function of the statistics of the timing offset. Reference 28 arbitrarily chooses W to be 0.1 in order that $D(t_o)$ and WJ are of the same order.

5.2.5.1 Eye contour maps[29]

It has been shown[29] that the eye diagram may provide misleading information of the system margins when the transmission path is designed to have a fairly abrupt cut-off. The conventional eye diagram assumes that the system is operating at precisely the Nyquist rate. However, the eye opening is also a function of the digit rate and the overall system cut-off frequency response. Usually the digit rate is controlled and is quite stable. The system cut-off frequency does not necessarily remain so rigorously controlled. If a second eye diagram is produced in which the digit rate or the cut-off frequency of the system was varied, the difference between the original diagram and the second one could be shown against the effective digit period. Ingram[29] has proposed that a better representation of the system margins can be obtained by an eye contour map, in which the eye opening is depicted by a series of contours as a function of both the timing offset and the digit period or digit rate. The contour map may be depicted as a three-dimensional roll-off surface map in which the three orthogonal axes are timing offset, digit period and eye opening. Figure 5.16 shows examples of both the contour map and the roll-off surface map.

Phase distortion has been shown to reduce significantly the eye opening and the eye width,[34] and for a very gradual full raised-cosine function as given by equation 5.18, produces 6.5 percent eye closure to 70 percent closure for the ($\alpha = 0.25$) cosine roll-off and complete eye closure for the ($\alpha = 0.125$) cosine roll-off factor using a

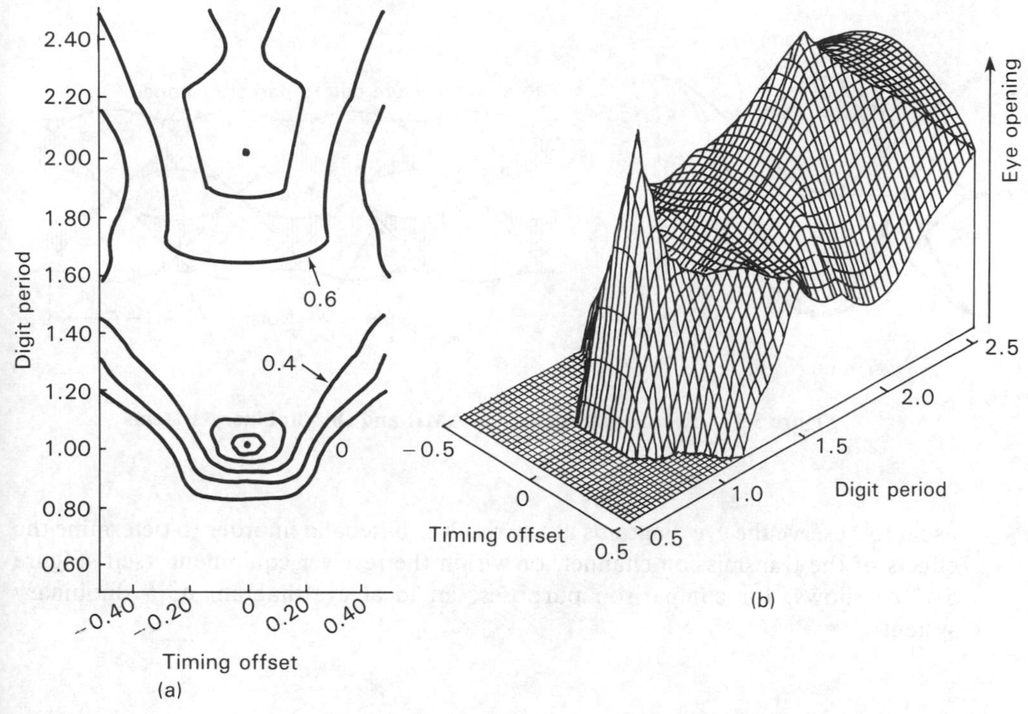

Figure 5.16 (a) Eye contour map, and (b) roll-off surface map (Courtesy of IEE, Reference 28)

15th-order rational function with an asymptotic rate of cut-off of 300 dB/decade. The Butterworth function eye diagram showed a 98 percent closure, and the Chebȳshev function resulted in a completely closed eye.

Eye diagams are not necessarily restricted to binary PAM. They may also be generated for duobinary and *M*-ary PAM systems as well as other modulation schemes as discussed in Chapter 6. The eye opening and structure is dependent on the number of code levels and the line code used. If an AMI coded signal is used, as in Section 5.1.1, there exists in the interval between the sampling instants three signal level states with the combination: 000, 00 + 1, 00 − 1, 0 + 10, 0 − 10, 0 + 1 − 1, 0 − 1 + 1, + 100, + 10 − 1, + 1 − 10, + 1 − 1 + 1, − 100, − 10 + 1, − 1 + 10 and − 1 + 1 − 1. These code combinations cause an eye diagram to be produced in which there is an open part of the pattern where the composite pattern created by the sequence of digital signals does not traverse. The 'open' area is thus not affected by all the possible combinations of the AMI coded signal (or any other coded signal for that matter) and is thus ideal for the detection process of the receiver. Figure 5.17(*a*) shows an ideal eye diagram of an AMI, HDB3 or ternary coded signal and it can be seen that the structure of the diagram due to the coding differs from that of the binary PAM signal of Figure 5.15. It can also be noticed that there does exist an opening in both cases in which the resultant signal trajectories do not traverse and these areas or eye openings are independent of the coded signal structure. It is

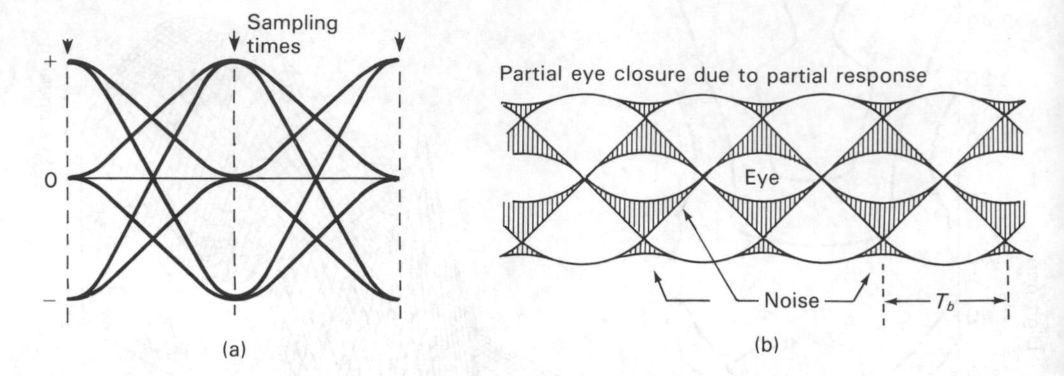

Figure 5.17 Eye diagrams for: (a) AMI, and (b) duobinary systems

usual to observe the eye diagrams at the receiver baseband in order to determine the effects of the transmission channel, or within the receiver equipment itself. Figure 5.17(*b*) shows, for comparison purposes, an ideal eye diagram of a duobinary system.

5.3 ERROR-CONTROL CODING[30-36]

As mentioned in Section 5.1.9, redundancy is the unused transmission capacity of a coded signal. This redundancy, however, is often used to permit the detection and sometimes the correction of noise-induced errors at the receiver, thus improving the system reliability. There are three kinds of error which may arise in a digital transmission system. These are: errors of substitution, in which the original digit is replaced by some other state, errors of omission, in which a symbol has been deleted from the bit stream; and errors in which a spurious symbol is inserted into the bit stream. The most frequent error is that of substitution. All kinds of error may occur in a binary bit stream or in a multilevel symbol. As the error probability in digital transmission is a direct function of the signal-to-noise ratio, a problem may arise, if for some reason or other, the signal power is limited to some maximum value which permits the occurrence of errors to arise with unacceptable frequency. These errors may be reduced with the aid of error-detection and -correcting circuits. The most extensive use of error-protecting coding is in the transmission of digital data, telegraphy and facsimile, whereas for speech it is not so critical. Since an integrated systems data network is now being phased in, it is becoming more difficult to determine those areas which require error protection and those which do not. The inevitable outcome is to provide error control coding on all systems to ensure their flexibility. Two distinct types of error conditions arise in transmission. The first kind are *random* errors which are uncorrelated between errored digits. The second kind involve error *bursts*, where a large number of consecutive digits are corrupted. The prevalence of either of these two kinds of error is a determining factor in the type of error code to be used. Error bursts occur from equipment such as scramblers or line coding equipment, where a single error is decoded as a word other than which originated. This in turn causes other errors to occur, thus multiplying or *extending* the error. The usual term for this type of error is, in fact, *extension error*. The effect of extension error introduces the term 'equivalent bit error rate' (EBER), in which a bit error rate measured under conditions known to be caused by extension error is reduced by an error extension factor to arrive at a figure of equivalent bit error rate. Error bursts are also typical of data transmission through the *switched* telephone network, in which impulsive noise rather than Gaussian noise predominates.

The most extensive use of error-protecting codes is to safeguard digital data and telegraphy circuits. Encryption systems used by the military and computer systems together with secure telephone voice communications are used to protect sensitive information by means of suitably chosen codes. These systems are used in the context of coding for data base communication as distinct from coding for error protection or detection. Error control coding is usually performed by modems using telephone connections, or a line communication channel, rather than in a digital line-of-sight trunk radio system. Various error-correcting codes have been evaluated and tested on HF, troposcatter and satellite systems.[36] As error control coding is of little relevance to the subject of this text, the reader is referred to some of the extensive treatments on this subject cited in the references.

5.4 EQUALIZATION

Due to imperfect filter design in practical systems and the changes in the communications channel characteristics, some amount of inter-symbol interference (ISI) is always present. To compensate for, and reduce the ISI to a level where it may be considered to be eliminated, we may construct a frequency-selective network, which has an amplitude and phase transfer function, that is the inverse of that of the channel and channel filter combination. Such a network is called an equalizer, and the process of compensation is often called 'mop-up equalization'. If the channel varies as the temperature changes; or as with a radio link, the climatic and environmental conditions change, which lead to multipath fading; or the equipment filters change, etc.; then the inverse characteristic, which may have been valid on initial setting, will no longer produce minimum ISI. The resultant variation from minimum ISI to maximum ISI under all conditions may be so large that it is unacceptable. This is especially so if the channel characteristics are time varying at a rate which allows significant changes to occur during the period of time that the channel is occupied. To overcome this situation, an automatic method of correcting the channel induced distortion, otherwise known as channel dispersion, is achieved using the *adaptive equalizer*. In this section we shall briefly describe the various types of equalizer and their uses.

5.4.1 Transversal equalizer

The transversal equalizer shown in Figure 5.18 consists of a delay line of $2M+1$ taps, each tap separated by the time interval T_s seconds, and connected through a variable gain device, designated $C_{-m}, \ldots, C_0, C_1, \ldots, C_m$ for the $2M+1$ taps, to a summing amplifier. If the known input pulse shape to the equalizer is $p_r(t)$ and the equalized output is $p_{eq}(t)$, the relationship between them both is

$$P_{eq}(t) = \sum_{m=-M}^{M} C_m \cdot p_r[t-(m+M)\cdot T_s] \tag{5.50}$$

Assuming $p_r(t)$ has its peak at $t=0$ and there is inter-symbol interference on both sides, the rate at which the output should be sampled is given by

$$t_k = (k+M)T_s \tag{5.51}$$

to give:

$$P_{eq}(t_k) = \sum_{m=-M}^{m} C_m \cdot p_r[(k-m)\cdot T_s] \tag{5.52}$$

To eliminate ISI, ideally we should have:

$$P_{eq}(t_k) = \begin{cases} 1 & \text{for } k=0 \\ 0 & \text{for } k\neq 0 \end{cases}$$

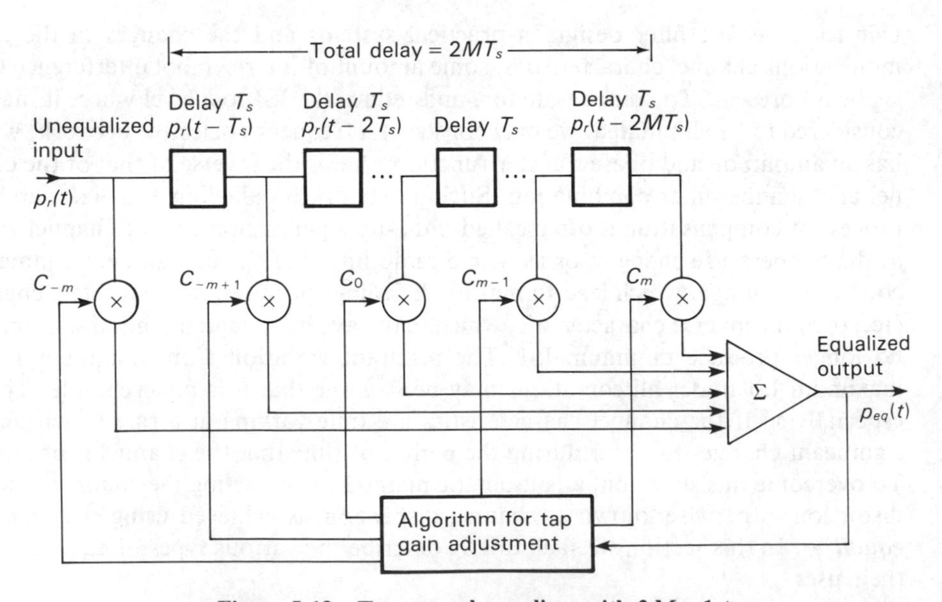

Figure 5.18 Transversal equalizer with $2M+1$ taps

As there are only $2M+1$ tap gains against the required infinite number, the value of $p_{eq}(t_k)$ may only be specified at the $2M+1$ points:

$$P_{eq}(t_k) = \begin{cases} 1 & \text{for } k = 0 \\ 0 & \text{for } k = \pm 1, \pm 2, \ldots, \pm M \end{cases} \tag{5.53}$$

From equations 5.52 and 5.53, a set of $(2M+1)$ simultaneous linear equations can be solved for each C_m. This set is given in matrix form as:

$$\begin{bmatrix} 0 \\ 0 \\ \vdots \\ 0 \\ 1 \\ 0 \\ \vdots \\ 0 \\ 0 \end{bmatrix} (2M+1) = \begin{bmatrix} p_r(0) & p_r(1) & \ldots & p_r(-2M) \\ p_r(1) & p_r(0) & \ldots & p_r(-2M+1) \\ . & & \ldots & . \\ . & & \ldots & . \\ . & & \ldots & . \\ & & & \\ p_r(2M) & & \ldots & p_r(0) \end{bmatrix} \begin{bmatrix} C_{-m} \\ C_{-m+1} \\ \vdots \\ C_0 \\ \vdots \\ C_{m-1} \\ C_m \end{bmatrix} \tag{5.54}$$

The matrix given in equation 5.54 is known as the 'zero forcing' matrix for equalizers, and selects the tap coefficients for the minimization of peak distortion by forcing the equalizer to zero at the M sample points either side of the required pulse.

Example

A transversal three-tap equalizer is to be adjusted to reduce the inter-symbol interference caused by the pulse $p_r(t)$ entering the equalizer. A diagram of the pulse shape entering the equalizer and leaving it is shown in Figure 5.19. Determine the tap gains and the values of the equalized pulse at each of the sampling instants.

Solution

With reference to Figure 5.19 and using equation 5.54, the matrix equation to be solved for the amplifier gains C_m is:

$$\begin{bmatrix} 0 \\ 1 \\ 0 \end{bmatrix} = \begin{bmatrix} 1.0 & -0.2 & 0.1 \\ 0.1 & 0.0 & -0.2 \\ 0.0 & 0.1 & 0.0 \end{bmatrix} \cdot \begin{bmatrix} C_{-1} \\ C_0 \\ C_1 \end{bmatrix}$$

giving:

$$\begin{bmatrix} C_{-1} \\ C_0 \\ C_1 \end{bmatrix} = \begin{bmatrix} 0.20 \\ 0 \\ -0.10 \end{bmatrix}$$

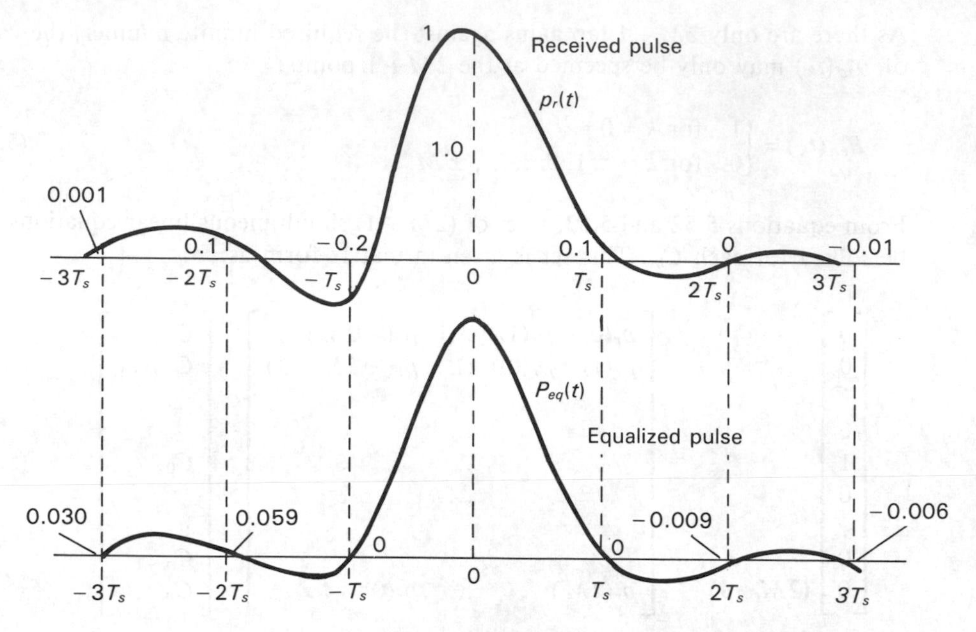

Figure 5.19 Input and output waveforms into a three-tap transversal equalizer

The values of the equalized pulse determined from equation 5.52 are in tabulated form:

k	−3	−2	−1	0	1	2	3
$P_{eq}(t_k)$	0.030	0.0587	0	1	0	0.009	−0.006

The equalized pulse is shown in Figure 5.19. Notice that on either side of $t = 0$ (that is $t = \pm T_s$), the output pulse crosses zero.

5.4.2 Automatic equalizers

Automatic equalizers can be divided into two groups: the *preset type*, which is dependent upon the transmission of a test pattern, and the *adaptive* type that adjusts itself continuously during the transmission of data according to the error between the actual received signal and the *a posteriori* estimate of the transmitted signal at the receiver. The error signal obtained is correlated with the data stream to obtain estimates of the gain coefficients. The process is iterative, so that the magnitude of the error signal is minimized.

5.4.2.1 Preset equalizers

To automatically preset the equalizer, a test pattern is transmitted and the *a priori* knowledge of the pattern at the receiver is used to compute the impulse response from the received signal. From the impulse response, either the tap coefficients are computed directly using equation 5.55, or an iterative technique is used to increment successively the coefficients until an optimum setting is obtained. This type of equalizer gives fast initial setting-up, but if the channel characteristics are changing, the test pattern has to be retransmitted at appropriate intervals to reset the coefficients. The iterative technique also uses equation 5.55 which may be written in terms of a $2M+1$ column vector I, a $2M+1$ square matrix X and a $2M+1$ column vector C, i.e.

$$I = XC \tag{5.55}$$

The iterations aim at attaining $I - XC = 0$ as given in equation 5.55 and at iteration k, and the result is an error vector e^k given by

$$e^k = XC^k - I \tag{5.56}$$

Figure 5.18 shows a feedback path which is used in automatic equalizers. The algorithm mentioned, which is used for optimizing the equalizer coefficients C_m has been the subject of considerable research. Two criteria have found widespread use. One is the peak distortion criterion and the other is the mean-square-error (MSE) criterion. The reader is referred to Chapter 6, Reference 26, for a detailed study of these criteria and algorithms.

5.4.2.2 Adaptive equalizers[26,37–43]

In digital radio systems the adaptive equalizer is used to overcome multipath effects; it is otherwise known as a *dispersive channel* because of its effect of spreading the signal power in time. When used with space diversity, it allows coherent addition of the signals since the equalizer in the diversity channel can be automatically adjusted to compensate for delay differences between channels. Alone, the adaptive equalizer also combats the dispersive effects as indicated by improvement in effective fade margin. Effective fade margin is discussed in Section 8.3.3.

The adaptive equalizer does not use a test pattern as does a preset equalizer, but assesses the error between the actual received signal and an *a posteriori* estimate of the transmitted signal, as discussed above. In order for an adaptive equalizer to work properly, the input bit sequence must be random. If the unequalized signal is badly distorted, the adaptive equalizer may make a large number of erroneous decisions and the iterative process will not converge. This of course means that equalization will not be achieved. In such a situation the start-up may be achieved by using a preset equalizer which converts to the adaptive mode after some measure of equalization has been attained.

In the simple adaptive transversal equalizer using iterative methods, the components of the error vector given by equation 5.56 are denoted by e_j^k, where j extends from $-M$ to $+M$. After the kth iteration the new adjusted value of the solution vector C^{k+1} is obtained from:

$$C^{k+1} = C^k - \Delta \operatorname{sgn}(e^k) \tag{5.57}$$

where

$$\text{sgn}(y) = \begin{cases} +1, & y > 0 \\ 0, & y = 0 \\ -1, & y < 0 \end{cases}$$

and Δ represents a positive increment. The iterations continue until C^k and C^{k+1} differ by less than some arbitrary small value.

In a digital radio system these filters can be implemented either at intermediate frequency (IF) or baseband, or both. When implemented at IF the delayed samples must feed complex weighting networks. Alternatively, after conversion to baseband, four separate real filter structures are required. The IF and baseband equalizer implementations are mathematically equivalent once carrier phase coherence has been established in the receiver; therefore in the design of these equalizers all computer simulation is done at baseband to ease the computing effort. Usually equalizers do not have more than eleven taps or ten discrete time delays. Although the IF equalizer has a lower complexity that its baseband counterpart, it is more difficult to implement digitally. A realistic practical design for an IF adaptive transversal equalizer is one with 5–9 taps.[40] Baseband equalization for modulated data signals is usually carried out after demodulation back into baseband. For quadrature amplitude and phase-modulated signals the demodulated signal can be resolved into two quadrature components. Each component can then be equalized separately. This is shown in Figure 5.20 together with a practical system block diagram. Since both signal components are operated on by similar channel characteristics, the two input signals are considered as the real and imaginary parts of a complex signal and likewise, the main and cross-modulation coefficients of the equalizer as the real and imaginary parts of an array of complex equalizer coefficients. Thus:

$$(I + iQ)(C + iK) = I + iQ \quad \text{where } i = \sqrt{-1} \tag{5.58}$$

The practical circuit shown also in Figure 5.20[41] consists of three sections. The first section is a two-dimensional equalizer comprising two analog delay lines that are tapped at symbol intervals, 5 taps per channel. For the I channel, five taps are weighted and summed together with weighted taps from the Q channel. This cross-correlation provides a means for reducing the crosstalk between the two. The Q channel is done in a similar manner. The algorithm used to adjust the weighted taps is a zero-forcing algorithm as given in its basic form by equation 5.55. This algorithm adjusts the tap weights according to the equations:

$$\left.\begin{aligned} C_j^{k+1}, I &= C_j^k, I - \text{sgn}(e_k) \cdot \text{sgn}(Y_k - j) \\ C_j^{k+1}, Q &= C_j^k, Q - \text{sgn}(e_k) \cdot \text{sgn}(Y_k - j) \\ d_j^{k+1}, I &= d_j^k, I - \text{sgn}(e_k) \cdot \text{sgn}(Y_k - j) \\ d_j^{k+1}, Q &= d_j^k, Q - \text{sgn}(e_k) \cdot \text{sgn}(Y_k - j) \end{aligned}\right\} \tag{5.59}$$

where $(C_j^{k+1}, I, \text{d}_j^{k+1}, I)$ are tap weights associated with the I-channel tapped delay

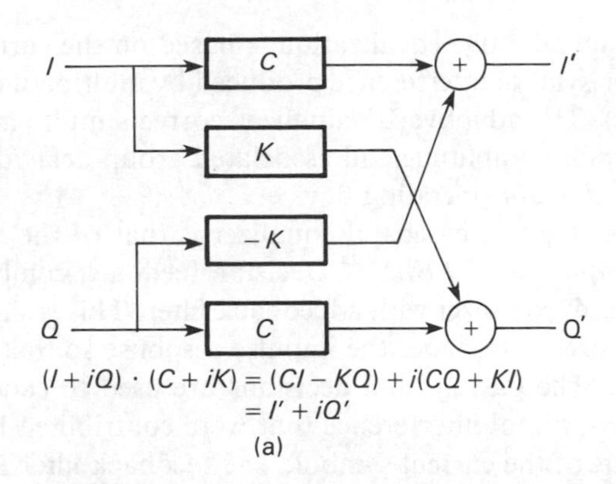

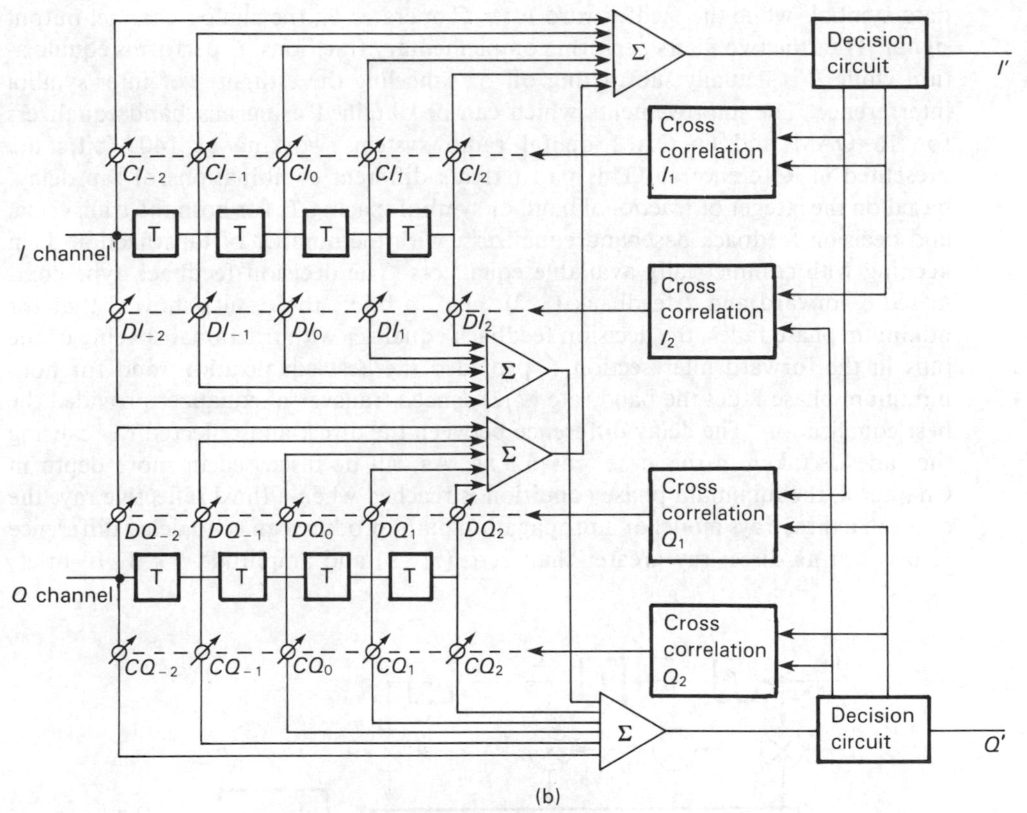

Figure 5.20 Baseband adaptive transversal equalizer

elements and (C_j^{k+1}, Q, d_j^{k+1}, Q) are with the Q-channel tapped delay elements for the ith tap weight at time $kT \cdot e_k = Y_k - A_k$, which represents the errors between the equalized and the transmitted samples for the I and Q channels.

The second section is the decision circuit, which provides the error polarity when inter-symbol interference is present. The third section is the cross-correlator, which correlates the error signal for updating the tap weights according to equation 5.59.

Baseband adaptive equalization is based on the correction, *in the time domain*, of the inter-symbol interference produced by multipath propagation effects (channel dispersion). The adaptive IF equalizer corrects multipath distortions by generating complementary amplitude and associated group delay distortions. It is effectively a *frequency domain* operating device.

Another type of baseband equalizer is that of the *linear forward plus decision feedback equalizer (DFE)*.[43,44] Decision feedback combines the use of a cascade or 'feedforward' equalizer with a feedback filter. This is shown in Figure 5.21. The forward equalizer C reshapes the impulse response so that the spectrum side lobes are eliminated. The past symbol decisions are used to cancel the effect of those parts of the inter-symbol interference that were contributed by previous data symbols to the estimate of the current symbol. The feedback filter B operates directly on digital data symbols while the feedforward filter C operates on the analog channel output signal. Thus the two filters perform complementary functions. C performs equalization while B is actually subtracting off or canceling the estimates of inter-symbol interference. The improvements which can be obtained using baseband equalizers for 16–QAM and 64–QAM digital radio systems working at 140 Mbit/s are presented in Reference 45. This paper treats different combinations of tap delay, based on the integer or fractional baud or symbol spacing T, for both the transversal and decision feedback baseband equalizers, with the number of taps fixed at 5, in keeping with commercially available equalizers. The decision feedback type comprised 3 forward and 2 feedback (3, 2) taps. In brief, the results showed that for minimum phase fades, the decision feedback equalizer with fractional spacing of the taps in the forward filter section C provided the best equalization, and for non-minimum phase fades the baud rate ($T/2$) spaced transversal equalizer provided the best equalization. The delay difference between the direct and reflected ray causing the fades is taken in this case[46] as 6.3 ns. As will be discussed in more depth in Chapter 8, the minimum phase condition is reached when a third reflective ray, the echo, in a three-rays model of a propagation path, produces an echo delay difference τ, between its direct ray greater than zero ($\tau > 0$) and amplitude $b < 1$. In brief,

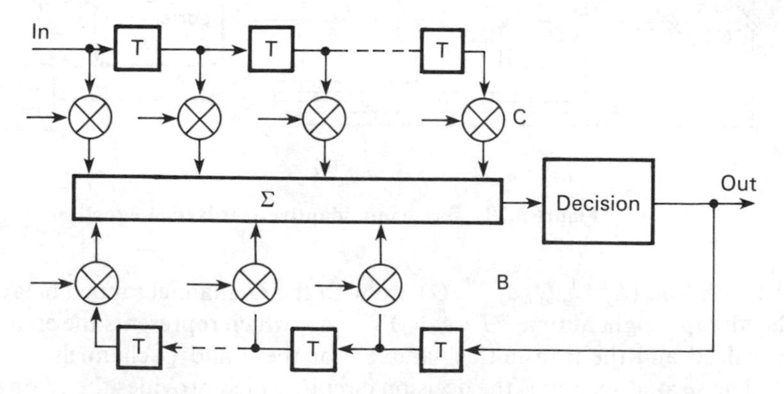

Figure 5.21 Baseband decision feedback equalizer (DFE) (Courtesy of IEEE, Reference 42)

then, there are two fade types; the minimum phase – which occurs where the direct ray is the stronger received signal and arrives first, and the nonminimum phase type – where the delayed echo provides the stronger signal and arrives after the direct ray, or arrives before the direct ray and is the weaker of the two.

The decision feedback equalizer (DFE) is the most widely used adaptive filter in communications. It requires a high carrier-to-noise ratio to avoid incorrect decisions perturbing the weight convergence. During conditions where fading changes between the minimum phase and nonminimum phase types, which may occur at 25 Hz, the equalizer must calculate a new set of weight values. This introduces a severe step discontinuity in long-delay multipath fading for most equalizers, but in the case of the DFE the new set of weight values are a time reversal of the set which is applicable to the previous fade type. Two-tap DFE equalizers are used for inter-symbol interference reduction in 4 and 6 GHz digital radios which use bandwidth-efficient modulation techniques (as described in Chapter 6) that confine a 140 Mbit/s quadrature phase-shift keying (QPSK) signal into 60 MHz channels spaced 66 MHz apart. These particular equalizer structures are only optimized for minimum-phase fades.

IF ADAPTIVE EQUALIZERS

The structure of an IF adaptive equalizer is shown in Figure 5.22.[42] The compensating network consists of a block capable of producing various distortion shapes of its amplitude-versus frequency characteristic and follows the commands of a fixed number of external control signals. The distortions produced by the compensating network should ideally be complementary to that presented by the transmission channel in the passband of the system, in order that equalization be achieved. The compensation block is usually followed by a variable gain flat amplifier through which the output level is held constant. This AGC circuit also compensates for the

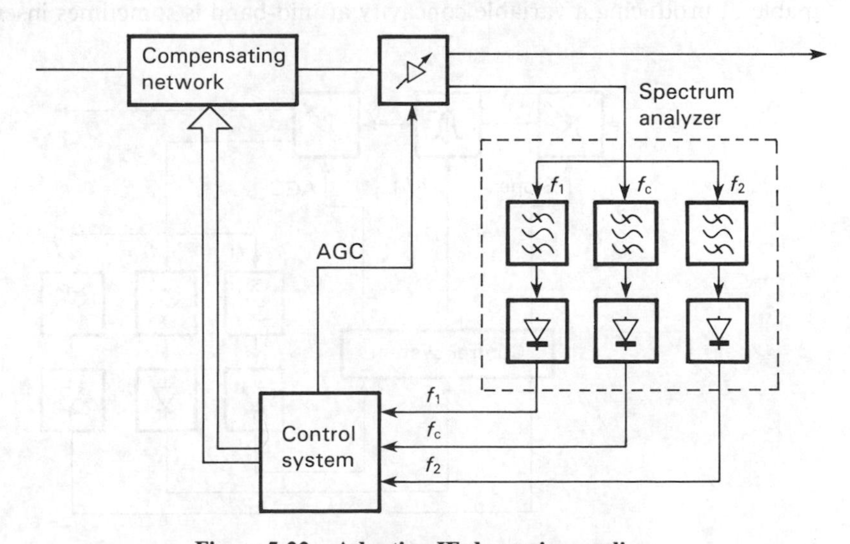

Figure 5.22 Adaptive IF dynamic equalizer

loss introduced by the compensating network. Because the transmitted signal power spectrum is constant, power measurements in different portions of the band symmetrically placed about the center frequency reflect what is happening to the overall frequency response, i.e. channel plus equalizer. Thus, included in the equalizer circuitry, are narrow-band analyzer filters, whose center frequencies are the power sampling frequencies. The AGC circuit reduces the requirements of the dynamic range of the detectors by keeping the average signal level constant at the analyzer circuit input. The control system processes the detector voltages provided by the analyzer and drives the compensating circuit in order to maintain them or restore them to their nominal levels corresponding to flat response conditions. It should be noted that frequency domain equalization acts generally on the frequency response only and is not very effective for compensating group delay distortion. In addition to this, the beneficial effects of the equalization of the amplitude are reduced or even canceled by the impairments on the overall phase distortion in the channel.

SIMPLE SLOPE EQUALIZER WITH MID-BAND RESONATORS

Figure 5.23 shows the circuit of this simple equalizer. This structure consists of a variable slope and notch equalizer with a central resonator. The compensating circuit consists of an AGC block, a variable-slope block and a variable-Q resonator fixed at mid-band frequency. Distortions are detected by a selective spectral detector which analyzes the received spectrum at different frequencies. The variable Q of the resonator provides the compensation of in-band deep notches. When used with the slope equalizer, it is able to realize a wide variety of symmetric as well as unsymmetric amplitude responses.

In order to compensate for the residual convexity that might result after slope equalization due to the convex quadratic component of both the channel and the compensating circuit, and also to equalize distortions corresponding to near-zero phase-shift values between the direct and reflected rays, a third compensating block capable of producing a variable concavity at mid-band is sometimes inserted. This

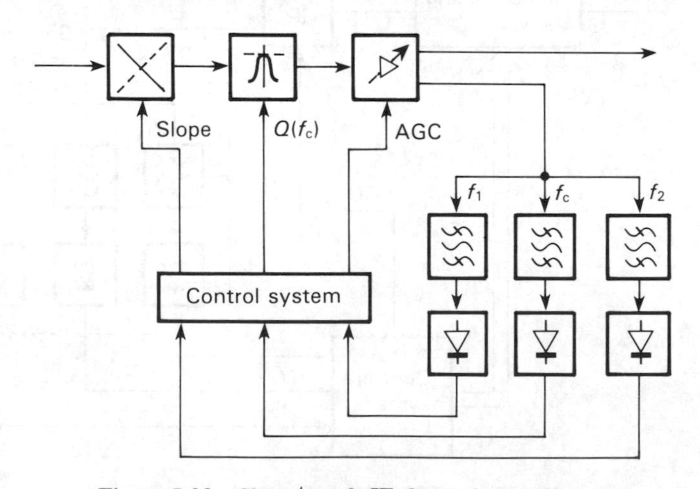

Figure 5.23 Slope/notch IF dynamic equalizer

can be done with the use of a band-stop-type variable-Q resonator introduced after the variable-Q band-pass resonator. The Q of both resonators can be easily varied by adjusting the equivalent load resistance using pin diodes. The control system consists simply of a few operational amplifiers, and automatic level regulation is derived directly from the center detector of the center frequency (f_c), circuit. Circuit simplicity is the major advantage of this type of equalizer, but the reduction in outage probability offered by its introduction into the receiver is not very high. The gain is about 3–4, depending on the type of modulation used, and for nonminimum phase distortions it is even less.

MOVABLE RESONATOR EQUALIZER

An alternative structure to that shown in Figure 5.23 is given in Figure 5.24, where the compensation circuit consists of a resonator in which both the Q and the resonator frequency are variable. This type of equalizer is known as a *movable resonator equalizer*. If the resonant frequency is exactly the same as the notch frequency of a multipath fade, a suitable value of Q will provide a distortion complementary to that of the channel. This equalizer also provides a group-delay distortion compensation with minimum-phase multipath. In a nonminumum phase condition the group-delay distortion is increased. As a consequence, a very high equalization efficiency can be obtained if the probability of nonminimum phase multipath is negligibly small. Even in this case, the group delay behavior of the equalizer is related to the amplitude response so that, while with minimum-phase fading there is also compensation of the channel group delay distortion, with non-minimum phase fading ($\tau < 0$) an impairment of such distortion occurs, causing the overall group delay to be roughly doubled. Since the control system has to perform complex operations in order to position the resonator frequency correctly and to optimize Q, microprocessor systems are normally used. Excellent performance is obtained for minimum phase fading ($\tau > 0$), but only modest gain with respect

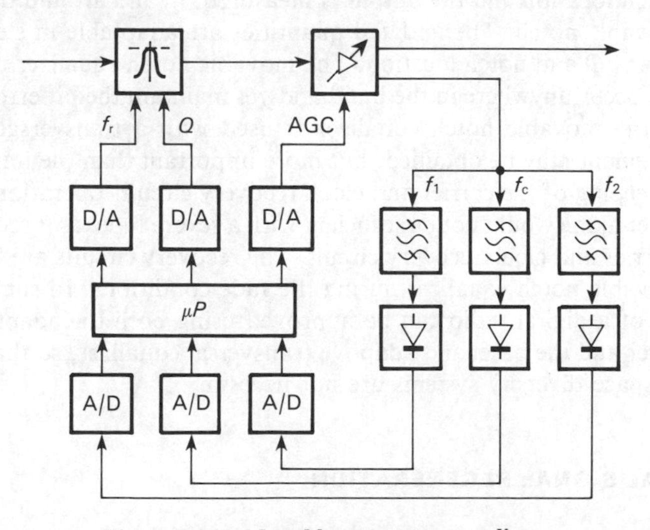

Figure 5.24 Movable resonator equalizer

to the unequalized performance is achieved in the case of nonminimum phase distortions.

The operation of the equalizer is such that it can accommodate a variety of fade shapes caused by variations in multipath fade depths ρ and delays τ. To do this the measurements taken by the selective spectral detector must be accurate enough to control the variety of equalization responses. In addition, the equalizer must be able to respond with an updated shape to be able to follow the fade. The IF signal is inserted into the movable resonator equalizer, or, as otherwise called, the movable notch equalizer. Since the circuit itself causes variation in amplitude, an AGC circuit is used to maintain a level of 0 dBm at the output. To determine what type of fade is present at the input to the circuit, the selective spectral detector is used. The controller directs this detector to measure the input and the output in a manner which gives it enough information to control the movable notch so that the output spectrum remains the shape of the transmitted spectrum during a no-fade condition. To take advantage of a movable notch circuit which can provide a wide range of responses, a microprocessor controller is normally used. This controller may permit 30 or so measurements to be made at 1 MHz intervals with a resolution of 0.25 dB. A single measurement may take less than one millisecond. The time required for a spectral reading is critical to overall system performance since the spectral detector is the weakest link in the chain. In order to follow a 100 dB/s fade to within 1 dB, the movable notch must be updated every 10 ms. In order to follow a fade of 30 MHz/s to within 1 MHz, the center frequency of the equalizer must be updated every 33 ms. The controller, which these days is a single-chip microprocessor (μP), provides the feedback from the spectral detector to the movable notch equalizer. In operation the controller takes nine measurements to characterize the input spectrum. Notch location and depth are obtained. From this information an open-loop jump to set the movable notch is made, based on microprocessor look-up tables. Once this jump has been made, the closed-loop algorithm begins: the input is measured to find the notch location and the output is measured, in and around the notch, to update the movable notch. The updated quantities are adaptable in step size and provide for depth, Q and notch location. The movable notch equalizer may permit a 23 dB fade to occur anywhere in the band and yet maintain the bit error rate at 1 in 1000. When the movable notch equalizer is used with a transversal equalizer, further improvement may be obtained, but more important than the improvement increase is the keeping of the carrier and clock recovery circuits operational. The transversal equalizer alone would not be sufficient with a severe fade, as it requires the operation of a carrier and clock recovery circuit. The recovery circuits are kept operational by the movable notch equalizer during the fade conditions. In summary, a multipath outage of a digital radio can be improved using both the adaptive movable notch equalizer and the baseband adaptive transversal equalizer, so that for most systems costly space diversity systems are not necessary.[41]

5.5 DIGITAL SIGNAL REGENERATION

The process which permits digital transmission to be superior to its analog counterpart is the process of digital regeneration. Regeneration is a process in which a digital signal which has been distorted and attenuated is regenerated to its correct amplitude and waveshape. This process may provide an alternative definition of digital transmission, since all digital transmission requires to be regenerated at some point in the transmission path. For a chain of digital transmission links, regenerators exist in baseband receiver circuits to permit the attenuated and dispersed incoming signal, which is perturbed by noise, to be preamplified and shaped before detection. Detection of the resultant waveform is accomplished by a threshold detector which produces an output that is a well-defined pulse. The detector circuit is able to operate on signal shapes which may be nearly sinusoidal. This implies that the baseband amplifier/filter bandwidth need not be as large as that required for the transmission of a pulse and thus the noise at the output of the baseband is reduced. The implication of reducing the bandwidth of the received baseband amplifier and thus widening the pulse spectrum is to permit a reduced tolerance to inter-symbol interference and thus an increase in the difficulty of performing the threshold decision. The result of course is the closing of the eye in the eye diagram. The regenerator design is thus a compromise between that of having a large bandwidth to reduce the inter-symbol interference and having an increase in the thermal noise which would also increase the effects of timing jitter due to the noise affecting the decision threshold. This has been discussed in brief in Sections 4.1.4 and 5.2.5. Figure 5.25(a) shows a general block diagram of a regenerator and Figure 5.25(b) shows the block diagram of a typical line regenerator. The line regenerator is considered because it provides an insight into the operation of practical regenerators.

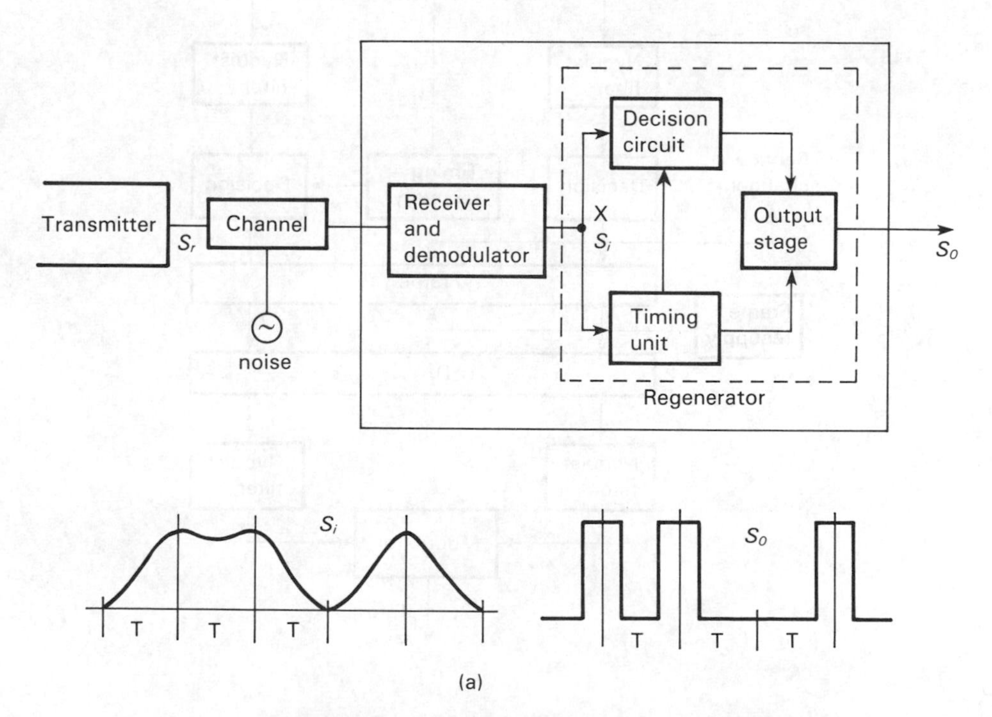

Figure 5.25 Digital regenerators: (a) general concept; (b) line repeater; (c) digital radio regenerator; (d) digital radio terminal

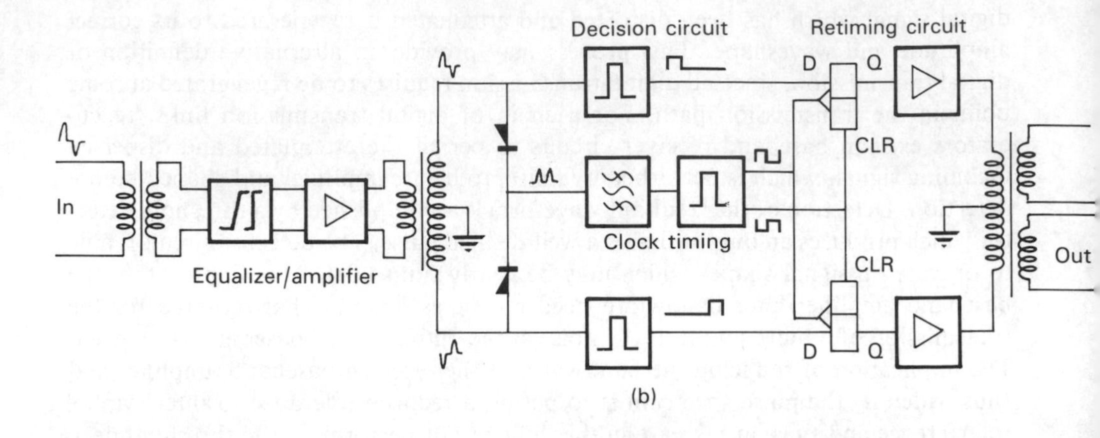

(b)

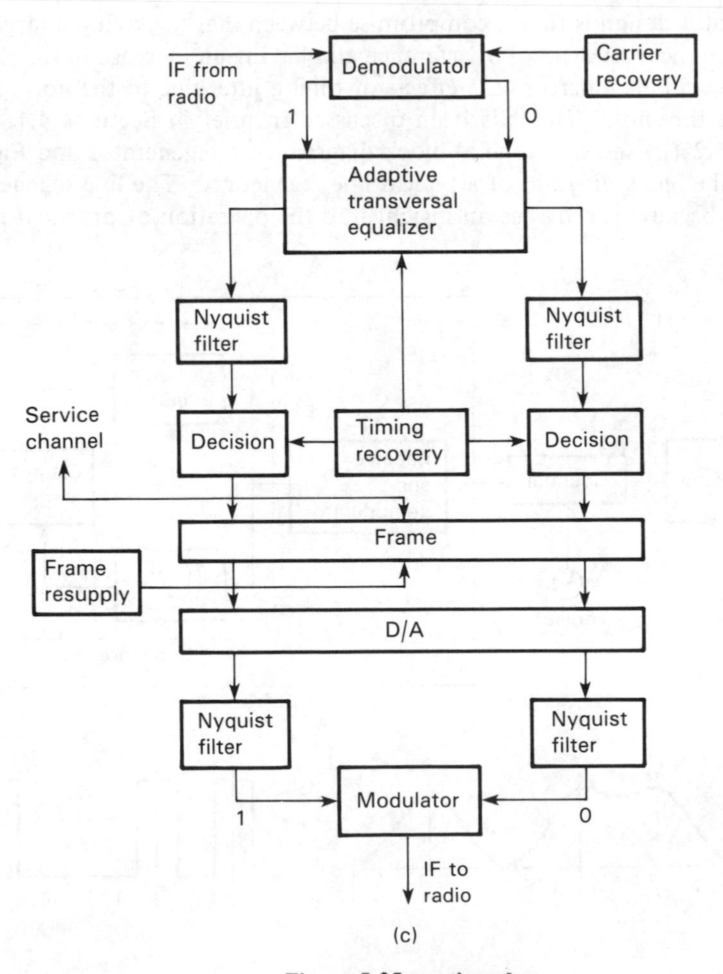

(c)

Figure 5.25 continued

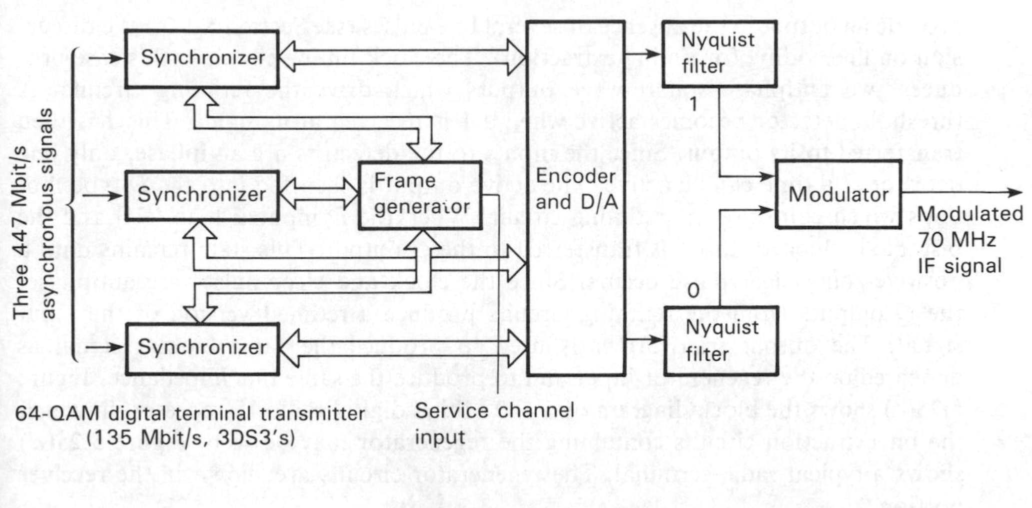

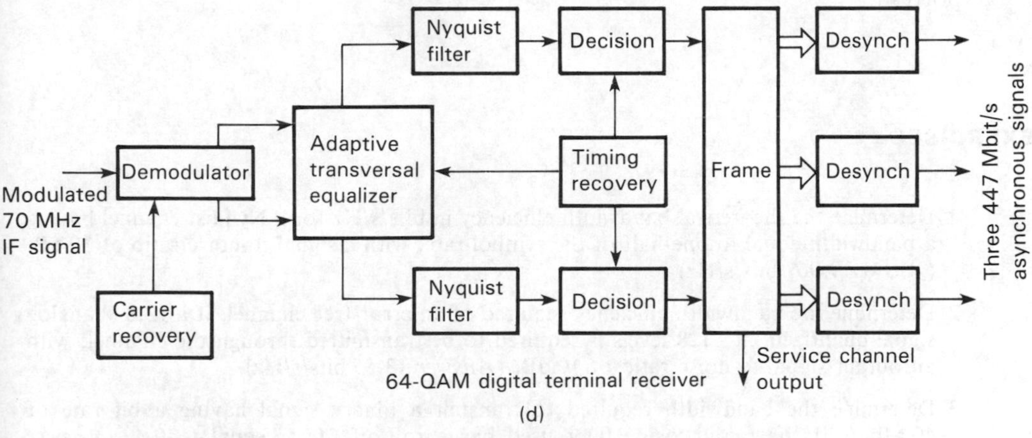

Figure 5.25 continued

In a digital radio the regenerator is usually located in the bit extraction circuits which come after the demodulator and transversal equalization stages. Referring to Figure 5.25(*b*), the incoming line signals pass from the line transformer and equalization circuits into an amplifier. Two antiphase outputs from this amplifier are fed to the threshold detectors. At the same time the two outputs are rectified and fed to the clock circuit.

5.5.1 Timing recovery and threshold detection

The timing recovery circuit or clock timing circuit, as shown in Figure 5.25(*b*), accepts input data bits at the line bit rate and either activates a series-tuned resonant circuit, or provides the input to a phase-locked loop.[47] Either circuit is designed to provide an output in the absence of several line pulses (see Section 5.1 for the discussion on line coding for timing extraction). The clock timing circuit in this case produces two antiphase square-wave outputs which drive the retiming circuits. A threshold detector becomes active when a 1 is received at its input. This 1 is then transferred to its output. Since the inputs to the detectors are antiphase, only one detector at a time can be active. The active output is then fed into the *D*-type flip-flops which comprise the retiming circuit. When the *D* input is high ('1') and the bistable is clocked, this 1 is transferred to the *Q* output. This state remains until a positive-going clear pulse occurs. Since the clock and clear pulses are antiphase, the *Q* outputs from the retiming circuits produce a retimed version of the input signal. The output transformer is used to produce the same bipolar signal as appeared on the regenerator input and to produce the same line impedance. Figure 5.25(*c*) shows the block diagram of a 140 Mbit/s digital radio IF repeater, in which the bit extraction circuits containing the regenerator may be seen. Figure 5.25(*d*) shows a typical radio terminal. The regenerator circuits are shown in the receiver portion.

EXERCISES

1 Determine the theoretical bandwidth efficiency in bits/s/Hz for a Nyquist channel having a bandwidth equal to one-half of the symbol rate, with a signal-to-noise ratio of 30 dB. (*Answer* 9.967 bits/s/Hz)

2 Determine the bandwidth efficiency required of an error-free channel, if a 12.8 V analog signal quantized into 128 levels is required to be transmitted through the channel, with an output signal-to-noise ratio of 10 dB. (*Answer* 13.55 bits/s/Hz)

3 Determine the bandwidth required to transmit a binary signal having a bit rate of 40 Mbit/s if the raised-cosine filter used has a roll-off factor equal to 0.4. (*Answer* 36 MHz)

4 Determine the bandwidth required to transmit a quaternary signal ($M = 4$) through a channel having a low-pass filter cut-off frequency equal to 50 percent excess of the Nyquist bandwidth $[f = r_s/2 + (0.5)r_s]$. (*Answer* $0.5 \times$ bit rate)

5 Determine the maximum bit rate of an $M = 8$ signal which can be transmitted through a channel having a bandwidth equal to 30 MHz and which uses a raised-cosine filter with a roll-off factor of 0.5. (*Answer* 90 Mbit/s)

6 Given $S/N = 15$ dB, determine the probability of error of a binary and an $M = 4$ baseband PAM signal. (*Answer* 9.65×10^{-9}, 8.93×10^{-3})

7 If results showed that for an $M = 8$ PAM baseband system, the probability of error was 6.39×10^{-9}, determine the upper limit of channel capacity, if this system was to be used over a Nyquist channel having a bandwidth of 20 MHz. (*Answer* 189.39 Mbit/s)

8 If the $M = 8$ PAM system is to be transmitted over a channel having a bit rate of 140 Mbit/s, what should the bandwidth be, if a raised-cosine filter of $\alpha = 0.4$ is used? (*Answer* 42 MHz)

9 If measurements made on the baseband of a receiver of a transmission system showed noise ratio had to be 15 dB, what percentage of eye opening would you expect to see on an eye diagram? (*Answer* 75.9 percent. Hint: use Figure 5.10)

10 Show that the spectral efficiency of an M-ary PAM system is given by $2 \log_2 M$.

11 Given that the spectral efficiency of an M-ary PAM system is $2 \log_2 M$, show that the theoretical signal-to-noise ratio is given by $10 \log(M^2 - 1)$ dB.

12 Determine the theoretical value of C/N for an $M = 4$ PAM for a P_e of 10^{-2}. (*Answer* 11.76 dB, 0.0625)

REFERENCES

1. Sipress, J. M., 'A New Class of Selected Ternary Pulse Transmission Plans for Digital Transmission Lines', *IEEE Transactions on Communications*, 1965, **COM–13**, No. 3, pp. 366–372.
2. Franaszek, P. A., 'Sequence State Coding for Digital Transmission', *Bell System Technical Journal*, 1968, **47**, pp. 143–157.
3. Lockstone, D. H., 'Field Measurements on MS43 Encoded 2 Mbit/s Digital Line System', *Telecommunication Transmission*, IEE Conference Publication No. 246, 1985, pp. 221–224.
4. Crosier, A., 'Introduction to Pseudo-Ternary Transmission Codes', *IBM Journal on Research and Development*, 1970, **14**, pp. 354–367.
5. Clothier, D. and Bilanski, P., 'A Digital Transmission System for Multiterminal Applications', *GEC Telecommunications Journal*, 1980, **41**, pp. 10–15.
6. Duc, N. Q. and Smith, B. M., 'Line Coding for Digital Data Transmission', *Australian Telecom Review*, 1977, **11**, pp. 14–27.
7. Walsh, J., 'A Closed Set of Orthogonal Functions', *American Journal of Mathematics*, 1923, **45**, pp. 5–24.
8. Davidson, I. A., 'An Investigation of Local area Networks', *Communications Equipment and Systems*, IEE Conference Publication No. 209, 1982, pp. 160–163.
9. Brewster, R., 'Data Transmission', *Local Telecommunications*, IEE Telecommunication Series 10 (Peter Peregrinus, 1983), pp. 89–110.
10. *Digital Networks – Transmission Systems and Multiplexing Equipments*, Recommendations G700–G956 (Study Groups XV and XVIII), CCITT Red Book, **III–3**, VIII Plenary Assembly, 1984.
11. Bylanski, P. and Ingram. D. G. W., *Digital Transmission Systems*, IEE Telecommunication Series 4, Rev. Ed. (Peter Peregrinus, 1980).
12. Gier, G. and Stummer, B., 'A 565 Mbit/s Line Equipment for Single-Mode Optical Waveguides', *Links for the Future*, IEEE International Conference on Communications, 1984, pp. 783–786.
13. *Fixed Service using Radio-Relay Systems*, CCIR Recommendations and Reports, 1986, **IX**, Part 1.
14. Taub, H. and Schilling, D. L., *Principles of Communication Systems* (McGraw-Hill, 1971).
15. Carlson, A. B., *Communication Systems– an Introduction to Signals and Noise in Electrical Communication*, 2nd Ed. (McGraw-Hill, 1975).
16. Shanmugam, K. S., *Digital and Analog Communication Systems*, (Wiley, 1979).
17. Shannon, C. E., 'A Mathematical Theory of Communications', *Bell System Technical Journal*, 1948, **27**, pp. 379–423 & 623–656. Also: Shannon, C. E., *Mathematical Theory of Communication* (University of Illinois Press, 1963).
18. Shannon, C. E., 'Communication in the Presence of Noise', *Proc. IRE*, 1949, **37**, p. 10.

19. Kretzmer, E. R., 'Generalization of a Technique for Binary Data Communication', *IEEE Transactions*, 1966, **COM-14**, pp. 67–68.
20. Kretzmer, E. R., 'Binary Data Communication by Partial Response Transmission', *IEEE Annual Communications Conference*, 1965, pp. 451–455.
21. Bennett, W. R. and Davey, J. R., *Data Transmission* (McGraw-Hill, 1965).
22. Lender, A., 'The Duobinary Technique for High Speed data Transmission', *IEEE Transactions on Communications & Electronics*, 1963, **82**, pp. 214–218.
23. Lender, A., 'Correlative Digital Communication Techniques', *IEEE Transactions on Communication Technology*, December 1964, pp. 128–135.
24. Lender, A., 'Correlative (Partial response) Techniques and Applications to Digital Radio Systems'; Chapter 7 of *Digital Communications – Microwave Applications* (Prentice-Hall, 1981) by Feher, K., pp. 144–182.
25. Coates, R. F. W., *Modern Communication Systems*, 2nd Ed. (Macmillan Press, 1983) p. 308.
26. Proakis, J. G., *Digital Communications* (McGraw-Hill, 1983).
27. Kawashima, M., 'A New Measurement for an Eye Opening Estimation on Equalized Digital Signals', *Measurements for Telecommunication Systems*, IEE Conference Publication No. 256, 1985, pp. 14–18.
28. Takawira, F. and Ingram, D. G. W., 'Significance of Phase Distortion in Digital Transmission Systems', *Proc. IEE*, 1986, **133**, Part F, No. 1, pp. 115–127.
29. Ingram, D. G. W., 'Analysis and Design of Digital Transmission Systems', *IEE J. Comput. & Digital Tech.*, 1979, **2**, No. 3, pp. 121–133.
30. Peterson, W. W. and Weldon, E. J. Jnr, *Error Correcting Codes*, 2nd Ed. (M.I.T. Press, 1972).
31. Berlekamp, E. R., *Algebraic Coding Theory* (McGraw-Hill, 1968).
32. Hamming, R. W., 'Error Detecting and Correcting Codes', *Bell System Technical Journal*, 1950, **29**, pp. 147–160.
33. Wosencraft, J. M. and Jacobs, I. M., *Principles of Communication Engineering* (Wiley, 1965).
34. Lin, S., *An Introduction to Error-Correcting Codes* (Prentice-Hall, 1970).
35. Lucky, R. W., Salz, J. and Weldon, E. J. Jnr, *Principles of Data Communications* (McGraw-Hill, 1968).
36. Brayer, K., 'Error Correction Code Performance on HF, Troposcatter, and Satellite Channels', IEEE Transactions on Communication Technology, 1971, **COM-19**, Part II, pp. 781–789.
37. Proakis, J. G. and Ling, F., 'Recursive Least Squares Algorithms for Adaptive Equalization of Time-Variant Multipath Channels', *Links for the Future*, IEEE International on Communications, 1984, pp. 1250–1254.
38. Picchi, G., Raheli, R. and Prati, G., 'Multiplication-Free Incremental Adaptive Equalizer using DM Coding', *ibid.*, pp. 1243–1246.
39. Zervos, N., Pasupathy, P. and Venetsanopoulos, A., 'Low Complexity Maximum Likelihood Equalizers', *ibid.*, pp. 1259–1262.
40. Wong, W. K., Williamson, B., Grant, P. M. and Cowan, C. F. N., 'Adaptive Transversal Filters for Multipath Compensation in Microwave Digital Radio', *ibid.*, pp. 989–992.
41. MacNally, D. E. and Huang, J. C. Y., 'Propagation Protection Techniques for High Capacity Digital Radio Systems', *ibid.*, pp. 1007–1010.
42. Moreno, L. and Salerno, M., 'Adaptive Equalization Structures in High Capacity Digital Radio: Predicted and Observed Outage Performance', *ibid.*, pp. 993–997.
43. Calandrino, L. and Bianconi, G., 'Adaptive Baseband Equalization in Digital Radio Links', *Telettra Review*, Special Edition, 1985.
44. Gersho, A. and Biglieri, E., 'Adaptive Cancellation of Channel Nonlinearities for Data Transmission', *Links for the Future*, IEEE International Conference on Communications, 1984, pp. 1239–1242.
45. Shafi, M. and Moore, D., 'Adaptive Equalizer Improvements for 16 QAM and 64 QAM

Digital Radio', *ibid.*, pp. 998–1002.
46. Rummler, W. D., 'A New Selective Fade Model: Application to Propagation Data', *Bell System Technical Journal*, May–June 1979, pp. 1037–1071.
47. Blanchard, A., *Phase-Locked Loops – Application to Coherent Receiver Design* (Wiley, 1976).

6 DIGITAL MODULATION AND DEMODULATION

Referring to Figure 1.1, we can see where, in the model of a digital system, this chapter is placed. The subject of this book implies that a digital modulator and demodulator is an integral part of a digital radio transmitter and receiver. This is so, since modulation techniques relevant to transverse screened cable, coaxial cable or optical fiber are not considered in this book, except by passing references. In this and the remaining chapters of the book, the main consideration is that of the digital radio and its design and planning. To lay the groundwork for a good understanding of later chapters, the various modulation techniques which have been developed will be reviewed in this chapter, with calculations of error probabilities for most cases, and methods of reducing the probability. The main objectives of a digital carrier link remain identical to those of a baseband system, in that a specified error probability and jitter amplitude must not be exceeded. These parameters have been discussed respectively in Section 5.2.4 and Chapter 4. As discussed under digital coding of the baseband system in Section 5.1, one of the aims of a baseband transmission code is to reduce the zero frequency or DC component to zero. This in effect produces a band-pass channel. The translation is one of the features considered in the 'matching' of the source information to the transmission channel. Similarly, if the only transmission channel which is available for the transmission of digital signals is the RF channel, then, in order to match this transmission channel, the digital signals must be made to modulate a carrier of appropriate frequency to permit it to be transmitted over the passband of this channel. The channel over which the signal is transmitted is limited in bandwidth to an interval of frequencies centered about the carrier as for double-sideband (DSB) modulation, or adjacent to the carrier as for single-sideband (SSB) modulation. If the bandwidth of the signals and channels is far smaller than the carrier frequency, they are known as *narrow band* band-pass signals. Digital modulation techniques which may be used to modulate the carrier may modify the amplitude, the phase, or the frequency of a carrier in discrete steps. In spite of the number of modulation schemes, performance analysis depends primarily on the type of demodulation or detection. There are two major classes of these types: *coherent* or synchronous detection which is prevalent in phase-shift keying (PSK) and assumes that, at the demodulator, local carrier waveforms have a fixed phase relationship with that of the transmitted carrier. This adds to the complexity of the receiver circuitry. The other class is that of *non-coherent* or envelope detection, which is used in conjunction with amplitude-shift keying (ASK) and frequency-shift keying (FSK). In the analysis of the performance of the various modulation and demodulation techniques, the band-pass signals and channels of some may be converted for mathematical simplicity to an equivalent low-pass configuration. This permits these analyses to be independent of carrier frequencies and channel frequency bands. The conversion of bandpass to the low pass equivalent has not been done for power spectral density functions given in this book.

6.1 REPRESENTATION OF DIGITALLY MODULATED SIGNALS

In a digital transmitter the modulator maps the sequence of binary digits into a corresponding set of M discrete carrier amplitudes, carrier phases or frequency deviations from the sinusoidal carrier frequency. The different forms which the mapping takes represents three main distinct classes of modulation. These are:

Amplitude-shift keying (ASK)
Phase-shift keying (PSK)
Frequency-shift keying (FSK)

Another form of modulation which is in common use in digital radio transmission is that in which a combination of PAM and PSK are used. The resulting modulation is known as:

Amplitude phase keying (APK)

6.2 AMPLITUDE-SHIFT KEYING (ASK)

Figure 6.1 shows the process of amplitude modulating a carrier with a binary signal 10101101. If the digital source has M states or levels and each of these levels is present for a period T, the modulated waveform corresponding to the ith state $s_i(t)$ is for pulse amplitude modulation (PAM) or amplitude-shift keying (ASK) equal to

$$s_i(t) = D_i(t) \cdot A_0 \cdot \cos \omega_0 t \tag{6.1}$$

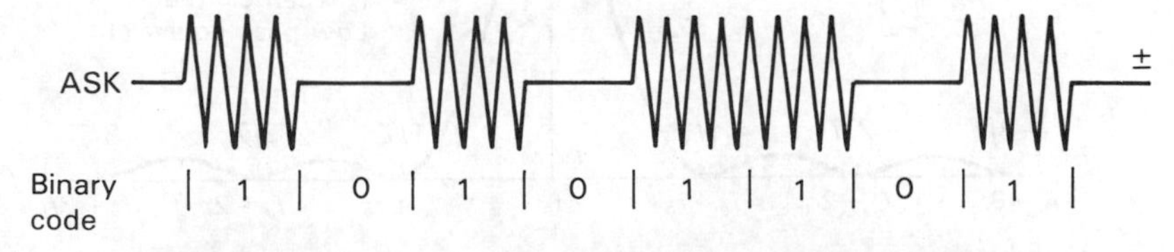

Figure 6.1 Digital modulation ASK for the binary signal 10101101

where $D_i(t)$ is the ith level of a multilevel waveform with a level duration of T. If it is assumed that the number of levels is restricted to two, as for a binary digital signal and that the carrier frequency is related to the rectangular binary waveform bit duration T by

$$\omega_0 = 2n\pi/T \tag{6.2}$$

then the power spectral density (psd), is given by the expression

$$\text{psd}_{\text{ASK}} = (A^2/16) \cdot \left[\delta(f - f_0) + \delta(f + f_0) + \frac{\sin^2 \pi T(f - f_0)}{\pi^2 T(f - f_0)} + \frac{\sin^2 \pi T(f + f_0)}{\pi^2 T(f + f_0)}\right] \tag{6.3}$$

Note that if an equivalent low-pass filter had been used, where f_0 was zero, the resulting spectrum without any loss of generality would be:

$$\text{psd}_{\text{ASK}} = (A^2/16) \cdot \left[\delta(f + f_0) + \frac{\sin^2 \pi T(f + f_0)}{\pi^2 T(f + f_0)}\right] \tag{6.4}$$

The spectrum for equations 6.3 and 6.4 is shown in Figure 6.2. Equations 6.3 and 6.4 are in two parts. The first is that term containing Dirac delta functions which implies discrete spectral components spaced at frequency intervals of $1/T$. These discrete frequency components vanish if the binary sequence has a zero mean, or with an M-level signal when each M level is equally likely. This permits the spectral characteristics of the digitally modulated signal to be controlled during the system design by proper selection of the information sequence to be transmitted. The second term is a continuous spectrum whose shape depends only on the spectral characteristic of the signal pulse. For the simple case of the binary digit as represented by equation 6.3, the impulse of the discrete spectral component exists only at the carrier frequency due to the presence of the spectral nulls spaced at frequency intervals of $1/T$.

The spectrum shown in Figure 6.2 contains 95 percent of its power in a bandwidth

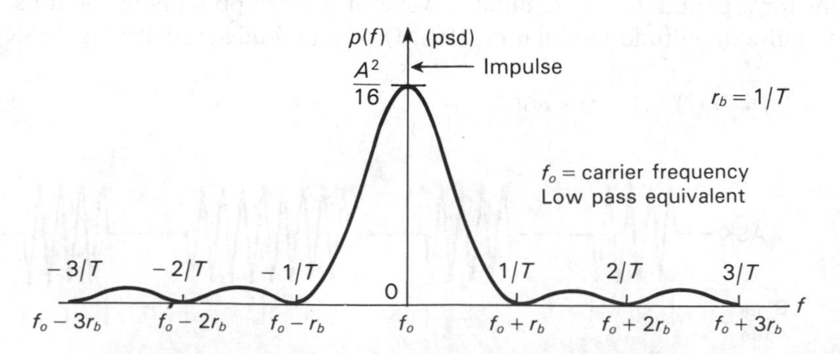

Figure 6.2 Power spectral density of the binary ASK signal

of $3/T$ or $3 \times$ (bit rate). This bandwidth can be reduced by using the raised-cosine pulse. The result is that the spectral nulls occur at intervals of $f_0 \pm n/T$ where $n = 2, 3, \ldots$. Consequently all the discrete spectral components vanish except at $f = f_0$ and $f = f_0 \pm 1/T$. The spectrum of the raised-cosine pulse has a broader main lobe making the ASK bandwidth equal to approximately $2/T$. The reception of the transmitted ASK signal may be achieved in two ways. The first is the coherent demodulation which uses complex circuits to maintain the phase coherence between the transmitted carrier and the local carrier, and the second is the noncoherent envelope demodulation process. In the discussion of these methods, the error probability will be given for each case.

6.2.1 Coherent ASK

With coherent detection, the receiver is synchronized to the transmitter. This means that the delay τ must be known at the receiver. This synchronization is derived from time measurements which are made on the received signal and is usually accurate to ± 5 percent of the signal bit period T. In addition to the time delay τ, the carrier phase $\phi = \omega_0 t$ must also be considered when the received signal is processed. As the delay τ varies with transmitter carrier frequency variations, the 5 percent estimation of T and the variations in propagation time for the carrier to reach the receiver mean that the value of ϕ cannot be determined with any certainty. For practical coherent detection systems, the carrier phase ϕ is an estimated quantity. Where M possible signal waveforms can be transmitted, the demodulator must decide which one has been actually transmitted. As noise is added to this signal, there is a definite probability that the ith signal state may be confused with that of its nearest neighbouring states. The probability of a decision error is minimized if the demodulator selects the received signal which has the largest probability of being the signal s_i and treating it as the signal which was transmitted. This decision strategy is called the *maximum a posteriori probability* (MAP) criterion and is shown to be optimum for additive zero-mean Gaussian noise and equiprobable states[1]. Two forms of the optimum demodulator can be derived.[2] The first is that of the cross-correlation-type demodulator and the second is that of the matched-filter type. Figure 6.3 shows both types. With a binary ASK signal, the receiver shown in Figure 6.4 may be used for coherent detection. This circuit is in effect a matched-filter demodulator which has a received input signal $s_i(t)$ together with white noise $n(t)$ that has been added during the transmission process. The receiver, after filtering away some of the noise and confining the signal to the required bandwidth ($2/T$ to $3/T$), is then multiplied by a local signal $A_c \cdot \cos \omega_0 t$. The local oscillator which can be represented by the difference in signaling waveform states $s_1(t) - s_0(t)$ is carefully synchronized with the frequency and phase of the received carrier. This product signal is then integrated by an 'integrate and dump' circuit, which is used because a perfect integrator is not possible to construct. The output of this integrator is compared with a threshold that is set half-way between the values u_1 and u_0, which are the levels into the decision circuit for an input of '1' or '0', that is $(u_1 + u_0)/2$. For the case when $s_1(t)$ is received in the absence of noise, the integrator computes and passes to the decision detector

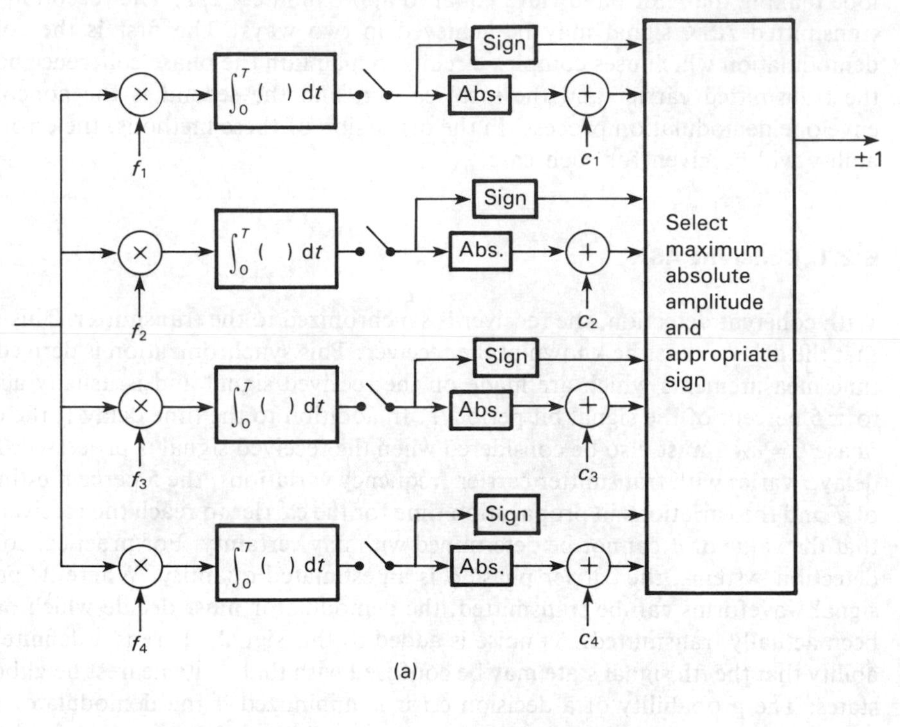

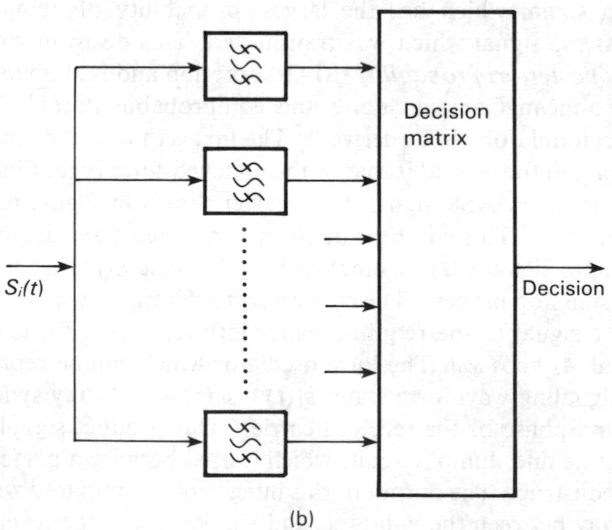

Figure 6.3 Optimum demodulators: (a) cross-correlation; (b) matched filter

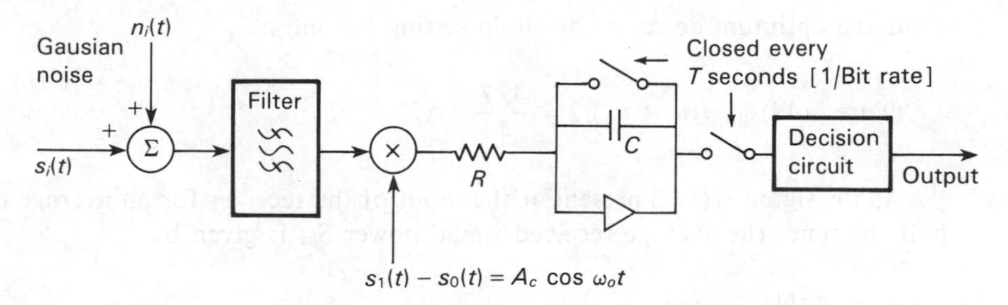

Figure 6.4 Binary coherent ASK demodulator

the value of u_1:

$$u_1 = \int_0^T s_1^2(t)\ dt - \int_0^T s_0(t) \cdot s_1(t)\ dt \tag{6.5a}$$

and when $s_0(t)$ is received:

$$u_0 = \int_0^T s_0(t) \cdot s_1(t)\ dt - \int_0^T s_0^2(t)\ dt \tag{6.5b}$$

If $u_1 > u_0$, i.e. the input exceeds the threshold level, the decision detector decides that $s_1(t)$ was transmitted. Similarly, if the input is less than the threshold level, the decision is that $s_0(t)$ was transmitted.

The two signaling waveforms of a binary ASK signal may be represented by:

$$\begin{aligned} s_1(t) &= A_1 \cos \omega_0 t \\ s_0(t) &= A_0 \cos \omega_0 t \end{aligned} \tag{6.6}$$

For these waveforms to be distinguished at the output of the integrator, the difference Δ in levels is defined in the same manner as the difference in quantization levels given in Section 2.2.1.2. Thus

$$\Delta = |u_1 - u_0| = \int_0^T [s_1(t) - s_0(t)]^2\ dt \tag{6.7}$$

The value of u_1 exceeds the threshold by $\Delta/2$ and the value of u_0 lies $\Delta/2$ below the threshold. By substituting equation 6.6 into 6.7, the value of Δ for a binary ASK signal may be found:

$$\begin{aligned} \Delta &= (A_1 - A_0)^2 \int_0^T \cos^2 \omega_0 t\ dt = (A_1 - A_0)^2 \cdot T/2 \\ &= A_c^2 T/2 \text{ for the case where no amplitude loss occurs.} \end{aligned} \tag{6.8}$$

Thus the optimum detector threshold setting becomes:

$$(\text{Threshold})_{\text{opt}} = (u_1 + u_0)/2 = \frac{A_c^2 T}{4} = \Delta/2 \tag{6.9}$$

As the signal $s_1(t)$ is present at the input of the receiver for an average of only half the time, the average received signal power S_{av} is given by

$$S_{av} = A_c^2/4 \tag{6.10}$$

6.2.1.1 Probability of error P_e

As Gaussian noise of variance σ^2 is introduced into the decision circuit it becomes probable that a false level will be detected. Equation 5.24 gives this probability as:

$$P_e = P(1) \cdot P(0 \mid 1) + P(0) \cdot P(1 \mid 0) \tag{5.24}$$

thus

$$P_e = P(1) \cdot P(n < -\Delta/2) + P(0) \cdot P(n > \Delta/2) \tag{6.11}$$

where n is the noise power. Assuming equiprobable digits, from equations 6.11 and 2.16, we have:

$$P_e = P(n > \Delta/2) = [1/(\sqrt{2\pi}\sigma)] \int_{\Delta/2}^{\infty} \exp(-n^2/2\sigma^2) \cdot dn$$

$$= (1/2)\text{erfc}\left[\frac{\Delta}{2\sqrt{2}\sigma}\right] \tag{6.12}$$

where σ^2 is the variance of the noise power distribution. This must be related to the optimum detector threshold in order to express the probability of error in terms of the unmodulated-input-carrier-to-noise ratio C/N. The noise power present at the input to the receiver is more suitably expressed as a power per unit frequency, for this insures that, whatever filters exist, the noise spectral density through them will not be affected. As in this treatment the noise is considered to be uniform throughout the spectrum, the double-sided spectral density $\eta/2$ watts/Hz is a constant value throughout the band considered. Thus the noise power passed by an ideal filter with unity gain and double-sided bandwidth $(2B = W)$ is ηB watts. This is equivalent to a single-sided bandwidth B, multiplied by a noise spectral density of η. The treatments in this chapter provide equations which produce C/N values based on double-sided noise. The diagrams also show the double-sided noise bandwidth since it is assumed that, on demodulation to the IF, the RF bandwidth will be centered on the IF center frequency. The case of the ASK probability of error requires special mention since the probability of error has been calculated on the basis that only half-power is being transmitted. To perform the comparisons with other modulation schemes this power has been doubled for the curves of Figure 6.5 and equation 6.17(a). The foregoing analysis is not rigorous and is presented only to

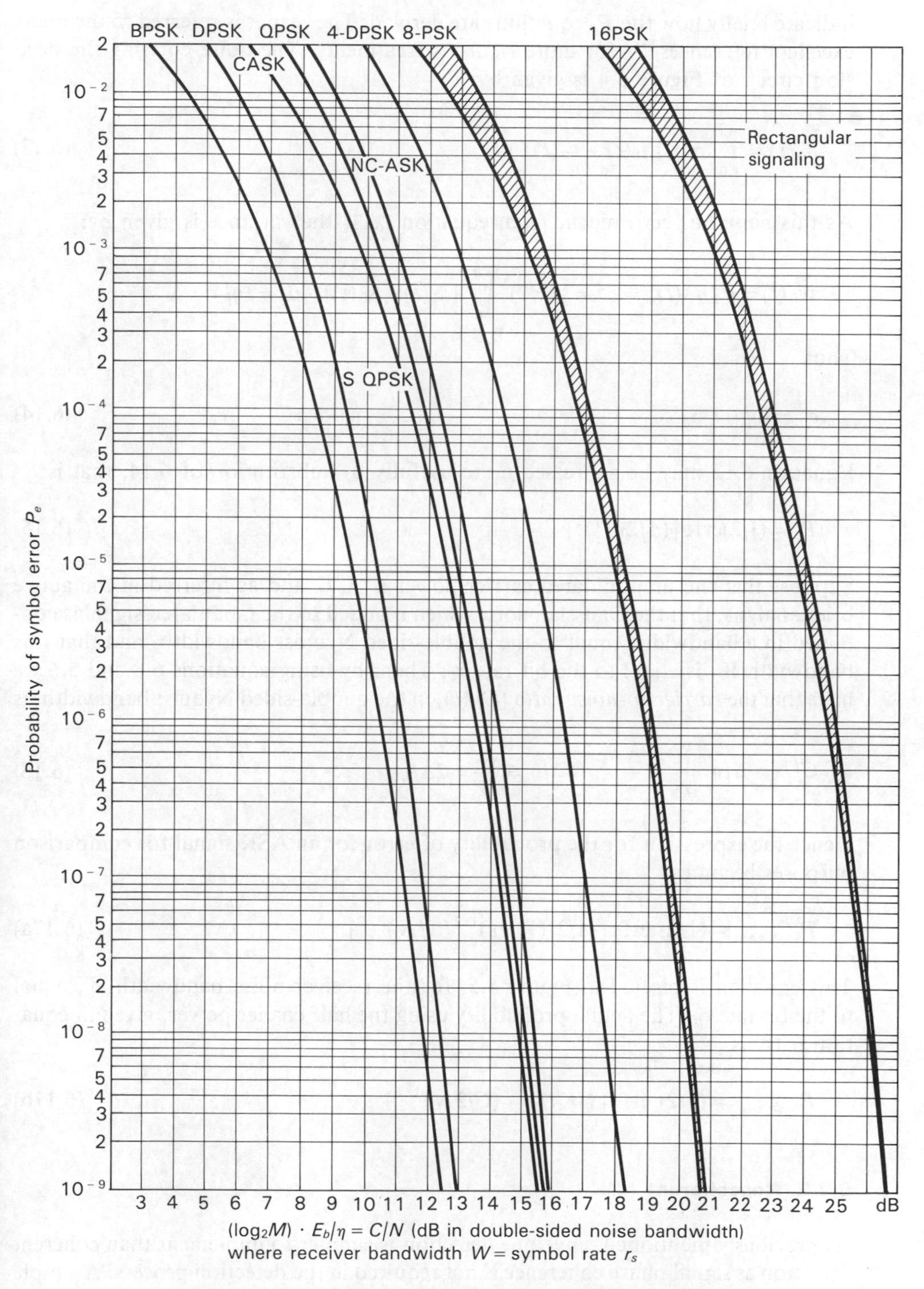

Figure 6.5 P_e performance of various modulation schemes

indicate briefly how the P_e equations are derived. The reader is referred to the many excellent references[1-5] for more rigorous treatments. The noise entering the decision circuit of Figure 6.4 is given by:

$$n_0(t) = \int_0^T n(t)[s_1(t) - s_0(t)]\, dt \tag{6.13}$$

As this noise has zero mean, from equation 2.11, the variance is given by:

$$N_0(t) = E[n_0^2(t)] = \sigma^2 = \tfrac{1}{2}(\eta/2) \int_0^T [s_1(t) - s_0(t)]^2\, dt = (\eta/4) \cdot \Delta$$

Thus

$$\sigma^2 = (\eta/4) \cdot \Delta \tag{6.14}$$

Equation 6.12 may be expressed more usefully by substitution of 6.14, that is

$$P_e = (1/2)\text{erfc}\,[(\Delta/2\eta)^{1/2}] \tag{6.15}$$

Suppose that the unmodulated carrier power is $A_c^2/2$ and as inferred in the above brief analysis, that the Gaussian noise which is added to the modulated signal is contained in a bandwidth equal to the double-sided Nyquist bandwidth, and that this bandwidth W, is equal to the bit rate r_b. Then, by using equations 6.8 and 5.6 we have that the *carrier-to-noise ratio* (C/N), in the double-sided Nyquist bandwidth, is

$$C/N = \Delta/\eta.\ \frac{W}{r_b} = \frac{\Delta}{\eta} \qquad 2\Delta/\eta \lesssim \frac{r_b}{W} = 2\Delta/\eta \tag{6.16}$$

Hence the expression for the probability of error for an ASK signal for comparison purposes becomes:

$$P_{e_{\text{ASK binary}}} = (1/2)\ \text{erfc}\,[\,(1/2)(W/r_b)^{1/2}(C/N)^{1/2}] \tag{6.17a}$$

This equation is plotted in Figure 6.5, for the receiver noise bandwidth W, equal to the bit rate r_b. The actual probability using the half carrier power, given in equation 6.10, is

$$P_{e_{\text{ASK binary}}} = (1/2)\ \text{erfc}\,[\,(W/r_b)^{1/2}(C/2N)^{1/2}] \tag{6.17b}$$

6.2.2 Noncoherent ASK

As previously mentioned, envelope detection is simpler to implement than coherent detection as signal phase coherence is not required in the detection process. A simple envelope detector is used following the IF amplifiers, or where there is no conversion stage, following the RF input band-pass filter. Consider the block diagram of a non-

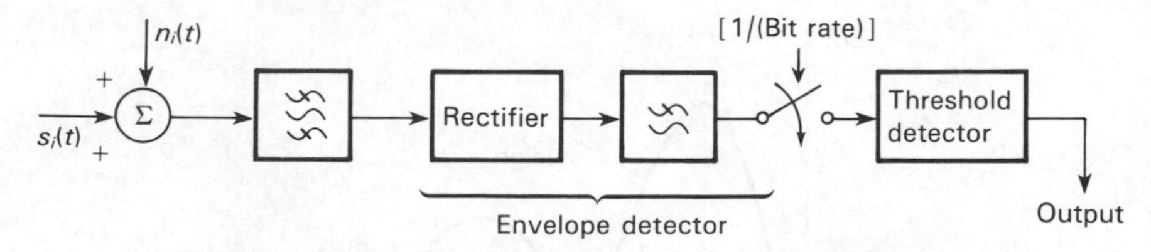

Figure 6.6 Block diagram of a noncoherent ASK demodulator

coherent ASK demodulator as shown in Figure 6.6. The detection system consists of a band-pass filter matched to the RF input binary 1 ASK waveform, as shown in Figure 6.1, followed by an envelope detector and a threshold detector (A/D convertor). Assuming that the filter has a bandwidth equal to twice the bit rate, that is $2/T$ and a center frequency of ω_o, then it passes the binary 1 ASK waveform without undue distortion. The noise power at the output of the filter is found from equation 5.38 to be:

$$n(t) = \sigma^2 = \eta B = 2\eta/T \tag{6.18}$$

Calculating the probability of error involves two pdfs. When an ASK zero is sent, the resultant envelope out of the envelope detector has a Rayleigh pdf (f_0) as given by equation 2.25:

$$f_0 = (x/\sigma^2)\exp(-x^2/2\sigma^2),\ x > 0 \tag{6.19}$$

The second pdf is when an ASK binary 1 is sent and is a Rice pdf (f_1). This is given by:

$$f_1 = [(x/\sigma^2) \cdot I_0 \cdot (xA_c/\sigma^2)]\exp\{-(x^2 + A_c^2)/(2\sigma^2)\},\ x > 0 \tag{6.20}$$

where $I_0 = I_0(u)$ is the modified Bessel function of the first kind and zero order defined by,

$$I_0(u) = (1/2\pi)\int_0^{2\pi} \exp[u\cos(v)]\ du \tag{6.21}$$

Figure 6.7 shows the two pdfs and the value of x which produces the lowest noise at the envelope detector output and consequently, the lowest probability of error. References 2–5 provide a detailed derivation of the probability of error of a noncoherent ASK detector and show that the minimum probability of error occurs when

$$x_{\min} = (A_c/2)[1 + 8\sigma^2/(A_c)^2]^{1/2} \tag{6.22}$$

The probability of error is given by the expression:

$$P_{e(\text{noncoherent ASK})} > (1/2) \cdot [1 + (1/\sigma A_c) \cdot (2/\pi)^{1/2}]\exp(-A_c^2/8\sigma^2)$$
$$> (1/2) \cdot \exp[-A_c^2/8\sigma^2] \text{ if } A_c \gg \sigma \text{ (Double-sided noise)}$$

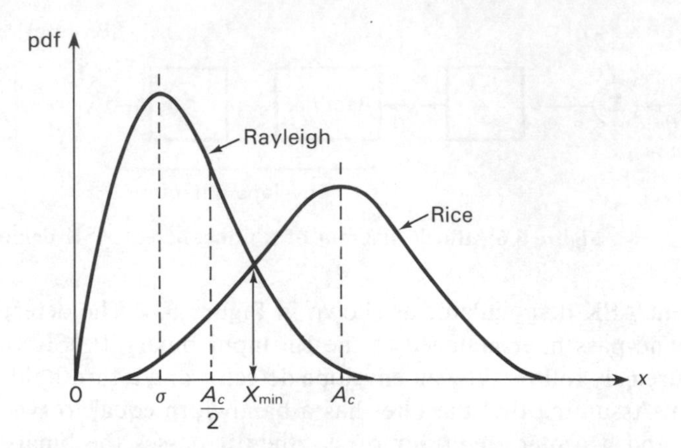

Figure 6.7 Rayleigh and Rice pdfs for noncoherent ASK demodulated noise and envelope plus noise

The lower bound may also be found[3] and therefore P_e is given for the double-sided noise case by

$$\exp[(-1/4)(W/r_b)(C/N)] \geqslant P_e > (1/2) \cdot \exp[(-1/4)(W/r_b)(C/N)] \quad \text{if } A_c \gg \sigma \tag{6.23}$$

Equation 6.23 is plotted in Figure 6.5, for the demodulator noise bandwidth W, equal to the bit rate r_b. Because the carrier is switched on and then off by the binary waveform, this form of modulation is known as *on-off keying* (OOK): either the carrier is switched on, or it is completely off. As described in section 6.3, if a bipolar binary signal produced an inversion in the carrier amplitude, so that a binary 1 produced a carrier with an amplitude of $+A_c$, and the binary zero produced a carrier with an amplitude of $-A_c$, the result would be another important form of modulation known as *phase-shift keying* (PSK). It can be seen from Figure 6.5 that both the coherent and noncoherent ASK detection methods produce almost the same results. The difference between the two values of C/N is no more than 1.5 dB, when $P_e \sim 10^{-3}$ and improves to about 0.5 dB at smaller values of P_e. The noncoherent or envelope detection method requires a higher C/N ratio for the same error rate as that of the coherent form of ASK, which is not a widely favoured modulation method because, as equation 6.10 shows, the average power of the modulated signal is reduced. When coherent ASK is compared with frequency- and phase-shift keying that becomes clear, since these modulation techniques use the full carrier. In addition to power considerations, the error probability is about three orders worse than a well-designed baseband system. Imperfect filtering, poor synchronization and the added costs and difficulties associated with the building of matched band-pass filters all lead to an unfavourable outcome when compared with other modulation systems.

To complete this section on ASK, multilevel (M-ary) ASK systems will be briefly discussed and the probability of error equations for both coherent and noncoherent receiver methods given.

6.2.3 *M*-ary ASK

For the reasons stated above, the multilevel ASK systems are not popular and are not usually found in manufacturers' stock items. The probability of error P_e for multilevel systems refers to symbol errors and not bit errors. As each symbol is comprised of $\log_2 M$ bits, the bit error rate will lie between $P_e/\log_2 M$ and P_e, the relation being dependent on the coding scheme used. Also, due to the higher bit rate of multilevel systems, the bandwidth is scaled down for comparison purposes and so the carrier-to-noise ratio and P_e are scaled down by the same amount.

For the *coherent* case:[3]

$$P_{e(\text{coherent ASK})} = [(M-1)/M] \cdot \operatorname{erfc}[(3/4) \cdot 1/(M-1) \cdot 1/(2M-1)(W/r_s)(C/N)]^{1/2} \tag{6.24}$$

where M is the number of levels of carrier amplitude which the digital signal is coded into. For a binary signal, $M=2$ and equation 6.24 reduces to that given by equation 6.17a. In equation 6.24, C/N is the ratio of unmodulated carrier to double-sided Nyquist-bandwidth noise and is not a logarithmic value. If a logarithmic value is given, it must be converted to a ratio, using the expression $C/N = \text{antilog}[(C/N)\ \text{dB}/10]$. As mentioned previously, in Figure 6.5, the double-sided noise power is used because it is expected that the receiver band-pass filter will be centered on the carrier and have a bandwidth equal to twice that of the information signal band, that is: $W = r_s = r_s/\log_2 M$.

For the *noncoherent* (or *incoherent*) case:[3]

$$\exp[(-3/4) \cdot 1/(2M-1) \cdot 1/(M-1)(W/r_b)(C/N)]^{1/2} \geqslant P_{e(\text{noncoherent ASK})}$$

$$P_{e(\text{noncoherent ASK})} > (1/M) \cdot \exp[(-3/4) \cdot 1/(M-1) \cdot 1/(2M-1)(W/r_s)(C/N)]^{1/2} \quad \text{for } A_c \gg \sigma \tag{6.25}$$

Again, for the binary case where $M=2$, equation 6.25 reduces to that given by equation 6.23.

6.3 PHASE-SHIFT KEYING (PSK)

Figure 6.8 shows the process of phase-modulating a carrier with a binary signal 10101101. In binary PSK the two signaling waveforms may be represented by:

$$\begin{aligned} s_1(t) &= \; = A \cos \omega_o t \\ s_0(t) &= -A \cos \omega_o t = \text{A} \cos(\omega_o t + \pi) \end{aligned} \tag{6.26}$$

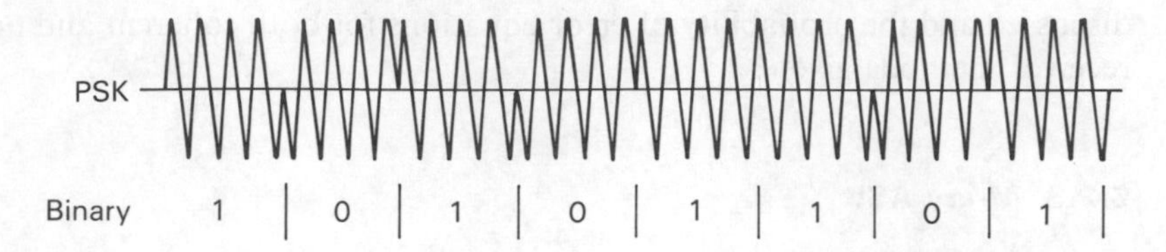

Figure 6.8 Digital modulation PSK for the binary signal 10101101

$s_1(t)$ represents the binary 1 and $s_0(t)$ the binary 0. As mentioned previously, the carrier amplitude of an ASK carrier is switched on and off. For the PSK carrier, the amplitude remains constant during transmission, but is switched between the $+A$ and $-A$ states and is thus *antipodal*. The $-A$ state may alternatively be considered as a phase change of 180° as shown in equation 6.26. The bandwidth requirement for both ASK and PSK, however, remains the same as shown by the power spectral density function. This is given by the expression

$$P(f)_{\mathrm{PSK}} = (A^2/4) \cdot \left\{ \frac{\sin^2 \pi T(f - f_0)}{\pi^2 T(f - f_0)^2} + \frac{\sin^2 \pi T(f + f_0)}{\pi^2 T(f + f_0)^2} \right\} \tag{6.27}$$

Comparison with equation 6.3 shows that the only difference between the power spectral density function $P(f)_{\mathrm{PSK}}$ and the power spectral density function for ASK modulation is that the PSK spectrum does not contain Dirac delta or impulse functions at the carrier frequency, and is thus a suppressed carrier modulation form. Figure 6.2 shows the power spectral density function for the ASK signal. The diagram may be considered to be that of the PSK spectrum if $P(f)$ is $A^2/4$ for f_0 and the impulse at f_0 is removed. As with ASK, the reception of the transmitted PSK signal may be achieved in two ways. The first is that of coherent demodulation which uses, typically, a circuit based on the block diagram shown in Figure 6.9, in which carrier recovery circuits insure that the local reference signal is synchronized in phase with the incoming signal. The second method is that of *differentially encoded PSK* (DPSK) in which, for binary DPSK, a binary 1 is transmitted by shifting the phase of the carrier by 180° relative to the carrier phase in the previous signaling interval. Demodulation is accomplished by comparing the phase of the received

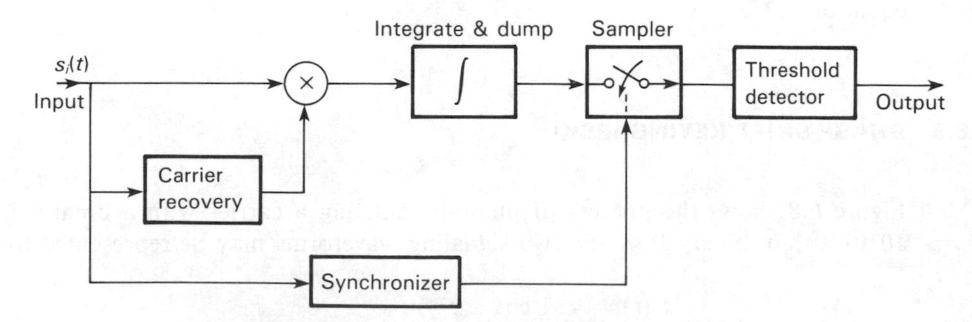

Figure 6.9 Coherent PSK demodulator

signal at two successive intervals. Again, the following sections on these two coherent demodulation methods provide expressions for the expected probability of error P_e.

6.3.1 Coherent PSK (CPSK)

Figure 6.9 shows the block diagram of the demodulator. The incoming signal $s_i(t)$ is used in the carrier recovery circuits, to derive a local oscillator signal which is in phase with the incoming signal. This oscillator signal may be given by the expression

$$s_1(t) - s_0(t) = 2A \cos \omega_0 t \tag{6.28}$$

This signal is then integrated using the 'integrate and dump' circuit shown in Figure 6.4. The signal components at the detector input may be given by the expression:

$$U_1 = \int_0^T s_1^2(t)\,\mathrm{d}t - \int_0^T s_0(t) \cdot s_1(t)\,\mathrm{d}t = +A^2T \tag{6.29a}$$

and for the reception of $s_0(t)$:

$$U_0 = \int_0^T s_0(t) \cdot s_1(t)\,\mathrm{d}t - \int_0^T s_0^2(t)\,\mathrm{d}t = -A^2T \tag{6.29b}$$

By using equation 6.7, the value of Δ for a binary PSK signal may be found to be:

$$\Delta = 2A^2T = 2A^2/r_b \tag{6.30}$$

The optimum threshold detector setting is given as

$$(\text{Threshold})_{\text{opt}} = (U_1 + U_0)/2 = 0 \tag{6.31}$$

which is independent of the carrier strength at the receiver input.

As the signal carrier is present at all times, the average received signal power is S_{av}

$$S_{av} = A^2/2 = C \tag{6.32}$$

6.3.1.1 Probability of error P_e[6]

The noise entering the decision circuit is given by $N = \eta B$, where $2B = W$, the double-sided noise bandwidth. Substituting η into equation 6.15 together with the value of Δ from equation 6.30, and C from equation 6.32 the probability of error becomes:

$$P_{e(\text{binary PSK})} = (1/2)\,\text{erfc}\,[\,(W/r_b)(C/N)]^{1/2} \tag{6.33}$$

This equation is shown in Figure 6.5, for $W = r_b$. Comparing binary coherent PSK with binary coherent ASK, for equal probability of error, we see that the average signal power of the ASK signal must be twice that of the PSK signal. This shows a 6 dB power advantage of PSK modulation over ASK modulation. If the local signal derived from the carrier recovery circuits is not in phase with the incoming signal, but has a fixed phase offset ϕ, the probability of error is increased, and is given by

$$P_{e_{CPSK\phi}} = (1/2)\text{erfc}\,[\,(C/N)(W/r_b)\,\cos^2\phi\,]^{\,1/2} \tag{6.34}$$

6.3.2 Differentially coherent PSK (DPSK)

Figure 6.10 shows the block diagram of a receiver used in the demodulation of DPSK. In this system the transmitter shifts the phase of the carrier by 180° relative to the phase in the previous signaling interval, once a binary digit 1 has been transmitted. The receiver demodulates the binary information by comparing the phase of the received signal with that previously found in the preceding interval. The advantage of this system is the decoding of the modulated carrier without the need for a coherent local oscillator signal. Therefore, the differentially coherent PSK modulation scheme may be considered as the noncoherent form of the previously discussed coherent PSK scheme. The generation of DPSK signals is shown in Figure 6.10. The binary signal to be transmitted is, for example, represented by the sequence $b'(t)$. Due to the delay circuit, the input to the level shifter is a modified binary sequence $b(t)$. This sequence has one more digit than $b'(t)$ and at the start of a sequence is arbitrarily chosen. Succeeding digits in $b(t)$ are determined using the following expression:

$$b_k' = b_{k-1}' \cdot b_k \oplus \bar{b}_{k-1}' \cdot \bar{b}_k \tag{6.35}$$

where $A \oplus B = A\bar{B} + \bar{A}B$, which represents the exclusive-or Boolean expression. By using the binary code 101011011, the encoded sequence $b(t)$ may be determined, as shown in Table 6.1. Before the differential sequence $b(t)$ appears at the input to the level shifter, the input level to this shifter is held at a constant voltage level, which corresponds to a binary logic state. In the example given, this arbitrary level corresponds to a binary 1. The differential sequence $b(t)$, after having its logic levels changed to + and − voltage values, phase-shift keys a carrier entering the balanced modulator. The modulator output is a carrier which changes phase when $b(t)$ changes.

Table 6.1 Differential encoding

Input sequence $b'(t)$	1	0	1	0	1	1	0	1	1	0
Differentially encoded sequence $b(t)$	0	1	1	0	0	1	0	0	1	0
Transmitted phase	π	0	0	π	π	0	π	π	0	π

The reason why the transmitted phase is shifted by 0°, when 1 is transmitted, instead of 180° is to prevent a carrier containing no phase shifts and having a relatively narrow spectrum from occurring, should a long sequence of '0's be sent. The transmission of $\lambda + \pi$ radians when a 1 is transmitted, with λ selected equal to π radians and the transmission of λ radians when a '0' is transmitted, permits the carrier phase to be shifted in every signaling interval even when a long stream of zeros is sent. This results in a signal spectrum with a width approximately equal to $1/T$. The spectral components about the carrier are used in some cases in maintaining time synchronization at the receiver.

The method of binary signal recovery of a DPSK carrier is shown in Figure 6.10(*b*). The input signal $s_i(t)$ plus noise $n(t)$ is first filtered to limit the noise power and then passed into a multiplier, or correlator, where it is multiplied by a one-signal-bit delayed version of the input signal. The output of this multiplier is passed through an integrator or low-pass filter to separate the twice-carrier-frequency term from the signal waveform, and the resultant signal is applied to the input of the decision circuit, where it is compared with a zero voltage. The decision on whether a '1' or '0' was received depends on whether the output voltage of the multiplier is positive or negative. The advantages of a DPSK system over the coherent PSK system is that the circuitry is not complicated by the need to generate a local carrier at the receiver. Noise occurring on the phase reference during transmission tends to

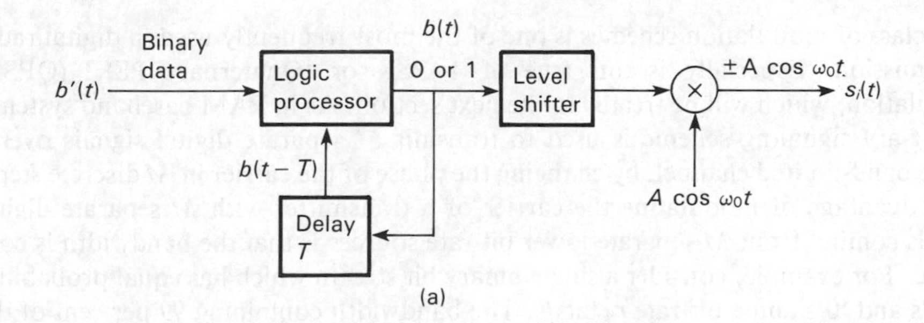

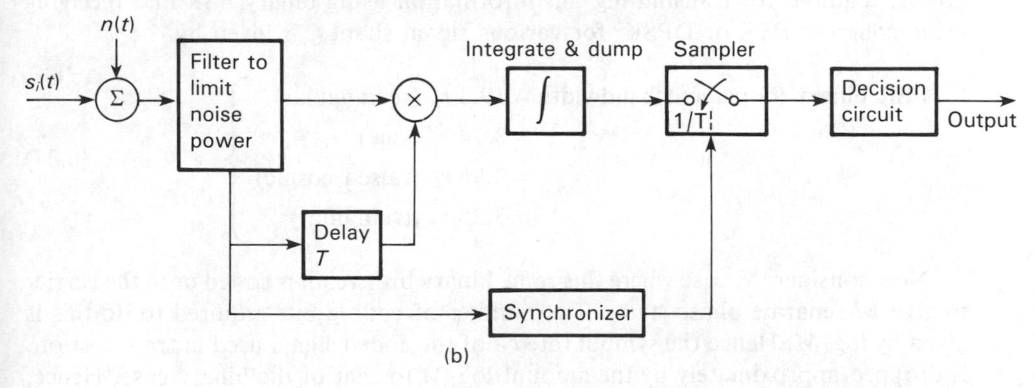

Figure 6.10 DPSK modulator and demodulator; (a) modulator; (b) demodulator

cancel out and the degradation in performance is not as great as may initially appear, but as the bit determination in the decision circuit may on the basis of the signal be received in two successive intervals, noise in a one-bit interval may cause errors to two-bit determinations. Thus there is an error extension factor where the bit errors tend to occur in pairs. The error rate of DPSK is therefore one or two dB greater than that of PSK for a given carrier-to-noise ratio.

6.3.2.1 Probability of error P_e

The probability of error in the presence of single-sided white Gaussian noise can be shown to be;[2,7]

$$P_{e_{\mathrm{DPSK}}} = (1/2)\exp[-(C/N)(W/r_b)] \tag{6.36}$$

In the presence of inter-symbol interference and Gaussian noise an expression for the probability of error in terms of modified Struve functions can be found in Reference 7. The expression will not be given here since its explanation is outside the scope of this book.

6.3.3 *M*-ary PSK[7]

This class of modulation schemes is one of the most frequently used in digital radio transmission. Especially is this true of 4–PSK or ‘Quaternary PSK’ (QPSK) modulation, which will be treated in the next section. As in PAM baseband systems, the M-ary signaling scheme is used to transmit M separate digital signals over a single band-limited channel, by changing the phase of the carrier in M discrete steps. The advantage of modulating the carrier of a transmitter with M separate digital signals coming from M separate lower bit rate sources is that the bandwidth is conserved. For example, consider a single binary bit stream which has equal probability of ‘1’s and ‘0’s and a bit rate r_b bits/s. The bandwidth containing 99 per cent of the power, required for transmitting this information using binary PSK and receiving using coherent PSK or DPSK, for various signal shapes, is given by[9]

$$\left.\begin{aligned} \text{PSK binary 99 percent bandwidth} &= 19.3\, r_b \text{ (rectangular)} \\ &= 3.74\, r_b \text{ (sine)} \\ &= 2.96\, r_b \text{ (raised cosine)} \\ &= 3.28\, r_b \text{ (triangular)} \end{aligned}\right\} \tag{6.37}$$

Now consider the case where this same binary bit stream is coded onto the carrier to give M separate phase states. The number of coding bits required to do this is given by $\log_2 M$. Hence the symbol rate r_s, of the coded signal used in transmission, is dropped approximately by the amount $\log_2 M$ to that of the binary case. Hence, as r_s is given by

$$r_s = r_b/(\log_2 M) \tag{6.38}$$

the M-ary PSK bandwidth becomes reduced by approximately the factor $\log_2 M$ to transmit the same information.

For the case where $M = 4$, the following 99 percent bandwidths are reduced accordingly for the different pulse shapes:

$$\left.\begin{aligned} \text{4-PSK or QPSK bandwidth} &= 19.30\, r_b/(\log_2 4) = 9.65\, r_b \text{ (rectangular)} \\ &= 3.84\, r_b/2 \quad = 1.92\, r_b \text{ (sine)} \\ &= 4.00\, r_b/2 \quad = 2.00\, r_b \text{ (raised cosine)} \\ &= 3.65\, r_b/2 \quad = 1.83\, r_b \text{ (triangular)} \end{aligned}\right\} \tag{6.39}$$

Usually the bandwidth is specified for a given bit error rate, rather than the 99 percent bandwidth and it is usually taken to be: $r_b/0.8$ for binary PSK, $r_b/1.9$ for 4–PSK, $r_b/2.6$ for 8–PSK and $r_b/2.9$ for 16–PSK. As the bandwidth efficiency is given as binary bit rate r_b divided by the IF bandwidth W, i.e. r_b/W, the practical spectral efficiency for binary PSK is 0.8 bits/s/Hz, for QPSK is 1.9 bits/s/Hz, for 8–PSK is 2.6 bits/s/Hz and for 16–PSK is 2.9 bits/s/Hz.[10] The reduction in bandwidth of an M-ary PSK system over its binary counterpart permits M separate signal sources to be transmitted for approximately the same bandwidth as that of a binary system. The alternative interpretation is that the M-ary PSK system permits a higher, (by the factor $\log_2 M$) input binary bit rate to the transmitter, to be transmitted into a binary PSK modulated system over a bandwidth that would just be sufficient for a single input binary bit rate. Band-limiting of a radio system is an important consideration as the radio spectrum is a shared resource that must be used efficiently to meet the increasing demands for transmission capacity. The band-limiting of the digital signal usually occurs at both the transmitter and receiver, and the filters must be designed to generate the desired band-limiting signaling pulse and to suppress out-of-band signal power. The composite filter function between the transmitter and receiver is so designed that the partitioning minimizes adjacent-channel interference and provides optimum detection at the receiver.

In a M-ary PSK system, the phase of the carrier is permitted to be in any of the phase states $\phi_k = 2\pi k/M$, for $k = 0, 1, 2, \ldots, M-1$ and each of these carrier states or signaling waveforms have equal energy. Thus the M possible signals that are transmitted in a symbol interval $T_s (T_s = 1/r_s)$ are given by:

$$s_k(t) = A\cos(\omega_0 t + 2\pi k/M + \lambda), \quad k = 0, 1, 2, \ldots, M-1, \text{ for } 0 \leqslant t \leqslant T_s \tag{6.40}$$

By expanding this cosine function, the k signaling waveforms may be expressed as:

$$s_k(t) = A_I \cos \omega_0 t - A_Q \sin \omega_0 t \tag{6.41}$$

where

$$A_I = A\cos(2\pi k/M + \lambda) \text{ and } A_Q = A\sin(2\pi k/M + \lambda) \quad k = 0, 1, 2, \ldots, M-1 \tag{6.42}$$

The signal in equation 6.41 may be considered as two quadrature carriers with amplitudes A_I and A_Q, which depend on the transmitted phase $2\pi k/M$, in any signaling interval T_s. Figure 6.11 shows the signal constellations for binary ASK, 4-ary ASK, binary PSK, 4-ary PSK or QPSK and 8-ary PSK, otherwise known as eight-phase PSK or 8–PSK. The phases of the PSK carrier relating to the signal given in equation 6.41 are represented as points in a plane at a distance A from the origin and separated in angle $2\pi k/M$. Decision thresholds in the receiver are centered between $2\pi k/M$, to enable correct decisions to be made if the received signal phase is within the threshold phases; $\pm\pi/M$ of the transmitted phase. For the case where $\lambda = 0$, the four possible waveforms, each of which is transmitted in the signaling interval T_s, are given by equation 6.41 as:

$$\left.\begin{aligned} s_0(t) &= \ \ A\cos\omega_0 t \\ s_1(t) &= -A\sin\omega_0 t \\ s_2(t) &= -A\cos\omega_0 t \\ s_3(t) &= \ \ A\sin\omega_0 t \quad \text{for } 0 < t < T_s \end{aligned}\right\} \tag{6.43}$$

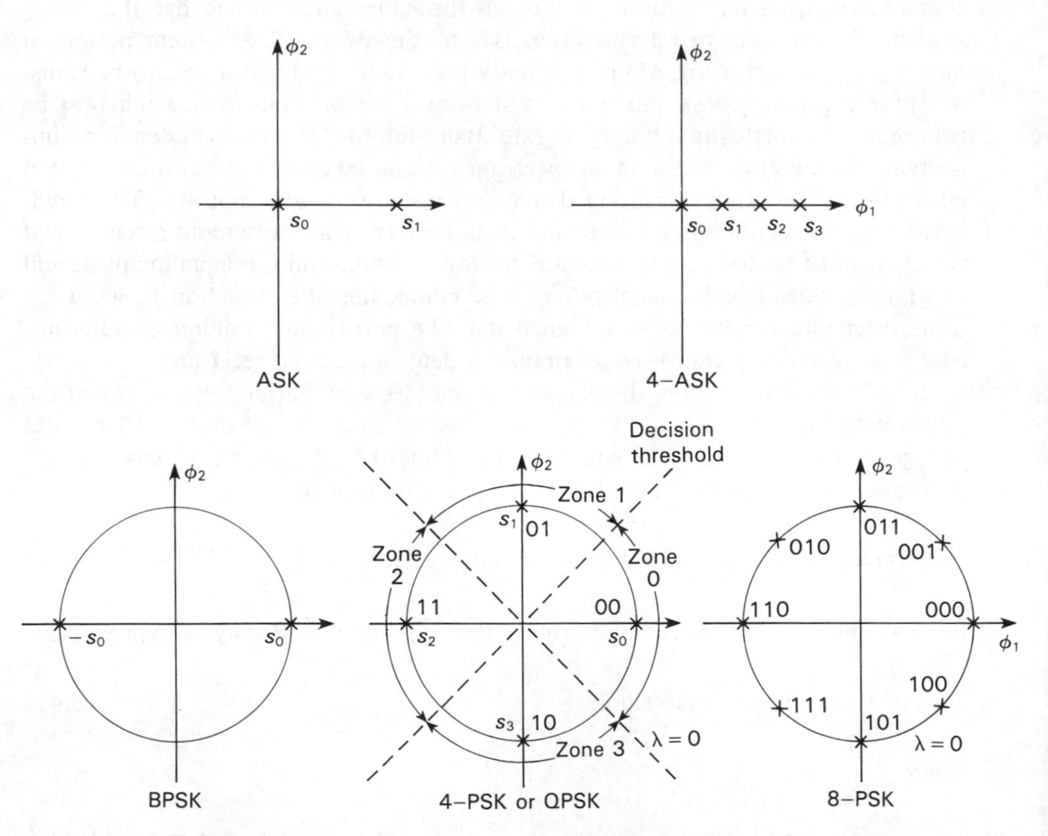

Figure 6.11 Constellation diagram for binary ASK, 4–ASK, PSK, 4–PSK and 8–PSK

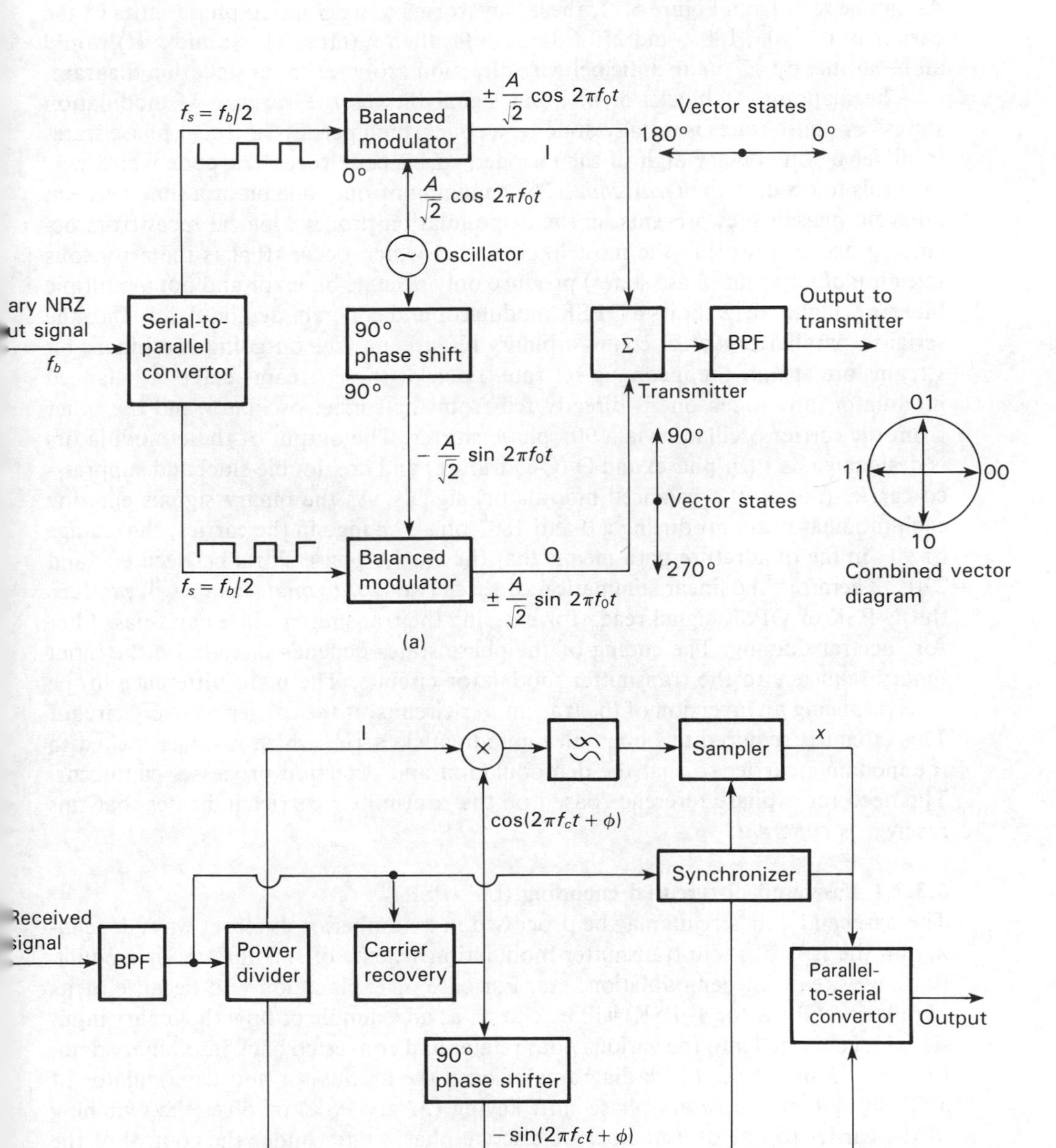

Figure 6.12 QPSK modulator and demodulator: (a) modulator; (b) demodulator

As can be seen from Figure 6.11, these waveforms correspond to phase shifts of the carrier of 0°, 90°, 180°, and 270°. If $\lambda = \pi/4$, then $s_0(t)$, $s_1(t)$, s_2t and $s_3(t)$ would all be shifted by 45° in an anticlockwise direction around the constellation diagram.

The mapping of blocks of m input signal bits to the various M modulation states[2] or phase states is usually done by a code which permits adjacent phase states to differ by one binary digit of the encoded M bit codeword. The code which permits this to occur is the *Gray code.* The changing of only one bit at a time between adjacent phase states prevents, in the demodulation process, logical *races* from occurring and insures that the most likely errors which occur (that is the erroneous selection of adjacent phase states) produce only a single bit error and not a multiple bit error. Figure 6.12 shows a QPSK modulator and coherent demodulator. Into the serial-to-parallel convertor comes a binary bit stream. The outgoing two binary bit streams are at half the incoming bit rate. Each of these streams enters a balanced modulator into which one is directly fed from the carrier oscillator and the other from the carrier oscillator via a 90° phase shifter. The output of these modulators is designated as I (In-phase) and Q (Quadrature) and are double-sideband suppressed carrier (due to the balanced modulator) signals. As the binary signals entering each modulator are producing a 0 and 180° phase change in the carrier, the change of 90° in the quadrature path means that the carrier phase shifts between 90° and 270°. Therefore the linear summation of these two *orthogonal* signals will produce the 4–PSK or QPSK signal ready for entering the transmitter via a band-pass filter for spectral shaping. The coding of the phase states depends directly on the input binary sequence to the transmitter modulator circuits. The main difference in the receiver, being an inversion of the transmitter circuits, is the carrier recovery circuit. This circuit is required to obtain an unmodulated carrier which is phase-locked to the incoming carrier so that the demodulation and detection processes can occur. The need for a phase reference based on the transmitter carrier indicates that this receiver is *coherent.*

6.3.3.1 Baseband differential encoding (DE–PSK)[10]

The baseband data stream may be processed in a number of different ways depending on the type of radio transmitter modulation scheme or the number of modulation levels, ease of demodulation, etc. For ease of explanation and because of its popularity, QPSK (or 4–PSK) will be chosen as an example of how the binary input signal is converted into the various phase states and converted back into binary data. Figure 6.12 shows the block diagrams of both the modulator and demodulator of a QPSK system. As M-ary phase-shift keying (M-ary PSK) involves the switching of the carrier to one of a number of discrete phase states under the control of the data stream, it is necessary to have a reference phase transmitted to detect these phase states at the receiver end. This can be done by sending phase 0° for a specified time before sending data but due to distortions in the RF waveform a change of phase may occur and the reference lost. That of course would result in the corruption of data. To satisfy the requirements for a reference phase without sending 0° the transmitted dibit (or block of two binary bits) is modified by an amount depending upon its relationship with the previous dibit. This is usually achieved by using a modulo-4 adder in which the incoming dibit is added to the sum of the previous dibits. The result of the signal out of the adder is then used to drive a binary coded (not a Gray coded) phase modulator. For this modulator, $s_0(t) = 00$, $s_1(t) = 01$, $s_2(t) = 11$, and $s_3(t) = 10$. The expression for the modulo-4 addition is

$$b'_k = b'_{k-1} + b_k \tag{6.44}$$

where b'_k is the output data from the modulo-4 adder, b'_{k-1} is the output delayed by one bit and b_k is the input to the adder. The addition is straight binary addition.

At the receiving end this process is reversed using a modulo-4 subtractor, given by the expression:

$$b'_k = b_k - b_{k-1} \tag{6.45}$$

where the subtraction is straight binary subtraction.

In order to demonstrate the principle, assume that the binary sequence is given by; 0100011100101011101. This sequence, blocked into dibits, becomes: 01 00 01 11 00 10 10 11 01. When the sequence is applied to the adder, the output from the adder is given by; 01 10 01 01 11 01 00 01. When applied to the binary coded modulator via the two-bit serial-to-parallel convertor, the sequence will produce the following output phases with respect to the reference phase; 90° 180° 90° 90° 270° 90° 0° 90°. On reception of the carrier, the receiver's demodulator recovers each phase shift in turn. Since the demodulator has no information regarding the reference phase, it will arbitrarily assign one of the four possible phase states as the reference (0°). Assuming that the first phase state received is designated as the reference, then the received phase states are; 0°90°0°0°180°0°270°0°. These phase states correspond to the binary code 00 01 00 00 10 00 11 00. Hence from equation 6.45, the demodulator output will be 01 11 00 10 10 11 01, which corresponds to the third dibit onwards. To start the sequence off, the first two dibits are lost out of the sequence, but there is no need to send directly a reference signal from which the demodulator can work. This system of modulation is known as differentially encoded PSK or DE–PSK (see below). The probability of error perfomance for different C/N, lies almost midway between that of the BPSK and DPSK as shown in Figure 6.5. For example, at 10^{-4}, the C/N of BPSK is around 8.4 dB, DE–PSK is 8.9 dB and DPSK is around 9.3 dB.

Digital phase modulation schemes usually require coherent detection. A PSK system requires a phase reference at the receiver,[12–14] which in practical communication systems is transmitted along with the keyed tone (and with DPSK is the same tone as that of the signal carrier). In order to maintain proper synchronization, the phase-keyed signal and the reference signal must be close to each other in frequency and in time so that fluctuations along the path of propagation affect both in the same way. There are four basic versions of PSK, of which all versions can be considered in M-ary PSK systems. The first version comprises binary PSK or BPSK and Quarternary PSK (QPSK) in which the phase of one quadrature component is keyed while the phase of the other remains essentially unkeyed. The integration over several keying intervals provided the unkeyed reference 'pilot' tone. The second version is differentially encoded PSK (DE–PSK), which has been described above. The third version is differential PSK (DPSK), in which, as mentioned above,

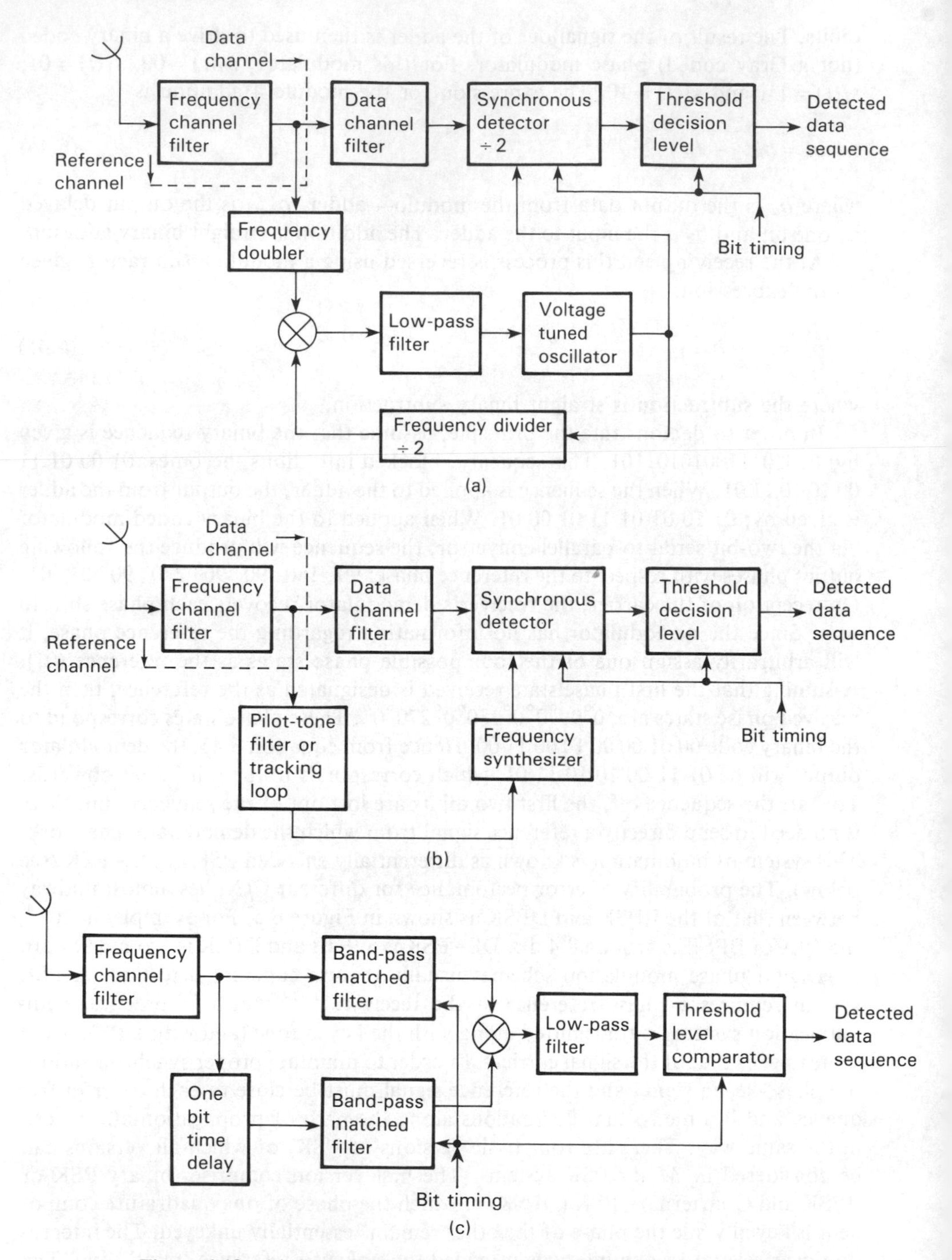

Figure 6.13 (a) Squaring-tracking loop receiver; (b) pilot-tone-tracking loop receiver; (c) DPSK receiver (Courtesy of IEEE, Reference 13)

the same tone serves as both the signal and the reference. The phase during one keying interval serves as a reference for the succeeding keying interval. The fourth version takes different names, but may be classed under the heading of adjacent-tone PSK (AT–PSK). This is a system in which a reference tone is transmitted at an adjacent frequency simultaneously with the keyed tone. At the receiver, the phase of the reference is adjusted to compensate for the frequency difference between the reference and the keyed tone. The adjacent-tone reference may be a residual carrier or pilot tone which designates the type of PSK as pilot-tone phase-shift keying (PT–PSK). The pilot tone frequency is near that of the data signal so as to undergo similar channel impairments. The phase reference for synchronous detection is obtained as the output of a tone filter or a phase-locked tone tracking loop. Many different methods have been investigated which are variations or combinations of the basic four forms listed above. The *decision-directed-measurement PSK* (DDM–PSK) is a system which reconstructs a reference tone by aligning the phases in successive keying intervals based on the already concluded decisions. The DDM–PSK can be regarded as a generalization of DPSK which employs more intervals than the immediately preceding interval. Another system which produces an error rate less than that of both the PT–ASK and DPSK is that of the 'squaring-tracking loop' system.[13] In this system the receiver has two signal processing channels which provide inputs to the detector section of the receiver. The data channel is simply a narrow band time-invariant filter which has a bandwidth large enough to avoid inter-symbol interference. Since the objective of the reference channel is to estimate the carrier phase, by averaging over more than a symbol interval, the phase-reversal keying must be removed from the suppressed-carrier data signal. This is achieved by a square-law device which removes the modulation and acts as a frequency doubler producing an unmodulated carrier having continuous phase at twice the intermediate frequency. A phase-locked loop can then be employed to track and smooth the double frequency output. By dividing this frequency by two, the carrier reference is recovered. Figure 6.13 shows block diagrams of (*a*) the squaring-tracking loop receiver, (*b*) the pilot-tone tracking loop receiver, and (*c*) the DPSK receiver, configurations. With DE–PSK, information is conveyed via transitions in carrier phase. Since a bit decision error on a detected bit will induce another error on the subsequent bit, the performance of DE–PSK is slightly inferior to that of coherent PSK. With DPSK, as with DE–PSK, the information is differentially encoded. The difference between these two is in the type of detector used. The DPSK detector makes no attempt to extract a coherent phase reference, since the signal from the previous bit interval is used as a phase reference for the current bit interval. Since the phase reference signal is not smoothed over many bit intervals, the performance of DPSK is not as good as that of DE–PSK.

6.3.3.2 Probability of error P_e for M-ary coherent PSK systems

In coherent detection, the best that the detector can do in the presence of additive stationary-white-zero mean-Gaussian noise is to make a guess at the transmitted message. Consequently, the measure of performance of the detector will be the number of times that it guesses wrong in a long typical sequence of messages. A reasonable rule which has been adopted[3] is that which assumes that the signal whose message point lies closest to the received signal level was the message which was actually transmitted. For coherent detection then, the decision rule which selects the message point closest to the received point minimizes the probability of error. Such a detector is known as a *maximum likelihood detector*. For QPSK, there are four possible transmitted signal points as shown in Figure 6.11. To realize the decision rule the signal space is partitioned into four zones which are equally spaced about the four signals $s_0, \ldots, s_3$. The decision rule is now simply to guess that $s_0(t)$ was transmitted if the received signal point falls within zone 0, etc. An erroneous decision will be made if, for example, $s_3(t)$ is transmitted and the noise is such that the received signal point falls outside zone 3. The probability of error for the coherent case and for *double-sided Gaussian noise* ($\sigma^2 = \eta B$), when the IF noise bandwidth W, where $W = 2B$, and the bit rate r_s* are given, is:

$$\tfrac{1}{2} \cdot \operatorname{erfc}[(C/N)(W/r_s) \cdot \sin^2(\pi/M)]^{1/2} < P_{e(M\text{-ary coherent PSK})} < \operatorname{erfc}[(C/N)(W/r_s) \sin^2 \pi/M]^{1/2} \quad (6.46a)$$

Figure 6.5 shows equation 6.46 plotted for $M = 4$ and $M = 8$ using the double-sided noise C/N, where $W = r_s = r_b/\log_2 M$. For the particular case of $M = 4$, the value of P_e can be given exactly:[3]

$$P_{e(\text{coherent QPSK})} = \operatorname{erfc}[(1/2) \cdot (C/N)(W/r_s)]^{1/2} - (1/4) \cdot \operatorname{erfc}^2[(1/2) \cdot (C/N)(W/r_s)]^{1/2} \quad (6.46b)$$

For constant P_e, increasing the value of M increases the difference between the required C/N to maintain this value of P_e. For a P_e of 10^{-5}, increasing M from 4 to 8 requires an additional 4 dB increase in C/N. Increasing M from 8 to 16 however requires an additional 5 dB of C/N to maintain the fixed P_e. For large values of M, doubling the value of M requires an additional 6 dB of C/N to achieve the same performance. Due to mapping of the modulator phase states k bit symbol errors contain only a single bit error; this permits the equivalent bit error rate (EBER), for M-ary PSK to be approximated by $(P_{e_{M\text{-ary PSK}}})/(\log_2 M)$. For a constant added noise density at high C/N and for the same error rate, 3–PSK requires 0.75 dB less energy per bit than 2–PSK or 4–PSK (QPSK). At low C/N, however, the 3-PSK requires 0.74 dB more.[15]

The operation of the demodulator shown in Figure 6.12 is similar to that proposed in 1956 by Costas[11] and its operation is as follows: the received signal enters the receiver band-pass filter which serves to eliminate out-of-band noise and any adjacent-channel interference. The power splitter splits the band-limited modulated signal into the I and Q paths. As can be seen there are two basic receivers in this structure with the same input signal but with local oscillator signals in phase quadrature to each other. As the name implies, the carrier recovery circuit is used to derive from the input signal a timing signal which is locked to the incoming signal. This recovered carrier is then split into two paths. In one path the carrier is applied directly to one of the two balanced mixers to demodulate the I signal and, in the other path, the carrier is shifted by 90° before being applied to the other mixer to demodulate the Q signal. The synchronizer also derived from the incoming signal

*See Sections 5.2.1, 6.2.1 and 6.3.5.

is used to provide the correct sampling instants for the detection of logic states on the baseband signal. Due to the transmission path and other causes, this recovered carrier wave may contain jitter, which will affect the optimum sampling instants and hence increase the bit error rate.[16,17]

6.3.3.3 8–PSK

The equation for the probability of error of an 8–PSK system is given by equation 6.46a, with $M = 8$. Figure 6.5 shows the theoretical expected probability of error for both the upper and lower bounds, for different input C/N in a receiver of bandwidth r_s. Note that some diagrams displaying the same curve may have given E_b/η as the energy/bit-to-noise spectral density. If this is so, then $(E_b/\eta) = (C/N)/\log_2 M = [(C/N) - 4.77]$ dB, for $M = 8$. This is discussed further in section 6.3.5. Figure 6.14 shows the block diagram of a typical 8–PSK transmitter and receiver. Figure 6.11 shows the expected constellation diagram for $\lambda = 0$. As can be seen from Figure 6.14, the input binary data are split into three parallel binary streams each having one-third the bit rate r_s, of the input bit stream. The modulator logic, or 2-state to 4-state convertor, provides four logic states from its two 2-level binary inputs A and B, or B and C. That is $+/+$, $+/-$, $-/+$ $-/-$. If the logic state on the input B is 1, the output of the top level convertor will have an output signal which is of a level greater than that of the bottom 2-level to 4-level convertor. The four-state baseband signals are then used as the input to a balanced modulator. The carrier side of the modulators is derived from an oscillator which may have a dual output. Each output is shifted in phase by $\pm 45°$, which permits the output constellation to be that shown in the diagram. The outputs of the suppressed carrier in-phase (I) and quadrature (Q) modulators are combined to give an IF which is carrier modulated with an 8–PSK signal. The 8–PSK spectrum is then filtered to have at the output of the band-pass filter a raised-cosine spectrum with a roll-off factor around 0.6. The demodulator consists usually of the following circuits:

A dynamic IF adaptive equalizer to compensate amplitude dispersions due to the effect of selective fading.
IF carrier recovery through digital baseband processing.
Clock recovery through IF envelope detection and narrow-band filtering.
Differential decoding (if differential encoding was used) and parallel-to-serial conversion to provide the original baseband input at the bit rate r_b.

8-PSK systems appear to find their use in digital radio applications up to 100 Mbit/s (70 Mbit/s–GTE; and 90 Mbit/s–Rockwell Collins).

6.3.4 Performance monitors

The estimation of the digital error rate of the receiver may be done by devices known as *performance monitoring units* (PMU). These devices will determine the error rate without the use of special transmissions, without a knowledge of the data being transmitted and without interrupting the traffic flow.[18,19] As an element of an adaptive communication system, the PMU could be used to determine when adaptive

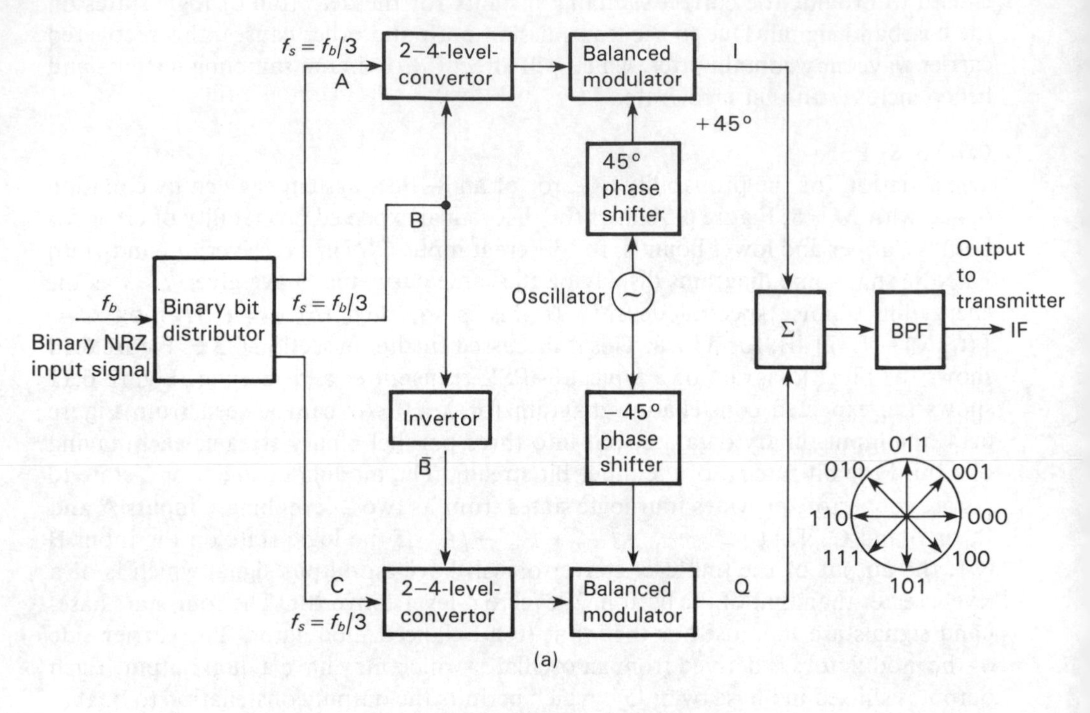

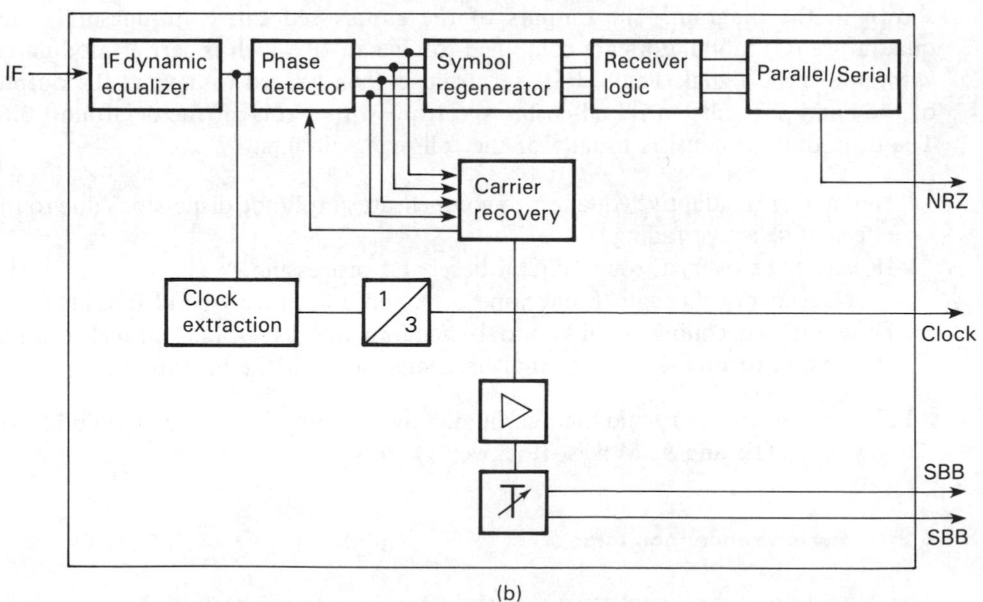

Figure 6.14 8–PSK modulator and demodulator: (a) modulator; (b) demodulator

change is needed and, by comparing the error rates which would result from the various available choices of adaptation, select the best change to be made at any time. Error rate estimates accurate to within a factor of 3 appear to be satisfactory for most applications. The most simple technique used is to count directly the errors. This requires that the transmissions contain redundancy either in the form of a test signal which is identified by the receiver, or by parity check digits in the error detection code. The disadvantage of this method is the inordinately long period of time required to count enough errors to give a reliable estimate of the actual error rates, if the error rate is very low. Another general technique used, which can be extended to most common forms of digital modulation (M-ary PSK, QPSK, FSK), is described in Reference 20. This technique involves the extrapolation of the error rate by the performance monitoring unit through employing two or more modified decision thresholds with a digital receiver, to produce a pseudo-error rate larger than that of the actual receiver, and to estimate this pseudo-error rate. This unit then extrapolates from the pseudo-error rates to obtain an estimate of that which would be obtained with an unmodified decision criterion. The PMU using this technique is applicable to both coherent PSK and differentially coherent PSK of arbitrary numbers of phase states M. The operation of the unit is to consider the phase difference between a received signal pulse and the phase reference and compare this difference with the modified decision threshold. From this comparison a pseudo-error is registered if the phase difference takes on certain values. Before considering the M-ary DPSK demodulator, we shall digress a little and discuss the relationship between the double-sided noise C/N which is being used in the probability of error equations and the average energy of a bit E_b, in relation to the noise power spectral density η, that is E_b/η.

6.3.5 Relationships between double-sided noise C/N, and E_b/η

The average energy of a bit E_b is defined as

$$E_b = S_{av}T = A^2T/2 \text{ (except for ASK where it} = A_c^2T/4) \qquad (6.47)$$
$$= S_{av}/r_b$$

where T is the bit duration = 1/bit rate = $1/r_b$ and S_{av} is the average carrier power. The double-sided noise power spectral density $\eta/2$, is given in terms of the double-sided noise power N or the variance of the Gaussian distribution σ^2 as:

$$N = \sigma^2 = \eta B = (\eta/2) \cdot (2B) = \eta/2 \cdot W \qquad (5.6)$$

where W is the double-sided symbol noise bandwidth under consideration. The noise power spectral density $\eta/2$ is, for a double-sided noise bandwidth, $2B$. Equation 5.6 shows that if a single-sided noise bandwidth is used, the noise spectral density is η, or twice that of the double-sided case. As the unmodulated carrier power C is considered only in the P_e equations, for the purposes of comparison between modulation schemes, S_{av} is redefined in terms of C:

$$C = S_{av} = A^2/2 = E_b r_b = E_s r_s \tag{6.48}$$

as $E_s = E_b \log_2 M$ and $r_b = r_s \log_2 M$ and $W_b = W \log_2 M$, where A is the voltage amplitude of the sinusoidal carrier. The ratio E_b/η for an *M-ary system* can be related to the C/N, using equations 5.6, 6.38 and 6.48, and is given by:

$$E_b/\eta = \left(\frac{C}{N}\right) \cdot \frac{W}{r_b} = \left(\frac{C}{N}\right) \frac{W}{r_s \log_2 M} = \frac{1}{\log_2 M} \left(\frac{E_s}{\eta}\right) \tag{6.49a}$$

where the subscript s, denotes symbol; r_s is the symbol or baud rate and W is the symbol bandwidth.

To accommodate the received carrier without distortion, the IF bandwidth is increased by the roll-off factor α (for raised-cosine filtering) over the Nyquist message bandwidth $r_s/2$. This increase is seen by the double-sided noise received, since its bandwidth must be the same as that of the receive IF filter. The IF filter is double the transmitted message bandwidth. The equations for P_e in this chapter are for the P_e of the transmitted C/N. The curves in Figure 6.5, however, reflect the expected P_e for the received signal, since ideally the double bandwidth of the receiver will introduce little distortion (ISI) to the single-bandwidth of the original transmitted source message signal. The receiver IF, noise bandwidth, or symbol bandwidth W, is given in terms of r_b and α (the raised cosine roll-off factor) as:

$$W_s = r_s(1+\alpha) = \frac{r_b(1+\alpha)}{\log_2 M} \tag{6.50}$$

Substituting equation 6.50 into 6.49a, the alternative expression for (C/N) for a raised-cosine filter is given as:

$$(C/N) = (Eb/\eta) \cdot \frac{\log_2 M}{(1+\alpha)} \tag{6.49b}$$

6.3.6 *M*-ary DPSK[21,22]

Consider an M-ary differential PSK signaling scheme where $M = 4$; from equation 6.41 the PSK signals may be considered to be two binary PSK signals using quadrature carriers. The four-phase differential encoding may be thought of as the differential encoding of two binary PSK carriers and the 4–DPSK receiver as consisting of two bi-phase detectors as shown in Figure 6.15. For $M = 8$, the same principle will apply in the construction and will be just an extension of the four-phase case; for this the receiver uses a delayed version of the received signal as its phase reference (discussed in Section 6.3.2 for the binary case). A distinguishing feature of the DPSK system is that there are no pre-assigned detection regions in the signal space which correspond to a particular transmitted signal. The decision, rather, is based on the phase angle between successively received signals. For 4–DPSK, the

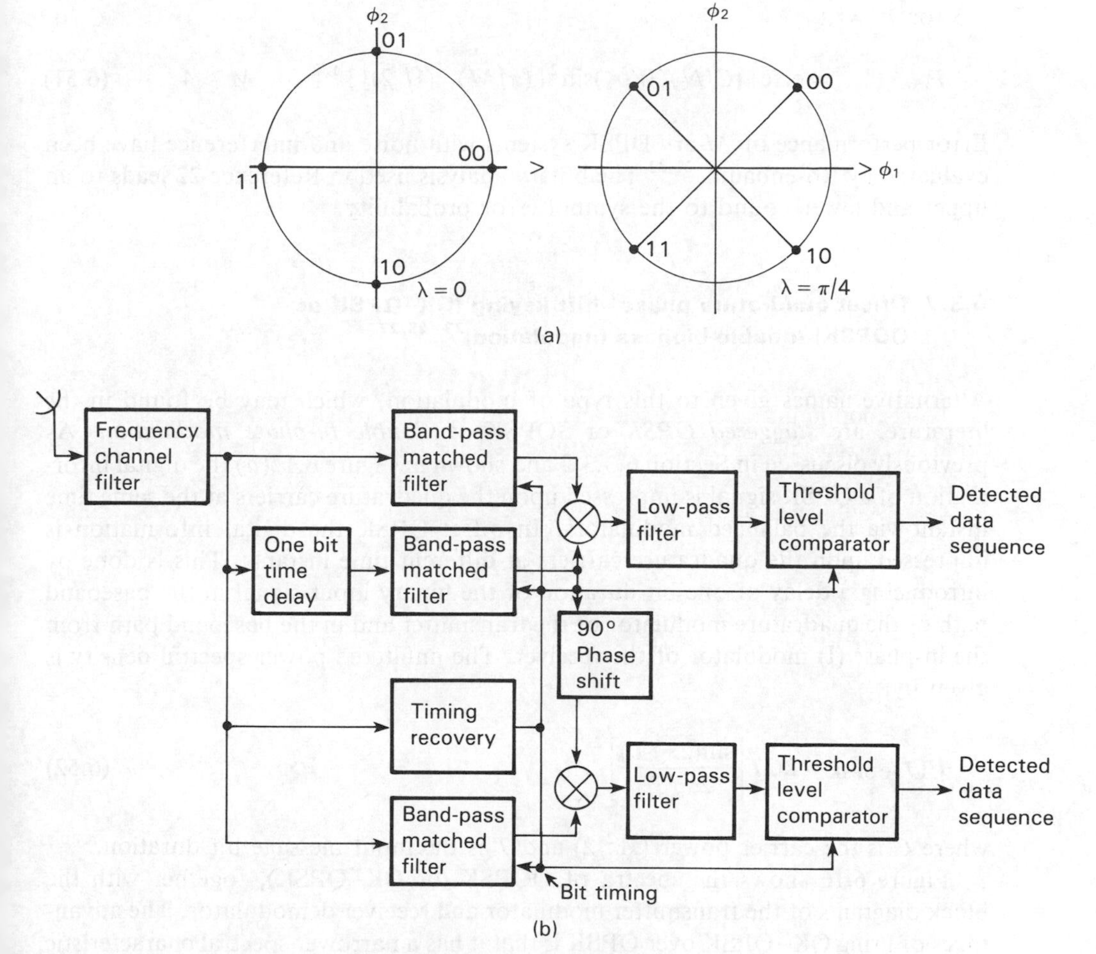

Figure 6.15 4–DPSK receiver: (a) constellation diagram; (b) 4–DPSK receiver

relative phase shifts between successive intervals are λ, $\lambda + \pi/2$, $\lambda + \pi$ and $\lambda + 3\pi/2$. Two commonly used signal constellations corresponding to $\lambda = 0$ and $\lambda = \pi/4$ are also shown in Figure 6.15. For $\lambda = 0$, the demodulator will have a 45° phase shifter located in each of the paths between the delay and the balanced multiplier. In comparison with coherent PSK, binary DPSK is a little inferior to binary PSK for large C/N. As DPSK does not require an elaborate method for estimating the carrier phase, it is often used in digital communication systems. Four-phase DPSK, however, is approximately 2.3 dB poorer in its performance than QPSK at large C/N and consequently the choice between the two systems is not so easily made. The 2.3 dB increase in C/N required for 4–DPSK must be considered in conjunction with the decrease in circuit complexity. The expression for the probability of error for M-ary DPSK is given in Reference 3 and for $M = 4$ is shown plotted in Figure

6.5 for $W = r_s$:

$$P_{e_{M\text{-ary DPSK}}} \simeq \operatorname{erfc}[(C/N)(W/r_s)\sin^2[(\pi/M)\cdot(1/\sqrt{2})]]^{1/2} \text{ for } M \geqslant 4 \tag{6.51}$$

Error performance of M-ary DPSK systems with noise and interference have been evaluated by Rosenbaum.[21,22] The binary analysis used in Reference 22 leads to an upper and lower bound to the symbol error probability.

6.3.7 Offset quadrature phase-shift keying (OK–QPSK or OQPSK) (double-biphase modulation)[23-25,27]

Alternative names given to this type of modulation, which may be found in the literature, are *staggered QPSK* or SQPSK or *double bi-phase modulation.* As previously discussed in Section 6.3.3.2 and shown in Figure 6.12(a) the digital information of a QPSK signal is impressed upon the quadrature carriers at the same time instant via the balanced modulators. In offset QPSK the digital information is impressed upon the quadrature carriers at different time instants. This is done by introducing a delay of one bit duration of the binary input signal in the baseband path of the quadrature modulator of the transmitter and in the baseband path from the in-phase (I) modulator of the receiver. The unfiltered power spectral density is given by:

$$P(f)_{\text{OQPSK}} = 2CT\left[\frac{\sin(2\pi fT)}{2\pi fT}\right]^2 \tag{6.52}$$

where C is the carrier power $(A^2/2)$ and T is the input message bit duration.

Figure 6.16 shows the spectra of OQPSK (or OK–QPSK), together with the block diagrams of the transmitter modulator and receiver demodulator. The advantages of using OK–QPSK over QPSK is that it has a narrower spectral characteristic and thus it is easier to derive the synchronization from the received signal. The output of the balanced modulators is twice the width of the baseband spectrum due to a double-sided spectrum being formed in the multiplication process. The filtering necessary to achieve the lowest C/N for a given P_e as well as maintaining the condition of zero inter-symbol interference is to use the raised-cosine filter with a roll-off factor of unity.[23] The spectral efficiency is 2 bits/s/Hz or $\log_2 M$. The error probability for detecting the information in a coherent OQPSK system using a raised-cosine filter, with a bandwidth of $2r_b$ or a roll-off factor equal to unity, is about 1 dB better than that of conventional QPSK. That is, to produce a given error rate, C/N needs to be 1 dB less. If a brick-wall filter is used, and the roll-off factor is zero, P_e for OK–QPSK is the same as that for QPSK.

Figure 6.16 Offset QPSK: (a) modulator and demodulator; (b) spectra; (c) typical data waveforms (Courtesy of IEEE, Reference 23)

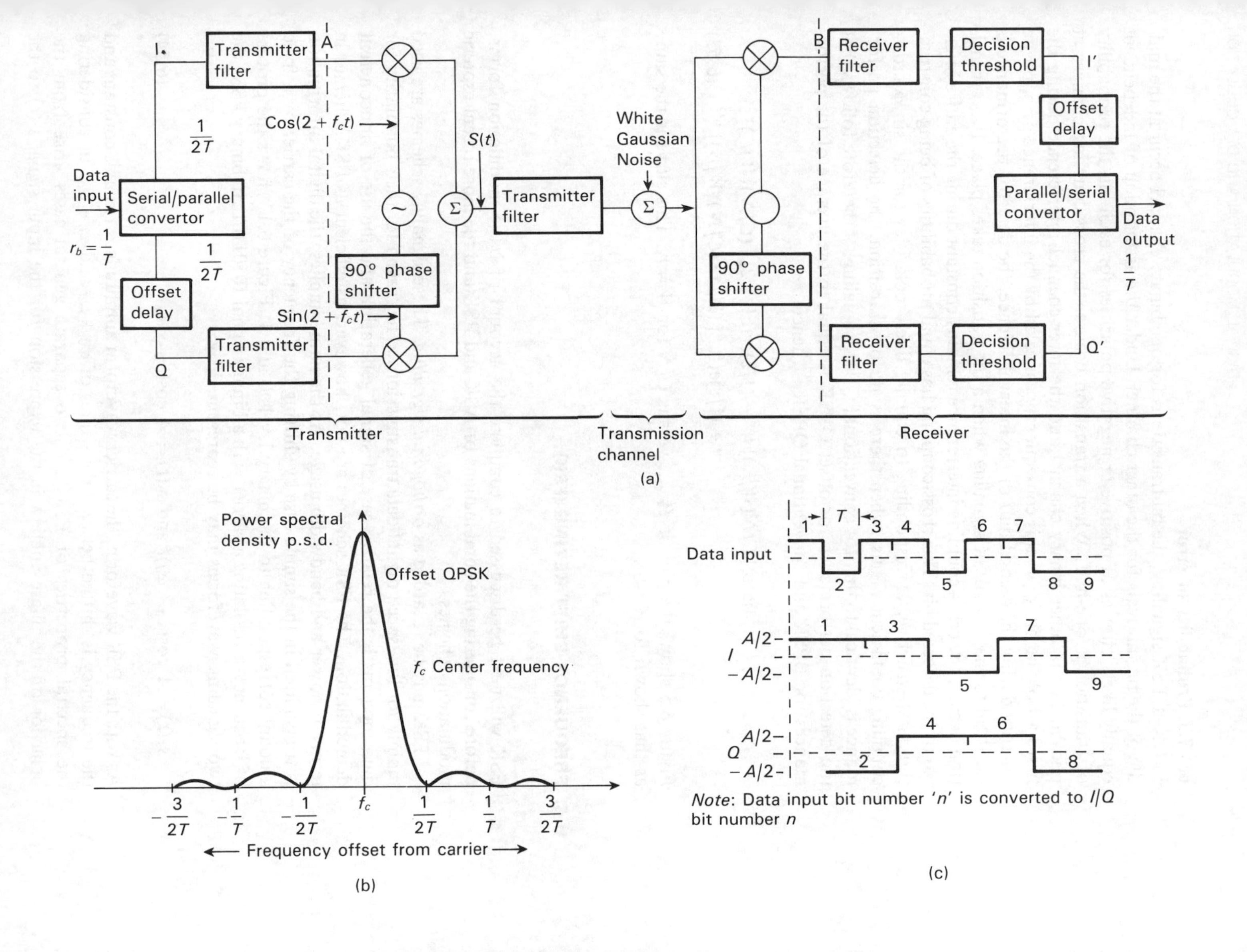

Data input
Serial/parallel convertor
$r_b = \frac{1}{T}$
I
Q
$\frac{1}{2T}$
$\frac{1}{2T}$
Offset delay
Transmitter filter
Transmitter filter
A
$\mathrm{Cos}(2 + f_c t)$
$\mathrm{Sin}(2 + f_c t)$
90° phase shifter
$S(t)$
Transmitter filter
White Gaussian Noise
B
90° phase shifter
Receiver filter
Receiver filter
Decision threshold
Decision threshold
I′
Q′
Offset delay
Parallel/serial convertor
Data output
$\frac{1}{T}$
Transmitter
Transmission channel
Receiver
(a)
Power spectral density p.s.d.
Offset QPSK
f_c Center frequency
$-\frac{3}{2T}$
$-\frac{1}{T}$
$-\frac{1}{2T}$
f_c
$\frac{1}{2T}$
$\frac{1}{T}$
$\frac{3}{2T}$
Frequency offset from carrier
(b)
Data input
T
1 2 3 4 5 6 7 8 9
$A/2$
I
$-A/2$
1 3 5 7 9
$A/2$
Q
$-A/2$
2 4 6 8
Note: Data input bit number 'n' is converted to I/Q bit number n
(c)

6.3.7.1 Probability of error

In OK–QPSK signaling, the bit transitions for one binary channel occur at the middle of the bit intervals for the other channel. Under the assumption of independent equally likely choices of positive or negative polarities for each bit, the probability of transition is one-half. When a transition occurs, the cross-coupling changes at mid-bit of the other binary channel and the inter-channel interference during the first half of the bit interval is consequently canceled by the interference of opposite polarity during the second half of the interval. Hence, the detector performance is identical to that for BPSK signaling when a bit transition takes place. If no transition occurs, the cross-coupling interference remains constant during the entire interval of the detected bit. This cross-coupling has equal probabilities of being constructive or destructive, corresponding to equally likely polarities for the sin ϕ cross-coupling coefficient. Thus, when there is no bit transition, the detection performance is identical to that for conventional QPSK signaling. It therefore follows that the detection performance for offset QPSK is equal to the average of the performances for BPSK and conventional QPSK,[37] hence:

$$P_{e(\mathrm{OK-QPSK})} = (1/4)\mathrm{erfc}\,[C/N(W/r_s)]^{1/2} + (1/2)\mathrm{erfc}\,[(1/2)\cdot(C/N)(W/r_s)]^{1/2} - (1/8)\mathrm{erfc}^2\,[(1/2)\cdot(C/N)(W/r_s)]^{1/2} \quad (6.53)$$

Figure 6.5 shows the plot of $P_{e\mathrm{SQPSK}}$ versus C/N for $W = r_s$. This curve is the same as that shown for CASK.

6.4. FREQUENCY-SHIFT KEYING (FSK)

FSK will next be discussed to complete the account of basic modulation forms, before considering the combinations of ASK and PSK and the more recent esoteric modulation schemes.

FSK may be regarded as *orthogonal signaling.* These signaling schemes are used mainly for low-speed digital data transmissions. The reason for their popularity for data modems is the relative ease of signal generation and the use of noncoherent demodulation. The FSK schemes are not, however, as efficient as PSK schemes in terms of power and bandwidth usage. As the name implies, the digital information is transmitted in the simple case by shifting the frequency of the carrier by a fixed amount corresponding to the binary levels 0 and 1. Figure 6.17 shows the process of frequency-modulating a carrier with a binary signal 10101101. In binary FSK, the two signaling waveforms may be represented by

$$s_1(t) = A\cos(\omega_0 + \omega_d)t \text{ and } s_0(t) = A\cos(\omega_0 - \omega_d)t \quad (6.54)$$

As with the PSK waveform, the carrier waveform amplitude A remains constant and the frequency is shifted between the values of $\omega_0 + \omega_d$ and $\omega_0 - \omega_d$. In considering the spectral properties of FSK[26–29] two separate general cases arise from the behavior of the phase angle λ in the expression for the input signal $s_i(t)$ to the

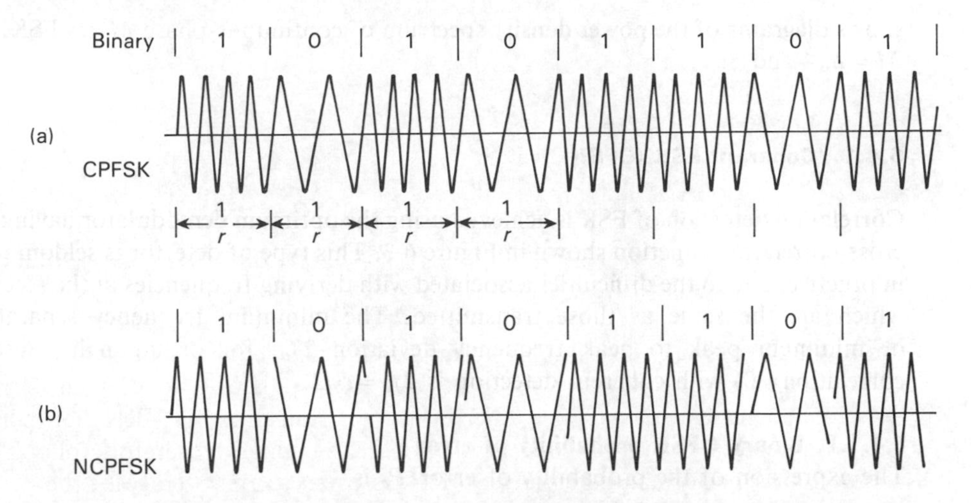

Figure 6.17 (a) Continuous-phase FSK (CPFSK); (b) noncontinuous-phase FSK (NCPFSK)

receiver, which is given by:

$$s_i(t) = A \cos\left\{\omega_0 t + \omega_d \sum_{\infty}^{\infty} a_k \int_0^t g(t - kT)\, dt + \lambda\right\} \tag{6.55}$$

where a_k are the digital weighting coefficients for the kth interval and are assumed to be discrete random variables. If λ is assumed to be random and uniformly distributed within 2π, then it bears no relation to the modulation and may at the signaling transitions take any random value. This leads to a possible phase discontinuity as shown in Figure 6.17(b) and the modulation is known as *phase-discontinuous FSK*. Phase-continuous FSK may be achieved by forcing λ to have a definite correlation with the modulating message. The transmission of binary data with a high frequency stability and negligible inter-symbol interference is difficult to achieve for the binary FM continuous-phase system. The reason is that binary FSK inherently requires two frequencies to represent two binary states, and to build a continuous-phase system using two separate oscillators requires much complicated circuitry. The alternative is to provide FM keying using a single voltage-controlled oscillator. While such a system has a continuous phase at the bit transition points, the frequency accuracy is relatively low and the bit rate is not locked to either of the two frequencies representing the 1 and 0 logic states. An ideal binary FM system has been postulated[30] in which the difference between the 1 and 0 frequencies, i.e. the peak to peak frequency deviation $2f_d$, equals the bit rate r_b, i.e. $2f_d = r_b$. In addition the 1 and 0 frequencies are locked to the bit rate. A system[31] has been proposed in which a single frequency source controls the system and provides output 1 and 0 signals locked to the bit rate. The expression for the FSK spectrum includes a series which is better given in diagrammatic form than mathematical form. References 26 and 28 provide the mathematical series and Reference 2, pp.126–130, shows diagrams of the power density spectrum of continuous-phase M-ary FSK, for $M = 2$, 4 and 8.

6.4.1. Coherent FSK (CFSK)

Correlation detection of FSK is achieved using the optimum demodulator having the cross-correlation function shown in Figure 6.3. This type of detector is seldom used in practice, due to the difficulties associated with deriving frequencies at the receiver which are the same as those transmitted. The minimum frequency separation or minimum peak to peak frequency deviation $2f_d$, for orthogonality (cross-correlation = 0) with coherent detection is $2f_d = r_b/2$.

6.4.1.1. Binary CFSK probability of error

The expression of the probability of error P_e is

$$P_{e_{\mathrm{FSK}}} = \tfrac{1}{2}\,\mathrm{erfc}\,[\tfrac{1}{2}(W/r_b)(C/N)]^{1/2} \tag{6.56}$$

Figure 6.18 shows a plot of this equation for the double-sided noise bandwidth.

Comparing the probability of error obtained for FSK as given by equation 6.56, against that obtained for PSK as given by equation 6.33, equal probability of error can be obtained in each system if the carrier power of the FSK signal is increased by 3 dB. The 99 percent transmission bandwidth required by a binary coherent FSK signal can be given by Carlson's rule, i.e. twice the peak to peak deviation plus twice the highest modulating frequency. If the highest modulating frequency extends from D.C., this can be considered as half the IF bandwidth W. The peak to peak deviation divided by the highest modulating frequency is defined as the modulation index m, and can be considered as the peak to peak deviation divided by the bandwidth W. The 99 percent transmission bandwidth equals $2(1+m)W$. Table 6.2 shows for different m, values of the 99 percent transmission bandwidth, normalized to the bit rate r_b, for *rectangular shaping filters*.

From Table 6.2, it appears that one should choose as small as possible deviation. However the error probability P_e is also a function of the deviation and reducing the deviation would increase P_e. The optimum value of deviation has been shown[32] to be approximately equal to 0.7 and reducing the deviation to 0.5 leads to a penalty of 1.6 dB in C/N when P_e equals 10^{-6} and $W = r_b$. The bandwidth $W = r_b$ is also optimal in this case.

Figure 6.19 shows the system block diagram of a coherent FSK receiver, where as described below, the output of each of the band-pass filters will contain a tone which is coherently related to the information-carrying frequencies. These tones will

Table 6.2 FSK 99 percent bandwidth for different modulation indices

Modulation index m	0.2	0.3	0.4	0.5	0.6	0.7	0.8	0.9
Bandwidth/r_b	0.78	1.00	1.10	1.17	1.25	1.80	1.94	2.05

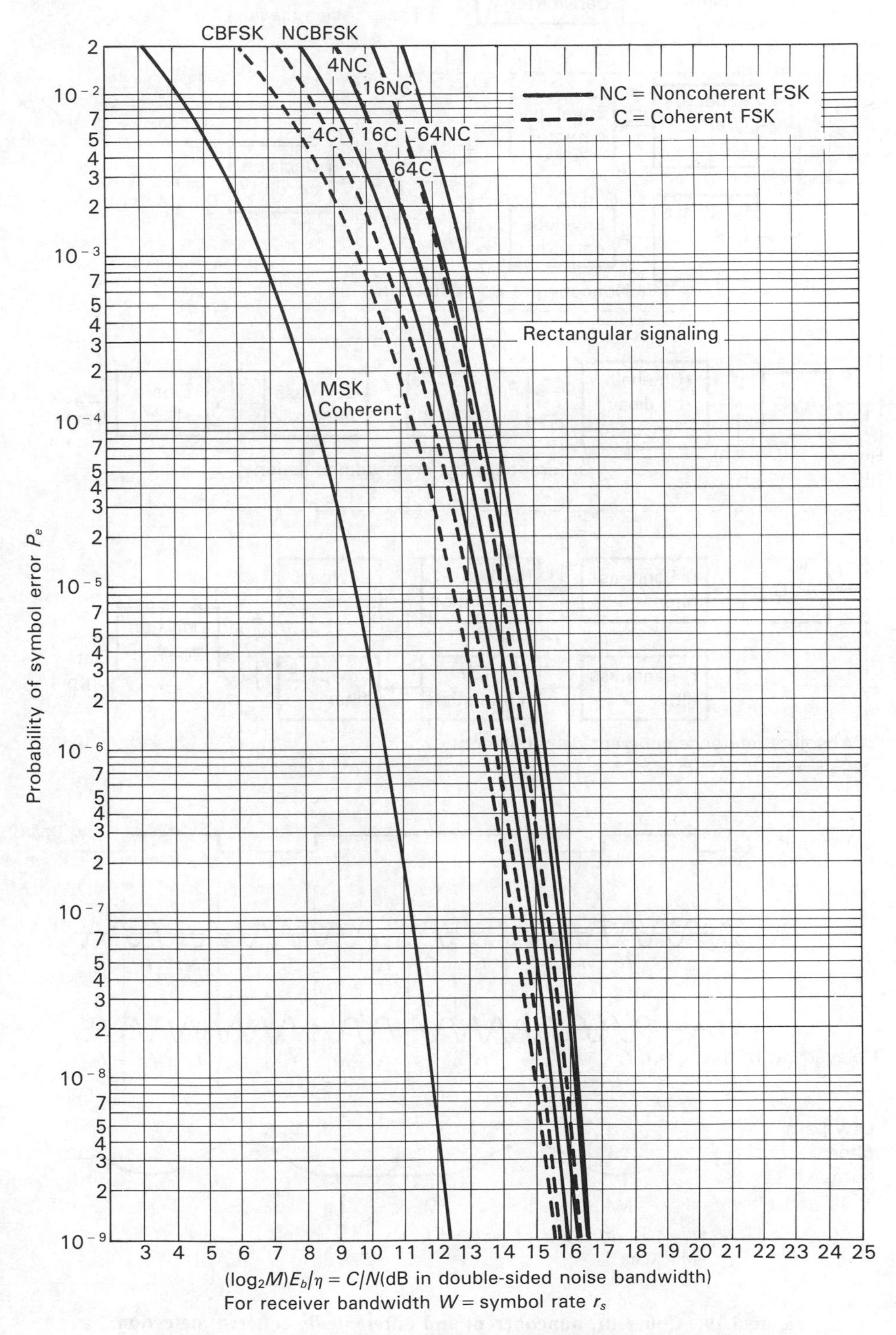

Figure 6.18 Probability of error curves for FSK modulation

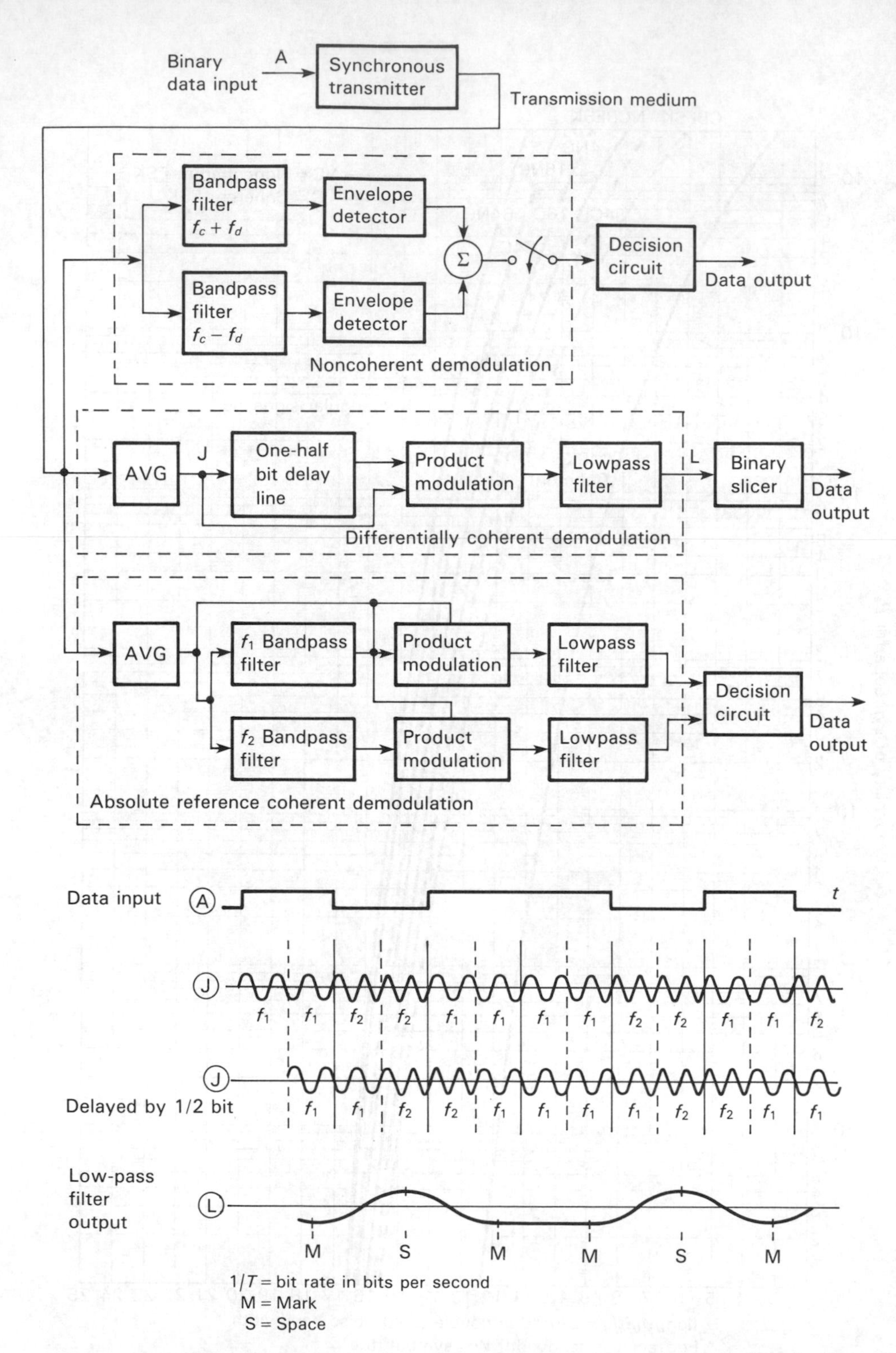

Figure 6.19 Coherent, noncoherent and differentially coherent detection systems

appear exactly at the 1 and 0 frequencies. In addition, they contain half the total power and their difference provides the bit clock frequency with a consistent phase.

6.4.2 Noncoherent FSK (NCFSK)

The frequency spectrum of FSK when the peak-to-peak deviation $2f_d = kr_b$, where k is an integer, appears as a two ASK signal spectra, having the carrier frequencies $f_0 - f_d$ and $f_0 + f_d$, each with a similar spectrum to that shown in Figure 6.2. This implies that the message signal under these conditions should be able to be detected by using two band-pass frequencies with their center frequencies at $f_0 - f_d$ and $f_0 + f_d$. A typical detection scheme is shown in Figure 6.19. When there exists a unique relationship between carrier frequency and bit rate, such that $f_0 = nr_b$, implying that the carrier is coherently related to the information-carrying bit speed, three types of detection processes are possible. The first which has been described already is that of coherent detection, the second is noncoherent detection and the third is differentially coherent detection using a delay line as shown in Figure 6.19.

6.4.2.1 Probability of error P_e of binary noncoherent FSK

The expression for the probability of error P_e is

$$P_e = (1/2)\exp[-(1/2)(W/r_b)(C/N)] \tag{6.57}$$

This equation is derived in some of the references provided.[2–4] Equation 6.57 is shown plotted in Figure 6.18 for double-sided noise bandwidth.

It may be worth while noting that where two band-pass filters are followed by an envelope detector and a decision device, the frequency spacing $2f_d$ must be at least $1/T$ (or $m \geqslant 1$), in order to prevent significant overlap of the passbands of the two filters. Alternatively, a discriminator can be used to convert the frequency variations to amplitude variations, so that amplitude-modulation envelope detection can be employed. This method eliminates the above-mentioned constraint on the modulation index m being equal to or greater than unity.

6.4.3 Differentially coherent FSK demodulation

The noncoherent detection process described in Section 6.4.2 suffers from changes in the frequency and phase shifts caused by the transmission medium. To overcome this problem, especially when C/N is low, the differentially coherent process with a delay line may be used. It is significant that the time delay of the delay line must be equal to *one-half* of the bit duration. For very low C/N this system is also dubious due to the relative carrier frequency reference being corrupted by noise. When C/N is low, advantage is taken of the discrete message components which are an integral part of the received waveform and which may provide an absolute coherent reference. These components carry half of the total power and are easily filtered out. In a practical system, simple band-pass filters with 3 dB bandwidths around 5 percent of the bit rate r_b, will suffice. The coherent clock frequency is obtained from the difference of the two transmitted carrier frequencies, which obviates the necessity to derive the clock from the data transitions. The differentially coherent detection process is shown in Figure 6.19 and is such that, whenever two frequencies are identical, they are either in phase or 180° out of phase, which provides a maximum or minimum at the output of the low-pass filter. If they differ during a half-bit interval, a transition between a 1 and 0 or vice versa takes place.

6.4.4 Comparison of FSK with ASK

From the curves plotted in Figures 6.18 and 6.5, FSK does not provide better error rates than ASK except for small values of C/N. Also the bandwidth required by FSK if a limiter discriminator is used is much larger than that of ASK. The advantages of FSK over ASK are:

The constant amplitude property of the carrier signal does not waste power and does produce some immunity to noise.

The optimum threshold level of the detector is independent of carrier amplitude A, and C/N. This means that the threshold need not be adjusted with changing transmission channel characteristics.

Binary FSK is mainly used for slow-speed data transmission but is finding applications also in frequency-hopping or spread spectrum communication systems using 1 and 0 tones which are not necessarily orthogonal. Discussion on these subjects may be found in Reference 2.

6.4.5 *M*-ary FSK

6.4.5.1 Coherent detection

The probability of error in multiple FSK systems with coherent detection is not given by a simple explicit function. A generally accepted expression[2,34,35] for the probability of error is given as:

$$P_{e(M\text{-ary coherent FSK})} = (1/\sqrt{2\pi}) \int_{-\infty}^{\infty} \{1 - [1 - (1/2)\mathrm{erfc}(x/\sqrt{2})]^{M-1}\} \exp\left\{-(1/2)\left[x - \left(\frac{W}{r_s}\right)(2C/N)^{1/2}\right]^2\right\} dx \quad (6.58)$$

where M is the number of keying frequencies and C/N is the carrier-to-noise ratio in the double-sided noise bandwidth. The values of P_e for various values of M are given in Figure 6.18. Since any of the M signaling waveforms each of a different frequency are equally likely, the expression given in equation 6.58 is the average probability of a symbol error. As shown by equation 6.49, the performance of various modulation schemes may be compared using E_b/η, rather than C/N. This permits an evaluation to be made between various M-ary modulation schemes using different values of M.

As mentioned previously, for M waveforms or states, each coded symbol requires $\log_2 M$ binary coding bits, hence from equation 6.49b, for a raised-cosine filter:

$$E_b/\eta = \frac{(1+\alpha)}{\log_2 M}(C/N) \tag{6.49b}$$

To convert the probability of a symbol error as given by equation 6.58 into an equivalent probability of a binary bit error, the way in which errors occur in an orthogonal system must be considered. The number of combinations ${}^{\log_2 M}C_n$, is the number of ways n binary bits out of $\log_2 M$ bits, may be in error. For equiprobable orthogonal signals, all symbol errors are equiprobable:

$$\text{Probability of symbol errors occurring} = P_e/(M-1) \tag{6.59}$$

Hence the number of bit errors per $\log_2 M$ is:

$$\sum_{n=1}^{\log_2 M} n[{}^{\log_2 M}C_n]P_e/(M-1) = [P_e/(M-1)]\,n(\log_2 M)!/[(\log_2 M - n)!n!]$$

$$= (P_e \text{ per bit})_{\text{FSK}} = P_{e_{\text{FSK}}} M/[2(M-1)] \tag{6.60}$$

$$= \text{BER (bit error rate)}$$

The frequency separation required for coherent FSK demodulation is given by $1/2T_s$. The bandwidth occupied by each signal is approximately $2f_d$; hence the channel bandwidth required for transmission of the M waveforms is given as:

$$\text{Bandwidth for coherent FSK} = 2Mf_d = M/(2T_s) \tag{6.61}$$

The bandwidth efficiency is given by the information rate in bits/s $[(\log_2 M)/T_s]$, divided by the required bandwidth. Thus

$$\text{Bandwidth efficiency of coherent FSK} = 2(\log_2 M)/M \tag{6.62}$$

6.4.5.2 Noncoherent detection

The probability of error in multiple FSK systems with noncoherent detection is given by:[35]

$$P_{e(\text{noncoherent FSK})} = \int_0^\infty I_0[2x^2(W/r_s)(C/N)]^{1/2} \cdot \{1 - [1-\exp(-x^2/2)]^{M-1}\} \cdot x \cdot \exp[-x^2/2 + (W/r_s)(C/N)] \cdot \mathrm{d}x \tag{6.63}$$

where $I_0(u)$ is the modified Bessel function of the first kind and of zero order as given in equation 6.21. Figure 6.18 shows the values of P_e for various values of M and double-sided noise carrier-to-noise ratio C/N. Comparing the probability of error between coherent and noncoherent multiplex FSK systems it is apparent that coherent detection is always the superior detection system for small values of M. The difference between the two systems becomes less apparent as the number of keying frequencies M increases. Orthogonality of the FSK waveforms which are noncoherently detected requires a frequency separation of $2f_d = 1/T_s$. Consequently the channel bandwidth required for transmission is:

$$\text{Bandwidth of } M\text{-ary noncoherent FSK} = M \cdot 2f_d = M/(T_s) \qquad (6.64)$$

This again shows that as the number of tones M increases the bandwidth increases; and yet from Figure 6.18 C/N approaches a limit. As the transmission rate is given by $(\log_2 M)/T_s$, we have:

$$\text{Bandwidth efficiency of noncoherent FSK} = (\log_2 M)/M \qquad (6.65)$$

This is half the value obtained for the coherent detection case.

From Figure 6.18 we see that, if the noise power remains constant, the transmitter power does not increase appreciably as M increases. The maximum error-free bit rate r_b, at which data can be transmitted using an M-ary orthogonal FSK signaling scheme is given by the channel capacity C' of a Gaussian channel of infinite bandwidth:

$$r_b = W \cdot (C/N) \cdot \log_2 e \qquad (5.7)$$

This means that if the bit rate r_b is less than the channel capacity the probability of error can be made arbitrarily small.

The constellation diagram for an M-ary FSK system may be represented by an M-orthogonal axis with vector magnitudes of $A/\sqrt{2}$. For $M = 3$, this is easily visualized as the three-dimensional coordinate axis with the positive axes x, y and z representing ϕ_1, ϕ_2 and ϕ_3.

Another FM technique which has received considerable interest and has resulted in commercially available equipment is that of *minimum-shift keying* (MSK), also called *fast frequency-shift keying* (FFSK).

6.4.6 Minimum-shift keying (MSK)[23,25,37]

MSK is a special case of continuous-phase FSK (CP–FSK), for which the deviation $2f_d$, equals 0.5 and coherent detection is used. This technique achieves a performance which is identical to coherent PSK and which exhibits the superior spectral properties of CP–FSK. In addition, MSK has the advantage of a relatively simple self-synchronizing implementation,[36] over coherent CP–FSK with a deviation of 0.7.

The uses for which MSK has been considered are for terrestrial digital microwave radio and satellite. If the pulses entering the transmitter circuits are filtered to produce 'full length' sinusoidal pulses prior to modulating with the carrier, MSK may be considered to be a modification of offset QPSK[23] (See Section 6.3.7). It has been shown[36] that an optimal and simple detector may be constructed with an ideal probability of error performance equal to that of a binary PSK receiver. Because of the P_e performance and the bandwidth efficiency (2 bits/s/Hz), this technique is being used in commercially available equipment, such as Telenokia 0.7, 2 and 8 Mbit/s digital radios. Coherent detection of MSK, or *FFSK*, as in the coherent detection of PSK signals, exhibits a degradation in P_e performance from the ideal due to a phase discrepancy between the carrier of the received signal and the locally generated noisy carrier reference. In conventional PSK systems, both BPSK and QPSK signaling exhibit almost the same P_e performance for the same value of C/N per bit, with a perfect phase reference. With a noisy phase reference the performance of these systems deteriorates, with the performance loss being greater for QPSK because of the coupling between the quadrature components. It has been shown[37] that offset QPSK produces a probability of error in detection that is equal to the average of the detection performances for BPSK and conventional QPSK. Because of the frequency instabilities in communications systems and the associated difficulty in obtaining carrier synchronization with sufficiently low jitter to preclude the detection losses, offset QPSK has the additional advantage of BPSK and QPSK by permitting the required value of C/N to be 3 dB lower than the level of the synchronizer phase reference for satisfying a specified value of allowable detection loss. While MSK is similar in structure to offset QPSK, its sensitivity to reference phase error is even less than that of offset QPSK, allowing another 2.5 dB to be taken off the level of the synchronizer phase reference to produce the same specified allowable detection loss.[38] The expression for the unfiltered power spectral density of MSK[25] is

$$P(f)_{\mathrm{MSK}} = [8CT(1 + \cos 4\pi fT)]/[\pi(1 - 16T^2f^2)]^2 \tag{6.66}$$

where f is the frequency offset from the carrier, C is the carrier power and T is the unit bit duration into the receiver. This spectrum is shown in Figure 6.20(*c*). Comparing the spectrum of MSK with that of offset QPSK, or OK–QPSK, in Figure 6.16, we note that the width of the main lobe of the MSK spectrum is 1.5 times larger than that of the main lobe of the OK–QPSK spectrum. It can be shown[23] that with the correct filtering the maximum bandwidth efficiency of MSK is the same as that of OK–QPSK, i.e. 2 bits/s/Hz. Figure 6.20 also shows a block diagram of the system modulator and demodulator, together with the expected data streams timing diagram. Frequency-shift keying (FSK) signals, like other FM systems, are a nonlinear process which makes the complete mathematical description very difficult. However, FSK signals which are designed to have the peak-to-peak deviation, or frequency shift, h (where $h = 2f_d$) equal to an integral multiple of the bit rate can be considered as a summation of two AM signals. This makes it easy to describe its frequency- and time-domain properties. The spectral density of such signals comprises both continuous and discrete components with the power equally divided among the two. As the discrete components do not contain any information, they are a waste of power. One of the advantages of MSK is that when the modulation index of FSK signals equals one-half (i.e. the peak-to-peak deviation h equals half

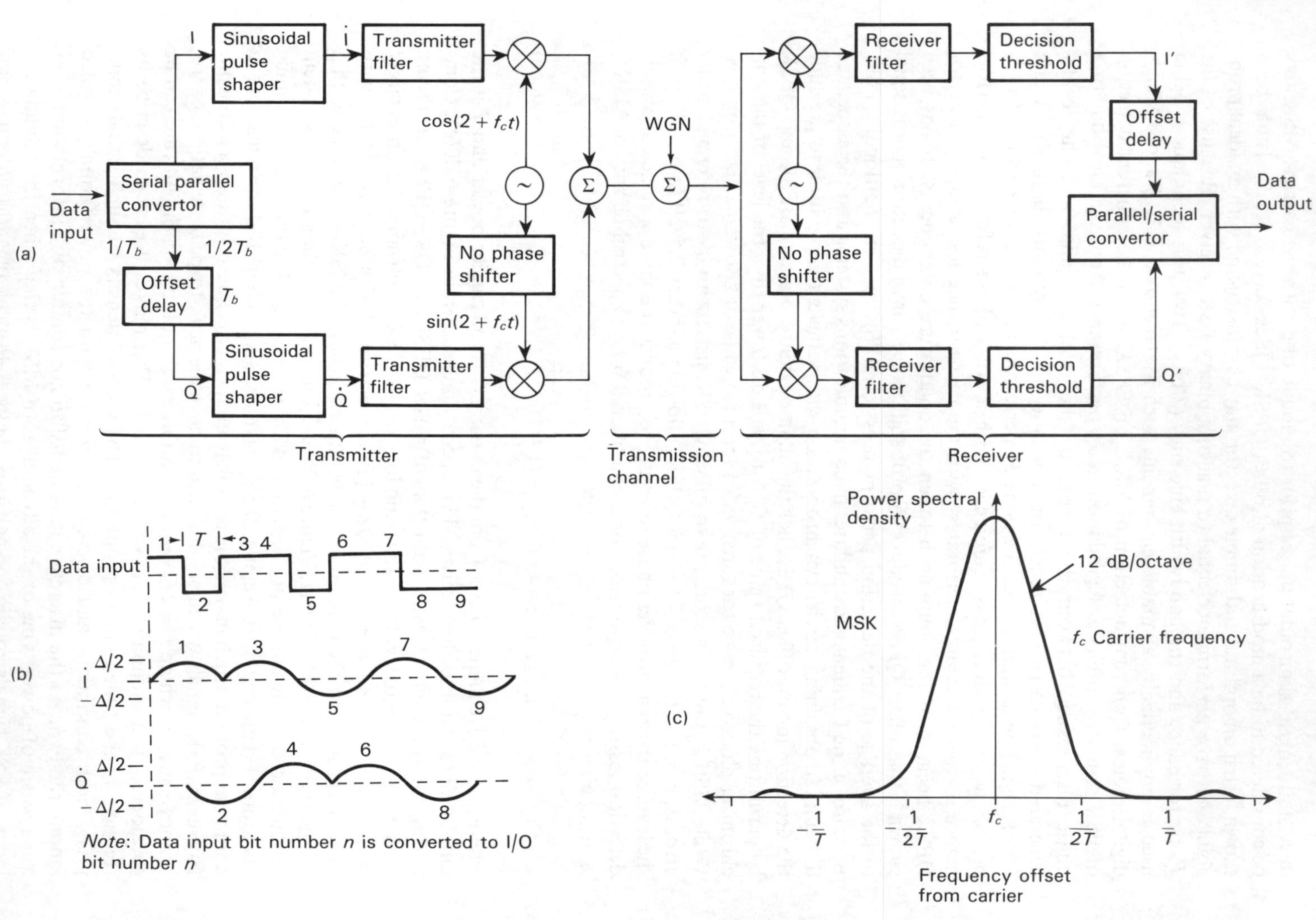

(a)
Data input
Serial parallel convertor
$1/T_b$
$1/2T_b$
Offset delay
T_b
I
Q
Sinusoidal pulse shaper
Sinusoidal pulse shaper
İ
Q̇
Transmitter filter
Transmitter filter
$\cos(2 + f_c t)$
$\sin(2 + f_c t)$
No phase shifter
WGN
No phase shifter
Receiver filter
Receiver filter
Decision threshold
Decision threshold
I′
Q′
Offset delay
Parallel/serial convertor
Data output
Transmitter
Transmission channel
Receiver
(b)
Data input
1 T 3 4 6 7
2 5 8 9
İ
$\Delta/2$
$-\Delta/2$
1 3 7
5 9
Q̇
$\Delta/2$
$-\Delta/2$
4 6
2 8
Note: Data input bit number n is converted to I/O bit number n
(c)
Power spectral density
12 dB/octave
MSK
f_c Carrier frequency
$-\frac{1}{T}$
$-\frac{1}{2T}$
f_c
$\frac{1}{2T}$
$\frac{1}{T}$
Frequency offset from carrier
Power spectral density of MSK

the bit rate), the spectral density contains only continuous components which carry the information. Another advantage of MSK over FSK with a deviation equal to unity, is that less bandwidth is required for the same bit rate, especially for *duobinary FM.*[39] Figure 6.20(*c*) shows that, for MSK, nearly all the signal energy is contained within a narrow frequency region equal to 1.5 times the bit rate and the spectrum skirts have an average slope of 12 dB/octave. In the case of duobinary FM, the bandwidth after demodulation is restricted by a raised-cosine filter to the point where no binary FSK signal can exist.[40] The resultant inter-symbol interference, however, can take such a form that the detection of the signal as a pseudo-ternary signal becomes possible. The subject of partial response detection will be dealt with in Section 6.6.3.

Another advantage of MSK is that the carrier, modulated by a random digital message, has a constant envelope and therefore does not encounter AM-to-PM conversion which would modify the spectrum. Because of these properties, MSK is very attractive for applications to nonlinear and power-limited systems such as satellite communication systems.[41]

Two generic techniques for the modulation and demodulation of MSK have evolved. These methods are referred to as parallel and serial methods. Both are completely equivalent in terms of bandwidth occupancy and probability of error performance. The parallel method amounts to quadrature multiplexing half-sinusoidal pulse-shaped data streams staggered by a one-half-symbol period on quadrature carriers as shown in Figure 6.20(*a*). Practical implementation of the modems using this approach requires that there is close balancing and alignment of the in-phase and quadrature channel data signals on carriers which are themselves balanced and whose phases are in quadrature. At the receiver, similarly, careful balancing and maintenance of the phase quadrature is required to minimize distortion and crosstalk. With the serial approach, the signal is produced from a bi-phase signal by filtering it with an appropriately designed conversion filter. The problems of balancing and maintaining phase quadrature carriers of the parallel approach are therefore replaced with the task of building a conversion filter with the appropriate sinc response. The demodulator comprises a filter matched to the transmitted signal spectrum, followed by coherent demodulation and bit detection. The implementation of a serial demodulator requires the synthesis of a band-pass filter which is closely matched to the MSK signal to ensure performance close to the ideal. MSK signals, like PSK signals, can be detected by either coherent or differential detection.[42] Differential detection is an attractive demodulation technique in burst-mode transmission, such as *time division multiple access* (TDMA) systems, because the circuit configuration is simple and carrier recovery is not necessary. This method uses noncoherent detection. The error rate performance, however, is inferior to that of coherent detection, as shown in Figure 6.18. MSK signals have the inherent properties that the absolute phases at any two moments are not independent and are a

---

Figure 6.20 MSK system: (a) quadrature modulation and demodulation; (b) modulator data timing; (c) power spectral density (Courtesy of IEEE, Reference 23)

function of the data transmitted between those two moments. The symbol detected from the difference in phase of two adjacent signaling intervals is the transmitted data in the noiseless situation. Similarly, the two symbols detected from the difference in phase of two alternate signaling intervals can be interpreted as the parity check sum of two successive transmitted data elements. The error rate performance can be improved by using the decoder for an error-correcting code which comprises the data and the parity bit.[42]

6.4.6.1 Probability of error of coherent MSK

The probability of error is the same as coherent antipodal phase-shift keying PSK as given by equation 6.34. Where the received reference is exactly in phase with that transmitted, $\phi = 0$ and equation 6.34 reduces to that of 6.33. Figure 6.5 shows the curve for $P_{e(\text{coherent MSK})}$ labeled as BPSK, and Figure 6.18 shows the same curve labeled as MSK coherent.

$$P_{e(\text{coherent MSK})} = (1/2)\text{erfc}\left[(W/r_b)(C/N)\cdot\cos^2\phi\right]^{1/2} \tag{6.67}$$

6.4.6.2 Probability of error of noncoherent or differentially detected MSK

Again, the probability of error is the same as that for noncoherent or DPSK and is also shown plotted against C/N in Figure 6.5 as DPSK.

$$P_{e(\text{differential detection MSK})} = (1/2)\exp\left[-(C/N)(W/r_b)\right] \tag{6.68}$$

6.4.6.3 Spectral shaping of the MSK system

As the requirement for higher bit rates increases, bandwidth-efficient digital radio systems are continually under development. Some such developments have been those of the different MSK-type modulation methods which are designed to achieve a compact signal spectrum.[43,44] The process involves the spectral shaping of the input data pulse, sinusoidal frequency-shift keying (SFSK) and multi-amplitude minimum-shift keying (MAMSK). SFSK has an excellent fractional out-of-band power characteristic.[46,47] Also SFSK has been investigated with respect to its crosstalk behavior[48] and found to be an excellent modulation scheme for packing many signals into a limited bandwidth when the signals are not synchronized in bit timing. Some applications require that signals are closely packed in frequency when no absolute phase reference is available at the receiver (noncoherent reception). Such applications resulted in studies being conducted on the crosstalk behavior of a phase-comparison version of SFSK, called *phase-comparison SFSK* (PCSFSK). The results show that PCSFSK allows a much closer packing of unsynchronized signals than DQPSK. The probability of error has been shown[49] to be slightly better than DQPSK when C/N is greater than 5 dB and slightly worse with C/N less than 5 dB. Another modification to the spectral skirts of the MSK modulation scheme is that produced by *double sine frequency-shift keying* (DSFSK), in which the spectrum tails have a 36 dB/octave average slope beyond $f = 4.75/T$, instead of 24 dB/octave for SFSK.[45] Finally, we finish this brief discussion on some different

FSK modulation methods by mentioning *multi-h phase-coded signaling.*[50-53] This type of signaling may be regarded as a generalization of continuous-phase frequency-shift keying (CPFSK) in which the modulation index or frequency deviation ratio is varied cyclically in successive T-second signaling intervals. Thus if $(h_0, h_1, \ldots, h_{i-1})$ is the set of modulation indices available to the modulator, the modulation index used in the kth signaling interval will equal that used in the $(k+i)$th interval.

6.5 COMBINED AMPLITUDE- AND PHASE-KEYING (CAPK) SCHEMES

Pulse amplitude modulation (PAM) was introduced in Section 5.2.4 as a method for directly transmitting digital information on a baseband system. The derivation of the probability of error was, however, derived for the Nyquist bandwidth, where it was assumed that the receiver had a cut-off frequency equal to half the bit rate r_b. The signal constellation for 8–PAM is shown in Figure 6.21. As M-ary PAM is a system where only the amplitude of the transmitted pulse changes in discrete steps, the various amplitude states about a reference level are shown as antipodal coded points along a straight line.

Combined amplitude- and phase-shift keying (CAPK) schemes are used in digital radio systems where bandwidth efficiencies of 3 bits/s/Hz and higher are required, without increasing excessively the required carrier power to maintain a reasonable error performance. In M-ary PSK the increase in carrier power at a P_e of 10^{-4} to go from BPSK to QPSK is 1 dB, from QPSK to 8–PSK is 2.5 dB, and from 8–PSK to 16–PSK is 5.5 dB. This shows that as M increases, so does the power. At large values of M, a 6 dB or fourfold increase is required to reach $2M$. Similarly, for PAM, if the value of M doubles, the carrier power to attain the same P_e is required to increase by 6.5 dB. This can be seen by referring to Figure 5.10. By combining M-ary PSK and M-ary PAM the increase in power required to maintain the same P_e, for a particular increase in M, is around 2.5 dB, against the individual modulation schemes requirement of 6.5 dB. The benefit of the bandwidth reduction by the amount $\log_2 M$ holds for CAPK as it does for PAM or PSK schemes. The general form of the combined M-ary amplitude and M-ary phase signal is given by;

$$s_i(t) = C_i \cos(\omega_0 t + \phi_i) = A_i \cos \omega_0 t + B_i \sin \omega_0 t \quad i = 1, 2, 3, \ldots, M \quad 0 < t < T_b \tag{6.69}$$

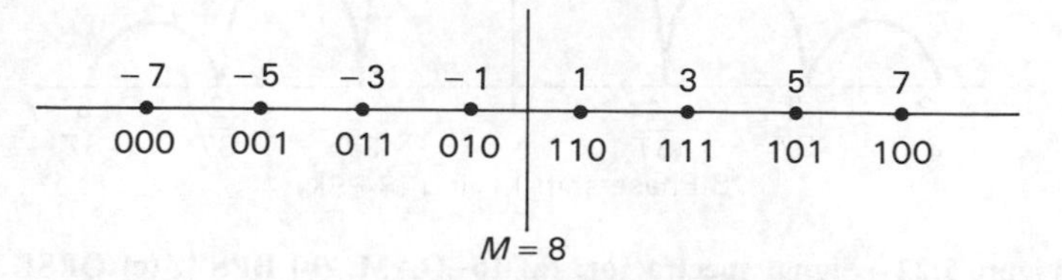

Figure 6.21 Signal constellation diagram for 8–PAM

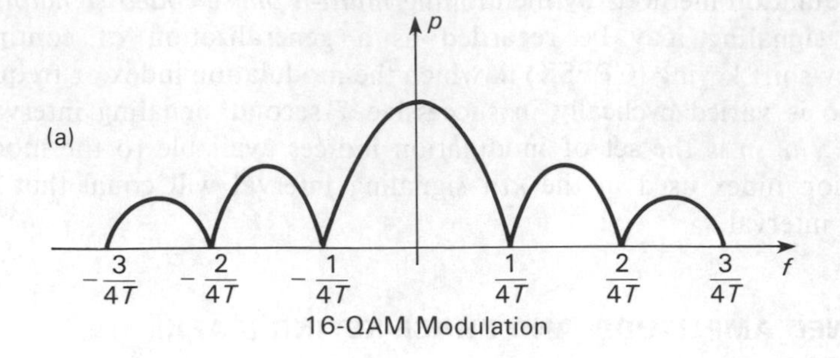

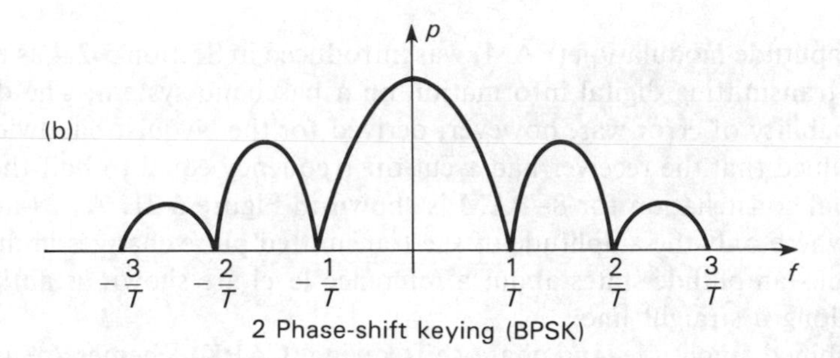

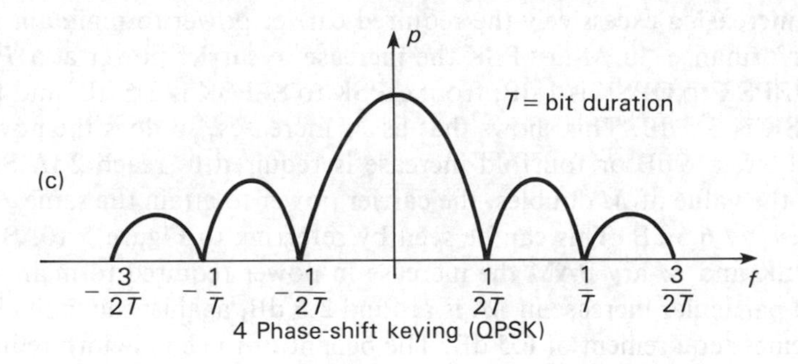

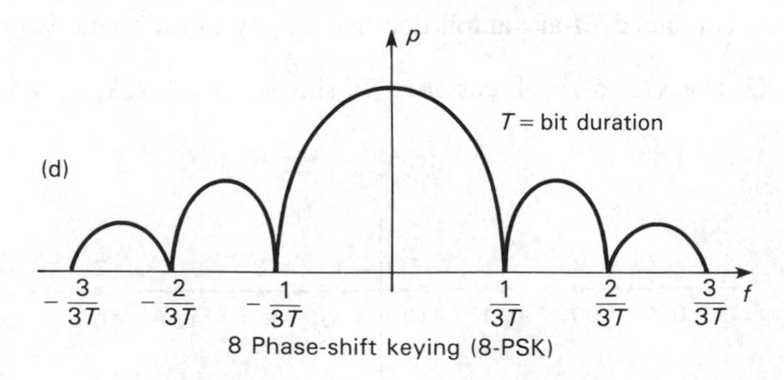

Figure 6.22 Signal spectra for: (a) 16–QAM, (b) BPSK, (c) QPSK, and (d) 8–PSK modulation

This expression shows that the combined waveform consists of two separate carriers which are in quadrature and that each is modulated by a set of discrete amplitudes (A_i and B_i). Consequently this CAPK scheme is commonly called *quadrature amplitude modulation* (QAM) rather than by its alternative name of *quadrature amplitude-shift keying* (QASK), which does not indicate the process defined in equation 6.69. QAM stands out as the ideal CAPK scheme because of its relatively low complexity and relatively good C/N performance in an ideal Gaussian channel. The amplitudes A_i and B_i in the QAM signal may be expressed in the form:

$$A_i = d_i \cdot A \qquad B_i = e_i \cdot A$$

where A is a fixed amplitude and the pairs (d_i, e_i) are defined to correspond to the desired signal points. Figure 6.22 shows the spectrum of a 16–QAM signal, together with that of BPSK, QPSK and 8–PSK for comparison purposes.

6.5.1 2-to-*L*-level convertor

In order to fully understand how the M-ary QAM, or M-ary PSK modulators and demodulators operate, a quick review of the operation of a baseband 2-to-L-level convertor is necessary. Consider a 2-to-4-level convertor, as shown in Figure 6.14 and again in Figure 6.24. Basically, what occurs is the grouping of the digital bit stream into two bit blocks, or dibits, as shown in Figure 6.23. The coding is such that dibit 10 represents a voltage level of $+v_2$, dibit 11 represents a level of $+v_1$, dibit 01 becomes $-v_1$ and dibit 00 becomes $-v_2$. Thus it can be seen that for four levels, only $\log_2 4 = 2$ binary bits or a dibit is required to produce four distinct baseband voltage levels. For L levels, blocks of $\log_2 L$ binary bits are required to assist in the binary coding of each of the L levels. For 16–QAM, the number of levels equals 4. If four baseband levels are required, the number of binary bits required to be grouped for coding purposes is 2 and the scheme shown in Figure 6.23 applies.

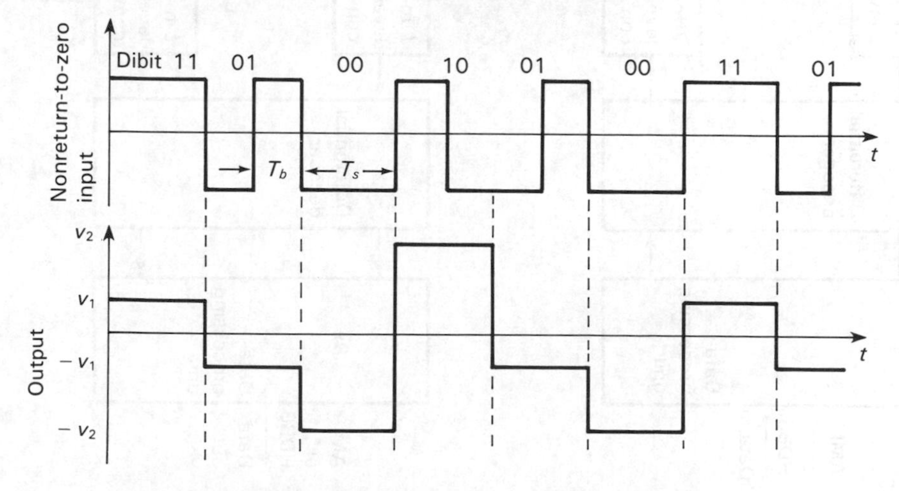

Figure 6.23 Principles of a 2-to-4-level convertor

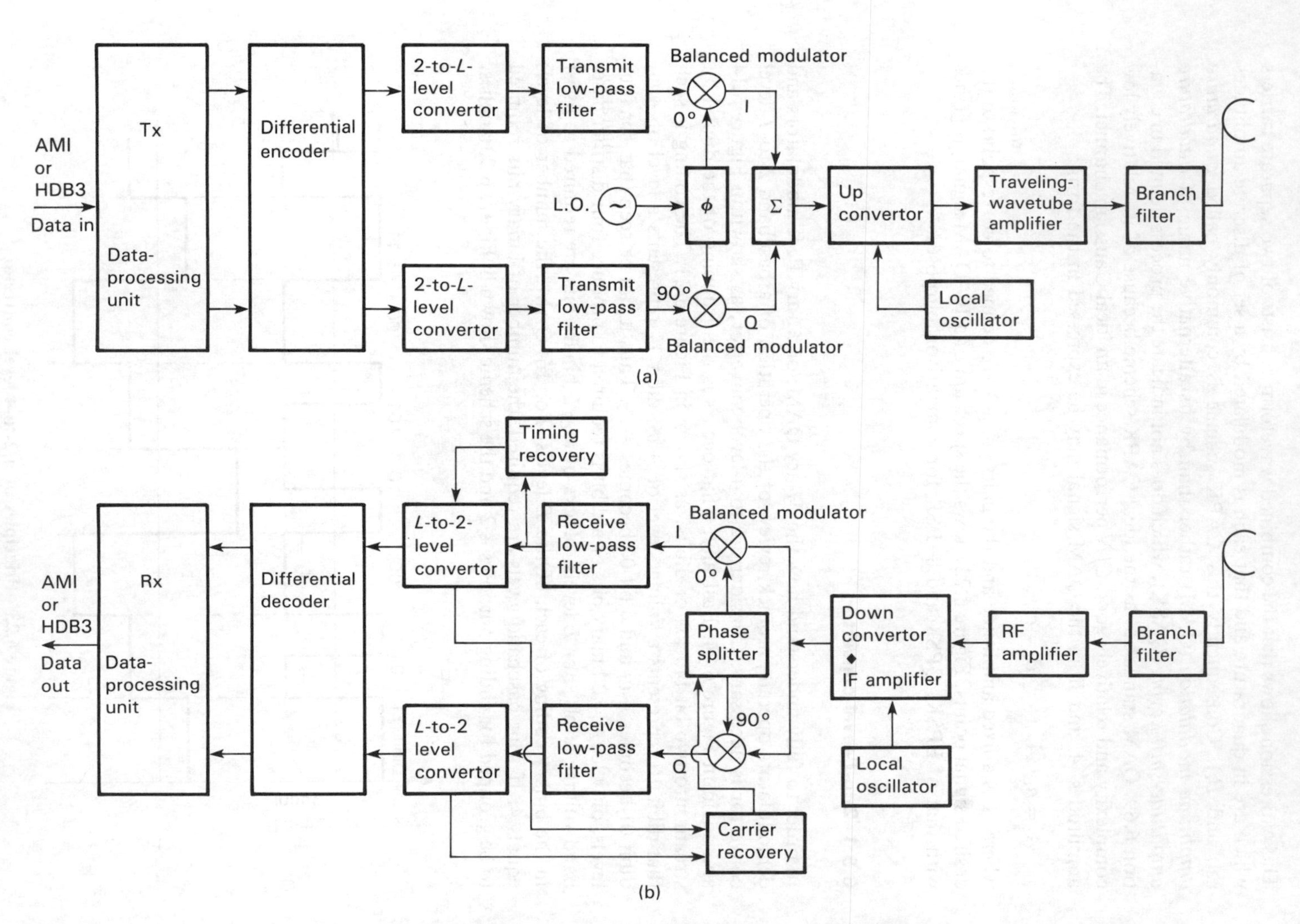

Figure 6.24 *M*-ary QAM modulator and demodulator: (a) modulator; (b) demodulator.

6.5.2. *M*-ary QAM modulator and demodulator

Figure 6.24 shows the block diagram of an *M*-ary QAM modulator and demodulator.

6.5.2.1 *M*-ary QAM modulator

In CAPK,[54–56] the phase and amplitude of the transmitted carrier are both used to transmit the digital information. For the duration of each symbol of the digital signal T_s, the phase and amplitude of the transmitted carrier signal assume values, chosen from a discrete set of possible amplitude levels and phases. Each combination of amplitude level and phase represents a transmitted symbol. In Figure 6.24 the incoming digital binary signal to the modulator, as with QPSK, is split into two separate bit streams having half the bit rate (that is $r_b/2$), of the incoming data stream bit rate r_b. This splitter also modifies the binary signal (which in practical systems is AMI or HDB3 coded), by changing the AMI or HDB3 code into non-return-to-zero (NRZ) bipolar digits and also performs the scrambling of the input bit stream as well as adding the overhead bits required for framing information, parity check bit, etc. The alternative name given to this block in practical systems is the DPU or data processing unit. The output of the DPU may in practical systems then pass into a differential encoder, as discussed below, which encodes the binary information and produces two output streams, whose bit rate is theoretically half that of the input bit rate r_b. In practice the bit rates will be slightly higher due to the addition of the overhead bits. The combination of the data processing unit (DPU) and the differential encoder is called the *transmit terminal interface unit* (TTIU). The two outputs of the TTIU are then passed into the input of the 2-to-L-level convertor. This convertor is in most schemes, a 2-to-4-level convertor, due to the construction of the transmitter. The resultant four-level signal coming out of the converter then enters into its own path transmit low-pass filter, which provides the necessary premodulation spectral shaping of the baseband signal. The symbol streams in each of the pre-I and -Q paths are still considered to be baseband signals. These spectrum shaped signals then modulate a 70 MHz (or usually some sub-multiple or multiple) local oscillator into a *M*-ary QAM signal. The transmitting and receiving filters are employed at the baseband, since two-dimensional modulation systems, such as 8–PSK, QPRS or 16–QAM, are sensitive to asymmetrical transmission responses and since it is difficult to realize the accuracy required by the filter at IF or RF.

6.5.2.2 Signal constellations of various examples of *M*-ary QAM systems[57,58]

$M = 4$

Figure 6.25[2] shows the constellation diagrams of both QPSK and QAM. With QAM, the four-point constellation comprises two different amplitudes a_1 and a_2. In order to provide the same average power in the two signal constellations, $(a_1^2 + a_2^2)/2 = 2A^2$. For a minimum separation of $2A$ between constellation points

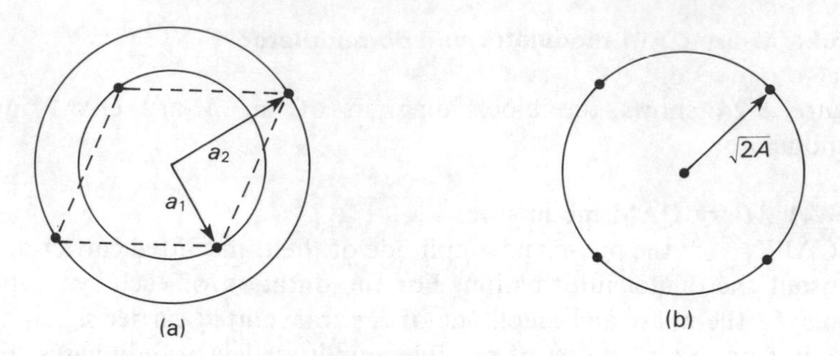

Figure 6.25 Constellation diagrams for 4–PSK and QAM: (a) 4–PSK; (b) QAM (Courtesy of McGraw-Hill, Reference 2)

$a_1 = A$ and $a_2 = \sqrt{3}\,A$. This also is the minimum distance between any two points in QPSK, and it implies that the error rate performance of the two types of modulation scheme are the same.

$M = 8$

Figure 6.26 shows the constellation diagrams for $M = 8$. For these four diagrams out of the many available sets, the minimum separation is $2A$ for two separate

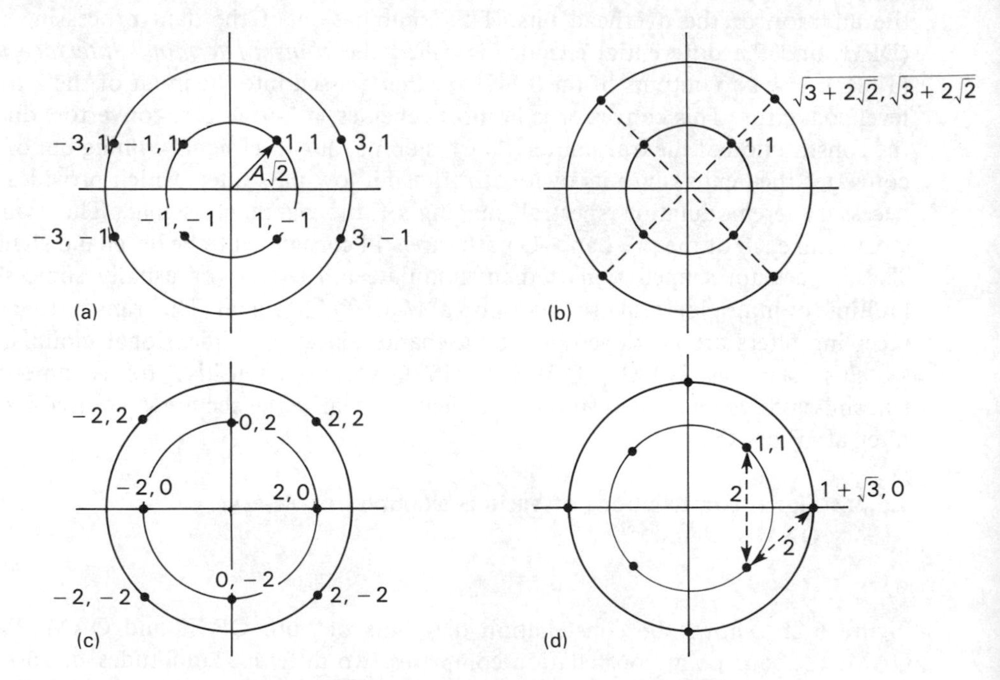

Figure 6.26 Constellation diagrams for 8–QAM (Courtesy of McGraw-Hill, Reference 2)

amplitudes and the error rate performance is the same. The coordinates (d_i, e_i) for each signal set are also shown. If it is assumed that the signal points are equally likely, the average transmitted power P_{av} is

$$P_{av} = (A^2/M) \sum_{i=1}^{M} (d_i^2 + e_i^2) \tag{6.70}$$

By using equation 6.70, the average power for each of the constellations can be computed. For Figure 6.26(a) and (b), the average power P_{av} is $6A^2$. For Figure 6.26(c), $P_{av} = 6.83A^2$ and for Figure 6.26(d), $P_{av} = 4.73A^2$. Comparing this with 8–PSK modulation indicates that 8–PSK requires 1.6 dB more power to achieve the same performance as that constellation shown in Figure 6.26(d). The difference between M-ary PSK and M-ary QAM as M increases is more significant. When M is greater than 8, the constellations usually take on a rectangular set, symmetrically distributed about the coordinate axis. The signal design problem is one involving many trade-offs of factors such as bandwidth reduction, power increase, complexity and suitability for the user's channel. This signal design, which is really a sphere-packing type of problem, dictates the use of symmetric patterns to reduce power requirements or, conversely, lower the probability of error. A problem arises at the receiver, however, in that the receiver can recognize the pattern of signal points, but cannot distinguish between the various symmetric phase orientations of the signal set. An L-fold rotational symmetry can be defined as a signal set for which the signal pattern remains unchanged after being rotated through $\pm I(2\pi/L)$ radians, where I and L are integers. Thus, given $M = 2^k$ signal points, the receiver cannot properly assign the k bit word to each detected signal point without first resolving the L-fold

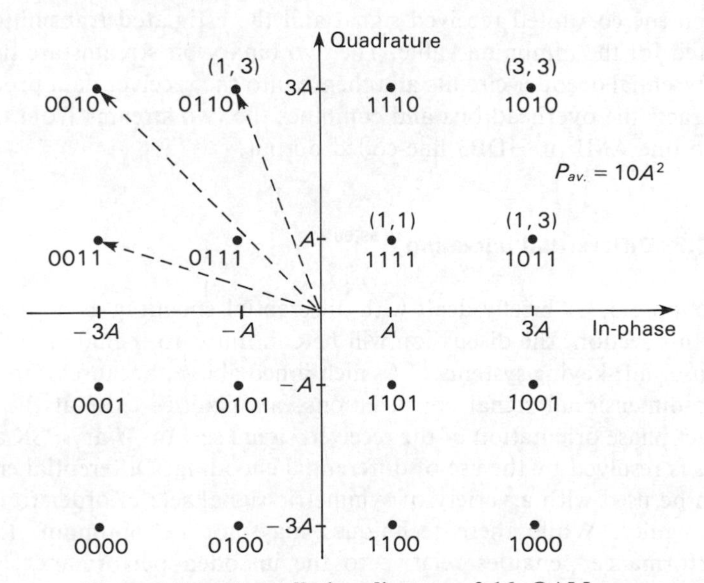

Figure 6.27 Constellation diagram of 16–QAM

ambiguity. For example, in BPSK as shown in Figure 6.11, the receiver needs some absolute way of distinguishing between the two equal and opposite signal points before it can assign a binary one to one phase and a zero to the other. This ambiguity can be resolved in three ways; the first is by the use of a constant reference signal which is transmitted with the modulated data signal; the second is by an acquisition signal and/or the periodic insertion of a synchronization sequence inserted into the data stream; and the third, which is described below, is by the use of differential encoding. Differential encoding has the advantage that by properly encoding and decoding the signal points, the proper bit detection takes place regardless of rotational phase ambiguities. Figure 6.27 shows the constellation diagram for a 16–QAM system on which the dibits (A_1, B_1 and A_2, B_2) are shown in parentheses. It is worth while mentioning at this point that the input binary signal recovered from the AMI or HDB3 code in the transmitter terminal interface unit is first converted into a Gray-coded binary bit stream, so that of the two binary bits which comprise the dibit, each one sequentially enters the I and Q paths. Thus the second dibit which enters the I and Q paths differs in each path by only one bit.

6.5.2.3 *M*-ary QAM demodulator

Figure 6.24 shows the block diagram of the *M*-ary QAM demodulator. The RF signal after being down converted to IF enters the QAM demodulator where it is split into two, for entering the I and Q balanced demodulators. The local oscillator for these demodulators is derived from the input signal using carrier recovery circuits, since all QAM demodulators are coherent in operation. The outputs from these demodulators then enter low-pass filters for spectral shaping before entering the 4-to-2-level convertor. Timing recovery circuits are used to reconstruct the digital signals from the converted 4-level signal and to insure that the distance between the corrupted received signal and the estimated transmitted signal are computed for the minimum value. The two binary bit streams are then passed into the differential decoder circuits and thence into the receiver data processing unit, which extracts the overhead bits and combines the two streams from the I and Q inputs, into one AMI or HDB3 line-coded output.

6.5.3 Differential encoding[10,59,60]

Section 6.3.3.1 briefly dealt with differential encoding as applied to PSK systems. In this section, the discussion will be continued to include multiple amplitude and phase shift keying systems.[59] As mentioned above, because of the symmetry in most two-dimensional signal constellations, ambiguities exist at the receiver as to the exact phase orientation of the received signal set. In *M*-ary PSK and QAM systems, this is resolved by the use of differential encoding. Differential encoding techniques can be used with a variety of symmetric signal sets in order to remove their phase ambiguity. While these techniques may not be optimum, they do have low-performance penalties relative to the uncoded performance, which are further reduced by the use of the minimum amount of differential encoding necessary to remove the ambiguity.

The differential encoding problem is how to assign blocks of k bits to the $2^k = M$, signal points. In the uncoded form, this assignment is a one-to-one mapping of the k bit blocks to the M signal points, which is usually done to minimize the number of bit errors when a symbol error occurs. As symbol errors are most likely to occur between the nearest neighbors (minimum distance), the probability of error is dominated by the minimum distance between signal points. Gray coding is an example in which each adjacent level or signal point differs by only one bit in a k bit block. *The one-bit difference between nearest neighbors is the goal for bit assignments for all signal sets.* This is an important aim.

Figure 6.27 shows for 16–QAM or 4–bit QASK an example of Gray coding, where the nearest neighbor differs by only one bit. In this case, a symbol error will give rise to a bit error. That is:

$$P_{e_{\text{bit}}} = (1/\log_2 M) \cdot P_{e_{\text{symbol}}} \quad \text{(Gray coding)} \tag{6.71}$$

If the signal set is such that it is not possible to code so that only single bit errors occur between adjacent signal points, the bit error equals the symbol error multiplied by some factor other than unity, i.e.

$$P_{e_{\text{bit}}} = K \cdot (1/\log_2 M) \cdot P_{e_{\text{symbol}}} \quad \text{(Non-single-bit change on error)} \tag{6.72}$$

The factor K is usually referred to as the Gray code penalty.

Differential encoding must remove the symmetric ambiguities but not at the expense of a large Gray code penalty, and may be considered as the definition of the bit-to-bit symbol encoding procedure, given a set of M points possessing L-fold symmetry, which:

1. Can be unambiguously decoded at the receiver, regardless of rotational equivalences.
2. Does so with a minimum increase in probability of bit error over that for the uncoded case.

Because of the wide variety of possible signal sets, it is not possible to present a general formula which applies to differentially encoding all sets. This is analogous to Gray coding, which must be handled on a case-by-case basis for two-dimensional signal sets. Reference 59 presents an efficient encoding algorithm for multi-amplitude and multi-phase signal sets, which is based on the differential encoding for PSK. Also is presented differential encoding for offset quadrature signal sets and MAMSK.

The key element to efficient encoding is the minimum use of differential encoding as applied to PSK, to resolve symmetric ambiguities. With most practical multi-amplitude and multi-phase signal sets, the performance penalty is less than two, which is the expected penalty for differentially encoding M-ary PSK.

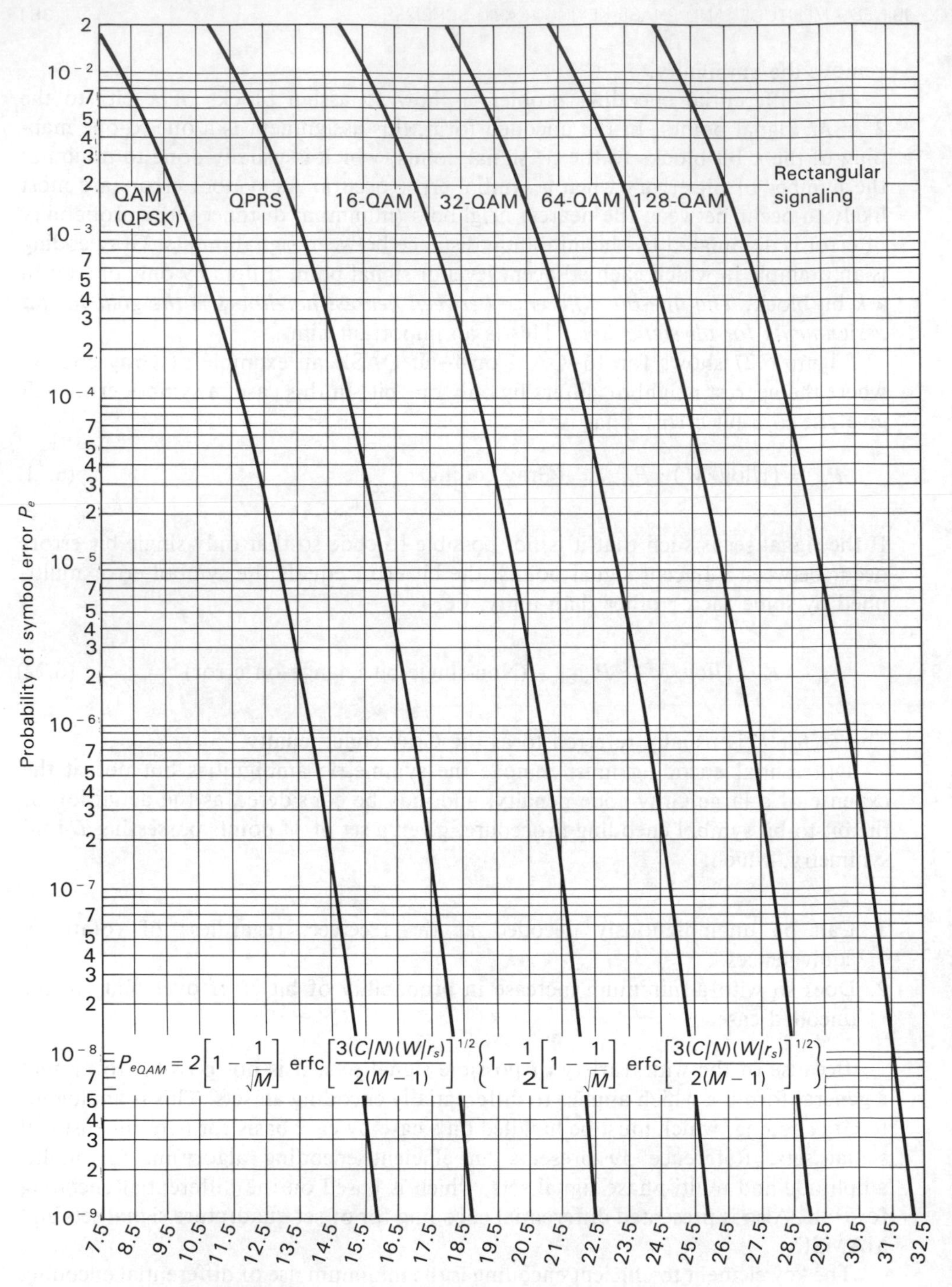

Figure 6.28 Probability of error versus double-sided noise C/N for QAM and QPRS systems

6.5.4 Probability of error of *M*-ary QAM systems

The upper bound for the probability of error is given by:

$$P_{e_{M\text{-ary QAM}}} \leqslant 2 \operatorname{erfc}[3(C/N)(W/r_s)/2(M-1)]^{1/2} \tag{6.73}$$

A more exact expression, when M is a multiple of 2, is given by:

$$P_{e_{M\text{-ary QAM}}} = 2[1 - 1/\sqrt{M}] \operatorname{erfc}[3(C/N)(W/r_s)/2(M-1)]^{1/2} \cdot \{1 - (1/2)[1 - 1/\sqrt{M}] \operatorname{erfc}[3(C/N)(W/r_s)/2(M-1)]^{1/2}\} \tag{6.74}$$

Figure 6.28 shows the probability of error plotted against the double-sided carrier-to-noise ratio (C/N) for various values of M, for the receiver bandwidth W equal to the symbol rate r_s. Table 6.3 below shows the required C/N for a P_e of 10^{-4}, together with general information on M-ary QAM systems.

The spectra for M-ary QAM and M-ary PSK are identical and are given by equation 6.27 for the same number of signal points. The ratio of the average power for

Table 6.3 *M*-ary QAM parameters

Symbol bits (*M*-ary QAM)	Signal constellation coordinates (d_i, e_i) (+ quadrant only)	Average power (where A^2 is given in equation 6.70)	Peak power	Required C/N for P_e of 10^{-4} Mean dB	Peak dB
2 (4–QAM)	(1,1)	$2A^2$	$2A^2$	11.8	11.8
4 (16–QAM)	(1,1), (1,3), (3,1), (3,3)	$10\,A^2$	$18\,A^2$	18.8	21.3
5 (32–QAM)	(1,1), (3,1), (5,1), (1,3) (3,3), (5,3), (1,5), (3,5)	$20\,A^2$	$34\,A^2$	21.8	24.1
6 (64–QAM)	(1,1), (3,1), (5,1), (7,1), (1,3), (3,3), (5,3), (7,3), (1,5), (3,5), (5,5), (7,5), (1,7), (3,7), (5,7), (7,7)	$42\,A^2$	$98\,A^2$	25.0	28.7
7 (128–QAM)	(1,1), (3,1), (5,1), (7,1), (9,1), (11,1), (1,3), (3,3), (5,3), (7,3), (9,3), (11,3), (1,5), (3,5), (5,5), (7,5), (9,5), (11,5), (1,7), (3,7), (5,7), (7,7), (9,7), (11,7), (1,9), (3,9), (5,9), (7,9), (1,11), (3,11), (5,11), (7,11)	$82\,A^2$	$170\,A^2$	27.9	31.1
8 (256–QAM)	(1,1), (3,1), (5,1), (7,1), (9,1), (11,1), (13,1), (15,1), (1,3), (3,3), (5,3), (7,3), (9,3), (11,3), (13,3), (15,3), (1,5), (3,5), (5,5), (7,5), (9,5), (11,5), (13,5), (15,5), (1,7), (3,7), (5,7), (7,7), (9,7), (11,7), (13,7), (15,7), (1,9), (3,9), (5,9), (7,9), (9,9), (11,9), (13,9), (15,9), (1,11), (3,11), (5,11), (7,11), (9,11), (11,11), (13,11), (15,11), (1,13), (3,13), (5,13), (7,13), (9,13), (11,13), (13,13), (15,13), (1,15), (3,15), (5,15), (7,15), (9,15), (11,15), (13,15), (15,15)	$170\,A^2$	$450\,A^2$	31.1	35.3

PSK to that of QAM is given by[2]

$$\frac{\text{Average } M\text{-ary PSK power}}{\text{Average } M\text{-ary QAM power}} = \frac{3M^2}{(2(M-1))\pi^2} \tag{6.75}$$

This ratio when expressed in decibels shows the difference between the two modulation schemes, with increased power required for PSK; the value of the ratio for various values of M is given in Table 6.4.

Table 6.4 shows that PSK systems require more power to transmit the same information for a given probability of error. As both QAM and PSK have the same bandwidth efficiency, that is, $\log_2 M$, it can be deduced that QAM is a superior modulation scheme. This is, however, at the expense of circuit complexity.

Table 6.4 Ratio of power requirements of *M*-ary PSK and of *M*-ary QAM (Courtesy of McGraw-Hill, Reference 2)

M	(Average M-ary PSK power)/(Average M-ary QAM power), dB
2	1
8	1.43
16	4.14
32	7.01
64	9.95

6.5.5 Offset QAM(OK–QAM, or OQAM) or staggered QAM (SQAM)

It has been shown[37] that the use of offset QPSK signaling allows almost a 3 dB relaxation in the limitation on the signal-to-noise ratio of the phase reference, compared with that for conventional QPSK. Offset QAM (SQAM), in the presence of residual phase jitter and additive Gaussian noise by the same token, also shows an improvement over conventional QAM, when the same bandwidth is used for comparison. This improvement, however, is proportional to the amount of excess bandwidth[61] available. Also with any Nyquist pulse shape ($\sin \omega t/\omega t$), the optimum staggered timing is half a symbol interval. Using a raised-cosine signal pulse with a half-bit-period staggered time delay is shown to be equivalent to a 3 dB gain in signal-to-noise ratio for a 100 percent excess bandwidth. Figure 6.29 shows the basic block diagram of the SQAM system. Reference 61 describes the generalized concept of parallel data transmission systems in which M-ary QAM and SQAM belong.

To add to the confusion which acronyms may cause, another SQAM system, which is an abbreviation of '*Superposed* Quadrature Amplitude Modulation', is reported in the literature.[63] In this system the non-return-to-zero input data stream is processed by the baseband SQAM processor to produce output signals with defined spectral characteristics. The outputs of the baseband signal processors then pass into a conventional offset QAM (or offset QPSK) modulator structure. The SQAM demodulator is the same as the demodulator of the conventional coherent offset QAM (or offset QPSK) system.

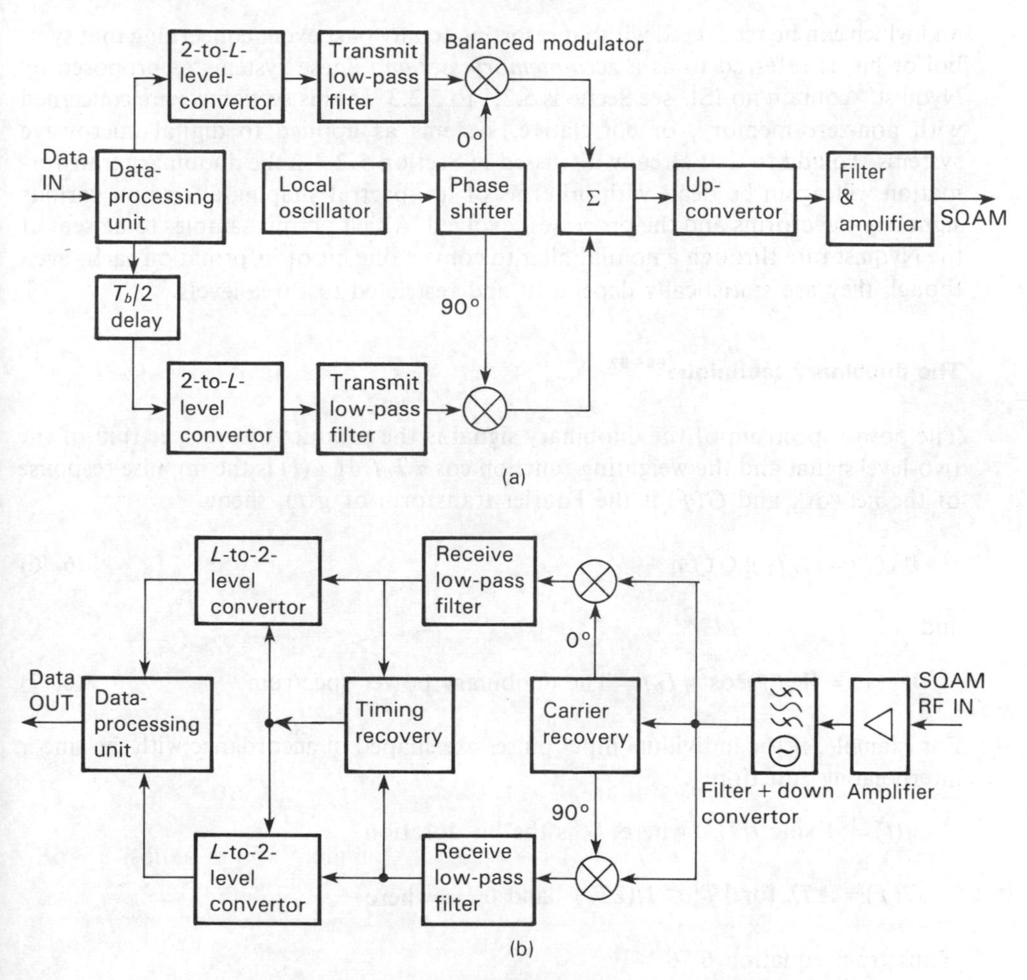

Figure 6.29 Staggered quadrature amplitude modulation (SQAM) system: (a) transmitter; (b) receiver

6.6 CORRELATIVE OR PARTIAL RESPONSE SYSTEMS (PRS)

These systems have been briefly mentioned in Section 5.2.4.2, under baseband duobinary systems, and in them, whether they are baseband or carrier, a deliberate amount of inter-symbol interference (ISI) is introduced over a span of one or more binary bits or symbols. This controlled amount of ISI permits an increase in the spectral efficiency by permitting data rates to reach and, in some cases, exceed the Nyquist rate. Due to the correlation between message digits from the introduction of ISI, distinctive patterns are formed which permit error detection without any special error-detecting overhead bits added to the transmitted message signal. A system in which any message symbol or bit, which is independent and uncorrelated and which can be recovered without resorting to any past event concerning that symbol or bit, is referred to as a *zero-memory system*. These systems as proposed by Nyquist[64] contain no ISI–see Sections 5.2.1 to 5.2.3. In this section we are concerned with non-zero-memory, or correlative, systems as applied to digital microwave systems. To add to that already discussed in Section 5.2.4.2 the duobinary transformation will again be dealt with in terms of its spectral shaping effects on various signaling waveforms and the processes involved, which permit samples to be sent at the Nyquist rate through a nonflat filter to convey one bit of information each, even though they are statistically dependent and restricted to three levels.[65]

The duobinary technique[65-67]

The power spectrum of the duobinary signal is the product of the spectrum of the two-level signal and the weighting function $\cos \pi T_b f$. If $g(t)$ is the impulse response of the network and $G(f)$ is the Fourier transform of $g(t)$, then:

$$W_x(f) = (1/T_b)|G(f)|^2 \tag{6.76}$$

and

$$W_y(f) = W_x(f)\cos^2 \pi T_b f = \text{The duobinary power spectrum} \tag{6.77}$$

For example, if the individual input pulses are shaped in accordance with the sinc x interpolating function:

$$g(t) = A \text{ sinc } t/T_b \quad \text{where } T_b \text{ is the bit duration}$$

$$G(f) = AT_b \text{ for } |f| \leqslant 1/(2T_b) \quad \text{and 0 elsewhere}$$

Thus from equation 6.76

$$W_x(f) = A^2 T_b \text{ for } |f| \leqslant 1/(2T_b) \quad \text{and 0 elsewhere}$$

and from equation 6.77

$$W_y(f) = A^2 T_b \cos^2 \pi T_b f \text{ for } |f| \leqslant 1/(2T_b) \quad \text{and 0 elsewhere} \tag{6.78}$$

Figure 6.30 shows the time function and its frequency spectrum, together with the spectrum after the duobinary spectral shaping for the sinc pulse, the raised-cosine pulse, the Gaussian pulse and the rectangular pulse.[65] From Figure 6.30, we see that the duobinary conversion has the following effects:

Sinc function
The spectrum of the sinc function before processing is strictly limited to a bandwidth of $1/2T_b$. The duobinary spectrum has the same bandwidth, but some of the energy has been shifted towards the low-frequency region.

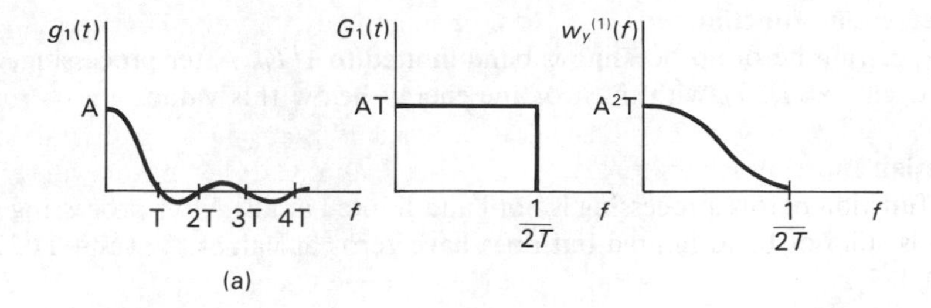

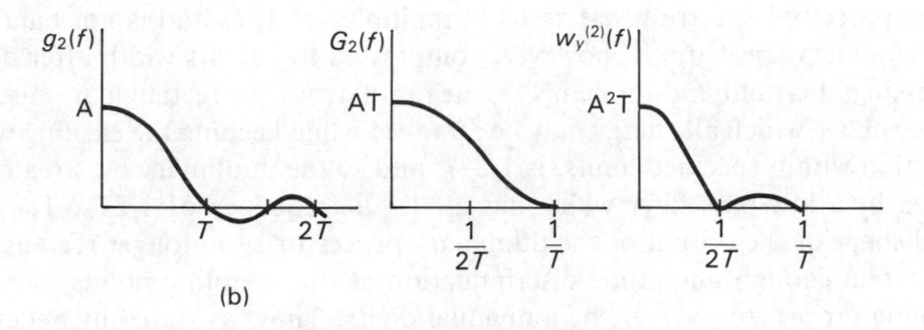

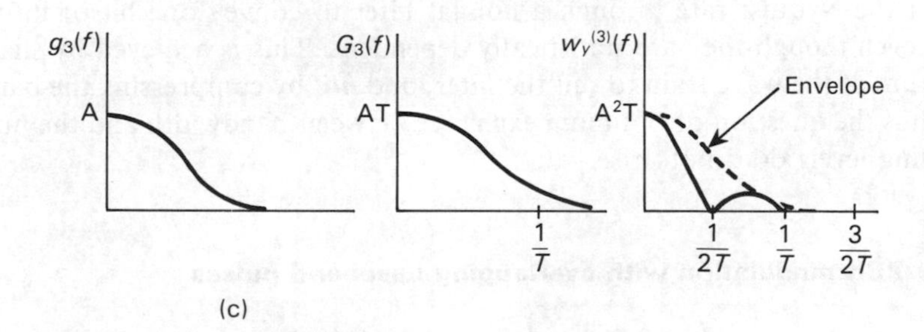

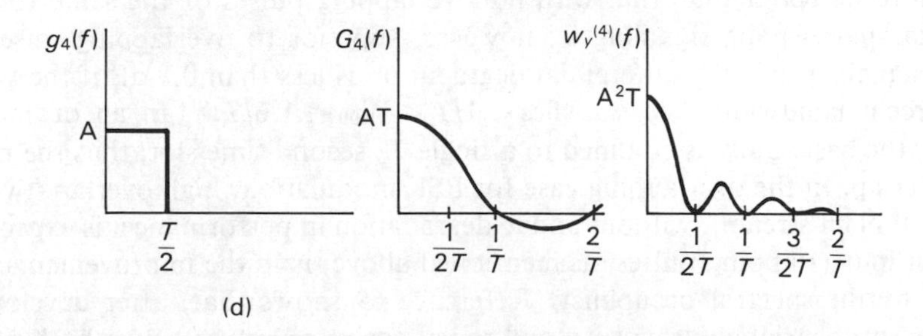

Figure 6.30 Normal and duobinary processed frequency spectra for different input signal waveshapes: (a) the sinc pulse $g_1(t)$ and its spectrum; (b) the raised-cosine spectrum $G_2(f)$; (c) the Gaussian pulse $g_3(t)$ and its spectrum; (d) the rectangular pulse $g_4(t)$ and its spectrum (Courtesy of IEEE, Reference 65)

Raised-cosine function
The spectrum before processing is band limited to $1/T_b$. After processing, a second zero occurs at $1/2T_b$ with most of the energy below this value.

Gaussian function
This function before processing is not band-limited at all. After processing the spectrum is still not band-limited but does have zeros at values $f=(2k+1)/2T_b$ where $k=0,1,2,\ldots$.

Rectangular pulse
The unprocessed spectrum has zeros at multiples of $1/T_b$ and is not band-limited. The duobinary spectrum is, however, compressed to half its width after duobinary processing; it is not band-limited. This means that, for the rectangular pulse, the frequency above which all energy may be removed while keeping the ensuing waveform distortion within specified limits, is $1/2T_b$; and so the duobinary bit stream is band-limited by a low-pass filter which has a cut-off frequency of $1/2T_b$. The resulting signal shape at the output of the duobinary processor is no longer rectangular, but does retain enough amplitude discrimination at the sampling points to enable the sampling process to recover the individual digits. Thus, as stated by Sekey:[65] 'The duobinary transformation is essentially a means for making it possible for samples sent at the Nyquist rate through a nonflat filter to convey one bit of information each, even though they are statistically dependent. This is achieved by shaping the spectrum of the wave train to suit the filter, and *not* by compressing the bandwidth, and thus the question of optimum exchange between bandwidth and the number of signaling levels does not arise.'

6.6.1 PSK modulation with overlapping baseband pulses

The spectral efficiency of PSK modulation with overlapping baseband pulses is known to be better than that with nonoverlapping pulses of the same form. Rectangular pulse-shape signaling is, however, superior to overlapping raised-cosine pulse signaling, but the differential degradation is less than 0.8 dB if the two-sided 99 percent bandwidth W_{99} satisfies $1.1/T < W_{99} < 1.6/T$.[68] In an ordinary PSK signal, the basic pulse is confined to a single T_b second time slot; thus the pulses do not overlap. In the overlapping case for PSK modulation, high overlap (two bits of the symbol bit stream), without undue degradation in performance, is expected. The interest in overlapping pulses, as mentioned above, is in the improvements brought about by the spectral occupancy. Reference 69 shows that, after developing the spectra for the triangular, cosine, and raised-cosine pulse types over both single and double intervals (T) together with Nyquist pulses and that produced from second- and third-order Butterworth filters, the overlapping baseband pulses do not improve the transmitted spectrum either by reducing the 3 dB bandwidth of the continuous spectrum or by reducing the bandwidth containing 99 per cent of the total power. The reader is referred to Reference 69 for the various spectral diagrams using the different pulse forms. For the overlapping PSK case, the degradation, or the increase in C/N, required to maintain the same probability of error as that of the rectangular-pulse nonoverlapping PSK (see equation 6.33), depends on the 99 percent bandwidth of the transmitting filter used, so that 99 percent of the power is contained in a given band. Also for a fixed probability of error, the signaling waveforms (raised cosine, triangular, etc.) will affect the amount of degradation over the nonoverlapping rectangular signaling. Figure 6.31(a) shows the overlapping raised-cosine pulses and Figure 6.31(b) shows the degradation in dB for an

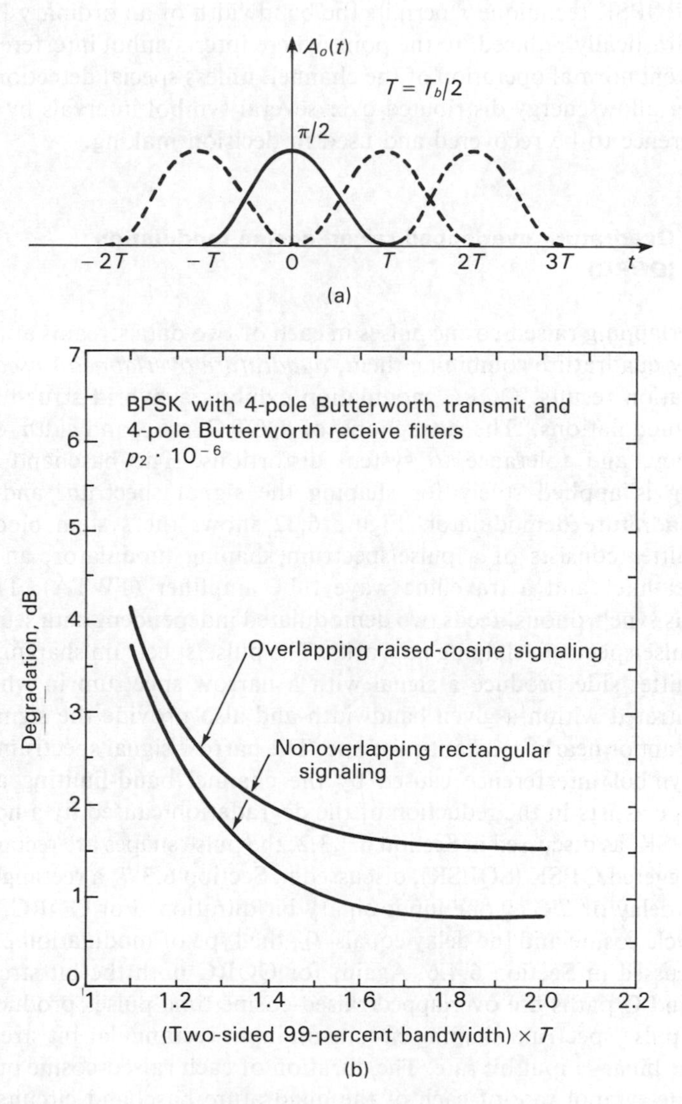

Figure 6.31 (a) Overlapping raised-cosine pulses; (b) degradation for a BPSK system with nonoverlapping rectangular and overlapping raised-cosine signaling
(Courtesy of IEEE, Reference 68)

overlapping raised-cosine pulse, when compared with the nonoverlapping rectangular signaling pulse. This means that the C/N has to be increased by the degradation factor in order to attain the ideal probability of error curve. As can be seen from Figure 6.31(b), if the transmitting filter 99 percent bandwidth is reduced, C/N requires to be increased to maintain the same probability of error. This exemplifies the power–bandwidth trade-off that is typical of PSK systems.

6.6.1.1 Reduced bandwidth QPSK (RBQPSK)

The RBQPSK technique[78] permits the bandwidth of an ordinary PSK transmission to be drastically reduced, to the point where inter-symbol interference is so great as to prevent normal operation of the channel, unless special detection processes in the receiver allow energy distributed over several symbol intervals by the inter-symbol interference to be recovered and used in decision making.

6.6.2 Quadrature overlapped raised-cosine modulation (QORC)

By overlapping raised-cosine pulses in each of two data streams at the baseband and then by quadrature combining them, *quadrature overlapped raised-cosine* (QORC) modulation results. QORC modulation exhibits a hybrid structure of QPSK and MSK modulations. The attractions of QORC are bandwidth efficiency, power efficiency, and tolerance to system distortions. The baseband pulse amplitude shaping is applied solely for shaping the signal spectrum and is discarded at the quadrature demodulator. Figure 6.32 shows the system block diagram. The transmitter consists of a pulse/spectrum shaping modulator, an up-convertor, a channel filter and a travelling wave tube amplifier (TWTA). The demodulator, which is synchronous, feeds two demodulated independent data sequences into identical pulse/spectrum shaping networks. The pulse/spectrum shaping networks on the transmitter side produce a signal with a narrow spectrum in which the power is concentrated within a given bandwidth and also provide the signal spectrum with a constant or nearly constant envelope. The narrow signal spectrum helps reduce the inter-symbol interference caused by the channel band-limiting and the constant envelope assists in the reduction of the degradation caused by a nonlinear channel. For QPSK, as discussed in Section 6.3.3.2, the pulse shapes are rectangular. Similarly, for staggered QPSK (SQPSK), discussed in Section 6.3.7, a rectangular pulse is used with a delay of T_b, or one input binary bit duration. For QORC, if the shaping is half-cycle cosine and the delay equals T_b, the type of modulation produced is MSK, as discussed in Section 6.4.6. Again, for QORC both the bit streams which form the I and Q paths are overlapped raised-cosine time pulses produced by the action of the pulse/spectrum shaping circuits on input rectangular bit streams operating at half the binary input bit rate. The duration of each raised-cosine pulse is $2T_s$, where T_s is the symbol rate of each of the quadrature baseband circuits ($T_b/2$).

Figure 6.33(a)[70] shows the modulator implementation, in which each of the I and Q channel data of data rate $r_s = 1/T_s$ is sampled and held or demultiplexed into odd and even data streams. The odd and even data streams are then weighted by

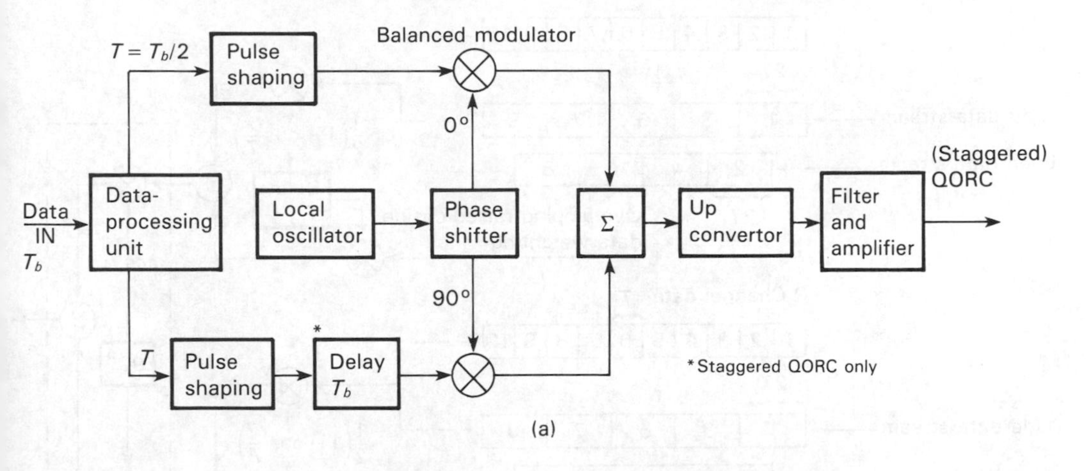

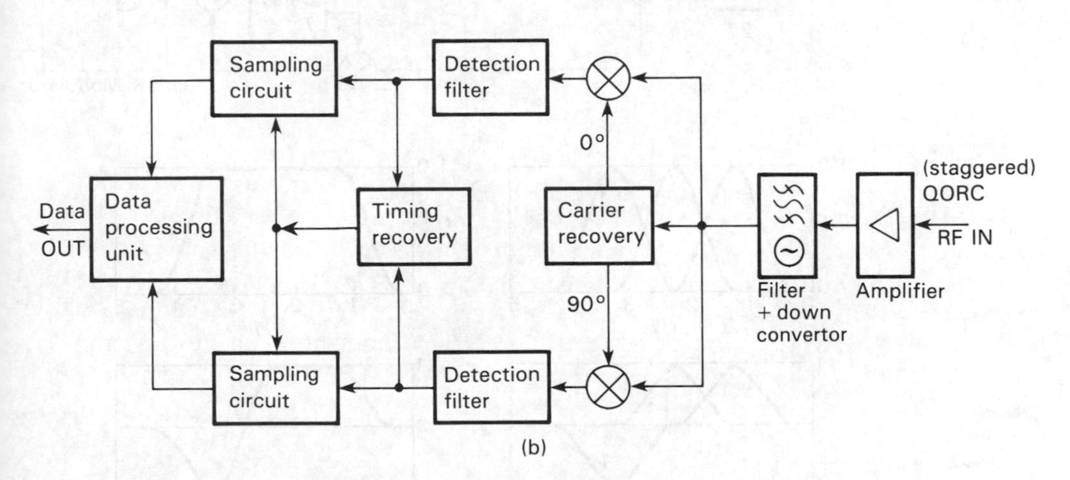

Figure 6.32 QORC modulator and demodulator: (a) transmitter; (b) receiver

a continuous raised-cosine waveform generator which has a period of $2T_s$, before being summed to produce the I baseband stream and the Q baseband stream. A conventional QPSK or SQPSK modulator then follows these circuits to modulate the shaped data streams. The spectrum of the QORC modulated data stream is given by;

$$P(f)_{\text{QORC}} = (A^2T/2)\cdot\left\{\left[\frac{\sin \pi T(f-f_0)}{\pi T(f-f_0)}\right]^2 + \left[\frac{\sin \pi T(f+f_0)}{\pi T(f+f_0)}\right]^2\right\}$$

$$\cdot\left\{\left[\frac{\cos \pi T(f-f_0)}{1-4T^2(f-f_0)^2}\right]^2 + \left[\frac{\cos \pi T(f+f_0)}{1-4T^2(f+f_0)^2}\right]^2\right\} \qquad (6.79)$$

Except for a normalization constant, the spectrum of QORC is the product of

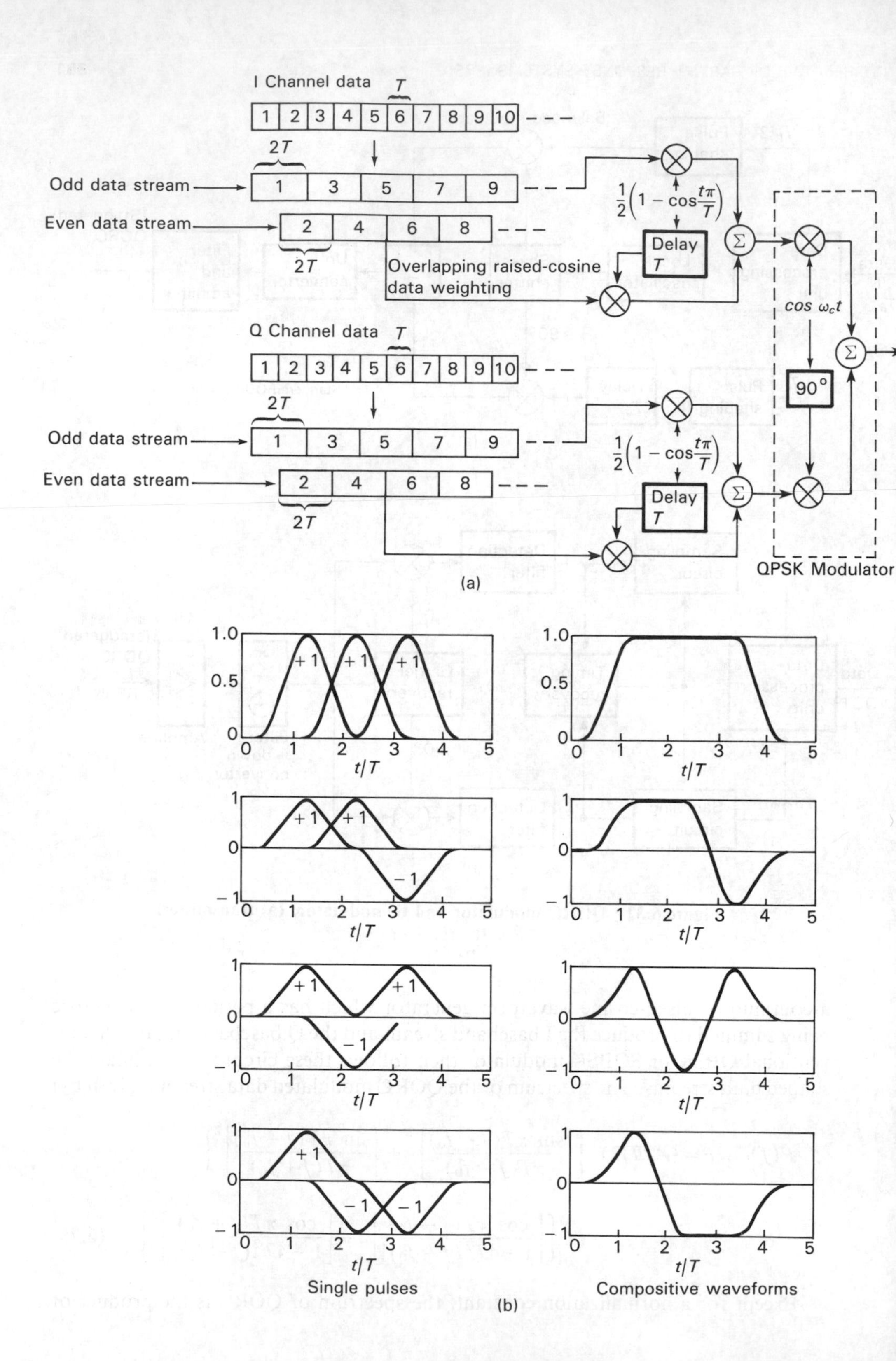

I Channel data
T
1 2 3 4 5 6 7 8 9 10
2T
Odd data stream
1 3 5 7 9
Even data stream
2 4 6 8
2T
$\frac{1}{2}\left(1 - \cos\frac{t\pi}{T}\right)$
Delay T
Overlapping raised-cosine data weighting
$\cos \omega_c t$
Q Channel data
T
1 2 3 4 5 6 7 8 9 10
2T
Odd data stream
1 3 5 7 9
Even data stream
2 4 6 8
2T
$\frac{1}{2}\left(1 - \cos\frac{t\pi}{T}\right)$
Delay T
90°
QPSK Modulator
(a)
+1
−1
t/T
Single pulses
Compositive waveforms
(b)

SQPSK and MSK spectra. In addition to the $1/f^6$ side-lobe roll-off, the power spectral density of QORC has the same primary null as QPSK and SQPSK. As the first null of MSK occurs at 1.5 times that of QPSK, QORC bears no similarity. For the remaining nulls of QORC, a plot of the spectra will show that they occur twice as often as those of QPSK, SQPSK and MSK. In Figure 6.33(*b*) are shown the isolated shaped pulses, together with their composite waveforms. Only four of the eight possible waveforms appear, since the missing four are only the negatives of the ones shown. No inter-symbol interference is introduced at the center of each pulse as indicated on these diagrams.

6.6.2.1 Offset QORC or staggered QORC (SQORC)

As with QPSK, another variation of QORC is staggered QORC, or SQORC. This form of modulation is implemented by having a delay of $T/2$ inserted in the quadrature baseband channel, as shown in Figure 6.32. One characteristic of QORC is the bit-to-bit correlation that is introduced by overlapping the pulses. This characteristic is similar to the partial response signaling techniques. The salient point in QORC modulation is that, only halves of each bit are overlapped successively and thus there is no inter-symbol interference at the center of each bit.

One of the more important considerations to be taken into account when comparing different modulation schemes is that of the power spectral density of the side-lobes. Low-level side lobes are a positive feature, because a transmitter needs to maintain low out-of-band transmitted signal power so as not to cause interference between other carriers which may be external to the system, or which may be part of a system that involves more than one carrier. At the modulator output the RF signal enters the channel filter which reduces the out-of-band side lobe power. This RF signal then passes through the transmission channel where it is exposed to channel nonlinearities. The effect of the transmission channel on the QSPK spectrum is to restore the side lobes to the level that they were before entering the transmit filter whilst, SQORC, SPQSK and MSK are restored to approximately the same low level. Both QORC and SQORC have excellent low-level side lobes at the output of a nonlinear TWT; SQORC is better than QORC, SQPSK, QPSK and MSK, while QORC is better than QPSK. As QORC is a nonconstant signal envelope type of modulation, it may at first appear that this form of modulation is not compatible with high-efficiency power amplifiers that are operating at or near saturation. Reference 70 shows that, for systems operating in real life, the performance of constant envelope modulation techniques, such as QPSK or MSK, also undergoes a nonconstant envelope transformation which is caused by the variation in group delay over the transmitting filter from the bandwidth limiting. For symbol rates which approach or exceed the transmit or channel bandwidth, signaling waveforms that have constant envelopes may perform no better than signaling waveforms with nonconstant envelopes. In the nonlinear channel where inter-symbol interference is

---

Figure 6.33 QORC modulator implementation: (a) modulator; (b) isolated pulses and composite waveforms (Courtesy of IEEE, Reference 70)

introduced and the pulse peak is reduced due to the transmit filtering, QORC provides a better probability of error performance than SQORC.

6.6.2.2 Staggered quadrature overlapped squared raised-cosine modulation (SQOSRC)

This form of modulation has been proposed[71] because of its desirable spectral property on a hard-limited satellite channel. A hard limiter is used at the input to a travelling wave tube when maximum output power is required. The pulse shape of SQOSRC is a squared raised-cosine pulse whose duration is the same as QORC, but the parts around the edges of the pulse are flatter than those of QORC. The rate of minimum to maximum value of the signal envelope is almost 0.9 for SQOSRC, whereas for SQORC it is almost 0.7. This indicates that the envelope of the former fluctuates less than that of the latter. The performance of SQOSRC in the presence of inter-symbol interference and pulse peak limiting caused by transmit filtering is better than that of SQORC.[71]

6.6.2.3 Correlative phase shift keying (CORPSK)[72]

CORPSK is a constant-envelope modulation technique with smooth transitions between fixed phase positions, and with successive transitions correlated. Main lobe bandwidth is the same as or less than in other constant envelope modulation techniques such as MSK, QPSK and CPFSK, but the out-of-band radiation is much lower. Coherent detection is used and the typical modulator or demodulator is similar in structure to those of Figure 6.32. The waveshaping network comprises a correlative encoder, a pre-modulation filter and an angle modulator in cascade. In addition, a precoder is often needed at the input to avoid error propagation. The modulator can be a linear phase modulator for those types of CORPSK for which the phase deviation is limited, but for those types for which the phase deviation is unlimited, a frequency modulator is required. To reach the proper phase positions at the end of each symbol interval, either the phase or frequency deviation of the modulator must be controlled in a feedback loop. There are many CORPSK modulation schemes, and they open up a new field. Reference 72 may be a good starting point for the reader who is interested in pursuing the subject.

6.6.3 Quadrature partial response (QPR)[73,74]

The quadrature partial response system (QPRS) is similar to two APK three-level duobinary signals in quadrature. Thus each signal in each of the baseband streams has three states, which on combining produce a state diagram of nine states. This is shown in Figure 6.34. The modulator and demodulator block diagrams are similar in construction to those of the QORC system, shown in Figure 6.32. The difference between the QORC signal and the QPRS system lies in the fact that the low-pass filters placed in the baseband pulse shaping network provide the duobinary filtering, using an analog cosine filter with the transfer function 2T cos $\pi f T$. Coherent carrier detection is normally used, and the carrier to be recovered is locked to one of the odd-numbered signal states (except state 9). Referring to the signal state or

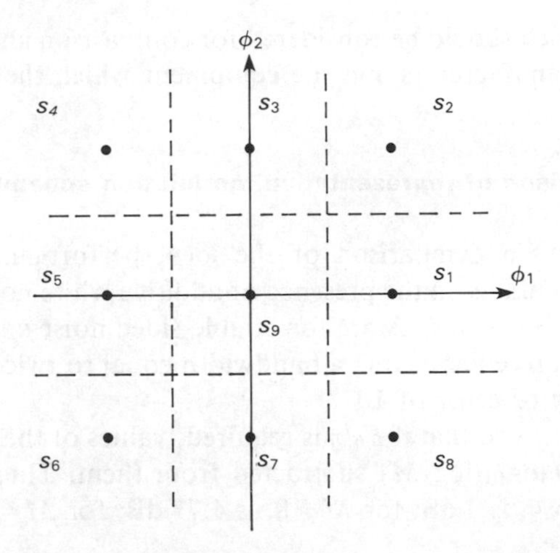

Figure 6.34 Constellation diagram of a QPRS signal

constellation diagram shown in Figure 6.34, the total number of signal states is nine, with the zero state being the states s_1 to s_8, and state s_9 being the 1 state. States s_2, s_4, s_6 and s_8 represent (0, 0) and states s_1 and s_5 represent (0, 1), whilst states s_3 and s_7 represent (1, 0). The absence of carrier state s_9 represents (1, 1). The QPRS system is therefore amplitude modulated on a noncontinuous carrier-type system. The probability of error is given as:[73]

$$P_{e\text{QPRS}} = \text{erfc}[(1/2)(W/r_s)^{1/2}(C/N)^{1/2} + (\ln 2)/(W/r_s)^{1/2}(4C/N)^{1/2}] + (1/2)\text{erfc}[(1/2)(W/r_s)^{1/2}(C/N)^{1/2} - (\ln 2)/(W/r_s)^{1/2}(4C/N)^{1/2}] \quad (6.80)$$

where C/N is the carrier-to-noise ratio in a double-sided noise bandwidth and is the received C/N if the receiver bandwidth is twice that of the message bandwidth, i.e. $W = r_s$. Figure 6.28 shows a plot of the QPRS probability of error against C/N for a rectangular input signal to the transmitter modulator, for $W = r_s$

6.7 COMPARISONS BETWEEN SOME COMMERCIALLY AVAILABLE SYSTEMS

As can be seen from brief descriptions in this chapter, there has been a rapid growth in the number of modulation schemes which are in use or will be coming into use in the near future. The schemes described, however, do not encompass all the different modulation schemes which have been reported in the literature. To attempt to complete a brief listing and description of the salient features of the different known schemes, which are pertinent to digital radio or satellite transmission, would require a full textbook. To assist the reader in choosing equipment which is commercially available, this section will be divided into two. The first will attempt to provide the features which should be considered for comparison and the second part will list some of the manufacturers and the equipment which they provide.

6.7.1 Comparison of representative modulation schemes

Table 6.5 shows a comparison of the ideal performance of the representative modulation techniques in the presence of additive white noise and using rectangular signaling. The values of C/N are for double-sided noise and for a signal that would be received in a receiver having a bandwidth equal to twice the message bandwidth at a probability of error of 10^{-6}.

It should be noted that if E_b/η is required, values of the C/N, given in Table 6.5, should have $10\log_{10}(\log_2 M)$ subtracted from them. Thus the amount to be subtracted for $M = 4$, is 3 dB; for $M = 8$, is 4.77 dB; for $M = 16$, is 6 dB; for $M = 64$,

Table 6.5 Ideal performance of representative modulation schemes

Type	M	Modulation scheme	C/N	Equation of P_e/Section with description
AM	2	Coherent ASK	13.5	Eqn 6.24/Section 6.2.1 and 6.2.3
	2	Noncoherent ASK	14.2	Eqn 6.25/Section 6.2.2 and 6.2.3
	4	Quadrature amplitude modulation (QAM or 4–APK or QASK)	13.8	Eqn 6.73 and 6.74/Section 6.5.4
	16	16–QAM or 16–APK	20.9	Eqn 6.74/Section 6.5.4
	32	32–QAM	24.1	Eqn 6.74/Section 6.5.4
	64	64–QAM	27.2	Eqn 6.74/Section 6.5.4
	128	128–QAM	30.3	Eqn 6.74/Section 6.5.4
	4	Quadrature partial response (QPRS)	16.95	Eqn 6.80/Section 6.6.3
FM	2	Coherent binary FSK	13.5	Eqn 6.56/Section 6.4.1.1
	2	Noncoherent binary FSK	14.2	Eqn 6.57/Section 6.4.2.1
	4	4–CFSK	13.9	Eqn 6.58/Section 6.4.5.1
	16	16–CFSK	14.5	Eqn 6.58/Section 6.4.5.1
	64	64–CFSK	14.7	Eqn 6.58/Section 6.4.5.1
	4	4–NCFSK	14.5	Eqn 6.63/Section 6.4.5.2
	16	16–NCFSK	15.1	Eqn 6.63/Section 6.4.5.2
	64	64–NCFSK	15.4	Eqn 6.63/Section 6.4.5.2
	4	MSK coherent detection (BPSK)	10.5	Eqn 6.33/Section 6.4.6.1
	4	MSK differential detection	11.15	Eqn 6.68/Section 6.4.6.2
PM	2	Coherent binary PSK	10.8	Eqn 6.33/Section 6.3.1.1
	2	Differentially encoded BPSK	10.8	$\frac{1}{2}(P_{e\text{BPSK}} + P_{e\text{DPSK}})$/Section 6.3.3.1
	2	Differential PSK (DPSK)	11.15	Eqn 6.36/Section 6.3.2.1
	4	4DPSK	16.3	Eqn 6.51/Section 6.3.6
	4	QPSK (Coherent detection)	13.75	Eqn 6.46b/Section 6.3.3.2
	8	8–PSK	19.0	Eqn 6.46a/Section 6.3.3.2
	16	16–PSK	24.8	Eqn 6.46a/Section 6.3.3.2
	4	Offset QPSK	13.5	Eqn 6.53/Section 6.3.7.1

is 7.8 dB; and for $M = 128$, is 8.45 dB. Also for all values the receiver IF bandwidth W is assumed to equal the symbol rate r_s.

6.7.2 Spectral characteristics[10]

There are many ways of comparing the spectral characteristics of the modulation schemes. Of particular interest is the effect which adjacent-channel interferers will have. A measure of this effect can be drawn from the attenuation of the power spectrum of a signal at some specified distance from the center frequency. If the attenuation at an arbitrary distance of $8/T_s$ is chosen, then, with AM schemes, the side lobes are about 25 dB down; with PM schemes the side lobes are about 33 dB down; and with continuous-phase FM the side lobes are 60 dB or more down. An MSK system at this distance of $8/T_s$ achieves a 95 dB attenuation down on the value at the center frequency. In general, for frequencies far from the center frequency, i.e. large $(f - f_0)T_s$, the spectra of AM and PM signals fall off at a rate proportional to the inverse of the frequency squared. For continuous-phase FSK (CP–FSK) and MSK the rate of fall-off is proportional to the inverse of the frequency to the fourth power. For PSK, the rate of fall-off is proportional to the inverse of the frequency squared. Initially it may appear that FM is the modulation scheme which offers the best advantage in this respect. However, the figures quoted above assume that PM systems suffer from abrupt phase transitions, and with the smoothing of these transitions improved spectral characteristics occur. For AM schemes such as that of a carrier being amplitude modulated by a biphase PCM code, the spectrum can also be made to fall off at the same rate as that for FM schemes.[75] As has often been mentioned in this chapter, another important spectral property is that of the bandwidth efficiency or spectral efficiency. The spectral efficiency is the bit rate r_b, divided by twice the Nyquist symbol or message bandwidth. If raised-cosine filtering is used, the message bandwidth is increased by the roll-off factor α so that the noise bandwidth or IF bandwidth becomes $2 \times \frac{1}{2} r_s(1 + \alpha) = r_s(1 + \alpha)$. Hence:

$$\text{Spectral efficiency} = \frac{r_b}{r_s(1+\alpha)} = \frac{r_s \log_2 M}{r_s(1+\alpha)} = \frac{\log_2 M}{1+\alpha} \tag{6.81}$$

As filters are not usually designed to be 'brickwall' in shape, the IF bandwidth may start to approach three times the message bandwidth depending on the roll-off factor. The smaller the roll-off factor, the higher the spectrum efficiency, but in the light of considerations such as the increase in waveform distortion due to multipath fading and the difficulty of accurate filter design, roll-off factors are usually in the range 0.5 to 0.7. When the roll-off factor is 0.5, more than 8-level modulation can satisfy the desired spectrum efficiency. For M-ary QAM systems usually a cosine roll-off factor of 0.5 is used for M more than 16. Figure 6.35 shows the spectrum efficiencies of various modulation methods. This diagram also indicates that more than 16-level modulation is necessary to realize a spectrum efficiency of more than 2 bits/s/Hz, when roll-off spectrum shaping is not applied.

In Table 6.6, the spectral efficiency is listed for some of the less well known

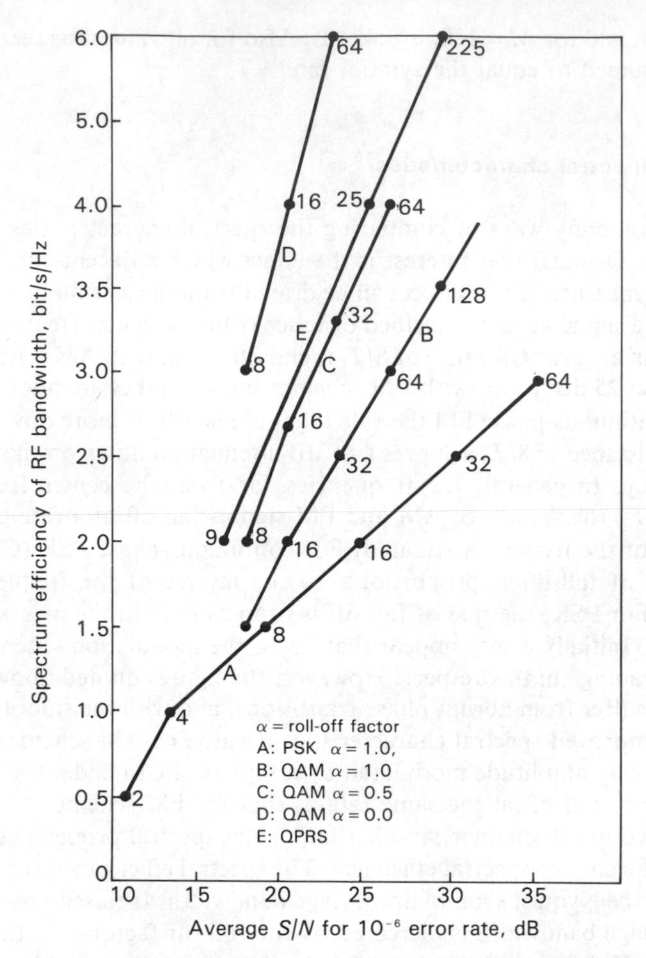

Figure 6.35 The ideal spectrum efficiencies of various modulation methods for different roll-off factors

techniques. The filters used are chosen for the ideal case to match the signaling pulse on the channel with the appropriate partitioning of the transmitting and receiving filters, to produce an optimum design so that the probability of error is minimized in the presence of impulse noise. The filter designs and partitioning are also optimum with respect to minimizing the effect of adjacent-channel interference where adjacent channels are assumed to be identical in form to the in-band channel. The material presented in this table has been taken from Reference 76, and the figures are indicative of the results to be expected. Included in this table, under modulator efficiency, is the ratio of the signal power output of a passive pulse-shaping network to the signal power output of the transmitter power amplifier, which is an important design consideration when the pulse shaping filter follows the power amplifier, and typical values of modulator efficiencies for the various modulator pulse forms and pulse-shaping networks.

Table 6.6 Relative spectral efficiencies of representative modulation schemes

Modulator efficiency	Modulation scheme	Efficiency	Ideal	C/N	Section	Signaling
0.62	MSK	1.8	2	10.5	6.4.6	Raised cosine($\alpha = 0.25$)
0.59	Offset QPSK/SQPSK	1.8	2	13.5	6.3.7	Raised cosine($\alpha = 0.25$)
0.59	QPSK	1.8	2	13.8	6.3.3	Raised cosine($\alpha = 0.25$)
0.59	8–PSK	2.7	3	19.0	6.3.3.3	Raised cosine($\alpha = 0.25$)
0.59	16–SQAM/16–offset–QASK	3.6	4	20.9	6.5.5	Raised cosine($\alpha = 0.25$)
0.52	MSK	2.1	2	12.8	6.4.6	Partial Response
0.64	Offset QPSK/SQPSK	2.1	2	15.8	6.6.1	Partial Response
0.64	QPSK	2.1	2	15.9	6.6.1	Partial Response
0.64	16–SQAM	4.2	4	22.8	6.6.1	Partial Response

6.7.3 Effects of interference

Another factor in evaluating potential modulation schemes for digital radio is the effect of co-channel and adjacent-channel interference. As discussed in Section 6.7.2, MSK and the other CP–FSK schemes have an advantage over the AM and PM schemes, when no post-modulation filtering or spectral shaping is employed. Another aspect of adjacent-channel interference is the amount of performance degradation caused by a specific level of interference. This subject will be dealt with in greater detail in Chapter 10. It suffices to say for the moment that an in-band carrier wave interferer, which is 10 to 15 dB down in power from the desired receive signal to produce a bit error rate of 10^{-4}, will have the least effect on noncoherent FSK and BPSK schemes, while the 8-ary and 16-ary schemes exhibit the most degradation from the ideal performance.

6.7.4 Effects of fading

Fading is another problem which is discussed in Chapter 8. if the fading is caused by two resolvable multipath components, using the two-ray path model, the secondary path may be considered as a carrier wave interferer and the results of Section 6.7.3 would apply. If the fading is flat Rayleigh fading, error-control coding may have to be used to overcome the degradation in bit error rate. Rain fading will also produce an effect on the BER. For the initial quick estimation, the results of the relative performance parameters indicated by Table 6.5 gives an indication of the performance to be expected for slow fading where the decision threshold of AM/PM schemes may have time to adjust, or the equalizers have time to compensate for the effects.

6.7.5 Effects of delay distortion[10,77]

Most of the delay distortion observed on line-of-sight radio links is introduced by the radios and not by the channel. If the distortion observed on the received signal

Table 6.7 Performance of representative modulation schemes in the presence of delay distortion ($m/T = 1$) for a bit error rate of 10^{-4}

Type	Modulation scheme	C/N, dB Quadratic	Linear
AM	Coherent ASK	12.8	12.4
	Noncoherent ASK	13.3	16.9
	QAM	12.8	18.8
	QPR	–	>28
FM	Coherent BFSK ($m = 0.7$)	*8.8	*8.6 (assumes 3 bit observation period)
	Noncoherent detection FSK ($m = 1$)	13.5	16 (interval)
	Noncoherent CP–FSK ($m = 0.7$)	*10.2	*12.7
	MSK ($m = 0.5$)	9.8	15.8
	MSK–differential encoding ($m = 0.5$)	10.8	16.8
PM	Coherent BPSK	9.8	9.6
	Differentially encoded BPSK	10.3	10.1
	Differential PSK (DPSK)	11.8	–
	QPSK	12.8	18.8
	DQPSK	19.3	–
	Offset QPSK	12.8	18.8
	8–PSK coherent detection	< 30	30

*Assumes an observation period over 3 bits.

can be modeled by passage of the transmitted signal through a linear filter, the delay distortion is the derivative of the phase-frequency characteristic of the filter. Table 6.7 is taken from Reference 10 and is based mainly on results derived from Sunde's comprehensive treatment of the effects of delay distortion on pulse transmission.[77] The performance of the representative systems is shown for quadratic and linear delay distortion for the case in which the maximum differential delay, which is relative to the mid-band delay, is equal to the symbol duration. It may be noted that some modulation schemes are severely affected by one type of distortion or the other. Differentially detected QPSK (DPSK) schemes suffer severe degradation from quadratic delay distortion, while QAM, MSK and variations of QPSK are degraded significantly by linear delay distortion.

6.7.6 Comparisons of some commercially available digital radios

Table 6.8 shows the modulation methods used in various types of commercially available radio equipment. To make the results more beneficial, the equipment type, frequency range, data rate, modulation method, power output, system gain, and receiver threshold for a BER of 10^{-3} have been included. The results are based on

Table 6.8 Comparison of commercially available equipment

Model	Frequency range, GHz	Data rate, Mbit/s	Modulation method	Power output, dBm	System gain, dB/10^{-4}	Receiver threshold, dBm at BER of 10^{-3} (unless stated otherwise)
Manufacturer: Ayadin Microwave Division, 75 East Trimble Road, San José, California, U.S.A.						
AMLD16/F2	1.7/1.9–2.3	1 × 34	16-QAM	32	114.6	−82.6
	1.9–2.3	2 × 34	16-QAM	32	111.6	−79.6
	1.9–2.3	3 × 34	16-QAM	32	108.8	−76.8
	2.5–2.69	1 × 34	16-QAM	32	114.6	−82.6
AMLD16/F3	3.8–4.2	1 × 34	16-QAM	37	118.6	−81.6
	3.8–4.2	2 × 34	16-QAM	37	115.6	−78.6
	3.8–4.2	3 × 34	16-QAM	37	112.8	−75.8
AMLD16/F4	4.4–5.0	1 × 34	16-QAM	33	113.6	−80.6
	4.4–5.0	2 × 34	16-QAM	33	110.5	−77.5
	4.4–5.0	3 × 34	16-QAM	33	108.8	−75.8
	6.43–7.11	1 × 34	16-QAM	35	115.6	−80.6
	6.43–7.11	2 × 34	16-QAM	35	112.6	−77.6
	6.43–7.11	3 × 34	16-QAM	35	110.8	−75.8
	6.43–7.11	4 × 34	16-QAM	35	109.6	−74.6
AMLD16/F7	7.125–8.5	1 × 34	16-QAM	30	110.1	−80.1
	7.125–8.5	2 × 34	16-QAM	30	107.1	−77.1
	7.125–8.5	3 × 34	16-QAM	30	104.3	−74.3
	7.725–8.275	1 × 34	QAM	30	110.1	−80.1
	7.725–8.275	2 × 34	QAM	30	107.1	−77.1
	7.725–8.275	3 × 34	QAM	30	104.3	−74.3
	7.725–8.275	4 × 34	QAM	30	104.1	−74.1
AMLD16/F11	10.7–11.7	1 × 34	QAM	36	116.1	−80.1
	10.7–11.7	2 × 34	QAM	36	113.1	−77.1
	10.7–11.7	3 × 34	QAM	36	110.3	−74.3
	10.7–11.7	4 × 34	QAM	36	110.1	−74.1
AMLD16/F12	12.75–13.25	1 × 34	QAM	30	110.1	−80.1
	12.75–13.25	2 × 34	QAM	30	107.1	−77.1
	12.75–13.25	3 × 34	QAM	30	104.3	−74.3
AMLD16/F14	14.4–15.3	1 × 34	QAM	33	111.1	−78.1
	14.4–15.3	2 × 34	QAM	33	108.1	−75.1
	14.4–15.3	3 × 34	QAM	33	105.3	−72.3
	14.4–15.3	4 × 34	QAM	33	105.1	−72.1
AMLD16/F18	17.7–19.7	1 × 34	QAM	27	105.1	−78.1
	17.7–19.7	2 × 34	QAM	27	102.1	−75.1
	17.7–19.7	2 × 34	QAM	27	99.3	−72.3
Note. See manufacturer's literature for other types and ranges as this is not an exhaustive list.						
Manufacturer: L. M. Ericson, Telefonaktiebolaget L. M. Ericson, Defence and Space Systems Division, P.O. Box 1001, 431 26 Molndal, Sweden.						
Mini Link 15	14.5–15.35	2 or 8	4-level FSK	15	–	−80 at BER of 10^{-4}
Mini Link 18	18.58–18.82	2 or 8	4-level FSK	10	–	−79 at BER of 10^{-4}
Note. See manufacturer's literature for other types and ranges at this is not an exhaustive list.						
Manufacturer: GEC, GEC Telecommunications Ltd, Transmission Division, P.O. Box 53, Coventry, CV1 2HJ, England.						
2 GHz	1.713–1.909	2 or 8	3-level FSK	24/33/36	–	−84.5 at BER of 10^{-4}
4 GHz	3.6–4.2	140	RBQPSK[78]	33/36	–	−70.5 at BER of 10^{-4}
6 GHz	5.85–6.425	140	RBQPSK[78]	33/36	–	−70.5 at BER of 10^{-4}
11 GHz	10.7–11.7	140	QPSK	33/40	–	−69 at BER of 10^{-5} −72
19 GHz	17.7–19.7	140	QPSK	20	–	−74
19DR2/8	18.58–18.82	2 or 8	3 level FSK	17	–	−81 or −77

Table 6.8 (*cont*)

Model	Frequency range, GHz	Data rate, Mbit/s	Modulation method	Power output, dBm	System gain, dB/10^{-4}	Receiver threshold, dBm at BER of 10^{-3} (unless stated otherwise)
Manufacturer: GTE, GTE Telecomunicaziona SpA, 20060 Cassina dei Pecchi, Milan, Italy.						
CTR 150	1.7–2.3/2.3–2.44	8.4/17/34	4–PSK (RF)	17/27/33		
CTR 1448	3.4–4.2	34	4–PSK (IF)	30/37		
CTR 1448–5	4.4–5.0	34	4–PSK (IF)	27/33		
CTR 147C	5.9–7.1	34	4–PSK (IF)	27/33		
CTR 147S	7.1–8.5	34	4–PSK (IF)	7G-29/33/39* 8G-28/39*		
CTR 177	7.1–8.5	70	8–PSK	28/37		
CTR 148S	10.7–11.7	34	4-PSK (IF)	27/30/41		
CTR 178	10.7–11.7	70	8–PSK	28/37		
CTR 188	10.7–11.7	140	16–QAM	25/33		
CTR 152	12.75–13.25	8/17/34	4–PSK	20/25		
CTR 155	14.5–15.35	8/17/34	4–PSK	17/23		−83/−80.5/−77.5 at BER of 10^{-4}
CTR 158	17.7–19.7	8/17/34	4–PSK	13		−78 for 34 Mbit/s
Note: 7G means 7 GHz radios, 8G means 8 GHz radios						
Manufacturer: Loral-Terrocom, Loral Microwave Communications, 9020 Balboa Avenue, San Diego, California, U.S.A.						
LMC 408D	7.125–8.500	Up to 20	FSK	28	102	−79 at BER of 10^{-6}
Hot standby	7.125–8.500	Up to 20	FSK	28	98	−79 at BER of 10^{-6}
Diversity	7.125–8.500	Up to 20	FSK	28	101	−79 at BER of 10^{-6}
LMC 412D	12.2–12.7	Up to 20	FSK	19	91	−74 at BER of 10^{-6}
Hot standby	12.2–12.7	Up to 20	FSK	19	87	
LMC 418D	17.7–19.7	Up to 20	FSK	23	91.5	−71 at BER of 10^{-6}
Hot standby	17.7–19.7	Up to 20	FSK	23	90	
Manufacturer: NERA, Elektrisk Bureau, Nera Divsion, P.O. Box 10, N5061, Tokstad, Norway.						
NL 104	1.7–2.1	2	2–PSK	27		−86
NL 123	2.1–2.3	8	2–PSK	27		−80
NL 111	1.7–2.3	8	4PSK	30		−86 or −89
NL 143	1.7–2.3	34	4PSK	30		−80 or −83
NL 146	3.6–4.2	34	4PSK	30		−83
NL 110	4.4–5.0	8	4PSK	30		−86
NL 142	4.4–5.0	34	4PSK	30		−80
NL 144	6.43–7.11	34	4PSK	20/28		−83
NL 175	6.43–7.11	140	16–QAM			
NL 112	7.125–7.725	8	4–PSK	20/28		−86
NL 141	7.125–7.725	34	4–PSK	20/28		−80
NL 145	7.7–8.5	34	4–PSK	28		−80 or −83
NL 102	12.75–13.25	2	2–PSK	20		−88
NL 133	12.75–13.25	8	2–PSK	20		−81
NL 132	12.75–13.25	34	4–PSK	20		−77
NL 148	12.75–13.25	34	4–PSK	20 or 27		−83
NL 106	14.5–15.35	2	2–PSK	20		−88 at BER of 10^{-5}
NL 134	14.5–15.35	8	2–PSK	20		−81 at BER of 10^{-5}
Manufacturer: Rockwell Collins, Collins Transmission Systems Division, Rockwell International, P.O. Box 10462, Dallas, Texas 75207, U.S.A.						
MDR 2204	3.7–4.2	45	16/64–QAM	33	102	
MDR 6	5.9–6.4	45	8–PSK	37	103 at BER of 10^{-6}	
MDR 2306	5.9–6.4	45	16/64–QAM	33	101	
MDR 8	7.13–8.5	45	8–PSK	37	105 at BER of 10^{-6}	
MDR8–5N	7.13–8.5	45	8–PSK	37	109 at BER of 10^{-6}	
MDR11–5	10.7–11.7	45	8–PSK	37	108 at BER of 10^{-6}	
MDR11–5N	10.7–11.7	45	8–PSK	37	109 at BER of 10^{-6}	

Table 6.8 (*cont*)

Model	Frequency range, GHz	Data rate, Mbit/s	Modulation method	Power output, dBm	System gain, dB/10^{-4}	Receiver threshold, dBm at BER of 10^{-3} (unless stated otherwise)
MDR 2311	10.7–11.7	45	16/64–QAM	36	107	
MDR 2411	10.7–11.7	90	16/64–QAM	33	101	
MDR12–5	12.2–12.7	90	8–PSK	37	109 at BER of 10^{-6}	
Manufacturer: SIAE Microelettronica SpA, Ferranti Industrial Electronics Ltd., Communications Systems Group, Bellesk House, Graton Road, Edinburgh EH5 1RD, U.K.						
RT 21	1.400–1.550	2 × 2	QPSK	29/35		−94 at BER of 10^{-4}
RT 22	2.300–2.440	2 × 2	QPSK	29/35		−94 at BER of 10^{-4}
RT 23	2.500–2.700	2 × 2	QPSK	29/35		−94 at BER of 10^{-4}
RT 26	14.50–14.62	2 × 2	QPSK	29/35		−94 at BER of 10^{-4}
Manufacturer: Telenokia, P.O. Box 33, 02601 Espoo 60, Finland.						
DR0.7–1500	1.429–1.525	0.7	MSK	25/33/38		−99
DR2–1500	1.429–1.525	2	MSK	25/33/38		−95
DR2 + 2–1500	1.429–1.525	2 or 2 + 2	MSK	25/33/38		−92
DR2 + 2–2600	2.486–2.686	2 or 2 + 2	MSK	25/33/38		−92
DR240–1800	1.7–2.3	8 + 8/8/2 + 2	4–PSK	26/35		−83/88
DR34 + 34.6800	6.430–7.110	34 + 34/34 + 8	QPSK	20/30		−77/80
Manufacturer: Telettra, Telefonia Elettronica e Radio SpA, 20092 Cinisello Balsamo (MI)-Viale F. Testi, 136, Italy.						
HA–2	1.7–2.3	8	QPSK	28		−88
HA–2	1.7–2.3	34	QPSK	33		−84.5
HTN-6u	6.4–7.1	140	16–QAM	27		−72
HA–6u	6.4–7.1	8/34	QPSK	24.5/28/33		−86/82.5
HMN–7	7.1–7.7	2 × 34	16–QAM	28		−74
HTN–11	10.7–11.7	140	16–QAM	33		−71
HMN–11	10.7–11.7	2 × 34	16–QAM	26		−74
HN–13	12.7–13.2	34	QPSK	25		−79 at BER of 10^{-4}
HD3–13	12.7–13.2	2/8	2–PSK	14		−89/84 at BER of 10^{-4}
HN–15	14.3–15.4	2/8	2–PSK	15/20		−89/84
HN–18	17.7–18.7	34	QPSK	18		−78 at BER of 10^{-4}
MP–18	18.82–18.92	2	QPSK	15		−85 at BER of 10^{-4}
MP–18	19.16–19.26	2	QPSK	15		−85 at BER of 10^{-4}
Manufacturer: Thorn-EMI Electronics Ltd, Communications Division, Wells, Somerset BA5 1AA, England						
Protected link						
ML 13–2	13	2		19		−81
ML 13–8	13	8		19		−74
ML 13–34	13	34		18		−70
Unprotected link						
ML 13–2	13	2		20		−85
ML 13–8	13	8		20		−79
Manufacturer: Plessey U.K. Ltd, Plessey Radio Systems, Martin Road, West Leigh, Havant, Hampshire PO9 5DH, England.						
DRS 1302	13	2		15		Unknown
DRS 1308	13	8		23		−77.5
DRS 1334	13	34		22		−75
Manufacturer: NEC Corporation, NEC Building, 33–1 Shiba 5-chome, Minato-ku, Tokyo 108, Japan						
TRP–13GD 34MB	13	34	16–QAM	25/30		−77
TRP–15G140MB	15	140	16–QAM	23		−72

a brief survey during the early part of 1986 and do not include, by any means, all of the commercially available equipment.

The above list of manufacturers and their various types of equipment is not complete. Further information can be obtained by writing to the manufacturers themselves. Omissions of manufacturers, of which there are possibly around one hundred, arises from lack of space and of information to hand.

6.7.7 CCIR comments on different modulation schemes

CCIR Report[79] 610–1, Section 2, deals with digital modulation methods for compatible operation with FDM–FM radio relay systems with respect to any special problems of interference between them. The comments made which are relevant to this chapter are listed under the various subsections of Section 6.7.7.

6.7.7.1 FM–DPSK

The attractiveness of this modulation method lies principally in that it permits the use of digital terminals on existing analog FM radio systems. This is because the signal is completely compatible with the properties of the analog FM channel and could operate through a number of FM repeaters before regeneration. In addition, the FM approach has an advantage, from the standpoint of the RF signal bandwidth, since it is possible to restrict, within limits, the transmitted spectrum by shaping the baseband pulses applied to the FM modulator.

6.7.7.2 CPSK modulation

Coherent detection of the CPSK modulated signal at the receiver yields a 2.3 dB E_b/η advantage over the FM–DPSK arrangement in a noise-limited environment. The coherent QPSK system using path-length rectangular-pulse switching requires all band shaping to be done by RF filters.

6.7.7.3 Quadrature partial response system (QPRS)

The PSK systems described above require the use of both polarizations to achieve the required transmission efficiency of 2 bits/s/Hz as recommended by the CCIR (Report 610–1, Section 2). Satisfactory digital transmission performance will probably require, however, that the cross-polarization discrimination (XPD) between the two PSK channels be maintained above about 15 dB during multipath fading. Even better XPD performance will be required for the FM–DPSK arrangement because its co-channel interference immunity is worse. The QPRS modulation scheme avoids the necessity for co-channel plans since, by using partial response coding, the second-order side lobes of the combined AM–PM signal at the transmitter can be attenuated to achieve about 2 bits/s/Hz on a single polarization without excessive inter-symbol interference or interference into adjacent RF channels. This is achieved at the expense of introducing envelope fluctuations in the carrier signal and of some 3–6 dB threshold degradation over coherent QPSK modulation (see Section 6.6.3). Such a plan would require signal regeneration at every repeater point since maintaining amplitude linearity would not be an easy task.

6.7.7.4 FSK-multi-level modulation (*M*-ary FSK)

This method has advantages over PSK in some circumstances, as previously mentioned. It is possible to restrict the transmitted spectrum by shaping the baseband signal and/or reducing the deviation of the carrier. Also, like FM–DPSK modulation it has the advantage of being able to use existing FDM–FM equipment. FSK is, however, particularly sensitive to interference as the number of levels is increased.

6.7.7.5 Coherent 8–PSK

This method is attractive due to its compatibility with existing FDM–FM equipment. Also it is possible to achieve good spectrum efficiency (see Table 6.6) while operating on a single polarization.

6.7.7.6 16–QAM

This method is attractive for high-capacity radio-relay systems, from a compatibility point of view, because it permits increased spectrum efficiency and should allow the analog channel arrangements to be used for digital systems of similar capacity.

6.7.8 CCIR modulation comparisons

CCIR Report 378–4 discusses the characteristics of digital radio-relay systems. Of particular relevance to this chapter is the comparison of the different modulation schemes. The report presents a table of comparisons of various modulation schemes along the lines of that presented in Table 6.5. The points of interest to note about the characteristics of the different modulation methods are:

1. Amplitude modulation, full carrier, double-sideband with envelope detection is simple but wasteful of bandwidth and power and is therefore not considered suitable for other than small-capacity requirements.
2. Frequency modulation, two-level with discriminator detection, is simple but wasteful of bandwidth and is also not considered suitable for other than very-small-capacity requirements.
3. Phase modulation, two-level, with coherent or differentially coherent detection, is fairly simple but wasteful of bandwidth. The method is considered the most suitable for small-capacity radio relay systems for the transmission of digital signals with gross bit rates up to and including 10 Mbit/s and possibly for medium-capacity radio relay systems with gross bit rates ranging from 10 to about 100 Mbit/s.
4. Phase modulation, four-level, with coherent or differentially coherent detection, is considered to be among the most suitable methods for medium- and high-capacity systems. It is also acceptable for small-capacity systems. A high-capacity system has a bit rate greater than 100 Mbit/s.
5. Frequency modulation, three-level (duo-binary), four-level or eight-level discriminator detection, is preferred for the simple adaptation to digital transmission of frequency-modulation radio-relay systems using frequency division multiplex (FDM). A large C/N will be available and the method of modulation need not be changed.

6. Phase modulation, eight level, with coherent detection, is particularly suitable for medium-capacity systems, operating below 12 GHz. This method is attractive when systems are to be mixed within existing analog channel arrangements because it offers good spectrum efficiency even when using a single polarization (see CCIR Report 610). Although theoretically more sensitive to distortion than coherent QPSK, comparable performance can be achieved with this method by optimal equipment filtering and equalization.
7. The use of partial response coding techniques may be very attractive to reduce bandwidth occupancy with only a small increase in equipment complexity. This reduction in bandwidth is achieved at the expense of a greater number of transmitted levels for a given number of input levels, or equivalent, at a greater value of C/N for a given error rate.
8. A 16–level QAM modulation technique is attractive for high-capacity digital radio-relay systems because it increases spectrum efficiency with only a small increase in equipment complexity.

EXERCISES

1 Determine the C/N per bit advantage of M-ary coherent PSK over M-ary DPSK for large C/N. (*Answer* 2.3 dB)

2 Determine the C/N advantage of 2–DPSK over NC–ASK for a probability of error of 10^{-6}. (*Answer* approx. 3 dB)

3 Raised-cosine signaling utilizes less bandwidth than rectangular signaling. Determine the BPSK 99 percent bandwidth ratio for raised-cosine signaling to rectangular signaling. (Answer $19.3r_b/2.96r_b = 6.52$)

4 Determine the theoretical bandwidth efficiency of 256–QAM. (*Answer* 8 bits/s/Hz)

5 How can a bandwidth efficiency of 2 bits/s/Hz be obtained using BPSK? (*Answer* cross-polarized channels)

6 To obtain offset QPSK, how much delay is used in the Q baseband path? (*Answer* 1 input data bit duration)

7 At what points do the spectral nulls of an 8–PSK modulated signal occur? (*Answer* $\pm k/3T$, where $k = 1, 2, \ldots$)

8 Determine the advantages of FSK over ASK. (*Answer* see Section 6.4.4)

9 What is the theoretical efficiency of coherent 8–FSK? (*Answer* 0.75 bits/s/Hz)

10 What is the theoretical bandwidth efficiency of BFSK? (*Answer* 1 bit/s/Hz)

11 What is the bandwidth efficiency of MSK using raised-cosine signaling, assuming a filter roll-off of 0.25? (*Answer* 1.8 bits/s/Hz)

12 Why is Gray coding used in say 16–QAM systems? (*Answer* see Section 6.5.3)

13 Determine the power increase for 256–QAM over 16–QAM for a bit error rate of 10^{-4}. (*Answer* 12.3 dB)

14 Determine the bandwidth efficiency ratio for 256–QAM and 16–QAM. (*Answer* 3 dB)

15 Determine the ratio of the power requirements of 64–PSK and 64–QAM. (*Answer* 9.95)

16 For offset QAM how much delay is required in the Q baseband path? (*Answer* half an input bit duration)

17 What would 8–PSK be used for? (*Answer* medium-capacity systems below 12 GHz – see Section 6.7.8)

18 Determine the double-sided Nyquist bandwidth of a 16–QAM signal operating at a bit rate of 140 Mbit/s. (*Answer* 35 MHz)

19 If the received C/N is 20.9 dB for the Nyquist filter case of Exercise 18, what would be the new C/N for the same P_e, if a filter with a roll-off of 0.50 was used instead of a Nyquist filter? (*Answer* double-sided noise bandwidth becomes 52.5 MHz, and new $C/N = 20.9 + 10 \log 52.5/35 = 22.66$ dB)

20 What effect does the duobinary conversion have on a rectangular signal waveform? (*Answer* see Section 6.6)

21 What is the fundamental difference between filters used in the baseband and those used in the RF output section? (*Answer* baseband are low-pass, RF are band-pass)

REFERENCES

1. Harman, W. W., *Principles of the Statistical Theory of Communications* (McGraw-Hill, 1963).
2. Proakis, J. G., *Digital Communications* (McGraw-Hill, 1983).
3. Arthurs, E. and Dym, H., On the Optimum Detection of Digital Signals in the presence of White Gaussian Noise – A Geometric Interpretation and a Study of three Basic Data Transmission Systems, *IRE Transactions on Communications Systems*, December 1962, pp. 336–372.
4. Shanmugam, K. S., *Digital and Analogue Communication Systems* (Wiley, 1979).
5. Carlson, A. B., *Communication Systems, 2nd Ed.* (McGraw-Hill, 1975).
6. Ekanayake, N. and Taylor, D. P., 'Binary CPSK Performance Analysis for Saturating Band-limited Channels', *IEEE Transactions on Communications,* 1979, **COM–27,** No. 3.
7. Tan, B. T. and Tjhung, T. T., 'On Binary DPSK Error Rates due to Noise and Intersymbol Interference', *ibid.*, 1983, **COM-31** No. 3, pp. 463–466.
8. Pawula, R. F., 'On M-ary DPSK Transmission over Terrestrial and Satellite Channels', *ibid.,* 1984, **COM–32,** No. 7, pp. 752–761.
9. Korn, I., 'Error Probability and Bandwidth of Digital Modulation, *ibid.*, 1980, **COM–28**, No. 2, pp. 287–290.
10. Oetting, J. D., 'A Comparison of Modulation Techniques for Digital Radio', *ibid.*, 1979, **COM–27**, No. 12, pp. 1752–1762.
11. Costas, J. P., 'Synchronous Communications', *Proc. IRE,* December, 1956, pp. 1713–1718.
12. Bussgang, J. J. and Leiter, M., 'Phase Shift Keying with a Transmitted Reference', *IEEE Transactions on Communication Technology,* 1966, **COM–14,** No. 1, pp. 14–20.
13. Fiestel, C. H. and Gregg, W. D., 'Performance Characteristics of Adaptive/Self-Synchronizing PSK Receivers under Common Power and Bandwidth Conditions', *ibid.*, 1970, **COM–18**, No. 5, pp. 527–536.
14. Oberst, J. F. and Schilling, D. I., 'Performance of Self-Synchronized Phase-Shift Keyed Systems', *ibid.*, 1969, **COM–17**, No. 6, pp. 664–669.
15. Pierce, J. R., 'Comparison of Three-phase Modulation with Two-phase and Four-phase Modulation', *IEEE Transactions on Communications,* Correspondence, 1980, **COM–28**, No. 7, pp. 1098–1099.

16. Chie, M. C. and Tsang, C-S., 'The Effects of Transmitter/Receiver Clock Time Base Instability on Coherent Communication System Performance', *ibid.*, 1982, **COM–30**, No. 3, pp. 510–516.
17. Polydoros, A., 'Effect of Data Asymmetry on Unbalanced QPSK Signals with Noise Phase Reference, *ibid.,* 1982, **COM–30**, No. 7, pp. 1735–1746.
18. Ashlock, J. C. and Posner, E. C., 'Application of the Statistical Theory of Extreme Values to Spacecraft Receivers', Jet Propulsion Laboratories, Pasadena, California, Technical Report 32–737, May 1965.
19. Robinson, W. G., 'Maintenance of Digital Radio using Remoted Performance Monitoring', *Links for the Future,* IEEE International Conference on Communications, 1984, pp. 456–461.
20. Gooding, D. J., 'Performance Monitor Techniques for Digital Receivers based on Extrapolation of Error Rate', *IEEE Transactions on Communication Technology,* 1968, **COM–16**, No. 3, pp. 380–387.
21. Rosenbaum, A. S., 'Binary PSK Error Probabilities with Multiple Co-channel Interferences', *ibid.,* 1970, **COM–18**, pp. 241–253.
22. Rosenbaum, A. S., 'Error Performance of Multiphase DPSK with Noise and Interference', *ibid.*, December 1970, pp. 821–824.
23. Morais, D. H. and Feher, K., 'Bandwidth Efficiency and Probability of Error Performance of MSK and Offset QPSK Systems', *IEEE Transactions on Communications,* 1979, **COM–27**, No. 12, pp. 1794–1801.
24. Morais, D. H., Sewerinson, A. and Feher, K., 'The Effects of the Amplitude and Delay Slope Components of Frequency Selective Fading on QPSK, Offset QPSK and 8 PSK Systems', *ibid.*, 1979, **COM–27**, No. 12, pp. 1849–1853.
25. Gronemeyer, S. and McBride, A., 'MSK and Offset QPSK Modulation', *ibid.*, 1976, **COM–24**, No. 8, pp. 809–820.
26. Anderson, R. R. and Salz, J., Spectra of Digital FM', *Bell System Technical Journal*, 1965, **44**, pp. 1165–1189.
27. Shimbo, O., Ohira, T. and Nitadori, K., 'A General Formula on Power Spectra of Pulse-Modulated Signals and its Applications to PCM Transmission Problems, *Journal of the Institution of Electrical Communication Engineering Japan*, 1963, **46**, pp. 9–17.
28. Shimbo, O., 'General Formula for Power Spectra of Digital FM Signals', *Proc. IEE,* 1966, **113**, pp. 1783–1789.
29. Marshall, G. J., Power Spectra of Digital PM signals, *ibid.,* 1970, **117**, pp. 1909–1914.
30. Sunde, E. D., 'Ideal Binary Pulse Transmission by AM and FM', *Bell System Technical Journal,* 1959, **38**, pp. 1357–1426.
31. Lender, A., 'Binary Orthogonal FM Technique with Multiple Properties', *IEEE Transactions on Communication Technology*, 1965, **COM–13**, No. 4, pp. 499–503.
32. Tjhung, T. T. and Wittke, P. H., 'Carrier Transmission of Binary Data in a Restricted Band', *ibid.,* 1970, **COM–18**, pp. 295–304.
33. Milstein, L. B., Pickholtz, R. L. and Schilling, D. L., 'Optimization of the Processing Gain of an FSK-FH System', *IEEE Transactions on Communications*, 1965, **COM–28**, No. 7, pp. 1062–1079.
34. Panter, P. F., *Modulation, Noise and Spectral Analysis* (McGraw-Hill, 1965).
35. Akima, H., 'The Error Rates in Multiple FSK Systems', National Bureau of Standards Technical Note 167, March 1963.
36. De Buda, R., 'Coherent Demodulation of Frequency-Shift Keying with Low Deviation Ratio', *IEEE Transactions on Communications,* Part I, 1972, **COM–20**, pp. 429–435.
37. Rhodes, S. A., 'Effect of Noisy Phase Reference on Coherent Detection of Offset-QPSK Signal', *ibid.*, 1974, **COM–22,** No. 8, pp. 1046–1055.
38. Matyas, R., 'Effect of Noisy Phase References on Coherent Detection of FFSK Signals', *ibid.,* 1978, **COM–26**, No. 6, pp. 807–815.
39. Lender, A., 'A Synchronous Signal with Dual Properties for Digital Communications', *IEEE Transactions on Communication Technology,* 1965, **COM–13**, pp. 202–208.

40. van den Elzen, H. C. and van der Wurf, P., 'A Simple Method of Calculating the Characteristics of FSK Signals with Modulation Index 0.5', *IEEE Transactions on Communications,* 1972, **COM–20**, No. 2, pp. 130–147.
41. Chan, H. C., Taylor, D. P. and Haykin, S. S., 'A Comparative Study of Digital Modulation Techniques for Single-Carrier Satellite Communications', 3rd International Conference on Digital Satellite Communications, Kyoto, Japan, November 1975.
42. Masamura, T., Samejima, S., Morihiro, T. and Fuketa, H., 'Differential Detection of MSK with Nonredundant Error Correction', *IEEE Transactions on Communications,* 1979, **COM–27**, No. 6, pp. 912–918.
43. Rabzel, M. and Pasupathy, S., 'Spectral Shaping in Minimum Shift Keying (MSK)-Type signals', *ibid.*, 1978, **COM–26**, No. 1, pp. 190–195.
44. Bayless, J. W. and Pedersen, R. D., 'Efficient Pulse Shaping using MSK or PSK Modulation', *ibid.,* 1979, **COM–27**, No. 6, pp. 927–931.
45. Bazin, B., 'A Class of MSK Baseband Pulse Formats with Sharp Spectral Roll-off', *ibid.,* 1979, **COM–27**, No. 5, pp. 826–829.
46. Simon, M. K., 'A Generalization of Minimum-Shift-Keying (MSK)-Type Signaling Based upon Input Data Symbol Pulse Shaping', *ibid.*, 1976, **COM–24**, No. 8, pp. 845–856.
47. Amoroso, F., 'Pulse and Spectrum Manipulation in the Minimum (Frequency) Shift Keying (MSK) Format', *ibid.,* 1976, **COM–24**, No. 3, pp. 381–384.
48. Kalet, I., 'A look at Crosstalk in Quadrature-Carrier Modulation Systems', *ibid.*, 1977, **COM–25**, No. 9, pp. 884–892.
49. Metzger, L. S., 'Performance of Phase Comparison Sinusoidal Frequency Shift Keying', *ibid.*, 1978, **COM–26**, No. 8, pp. 1250–1253.
50. Anderson, J. B. and Taylor, D. P., 'A Bandwidth-Efficient Class of Signal-Space Codes', *IEEE Transactions on Information Theory,* 1978, **IT–24**, pp. 703–712.
51. Anderson, J. B. and de Buda, R., 'Better Phase-Modulation Error Performance using Trellis Phase Codes', *Electronics Letters,* 1976, **12**, pp. 587–588.
52. Mazur, B. A. and Taylor, D. P., 'Demodulation and Carrier Synchronization of Multi-*h* Phase Codes', *IEEE Transactions on Communications,* 1981, **COM–29**, No. 3, pp. 257–266.
53. Wilson, S. G. and Gaus, R. C., 'Power Spectra of Multi-*h* Phase Codes', *ibid.,* 1981, **COM–29**, No. 3, pp. 250–256.
54. Cahn, C. R., 'Combined Digital Phase and Amplitude Modulation Communication Systems', *IRE Transactions on Communication Systems,* 1960, **CS–8**, No. 3, pp. 150–155.
55. Hancock, J. C. and Lucky, R. W., 'Performance of Combined Amplitude and Phase-Modulated Communication Systems', *ibid.*, 1960, **CS–8**, No. 4, pp. 232–237.
56. Campopiano, C. N. and Glazier, B. G., 'A Coherent Digital Amplitude and Phase Modulation Scheme', *ibid.,* 1962, **CS–10**, pp. 90–95.
57. Horikawa, I., Murase, T. and Saito, Y., 'Design and Performances of a 200 Mbit/s 16 QAM Digital Radio System', *IEEE Transactions on Communications*, 1979, **COM–27**, No. 12, pp. 1953–1958.
58. Dupuis, P., Joindot, M., Leclert, A. and Soufflet, D., '16 QAM Modulation for High Capacity Digital Radio System', *ibid.,* 1979, **COM–27**, No. 12, pp. 1771–1781.
59. Weber, W. J., 'Differential Encoding for Multiple Amplitude and Phase Shift Keying Systems', *ibid.,* 1978, **COM–26**, No. 3, pp. 385–391.
60. Lindsey, W. C. and Simon, M. K., 'On the Detection of Differentially Encoded Polyphase Signals', *ibid.*, 1972, **COM–20**, No. 6, pp. 1121–1128.
61. Gitlin, R. D. and Ho, E. Y., 'The performance of Staggered Quadrature Amplitude Modulation in the presence of Phase Jitter', *ibid.*, 1975, **COM–23**, No. 3, pp. 348–352.
62. Saltzberg, B. R., 'Performance of an Efficient Parallel Data Transmission System', *IEEE Transactions on Communication Technology,* 1967, **COM–15**, No. 6, pp. 805–811.
63. Seo, J. S. and Feher, K., 'Performance of SQAM Systems in a Nonlinearly Amplified Multichannel Interference Environment', *Proc. IEE*, 1985, **132**, Part F, No. 3, pp. 175–180.

64. Nyquist, H., 'Certain Factors Affecting Telegraph Speed', *Bell System Technical Journal,* 1924, **3**, pp. 324–326.
65. Sekey, A., 'An Analysis of the Duobinary Technique', *IEEE Transactions on Communication Technology*, 1966, **COM–14**, No. 2, pp. 126–130.
66. Lender, A., 'The Duobinary Technique for High-Speed Data Transmission', *IEEE Transactions on Communications and Electronics*, 1963, **82**, pp. 214–218.
67. Lender, A., 'Faster Digital Communication with Duobinary Techniques', *Electronics,* 1963, **36**, pp. 61–65.
68. Prabhu, V. K., 'PSK-Type Modulation with Overlapping Baseband Pulses', *IEEE Transactions on Communications,* 1977, **COM–25**, No. 9, pp. 980–990.
69. Greenstein, L. J., 'Spectra of PSK Signals with Overlapping Baseband Pulses', *ibid.*, 1977, **COM–25,** No. 5, pp. 523–530.
70. Austin, M. C. and Chang, M. U., 'Quadrature Overlapped Raised-Cosine Modulation', *ibid.,* 1981, **COM–29**, No. 3, pp. 237–249.
71. Sasase, I. and Mori, S., 'Performance of Staggered QOSRC Modulation on a Hard-limited Satellite Channel in the presence of Uplink and Downlink Noise and Intersymbol Interference', *Links for the Future,* IEEE International Conference on Communications, 1984, pp. 941–945.
72. Muilwijk, D., 'Correlative Phase Shift Keying – A class of Constant Envelope Modulation Techniques', *IEEE Transactions on Communications*, 1981, **COM–29**, No. 3, pp. 226–236.
73. Sundal, G. J., Error Rate of QPRS evaluated in Amplitude-Phase Space', *ibid.*, 1979, **COM–27**, No. 12, pp. 1802–1805.
74. Lender, A., 'Correlative (Partial Response) Techniques and Applications to Digital Radio Systems', Chapter 7 of *Digital Communications – Microwave Applications* (Prentice-Hall, 1981) by Feher, K.
75. Shehadeh, N. M. and Chiu, R. F., 'Spectral Density Functions of a Carrier Amplitude Modulated by a Biphase PCM Code', *IEEE Transactions on Communications,* 1970, **COM–18,** No. 6, pp. 263–265.
76. Bayless, J. W., Collins, A. A. and Pedersen, R. D., 'The Specification and Design of Bandlimited Digital Radio Systems', *ibid.*, 1979, **COM–27**, No. 12, pp. 1763–1770.
77. Sunde, E. D., 'Pulse Transmission by AM, FM and PM in the presence of Phase Distortion', *Bell System Technical Journal,* 1961, **40**, pp. 353–422.
78. Chamberlain, J. K., Clayton, F. M. and Collins, R. V., 'Reduced Bandwith Quaternary Phase-shift Keying (RBQPSK) – an Evolutionary Approach to Bandwidth Efficient Digital Microwave Transmission', *Proceedings of the IEEE International Seminar on Digital Communications,* Zurich, March 1980, A3.1–A3.6.
79. 'Fixed Service using Radio-Relay Systems', Recommendations and Reports of the CCIR, **IX–1**, XV Plenary Assembly, 1982.

7 DIGITAL RADIO RELAY SYSTEMS AND PERFORMANCE OBJECTIVES

7.1 COMPARISON WITH FDM–FM SYSTEMS

Some of the advantages of digital radio systems are:

1. Digital equipment is transparent to the type of traffic which it carries, i.e. the traffic may originate from telephone, computer, facsimile, telex, etc., may be integrated into one bit stream for transmission over the same radio bearer (or other transmission system), and may be switched together. As the data stream is always present, whether information is contained in it or not, the load on the radio system is therefore constant. This is not true for analog FDM–FM radio systems, where the loading depends upon the amount and the type of traffic being carried at any one time.
2. The accumulated noise of FDM–FM systems which posed serious constraints on equipment design is avoided in a digital system by regenerating the data stream. This regeneration may be done at the highest baseband bit rate without having to demultiplex down to the original bit rate. The regenerator equipment design is thus kept at a low level of complexity (relative to what it would be by 'mixing down' to origin).
3. Digital radio systems are less prone to interference and can operate satisfactorily with a carrier-to-interference ratio of 15 to 30 dB. This permits the same frequency to be re-used on the orthogonal polarization, thus conserving bandwidth and effectively doubling the spectrum efficiency or doubling the capacity which the RF channel can carry. This feature is one of the best advantages of digital radio over analog, for in the analog case the requirement for carrier-to-interference ratio is much higher (CCIR Report 780–1).[1]
4. The output power of a digital radio transmitter can be less than that of the analog radio for a given transmission quality. This lowers the cost of the equipment, increases reliability and saves on power and air-conditioning costs. Smaller output power also has the advantage that smaller interference levels are produced, making it possible for higher co-channel frequency re-use within a given geographical area and coexistence of digital systems with existing analog systems. CCIR Report 378–4 deals with this problem and concludes that for medium- and

small-capacity systems the use of multilevel PSK digital systems existing with analog systems is feasible. For high-capacity systems (greater 100 Mbit/s), multilevel systems (with many levels) should allow acceptable compatibility with FDM–FM systems if the system RF output power is kept to a value which is just sufficient to insure acceptable digital performance. Since in most commercially available equipment, the emitted spectrum of a digital transmitter is continuous without any discrete components or carrier, the interference which is induced in another nearby system is similar to thermal noise.

5. Depending on the modulation system in use, digital systems are able to produce a voice channel of acceptable quality with a carrier-to-noise ratio of as little as 15 dB, whereas for an analog system this may be designated as 30 dB or more, again depending upon the system.

Figure 7.1 shows the comparison of the signal-to-noise ratio versus input receive level (receiver quieting curve) of an analog and a digital radio system. Under flat fading conditions, the analog system shows a gradual decline of signal-to-noise ratio (dB for dB) for a decrease in received signal level. The digital system is unaffected until a threshold is reached. This characteristic is due to the regeneration process, in which the digital signal can be regenerated to its pristine form as long as the system is operated above a predetermined threshold (C/N of approx. 15 dB). This threshold behavior thus permits a constant transmission quality to be achieved, which is independent of the received field variations, path

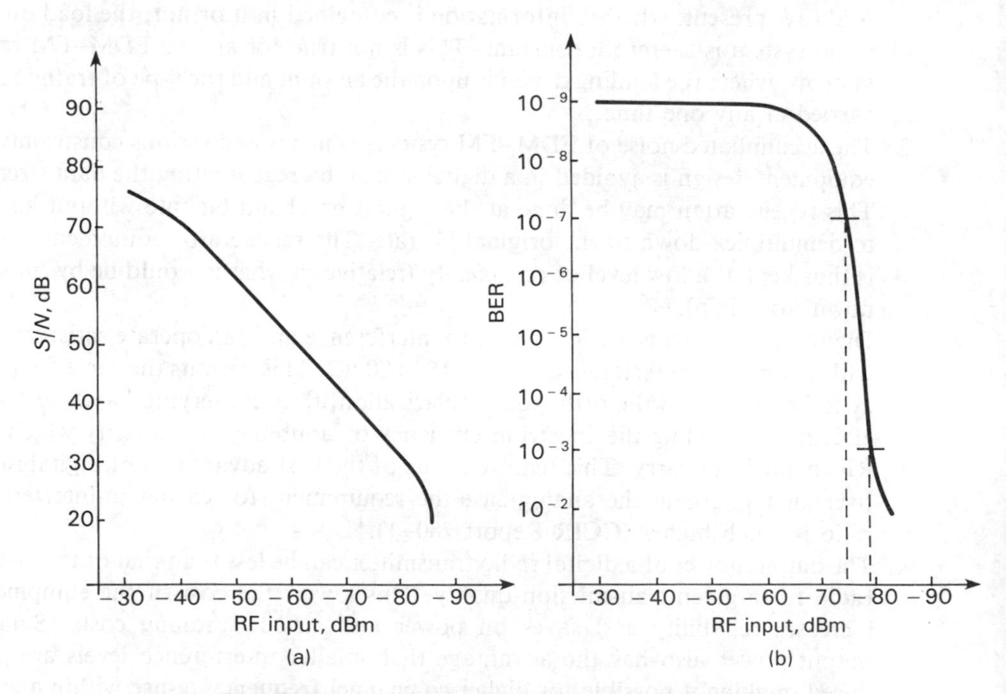

Figure 7.1 Comparison of the output signal-to-noise behavior of an analog and digital system with the reduction in receive signal level: (a) analog system (960 channels); (b) digital system (34 Mbit/s, 4-PSK)

length and the number of repeaters. It also permits a constant transmission quality to be achieved over the dynamic range of the message signal. Another feature is the ability of a low quality bearer to produce a high performance circuit and maintain, in a high interference environment, a relatively unaffected output signal.

6. The interconnection of metropolitan areas, which are these days sources of digital transmission activity, provides the digital connectivity and reduces the cost of inter-city trunking by providing a facility which is directly compatible with data transmission and other digital services. These points together with the fact that a digital radio network can be established by using existing towers and antennas make digital systems attractive. With the reduction in large scale integrated (LSI) circuit costs, and the large scale production of digital integrated circuits, equipment costs are reduced, making digital systems cost competitive with that of analog systems. Indeed the total cost for a digital system including muldexes and radio equipment is at present generally lower than for an analog system. This is especially so for short-haul systems or in systems where a significant number of drop and insert circuits are required.

Some disadvantages of digital radio systems are:

1. Digital radio and digital systems have to be integrated in the early stages with an existing analog network. The associated interfacing problems, costs and interference problems make the situation difficult. This may not be alleviated with the introduction of the ISDN due to the rapid advances in technology and the decreasing buffer time to adapt to it. In order to solve problems which are introduced by new technologies, a buffer time is required. This buffer time is becoming shorter and yet the problems are becoming increasingly more complex.
2. Present digital radio systems are not as RF bandwidth efficient as analog systems of 600 voice channels or more. It is felt however, that in time these problems will be resolved or not be that important, as the operating carrier frequencies are pushed higher and optical fiber networks carry the bulk of the large-capacity requirements of trunk networks. At the time of writing, optical fibers are in operation which are carrying 565 Mbit/s, whereas digital radio is only able to carry 140 Mbit/s.
3. What causes optical fibers to be a more realistic alternative to digital radio at the present time is the traffic capacity which they can carry without impairments from natural causes (a man with a jack-hammer cutting through the cable is not considered, in this context, as a natural cause). To carry the same traffic as an optical fiber, a digital radio requires to operate with a large RF bandwidth. This large bandwidth makes the radio link susceptible to selective fading which results in the consequent loss of all traffic. A wideband analog system by the same token would suffer only an increase in noise.
4. The introduction of new and many varied techniques over those required in the design, testing and maintenance of analog systems poses a problem in the retraining of personnel and in the training of new engineering staff. There is in reality a threefold problem. The first aspect is that there is heavy capital investment in

analog systems, which have proved to be reliable. Personnel are thus still heavily committed in this area and will be for some years to come. Secondly, digital radio or systems require that new skills be learnt and acquired, which are not easily accepted in general. Thirdly, transmission technology has strong competitors in the information technology and computer technology areas, which may be more immediately appealing to prospective new engineering staff.

The decisive advantages of digital radio over analog are: economic, better interference immunity, and better quality circuits over long distances. With the increasing quantity of digital switches being installed throughout the world, even greater cost savings are achieved, since there is no need to aquire complex and costly interfacing circuits.

7.2 PERFORMANCE OBJECTIVES FOR DIGITAL RADIO RELAY SYSTEMS

A measure of the transmission quality of a digital section is the quantity of incorrect symbols, or errors which occur. These errors may arise through thermal noise, intersymbol interference or from some external interference which produces impulse noise, etc. Whatever the sources are, there must be an objective which specifies the number of errors permitted to occur for a system which is considered to be operating properly. This section is concerned with providing those objectives and the conditions under which they are to be met. Errors tend to occur in 'bursts' and may not be randomly distributed. This arises because of the use of self-synchronizing scramblers and 'descramblers' in the transmitter and receiver baseband circuits. An error which occurs is multiplied by the descrambling circuit before the circuit locks back onto the nonerrored incoming unscrambled baseband bit stream. This is known as the error extension effect. In addition to the error extension effect, bursts of errors may occur due to impulsive noise, which is frequently a natural source of interference. The occurrence of intense error bursts, widely spaced in time, can have a significant effect on the number of errors counted over a period and thus, the mean-bit-error rate, without really seriously affecting operational circuits. As a consequence of this situation, it is necessary to establish an alternative means of defining error performance, which will directly relate to the effects of the different ways in which errors occur, on connected services. For an example, one service may be insensitive to a low background error ratio with a random distribution, but sensitive to error bursts, such as the case of video services. Error-correcting codes can be used to deal with single or small error bursts. An example of the opposite situation is with typical data services in which information at the source is assembled and transmitted in blocks. The systematic inclusion of error-control bits in each block permits the receiving data terminal to check the validity of the received block of data and if errors occur, ask the source to retransmit the entire block. With this type of arrangement the actual number of errors is not important and therefore if errors are to appear at all, error bursts are preferred due to the resultant reduction in the number of times a block has to be retransmitted. This of course increases the throughput efficiency. Error sources may be classified into those which are predictable and those which are not. *Predictable sources* may be error-producing sources arising from thermal noise, crosstalk and propagation effects, whilst *unpredictable sources* can be equipment imperfections, electrical interference and human interference.

The CCIR quality performance objectives (Report 930,[1]) consider the performance of the 2500 km hypothetical digital reference path (HDRP) or any shorter digital section which may be defined. This could be expressed in terms of two error ratios which are the two error-free second (EFS) objectives and an objective based upon error burst criteria. The higher error ratio, the short-term EFS or the error burst objective would in many systems be the most significant and would determine repeater spacing. The lower error-ratio objective would serve to control performance for the majority of the time. The quality objectives only become valid when the system is considered available (as discussed below). The HRDP may form part of a longer international digital connection, which defines the performance requirements for a hypothetical reference connection forming part of an integrated service digital network (ISDN). This is taken into account in the modified error rate performance objectives discussed in Section 7.3.3.

7.2.1 Low error ratio

The CCITT has proposed[1] an error-ratio design objective of 1 in 10^{10} per km for the transmission system in a 25000 km hypothetical reference digital connection. For the 2500 km HRDP this gives an error ratio of 2.5×10^{-7}, which excludes the contributions due to the multiplexing equipment. In digital radio relay systems, this error ratio will, according to present design criteria, be achieved for at least 99 percent of the time. Therefore a low value of error ratio of about 10^{-7} is appropriate for a 2500 km hypothetical reference digital path. The CCITT has *recommended*[4] (G821) that the bit-error-ratio on the 25000 km high-grade portion in the longest hypothetical reference connection should not exceed 1×10^{-6} during more than 4 percent of minutes within any month.

7.2.2 High error ratio

The proportion of time for which the high error ratio may be exceeded has a very great influence on the design of a system. A requirement that the higher error ratio may not be exceeded for more than 0.01 percent of any month for a 2500 km circuit would have severe repercussions on the economics of a practical system. However, a requirement that this error ratio should not be exceeded for more than 0.1 percent of any month would result in a more economic system, but one which may not compare favorably with existing FDM–FM radio relay systems. The figure which is adopted is, as expected, somewhere within these two extremes and is 0.054 percent of any month using an integration time of one-second.

7.2.3 Long-term error-free-second (EFS) objective[1]

The bit error ratio (BER) does not give an accurate estimate of the data circuit performance except when the error distribution is known. The more desirable alternative block error-ratio criterion unfortunately requires a knowledge of data block size and data rate which vary from user to user. The EFS objectives are one way in which the above limitations can be resolved. The CCITT has proposed that for data transmission a 95 percent EFS objective should be attained on the local exchange portion of the 25000 km hypothetical reference connection. A value of 99.5 percent EFS for the 2500 km HRDP would then seem appropriate. There is really no simple relationship between BER performance and the long-term EFS performance for radio systems during multipath fading, as EFS is a function of the rate of fading, whereas BER is independent of fading rate.

7.2.4 Error burst objective

With digital demultiplex equipments that are in common use, the occurrence of error bursts on large-capacity digital radio-relay systems will cause undesirable effects in the network. A limit has been proposed therefore, by the CCIR, on the permitted size and duration of these error bursts, primarily to control the incidence of multiple call drop-out in the telephony network. Figure 7.2 shows the possible error burst objective as shown proposed in CCIR Report 930.[1]

7.3 ERROR PERFORMANCE OBJECTIVES

7.3.1 Hypothetical reference digital path (HRDP) (CCIR Recommendation 556, Reference 1)

To determine error performance standards for a digital radio relay system it is necessary that the definition of the hypothetical reference digital path be understood. This will provide guidance to the design of equipment and systems for use in international telecommunication networks. The length of the path for systems having a bit rate above second-order digital multiplex (2048 kbit/s or 1554 kbit/s) is 2500 km in length. The HRDP is shown in Figure 7.3 and, as shown, allows for each direction of transmission, nine sets of digital multiplexing equipment at the CCITT recommended hierarchical levels and includes nine consecutive identical digital radio sections of equal length.

7.3.2 Digital radio sections (CCIR Recommendation 556, Reference 1)

A digital radio section consists of two consecutive radio terminal equipments and their interconnecting transmission medium, which together provide the whole of the

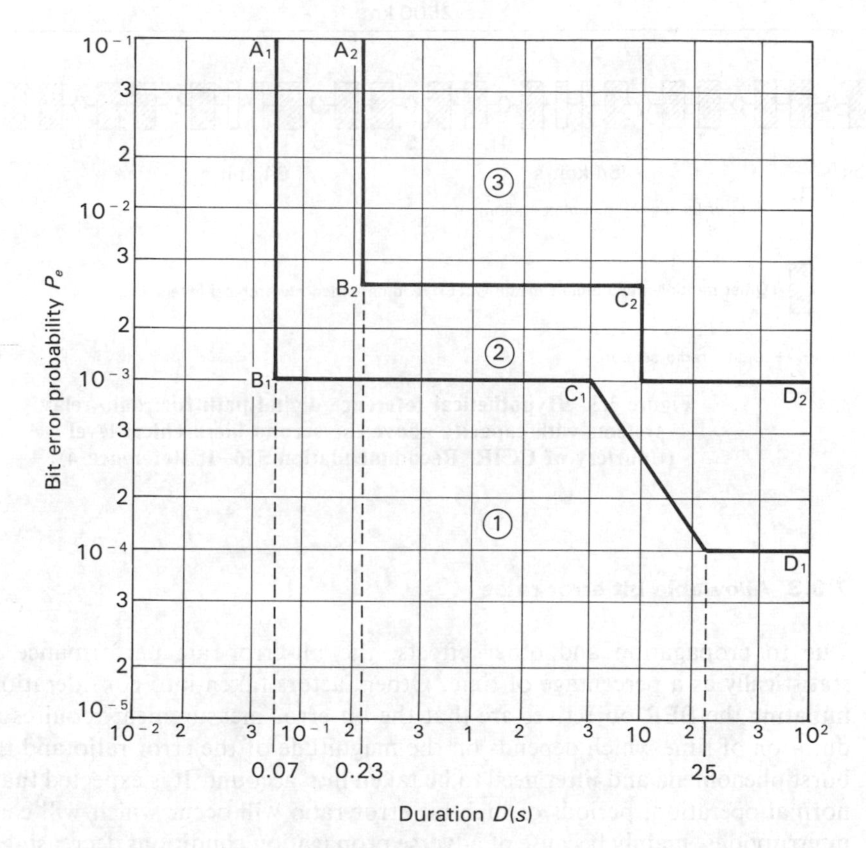

Area ① Error bursts confined within this area will have a negligible probability of causing multiple call drop-out

Area ② Error bursts extending into this area will have an increased probability of causing multiple call drop-out and would probably be unacceptable

Area ③ Error bursts extending into this area are likely to have an unacceptably high probability of causing multiple call drop-out

A_1, B_1, C_1; A_2, B_2, C_2 } Boundaries related to signaling performance criteria

C_1, D_1; C_2, D_2 } Boundaries related to subjective performance criteria

Figure 7.2 Possible error burst objective (Courtesy of CCIR, Report 930–1, Reference 4)

means of transmitting and receiving between two consecutive digital distribution frames. Due to the addition of overhead bits included in the radio equipment, the bit rate within the digital radio section may be different from the CCITT recommended hierarchical level or from an integral multiple of it.

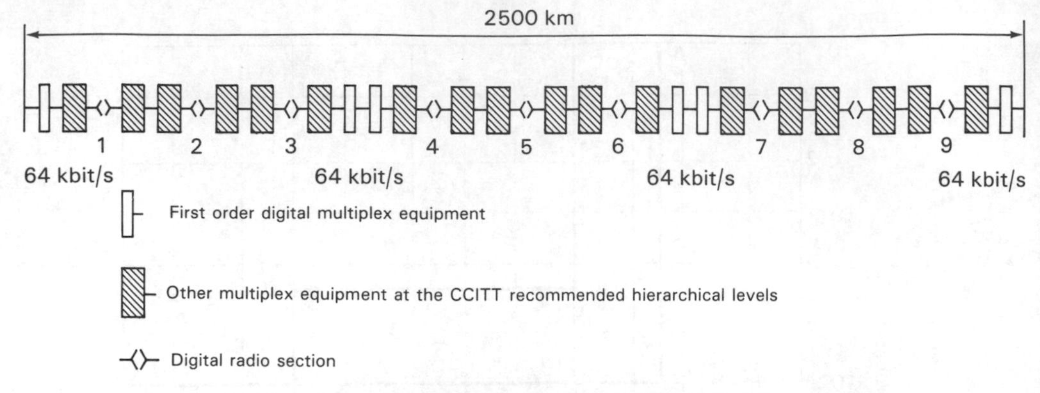

Figure 7.3 Hypothetical reference digital path for radio-relay systems with capacity above the second hierarchical level (Courtesy of CCIR, Recommendation 556–1, Reference 4)

7.3.3 Allowable bit error rates

Due to propagation and other effects, the bit-error-rate performance is stated statistically as a percentage of time. Other factors taken into consideration in formulating the BER objectives are that the bit-error measurement requires a certain duration of time which depends on the magnitude of the error ratio and that error burst phenomena and jitter need to be taken into account. It is expected that, during normal operation, periods of high bit error ratio will occur which will cause short interruptions, mainly because of adverse propagation conditions decreasing the level of wanted signal and/or increasing the level of interfering signals. It should be pointed out that the performance objectives for an international digital connection forming part of the Integrated Services Digital Network (ISDN) has been specified by the CCITT (Recommendation G821).[2] The HRDP corresponds to the 'high grade performance' circuit as specified in CCITT Recommendation G821, which also may form part of the ISDN and for which the CCITT has established a rule of apportionment of the total impairment allowance of a hypothetical digital reference circuit. This apportionment is:

1. Fewer than 0.00016 percent per km up to 25000 km of one-minute intervals are to have a bit error ratio worse than 10^{-6}. (Degraded minutes objective*)
2. Fewer than 0.000128 percent per km up to 25000 km of one-second intervals to have any errors (equivalent to (100 – 0.000128/km)) percent error-free seconds. (Severely errored seconds objective*)
3. Fewer than 0.05 percent of one-second intervals to have a bit-error ratio worse than 10^{-3} for a 2500 km HRDP. (Errored seconds objective*)

*See Section 7.4.1

For example, for the HRDP of 2500 km, the following will apply:

1. Fewer than 0.4 percent of one-minute intervals are to have a bit error ratio worse than 10^{-6}.
2. Fewer than 0.32 percent of one-second intervals are to have any errors (equivalent to 99.68 percent error-free seconds).
3. Fewer than 0.05 percent of one-second intervals to have a bit error ratio worse than 10^{-3}.

7.3.3.1 CCIR Recommendation 594–1* quality performance objectives

The CCIR recommends that the following quality performance objectives for each direction of the 64 kbit/s HRDP in which fading, interference and all other sources of performance degradation are taken into account:

1. That the bit error ratio should not exceed the following values:
   10^{-6} during more than 0.4 percent of any month, integration time being one minute (see Notes 1 and 2 below).
   10^{-3} during more than 0.054 percent of any month, integration time being one second.
2. That the total of errored seconds should not exceed 0.32 percent of any month.

Notes

1. Measurements of BER are normally made at a much higher bit rate than 64 kbit/s, i.e. usually at the bit rate of the radio relay system. Practical and theoretical considerations of the BER measurements are included in CCIR Report 613.
2. The seconds during which the BER exceeds 10^{-3} should not be taken into account in the integration time.
3. Contributions to the BER from multiplex equipment are not included in the above stated objectives.
4. The recommendation applies only when the system is considered to be available. The system is considered to be unavailable if one or both of the two following conditions occur for at least 10 consecutive seconds: the digital signal is interrupted (i.e. alignment or timing is lost); the error ratio is greater than 10^{-3}.

   This recommendation includes periods of high BER exceeding 10^{-3} which persist for periods of less than 10 consecutive seconds.

7.3.3.2 Real systems

The CCIR does not have any objective for real circuits, but does recognize that paths will differ in both length and composition from the HRDP and that it is desirable to provide planning objectives for the allowable BER of such paths, par-

*Modified from the recommendation given in Reference 1 to include that HRDP which may form part of the ISDN.

ticularly those which are shorter in length than the HRDP. Also it recognizes that in general for real paths there are two problems:

1. To find a method of apportioning the error objectives for systems which are intended to be connected in tandem in order to form paths similar to the HRDP
2. To take into account the CCITT hypothetical reference connection which includes the ISDN

For the first problem, the assumptions made in the method of apportioning these objectives are based on the statistical dependence of any section of the HRDP when subjected to fading, thus permitting the HRDP performance to be determined by the convolution of the probability density functions of all sections. On the assumption that the HRDP has a background BER of 10^{-7}, the background BER of any shorter section should be better than 10^{-7}. If the background BER of the HRDP is considerably better than 10^{-7}, it may be satisfactory to define the performance of a shorter section by apportioning the small percentage of the time that the BER of 10^{-7} can be exceeded. These objectives may be decided on the basis of the philosophy followed by the CCIR in laying down quality performance objectives of the real radio relay links for FDM telephony. Considering that the HRDP comprises nine radio sections, each of length 280 km, the following objectives could be adopted for the real radio relay links of length L:

1. For L less than 280 km:
   (a) The BER should not equal or exceed 10^{-6} during more than $(280/2500) \times 0.4$ percent of any month, integration time being one minute.
   That is; BER $\geqslant 10^{-6}$ for 0.045 percent of any month.
   (b) The BER should not equal or exceed 10^{-3} during more than $(280/2500) \times 0.054$ percent of any month, integration time being one second.
   That is; BER $\geqslant 10^{-3}$ for 0.006 percent of any month.
   (c) The total of errored seconds should not exceed $(280/2500) \times 0.32$ percent = 0.036 percent of any month.
2. For L greater than 280 km but less than 2500 km:
   (a) The BER should not equal or exceed 10^{-6} during more than $(L/2500) \times 0.4$ percent of any month, integration time being one mintue.
   (b) The BER should not equal or exceed 10^{-3} during more than $(L/2500) \times 0.054$ percent of any month, integration time being one second.
   (c) The total of errored seconds should not exceed $(L/2500) \times 0.32$ percent of any month.

7.4 ERROR PERFORMANCE OF AN INTERNATIONAL DIGITAL CONNECTION FORMING PART OF AN ISDN (CCITT RECOMMENDATION G821: DEFINITIONS)[2]

CCITT Recommendation G821 deals with the error performance of an international digital connection forming part of an Integrated Services Digital Network (ISDN).

It is included here, in order that the HRDP and CCIR objectives can be placed in their proper perspective with regard to international circuits. This recommendation has been made because services in the future may expect to be based on the concept of the ISDN, which is required to transport a wide range of services as well as voice services, which are considered to be most likely to predominate in the ISDN. Because of the concept of the ISDN and because errors are a major source of degradation affecting voice services in terms of voice distortion and affecting data-type services in terms of inaccurate information or reduced throughput, performance objectives are required. These objectives are stated in Recommendation G821 for each direction of a 64 kbit/s circuit-switched connection used for voice traffic or as a 'bearer channel' for data-type services. The 64 kbit/s circuit-switched connection referred to is an all-digital hypothetical reference connection (HRX) and is shown in Figure 7.4. It encompasses a total length of 27500 km and is a derivative of the standard hypothetical reference connection (HRX) shown in Figure 7.5. As

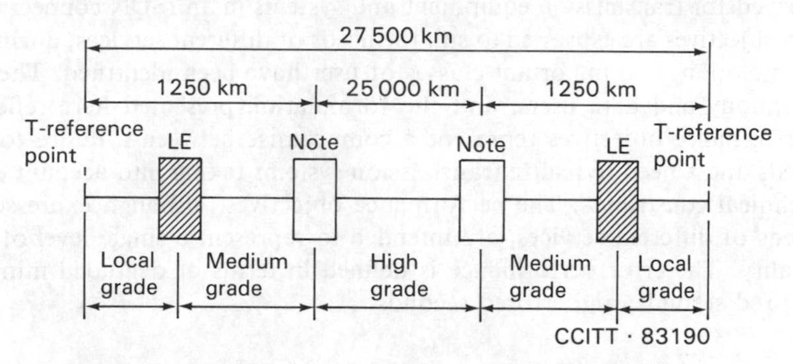

Figure 7.4 Circuit quality demarcation of longest HRX (Courtesy of CCITT, Recommendation G821, Reference 2)

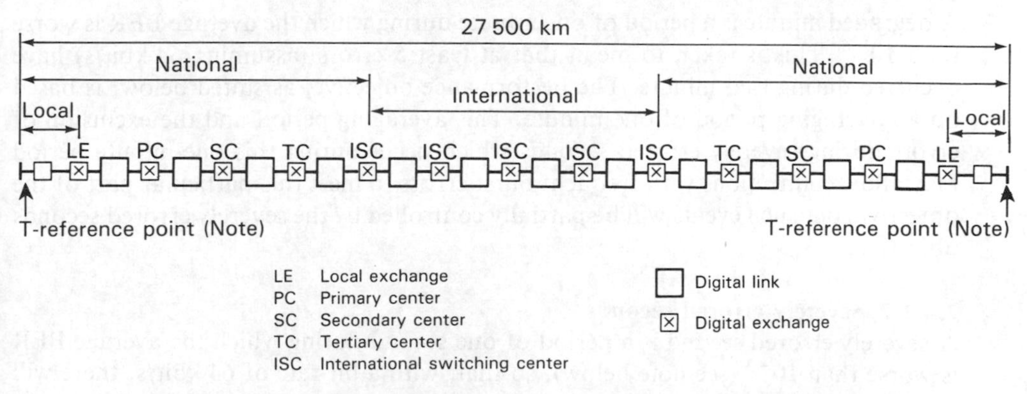

Figure 7.5 Standard digital hypothetical reference connection (longest length) (HRX) (Courtesy of CCITT, Recommendation G801, Reference 2)

the objectives are aimed at network users, the performance is expressed in a way that gives the most information about the effects of the performance degradations on services. In terms of error performance parameters, the quality performance objective is defined as:

> 'The percentage of averaging periods each of time interval T_0 during which the bit error ratio (BER) exceeds a threshold value. The percentage is assessed over a much longer time interval T_L'

T_L is split into two parts, namely, time for which the connection is deemed to be available and time when it is unavailable. The performance objectives aim to serve two main functions. These functions are to give the user of future national and international digital networks an indication of the expected error performance under real operating conditions, thus facilitating service planning and terminal equipment design; and, secondly, to form the basis upon which performance standards are derived for transmission equipment and systems in an ISDN connection. Although the objectives are expressed to suit the needs of different services, during the original formulation, two important classes of user have been identified. These classes are telephony and data users, and the formulation presented here reflects this. The performance objectives represent a compromise between a desire to meet service needs and a need to realize transmission systems taking into account economic and technical constraints. The performance objectives, although expressed to suit the needs of different services, are intended to represent a single level of transmission quality. The error performance is defined in terms of degraded minutes; severely errored seconds; and errored seconds.

7.4.1 Performance objectives

7.4.1.1 Degraded minutes

A degraded minute is a period of one minute during which the average BER is worse than 10^{-6}. This is taken to mean that at least 5 errors (assuming 64 kbit/s) have occurred during that minute. The performance objective, as stated below, is based on an averaging period of one minute. This averaging period and the exclusion of errors during severely errored seconds which occur during this one-minute period may allow connections with frequent burst errors to meet this particular part of the objective, but such events will be partially controlled by the severely errored seconds objective.

7.4.1.2 Severely errored seconds

A severely errored second is a period of one second during which the average BER is worse than 10^{-3} (see note below), so that, with a bit rate of 64 kbit/s, there will be more than 64 errors in that second.

Note. CCITT Study Group XVIII Question 10 propose 10^{-4} as a threshold BER for failure for an ISDN connection.

7.4.1.3 Errored seconds

An errored second is a period of one second during which at least one bit error has occurred.

The performance objectives for an international ISDN connection are shown in Table 7.1. It is intended that the international ISDN connections should meet all the requirements of Table 7.1 concurrently. The connection fails to satisfy the objective if any one of the requirements is not met.

An alternative interpretation of Table 7.1 is given in Table 7.2.

Table 7.1 Error performance objectives for international ISDN connections

Performance classification	Objective
(*a*) (Degraded minutes) (Notes 1 & 2)	Fewer than 10 percent of one-minute intervals to have a bit error ratio worse than 10^{-6}. (Note 4)
(*b*) (Severely errored seconds) (Note 1)	Fewer than 0.2 percent of one-second intervals to have a bit error ratio worse than 10^{-3}.
(*c*) (Errored seconds) (Note 1)	Fewer than 8 percent of one-second intervals to have any errors (equivalent to 92 percent error-free seconds)

Notes

1. The terms 'degraded minutes', 'severely errored seconds' and 'errored seconds' are used as a convenient and concise performance objective 'identifier'. Their usage is not intended to imply the acceptability, or otherwise, of this level of performance.
2. The one-minute intervals are derived by removing unavailable time and severely errored seconds from the total time and then consecutively grouping the remaining seconds into blocks of 60. The basic one-second intervals are derived from a fixed time pattern.
3. The time interval T_L over which the percentages are to be assessed has not been specified, since the period may depend upon the application. A period of the order of any one month is suggested as a reference.
4. For practical reasons, at 64 kbit/s, a minute containing four errors (equivalent to an error ratio of 1.04×10^{-6}) is not considered degraded. However, this does not imply relaxation of the error ratio objective of 10^{-6}.

Table 7.2 Alternatively expressed objectives of Table 7.1

(*a*) Degraded minutes	At least 90 percent of one-minute intervals to have 4 or fewer errors (Note 1)
(*b*) Severely errored seconds	At least 99.8 percent of one-second intervals to have less than 64 errors (Note 2)
(*c*) Errored seconds	At least 92 percent of one-second intervals to have zero errors (Note 3)

Notes

1. Four errors over one minute is equivalent to a BER of approximately 1×10^{-6}.
2. Sixty-four errors over one second is equivalent to a BER of 1×10^{-3}.
3. The one-second integration period was chosen in the belief that it represented a short enough time interval to display sufficient detail of the error structure, yet long enough to encompass most block lengths that data users may choose for error control.

7.4.2 Allocation of overall objectives

The objectives provided in Table 7.1 and Table 7.2 relate to an overall connection. These objectives need to be subdivided into their constituent parts to form the basis on which the performance objectives are derived for transmission equipment and systems used to support connections. This subdivision involves the use of two slightly different strategies. One is applicable to the degraded minutes objective and to the errored seconds objective and the other is applicable to the severely errored seconds objective.

7.4.2.1 Principles of apportioning error objectives

The apportionment is based on the assumed use of transmission systems having qualities falling into any one of three different classifications, namely, local-, medium- and high-grade systems, which are independent of the transmission systems used. Their usage generally depends upon their location within a network. Figure 7.4 shows the circuit quality demarcation of the longest HRX. It is not possible to provide a definition, however, of the location of the boundary between the medium- and high-grade portions of the HRX. Note 4 of Table 7.3 does provide some clarification of this point.

The three circuit quality classifications for transmission systems are as follows:

LOCAL GRADE

This embraces systems assumed to be operating between the user's premises and local exchanges (LE) and will typically be low-digit-rate transmission systems operating at a bit rate below the primary levels of 2048 kbit/s or 1544 kbit/s, over a range of cable types and other transmission media.

MEDIUM GRADE

These types of system are assumed to be operating between local exchanges (LE) and beyond into the national part of the connection. The actual distance covered by such systems will vary considerably from one country to another, but in most circumstances will not exceed 1250 km. In large countries, this distance may only reach the primary switching center (PC), whilst in smaller countries it may go as far as the secondary center, tertiary center or the international switching center. For this reason, transmission systems assigned to this classification will vary in quality, falling between the local- and high-grade classifications. In many systems the variation in quality relates more to network planning and maintenance than to intrinsic differences in system design. Typically these systems will operate at low, medium or high bit rates and use a variety of different transmission media.

HIGH GRADE

These systems are used for realizing transmission over long-haul national and international parts of a connection. Typically such systems operate at medium or high digit rates utilizing all types of transmission media, such as satellite, terrestial line-of-sight radio relay systems, optical fiber, etc. Of the total HRX length (27500 km), a significant proportion will be made up of high-grade systems.

The following general assumptions apply to the apportionment strategy that follows:

1. In apportioning the overall objective to the various constituent parts of a connection, it is the 'percentage of time' or 'percentage of averaging periods that are permitted to exceed the threshold values' that is subdivided.
2. An equal apportionment of the objectives applies for both the 'percentage of degraded minutes' and 'percentage of errored seconds'.
3. The error ratio threshold is not subdivided, for it is expected that the performance of real circuits forming the parts of the HRX will normally be significantly better than the degraded minute threshold.
4. The errror contributions from both digital switching and digital multiplex equipments are assumed to be negligible in comparison with the contributions from transmission systems.

For terrestrial line-of-sight radio relay systems, the apportionment of the 'degraded minute' performance classification to smaller entities may require subdivision of the error ratio objective as well as the subdivision of the 'percentage of time', with distance. In Section 7.3.3.2, it was suggested that only the 'percentage of time', should be subdivided. The CCITT state that this is the subject of further study (CCITT Recommendation G821).[2]

7.4.2.2 Apportionment strategy for the degraded minutes and errored seconds requirements

The apportionment of the permitted degradation, i.e. 10 percent degraded minutes and 8 percent errored seconds, is given in Table 7.3. The derived network performance objectives are given in Table 7.5.

7.4.2.3 Apportionment strategy for severely errored seconds

The total allocation of 0.2 percent severely errored seconds is subdivided into each circuit classification as follows:

(*a*) 0.1 percent is divided between the three circuit classifications in the same proportions as adopted for the other two objectives. The results of this allocation is shown in Table 7.4.

(*b*) The remaining 0.1 percent is a block allowance to the medium- and high-grade classifications to accommodate the occurrence of adverse network conditions occasionally experienced (intended to mean the worst month of the year) on transmission systems. Because the occurrence of worst-month effects in a worldwide connection is determined statistically, it is considered that the following allowances are consistent with the total 0.1 percent figure: 0.05 percent to a 2500 km HRDP for radio relay systems which can be used in the high- and medium-grade portions of the connection; and 0.01 percent to a satellite HRDP.

The overall high-grade objective is subdivided on the basis of length, resulting in a conceptual per kilometer allocation which can be used to derive a block allowance for particular system lengths. Based on Table 7.5, the derived network performance objectives are given in Table 7.6.

Table 7.3 Allocation of the degraded minutes and errored seconds objectives for the three circuit classifications

Circuit classification	Allocation of the degraded minutes and errored seconds objectives for Table 7.2
Local grade (2 ends)	15 percent block allowance to each end (Notes 1, 4 and 5)
Medium grade (2 ends)	15 percent block allowance to each end (Notes 2, 4 and 5)
High grade	40 percent (equivalent to conceptual quality of 0.0016 percent per km for 25000 km (Notes 3, 6 and 7)

Notes

1. The local-grade apportionment is considered to be a block allowance, i.e. an allowance to that part of the connection regardless of length.
2. The medium-grade apportionment is considered to be a block allowance, i.e. an allowance to that part of the connection regardless of length. The actual length covered by the medium-grade part of the connection will vary considerably from one country to another. Transmission systems in this classification exhibit a variation in quality falling between the other classifications.
3. The high-grade apportionment is divided on the basis of length resulting in a conceptual per kilometer allocation which can be used to derive a block allowance for a defined network model (e.g. hypothetical reference digital link).
4. The local-grade and medium-grade portions are permitted to cover up to the first 1250 km of the circuit from the *T*-reference point extending into the network.
5. Administrations may allocate the block allowances for the local- and medium-grade portions of the connection as necessary within the total allowance of 30 percent for any one end of the connection. This also applies to the objectives given for local and medium grades in Table 7.4.
6. Based on the understanding that satellite error performance is largely independent of distance, a block allowance of 20 percent of the permitted degraded minutes and errored second objectives is allocated to a single satellite HRDP employed in the high-grade portion of the HRX.
7. If the high-grade portion of a connection includes a satellite system and the remaining distance included in this category exceeds 12500 km or if the high-grade portion of a nonsatellite connection exceeds 25000 km, the objectives of Recommendation G821 may be exceeded. The occurrence of such connections is thought to be relatively rare and studies are continuing in order to investigate this. The concept of satellite equivalent distance (the length of an equivalent terrestrial path) is useful in this respect and it has been noted that a value in the range 10000–13000 km might be expected.

Table 7.4 Allocation of severely errored seconds

Circuit classification	Allocation of severely errored seconds objectives
Local grade	0.015 percent block allowance to each end (See Note 5 to Table 7.3)
Medium grade	0.015 percent block allowance to each end (See Note 5 to Table 7.3)
High grade	0.04 percent (Note 1)

Notes

1. For transmission systems covering the high-grade classification, each 2500 km portion may contribute not more than 0.004 percent.
2. For a satellite HRDP operating in the high-grade portion there is a block allowance of 0.02 percent severely errored seconds (see Note 6 to Table 7.3).

Table 7.5 Allocation of degraded minute intervals and errored seconds objectives

Circuit classification	Network performance objective at 64 kbit/s	
	Percentage of one-minute intervals with > 4 errors (% degraded minutes)	Percentage of seconds with one or more errors (% errored seconds)
Local grade: block allowance for each end	<1.5	<1.2
Medium grade: block allowance for each end	<1.5	<1.2
High grade for 25000 km	<4.0	<3.2

Table 7.6 Allocation of percentage of one-minute and errored seconds objectives for high-grade systems

High-grade system length, km	Network performance objective at 64 kbit/s	
	Percentage of one-minute intervals with > 4 errors (% degraded minutes)	Percentage of seconds with one or more errors (% errored seconds)
280	<0.0448	<0.03584
420	<0.0672	<0.05376
2500	<0.4	<0.32

7.4.3 Interpretation of the error performance objectives for international ISDN connections

Table 7.1 provides the details of these objectives for a 27500 km HRX. Annex B to the CCITT Recommendation G821 provides an algorithm to assist in the interpretation of the objectives of Table 7.1. An alternative to this algorithm is the following explanation.

The degraded minutes, severely errored seconds and errored seconds are monitored over a total time T_L. The value of T_L, as mentioned previously, is not specified, but suggested to be one month, which is in accordance with CCIR Recommendation 594 (error performance of the HRDP). If T_u is the total unavailable time, as defined above, the available time T_a is given by

$$T_a = T_L - T_u \tag{7.1}$$

where T_a is recorded in seconds. If S is the number of one-second intervals counted which contain more than 64 errors (severely errored seconds), then S must be less than 0.2 percent of T_a. That is

$$100\ S/T_a < 0.2 \tag{7.2}$$

The remaining $T_a - S$ seconds are grouped into packets of 60. The number of these packets M, which contain more than four errors (degraded minutes), must be less than 10 percent:

$$100\ M < 10 \tag{7.3}$$

Over an interval of T_L, count the number of one-second intervals that contain at least 1, but less than 64, errors (errored seconds). The total number of these errored seconds E must be less than 8 percent of T_a:

$$100\ E/T_a < 8 \tag{7.4}$$

7.4.4 Worked example on material presented in Section 7.4

Consider a 1200 km hypothetical reference circuit in which all of the trunk sections are to be high grade. The circuits employed in this system consist of local distribution, back-haul systems from a customer data distribution node (DN) to a main trunk access node (TAN), and trunk routes between TANs. Determine the error performance objectives for this system based on CCITT Recommendation G821 and measured at 64 kbit/s. The objectives may be worked out as in Sections 7.4.4.1 to 7.4.4.4.

7.4.4.1 Degraded minutes

From Table 7.5, the following allowances are made:

LOCAL CIRCUITS

These include local distribution circuits

Allowance for each end < 1.5 percent (block allowance)

MEDIUM-GRADE CIRCUITS

These include back-haul systems

Allowance for each end < 1.5 percent (block allowance)

HIGH-GRADE CIRCUITS

These include trunk and radio circuits and the allowance is based upon the circuit length. As Table 7.5 shows, the 4 percent allowance is for 25000 km (see Figure 7.4). Thus for 1200 km, the apportioned allowance is given as less than (1200/25000) × 4 percent = 0.192 percent.

7.4.4.2 Errored seconds

LOCAL CIRCUITS

Allowance for each end < 1.2 percent (block allowance)

MEDIUM-GRADE CIRCUITS

Allowance for each end < 1.2 percent (block allowance)

HIGH-GRADE CIRCUITS

The apportionment allowance in this case is given in Table 7.5 as being less than 3.2 percent. Thus for 1200 km, the apportionment becomes less than (1200/25000) × 3.2 percent = 0.154 percent.

7.4.4.3 Severely errored seconds

From Table 7.4, the following apportionments are given:

LOCAL CIRCUITS

Allowance < 0.015 percent (block allowance)

MEDIUM-GRADE CIRCUITS

Allowance < 0.015 percent (block allowance)

HIGH-GRADE CIRCUITS

The apportionment allowance in this case, given in Table 7.4, is based on a HRX trunk section of 25000 km. For 1200 km the allowance is taken as *pro rata*, i.e. (1200/25000) × 0.04 = 0.00192 percent. If the trunk route was a radio-relay system, the apportionment for 25000 km would be given as 0.05 percent and the allowance would be less than (1200/25000) × 0.05 = 0.0024 percent.

Table 7.7 Error performance objectives for example of Section 7.4.4

	Local distribution	Back-haul	Primary Trunk (cable)	Primary trunk (μ-wave)
Reference path length, km	Block allowance	Block allowance	600	600
% Degraded minutes	<1.5	<1.5	<0.096	<0.096
% Errored seconds	<1.2	<1.2	<0.077	<0.077
% Severely errored seconds	<0.015	<0.015	<0.00096	<0.0012

Note. Note 5 of Table 7.3 states that administrations may allocate the block allowances for the local- and medium-grade portions of the connection as necessary within the total allowance of 30 percent for any one end of the connection. That note also applies to Table 7.4. This means that the block allowances worked out in the example of Section 7.4.4 and given in Table 7.7 are maximum values for the degraded minutes and errored seconds.

7.4.4.4 Summary

Table 7.7 shows a summary of the above results. A small change has been made to bring out a simple point. The high-grade circuits are now assumed to be 600 km of optical fiber cable joined to 600 km of microwave radio-relay systems. This means that for the high-grade circuits the figures calculated above are halved.

7.5 AVAILABILITY

Availability is the ability of a component, in the light of its reliability, maintainability and maintenance/logistic support, to perform its required function: availability is expressed as a percentage of time. In the analog network, availability objectives are used mainly as system design parameters. In contrast, in a digital network, availability objectives are used as network performance parameters, and together with the quality objectives discussed in Sections 7.2 to 7.4 comprise total performance objectives of a system. To establish availability objectives, conditions which make a service and/or a connection unavailable must be defined. A service can be considered as unavailable to the user when its level of performance is insufficient for the intelligible transfer of information. This performance level will vary from user to user and hence it is necessary to evaluate the effects on the service of the various digital transmission impairments. There are six transmission parameters which may be used as a measure of unsatisfactory *quality* performance. These are BER, short interruptions, delay, jitter, slip and quantizing noise. BER and short interruptions are the main indicators of unavailability. This is because jitter and slip will cause bit errors and short interruptions in the network and that delay and quantizing noise are relatively fixed quantities in any connection. At the present time, voice services represent the largest percentage of all communication services. They are considered to be more *robust* than other services, or more tolerant to bit errors and short interruptions. Because of this, voice services can be used to indicate the minimum level of performance for the availability of a connection. The availability of a connection varies from administration to administration, but if a minimum of 99.8 percent of the time in any three-month interval with no more than 30 outages during that time is chosen as the availability objective, the network is expected to provide a very good grade of service. When measuring network error performance it is assumed that the test signal is present during the whole measurement period T_L. During periods when the test signal is not present, or the test signal becomes so degraded as to make the circuit unusable for traffic, the measurement is suspended and the circuit is said to be unavailable. Periods of unavailability are therefore discounted in the percentage calculations. There are two distinct types of unavailability. These are:

1. The test circuit becomes unavailable immediately the error detection equipment loses synchronism for whatever reason and becomes available immediately synchronism is restored.
2. A period of unavailable time begins when the BER in each second is worse than

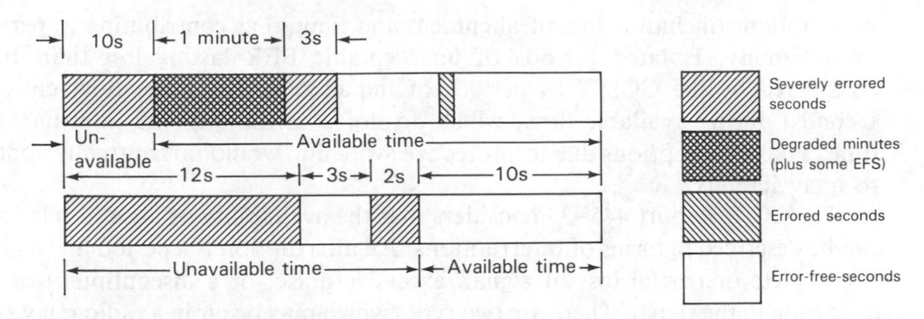

Figure 7.6 Examples of sequences containing unavailable time

10^{-3} for a period of ten consecutive seconds; these ten seconds are considered to be *unavailable time*. The period of unavailable time ends when the BER in each second is better than 10^{-3} for a period of ten consecutive seconds; these ten seconds are considered to be *available time* (CCITT Recommendation G821).[2] Figure 7.6 helps clarify this definition, for there are several ways in which a period of unavailable time may be recorded. Note that degraded minutes will normally contain a number of error-free seconds.

The CCIR concept of unavailability of a hypothetical reference digital path (HRDP) is (CCIR Recommendation 557)[1] that, in at least one direction of transmission, one or both of the two following conditions occur for at least 10 consecutive seconds: the digital signal is interrupted (i.e. alignment or timing is lost); the error ratio is greater than 10^{-3}. The same recommendation goes on to state that, in the estimate of unavailability, all causes must be included which are statistically predictable, unintentional and resulting from the radio equipment, power supplies, propagation, interference and from auxiliary equipment and human activity. The estimate of unavailability includes consideration of the mean time to restore.

The availability objective appropriate to a 2500 km HRDP should be 99.7 percent of the time, the percentage being over a period of time sufficiently long to be statistically valid; this period is probably greater than one year. It is recognized by the CCIR (Recommendation 557)[1] that the value of 99.7 percent is a provisional one and that, in practice, the objectives selected may fall into the range 99.5–99.9 percent. The choice of a specific value in this range depends on the optimum allocation of outage time among the various causes which may not be the same when local conditions such as propagation, geographical size, population distribution, organization and maintenance are taken into account. The availability of radio-relay systems is only one of the many aspects that insure the acceptable grade of service of the telephony traffic; the choice of an optimum value for this particular aspect can be made only by considering all transmission systems either existing or planned in the network under study. Because of this, administrations may select different values of availability objective for use by their planning organizations, in the above range 99.5–99.9 percent. Availability of multiplex is not included in the range of availability figures quoted. The only difference between the CCIR and CCITT definitions of availability is that CCIR Recommendation 557 specifically includes interruptions (including loss of alignment and timing) as contributing to periods of unavailability. Isolated periods of unacceptable BER lasting less than 10 s are considered by the CCITT as periods of impaired performance (severely errored seconds) during available time, which do not contribute to the total unavailable time. Thus interruptions due to protective switching would not normally contribute to unavailability.

The CCIR Report 445–3[4] considers that the availability of a radio relay system can be described in terms of interruptions. An interruption is a period in which there is complete or partial loss of signal, excessive noise, or a discontinuity or severe distortion in the signal. There are two types which may occur in a radio relay system. These are long interruptions, which may last from 10 s to an hour, occurring once per week or perhaps once per year and, short interruptions, which are fairly frequent. These may last from a number of seconds down to very small fractions of a second, but which may occur many times in an hour on some occasions. These two types of interruption have separate significance when considering the availability of the system. For long interruptions, the parameters which require to be defined are the mean time between interruptions, so that they do not occur too frequently; and the total outage over a period of, say, one year. This insures that the availability of the system is adequate and that the mean duration of an interruption if it occurred would not be too long. For short interruptions, the parameters required to be defined are: the rate of occurrence, measured on an hourly basis; the total outage time over, say, the worst month; and the duration of these short interruptions. These should be defined statistically. It might be more desirable to specify several points of an objective distribution, as discussed in Section 7.6.2.

7.5.1 Causes of unavailability

The large number of factors affecting the availability of the ISDN and the inclusion of an international connection in the HRX are possibly two reasons why the CCITT has not yet provided a general availability objective. The CCIR Recommendation, which does not include the availability of the multiplex equipment, is restricted to fixed radio-relay services and therefore is able to provide a value which is more easily shown to be realistic. The causes of unavailability in radio relay systems can be divided into three categories. These are:

1. Deep fades or unacceptable interference due to adverse propagation conditions on radio paths
2. Equipment failures or faults
3. Human interference, which may be an order greater than the above-mentioned other two, put together

Deep fades cause noise to exceed a certain limit. This may be due to ducting and usually lasts a fairly long time as does excessive precipitation attenuation, which is caused mainly by heavy rainfall and in some cases by heavy snowfall. Fading causes short interruptions. Interruptions caused by equipment failures may be due to the degradation of the radio equipment, including the modulators and demodulators, as well as auxiliary equipment such as the switchover equipment, the primary power supply equipment and the failure of an antenna or feeder.

7.5.2 A proposed end-to-end availability objective

Consider the range of availability objectives recommended in CCIR Recommendation 557[1] of 99.5–99.9 percent. If 99.5 percent is chosen for a system and a margin of 0.3 percent allowed for outages caused by human error, the availability objective for system design purposes is 99.8 percent end-to-end. If it is possible to minimize the scope of human interference by providing automated network management, without any possible adjustments to equipment installed in the field, then this 0.3 percent margin is left for the unknown causes of outages due to human interference. The availability A percent is given by CCIR Report 445–3,[4] as

$$A = 100 - (2500 \times 100/L) \times [(T_1 + T_2 - T_b)/T_s] \tag{7.5}$$

where A is the percentage availability based on 2500 km,
L is the length of a one-way radio channel under consideration,
T_1 is the total interruption time for at least 10 consecutive seconds for one direction of transmission,
T_2 is the total interruption time for at least ten consecutive seconds for the other direction of transmission,
T_b is the bidirectional interruptions for at least ten consecutive seconds and
T_s is the period of study

For unidirectional transmission, $T_2 = 0$ and $T_b = 0$. Availability and unavailability should be evaluated on the same basis. It should be noted that for short circuits, the unavailability is unlikely to add in a linear manner and it may be necessary to define a minimum value for the parameter L. For higher-grade circuits, the value of 280 km quoted in CCIR Report 930[1] might be appropriate.

The measured data on availability for real circuits show a wide distribution; a reliability figure on the actual availability can be estimated only as an average of a large amount of data collected from many radio relay routes for a sufficiently long period of time.

7.5.3 Apportionment of unavailability

As for the apportionment of the error rates described in Section 7.4, it is possible to apportion unavailability to particular network elements. With the 99.8 percent figure chosen as an example in Section 7.5.2, the value permitted for the unavailability, excluding human interference, is given as 0.2 percent of the time. This figure will represent the sum of all of the unavailability apportionments to local distribution circuits, back-haul and various forms of primary trunk systems. As the trunk routes carry the most traffic, it could be expected that the unavailability figures for these circuits are lower than those for the back-haul or local circuits. However, since their circuit lengths are usually much longer than the back-haul or local systems, the unavailability figures may be increased to the point where their values are equal or greater. Optical fiber and cable trunk systems would be expected to have a lower unavailability than microwave radio relay systems due to the absence of propagation effects such as multipath fading, etc. Of the primary trunk systems, unavailability allocations for a 600 km optical fiber cable may be 0.030 percent, whereas, for a microwave radio relay system of 600 km, it may be chosen to be 0.060 percent. This figure depends of course on such factors as climate, terrain, operating frequency, etc. Back-haul systems would be expected to have an unavailability allocation of 0.0225 percent for each end, and local distribution networks, as short as they may be, to have an allocation as high as 0.0325 percent. Usually the local distribution network is a main source of trouble due to the direct interfacing with the user. Both the back-haul and local network allocations are independent of their length or of factors such as the number of users being served by the local distribution. The purpose of the unavailability allocation is to permit the network elements to be specified so that the end-to-end availability objective will be met for the complete system. The allocation of 0.060 percent, given as an example for the 600 km primary trunk radio relay circuit, may again be subdivided into two parts. One part is for propagation outages and the other for equipment failures. The ratio of the subdivision is either arbitrary or based on expected reliability figures for the equipment. A reliability analysis may suggest that redundancy protection is necessary in order to meet the total amount allocated to the radio system. In a case such as that, the unavailability allocation due to equipment failure may become a small part of the radio relay unavailability allocation, leaving a greater margin for propagation outages. On the other hand, it may suggest that a more reliable equipment should be sought as there is not enough allowance for propagation effects.

7.5.4 Error performance and availability objectives for local-grade digital radio-relay systems

CCIR Report 1053 ANNEX 1[4] considers the common use of local network transmission facilities as part of radio-relay systems under the ISDN, and it has recommended error performance and availability objectives for the portion of the HRX between the subscriber and the local exchange. These recommendations in part are:

1. The BER should not exceed 10^{-3} for more than A percent of any month, with an integration time of 1 second. Values of A in the range 0.005–0.015 have been proposed.
2. The BER should not exceed 10^{-6} for more than B percent of any month, with an integration time of 1 minute. Values of B in the range 0.5–1.5 have been proposed.
3. The total errored seconds should not exceed C percent of any month. Values of C in the range 0.4 to 1.2 have been proposed.

4. The residual bit error ratio (RBER), which is the bit error ratio of the system in the absence of fading and short-term interference, but which includes the presence of long-term interference that arises from sources within line-of-sight of the victim receiver and which is typically low in level and constant in value, shall not exceed D. Values of D in the range 10^{-7} to 10^{-10} have been proposed.
5. The total unavailability due to all causes of a bi-directional digital radio system used in the local grade portion of an ISDN, shall not exceed E percent averaged over F years. The value of E is probably in the range 0.08 to 1 and F should be greater than one.

The residual BER or RBER is the bit error ratio understood to mean the error ratio in the absence of fading and includes allowance for system inherent errors, environmental and aging effects and long-term interference. A value of 5×10^{-9} for the RBER is specified as appropriate in CCIR Recommendation 634 for a 2500 km high-grade real circuit. Such a value is consistent with a range of values that can be derived theoretically from the performance objectives on the assumption that the various error statistics are known. For L, less than 2500 km RBER $\leqslant (5 \times 10^{-9}) L/2500$, for $280 \leqslant L < 2500$ km.

7.6 ERROR-FREE SECOND PERFORMANCE

The error performance objectives for an errored second as specified by the CCITT Recommendation G821[2] for a 27500 km HRX, is that fewer than 8 percent of one-second intervals are to have any errors, where an errored second is a period of one second during which at least one bit error has occurred. This leads to the concept of *error-free seconds* (EFS), as being the situation where for 98 percent of the time there are no errors or errored seconds. If the equivalent HRDP is derived, as discussed in Section 7.3.3, then the figure is decreased to 95 percent. The CCIR Recommendation 594[4], which takes into account only the radio system, provides a figure of errored seconds as 0.32 percent or less in any month. This has been discussed in Section 7.3.3.1. This means that the error-free second objective is for any month, at least 99.68 percent for the HRDP of 2500 km. In this section we will attempt to provide an equivalent set of end-to-end BER objectives which provide this error-free performance for a two-way 64 kbit/s connection. The BER objectives will take the form of:

Better than 10^{-6} for x percent of the time
Better than 10^{-5} for y percent of the time
Better than 10^{-3} for z percent of the time

Conversion between BER and percentage of error-free seconds (%EFS) requires consideration of the distribution of errors, which is facility dependent. There is no universal, simple formula which is valid always. To understand the processes involved, some assumed error distributions will be considered together with different interpolations.

7.6.1 Binomial distribution[3]

The binomial distribution is one which describes a situation where an information bit either is in error or is not. If the number of errors is given by the BER or probability of error P_e, the probability of an error not occurring is given by $(1 - P_e)$. If a communications channel is carrying information with a bit rate equal to r_b, then, after a time t, the number of received bits should be $r_b t$ and the probability of n of these bits being in error $P(n)$ is given by the binomial distribution:

$$P(n) = {}^{r_b t}C_n P_e^n (1 - P_e)^{r_b t - n} \tag{7.6}$$

where

$$C = (r_b t)!/[n!(r_b t - n)!]$$

The probability of getting *at least* E errors in $r_b t$ bits is given by:

$$P(E) = \sum_{n=E}^{r_b t} P(n) \tag{7.7}$$

For zero errors occurring in time $t = 1$ second, n is set to zero and equation 7.6 becomes:

$$P(0) = (1 - P_e)^{r_b} \tag{7.8}$$

Thus the percentage error-free seconds is given by:

$$\%\text{EFS} = 100(1 - P_e)^{r_b} \tag{7.9}$$

The Poisson distribution can be said to describe the occurrence of isolated events in a continuum. Thus, if r_b is large and P_e is small, the errors can be considered as isolated events and the binomial distribution reduces to that of the Poisson distribution, where:

$$P(n) = (r_b \cdot P_e)^n \cdot e^{-r_b P_e}/n! \tag{7.10}$$

For zero errors

$$P(0) = e^{-r_b P_e} \tag{7.11}$$

and when $r_b \cdot P_e$ is small this approximates again to produce the percentage of error-free seconds:

$$\%\text{EFS} = 100(1 - r_b \cdot P_e) \tag{7.12}$$

Channel error performance is not usually expressed in terms of an error probability density function, but is more generally expressed in terms of BER targets to be met, as discussed in Section 7.6. These BER targets may be considered as points on the cumulative distribution function $F(E)$, which is the probability that P_e exceeds the value E. The complementary cumulative distribution function $\overline{F(E)}$, which is equal to $1 - F(E)$, is more useful for expressing the BER targets, for it determines the probability of P_e not exceeding E. Consider Figure 7.7, which is a plot of $\overline{F(E)}$ against E. Using elementary calculus, the probability of P_e lying within the interval $(E - \Delta E/2)$ to $(E + \Delta E/2)$ is given in the limit for lim $\Delta E/2$, as $\Delta E \to 0$, as:

$$F(-\Delta E/2 < P_e < \Delta E/2) = \frac{d\overline{F(E)}}{dE}\,\Delta E$$

$$\text{as } \Delta E \to 0 \tag{7.13}$$

From equation 7.9 the incremental error-free second contribution of the error probability contained in the limited interval ΔE is given by:

$$d(\%\text{EFS})/dE = 100.(1 - E)^{r_b} \cdot d\overline{F(E)}/dE \cdot \Delta E \tag{7.14}$$

and the total error-free second contribution of the error probability E is, for the discrete case, given by:

$$\%\text{EFS} = -100\Sigma(1 - E)^{r_b} \cdot dF(E)/dE \cdot \Delta E \tag{7.15}$$

and for the continuous case it is given by:

$$\%\text{EFS} = -100 \cdot \int_0^1 (1 - E)^{r_b} \cdot dF(E)/dE \cdot dE \tag{7.16}$$

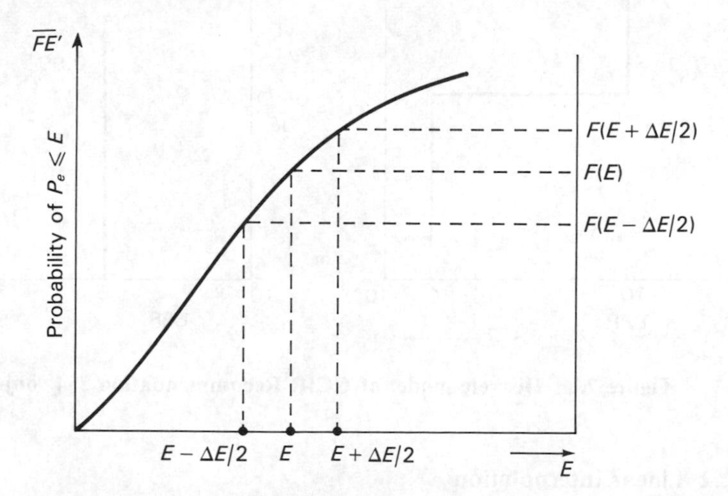

Figure 7.7 Complementary probability distribution function $\overline{F(E)}'$

where

$$\mathrm{d}\overline{F(E)}/\mathrm{d}E = -\mathrm{d}F(E)/\mathrm{d}E \tag{7.17}$$

7.6.2 Forms of *F(E)*

CCIR Recommendation 594[1] states that the bit error ratio for the 64 kbit/s HRDP should not exceed the following provisional values given below, which take account of fading, interference and all other sources of degradation of the performance:

10^{-7} for more than 1 percent of any month; integration time under study
10^{-3} for more than 0.05 percent of any month; integration time 1 second.

This error performance objective will be used to demonstrate how the different percentage EFS objective may be deduced for different distribution forms of $F(E)$.

7.6.2.1 Discrete model

Consider Figure 7.8. This figure results by considering that the two points given in the CCIR Recommendation discussed in Section 7.6.2 are points on a fixed value $F(E)$. Figure 7.8 also shows the negative differential coefficients.

From equation 7.15:

$$\%\mathrm{EFS} = 100[(1-10^{-7})^{r_b}\cdot(0.99) + (1-10^{-3})^{r_b}\cdot(0.0095)]$$

which, if $r_b = 64$ kbit/s, gives %EFS = 98.37.

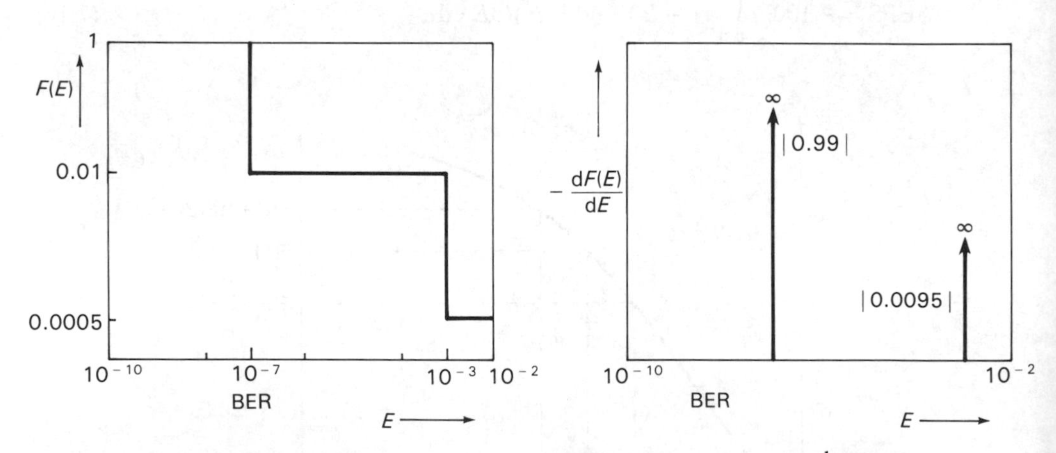

Figure 7.8 Discrete model of CCIR Recommendation 594[1] objectives

7.6.2.2 Linear interpolation

Figure 7.9 shows the case where the CCIR Recommendation 594 objectives are

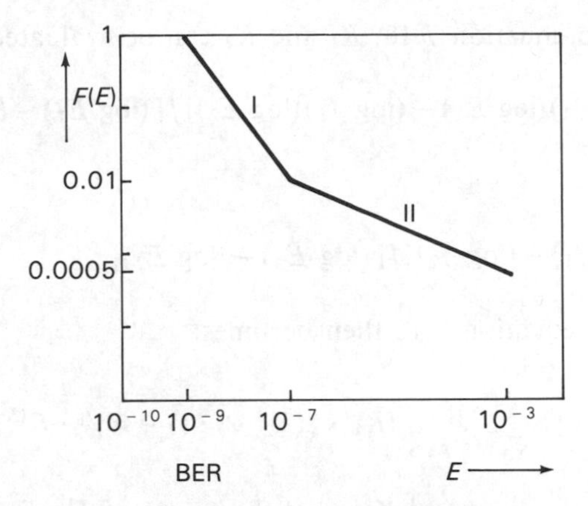

Figure 7.9 Linear interpolation of CCIR Recommendation 594[1]

linearly interpolated between the objective points. To obtain the straight line interpolation for line I, the lowest BER assumed to be achieved is 10^{-9}. The value of $\mathrm{d}F(E)/\mathrm{d}E$ is the slope of the straight line sections, i.e.

Slope of straight line section I $= -9 \times 10^5$

Slope of straight line section II $= -9.5$

Using equation 7.16, the %EFS is given by:

$$\%\mathrm{EFS} = 9 \times 10^7 \int_{10^{-10}}^{10^{-7}} (1-E)^{r_b}\,\mathrm{d}E + 9.5 \times 10^2 \int_{10^{-7}}^{10^{-3}} (1-E)^{r_b}\,\mathrm{d}E$$

For $r_b = 64$ kbit/s, the %EFS = 98.69

7.6.2.3 Logarithmic interpolation of *F*(*E*)

Assuming that $F(E)$ and E are plotted using log scales, i.e. $\log F(E)$ against $\log E$, and that logarithmic interpolation is used, straight line plots will result and be of the form

$$\log F(E) = \log K_1 + K_2 \cdot \log E = \log(K_1 \cdot E^{K_2}) \tag{7.18}$$

or

$$F(E) = K_1 \cdot E^{K_2} \tag{7.19}$$

thus

$$-\mathrm{d}F(E)/\mathrm{d}E = -K_1 \cdot K_2 \cdot E^{(K_2-1)} \tag{7.20}$$

If the terminal points $(E_1, F(E)_1)$, $(E_2, F(E)_2)$ of each line section are substituted into equation 7.18, K_1 and K_2 can be evaluated:

$$K_1 = [(\log f_2)(\log E_1) - (\log f_1)(\log E_2)]/[(\log E_1) - (\log E_2)] \tag{7.21}$$

and

$$K_2 = [(\log f_1) - (\log f_2)]/[(\log E_1) - (\log E_2)] \tag{7.22}$$

The integral in equation 7.16 then becomes:

$$\%\text{EFS} = -100 \sum_{N=1}^{N=N} \int_{E_{1,N}}^{E_{2,N}} (K_{1,N})(K_{2,N}) \cdot (1-E)^{r_b} \cdot E^{(K_{2,N}-1)} \cdot \mathrm{d}E \tag{7.23}$$

where $E_{1,N}$, $E_{2,N}$, $K_{1,N}$ and $K_{2,N}$ are the values of E_1, E_2, K_1 and K_2 appropriate to the Nth approximating line section. Table 7.8 shows values of %EFS for minimum values of BER assumed.

Table 7.8 % EFS for CCIR Recommendation 594[1,4]

Minimum BER assumed	10^{-9}	2×10^{-9}	5×10^{-9}	10^{-8}	2×10^{-8}	5×10^{-8}
% EFS	99.716	99.702	99.671	99.629	99.558	99.376

7.6.3 %EFS performance in multi-section links

Consider an N-section link comprising sections $S_1, S_2, \ldots, S_N$. If the error-free second performance of each section is $\text{EFS}_1, \text{EFS}_2, \ldots, \text{EFS}_N$ and the direction of bit propagation is from S_1 to S_N, then the errored seconds are propagated in the direction S_1 to S_N, it being assumed that a bit error in one section is not corrected in a succeeding section. Furthermore, as it is only error-free seconds which are susceptible to errors, the ratio of error-free seconds leaving the ith section to the error-free seconds entering the ith section is EFS_i and for concatenated links, the overall %EFS is given by:

$$\%\text{EFS (multi-section)} = 100 \cdot \prod_{i=1}^{N} \text{EFS}_i \tag{7.24}$$

A more thorough approach would require a knowledge of the distribution of error probability P_e in each section of the link. If this information was known, the overall distribution could be determined by the successive convolution of the individual distributions. This overall distribution would then be used in conjunction with equation 7.23 to give the %EFS performance of the link.

7.7 ERROR MEASUREMENT

This section deals with the CCIR report 613–3[4] for bit error performance measurements in digital radio-relay systems, both under out-of-service conditions and under normal traffic loading.

7.7.1 Basic criteria for bit-error performance evaluation

As already mentioned, the parameter used to describe the performance of a digital system is the bit error probability, i.e. the incorrect reception of a single bit. Experimentally, the most used parameter is the bit error ratio BER. This is defined as

$$\text{Bit error ratio} = N_e/N_t = N_e/r_b t_0 \tag{7.25}$$

where N_e = number of bit errors in time interval t_0

N_t = number of transmitted bits in the time interval t_0

r_b = bit rate of a binary signal at the point where the measurement is performed

t_0 = measuring time interval or error counting time.

Where the error generation process is random and stationary and the errors are counted in a sufficiently long interval t_0, equation 7.25 can give an estimate of the error probability. The accuracy of the estimate increases as N_e increases. For small error occurrence in the time interval t_0, i.e. for small N_e, the measuring time interval t_0 becomes inordinately large if N_e is to be sufficiently large for reasonably good estimation. Practical requirements on the measuring time interval usually limit, however, the values which can be obtained for N_e. The minimum acceptable value of N_e seems to be about 10 and then the true error probability is contained in a range equal to ± 50 percent of N_e/N_t with a confidence coefficient of 90 percent.

The data required in equation 7.25 may be obtained by means of different procedures. In particular, the measurement could be made of the number of errors detected in a fixed time interval t_0.

This method presents an advantage for extensive statistical measurements of performance indicators and availability. Measurements at 64 kbit/s in accordance with the performance parameters given throughout this chapter should be performed by counting the number of errors in each one-second interval and then processing the results using the algorithm of Figure 7.10. The performance indicators of digital radio-relay circuits and sections are normally measured at the system bit rate. The relationship between measurement results obtained at the system bit rate and at 64 kbit/s has not as yet been definitely established by the CCIR, but it is thought that the integration time required to achieve the same confidence level will be dependent upon the capacity of the system. Measurements at the system bit rate indicate that the percentage of time for which the BER exceeds 10^{-6} is 4 to 7 times lower

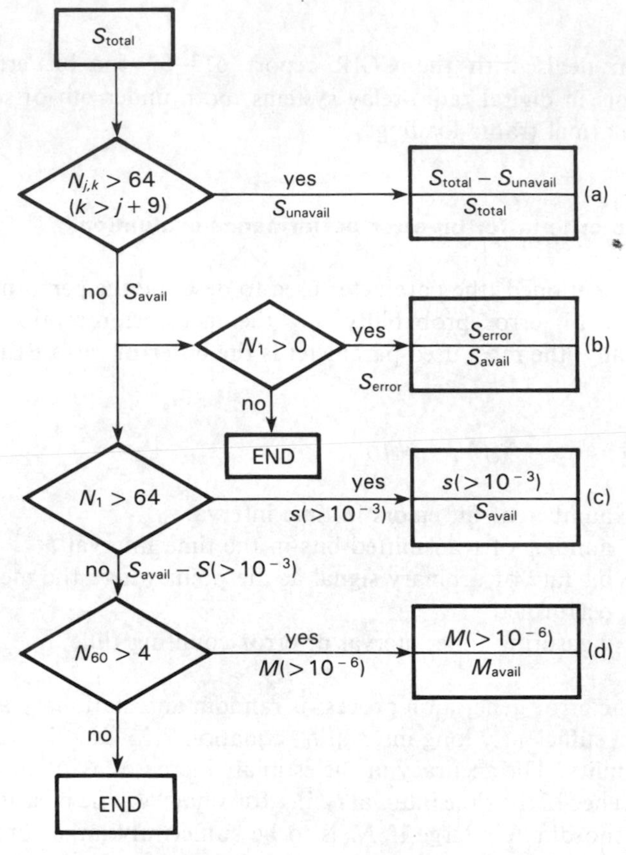

Key

S_{total} : Total measured seconds = one month

N_j : Number of bit errors in j second intervals

$N_{j,k}$: Number of bit errors in each second interval between jth second and kth second

$S(>10^{-3})$: Total time during which the BER exceeds 10^{-3} in each second interval, seconds

$S_{unavail}$: Unavailable time, seconds

S_{avail} : Available time, seconds

S_{error} : Total time where the number of errors is one or more in each second interval, seconds

M_{avail} : Available time, minutes = $\dfrac{S_{total} - S_{unavail}}{60}$

$M(>10^{-6})$: Total time that the BER exceeds 10^{-6}, measured in blocks of 60 consecutive one-second intervals derived by excluding any one-second intervals during which the BER exceeds 10^{-3} minutes

Figure 7.10 CCIR BER measurement algorithm: (a) availability; (b) errored seconds; (c) severely errored seconds; (d) minutes of degraded performance (Courtesy of CCIR, Report 613–3, Reference 4)

if evaluated with a one-second, instead of a one-minute, integration time (CCIR Report 930–1).[4]

Another procedure which may be used to provide the data for equation 7.25 could be followed by the measurement of the time interval required to detect exactly N_e errors, or more conveniently, the number N_t of bits transmitted in such an interval. This method has the advantages of nearly constant measurement accuracy for any value of the error ratio and minimum time to perform the measurement with the required accuracy. In certain cases it seems more convenient to measure the bit error ratio over a sufficiently large number of one-second time intervals following each other. During in-service measurements, the true bit error ratio may differ from the measured value and an allowance must be made for this possibility in the calculation. Transmission burst errors may also occur which require to be taken into account. These burst mode errors are groups of errors of short duration with relatively high error frequency, separated by much longer intervals with a much lower error rate. The occurrence probability of such a phenomenon, the statistical characteristics of it (burst duration, error density during the burst, low-error interval duration, etc.), and the effect on the services carried by the digital link, in particular any framing structures which might be necessary, require further study to arrive at a definition of a suitable performance parameter.

As mentioned in Section 6.3.3.2, for M-ary PSK, Gray coding is used in the mapping of the modulator states, permitting adjacent PSK symbols to differ in only one bit position. Any symbol error which occurs is then related only to a single bit error. The symbol error rate P_e or probability of symbol error for M-ary PSK schemes can be related to the bit-error-rate (BER) by:

$$\text{BER}_{\text{PSK}} = \frac{P_{e\text{PSK}}}{\log_2 M} \text{ (Gray coding)} \tag{6.71}$$

As given by equation 6.60, for equiprobable orthogonal signals, such as M-ary FSK, the symbol error rate P_e can be related to the bit-error-rate by:

$$\text{BER}_{\text{FSK}} = \frac{M}{2(M-1)} P_{e\text{FSK}} \tag{6.60}$$

Section 6.5.3 considers the case where the signal set is such that it is not possible to code so that only single bit errors occur between adjacent signal points, and the BER equals the symbol error multiplied by some factor other than $1/\log_2 M$, i.e.

$$\text{BER} = \frac{KP_e}{\log_2 M} \quad \text{(Non Gray coding)} \tag{6.72}$$

where K is known as the Gray code penalty factor. For M-ary QAM

$$\text{BER}_{\text{QAM}} = \frac{P_{e\text{QAM}}}{\log_2 \sqrt{M}} \quad \text{(Gray coding)} \tag{7.26}$$

Figures 6.5, 6.18, and 6.28 all show the probability of symbol error P_e, except for binary schemes.

7.7.1.1 Residual bit error ratio (RBER)

The RBER is the bit error ratio understood to mean the error ratio in the absence of fading, and it includes allowance for system inherent errors, environmental and aging effects and long-term interference, but excludes short-term interference. Short-term interference is the interference due to anomalous propagation conditions and typically consists of very high levels of interference which occur only rarely and for short periods of time. Long-term interference is that which arises from sources within line-of-sight of the victim receiver and is typically low in level and constant in value. For a 2500 km high-grade real circuit, the RBER value, which may be linearly apportioned according to the circuit length L km, is given by CCIR Draft Recommendation 634,[4] as

$$\text{RBER (high-grade circuit)} \leqslant (L \times 5.0 \times 10^{-9})/2500 \tag{7.27}$$

The measurement of the RBER may be obtained by taking BER measurements over a period of one month using a 15-minute integration time, by discarding the 50 per cent of 15-minute intervals which contain the worst BER measurements and by taking the worst of the remaining measurements. The CCIR do specify that this measurement method is, for the moment, provisional and that other measurement integration periods may also be suitable, particularly with small-capacity systems. A difficulty arises when defining a measurement procedure for RBER, because it cannot be easily verified that fading is absent on all hops within a section during the measurement period. In the CCIR provisional testing method just discussed, it should be noted that periods containing errors due to effects other than fading may also be discarded by the procedure, and, in some circumstances, periods affected by fading may be included. In determining the percentage of intervals to be discarded, account needs to be taken of the need to insure that the errors in concatenated digital sections do not accumulate to form additional degraded minutes within the overall HRDP. If measurements are made over a shorter period than one month, it may be appropriate to discard a higher percentage of intervals. For the local-grade digital radio relay systems, which may form part of the integrated services digital network (ISDN), the residual bit error ratio (RBER) shall not exceed D, where $10^{-7} < D < 10^{-10}$.

7.7.2 Error performance measurement under out-of-service conditions

An out-of-service test or intrusive measurement implies that a digital transmission link has been taken out of revenue or monitoring service. This form of measurement is therefore more suitable for design, production testing and initial installation and commissioning tests. This is particularly so for high-capacity long-haul transmission systems. The test is done by inserting a test pattern into the input of the link, which simulates the normal traffic data stream. The output of the link is compared bit by bit with a locally generated error-free reference pattern in an error detector. The choice of test pattern is usually made between a pseudo-random binary sequence (PRBS), sometimes called an 'm' sequence, to simulate the traffic, or specific word patterns to examine the transmission equipment for pattern-dependent tendencies or critical timing effects. The pattern is normally encoded into some standard line interface code to allow single port connections of the data signals between terminals, test equipment and transmission equipments. The test pattern which is a pseudo-random test sequence has a repetition period equal to $2^n - 1$. Also the shift-register generator which produces, from a register of n bits, the test sequence is designed to produce within the repetition period, at least once, every combination or sub-sequence of n bits. The combination comprising n zeros is prohibited. The sub-sequences lower than n bits should also be balanced in their occurrence throughout the repetition period. To have a reliable measurement on the inter-symbol interference effects, the register length n should be greater than, or equal to, the number of pulses which are affected by the channel response to a single pulse. As this requirement depends on characteristics of the transmission channel which are often unknown, a conservative value of n is usually chosen. To check the suitability of the test sequence, the error rates observed with different register lengths can be compared. Another aspect to be taken into account when evaluating the measurement accuracy is the coding operation. The coding operations are likely to modify the symbol sequence in the modulated signal and may cause an increase in the detected errors during the decoding and self-descrambling operations. The test sequence period, equal to $(2^n - 1)$, gives rise to a frequency spectrum with discrete components and with a fundamental frequency $r_s/(2^n - 1)$, where r_s is the line symbol rate. Care must be taken to insure that the test sequences applied to the input of the digital modulator do not generate undesirable RF spectra, especially when taking measurements on interference. Table 4.10 in Section 4.4.3.1 shows the required pattern length for the various hierarchical bit rates as per CCITT Recommendations O151 and O152.[5] Table 7.9 provides a summary of the test pattern length against hierarchical bit rate, which is required for test purposes.

Table 7.9 Hierarchical bit rates and test pattern lengths

Bit rate, kbit/s	Test pattern length
64	$2^{11} - 1$
2048, 8448, 1544, 6312, 32064, 447346	$2^{15} - 1$
34368 and 139264	$2^{23} - 1$

7.7.3 Error performance measurement under in-service conditions

The main problem in this type of measurement is the detection of bit errors in the unknown digital pulse stream. Methods which may be used make use of *a priori* knowledge of some characteristics of the received signal. Other problems associated with this type of measurement are the duration of measurement which may be limited by the necessity for immediate action and the nonstationary behavior of the transmission channel which varies the readings of the error ratio obtained from one measurement to the next. Most digital transmission systems have some built-in error-monitoring facilities. The form which it takes may be based on some property of the traffic which is being carried. For terminal equipment it may be the frame alignment signal, for optical line equipment it may be light-emitting diodes indicating the bit error rate, etc. The most common method of in-service error monitoring is detection of line code violation. A specification for a low-bit-rate instrument to do this type of monitoring is specified in CCITT Recommendation O161.[5] This is possible for systems where the interface code signals to the radio link have the same format as the line code, e.g. AMI, HDB3, B6ZS or B8ZS. Another common method of in-service error monitoring is to check some fixed pattern always present in the transmitted signal. This overcomes some limitations of code violation monitoring and can take several forms, the most common being that of the frame alignment signal (FAS) of the terminal multiplex connected to the digital radio link. The approach has the advantage of checking a fixed pattern at various points along the route, where the overall digital connection comprises multiplexing and demultiplexing equipments and a number of digital line sections which operate at various bit rates. The checking of the fixed pattern can be done wherever there is a digital multiplex breakout. A problem arising with higher-order multiplex systems is the translation of errors down to the 2 Mbit/s and 64 kbit/s levels. This can be done by measuring the errors on an out-of-service 64 kbit/s channel or 2048 kbit/s tributary which is contained within the higher-order system. The implications of this are that the error detector must itself contain demultiplexers so that the 64 kbit/s or 2 Mbit/s test stream may be extracted from a system carrying traffic at the system monitor points. This demultiplexing capability of the measuring equipment is becoming more feasible as component costs continue to fall. The cost of taking a couple of 64 kbit/s circuits out of service is outweighed by the increase in the value of the BER measurement made under real traffic conditions. This type of measurement provides a more accurate estimate of the average error rate in the traffic because it can be based on a bit-by-bit comparison of the decoded binary signal including any error extension which will be seen at the system terminal.

The methods recommended by the CCIR (Report 613–3)[4] are given in Sections 7.7.3.1 to 7.7.3.3.

7.7.3.1 Test sequence interleaving

The error performance may be estimated by inserting sample pulses into the digital pulse stream at a rate of $1/N$ of the clock rate. At the receiver these pulses are extracted and their errors counted. The error performance of the system is estimated from this count of the errors within the sample pulses. Using this method the time taken to recognize a specific error ratio is N times that required for counting the errors of all transmitted pulses. The error performance of the radio relay circuit can be estimated independently of the signal content at the line interconnection interface TT', as shown in Figure 5.1. A separate synchronizing frame is not always needed to extract the sample pulses at the receiver. The interleaved pulses should be independent of the digital multiplexer frame.

7.7.3.2 Parity-check coding

For this method a parity-check symbol is added to each group of s symbols at the transmitter end. The value of the added symbol equals the sum of the s symbol modulo M where M is the number of possible symbol values, i.e. the number of levels being transmitted. That is:

$$\text{Value of parity symbol} = \sum_{n=1}^{s} G_n \text{ modulo } M$$

where G_n is the value of the nth symbol of the group of s symbols.

At the receiver the modulo M sum of the s group of received signals is compared with the value of the received parity-check symbol. It is assumed that if there is agreement, no errors have occurred, and vice versa. For a low error probability (less than 10^{-3}) the number of disagreements obtained from the parity check comparison is approximately equal to the total errors of all the pulses in the digital signal. This method requires a synchronizing frame.

7.7.3.3 Cyclic code error detection

The error performance of a digital radio relay system can be estimated by using a technique called *cyclic code error detection*. CCITT Recommendation V41,[6] Appendix I, shows the encoder/decoder pair and describes the operation. This technique is used in data transmission because of its ability to detect multiple errors. A four-bit synchronizing pattern (0101) is required so that the 16 check bits may be examined in successive bit intervals. When the pattern is recognized the register and bit counter in the decoder are set to zero and decoding proceeds normally.

7.7.3.4 Summary

The digital pulse stream from a digital multiplexer usually contains added bits as well as the information bits. These are, for example, used for framing. In general, error performance should be estimated independently of the format of the information signal. It follows that it is considered inadvisable in a digital radio relay system to use such added bits for measuring the bit error performance.

7.7.4 Performance monitoring units (PMU)[7-9]

In Section 6.3.4 a performance monitoring unit was described[9] which used an extrapolated error rate technique obtained by modifying the decision thresholds in conjunction with a digital receiver, to produce a pseudo-error rate larger than that of the receiver. A method of estimating the pseudo-error rates corresponding to two or more modified decision thresholds was developed. By extrapolating the results obtained from the estimates of two or more of the resulting pseudo-error rates, an estimate of the error rate was determined, which is equivalent to that which would be obtained with an unmodified decision criterion, i.e. an estimate of the actual error rate of the receiver. The major advantage of using this method over the direct error-counting method is that the measuring time is less to obtain a statistically significant number of actual errors. This is because the pseudo-error rate is larger than the actual error rate. The CCIR mentions in Report 613–3, as part of its in-service bit error ratio measurements, this technique and quotes References 10, 11 and 12.

Another method of in-service detecting of transmission performance is to provide some unique time sequence of symbols to be transmitted. If that unique sequence is not received, a transmission problem is indicated. Due to the additional requirements of digital radio transmission systems, such as auxiliary channels and multiplexing of multiple digital input signals, a frame format is often required which occurs often enough to be used together with parity bits generated from the information bits. This combination permits an accurate determination of performance by verifying the receipt of the expected sequence of frame and parity bits. Both logical and symbol state faults are detectable by this method, but logical faults cannot be detected using the pseudo-error rate method.

A technique for monitoring the error performance of a pilot-tone PSK system is described in Reference 13. This method works well with Gaussian noise, impulse noise, inter-channel or inter-symbol interference, or with interference from other transmissions as well as when the system is operating over a fading channel. The measurement time required for reliable estimates of the error rate is dependent on the statistics of the noise, the order of diversity and the actual error rate.

The information obtained by continual monitoring of the performance of digital radio whilst in-service and without disrupting traffic may be passed to a centralized maintenance operations center using either telemetry networks or the digital radio auxiliary channels. The information transferred to this center may often include, with the error performance information, all alarm information.

7.7.5 Centralized maintenance operations[14–16]

For maximum effectiveness of centralized maintenance operations, one important point is that a transmission system, like a digital radio, should be in a situation where only *corrective* maintenance is applied, i.e. no routine maintenance scheme is permitted. Today that is possible due to improved reliability of components and improved circuit designs. When a failure occurs which requires attention, the built-in test equipment must reliably and accurately indicate the failure to the centralized maintenance center. The combined effects of a need for corrective maintenance procedures only and the demanding performance requirements of some digital transmission services make it necessary to be able to detect and report intermittent and very low error-rate failures. Such values of error rate may be as low as 10^{-9}. If an error occurs, the maintenance center operator is required to take the necessary action based upon the information at hand. This information may arise not only from performance monitor alarms but from the complete loss of transmission alarms. That places the additional requirement on the system that the complete loss of system alarms should be separated from the performance alarms and data. The categories in which transmission impairments may be classed are:

A low but unacceptable error rate.
Short-duration error bursts that cause momentary loss of the digital radio frame.
Longer-duration error bursts that cause error rates worse than 10^{-6}.

The characteristics of low but unacceptable error-rate problems are:

A meaningful average error rate cannot be determined because errors which occur are not significantly correlated in time.
There are no periods of high error rate or loss of frame.

Short duration error bursts that cause momentary loss of frame may be considered as a problem which is the other extreme of the low-error-rate problem. This category of problem has the characteristics of:

Short periods of high error rate indicated by a loss of frame synchronization.
Loss-of-frame duration which may be between 5 and 25 ms.
Significant time correlation of events, which means that errors of 10^{-6} occur in bunches lasting several seconds and then periods of several hours where no errors occur.

These three categories of transmission impairment are difficult to identify because multipath and other propagation phenomena also cause similar parity errors and frame losses. Typically, multipath fading events usually occur with a 1–2 hour period of a day where error rates vary slowly over seconds. In addition to this there are, during the fading, periods of high error rate worse than 10^{-6}, which may last for several seconds after loss of frame alignment occurs.

The methods of analyzing the performance data at the local site are; displaying from the built-in test equipment; the impairment category that has been detected; the current error-rate performance and an indication of any short error bursts or frame losses. This facility reduces the amount of test equipment required to be transported to site. In addition to this local display of transmission performance, the performance data may be analyzed using microprocessors which are designed to recognize any of the transmission impairment categories and report a performance alarm to the maintenance center. This method assists in the prevention of false alarms and from propagation-induced data. Since every transmitter-to-receiver pair is continuously monitored, direct alarm indications of unacceptable performance can be provided to the maintenance center. This reduces the fault location time. The performance analysis facilities offered to the local center may be also performed at the maintenance center by transmitting the information from the built-in test equipment at the local site or by creating a microprocessor-to-microprocessor link between the site and the maintenance center.

7.8 RELIABILITY OBJECTIVES[4,17]

Some administrations consider that there should be two separate recommended objectives for unavailability, one associated with the failure of equipment, and the other with fading, (which would primarily apply to systems operating above about 10 GHz). The rationale is that these two causes of unavailability are seen quite differently by the user and make quite different demands upon the operator. Unavailability due to equipment failure occurs less frequently (about every three years on average for typical mean-time-between-failures (MTBF), which is discussed later in this section), but results in longer interruptions. Unavailability due to equipment unreliability is given by the expression:

$$\begin{aligned}\text{Unavailability} &= [1 - (\text{MTBF})/(\text{MTBF} + \text{MTTR})] \times 100 \text{ percent} \\ &= 100 \times [(\text{MTTR})/(\text{MTBF} + \text{MTTR})] \text{ percent} \qquad (7.28)\end{aligned}$$

and therefore

$$\text{Availability} = 100 \times [(\text{MTBF})/(\text{MTBF} + \text{MTTR})] \text{ percent} \qquad (7.29)$$

where MTBF is the equipment mean-time-between-failures in hours, and MTTR is the mean-time-to-restore-service in hours. This includes the response time to detect the fault and to start the repair.

Figure 7.11 shows the resulting unavailability as a function of MTTR, given that the MTBF is 25000 hours for a link comprising two transceivers, each with an MTBF of 50000 hours. Due to differing operational environments, the MTTR will vary over a wide range. Values of from 6 to 48 hours have been suggested, which correspond to unavailabilities of between 0.025 and 0.200 percent. It should be noted that the CCIR recommendation on availability (CCIR Recommendation 557)[4] currently applies to both directions of transmission, simultaneously, whereas the CCITT Recommendation G602, and the error performance recommendations, apply to each direction independently.

Reliability in the field of communications deals with the preferred or unpreferred events in the service provided by physical equipment. Since it is not possible to predict with certainty when these events will occur, probability concepts have to be employed to give reliability a quantitative meaning. One of the more important practical probability concepts is that of the mean-time-between-failures (MTBF). Manufacturers usually provide estimates of the MTBF for various items of equipment, so that availability may be determined for a complete system. One common denominator which must be considered is the confidence limit placed on the MTBF provided. This may not always be readily procurable and when it is not, most calculations will assume unscientifically that the availability calculations are reasonably probable. Table 7.10 lists values of the MTBF for various items of equipment, which may be used as a guide. These values will of course vary from manufacturer to manufacturer and are presented here to give some idea of what to expect and to assist in design of systems.

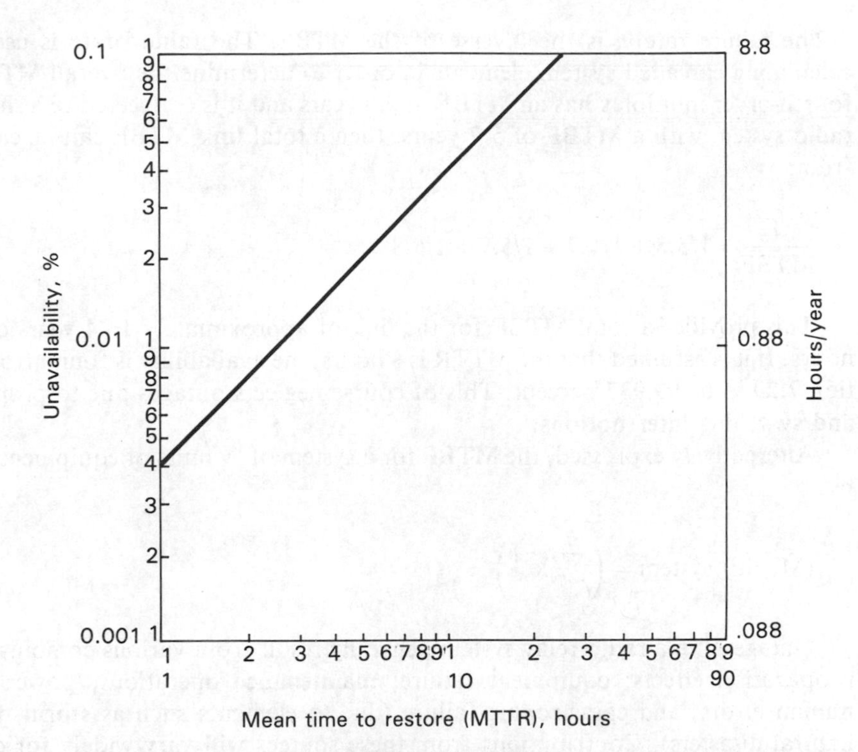

Figure 7.11 Equipment reliability (bidirectional link comprising two 50000 hour MTBF transceivers (Courtesy of CCIR, Report 1053, Reference 4)

Table 7.10 Typical values of the MTBF for various items of equipment

Equipment	MTBF, years	Failures/ year	Equipment	MTBF, years	Failures/ year
Multiplex equipment					
Primary multiplex	4.5	0.22	Second-order MUX	9.4	0.11
Third-order MUX	8.2	0.12	Fourth-order MUX	5.8	0.17
Radio transceivers					
2 Mbit/s unprotected	1.0		34 Mbit/s protected	53.5	0.019
140 Mbit/s unprotected	5.7	0.18	140 Mbit/s protected	540	0.002
Ancilliary equipment					
Selection switch	250000	4×10^{-6}	Hot-standby switch	83333	1.2×10^{-5}
Power	10^7	10^{-7}			
Optical fiber link equipment (per 100 km of path)	2.8	0.36			

The failure rate λ is the inverse of the MTBF. The failure rate is used when calculating cascaded system elements in order to determine the overall MTBF. If a fourth-order multiplex has an MTBF of 5.8 years and it is connected to a short-haul radio system with a MTBF of 5.7 years, then a total link MTBF can be calculated from:

$$\frac{1}{\text{MTBF}} = 1/5.8 + 1/5.7 + 1/5.7 + 1/5.8$$

This provides a total MTBF for the link of approximately 1.44 years or 12600 hours. If it is assumed that the MTTR is 8 hours, the availability is found from equation 7.29 to be 99.937 percent. This of course neglects outages due to propagation and switching interruptions.

Alternatively expressed, the MTBF for a system of N units of equipment is given by:

$$(\text{MTBF})\ \text{system} = \left(\sum_{i=1}^{N} \lambda_i^{-1} \right) \tag{7.30}$$

Outages[17] in a radio relay system generally result from various combinations of propagation effects; equipment failure; maintenance operations; power failure; human errors; and catastrophic failure (due to elements such as storm, war and natural disasters). Contributions from these sources will vary widely for different facilities. There are various methods of minimizing the outage of microwave radio relay systems. These include design of quality equipment, good path engineering, and an adequate operational environment, which provide the basic reliability of microwave systems. That alone is not sufficient for high-capacity common-carrier systems, particularly with respect to equipment and propagation failures, if the availability objectives for these systems are to be attained. Any technique used to reduce outage or unavailability time must protect the service and permit maintenance without unduly increasing the system cost and complexity, and without decreasing the performance. No method other than complete alternative routing can provide an effective counter-measure against multiple-channel outages caused by multipath fades or, where the frequencies are applicable, rain fades. Frequency and space diversity in their various combinations together with hot-standby systems and polarization diversity are some of the usual protection measures. These techniques, together with a more detailed discussion on fading, will be discussed in Chapter 8. It suffices to say that for outages caused by equipment failures, some redundancy must be provided if the availability objectives set for the system cannot be attained by an unprotected system.

7.8.1 Simple redundancy

The most simple form of redundancy which can be employed is that of a standby system operating in parallel with the active system. This standby system takes over whenever the on-line system fails. Thus a failure on the on-line system constitutes a complete system failure only if the standby system is unavailable. If the MTBF of the on-line system is given as T, its failure rate is given as $1/T$. If the standby system is identical to that on-line, it will also have a failure rate of $1/T$, so that the probability of either channel failing within some unit time which includes the MTTR t, is $2/(T+t)$. The MTTR is included in this unit time to take account of the case where one of the two systems may be under repair. This is the same as saying that, as two systems are operating together at the same time, the probability of both systems failing at different times (or together) within the same year, say, is twice that of a single system failing. Of these failures, the only time a *complete* system failure occurs, is when the standby system is unavailable and under repair. If the mean time to repair a fault (i.e. the MTTR) is t, the time a system is unavailable will be when it is being repaired; hence the probability of either channel being unavailable at any instant is the ratio of the MTTR to the total time (MTBF + MTTR), or $t/(T+t)$, and so the effective failure rate of the systems is the probability of either system failing *and* the standby system being under repair:

$$\text{Dual system failure rate } 1/T' = [2/(t+T)]\cdot[t/(t+T)] = 2t/(T+t)^2 \qquad (7.31)$$

The effective system mean-time-between-failures T' is then given by

$$T' + t = (T+t)^2/2t \qquad (7.32)$$

and the availability is given from equations 7.29 and 7.32 as:

$$\begin{aligned}\text{System availability} &= 100\times[((T+t)^2/2t) - t]/[(T+t)^2/2t] \text{ percent}\\ &= 100\times[(T+t)^2 - 2t^2/(T+t)^2] \text{ per cent} \qquad (7.33)\end{aligned}$$

Equation 7.33 applies to a monitored hot-standby (MHSB) radio terminal comprising two transmitters and two receivers. Only one transmitter, however, is 'on-line'. The other transmitter feeds a dummy load although it shares the same baseband information. On detection of a fault the RF switch automatically switches the off-line transmitter to 'on-line'. The two receivers of the MHSB terminal are both fed to a baseband switch which switches either receiver 'out' should one go faulty.

Example 1

Determine the system MTBF T' and the availability for a hot-standby system in which each system has a MTBF of 10000 hours and it is assumed that the MTTR is 12 hours if either system fails. It is also assumed that the switchover network is perfectly reliable and that it takes zero time to switch over from the faulty system to the standby system.

Solution

Using equation 7.32, the effective system MTBF is found 4 176 661 hours (approx. 476 years). This is a redundancy improvement factor of 4 176 661/10000 = 417.66. Using equation 7.33, the availability is determined to be 99.99971 percent. Note that the availability of a nonredundant system would be 99.88 percent.

To perform a complete analysis, the system may be considered as a chain of components comprising that system. Thus a series analysis in terms of failure rates may be made, in which each component failure rate is determined from the MTBF of that component. The final system failure rate is then determined by the summation of all of the failure rates and the system MTBF determined as given by equation 7.30. The availability can then be determined by using equation 7.29.

The weakest link in the chain for a hot-standby radio relay system are the hot-standby switches (Baseband or RF). If these switches do not do their job, the advantage of having a hot-standby system is lost.

7.9 PROTECTION SWITCHING

7.9.1 Factors influencing the choice of switching criteria

Continuous monitoring of a digital radio-relay system is necessary for initiating protection switching under conditions of channel failure. The switching of the protection switch in a multi-line protection switching system, however, is not always advantageous or practical if there exists only an intermittent failure mode condition which produces a high error ratio, or frame loss persisting for a few microseconds. The capability of a system to detect interruptions and inform maintenance personnel of the adverse condition is however, necessary and has been described in Sections 7.7.4 and 7.7.5. If switching is to be used to protect against equipment failure, relatively slow recognition of the switching criteria, transmission of the switching instructions and switch changeover can be tolerated. This slow changeover will cause a loss in synchronism at the remote terminal and provision may be required to minimize the number of interruptions which are caused. This in turn means that the switchover time is limited. If switching is used to improve performance during poor propagation conditions, the opposite situation exists, since rapid recognition of the switching criteria is required and switching to the standby channel without loss of synchronism is desirable. The loss or addition of a bit, due to the switching to the prior channel without prior arrangement with the distant terminal to insure coincidence, may cause the downstream transmission chain to be completely desynchronized. This affects system performance and cannot thus be considered as a single isolated error.

7.9.2 Protection arrangements

The type of switching equipment used and its characterization is determined by the service which it is to provide, such as an improvement in availability, switching speed, etc., the point in the circuit or path where the switching is to be done, and the switching criteria and mode of transmission of the switching instructions. As described in Section 7.8, the main purpose of protection switching is to improve the system availability by switching to available standby channels in the event of equipment failure. The additional benefit which also can be gained is when the equipment protects against outages caused by frequency-selective fading by switching in different systems in a multi-line protection-switched system. Space diversity protection may also be used to improve system performance during poor propagation conditions. This will be discussed in more detail in Chapter 8. Dual route diversity protection facilitates the use of greater hop lengths in frequency bands where attenuation due to precipitation becomes appreciable. This form of protection involves the use of switching at the terminal receiver and consequently it may be necessary to equalize the difference in transmission time between the path lengths of each route, in order that the signals on both routes may be aligned at the instant of changeover. Single-route systems may be protected using multi-line and single-line switching with either a dedicated or nondedicated protection channel, or by space diversity operation. Theoretically, the switch may operate at RF, IF or baseband, but in practice, baseband switching is preferred since this may protect the complete channel from the input port to the output port with minimum duplication of equipment outside the switched path.

EXERCISES

1 In a telephone network, an error burst which lasts approximately one second occurs. An error-rate measuring set monitoring a 64 kbit/s circuit shows that this error burst gave rise to seven errors. Would this error burst be large enough to cause multiple call drop-out, if it occurred on all 64 kbit/s circuits in the network? (*Answer* see Section 7.2.4, which indicates that there is a negligible probability of multiple call drop-out being caused.)

2 For a hypothetical reference digital path of 1000 km used for digital radio relay systems determine the degraded minutes objective, errored seconds objective and the severely errored seconds objective.

(*Answer* Degraded minutes: the BER should not exceed 10^{-6} during more than 0.16 percent of any month
Errored seconds; the BER should not exceed 10^{-3} during more than 0.128 percent of any month
Severely errored seconds; fewer than 0.054 percent of one-second intervals are to have a BER worse than 10^{-3} during any one-month interval.)

3 If the system considered in Question 2 records seconds during which the BER exceeds 10^{-3}, should this be taken account of in determining the errored seconds objective? (*Answer* the severely errored seconds must be subtracted from the value of the errored seconds objective.)

4 If the system considered in Question 2 records a BER greater than 10^{-3} for greater than ten consecutive seconds, should the severely errored seconds be discounted from the errored seconds measurements? (*Answer* the system under these conditions is considered as 'unavailable' and none of the three measurements is valid during this period.)

5 What does the statement 'at least 99.944 percent of one-second intervals are to have zero errors', refer to? (*Answer* a hypothetical reference digital path of 2500 km using digital radio-relay systems: errored seconds objective.)

6 Express the severely errored seconds objectives for a digital radio-relay system HRDP, in the same form as that given in the statement of Question 5. (*Answer* at least 99.944 percent of one-second intervals are to have a BER less than 64 errors when measured at 64 kbit/s.)

7 Express the degraded minutes objectives for a digital radio-relay system of length L, where 280 km $< L <$ 2500 km, in the same form as that given in the statement of Question 5. (*Answer* at least $(100 - 0.4L/2500)$ percent of one-minute intervals are to have four or fewer errors, when measured at 64 kbit/s.)

8 For a 42 km radio link, determine the percentage of one-minute intervals with greater than four errors permitted for a high-grade system. (*Answer* 0.045 percent of any month)

9 For a 28 km radio link, determine the percentage of seconds with one or more errors which is permitted for a high-grade circuit. (*Answer* 0.036 percent of any month)

10 For a 35 km radio link, determine the percentage of severely errored seconds objective allowed at each end of the link for a medium-grade circuit (*Answer* 0.015 percent block allowance. Note that this is also the same for a local-grade circuit.)

11 A high-grade circuit of 280 km was found to have a both-way interruption of 125 s during a heavy rain storm in February. Determine the percentage availability of the circuit during this month and whether the link would satisfy the availability objective of 99.8 percent if no equipment failures occurred during this interval. (*Answer* 99.96 percent, yes)

12 Determine the residual bit error ratio (RBER) objective for a 280 km high-grade circuit. (Answer 5.6×10^{-10})

13 Would a radio system with an MTBF of 20000 hours and an MTTR of 48 hours be able to satisfy the availability objective of 99.98 percent? (*Answer* availability = 99.76 percent, no)

14 Would a hot-standby radio system with an ideal changeover switch and the same system MTBF and MTTR as in Question 13, be able to satisfy the availability objective of 99.98 percent? (*Answer* availability = 99.999 percent, yes)

15 You are given that each end of a system comprises a radio terminal with an MTBF of 5.7 years, one fourth-order multiplex with an MTBF of 5.8 years, two third-order multiplexes, each with an MTBF of 8.2 years, two second-order multiplexes, each with an MTBF of 9.4 years, and four primary multiplexes, each with an MTBF of 4.5 years. Determine (*a*) the system MTBF and (*b*) the availability of the system if an MTTR of 48 hours total is expected for one year and that any unit failing in any multiplex configuration is considered as a failure. (*Answer* expected failure rate for complete system is 3.38/year, and therefore system MTBF is 0.296 year or 2593 hours. Availability = 98.18 percent)

16 If a hot-standby radio link is used with an ideal changeover switch and MTTR of 8 out of the 48 hours given for the system MTTR over the year, determine the new availability figure for Question 15. (*Answer* new failure rate = 1.510225/year, MTBF = 0.662 year, availability = 99.32 percent)

REFERENCES

1. 'Fixed Service using Radio-Relay Systems', Recommendations and Reports of the CCIR, **IX–1**, XV Plenary Assembly, 1982.
2. 'Digital Networks. Transmission Systems and Multiplexing Equipment', Recommendations of the CCITT, G700–G956, **III–3**, VIII Plenary Assembly, 1984.

3. King, T., Private communication, 1986.
4. 'Fixed Service using Radio-Relay Systems', Recommendations and Reports of the CCIR, **IX–1**, XVI Plenary Assembly, 1986.
5. 'Specifications of Measuring Equipment', Recommendations of the CCITT, O Series, **IV–4**, VIII Plenary Assembly, 1984.
6. Recommendations of the CCITT, V series, 'Data Communication over the Telephone Network', **VIII–1**, VIII Plenary Assembly, 1984.
7. Ashlock, J. C. and Posner, E. C., 'Application of the Statistical Theory of Extreme Values to Spacecraft Receivers', Jet Propulsion Laboratories, Pasadena, California, Technical Report 32–737, May 1965.
8. Robinson, W. G., 'Maintenance of Digital Radio using Remoted Performance Monitoring', *Links for the Future*, IEEE International Conference on Communications, 1984, pp. 456–461.
9. Gooding, D. J., 'Performance Monitor Techniques for Digital Receivers based on Extrapolation of Error Rate', *IEEE Transactions on Communication Technology*, 1968, **COM–16**, No. 3, pp. 380–387.
10. Calindrino, L. and Crippa, G., 'Indirect Method for Error-Rate Measurement in Radio-Relay Systems', Telettra Internal Report, 1975.
11. CCIR Documents, 9/81 (Netherlands), 1974–1978.
12. Sant'Agostino, M., 'Performance Monitoring Devices for Digital Radio-Relay Links', *CSELT Rapporti Tecnici*, 1975, **3**, No. 4, pp. 13–29.
13. Hingorani, G. D. and Chesler, D. A., 'A Performance Monitoring Technique for Arbitrary Noise Statistics', *IEEE Transactions on Communication Technology*, 1968, **COM–16**, No. 3, pp. 430–435.
14. Robinson, W. G. See Reference 8 above.
15. Guenther, R. P., 'Planning for Centralized Maintenance Control of Digital Facility Networks', *Links for the Future*, IEEE International Conference on Communications, 1984, pp. 462–465.
16. Mines, J. A. and Zuckerman, D. N., 'Integrating Performance Monitoring Data with other Alarm Information to enhance Digital Transmission System Maintenance', *ibid.*, pp. 471–475.
17. Barnett, W. T., 'Microwave Radio-Relay: Attainment of Reliability Objectives', *IEEE Transactions on Communication Technology*, 1966, **COM–14**, No. 1, pp. 39–46.

8 PROPAGATION AND FADING CHANNELS

8.1 INTRODUCTION

The most significant parameter for evaluating the quality of a digital radio system is the outage time, or the probability of exceeding a stated BER. The outage time can be evaluated starting from the fade margin, which is the difference between the RF received power and the threshold level of the receiver, for a specified BER.

From the fade margin, determined from the system data for each path, the time percentages for the performance and availability criteria can be calculated. In this chapter, information on calculating the fading parameters due to multipath propagation and rain, etc., is presented so that the fade margin may be calculated. With the value of fade margin obtained, calculations on the expected outage time will be given to permit a calculation on the expected percentage availability for the link or system. Also methods used to reduce the outage time will be considered. The figures so obtained can then be compared with the objectives given in Chapter 7, to determine if link or system redesign is necessary to meet these objectives. To insure that the processes of fading can be placed into proper perspective and to lay the groundwork for line-of-sight radio link engineering, to be discussed in Chapter 9, the fundamentals of radio wave propagation including the more important equations will first be presented.

8.2 FUNDAMENTALS OF RADIO WAVES

The two classes of wave which are readily found in the real world are longitudinal waves and transverse waves. In longitudinal waves, the particles comprising the medium are compressed or expanded in the direction of the wave motion. An obvious example of this type of propagation is that of sound waves in the air. For electromagnetic radiation, the waves are transverse, i.e. waves are propagated in a direction which is perpendicular to their motion. The movement of a snake is similar, in that its body movement is almost perpendicular to the direction in which it is traveling. An electromagnetic wave in three-dimensional space consists of two component waves. These are the electrostatic or electric component E and the magnetic component B or $H(B = \mu H)$. These components are orthogonal. A radio wave is polarized in the direction of its *electric* component. The dipole radiator, whether alone, used in a dish antenna or used in other types of antenna, aligns itself to the E field and thus indicates the direction of polarization. That is a vertically orientated dipole indicates vertical polarization, however a rectangular waveguide feed has its E field perpendicular to the longest side of its cross-section. Radio waves may be propagated about the earth differently. The two main routes which they may travel from the transmitting antenna to the receiving antenna are either by the ionosphere (sky wave), or by hugging the ground (ground wave). The ground wave may itself be divided into two types: the surface wave and the space wave. For the space wave, three paths may be used to traverse the distance between the transmitting antenna and the receiving antenna. These are the direct wave, the ground reflected wave, and the tropospherically reflected wave.

8.2.1 Ground waves

Ground waves cover those waves which are not influenced by the ionosphere, i.e. the *surface* and the *space wave*.

8.2.1.1 Surface wave

In surface wave propagation the wave energy glides over the Earth's surface, more or less in the same manner as a wave travels along a wire. This mode of propagation is useful only for the lower radio frequencies, i.e. below 30 MHz. The Earth's attenuation increases rapidly with the increase in frequency, since the attenuation depends upon the effective conductivity and dielectric constant of the Earth. Above 30 MHz the Earth acts as a poor conductor and the attenuation is too high for VHF, UHF and SHF propagation.

8.2.1.2 Space wave

This mode of propagation is that which is mostly used by systems considered in this book. The wave travels in the troposphere, which extends to ten miles above the Earth's surface. The wave energy travels from the transmitting to receiving antenna either in a straight line (line of sight) or is reflected at the ground or from the troposphere. The space wave is the one which is of importance in VHF, UHF and SHF communications.

THE DIRECT WAVE

This wave proceeds directly from the transmitter antenna to the receiver antenna, without being refleced in any way. Under normal propagation conditions, it has a greater magnitude than any reflected wave arriving at the receiver.

THE GROUND REFLECTED WAVE

This wave may arrive at the receiver antenna after being reflected one or more times from the ground or surrounding objects. The reflection does not have to occur only in the vertical plane, but may occur in the horizontal plane, such as that reflected from an obstruction, offset from the main path. The reflected wave will have a different phase and amplitude than those of the direct wave and if the distance traversed is greater by an odd number of wavelengths, the phase will be 180° with respect to the direct wave at the receiver antenna and the effect will be to cancel to some extent the direct wave signal. The extent to which the direct wave is canceled depends on the amplitude of the reflected wave.

THE TROPOSPHERICALLY REFLECTED WAVE
Due to changes in the refractive index of the air at varying heights above the Earth's surface, a wave may be diffracted and, depending on the angle at which it was launched, totally reflected from the troposphere. In this event, it would appear that there is a boundary which acts as a reflecting surface, which sends a wave back down to Earth. One of the rays may reach the receiving antenna, where it may subtract from the direct wave due to the change in phase and amplitude caused by the reflection.

8.2.2 Propagation mechanisms

There are three types of propagation mechanism to be considered in clear weather conditions. These are refraction, diffraction and reflection. These will be briefly discussed before we consider the different forms of fading mechanisms which affect the design of a radio link for real world conditions. All propagation mechanisms may be classified in two groups: surface mechanisms which take account of line-of-sight and ground-wave propagation; and volume mechanisms.

Surface mechanisms
The Earth's surface is the major instrument in the reflection and diffraction of a radio wave, whereas the atmosphere above the Earth's surface contributes mainly to refraction of the wave. Ducting, which is likely to cause interference on both line-of-sight and trans-horizon (troposcatter) paths, can occur at both the surface and at elevated positions up to 200 meters or so, above the surface of the Earth. In all surface mechanisms, propagation takes place approximately on the great circle path.

Volume mechanisms
These arise from scatter in the fine structure of the troposphere and include tropospheric scattering from fluctuations in the refractive index of the atmosphere due to its gaseous composition. Scatter may also be caused by solid or liquid particles such as rain or hail. Scatter from particles is not confined to the great circle path but is highly dependent on the radiation characteristics and relative geometry of the antennae.

8.2.2.1 Refraction

All electromagnetic waves are refracted when they pass from a material of one refractive index to another. In the atmosphere, the changes in refractive index are at most times gradual since the density of the air decreases with height, theoretically, at a uniform rate. Above 30 MHz, the water content of the air plays a predominant part in refractivity changes, for the dielectric constant of water is approximately eight times that of air. In a normal atmosphere, the specific humidity, or the mass of water vapor per unit mass of air, is constant. This means that the dielectric constant and the refractive index both decrease continuously with increasing height. The general effect of the vertical changes in refractive index of the atmosphere is to produce a curving of the waves in the vertical plane, as they travel from the transmitter to the receiver. The amount of path curvature varies with the time due to changes in temperature, pressure and humidity. Under normal propagation conditions, the path curves away from the true Earth's surface so that the radio horizon is effectively extended. When the vertical refractivity gradient increases, however, the path of the radio wave is bent towards the true Earth's surface, reducing the clearance over the underlying terrain. This situation may produce an effect where the Earth itself, or the vegetation, causes an obstruction. Under conditions such as these, obstruction fading may occur. Good radio link design prevents this from occurring, by insuring that the line of sight from the transmitter and receiver, by proper antenna height placement, is rarely lost under the worst refractivity conditions. The changes in the refractive index n of the air of only a few parts per million can have an effect on the propagation of the radio wave. The values of n are very close to unity (typically 1.00035) and so it is more usual to deal with the *refractivity* N, where:

$$N = (n - 1) \times 10^6 \tag{8.1}$$

The variability of the refractive index n with elevation arises from the fact that n depends on pressure P (in millibars), temperature T (in kelvins), together with the water vapor pressure e (in millibars), which in turn, change with height. For frequencies up to 30 GHz, the refractivity N is given by

$$N = 77.6\ P/T + 3.73 \times 10^5\ e/T \quad N\text{-units} \tag{8.2}$$

This equation is correct to within 0.5 percent for atmospheric pressures between 200 and 1100 mbar for air temperatures between 240 and 310 K, and for radio frequencies less than 30 GHz. The term $77.6\ P/T = N_{\text{dry}}$ and the term $3.73 \times 10^5\ e/T^2 = N_{\text{wet}}$.

One of the most significant factors in the influence of radio wave propagation is the large-scale variation of refractive index with height, and the extent to which this changes with time. In practice, the measured median of the refractivity gradient in the first kilometer above ground in most temperate regions is about -40 N-units/km. Under the assumption of a constant refractivity gradient, the radio wave is an arc of circle of radius r, related to the refractive index n, by

$$1/r = -\mathrm{d}n/\mathrm{d}h \tag{8.3}$$

where h is the height above the Earth's surface in the same units as r. The effective Earth's radius a_e due to the change in refractive index is given by

$$1/a_e = 1/a - 1/r = 1/a + \mathrm{d}n/\mathrm{d}h \tag{8.4}$$

where a is the Earth's radius (6.37×10^3 km). If the effective Earth's radius a_e is given by

$$a_e = ka \tag{8.5}$$

then

$$k = a_e/a = 1/(1 + a \cdot \mathrm{d}n/\mathrm{d}h) = 1/[1 + a \cdot (\mathrm{d}N/\mathrm{d}h) \times 10^{-6}] \tag{8.6}$$

where k is the k-factor, or the effective Earth-radius factor.

$\frac{\mathrm{d}n}{\mathrm{d}h}$ is the gradient of the radio refractive index with respect to height.

$\frac{\mathrm{d}N}{\mathrm{d}h}$ is the gradient of refractivity per kilometer and is expressed in N-units/km.

When $\mathrm{d}N/\mathrm{d}h > -39$ N-units/km, the wave is said to be 'subrefracted' or bent down *less than normal*, i.e. it is bent towards the sky ($0 < k < 4/3$). When $\mathrm{d}N/\mathrm{d}h < -39$ N-units/km, it is said to be 'superrefracted', or bent down *more than normal*; in other words the radio wave is bent towards the ground ($4/3 < k < \infty$). When $\mathrm{d}N/\mathrm{d}h = -39$ N-units/km, it is said to be in a $k = 4/3$ condition. As a large number of experimental values have been obtained for the rate of change of refractivity with height and the average value has been found to be -39 N-units/km, this value is used to determine the normal refractive path of the radio beam through the atmosphere within 1 km of the Earth's surface. Such an atmosphere is usually referred to as 'standard'.

If the Earth's radius ($a = 6.37 \times 10^3$ km) is substituted into equation 8.6, the expression becomes:

$$k = 157/(157 + \mathrm{d}N/\mathrm{d}h) \tag{8.7}$$

for $\mathrm{d}N/\mathrm{d}h = -39$ N-units/km, $k = 4/3$. Equation 8.7 is shown plotted in Figure 8.1.

The k-factor is an alternative and more often used method of describing the amount of bending undergone by the microwave beam. This factor, multiplied by the actual Earth's radius, gives the fictitious radius of the Earth's curvature. During standard atmospheric conditions, the range of k is from 1.2 in dry elevated areas and 4/3 in typical inland areas, to 2 or 3 in humid coastal areas. Reference 2, Section 8.7, provides details of the k-factor in different parts of the world. When k equals infinity, the Earth appears to the microwave beam, to be perfectly flat, since the beam curves at exactly the same rate as the curvature of the Earth. If the value of k becomes less than unity, the beam curves upwards away from the Earth's surface. This appears to the radio wave as if the Earth is 'bulging' and is going to obstruct the transmission path. This phenomenon is due to the atmospheric density increasing with height instead of decreasing with height. A substandard or subrefracting atmosphere is caused by the formation of fog created with the passage of warm air

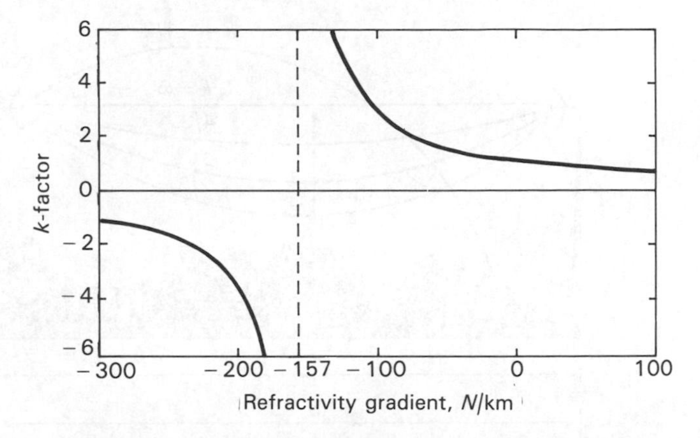

Figure 8.1 Effective Earth's radius factor k for different refractivity gradients dN/dh N-units/km

over a cool or moist surface. The result of the fog is to cause the atomospheric density to be lower near the ground than at higher elevations, causing an upward bending of the beam. Superrefraction, or superstandard refraction, results from conditions where there is a rise in temperature with increasing height (temperature inversion), or a marked decrease in total moisture content in the air with increasing height. Either condition will cause a reduction in the atmospheric density with height. In this case, k increases and the Earth's curvature appears to the radio wave to be flattened. As the k-factor is increased towards infinity, the radio wave will tend to move parallel with the surface of the Earth. An extreme case of superrefraction will bend the radio wave downward with a radius smaller than that of the Earth (negative k) and cause blackout fading if the receiver is beyond the point where the wave dips into the earth. This is not a normal condition by any means and comes under the heading of anomalous propagation. Superrefraction may be caused by the passage of cool air over a warm body of water. Evaporation of the water will cause an increase in humidity and the low temperatures near the surface are a sign of temperature inversion. Low temperatures and high humidities greatly increase the atmospheric density near the surface of the Earth, causing an abnormally high downward bending of the radio wave. Figure 8.2 shows the effects of differing k-factor on a radio beam in which the true Earth's radius is drawn.

The 'Earth bulge' at any point in a radio path is given by:

$$h = (4/51) \cdot d_1 \cdot d_2/k \text{ meters} \tag{8.8}$$

where d_1 and d_2 are the distances in kilometers to the near end and far end of the path, respectively. Unless special path profiling paper is used with the earth bulge plotted for a specific k-factor, the actual Earth bulge must first be plotted from the horizontal axis, so that the microwave beam can be drawn as a straight line (opposite to that shown in Figure 8.2).

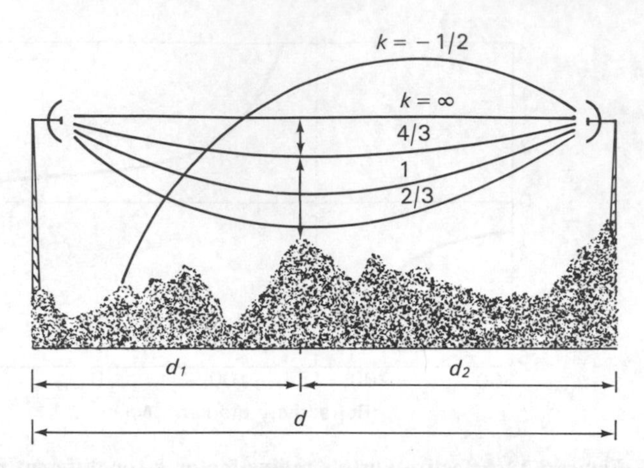

Figure 8.2 Variation of radio wave path for different *k*-factors

MINIMUM VALUE OF *k*-FACTOR

The gradient of refractivity dN/dh, does not remain constant over the entire length of the propagation path, especially on long hops. To establish path-clearance criteria, the following factors must be considered:[3,4]

1. The actual radio path is still approximately an arc of a circle with a curvature of

$$1/r = -dN_e/dh \times 10^{-6} \tag{8.9}$$

   where dN_e/dh is the 'equivalent' refractivity gradient.
2. At any given time, dN_e/dh equals the mean value of the refractivity gradients dN/dh encountered along the hop path.
3. dN_e/dh is a random variable which tends to have a Gaussian distribution on long hops. If μ and σ are the mean and standard deviation of dN/dh, the corresponding results apply:

$$\mu_e = \mu \tag{8.10}$$
$$\sigma_e = -\sigma[1 + (d/d_0)]^{-1/2} \tag{8.11}$$

   where d is the hop length in km and $d_0 = 13.5$ km. The above model for dN_e/dh holds for hops longer than about 20 km. For hops shorter than about 20 km, the distribution of dN_e/dh approximately coincides with that of dN/dh at an arbitrary point along the path.
4. For hops greater than 20 km, the value of dN_e/dh which is exceeded for a given percentage of time, is computed by determining the area under the Gaussian probability density function. In particular:

$$dN_e/dh \text{ (0.1 percent)} \simeq \mu_e + 3.1\sigma_e \tag{8.12}$$
$$dN_e/dh \text{ (0.01 percent)} \simeq \mu_e + 3.7\sigma_e \tag{8.13}$$

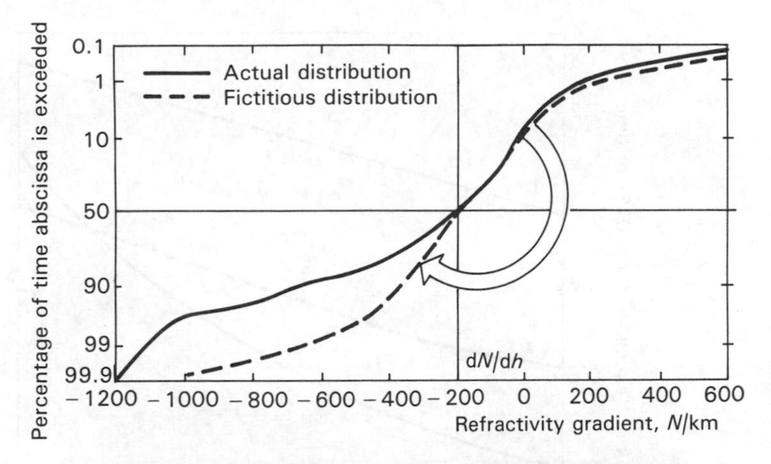

Figure 8.3 Actual and fictitious distributions of dN/dh (Courtesy of Telettra S.p.A., Milan, Reference 4)

5. In computing μ and σ, attention must be paid to the propagation conditions which characterize the conditions of reduced path clearance as gradients greater than the median value dN_m/dh occur. This may be taken account of by letting $\mu = dN_m/dh$ and by computing σ from a fictitious gradient distribution, which coincides with the actual distribution for $dN/dh > dN_m/dh$. Where $dN/dh < dN_m/dh$, the value of σ is obtained by folding back the actual distribution around the point of abscissa dN_m/dh, as shown in Figure 8.3.

In the determination of the minimum value of k, equation 8.7 becomes:

$$k_{\min} = 157/(157 + dN_e/dh) \tag{8.14}$$

Equation 8.11 shows that, the longer the hop, the less disperse is dN_e/dh compared with dN/dh. The physical reason for this is, that partial refractions suffered by the wave on the various segments of the hop tend to compensate because of the different meteorological conditions existing along the hop path. This in turn means that the dispersion of the k-factor decreases also as the hop length increases and thus the probability that k attains values so low as to produce obstruction fading is also reduced. In the calculation of tower heights, usually the minimum value of k is used to ensure that obstruction fading does not become a problem. This means that if k (0.01 percent) is used, it is expected that the value of k which would be less than this would occur for only 0.01 percent of the time. Thus, for design purposes, the minimum value of k should be:

$$k_{\min(0.01\%)} = 157/[157 + dN_e/dh\ (0.01\%)] \tag{8.15}$$

Figure 8.4 shows a plot of $k_{\min(0.01\%)}$ for different path lengths for three different climates.[4]

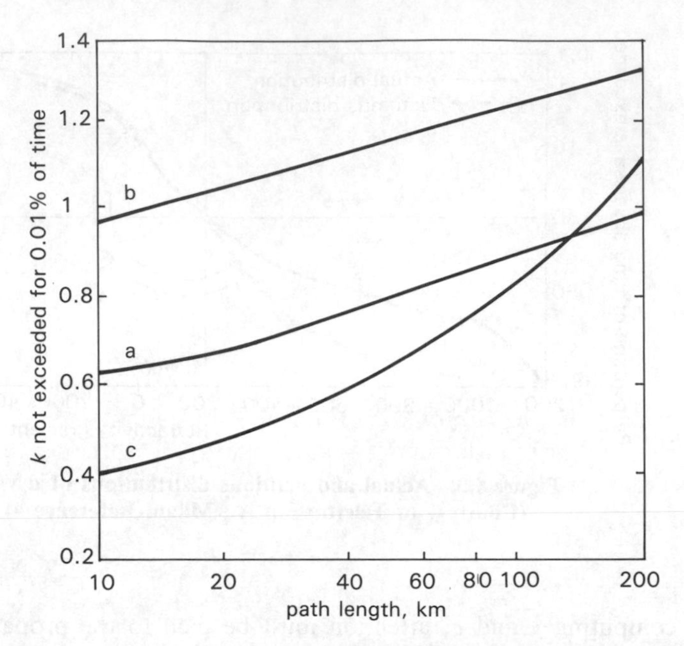

Figure 8.4 ***k*-factor not less than specified for 0.01 percent of the time as a function of path length (a) Trappes (France), temperate climate; (b) Ostersund (Sweden), northern climate; (c) Dakar (Senegal), tropical climate**
(Courtesy of Telettra S.p.A., Milan, Reference 4)

MODIFIED REFRACTIVE INDEX

When dealing with superrefractive propagation conditions, for convenience, a modified refractive index M is used. Its relationship to the refractivity N is given by:

$$M = (n - 1 + h/a) \times 10^6 = N + h/a \times 10^6 \quad (8.16)$$

If 6.37×10^3 km is used for the Earth's radius a, the rate of increase of M with height is given by

$$\mathrm{d}M/\mathrm{d}h = (\mathrm{d}n/\mathrm{d}h + 1/a) \times 10^6 = (\mathrm{d}N/\mathrm{d}h + 157)M\text{-units/km} \quad (8.17)$$

This shows that conventional gradients less than -157 N-units/km correspond to negative modified gradients and positive gradients or gradients greater than -157 N-units/km, correspond to positive modified gradients. Blackouts occur when the atmosphere has an extreme drop in density with increasing height due to temperature inversions or when a marked decrease in humidity with increasing height occurs. If the density drop is so sharp as to produce refractivity gradients less than -157 N-units/km, the radio wave is bent downwards with a radius smaller than that of the Earth (negative k), causing a blackout fade if the receiver is beyond the point where the wave enters the Earth. When the radio horizon becomes short, because of the very negative refractivity gradients, either the signal reaches the receiver or it does not. If the signal does not reach the receiver, the fade suffered at the receiver is so great that increasing the fade margins or adding diversity will not produce any improvement.

DUCTING

The first necessary condition for a duct to occur is that the refractive index gradient shall be equal to or more negative than -157 N-units/km. This causes the rays to remain close to the Earth's surface beyond the normal horizon. The second necessary condition is that this gradient should be maintained over a height of many wavelengths. Figure 8.5(a) shows how rays entering a duct at small angles less than the critical angle θ_c are bent downwards and are trapped in a surface duct. Where the transmitter is located in the duct, the angle of radiation θ becomes critical. For any angle less than that shown by ray A in Figure 8.5(a), the signal is trapped within the duct and may be propagated far beyond the horizon. In extreme conditions, it may be bent down so much that it is reflected and propagated down the duct in a series of hops as shown by ray 4 in Figure 8.5(d). Ray 1 illustrates a leakage signal escaping from the duct. There are two rays that define the blackout region. They are labeled GOL (grazing over the layer) and GBL (grazing below the layer). Both rays originate as one, ray A, from the transmitter and on reaching the top of the duct separate into two. Any ray leaving the antenna is either above or below ray A; hence a zone between the GOL and GBL rays is a shadow region in which a blackout will occur for any receiver located in this region. Ray A is launched with the critical angle θ_c, which is given by the expression;

$$\theta_c = \arcsin(2\,\Delta M \times 10^{-6})^{1/2} \tag{8.18}$$

where ΔM is the difference between the value of the modified refractive index at the transmitter height and the value of the modified refractive index at the top of the duct. In practice, the value of θ_c never exceeds about 0.5°, which corresponds to $\Delta M = 30$ M-units. For efficient duct propagation, the transmitter carrier wavelength λ must be less than a critical value λ_c;

$$\lambda < \lambda_c = 1.9 \times t^{1.8} \times 10^{-4} \tag{8.19}$$

where t is the thickness of the duct in the same units as the wavelength λ. For the case where the transmitter is placed above the duct [Figure 8.5(b)], the splitting of ray A into GOL and GBL will also occur with the same type of shadow region. For the case shown in Figure 8.5(c), $\mathrm{d}M/\mathrm{d}h$ is very small and the duct can be responsible for the transmission signals with large increases in the received signal level to a receiver whose antenna is placed at the same height. This is because the energy within a duct spreads with distance in the horizontal, but is constrained in the vertical direction, allowing the field strength within a duct to be much greater than the free-space field strength for the same distance. With no spread in the vertical but normal spread of energy in the horizontal, the transmission loss is expected to be inversely proportional to the distance d, instead of d^2 as in the free-space law. Finally, Figure 8.5(d) shows the case where a signal may be received in a blackout region

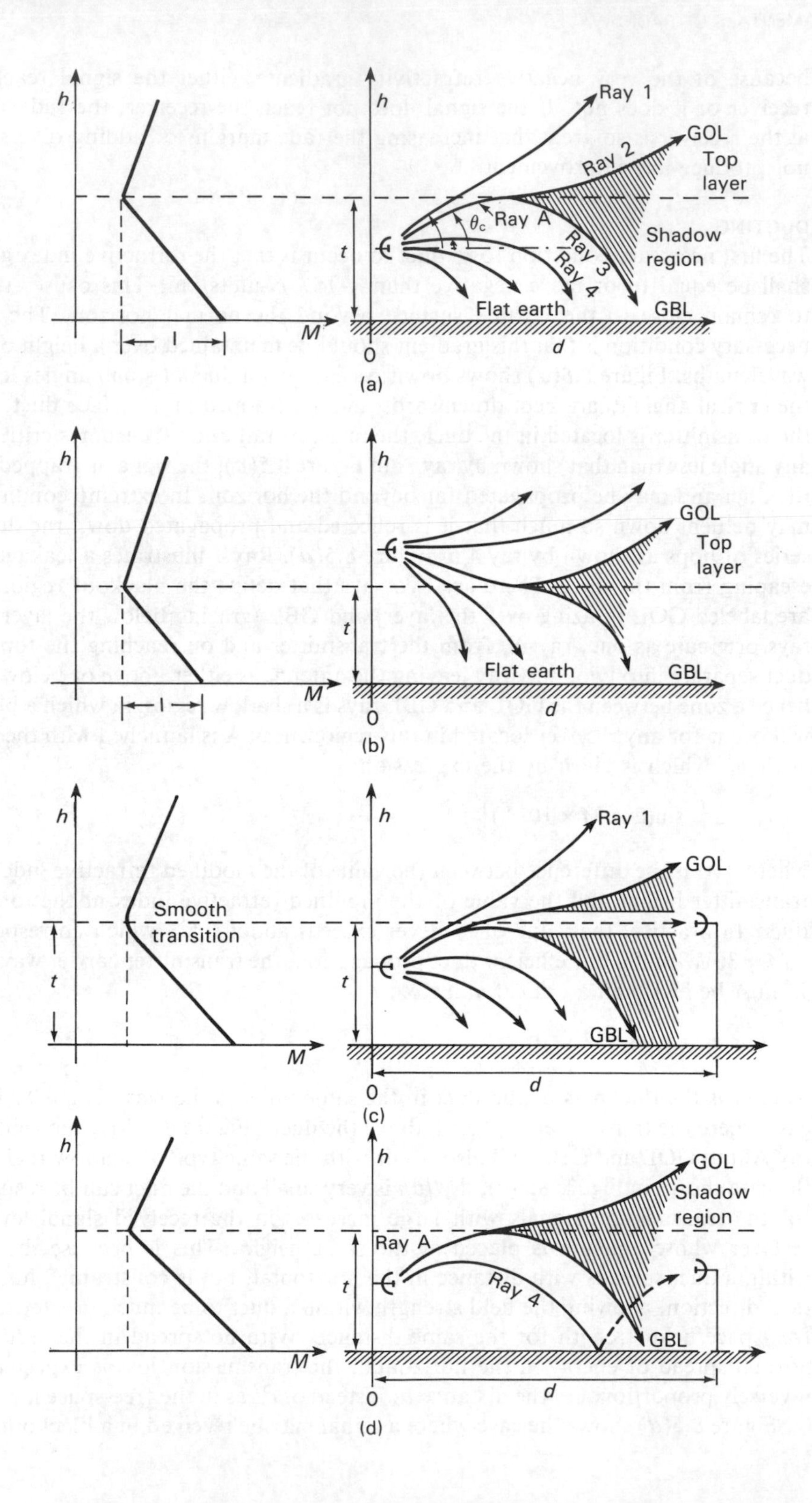

h
M
t
0
d
Ray 1
GOL
Ray 2
Top layer
θc
Ray A
Shadow region
Ray 3
Ray 4
Flat earth
GBL
(a)
(b)
Smooth transition
(c)
(d)

through reflections from the ground. These reflections should be considered more from the standpoint of interfering with other co-channel operating receivers than for reception of a wanted signal by a receiver which has its antenna in the blackout region. Ground-based ducts may be formed by an unusually rapid decrease in the water vapor density with height or an increase in temperature with height or both effects together. The causes which are associated with expanses of water are evaporation and advection. Evaporation of water vapor from the sea or a lake, may cause a high humidity zone (high refractive index n) below a region of drier air. These conditions are likely to be afternoon occurrences due to the prolonged solar heating during the day. The duct thicknesses are of the order of 15 m. In the tropics, high humidity which exists over the surface of water, or swamps produces almost permanent ducts with a refractivity change of 40 N-units. Advection, which is the movement of one type of air over another, may cause hot dry air from the land to be blown over cold wet air from the sea, producing a region of low refractive index located above a region of high refractive index. This occurs at the evening times when the land breeze occurs. The duct thickness is of the order of 25 m. Reference 5 describes the formation of the various M profiles for different topological conditions.

CCIR Report 569–3 [10] describes a prediction method for calculating the path loss contribution or specific attenuation (γ_d dB/km) due to superrefraction and ducting. This method is evaluated for the specific attenuation occurring for less than, or equal to, one percent of the time and is based on the calculation of the propagation of energy in a tropospheric duct. The method is also valid for the cases of high superrefraction without ducting and of reflections from thin layers. To determine the value pertinent to a particular radio-climatic zone, three main zones are defined. These are:

Zone A_1 Coastal land and shore areas, i.e. land which is adjacent to a Zone B or a Zone C as defined below, up to an altitude of 100 m relative to mean sea level, but limited to a distance of 50 km from the nearest Zone B or Zone C as the case may be.

Zone A_2 All land, other than coastal land and shore defined as Zone A_1 above.

Zone B 'Cold' seas, oceans and other large bodies of water which cover a circle at least 100 km in diameter, and which are situated in latitudes above 30°, with the exception of the Mediterranean and the Black Sea.

Zone C 'Warm' seas, oceans and other large bodies of water which cover a circle at least 100 km in diameter, situated at latitudes below 30° and including the Mediterranean and the Black Sea.

The specific attenuation γ_d, is given by:

$$\gamma_d \text{ (ducting)} = [C_1 + C_2 \cdot \log(f + C_3)]\, p^{C_4} \text{ dB/km} \tag{8.20}$$

Figure 8.5 Propagation in a surface duct: (a) transmitter antenna within duct; (b) transmitter antenna above duct; (c) anomalous propagation; (d) reflection of refracted rays

Table 8.1 Constants used to calculate attenuation due to ducting for different zones

Zone	C_1	C_2	C_3	C_4
A_1	0.109	0.100	−0.10	0.16
A_2	0.146	0.148	−0.15	0.12
B	0.050	0.096	0.25	0.19
C	0.040	0.078	0.25	0.16

The actual attenuation is given by:

$$L(\text{ducting}) = \gamma_d \cdot d \text{ dB} \tag{8.21}$$

where d = the path length in km
f = the frequency of the carrier in GHz
p = the annual time percentage ($\leqslant 1$ percent)
C_1, C_2, C_3 and C_4 are the constants as given in Table 8.1.

The values for γ_d applicable to land paths are based on data for regions where the highest monthly mean value of ΔN is below 70 N-units/km. For regions where the highest monthly mean value of ΔN is significantly greater than this, the value of γ_d given by equation 8.21 for Zone A_2 should be multiplied by 0.6, 0.7, 0.8 and 0.8 for time percentages of 1, 0.1, 0.01, and 0.001 percent respectively. Equation 8.21 relates only to paths within one radio-climatic zone. Should the radio path cross climatic zones, then the sum of the individual specific attenuations multiplied by the individual path lengths within each individual zone, must be used.

SCINTILLATION[6–9]

Scintillation on a line-of-sight microwave radio signal is caused by fluctuations in the refractive index of the region of the atmosphere through which the signal is propagated. These fluctuations produce a slight focusing and defocusing in the radio beam. The received signal is then the sum of a large number of components that arrive from different directions with continually changing amplitude and path length, so that the resultant shows a rapid and random variation of amplitude and phase. Four main types of short-term fluctuation may be identified. These are:

1. Pure scattering associated with a steady mean level with a relatively small fluctuation of about 0.5 dB
2. Scattering with reflection which is associated with short-term fluctuations superimposed on a mean level, changing in an almost cyclic manner over a period of some minutes
3. Scattering with intense patches possibly due to the signal passing through smoke or hot air
4. Reflection from a layered structure which is associated with a much steadier signal having a cyclic type of variation in mean level.

An important application of the scintillation effect is for high-capacity microwave links operating at frequencies greater than 10 GHz in an urban environment.

The effects of this type of refractive fading are reduced if large antennas are used, since the range of angles from different receive paths is then reduced, and there is some spatial averaging across the aperture of the incident wave front. If for the various contributions to a given signal level there is a range of delay times Δt, there will be only a limited bandwidth W Hz that can be transmitted without distortion, such that

$$W = 1/\Delta t \tag{8.22}$$

Satellite paths also show rapid fading of the signal amplitude over a narrow peak-to-peak range similar to that described above. CCIR Report 718–2[10], Section 5 deals at length with this subject.

8.2.2.2 Diffraction

Diffraction of radio waves is, in practice, the bending of the waves around objects. The amount of bending increases as the object thickness is reduced and also as the wavelength of the wave increases. The amount of bending or diffraction by radio waves is therefore much greater than that of light around the same sized object. The reason why diffraction concerns the radio link design engineer is that obstructions which are not necessarily in the direct path of the radio beam may cause the beam to become attenuated. This attenuation is due to 'diffraction losses' or 'diffraction attenuation'. This arises from the concept of Fresnel zones which are ellipsoids surrounding the direct path beam from the transmitter antenna to the receiver antenna. The first ellipsoid has the locus of its outer shell such that any signal reaching the receiver antenna via this path will travel half the carrier frequency wavelength ($\lambda/2$) more than a signal traveling via the direct path. The region inside this first ellipsoid is called the *first Fresnel zone*. The first Fresnel zone ellipsoid contains most of the power which reaches the receiver. If an obstruction exists at the edge of the first Fresnel zone, the reflected wave will tend to cancel the direct wave. The extent of this cancelation will depend on the relative amplitudes of the reflected and direct waves. In practice, propagation is assumed to be pure line-of-sight, i.e. without any diffraction attenuation occurring, if there is no obstacle within the first Fresnel zone. Besides the first Fresnel zone, there is a family of ellipsoids surrounding this first shell. These are the second, third, fourth, etc., Fresnel zones. They have little effect in producing noticeable diffraction attenuation due to the signal power contained within them, being so small. The radius of the family of ellipsoids about the direct path varies along the propagation path and is given by:

$$F_n = [n \cdot \lambda \cdot d_1 \cdot d_2/(d_1 + d_2)]^{1/2} = [n \cdot \lambda \cdot d_1 \cdot d_2/d]^{1/2} \tag{8.23}$$

where n = the Fresnel zone number (1, 2, 3, ...).
λ = the carrier wavelength.
d_1 = the distance from one terminal to the point where the Fresnel zone radius is being calculated.

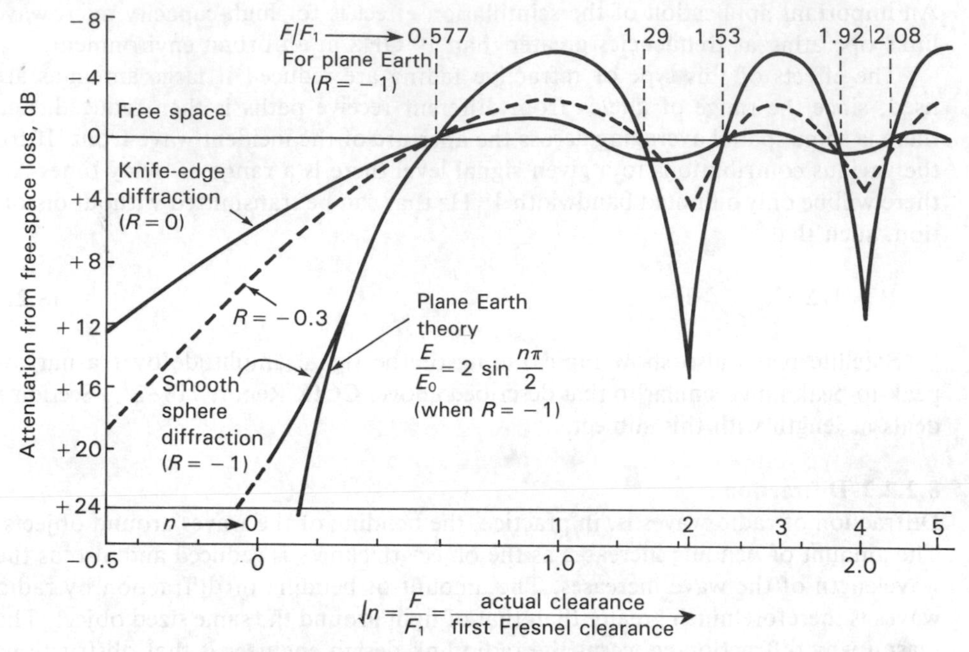

Figure 8.6 Effect of path clearance on radio transmission

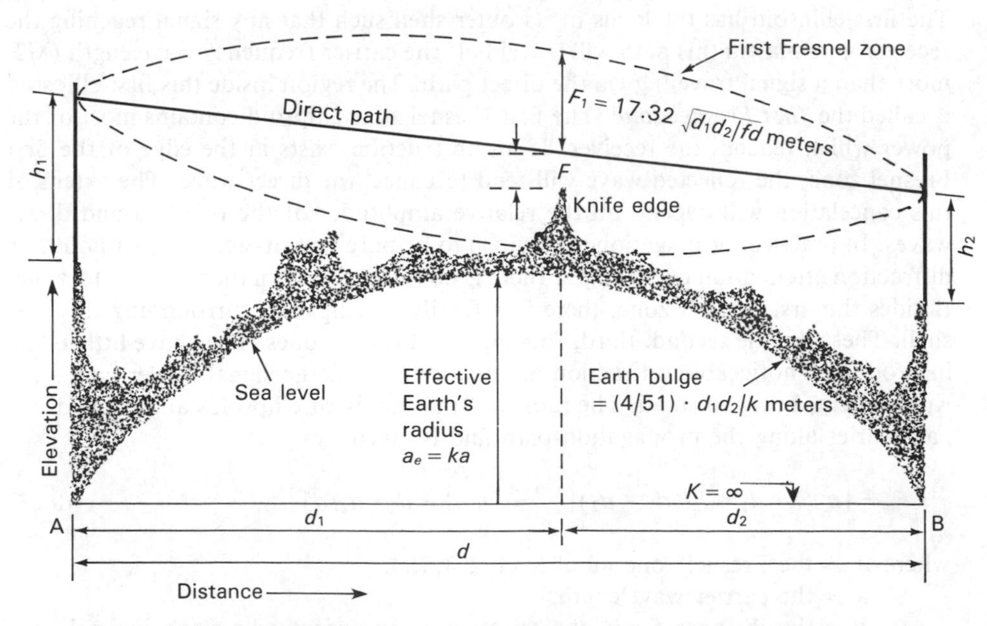

Figure 8.7 Typical path profile showing earth bulge and first Fresnel zone

d_2 = the distance from the other terminal to the point where the Fresnel zone radius is being calculated.
$d = d_1 + d_2$.

All quantities are in the same units. The radius of the first Fresnel zone F_1 is given by:

$$F_1 = 17.32 [d_1 \cdot d_2/(f \cdot d)]^{1/2} \text{ meters} \quad (8.24)$$

where d_1, d_2 and d are in km and f is the frequency of the carrier in GHz.

Diffraction losses differ for different types of terrain which may obstruct the first Fresnel zone. These types of terrain are spherical earth, knife-edge and one rounded obstacle. Figure 8.6 shows the attenuation in dB with respect to free-space propagation loss, as a function of normalized clearance F/F_1, where F is the actual clearance and F_1 is the value of the first Fresnel zone. Figure 8.7 shows the representation of F and F_1 on a typical path profile which has taken into account the Earth bulge as given by equation 8.8.

DIFFRACTION LOSS DUE TO GRAZING OVER SMOOTH EARTH

The electrical characteristics of any medium may be expressed by three parameters; the permeability μ (henrys/meter, abbreviated H/m), the permittivity ε (farads/meter, abbreviated F/m) and the conductivity σ (siemens/meter, abbreviated S/m). The permeability of the ground and the sea can normally be regarded as equal to the free-space permeability so that in most propagation problems only permittivity and conductivity are considered. At frequencies above 100 MHz account must be taken of the variation of permittivity and conductivity with frequency. A value of 80 has often been used for the relative permittivity ε_r of sea water at low temperatures and frequencies below 1 GHz. The relative permittivity ε_r, is relative to that of a vacuum of free-space. The actual value, even at low frequencies, depends on the temperature and composition of the sea water. Typical values of conductivity and permittivity for different types of ground as a function of frequency are given below (CCIR Recommendation 527–1).[10] In the absence of data on permittivity, it is normally sufficient to use the values of ε_r associated with the conductivity σ by the empirical expression:[11]

$$\varepsilon_r = 50\sigma^{1/5} \quad (8.25)$$

where conductivity is expressed in S/m.

Graphs of relative permittivity and conductivity, plotted against frequency, are given in Figure 8.8. The values of relative permittivity ε_r indicate the approximate range for different types of ground, but in extreme cases, values outside this range may occur. In wet fertile areas higher values of conductivity will be found, while in mountainous and arctic regions the conductivity of frequencies below 100 MHz may be as low as 10^{-5} S/m. With snow-covered ground, lower values of permittivity than shown in curve E may be met. The water in lakes and rivers has a conductivity that increases with the concentration of impurities. The expected ranges of ground

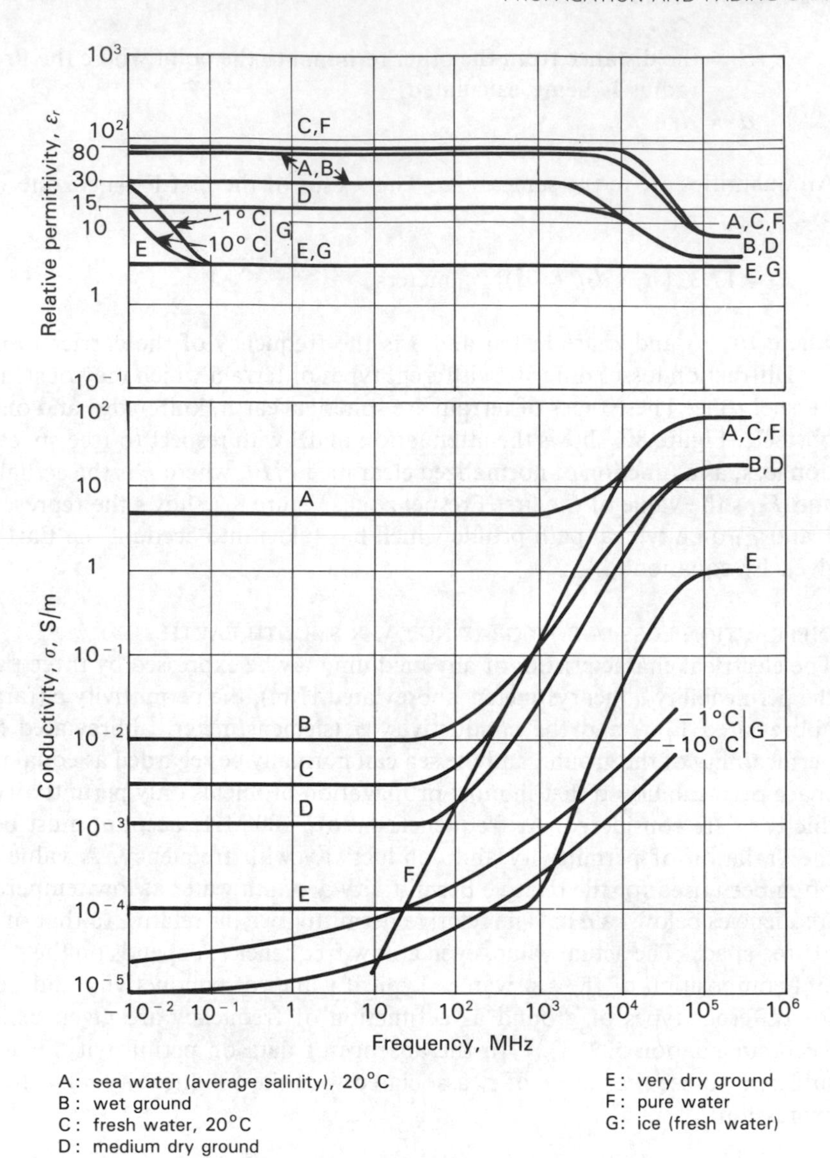

Figure 8.8 Relative permittivity ε_r, and conductivity σ as a function of frequency
(Courtesy of CCIR, Recommendation 527–1, Reference 10)

conductivity for different terrains are given in Table 8.2.

To determine the diffraction loss caused by the radio path being obstructed by the Earth, the surface admittance K of the Earth must first be determined. This can be done from equation 8.26 or 8.27, depending on whether the radio wave is horizontally or vertically polarized respectively,

$$K_h = 1.942 \times 10^{-3} \cdot (k \cdot f)^{-1/3} \cdot [(\varepsilon_r - 1)^2 + (18\sigma/f)^2]^{-1/4} \quad (8.26)$$

Table 8.2 Ranges of effective ground conductivity

Type of terrain	Range of ground conductivity, S/m
Sea water	4 to 5
Very moist soil; moist cultivated land	10^{-2} to 10^{-1}
Marshes (fresh water); cultivated land	3×10^{-3} to 3×10^{-2}
Fresh water; sandy loam, hilly country in temperate climates	10^{-3} to 10^{-2}
Medium dry ground; rocks; sand; medium-sized towns	3×10^{-4} to 3×10^{-3}
Dry ground; desert	10^{-4} to 10^{-3}
Very dry ground; granite mountains in cold regions; industrial areas	3×10^{-5} to 3×10^{-4}
Dry glaciers in mountainous areas, permafrost, polar ice, high mountains	10^{-5} to 10^{-4}

$$K_v = K_h \cdot [\varepsilon_r^2 + (18\sigma/f)^2]^{1/2} \tag{8.27}$$

where k is the k-factor for the link, f is the frequency in GHz, ε_r is the relative permittivity as determined from Figure 8.8, and σ is the conductivity of the earth in S/m, also determined from Figure 8.8. If K is less than 0.001, the electrical characteristics of the Earth are not important. For values of K greater than 0.001, the appropriate formulae given below should be used. The attenuation L_{diff} due to diffraction is found from

$$L_{\text{diff}} = F(X) + G(Y_1) + G(Y_2) \text{ dB} \tag{8.28}$$

where F and G denote functions of X, Y_1 and Y_2 respectively and X is the normalized length of the path between the antennae at normalized heights Y_1 and Y_2, given by

$$X = 0.064\beta \cdot d \cdot [f/k^2]^{1/3} \tag{8.29}$$

$$\beta = [1 + 1.6K^2 + 0.75\,K^4]/[1 + 4.5\,K^2 + 1.35\,K^4] \tag{8.30}$$

For horizontal polarization at all frequencies, and for vertical polarization above 20 MHz over land, or above 300 MHz over sea, β may be taken as equal to 1. $F(X)$ is given by:

$$F(X) = 11 + 10 \log X - 17.6X \tag{8.31}$$

and

$$Y = 0.05179h \cdot \beta \cdot [f^2/k]^{1/3} \tag{8.32}$$

where d is the path length in km, k is the k-factor, h is the antenna height (h_1 or h_2) in meters and f is the frequency in GHz.

The values of $G(Y)$ may be calculated from the following formulae:

For $Y > 2$

$$G(Y) \simeq 17.6[Y - 1.1]^{1/2} - 5\log(Y - 1.1) - 8 \tag{8.33a}$$

For $10K < Y < 2$

$$G(Y) \simeq 20\log(Y + 0.1Y^3) \tag{8.33b}$$

For $K/10 < Y < 10K$

$$G(Y) \simeq 2 + 20\log K + 9\log(Y/K)\cdot[(\log(Y/K) + 1)] \tag{8.33c}$$

For $Y < K/10$

$$G(Y) \simeq 2 + 20\log K \tag{8.33d}$$

Example

Determine the diffraction loss for a 100 km vertically polarized link over sea water in a temperate climate. The antenna heights are 60 meters above sea level at one end and 120 meters above sea level at the other end. Assume that the carrier frequency is 4 GHz.

Suggested solution
From Figure 8.8, the value of relative permittivity ε_r and conductivity σ are, for a frequency of 4 GHz and for sea water of average salinity at 20°C, 69 F/m and 8.5 S/m respectively. Assuming that the k-factor is 4/3, then, from equations 8.26 and 8.27, K_v is found to be 9.928×10^{-3}. From equation 8.29, $X = 8.389$ and from equation 8.31, $F(X) = -127.4$ dB. From equation 8.32, Y is found to be $h \times 118.57 \times 10^{-3}$. For $h_1 = 60$ meters, $Y_1 = 7.114$ and for $h_2 = 120$ meters, $Y_2 = 14.228$. From equation 8.33a, $G(Y_1) = 31.27$ dB and $G(Y_2) = 50.18$ dB. Thus the diffraction loss as determined from equation 8.28 is $L_{\text{diff}} = -45.95$ dB. This means that as well as the free-space loss there will be an additional 46 dB of loss. This is quite significant and shows one of the difficulties of designing long hops over sea. CCIR Report 715–2[10] gives nomograms which may be used to determine quickly the diffraction loss over various terrains and with different polarizations.

DIFFRACTION LOSS DUE TO KNIFE-EDGE OBSTRUCTION

Figure 8.9 shows a diagram of the extremely idealized case of a single knife-edge obstruction. This seldom occurs in nature and thus the diffraction attenuation equation given in equation 8.35 below can be only approximate for most real-life situations. Also the equation applies when the carrier wavelength is small in relation to the size of the obstacles. For the obstruction considered, all the geometrical parameters are lumped together in a single dimensionless parameter normally denoted by ν. This parameter may assume a variety of equivalent forms according to the geometrical parameters selected. The equations relating ν to the geometrical parameters are:

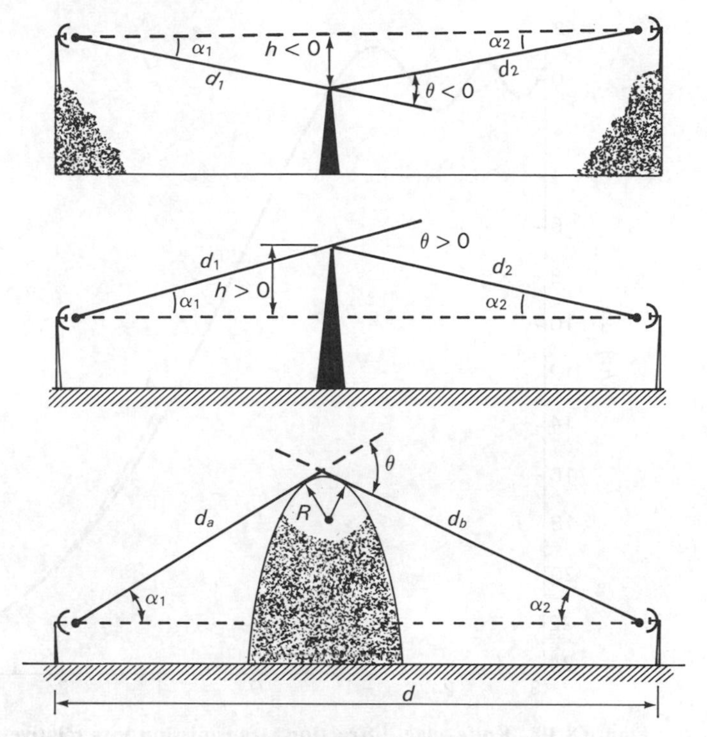

Figure 8.9 Knife edge and rounded obstacles

$$\nu = h[(2/\lambda) \cdot (1/d_1 + 1/d_2)]^{1/2} \tag{8.34a}$$

$$\nu = \theta \cdot [2/[\lambda(1/d_1 + 1/d_2)]]^{1/2} \tag{8.34b}$$

$$\nu = [2h \cdot \theta/\lambda]^{1/2} \text{ where } \nu \text{ has the sign of } h \text{ and } \theta \tag{8.34c}$$

$$\nu = [2d \cdot \alpha_1 \cdot \alpha_2/\lambda]^{1/2} \text{ where } \nu \text{ has the sign of } \alpha_1 \text{ and } \alpha_2 \tag{8.34d}$$

where h = height of the top of the obstacle above the straight line joining two ends of the path. If the height is below this line, h is negative.

d_1 and d_2 = distances of the two ends of the path from the obstacle.

d = length of the path, $d = d_1 + d_2$, as the height is much smaller than the path length d.

θ = the angle of diffraction in radians: its sign is the same as that of h. The angle θ is assumed to be less than 0.2 radian (roughly 12°).

α_1 and α_2 are angles between the top of the obstacle and one end as seen from the other end. α_1 and α_2 have the same sign as that of h in the above equation and h, d, d_1, d_2 and the carrier wavelength λ, are expressed in the same unit.

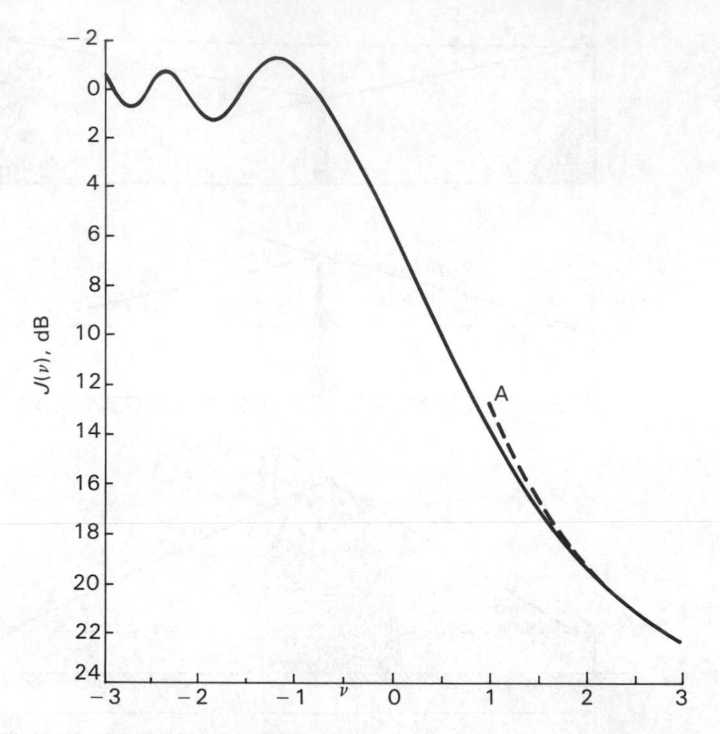

Figure 8.10 Knife-edge diffraction, transmission loss relative to free space. Curve A: asymptote: $L(\nu) = 13 + 20 \log \nu$ (Courtesy of CCIR, Report 715–2, Reference 10)

Figure 8.10 shows, as a function of ν, the loss in decibels caused by the presence of the obstacle. For ν more positive than -1 an approximate value which is within 0.5 dB of the nominal value, can be obtained from the expression:

$$L(\nu) = 6.4 + 20 \log[(\nu + 1)^{1/2} + \nu] \text{ dB} \tag{8.35}$$

DIFFRACTION LOSS DUE TO A SINGLE ROUNDED OBSTACLE

For a single large obstacle in the path of a radio beam, the diffraction attenuation $L(r)$ may be evaluated from the equation:

$$L(r) = L(\nu) + T(\rho) + Q(X) \tag{8.36}$$

The Fresnel–Kirchhoff loss $L(\nu)$ is obtained from Figure 8.10 after the value of ν has been obtained from the equation

$$\nu = [2 \sin(\theta/2)] \cdot [2(d_a + R \cdot \theta/2) \cdot (d_b + R \cdot \theta/2)/(\lambda \cdot d)]^{1/2} \tag{8.37}$$

where λ is the radio carrier wavelength and d_a and d_b are the distances from each terminal to their horizons on the terrain. As the obstacle can be a large percentage of the path length, the values of d_a and d_b may not be the same as the path distances

d_1 and d_2 to the center of the obstacle. R is the effective radius of curvature for the terrain between horizons as determined by the product of the geometrical radius and the k-factor. All distances and lengths are in the same units.

$T(\rho)$, the loss in decibels for incidence upon the curved surface, may be determined[13] from:

$$T(\rho) = 7.2\rho - 2\rho^2 + 3.6\rho^3 - 0.8\rho^4 \tag{8.38}$$

where ρ is given by

$$\rho^2 = [(d_a + d_b)/(d_a \cdot d_b)] \cdot [[\pi \cdot R/\lambda]^{1/3} \cdot 1/R]^{-1} \tag{8.39}$$

$Q(X)$, the decibel loss for propagation along the surface between the horizons, is given by:

$$Q(X) = \begin{cases} T(\rho) \cdot X/\rho & \text{for } -\rho \leqslant X < 0 \\ 12.5X & \text{for } 0 \leqslant X < 4 \\ 17X - 6 - 20 \log X & \text{for } X > 4 \end{cases} \tag{8.40}$$

and

$$X = [\pi R/\lambda]^{1/3} \cdot \theta \simeq [(\pi/2)^{1/2} \cdot \nu\rho] \text{ if } \theta \ll 1 \tag{8.41}$$

Example

A 50 km path has a mountain range running across it. After the earth bulge has been accounted for, it is found that the peak of the range which appears like a single knife edge, protrudes 100 meters past the direct beam at a distance of 30 km from one of the radio terminals. If the operating frequency of the radio link is 4 GHz, determine the expected path loss due to this obstruction.

Solution

From the above information, $d_1 = 30$ km and $d_2 = 20$ km. The wavelength of the carrier is found to be 0.75×10^{-4} km and $h = 0.1$ km. Using equation 8.34a, ν is found to be 4.714, and the loss is calculated from $L(\nu) = 13 + 20 \log \nu$ giving $L(\nu) = 26.46$ dB, for the asymptotic value, for large ν, or from equation 8.35, $L(\nu)$ is found to be 23.3 dB, which is the more accurate value.

The diffraction loss L_{diff} over average terrain can be approximated for losses greater than about 15 dB (CCIR Report 338–5[10] by

$$L_{\text{diff}} = -20\, h/F_1 + 10 \text{ dB} \tag{8.42}$$

A rough plot of this equation is shown in Figure 8.6 by the curve with the reflection coefficient $R = -0.3$. Equation 8.42 is based on measurements over average terrain in the United States.[14] Data obtained in the USSR over average terrain show that diffraction loss may be approximated by equation 8.42 for about half the radio relay links.

8.2.2.3 Planning criteria for path clearance

Diffraction fading as described above can be designed out of a system by insuring that the antenna heights are sufficient, so that in the worst case of earth bulge (k-factor minimum), the receiver antenna is not placed in the diffraction region. As can be seen from Figure 8.6, for zero additional loss over that of free space, the clearance is to be 0.577 F_1. That is, the direct path between the transmitter and the receiver needs a clearance above ground or any obstruction of at least 60 percent of the radius of the first Fresnel zone (equation 8.24) to achieve free-space propagation conditions. The CCIR Report 338–5 [10] advises that for temperate and tropical climates, the following procedure should be used:

1. Determine the antenna heights required for the appropriate median value of the point (i.e. local) k-factor as provided in CCIR Report 563–2. [10] For a standard atmosphere this is 4/3 and may be used in the absence of any available data, and a clearance factor of $1.0F_1$ over the highest obstacle.
2. Obtain the value of k-factor exceeded for approximately 99.9 percent of the worst month in a continental temperate climate from Figure 8.11, for the path length in question.
3. Calculate the antenna heights required for the value of k-factor exceeded for 99.9 percent of the worst month, obtained from 2 above and the following Fresnel zone clearance radii:
   (a) 0.0 F_1 (i.e. grazing) if there is a single isolated path obstruction – temperate climate.
   0.6 F_1 for path lengths greater than about 30 km – tropical climate.
   (b) 0.3 F_1 if the path obstruction is extended along a portion of the path.

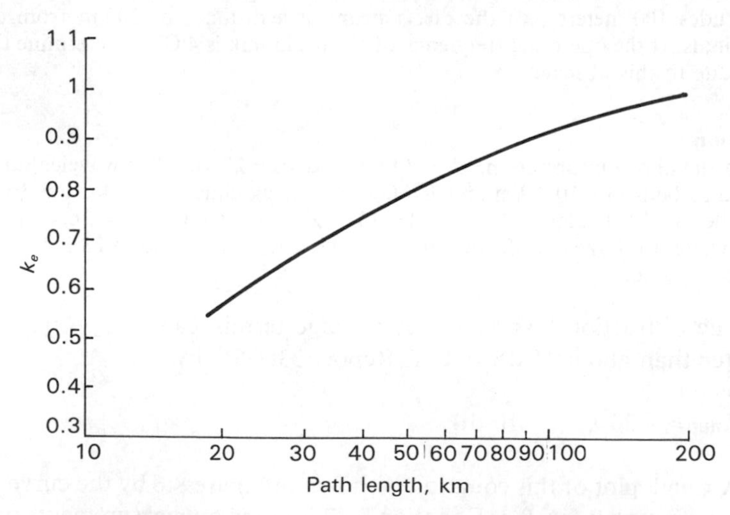

Figure 8.11 Value of k exceeded for approximately 99.9 percent of the worst month (continental temperate climate) (Courtesy of CCIR, Report 338–5, Reference 10)

4. Use the larger antenna heights obtained by steps 1 and 3.

In the U.S.A. for various propagation climates the maximum and minimum k-factors together with their corresponding clearance factors (proportion of first Fresnel zone radius) have been used:

Good propagation climate k_{min}/clearance = 0.66/ > 0.0, k_{max}/clearance = 1.0/0.6

Average propagation climate k_{min}/clearance = 0.66/0.3, k_{max}/clearance = 1.33/1.0

Difficult propagation climate k_{min}/clearance = 0.50/ > 0.0, k_{max}/clearance = 0.50/ > 0.0

In France k is determined from Figure 8.11:

k_{min}/clearance = k/0.0, k_{max}/clearance = 1.33/1.0

In the U.K. no diffraction fading has been observed from 300 000 hours of records from 21 radio relay stations:

k-factor/clearance factor = 0.7/0.6

In the Federal Republic of Germany it is found that:

For 50 percent of the time, k_{max}/clearance = 1.33/1.0
For 99 percent of the time to protect against diffraction fading for hops longer than 100 km, k_{min}/clearance = 1.0/0.0

For various other countries k-factor/clearance factor = 1.33/1.0

Majoli[4] states that the designer's task is to guarantee that diffraction losses do not exceed the system fade margin except for a negligibly small percentage of the time. His method is to keep in mind the two-way unavailability limit of the CCIR HRDP, which is:

$$\text{Unavailability} \leqslant 0.01 \times d/2500 \text{ percent} \tag{8.43}$$

where d is in km.

The first step in the design procedure is to determine the value of k which is not exceeded for more than 99.9 percent of the time as a function of the path length. This can be determined from Figure 8.4 or from the method outlined in Section 8.2.2.1. The corresponding diffraction loss is then determined from either of the methods outlined in Section 8.2.2.2, from equation 8.42 for average terrain, or from equation 8.44 below for flat terrain:

$$L_{diff} = -25\ h/F + 15 \text{ dB} \tag{8.44}$$

Comparing the required fade margin F_m, with the value of this diffraction loss L_{diff}, provides an indication of the tower height adjustments required to make the fade margin equal to the absolute value of the diffraction loss ($F_m = L_{diff}$). When this condition is met, additional tower height is provided to meet the CCIR limit provided in equation 8.43 and to allow for the growth of vegetation and errors in terrain elevation, etc. The tower heights can be increased to reduce the estimated unavailability to a specified fraction (0.1) of that time during which diffraction attenuation exceeds F_m, i.e. a fraction of that given by equation 8.43. Majoli[4] suggests that a clearance of about 0.6 at the median value of k for the site considered be used.

Example

Determine the minimum fade margin for a 100 km radio link operating at 4 GHz in a temperate climate, if the tower heights at either end of the link are 60 m and 120 m above sea level.

Solution
From Figure 8.11, the minimum value of the k-factor is found to be 0.92 for a 100 km hop length. This compares favorably with that in Figure 8.4. As the difference between the antenna heights is 60 m, at a distance of 50 km, where the first Fresnel zone is at a maximum, the direct beam is 30 m higher than it would have been if the antenna heights were equal. This means that the direct beam is 90 m above sea level. The first Fresnel zone radius at mid-path is calculated, from equation 8.24, to be 43.3 m. This means that the edge of the first Fresnel zone is 46.7 m above the Earth. As the k-factor is 0.92, the Earth bulge at mid-path is calculated from equation 8.8, to be 213.1 m. Thus, the Earth protrudes $(213.1-46.7) = 166.4$ m into the first Fresnel zone. From equation 8.44 the attenuation due to a flat terrain is found to be 81 dB. This means that the fade margin required to equal this loss must be 81 dB. As this is impractical, the tower heights must be raised and the calculations redone.

8.2.2.4 Reflection from the surface of the Earth

The variation in transmission loss for many paths which have been studied arise from two relatively simple propagation mechanisms: the refraction associated with the time-varying vertical gradient of refractive index, as discussed in Section 8.2.2.1, and the phase-interference patterns due to diffraction and reflection by the Earth's surface and atmospheric discontinuities in refractive index. In certain circumstances, the direct beam will be interfered with by the ground-reflected ray or other multipath rays. The most severe fading occurs when the reflected or multipath ray is of the same amplitude as, and opposite in phase to, the direct beam. Before considering multipath fading due to changes in the refractive index of the atmosphere, we shall discuss ground reflections. The strength of the reflected signal at the receiving antenna depends to a large extent on the directivity of the antennas, the height of the antennas above the Earth and the nature of the surface and length of the path.

Figure 8.12 shows the geometrical relationships for a radio path within the horizon. For a perfectly conducting surface, the attenuation relative to free space

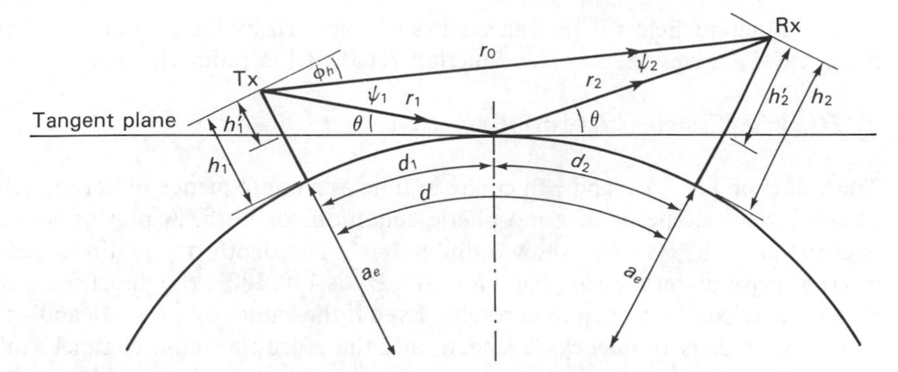

Figure 8.12 Geometrical relationships for a radio path within the horizon

of the received signal, as a result of the sum of the direct and ground reflected waves, is:

$$L_{\text{ref}} = -6 - 10 \log[\sin^2(\pi \Delta r/\lambda)] \text{ dB} \tag{8.45}$$

where λ is the radio carrier wavelength and $\Delta r = r_1 + r_2 - r_0$ is the difference in the path lengths. For small grazing angles, the reflected wave will generally have some phase delay ϕ_Δ due to the difference between the ray paths Δr, such that

$$\phi_\Delta = 2\pi \Delta r/\lambda = 4\pi h_1' \cdot h_2'/\lambda \text{d} \tag{8.46}$$

where h_1' and h_2' are the heights of the antennas above the tangent plane, as shown in Figure 8.12, and d is the path length. There will also be some phase delay ϕ_R associated with the reflection coefficient R of the surface. The reflection coefficient is given by:

$$R = |R| \cdot \exp(j\phi_R) \tag{8.47}$$

The resultant field E for the direct and reflected rays is then

$$E = E_d[1 + R \cdot \exp(j\phi_\Delta)] = E_d[1 + |R| \exp(j(\phi_\Delta + \phi_R))] \tag{8.48}$$

where E_d is the direct wave. The magnitude of the reflection coefficient $|R|$ and its phase ϕ_R depends on the nature of the reflecting surface, the radio carrier wavelength λ, the curvature of the surface and the degree of surface roughness, as well as the grazing angle θ, i.e. the angle between the tangent plane and the incident or reflected ray. The difference between the phase of the direct ray and that of the reflected ray ϕ_Δ can also be expressed as a time difference, known as the difference in propagation delays τ. As $\text{d}\phi/\text{d}t = \omega$, where ω is the angular frequency of the carrier (in radians/s), we have:

$$\phi_\Delta = -\omega\tau \tag{8.49}$$

The resultant field for the reflected ray E, divided by the resultant field for the direct ray E_d, gives the transfer function $H(\omega)$ of the radio channel:

$$H(\omega) = E/E_d = 1 + |R| \exp[j(\phi_R - \omega\tau)] \quad (8.50)$$

The values of $|R|$, ϕ_R and τ fluctuate in time as a consequence of variations either of the surface shape or of atmospheric conditions or both. A plot of $\phi_R - \omega\tau$ (in degrees) against $H(\omega)$ will show a dip at 180°. The depth of the dip or reduction in $H(\omega)$ depends on the value of $|R|$. If $|R|$ is 1 at 180°, the direct wave is completely canceled and a deep fade results. Even if the values of $|R|$, ϕ_R and τ remain constant, $H(\omega)$ is frequency-dependent and therefore the radio channel exhibits a frequency-selective characteristic. This can be intuitively understood by considering the nature of the surface. Instead of a smooth, perfectly conducting surface as described above, assume a more practical surface in which irregularities appear. At short wavelengths a small irregularity in the surface will appear larger to a high-frequency carrier wavelength than it would to a lower-frequency carrier wavelength. This will directly affect the reflection coefficient. The reflection coefficient is no longer $|R|$ in this case, as will be discussed below. The field due to a scattering from a moderately rough surface is the sum of a *specular component* and a *diffuse component*. For the specular component, the scattered energy is contained within a cone, close to the direction for which the angle of reflection is equal to the angle of incidence, and its phase is coherent, which means that its mean value can be determined for any point in space. The specular component results from the reradiation of energy from points on a Fresnel ellipsoid which give rise to equal phase at the receiver. The diffuse component has little directivity and originates over a larger area of the scattering surface than that of the specular component, but its fluctuations are Rayleigh distributed and have a large amplitude. Its phase, however, unlike the specular component, is incoherent and has equal probability of having any value at any point in space. If a surface is very rough, there will be no specular component and the diffuse component will be uniformly distributed with angle.[5]

SPECULAR REFLECTION FROM A PLANE SURFACE

The Fresnel reflection coefficient R_o, of a plane surface is given by the expression

$$R_o = (\sin\theta - \sqrt{C})/(\sin\theta + \sqrt{C}) \quad (8.51)$$

where θ = the angle of grazing as shown in Figure 8.12
$C = \eta - \cos^2\theta$ for horizontal polarization
$C = (\eta - \cos^2\theta)/\eta^2$ for vertical polarization
with the refractive index $\eta = \varepsilon_r(f) - j \cdot 60\lambda\sigma(f)$ where $\varepsilon_r(f)$ is the relative permittivity of the surface with conductivity $\sigma(f)$, in S/m and λ is the free-space wavelength in meters of the carrier frequency f. The value of the Fresnel reflection coefficient R_o, is that of a plane surface of infinite extent.

SPECULAR REFLECTION FROM A SMOOTH SPHERICAL EARTH

The reflection coefficient R, given in equation 8.47, may be considered to comprise;

the electrical characteristics of the ground, the roughness and curvature of the reflecting surface, and the transmitting and receiving antenna patterns,[4] so that

$$R = |R| \exp(j\phi_R) = R_o \cdot D \cdot A_T \cdot A_R \tag{8.52}$$

where D is the divergence factor due to the Earth's curvature and A_T and A_R are the attenuations due to the directivity of the transmitting and receiving antennas in the direction of the reflection point. The combination $A_T \cdot A_R$ may also be considered as the transmission factor for the aperture or reflection factor for the surface. For practical purposes it is often sufficient to assume the ground to be approximate to a large flat surface, so that $A_T \cdot A_R = 1$. In practice and in most cases $A_T \cdot A_R$ is found to have a value between 0.1 and 1.2. For a smooth spherical Earth it is practical to assume that $R = -1$ as shown in Figure 8.6. To evaluate equation 8.51, the curves of Figure 8.8 may be used. The determination of R is based on the condition that the reflecting surface is smooth and that a high specular component exists. This assumption, as previously discussed, is valid only when the surface irregularities are negligible in comparison to the wavelength.

REFLECTION FROM ROUGH SURFACES

To give a quantitive definition of roughness, the Rayleigh roughness criterion is often used. This is:

$$g = 4\pi (S_h/\lambda) \cdot \sin\theta \tag{8.53}$$

where S_h is the standard deviation of the surface height about a local mean value within the first Fresnel zone, λ is the free-space wavelength and θ is the grazing angle. In general, a surface can be considered smooth for $g < 0.3$. When the surface is rough, the reflected signal has two components: one specular and the other diffuse.

DIVERGENCE FACTOR

When rays are specularly reflected from a spherical surface, there is a reduction in the reflection coefficient. To account for the curvature of the Earth which produces this reduction, a 'divergence factor' D is introduced such that the ratio of the reflected ray to the incident ray amplitudes is the product of D by the reflection coefficient R_o as computed for plane spherical Earth reflection.

$$D = [1 + (2d_1 \cdot d_2)/[(a_e \cdot \sin\theta) \cdot (d_1 + d_2)]]^{-1/2} \tag{8.54}$$

Equation 8.52 shows how the divergence factor is used in determining the effective reflection coefficient of a spherical Earth which includes also the effect of partial reflection.

CALCULATION OF REFLECTION POINT (CCIR REPORT 1008)[10]

An analysis of surface reflections requires a determination of the geometrical specular reflection point located at some distance d_1 from one of the terminals. An exact value to the location of the reflection point exists only for a flat earth. Approximate solutions are available for small angular distances for terminals near the surface of the Earth. To derive a value for the reflection point, it is convenient to define two intermediate quantities, m and c and refer to Figure 8.12:

$$m = d^2/[4a_e(h_1 + h_2)] \tag{8.55}$$

$$c = (h_2 - h_1)/(h_1 + h_2) \quad h_2 > h_1 \tag{8.56}$$

where a_e is the effective Earth's radius and other quantities are those shown in Figure 8.12. The third quantity b is then found to be:

$$b = 2[(m+1)/3m]^{1/2} \cdot \cos[(\pi/3) + (1/3) \cdot \text{arc}\cos\{(3c/2) \cdot [3m/(m+1)^3]^{1/2}\}] \tag{8.57}$$

This permits the determination of the distance d_2 from the terminal with a height above sea level h_2 to be determined, together with the path length difference Δr, and the grazing angle θ:

$$d_2 = (d/2) \cdot (1 + b) \tag{8.58}$$

$$\Delta r = (2d_1 \cdot d_2/d) \cdot \theta^2 \tag{8.59}$$

$$\theta = [(h_1 + h_2)/d] \cdot [1 - m(1 + b^2)] \tag{8.60}$$

The method used to reduce the effect of reflections is space diversity. This will be discussed a little later in this chapter, when we consider the techniques used to reduce fading effects.

8.2.3 Propagation fundamentals

This section is concerned with only fundamental concepts affecting the development of equations in common use.

8.2.3.1 Field strength

The field strength is defined as the dielectric stress existing between two points which are unit distance apart and lying on a line parallel to the electric field. Alternatively, it may be defined as 'The electric field strength at a point (measured in V/m) is the force acting upon a positive stationary charge per unit charge'. Being a force, it may be expressed in newtons per unit charge. Field strengths may be given in units of megavolts/meter (MV/m) down to microvolts/meter (μV/m). Unless otherwise stated, the figures are RMS values.

8.2.3.2 Radiation from an isotropic radiator

An isotropic radiator is a figment of the imagination, for it is a point radiator in free space which will radiate in all directions, so that the wavefront produced is the surface of a sphere. If the power radiated at the center of the sphere is P_1 watts, then this power, which will be evenly distributed over the spherical surface of the wavefront, will have a power density of $P_1/(4\pi d^2)$ W/m^2, where d is the distance of the wavefront from the center of the sphere, i.e. the radiator. If d is greater than 100 meters, the surface area equal to one square meter will be a very small portion of the total surface area of the sphere, and may be considered flat. As the power contained in one square meter of the wavefront is the amount of energy passing through it in one second, it is convenient to consider that this energy passes through a surface of one square meter after traveling through one meter. This permits the amount of energy stored in an equivalent capacitor to be calculated. The capacitor has a surface area of 1 m^2 and its plates (wavefronts) are separated by 1 m. The amount of energy stored in a capacitor is given by:

$$W = C \cdot V^2/2 \text{ joules} \tag{8.61}$$

where W is the energy stored in the capacitor, C is the capacitance in farads, and V is the potential across the capacitor in volts.

As the capacitance of a parallel plate capacitor is given by

$$C = \varepsilon_0 A/s \text{ farads} \tag{8.62}$$

where ε_0 is the permittivity of free space (8.85419×10^{-12} F/m), A is the area of the plates (m^2) and s is the distance that the plates are apart (m). For one cubic meter of air, the capacitance C is:

$$C = 8.85 \times 10^{-12} \text{ farads}$$

From equation 8.61, the energy stored in a capacitor is:

$$W = (1/2) \cdot \varepsilon_0 \cdot (A/s) \cdot (V/s)^2 \cdot s^2 = (1/2) \cdot \varepsilon_0 \cdot A \cdot E_0^2 \cdot s \tag{8.63}$$

For $A = 1$ m^2, and $s = 1$ m, the energy stored in one cubic meter is given by:

$$W = E_0^2 \times 4.425 \times 10^{-12} \text{ joules}$$

where E_0 is the field strength in V/m.

This electrostatic energy passes through the spherical wavefront at near the velocity of light, so that the time taken for it to travel through 1 meter will take $1/c = 1/(3 \times 10^8)$ s. Thus, the energy passing through the one cubic meter in one second, or the power across one square meter of the wavefront is:

$$1.326 \times 10^{-3} \times E_0^2 \text{ watts}$$

The power/square meter originating from the isotropic radiator contains both the magnetic and electric power components. Thus, for the electric component alone, the power/square meter is given by $P_1/8\,\pi d^2$. If a surface area of the wavefront of one square meter is taken, then

$$P_1/(8\pi d^2) = 1.326 \times 10^{-3} \times E_0^2$$

Thus

$$E_0 = [30P_1/d^2]^{1/2} \text{ V/m} \quad (8.64)$$

In practical antennas, the power of an isotropic antenna is concentrated in a certain direction. The more the power is concentrated, the higher is the gain of the antenna. This is because this concentrated power would require an isotropic antenna with a power equivalent to the concentration factor or the practical antenna gain. Thus equation 8.64 would become:

$$E_{0(\text{practical})} = [30G_1 \cdot P_1/d^2]^{1/2} \text{ V/m} \quad (8.65)$$

which means that the radiated power P_1 of the isotropic radiator has been increased by the factor G_1. Thus, the gain of a practical antenna, if defined as the gain relative to an isotropic antenna, is

$$\begin{aligned} &20 \log [E_{0(\text{practical})}/E_0] \\ &= 10 \log [(30G_1 \cdot P_1/d^2)/(30P_1/d^2)] \\ &= 10 \log G_1 = G_1 \text{ dBi} \end{aligned} \quad (8.66)$$

where dBi denotes gain in decibels over an isotropic antenna.

8.2.3.3 Transmission within line-of-sight

From simple geometry, the difference in length between a direct ray and the reflected ray, Δr, can be found for the plane Earth case to be:

$$\Delta r_{(\text{plane})} = 2h_1' \cdot h_2'/d \quad (8.67)$$

where h_1' and h_2' are the antenna heights above the plane earth or tangent plane, as shown in Figure 8.12, and d is the distance between the terminals. If $\Delta r_{(\text{plane})}$ is expressed in terms of the carrier frequency phase, the distance difference becomes a phase difference, equal to:

$$\Delta\phi_r = \Delta r_{(\text{plane})} \cdot (2\pi/\lambda) = 4\pi h_1' \cdot h_2'/(\lambda \cdot d) \text{ radians} \quad (8.68)$$

Although equation 8.68 provides the difference in phase between the direct and reflected ray, it has not taken into account the phase change at the point of reflection. This phase change is dependent on the grazing angle, nature of the surface, polarization and frequency, as discussed in Section 8.2.2.4. However, if the grazing angle is very small, the phase change is approximately 180° or π radians. The total phase difference $\Delta\phi$ at the receiver between the direct and reflected ray will be the sum of $\Delta\phi_r$ and the phase change on reflection $\Delta\phi_p$, i.e.

$$\Delta\phi = 4\pi h_1' \cdot h_2'(\lambda \cdot d) + \pi \text{ radians} \tag{8.69}$$

If it is assumed that the magnitude of the reflection coefficient is unity, the field strength at the receiver will be equal to the vector sum of two components, each equal to the free-space field strength E_0, with a phase difference of $\Delta\phi$ radians. The resultant vector is the resultant field strength E at the receiver and is given by

$$E = 2E_0 \cdot \sin[2\pi h_1' \cdot h_2'/(\lambda \cdot d)] \text{ V/m (over plane Earth)} \tag{8.70}$$

Equation 8.70 may be derived also from the propagation equation for radio waves over a plane Earth, given in equation 8.71[15] when $R = -1$ for near grazing paths and the 'surface wave' attenuation factor A is neglected. This condition applies as long as both antennas are elevated more than a wavelength above the ground, or more than 5 to 10 wavelengths above sea water. Under these conditions, the effect of the Earth is independent of polarization and ground constants.

$$(E/E_0) = \underbrace{1}_{\text{Direct Wave}} + \underbrace{R \exp(j\,\Delta r)}_{\text{Reflected Wave}} + \underbrace{(1-R)A \cdot \exp(j\,\Delta r)}_{\text{'Surface Wave'}} + \underbrace{\ldots}_{\text{Induction field and secondary ground effects}} \tag{8.71}$$

HEIGHT/GAIN PATTERNS

Equation 8.70 shows that the lobe structure of the signal oscillates around the free-space value E_0, i.e. $2E_0$ and zero, as the expression $2\pi h_1' \cdot h_2'/(\lambda \cdot d)$ varies. If one of the antennas' heights (say h_2'), is varied with all other terms remaining constant, E will have a maximum value of $2E_0$ whenever

$$2\pi h_1' \cdot h_2'/(\lambda \cdot d) = (2n+1)\pi/2 \text{ radians}$$

or when

$$h_2' = (2n+1) \cdot \lambda d/(4h_1') \tag{8.72}$$

Similarly, E will have a minimum value of zero whenever

$$2\pi h_1' \cdot h_2'/(\lambda d) = n\pi$$

or when

$$h_2' = n\lambda \cdot d/(2h_1') \tag{8.73}$$

Note that all distances are measured in the same units.

Thus, as the height of the aerial is increased, the field strength E varies in a sinusoidal manner, setting up a pattern of interference fringes. The variation of the field strength with the antenna height is known as the height/gain pattern. The spacing of two antennas feeding a common receiver, one at a null and the other at a peak ($2E_0$), is the principle of space diversity protection.

MAXIMUM POWER DELIVERED TO A RECEIVER

The maximum useful power P_2 that can be delivered to a matched receiver is given by:

$$P_2 = [E \cdot \lambda/2\pi]^2 \cdot (G_2/120) \text{ W} \tag{8.74}$$

where E is the received field intensity V/m, λ is the wavelength in meters and G_2 is the receiving antenna gain ratio over an isotropic antenna. This equation can be derived by considering that the power flux density in watts/square-meter, as derived above for the electric field, is given by:

$$S_e = E^2/(240\pi) = 1.326 \times 10^{-3} \times E_0^2 \text{ W/m}^2 \tag{8.75}$$

As the magnetic component also is required to be taken into account, the available flux density at the receiver antenna is twice that given by equation 8.75. Thus available power flux density at the receiver antenna is:

$$S_r = E^2/(120\pi) \text{ W/m}^2 \tag{8.76}$$

The effective area of an isotropic antenna is given by $\lambda^2/4\pi$, and the flux density at the antenna is the received power divided by the effective area. This flux density is the same as that given by equation 8.76, thus

$$S_r = E^2/(120\pi) = P_2 \cdot (4\pi/\lambda^2)$$

Hence the received power P_2 for an isotropic receiver antenna, is found to be:

$$P_2 = E^2 \cdot \lambda^2/(480\pi^2) \tag{8.77}$$

When the antenna is not isotropic, the gain is given by:

$$G_2 = (4\pi) \cdot A_e/\lambda^2 \tag{8.78}$$

where A_e is the effective area in the direction of transmission. Using equation 8.78 in the derivation for the flux density at the receiver antenna, we have:

$$S_r = E^2/(120\pi) = P_2 \cdot (4\pi/G_2 \cdot \lambda^2)$$

Solving for P_2, we get equation 8.74.

TRANSMISSION LOSS

The CCIR Recommendation 341–2[10] defines the transmission loss of a radio link (symbols L or A) as: 'The ratio, usually expressed in decibels, for a radio link between the power radiated by the transmitting antenna and the power that would be available at the receiving antenna output if there were no loss in the radio frequency circuits, assuming that the antenna radiation diagrams are retained'. The notes which accompany this definition are:

Note 1 The transmission loss may be expressed by $L = L_s - L_{tc} - L_{rc}$ dB, where L_{tc} and L_{rc} are the losses, expressed in dB, in the transmitting and receiving antenna circuits respectively, excluding the dissipation associated with the antenna's radiation, i.e. the definitions of L_{tc} and L_{rc} are $10 \log(r'/r)$, where r' is the resistive component of the antenna circuit and r is the radiation resistance.

Note 2 Transmission loss is equal to system loss minus the loss in the radio frequency circuits associated with the antennas.

The *free-space basic transmission loss (symbols: L_{bf} or A_0)* is the transmission loss that would occur if the antennas were replaced by isotropic antennas located in a perfectly dielectric, homogeneous, isotropic and unlimited environment, the distance between the antennas being retained. If the distance d between the antennas is much greater than the wavelength λ, the free-space transmission loss A_0 can be derived from equation 8.65 and equation 8.77, as:

$$A_0 = 10 \log(P_1/P_2) = 20 \log((4\pi d/\lambda) \cdot (E_0/E)) \text{ dB}$$

In free-space conditions where no reflected wave is present, the received field strength E is equal to the free-space field strength E_0, hence

$$\begin{aligned} A_0 &= 20 \log(4\pi d/\lambda) \text{ dB} \\ &= 32.5 + 20 \log f(\text{MHz}) + 20 \log d(\text{km}) \text{ dB} \end{aligned} \tag{8.79}$$

where f is the carrier frequency in MHz, and d is the distance in km, between the two antennas at either end of the path. The *transmission loss A* may be derived from equation 8.65 and equation 8.74 to give:

$$\begin{aligned} A &= 10 \log(P_1/P_2) = A_0 - G_1 - G_2 \text{ dB} \\ &= 32.5 + 20 \log f \text{ MHz} + 20 \log d \text{ km} - G_1 - G_2 \text{ dB} \end{aligned} \tag{8.80}$$

Total loss of a radio link (symbols: L_l or A_l) is the ratio, usually expressed in decibels, of the power supplied by the transmitter of a radio link and the power supplied to the corresponding receiver in real installation, propagation and operational conditions.

Note. It is necessary to specify in each case the points at which the power supplied by the transmitter and the power supplied to the receiver are determined, for example: before or after the radio frequency filter or multiplexers that may be employed at the sending or the receiving end; and at the input or at the output of the transmitting and receiving antenna feed lines.

System loss (symbols: L_s or A_s) is the ratio, usually expressed in dB, for a radio link, of the radio frequency power input to the terminals of the transmitting antenna and the resultant radio frequency signal power available at the terminals of the receiving antenna.

Note 1. The available power is the maximum real power which a source can deliver to a load, i.e. the power which would be delivered to the load if the impedances were conjugately matched.
Note 2. The system loss may be expressed by:

$$L_s = 10 \log(P_t/P_a) = P_t - P_a \text{ dB} \tag{8.81}$$

where P_t is the radio frequency power input to the terminals of the transmitting antenna and P_a is the resultant radio frequency original power available at the terminals of the receiving antenna.
Note 3. The system loss excludes losses in feeder lines but includes all losses in radio frequency circuits associated with the antenna, such as ground losses, dielectric losses, antenna loading coil losses, and terminating resistor losses.

Basic transmission loss of a radio link (symbols: L_b or A_i) is the transmission loss that would occur if the antennas were replaced by isotropic antennas with the same polarization as the real antennas, the propagation path being retained, but the effects of obstacles close to the antennas being disregarded.

Note 1. The basic transmission loss is equal to the ratio of the equivalent isotropically radiated power of the transmitter system to the power available from an isotropic receiving antenna.
Note 2. The effect of the local ground close to the antenna is included in computing antenna gain, but not in the basic antenna loss.

Loss relative to free space (symbols: L_m or A_m) is the difference between the basic transmission loss and the free-space basic transmission loss, expressed in decibels.

Note 1. The loss relative to free space may be expressed by:

$$L_m = L_b - L_{bf} \text{ dB} \tag{8.82}$$

Note 2. Loss relative to free space (L_m) may be divided into losses of different types, such as:

Absorption loss (ionospheric, atmospheric gases or precipitation)
Diffraction loss, as for ground waves
Effective reflection or scattering loss, as in the ionospheric case including the results of any focusing or defocusing due to curvature of a reflecting layer
Polarization coupling loss; this can arise from any polarization mismatch between the antennas for the particular ray path considered
Aperture-to-medium coupling loss or antenna gain degradation, which may be due to the presence of substantial scatter phenomena on the path

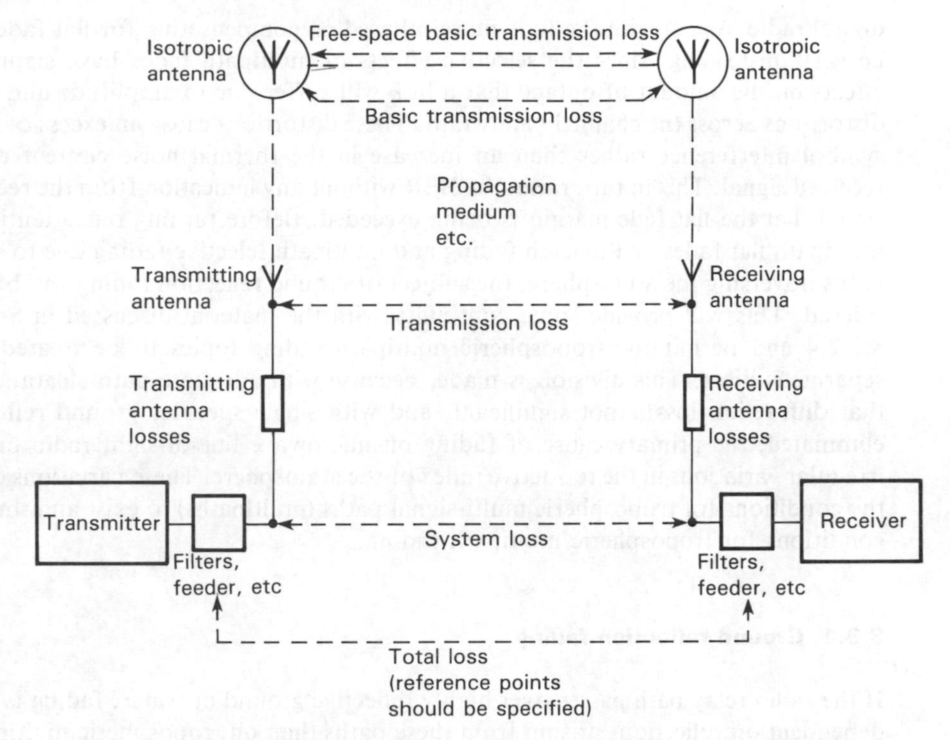

Figure 8.13 Graphical depiction of terms used in the transmission loss concept (Courtesy of CCIR, Recommendation 341–2, Reference 10)

Effect of wave interference between the direct ray and rays reflected from the ground, other obstacles or atmospheric layers

Figure 8.13 shows in graphical form the CCIR Recommendation 341–2[10] definitions presented in this section.

8.3 FADING

Fading is the variation in the strength of a received microwave radio carrier signal due to atmospheric changes and ground and water reflections in the propagation path. There are two main types of fade which are of concern to radio system designers. These are *flat fades*, which mainly affect the small-capacity digital radio system designer and which attenuate the carrier signal uniformly across the band, and *selective fades*, which as the name implies are frequency selective and are of major importance to the high-capacity digital radio system designer. These two classes of fading may occur individually or together. To design a high-capacity digital radio system with the fade margin based on compensating for flat fades can be very misleading, since the selective effects of multipath fades have significant effects on the amount of outage that a link will suffer due to amplitude and delay distortions across the channel bandwidth. These distortions cause an excess of intersymbol interference rather than an increase in the thermal noise content of the received signal. This in turn raises the BER without any indication from the received signal that the flat fade margin is being exceeded. Before turning our attention to multipath flat fades or Rayleigh fading and multipath selective fading due to signal paths traversing the atmosphere, the subject of ground reflection fading will be considered. This will provide some continuity with the material discussed in Section 8.2.2.4 and permit the tropospheric multipath fading topics to be treated as a separate subject. This division is made, because with adequate path clearance, so that diffraction loss is not significant, and with single specular ground reflection eliminated, the primary cause of fading on microwave line-of-sight radio links is irregular variations in the refractive index of the atmosphere. These variations create the conditions for tropospheric multi-signal paths (multipaths) to exist and thus the conditions for tropospheric multipath fading.

8.3.1 Ground reflection fading

If the radio relay path passes over highly reflective ground or water, fading is more dependent on reflections arising from these paths than on tropospheric multipaths, especially if the paths are short. This is because reflections from the ground make the received signal strength fluctuate randomly owing to the propagation parameter variations caused by meteorological conditions. Reflected signal levels which are less than 10 dB down on the direct path signal level can occur if the link crosses areas such as sea, lake, flat and humid areas, wet ground and moist river valleys or moors, where there may be total reflection in an atmospheric layer near the ground when mist or ground fog is present. Reflection fade characteristics are different from those characteristics used to describe multipath fading. In particular, the techniques used to predict multipath fading do not apply to reflection fading, where smaller or larger fade depths occur, depending on the height of the antennas above the reflective surface and the surface coefficient of reflection. Although there is little information in general on prediction methods for paths with strong surface reflections some methods have been described.[16,4] Majoli's method[4] is to consider that for a given carrier angular frequency ω, the fade depth due to reflection FAD_r, may be computed from equation 8.50 as

$$FAD_r = -20 \log(H(\omega)) \tag{8.83}$$

Using de Moivre's theorem and letting $(\phi_R - \omega\tau) = \Phi$ and $|R| = b$, equation 8.83 becomes:

$$FAD_r = -10 \log(1 + b^2 + 2b \cos \Phi) \tag{8.84}$$

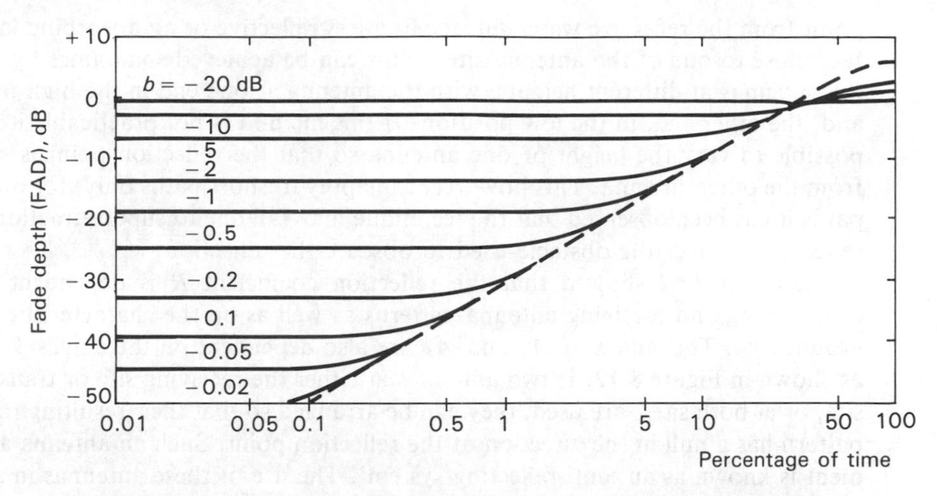

Figure 8.14 Percentage of the time that fade depth indicated in ordinate is exceeded (Courtesy of Telettra S.p.A., Reference 4)

If it is considered that the reflected path delay is far greater than one period of the carrier wave, it may be assumed that the fluctuations of τ in time are large enough to be taken as a random variable, uniformly distributed about half of the carrier wave's period. Assuming that the phase shift associated with the reflection is constant, then, from $(\phi_R - \omega\tau) = \Phi$, the same model applies to Φ. Therefore, taking b as constant, the distribution of FAD_r may be computed from equation 8.84 with the results shown in Figure 8.14 for various values of b. From Figure 8.14 it can be seen that for a reflection coefficient of 0.96 ($b = -0.175$ dB), which indicates a very strong reflection, the attenuation will be greater than about 35 dB for about 0.5 percent of the time. This indicates that if the receiver had a 35 dB fade margin and the link was to have an availability of 99.98 percent, then without adequate protection from reflection fading the link would not be able to meet the availability objective. The assumptions made to derive the curves of Figure 8.14 are not always valid, since τ may not be a uniformly distributed random variable and since b is not always constant but varies with changing propagation conditions as does ϕ_R. The CCIR Report 338–5[10] states that over sufficiently flat paths the number of fades (except subrefractive) rises with increased path clearance. On coastal paths, the number of fades is substantially higher than over land paths. For fades measured on over-sea paths it has been found that there were five times as many as that on overland paths with a flat region in the vicinity of the ground-reflection point.

8.3.1.1 Techniques for reducing the effects of surface multipath fading (ground reflection fading)

Usually only space diversity reception and transmission with vertically spaced antennas result in an efficient counter-measure against surface-reflected rays, while the usage of frequency diversity alone is effective only if cross-band diversity is applied. Sometimes, especially on paths over water, it is appropriate to shift the reflection point from the reflective water surface to a less reflective or an absorbing land surface close to one of the antenna sites. This can be achieved sometimes by making the antennas at different heights, with the antenna at one end in the high position, and, the other end, in the low position. If this method is not practicable it may be possible to vary the height of one antenna so that the reflection point is shielded from the other antenna. This however, can apply to short paths only, for on longer paths it has been observed that the technique may fail due to superrefraction of the reflected ray over the obstacle used to obscure the reflection.

Equation 8.52 showed that the reflection coefficient R is dependent on the transmitting and receiving antenna patterns as well as on the characteristics of the ground, etc. The values of A_T and A_R are also dependent on the angles ψ_1 and ψ_2 as shown in Figure 8.12. If two antennas at either the receiving site or transmitting site, or at both sites, are used, they can be arranged so that their resulting radiation pattern has a null in the direction of the reflection point. Such an antenna arrangement is known as an 'anti-reflecting system'. The use of these antennas in such an array also permits a 3 dB increase in antenna gain.[17] Figure 8.15 shows such an arrangement.[17,18] If the direct ray arrives horizontally, the two antennas are set in the vertical plane with a spacing of s metres between them. The output of each antenna is then connected into a hybrid through two feeders, which are cut to length, in order that the direct received signals add in phase at the hybrid output. To trim the phase for optimum performance, a phase shifter is used in one of the feeders. Because of the different path lengths traversed by the reflected signals in reaching each of the two antennas, their phases will not add in phase. From Figure 8.15 it can be seen that the path difference is $s \cdot \sin \psi_2$, which produces a phase shift of:

$$\delta = (2\pi s \cdot \sin \psi_2)/\lambda \tag{8.85}$$

By choosing the value of s such that $\delta = \pi$, i.e.

$$s = \lambda/(2 \sin \psi_2) \tag{8.86}$$

a phase opposition is produced and the reflected signals cancel each other.

The practical constraints on obtaining exact reflected signal cancelation are: the variability of atmospheric refractive index changing the angle ψ_2, the movement of the antenna due to wind, length variations in feeders due to temperature changes, change in the carrier frequency over that used in the design, and difficulty in implementing and calibrating. This system cannot be used when ψ_2 is very small (less than or equal to 20′). For a 2 GHz system with $\psi_2 = 1^\circ$, Majoli[4] has shown that for 15 percent of the time a 28.6 dB attenuation of the reflected ray is obtained, whilst in 80 percent of the cases the actual attenuation is barely 15 dB. It is expected that as the angle ψ_2 is increased improvement will result. Another practical consideration is that the spacing of the antennas is limited as the frequency of the carrier rises. Assuming a standard 1.2 m antenna which is often used at 13 GHz or so, the minimum separation of the antennas is limited to 1.2 m. From equation 8.86, for $\psi_2 = 2^\circ$, the minimum value of λ is 0.08376 m, which corresponds to a maximum frequency of 3.58 GHz, or 7.16 GHz if $\psi_2 = 1^\circ$.

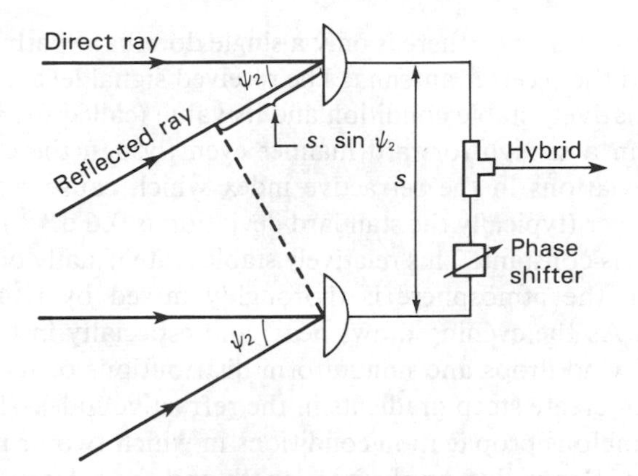

Figure 8.15 Anti-reflecting system
(Courtesy of Telettra S.p.A., Reference 4)

8.3.2 Multipath or Rayleigh fading

For a well-designed path which is not subject to diffraction fading or surface reflections, multipath propagation is the dominant factor in fading below 8 GHz. Above this frequency, the effects of precipitation tend increasingly to determine the permissible path length through the system outage objectives (see Section 7.5.3). 'It is generally agreed that, with adequate path clearance and in the absence of a single specular reflection on a path, the very deep fading is due to multipath propagation through the atmosphere, which, over a single section, gives rise to a Rayleigh distribution of the received signal amplitude against time'.[15] Such a distribution is characterized by a 10 dB-per-decade slope for fading depths greater than 10 dB[19] in normal hops and 20–30 dB on long hops (>60 km). For short hops, this asymptotic decrease starts for fade depths exceeding only a few dB. If multipath fading occurs for a proportion of the month, during the rest of the time, the fading depth has a much less severe distribution, which follows a log-normal distribution.* These fluctuations are known as *shallow fades.*

For depths of fade greater than or equal to 20 dB, on a 4 GHz and 6 GHz path, Barnett[20] showed that

$$P(FAD_m > F) = P_0 10^{-F/10} \tag{8.87}$$

where $P(FAD_m > F)$ is the probability that the multipath fading depth is greater than the fade depth F dB, and where the constant P_0 varies from about 0.001 to 15 and is dependent on the various link parameters such as operating frequency, hop length, climate and terrain roughness, as well as the time of the day and month of the year. Under normal propagation conditions with the atmosphere along the link relatively homogeneous (that is, there are little or no vertical or horizontal variations

*See Section 2.1.3.1

in the refractive index), there is only a single dominant path between the transmitter antenna and the receiver antenna. The received signal level exhibits under these conditions a relatively stable condition and its value (called the free-space value) can be calculated in a straightforward manner even though there are small time-varying random deviations in the refractive index which cause small scintillations in the received power (typically the standard deviation is 0.6 dB[20]), even when the average value remains constant. This relatively stable state usually occurs during the midday hours when the atmosphere is thoroughly mixed by rising convection currents and winds. As the evening draws near, and especially in the summer months, the amount of wind drops and nonuniform distributions of temperature and humidity form, which create steep gradients in the refractive index. This type of atmosphere brings anomalous propagation conditions in which two or more propagation paths are possible. Depending on the hop length and the refractive index gradient value, thickness and height, up to three propagation paths are possible.[21] It is thought that multipath propagation occurs in the two-ray mode for 80–90 percent of the time, and that the three-ray mode is an exception rather than a usual occurrence[22] during conditions of multipath fading.

As mentioned above, multipath fading does not normally produce inter-symbol interference to the same extent with a small-capacity digital system as it does with a large-capacity system. This is true provided that the bandwidth of the small-capacity system is much less than $1/\tau$, where $1/\tau$ is the inverse of the largest multipath delay. Under these conditions the flat fade margin is a valid design parameter. In high-capacity systems, where the IF bandwidth is far greater, in-band amplitude dispersions may be sufficient to cause outages which are not immediately apparent when considering nonselective multipath fading. It is perhaps worth while mentioning at this point that multipath fading is frequency selective, but here only the nonselective properties are being considered.

8.3.2.1 Multipath fade occurrence factor P_o

Equation 8.87 above, may be used to predict the probability of a multipath fade occurring at a depth greater than some designated fade depth F, if the parameter P_o could be determined. An expression for the multipath fade occurrence factor for the worst month P_o, which equals the multipath activity factor $\eta \times$ probability of Rayleigh fading P_R is given as:[23]

$$P_o = C(f/4) \cdot d^3 \times 10^{-5} = \eta P_R \tag{8.88}$$

where C is the terrain factor and has the values:

$$C = \begin{cases} 1 & \text{Average terrain in temperate climates} \\ 4 & \text{Over water and Gulf Coast or in damp climates or climates} \\ & \text{exhibiting strong thermal excursions such as that found in deserts} \\ 0.25 & \text{Mountains and dry climates} \end{cases}$$

and f is in GHz and d is in km.

To allow for the roughness of the terrain Majoli[4] uses a modified form of equation 8.88, which is:

$$P_o = 0.3aC(f/4) \cdot (d/50)^3 = \eta P_R \qquad (8.89)$$

where C remains as above, but a takes into account the roughness of the terrain and varies between 0.25 and 4 as the roughness decreases. When $a = 0.24$ equations 8.88 and 8.89 are identical. For rolling terrains with a roughness u, in the range 5 to 100 m, the following equation applies:

$$a = (u/15)^{-1.3} \qquad (8.90)$$

8.3.2.2 Flat fade margin and system outage

The necessary reduction in path length with increase in carrier frequency reduces the severity of multipath fading. As fading due to precipitation and fading due to multipath effects are mutually exclusive events, the respective outage times predicted for each fading cause therefore have to be added. Given the split between availability and performance in the performance objectives, precipitation effects contribute mainly to unavailability (outage greater than 10 s) and multipath propagation mainly to system quality performance. The fade margin is the level to which the received signal can drop from the unfaded received signal level, before the system fails to operate correctly. This definition is applicable to analog radio systems and to low-capacity digital systems. The fade margin F_m dB is related to the unfaded received signal level W_o dBm, and the lowest practical received signal level W dBm, before the system fails to operate correctly, by

$$F_m = 10 \log(w_0/w) \text{ dB} = (W_o \text{ dBm} - W \text{ dBm}) \text{ dB} \qquad (8.91)$$

Equation 8.87 provides the equation for the system outage probability $P(FAD_m > F_m)$:

$$\text{Probability of outage} = P(FAD_m > F_m) = P_o 10^{-F_m/10} \qquad (8.92)$$

where P_o is given by equation 8.89. The short-term outage objective (see Section 7.3.3.2) during fading is given as:

$$\text{Maximum outage time} = (0.054d/2500) \text{ percent of any month, for } 280 < d < 2500 \text{ km} \qquad (8.93a)$$

$$= 0.006 \text{ percent of any month, for } d < 280 \text{ km} \qquad (8.93b)$$

where an outage is considered as a BER equal to or greater than 10^{-3} with an integration time of one second. If η is the fraction of time in which Rayleigh multipath propagation occurs (the multipath activity factor), then $(1 - \eta)$ is that period of time in which non-Rayleigh fading or log-normally-distributed fading occurs. Hence during the worst month, η will be the period in which the worst multipath fading activity will occur. Majoli[24] has provided a relationship between η and P_o:

$$\eta = \begin{cases} 1 & \text{for } P_o > 10 \\ 0.812\ P_o^{0.7} & \text{for } 0.1 < P_o < 2 \\ 1.44\ P_o & \text{for } P_o < 10^{-2} \end{cases} \tag{8.94}$$

This value of η so computed does not by itself indicate the multipath fading which is greater in depth than that of the fade margin, but does indicate the amount of multipath activity during the worst month. Hence η is known as the multipath activity factor.

Example 1

Determine the amount of multipath activity as a percentage of the worst month that can be expected on a 6 GHz link of length 60 km over an average terrain of surface roughness 10 m.

Solution
From equation 8.89, the value of P_o is computed as 1.317, hence from equation 8.94, $\eta = 0.221$ or 22 percent. This indicates that multipath activity occurs for 580 369 seconds during the worst month. This does not however mean that outages will occur for this length of time. If η had been determined to be 1, this would have meant that multipath activity would have occurred throughout the whole of the worst month.

To determine the percentage of the worst month in which outages will occur, the period of multipath fading activity during the worst month is required to be multiplied by the probability of the Rayleigh multipath fade which exceeds the fade margin. That is, from equation 8.92:

$$\text{Percentage outage time} = 100\ P(FAD_m > F_m) = 100\ P_o 10^{-F_m/10}\ \text{percent} \tag{8.95}$$

If the maximum outage time objective given in equation 8.93 is used, a figure for the minimum fade margin can be determined. By equating equation 8.93 with 8.95, we have

$$F_m \geqslant 10 \log[(4.63 \cdot P_o/d) \times 10^6]\,\text{dB} \qquad \text{for } 280 < d < 2500 \tag{8.96a}$$

$$\geqslant 10 \log[(16.54 \cdot P_o) \times 10^3]\,\text{dB} \qquad \text{for } d < 280\ \text{km} \tag{8.96b}$$

Example 2

Determine the flat fade margin for multipath effects only, required to satisfy the CCIR performance objective, if a 50 km link in a temperate climate, over rolling terrain with a roughness of 30 m and operating at 6 GHz, is to be designed.

Solution
From equation 8.89, P_o is determined to be 0.183. From equation 8.94, the value of η is determined to be 0.055, and from equation 8.96b the minimum value of fade margin F_m is determined to be 34.79 dB.

CCIR METHOD OF DETERMINING THE FLAT FADE MARGIN Fm

The equation provided by the CCIR (CCIR Report 338–5,[10] which is used for determining the average worst month fading time fraction for less than 1 percent of the time is approximated by:

$$\text{Percentage outage time} = KQ \cdot f^B \cdot d^C \cdot 10^{-F_m/10} \times 100 \text{ percent} \qquad (8.97)$$

where Table 8.3 provides values for the constants KQ, B and C. If equation 8.97 is equated to the outage objective given in equation 8.93, the equation for the minimum fade margin can be determined as before:

$$F_m(\text{CCIR}) \geqslant 10 \log[4.63 \times 10^6 \times KQ \cdot f^B \cdot d^C] \text{ dB for } 280 < d < 2500 \text{ km} \qquad (8.98a)$$

$$\geqslant 10 \log[16.54 \times 10^3 \times KQ \cdot f^B \cdot d^C] \text{ dB for } d < 280 \text{ km} \qquad (8.98b)$$

Example 3

Using the same figures as given in Example 2, determine the minimum value range for the fade margin, considering Rayleigh multipath flat fading activity only. Assume that the link is in the U.K.

Solution

Table 8.3 gives two values of KQ and values for B and C. From the table the higher value of KQ is found to be $8.1 \times (1/30^{1.4}) \times 10^{-7} = 6.926 \times 10^{-9}$; $B = 0.85$; $C = 3.5$. Hence from equation 8.98b, the value of F_m is calculated to be 26.67 dB. The lower value of KQ is found to be $4.0 \times (1/30^{1.4}) \times 10^{-6} = 3.421 \times 10^{-8}$; $B = 0.85$; $C = 3.5$. Hence from equation 8.98b, the value of F_m is calculated to be 33.60 dB. Therefore the range of minimum fade margin is from 26.67 dB to 33.60 dB. For North-west Europe the result would be 38.4 dB, which is in the range expected. This result is less conservative than that obtained by using Majoli's equations (34.79 dB).

The values of the parameters given in Table 8.3 may be distorted by the inclusion of some data for paths with strong surface reflections and/or diffraction fading. These values may have been influenced by the average path clearance, the path elevation angle, and the beam widths of the antennas. In spite of this, the use of either equation 8.96 or equation 8.98 does provide an indication of the flat fade margin required if the CCIR severely errored second performance is to be met.

8.3.3 Selective fading

Selective fading affects mainly medium-capacity (32 Mbit/s, 34 Mbit/s, 45 Mbit/s) and high-capacity (98 Mbit/s, 140 Mbit/s and eventually above) digital radio-relay systems. The outages which occur in these systems below the flat fade margin are mainly due to amplitude and group delay distortions across the channel bandwidth. Experimental results have shown[25] that the frequency selectivity accompanying multipath fading has a dramatic effect on the transmission performance of the link. This impact has been on a system with a flat fade margin of 46.3 dB for a BER of

Table 8.3 Empirical values of parameters for equation 8.97 (Courtesy of CCIR, Report 338–5, Reference 10)

Proposed for	Japan[1]	N.W. Europe	United Kingdom	United States	U.S.S.R.	Northern Europe[2]
Reference	[Morita, 1970]		[Doble, 1979]	[Barnett, 1972 Vigants, 1975]	[Nadenenko, 1980]	[Laine, 1979, Blomquist *et al.*, 1980; Danielson, 1983; Tanem, 1985]
B	1.2	1.0	0.85	1.0	1.5	1.0
C	3.5	3.5	3.5	3.0	2.0	3.0
KQ for maritime temperature, mediterranean, coastal or high-humidity-and-temperature climatic regions				$\frac{4.1 \times 10^{-5}}{S_1^{1.3}}$	2×10^{-5}	
KQ for maritime subtropical climatic regions				$\frac{3.1 \times 10^{-5}}{S_1^{1.3}}$		
KQ for continental temperate climates or mid-latitude inland climatic regions with average rolling terrain	10^{-9}	1.4×10^{-8}	$\frac{8.1 \times 10^{-7}}{S_2^{1.4}}$ to $\frac{4.0 \times 10^{-6}}{S_2^{1.4}}$	$\frac{2.1 \times 10^{-5}}{S_1^{1.3}}$	4.1×10^{-6}	$\frac{2.3 \times 10^{-5}}{S_1^{1.3}}$
KQ for temperate climates, coastal regions with fairly flat terrain	$\frac{9.9 \times 10^{-8}}{\sqrt{h_1 + h_2}}$				2.3×10^{-5} to 4.9×10^{-5}	$\frac{6.5 \times 10^{-5}}{S_1^{1.3}}$

KQ for high dry mountainous climatic regions	3.9×10^{-10}			$\frac{10^{-5}}{S_1^{1.3}}$		10^{-8}
KQ for temperate climates, inland regions with fairly flat terrain					7.6×10^{-6} to 2×10^{-5}	$\frac{3.3 \times 10^{-5}}{S_1^{1.3}}$
Accuracy of equation 8.97 with coefficients indicated		$\bar{e} = -2.9$ dB, $\sigma = 6.6$ dB 47 links in U.K./France, 7.5–95 km, 2–37 GHz	$\bar{e} = 1.3$ dB, $\sigma = 5.0$ dB 28 links, 7.5–75 km 2–37 GHz		$\bar{e} = -0.04$ dB, $\sigma = 2.0$ dB 26 links, 36–75 km 3.7–8 GHz	

[1]The Japanese data apply for the worst season.
[2]The coefficients are based on data from Finland and Sweden for nonmountainous regions, and data from Norway for mountainous regions.

Notes

h_1 and h_2 are antenna heights in meters.
S_1 is the terrain roughness measured in meters by the standard deviation of terrain elevations at 1 km intervals ($6 \text{ m} \leqslant S_1 \leqslant 42$ m). The height of the radio sites has to be excluded.
S_2 is defined as the RMS value of the slopes (in milli-radians) measured between points separated by 1 km along the path, but excluding the first and the last complete km interval ($1 < S_2 < 80$).
It is possible that some of these parameters may be distorted by the inclusion of data from paths which experience surface reflection problems and/or diffraction fading.
$\bar{e}$ is the mean value of the errors (predicted – measured fade depths) in dB.
σ is the standard deviation of the errors in dB.

10^{-3}; experimental results have shown that no fades have exceeded this flat fade depth, yet instances where the BER is greater than 10^{-3} have occurred on numerous occasions. Due to the misleading concept of a 'flat fade margin' in the higher-capacity systems, a better way to predict the bit-error-rate performance has been found, which involves the measurement of the peak-to-peak amplitude dispersion (frequency response) at the input to the digital demodulator. Fades which do not exhibit channel dispersion do correspond to the BER predicted by the 'flat fade margin'. Barnett[25] has shown that average power fade depths corresponding to BER $\geqslant 1.2 \times 10^{-3}$ ranged from 15 to 40 dB with a median value of 28 dB. The results also indicated that the probability of experiencing dispersion with a faded channel grows significantly with fade depth. Giger and Barnett[26] showed also that a 10^{-3} BER was obtained long before a fade depth of 41 dB, corresponding to the flat fade margin for a BER of 10^{-3}, was reached. In some cases the 10^{-3} BER was reached after a 19 dB fade occurred. From this paper it is concluded that the main source of degradation is inter-symbol interference due to the effects of the dispersive channel. Much work has been carried out since the early sixties on the development of a suitable channel model which can assist in the understanding and prediction of the effects of multipath selective fading.

8.3.3.1 Multipath propagation models

The following is a list of the various channel models which can be constructed from commercially available equipment (CCIR Report 784–2).[97]

Two-ray model[27–29]
Two-ray model with flat attenuation (also referred to as 'simplified three-ray model')[30,31]
Three-ray model[28,32]
Polynomial model for amplitude and group delay versus frequency characteristics[33–35]
Single *S*-plane zero-plus attenuation[36]
Multi-echo time-domain model[28,37]
Joint characterization of a space diversity channel pair with simplified three-ray model.[38–40]

The CCIR state that it may be possible to show that these approaches can be simply related to one another, leading to a single preferred theoretical formulation of the channel distortion.

Before concentrating on the two-ray model, which adequately describes a limited bandwidth channel for 80–90 percent of all multipath propagation conditions,[22] a brief review of the models listed above will be given (CCIR Report 718–2).[10]

MULTIPLE-ECHO MODEL

In reality, the number of physical rays reaching the receiver is much larger than 3, the number which is used to produce a model that accurately describes the processes of the channel transmission. If a model of N rays is considered, $H_A(\omega)$, the transfer function of this model is expressed by

$$H_A(\omega) = \sum_{i=1}^{N} A_i \cdot \exp(-j\omega T_i) \tag{8.99}$$

where A_i and T_i are the absolute amplitude and delay of the ith ray. However, since it is difficult to measure A_i and T_i, a normalized N-ray model has been proposed:

$$H_A(\omega) = H_0 + \sum_{i=1}^{N} a_i \cdot \exp(-j\omega\tau_1) \tag{8.100}$$

Here one ray H_0, has been taken as a reference and a_i and τ_i are respectively the relative amplitude and delay of the ith ray.

GENERAL THREE-RAY MODEL
It has been shown that equation 8.100 with $N = 2$ can be fitted to experimental results, taking a reference ray $H_0 = 1$[28] or a reference ray smaller than 1.[32]

PHASE-SHIFTED TWO-RAY MODEL WITH FLAT ATTENUATION (SIMPLIFIED THREE-RAY MODEL)
The transfer function of this model is:

$$H(\omega) = a[1 - b \cdot \exp[\pm j(\omega - \omega_M)\tau]] \tag{8.101}$$

In the model the notch position can be shifted with respect to the center frequency ω_M. The model has four parameters (a, b, ω, τ) which can be reduced to three parameters if τ is made constant. This model with a fixed τ represents correctly the measured transfer functions for *bandwidths up to 55 MHz*, providing that $\tau = 1/(6B)$, where B is the bandwidth. Statistics of parameters of this model have been obtained for a 42 km link at 6 GHz and, using a 26.6 MHz bandwidth, for which τ was fixed at 6.31 ns.[31] On 37 km and 50 km links at 11 GHz, the same statistics have been produced for a bandwidth of 55 MHz, using a value of τ, fixed at 3.03 ns.[41] It was found that, on these links, ω_M is statistically independent of the other two parameters and its distribution is uniform across the 55 MHz band.

NORMALIZED TWO-RAY MODEL
The transfer function of the normalized two-ray model is expressed by equation 8.101, with $a = 1$. This model has three parameters $(b, \omega$ and $\tau)$. The model represents correctly the measured transfer function for a bandwidth of 55 MHz. The three parameters are statistically independent and b and $\omega_M\tau = \phi_R$ are uniformly distributed. Equation 8.50 shows the final equation where $|R| = b$ and $\phi_R = \omega_M\tau - \pi$. If $\omega_M = 0$, the normalized two-ray model can be reduced to two parameters. The notch position is then fixed and centered in the channel band.

POLYNOMIAL METHODS
The polynomial expression of amplitude and group delay as a function of frequency is given in the general form as:[33-35]

$$p(\omega) = C_0 + C_1 \cdot \omega + C_2 \cdot \omega^2 + \cdots + C_N \cdot \omega^N \tag{8.102}$$

The complex polynomial for the transfer function with the general form[36,41] is

$$H(\omega) = A_0 + \sum_{k=1}^{N} (A_k + jB_k) \cdot (j\omega)^k \tag{8.103}$$

The complex polynomial developed in the first order is in good agreement with experimental data obtained on a 42.5 km hop at 6 GHz in a 26 MHz band[36] and with data on a 37 and 50 km hop at 11 GHz in a 55 MHz band.[41]

8.3.3.2 The two-ray model[31]

The multipath fading (MPF) channel transfer function given in equation 8.100 possibly represents the real-life situation. However, with $N = 2$ and the direct or reference ray H_0 taken as unity and $\tau_2 = \tau$, the general three-ray model results. This is given by:

$$H(\omega) = 1 + a_1 \cdot \exp(-j\omega\tau_1) + a_2 \cdot \exp(-j\omega \cdot \tau) \tag{8.100}$$

As $a_1 \cdot \exp(-j\omega\tau_1)$ is a vector rotating clockwise with radius a_1, for ω fixed (at an instant) and τ_1 fixed, this transfer function can be represented by the resultant of 1 and $a_1 \cdot \exp(-j\omega\tau_1)$, and this resultant combining with the vector $a_2 \cdot \exp(-j\omega\tau)$ will then form the overall resultant. This is shown in Figure 8.16(c). If τ_1 is made sufficiently small, the phase $\omega\tau_1$ changes by only a small amount when the fade minimum or notch angular frequency ω changes and moves over the bandwidth of the carrier. This in turn means that the vector $1 + a_1 \cdot \exp(-j\omega\tau_1)$ will not vary appreciably. The resultant vector for small τ_1 is shown in Figure 8.16(a). Thus

$$1 + a_1 \cdot \exp(-j\omega\tau_1) \simeq a \cdot \exp(-j\phi) \tag{8.104}$$

If the notch frequency is given by ω_0, and τ is the delay difference in the channel between the indirect ray path and the direct ray path, then $\omega_0\tau$ could be considered as the change in phase caused to the resultant transfer function by the frequency dependence of the indirect ray. This change in phase, however, is in opposition to the direct ray and has a phase ϕ at the center of the channel, given by

$$\phi = \omega_0\tau + \pi \tag{8.105}$$

By substituting equations 8.104 and 8.105 in equation 8.103, and letting $b = a_2/a$, the transfer function becomes (remembering that $\exp(j\pi) = -1$):

$$H(\omega) = a \cdot \exp(-j\phi) \cdot [1 - b \cdot \exp\{-j(\omega - \omega_0)\tau\}]$$

Since $\omega\tau_1$ is very small, as shown in Figure 8.16(a), the resultant angle ϕ is also very small. Hence $\exp(j\phi)$ is approximately equal to one. The final multipath fading transfer function $H(\omega)$ is given by

$$H(\omega) = a[1 - b \cdot \exp[-j(\omega - \omega_0)\tau]] \tag{8.106}$$

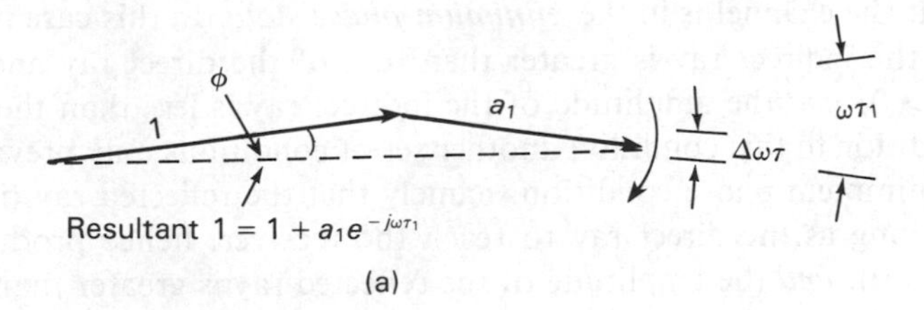

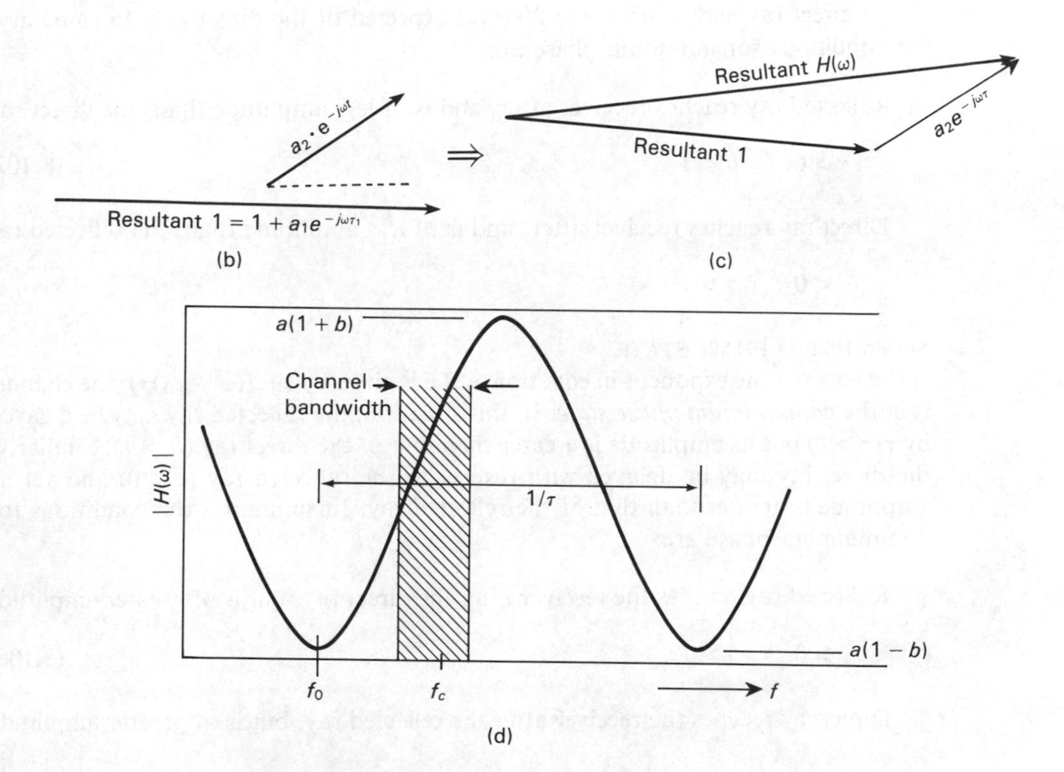

Figure 8.16 Three-path fade vectors and frequency response of the multipath fading channel transfer function; (a) Formation of $1 + ae^{-j\omega t}$, (b) and (c) formation of $H(\omega)$; (d) frequency response

Equation 8.106 shows that $H(\omega)$ is a periodic function of f, where $\omega = 2\pi f$, with a period equal to $1/\tau$. Its amplitude varies between the minimum value of $a(1 - b)$, which occurs when $f = f_0 + k/\tau$, where $k = \ldots, -1, 0, 1, \ldots$, and the maximum value of $a(1 + b)$, which occurs at values of $f = f_0 + 1/(2\tau) + k/\tau$. If it is assumed that the carrier frequency f_c is fixed, then the variable f_0 is the frequency nearest the carrier at which a notch occurs and which can vary in the interval $(-1/2|\tau|, 1/2|\tau|)$. Figure 8.16(d) shows the frequency response of $|H(\omega)|$.

MINIMUM PHASE STATE

The value of the exponent in equation 8.106 is given by $(-j(\omega - \omega_0)\tau)$. This means that the channel is in the *minimum phase state*. In this case it is seen that the delay of the indirect ray is greater than that of the direct ray and a positive τ is taken ($\tau > 0$) *and* the amplitude of the indirect ray is less than the direct ray ($b < 1$). In addition to this condition another set of conditions may prevail, which still produce a minimum phase condition, namely that the reflected ray (indirect) does not take as long as the direct ray to reach the receiver, hence producing a negative delay ($\tau < 0$), *and* the amplitude of the reflected ray is greater than that of the direct ray ($b > 1$). In other words, for these conditions, the reflected ray takes over the role of the direct ray and exhibits the features expected of the direct ray. In summary, the conditions for minimum phase are:

Reflected ray reaches receiver after, and is of less amplitude than, the direct ray

$$\tau > 0 \qquad b < 1 \tag{8.107}$$

Direct ray reaches receiver after, and is of less amplitude than, the reflected ray

$$\tau < 0 \qquad b > 1$$

NONMINIMUM PHASE STATE

If the value of the exponent in equation 8.106 is given by $(+j(\omega - \omega_0)\tau)$, the channel is in the *nonminimum phase state*. In this situation, the reflected ray may be delayed by $\tau(\tau > 0)$ but its amplitude is greater than that of the direct ray ($b > 1$). Similarly, the direct ray may be delayed with respect to the reflected ray ($\tau < 0$) and yet its amplitude is greater than that of the reflected ray. In summary, the conditions for nonminimum phase are:

Reflected ray reaches the receiver after the direct ray, but is of greater amplitude

$$\tau > 0 \qquad b > 1 \tag{8.108}$$

Direct ray reaches the receiver after the reflected ray, but is of greater amplitude

$$\tau < 0 \qquad b < 1$$

GROUP DELAY DISTORTION

If the set of parameters (a, b, ω_0, τ) is replaced by the set $(-ab, 1/b, \omega_0, -\tau)$ in equation 8.106,[42] except for the muliplicative factor $\exp(-j(\omega - \omega_0)\tau)$, the same transfer function is obtained. This permits the sign of τ to change together with the parameters a and b, and still have the same propagation phenomena represented in the same way. If it is assumed for what follows that $\tau > 0$, then the minimum phase condition appears when $b < 1$ and the nonminimum phase condition appears when $b > 1$. By substituting $1/b$ into equation 8.106, the following new transfer function $H'(\omega)$ is formed with a complex conjugate relationship to the original $H(\omega)^*$:

$$\begin{aligned} H'(\omega) &= a[1 - (1/b) \cdot \exp\{-j(\omega - \omega_0)\tau\}] \\ &= -(1/b) \cdot \exp\{-j(\omega - \omega_0)\tau\} \cdot H(\omega)^* \end{aligned} \tag{8.109}$$

The amplitude characteristic of the propagation medium varies only by the multiplicative factor $(1/b)$, while the group delay characteristic $\tau_{g(\omega)}$, changes because of the sign and addition of the constant value τ, i.e.

$$\tau'_{g(\omega)} = -\tau_{g(\omega)} + \tau \tag{8.110}$$

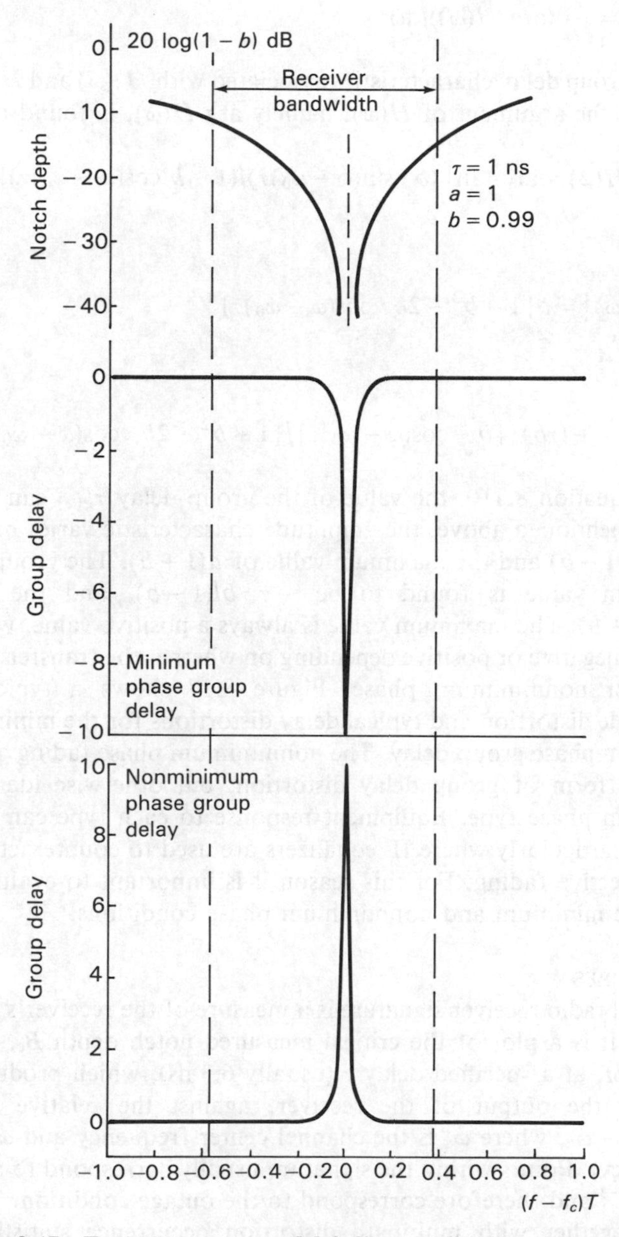

Figure 8.17 Typical two-ray model fading amplitude and group delay distortions

where

$$\tau'_{g(\omega)} = -\mathrm{d}(\arg H'(\omega)/\mathrm{d}\omega$$

and (8.111)

$$\tau_{g(\omega)} = -\mathrm{d}(\arg H(\omega))/\mathrm{d}\omega$$

are the group delay characteristics associated with $H'(\omega)$ and $H(\omega)$ respectively. The value of the argument of $H(\omega)$, namely arg $H(\omega)$, is found to be

$$\arg H(\omega) = \arctan[(b \cdot \sin(\omega - \omega_0)\tau)/(1 - b\cos(\omega - \omega_0)\tau)]$$

and

$$|H(\omega)| = a[1 + b^2 - 2b \cdot \cos(\omega - \omega_0)\tau]^{1/2} \tag{8.112}$$

Thus

$$\tau_{g(\omega)} = -(\tau b) \cdot [b - \cos(\omega - \omega_0)\tau]/[1 + b^2 - 2b \cdot \cos(\omega - \omega_0)\tau] \tag{8.113}$$

Using equation 8.110, the value of the group delay $\tau'_{g(\omega)}$ can be found.

As mentioned above, the amplitude characteristic varies between the minimum value $a(1 - b)$ and the maximum value of $a(1 + b)$. The group delay characteristic minimum value is found to be $-\tau \cdot b/(1 - b)$, and the maximum value is $\tau \cdot b/(1 + b)$. The maximum value is always a positive value, whereas the minimum value is negative or positive depending on whether the transfer function is minimum phase or nonminimum phase. Figure 8.17 shows a typical two-path fading amplitude distortion and typical delay distortions for the minimum phase and non-minimum phase group delay. The nonminimum phase fading group distortion is an inverted form of group delay distortion, but otherwise identical to that of the minimum phase type. Equipment response to each type can be considerably different, particularly where IF equalizers are used to counteract the effects of multipath selective fading. For this reason it is important to evaluate the response for both the minimum and nonminimum phase conditions.

SIGNATURES

A digital radio receiver signature is a measure of the receiver's tolerance to selective fading. It is a plot of the critical measured notch depth B_c, generated by a path simulator, at a specified delay τ (usually 6.3 ns), which produces a specified error ratio at the output of the receiver, against the relative notch frequency ω_0 ($\omega_0 = \omega_c - \omega_n$, where ω_c is the channel center frequency and ω_n is the actual notch frequency). Points within the signature usually correspond to an error ratio greater than 10^{-3} and therefore correspond to the outage condition. The signature can be used, together with multipath distortion occurrence statistics, to compute the amount of time an outage is expected to last. On considering the shape of the equipment signature, the following may be noted:

The smaller the area of the signature, the less sensitive that equipment will be to multipath distortion.
The width W of the signature is approximately independent of the delay τ used.
The height of the signature $\lambda(\tau)$, is sensitive to multipath delay τ.
The system signatures are degraded in the presence of noise and/or interference, which increase its width and the height. Therefore the area inside the signature increases.
The idealized signature shape is a rectangle, but in practice occurs as M- or W-curves.
The greater the signature area, the greater is the outage region and the more prone the receiver will be to outages caused by selective fading.
A radio system whose signature falls entirely below, or is contained entirely within, that of a second system (both at a relative echo delay of 6.3 ns) will outperform that second system.

Usually the notch depth $(1-b)$, which is the direct ray amplitude minus the reflected ray amplitude, is described by its logarithmic value and for the critical value where the radio system produces a BER of 10^{-3}, is denoted by B_c, i.e.

$$B_c = -20 \log(1-b) \tag{8.114}$$

As mentioned above, a value of the delay $\tau = 6.3$ ns is often used. No loss of generality occurs by fixing τ, when the ratio of τ to the symbol period is small. Sometimes values of τ other than 6.3 ns are used in the simulator. When this occurs a simple scaling rule can be applied within a limited range to get the critical curve $B_c(\tau = 6.3 \text{ ns})$. This scaling rule is applied, since only the height of the signature is dependent on τ and not the width. Two types of signature are usually produced for a system, which depend on whether the selective fading is of the minimum or non-minimum phase type. These signatures are usually designated; M-type signature for minimum phase conditions; and W-type signature for nonminimum phase conditions.

In most signatures, a dip is found in the center of the band. This indicates that the receiving equipment has less susceptibility for band-centered fades than for fades which occur off-center. The band-centered fades cause maximum pulse spreading, which leads to in-phase inter-symbol interference, whereas off-centered fades do not cause the same amount of pulse spreading, but cause considerable crosstalk inter-symbol interference. The crosstalk inter-symbol interference is the worse of the two, and it increases the susceptibility of the equipment to off-center fading. As discussed below, signatures may be used to assist in the prediction of outages.[29,43–46] In addition to this, they may be also be used to compare the improved performance of a radio receiver with and without the inclusion of adaptive IF and baseband equalizers, especially since equalizers are usually designed specifically for minimum-phase or for nonminimum-phase conditions. Figure 8.18 shows typical signatures for different modulations.

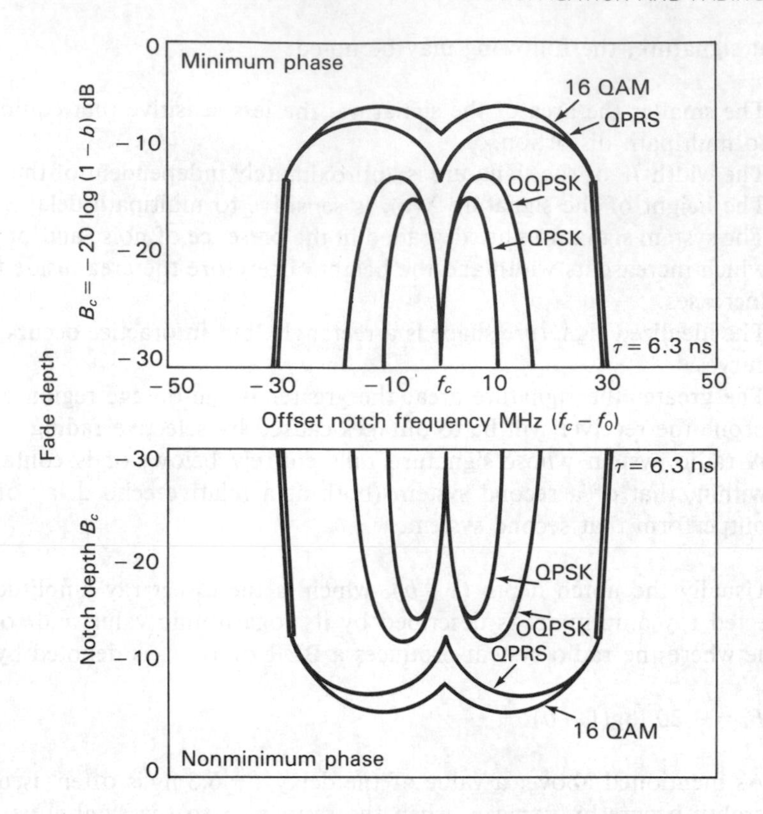

Figure 8.18 Typical signatures for different modulations

To scale a signature measurement height (signature 2), produced with a given delay (τ_2 ns) in the simulator path, to another signature height (signature 1), with a different delay (τ_1 ns, where τ_1 may have a value such as 6.2 ns), the following equation may be used:

$$\text{Signature 1} = \text{signature 2} - 20 \log(\tau_1/\tau_2) \text{ dB} \tag{8.115}$$

Equation 8.115 shows that the signature height is reduced as the delay is increased. This is because $B_c = -20 \log(1-b)$, which provides the limits of B_c as: $B_c = \infty$, when the reflected ray equals the direct ray at zero delay (worst case) and $B_c = 0$ dB when $b = 0$ or the delay is infinite (best case). The best signature to have is one which has almost zero area under the M- or W-curves. Consider for the sake of clarity, only the M-curve (minimum phase condition): a signature with almost zero area under the M-curve would indicate that the receiver would not be affected by a reflected wave which had almost zero delay and was nearly of equal amplitude to the direct wave. The worst sort of signature to have is one which extends nearly all the way up to the $B_c = 0$ dB axis, for this would show that, when the reflected ray was almost zero and had a very long delay time, it would still cause a receiver outage.

Before applying the scaling process according to equation 8.115, two important aspects must be considered. These are:

1. Equation 8.115 is valid only for deep notches which have only a small portion of the interference pattern falling in the channel. This means that there must be a linear relationship between changes to the delay and notch depth and the amount of distortion produced in the radio channel,
2. The channel is operating in a noise- and interference-free environment.

During multipath fading not only is there a reduction in the received signal level, but there is an equivalent loss of cross-polar discrimination (see later in this chapter for multipath effects on cross-polarization). This effect is most pronounced when the channel delays are shortest. To add to this, short delays also invalidate the scaling rule. The effects of different delay times on a channel which experiences adjacent channel interference are:

Short delays: degradation from cross-polarization discrimination predominates.
Medium delays: both cross-polarization discrimination (XPD) and signal distortion occur.
Long delays: distortion in the channel predominates.

These effects must be considered in predicting the performance of a radio system from its signature measurement.

The relationship between fade depth and percentage of minimum phase fades is as follows:[49]

Notch depths to 19 dB below unfaded signal level occur for nearly 100 percent of the time.
Notch depths between 19 dB and 37 dB below unfaded signal level occur at a decreasing linear rate of 2.8 percent per dB from almost 100 percent at 19 dB, down to 50 percent of the time at 37 dB.
Notch depths greater than 37 dB occur for only 50 percent of the time.

Consider deep fading where the direct and reflected path components are close in amplitude. Any slight change of the relative amplitudes will produce a high probability of causing an alternation between the two phase states. Therefore, for deep fades, each of the two phase states must have equal probability. For light fading, the possibility of changing state is not so great and the minimum phase condition is the dominant phase state. There is an additional likelihood of the nonminimum phase state occurring during shallow fades, if the system has locked onto a nonminimum phase state during a deep fade and remains locked to this state as the fade depth decreases.

MULTIPATH SIMULATORS[47,48]

Figure 8.19 shows a typical IF fading simulator. Although an RF fade simulator could be used, the IF type permits signals and noise levels to be easily adjusted for

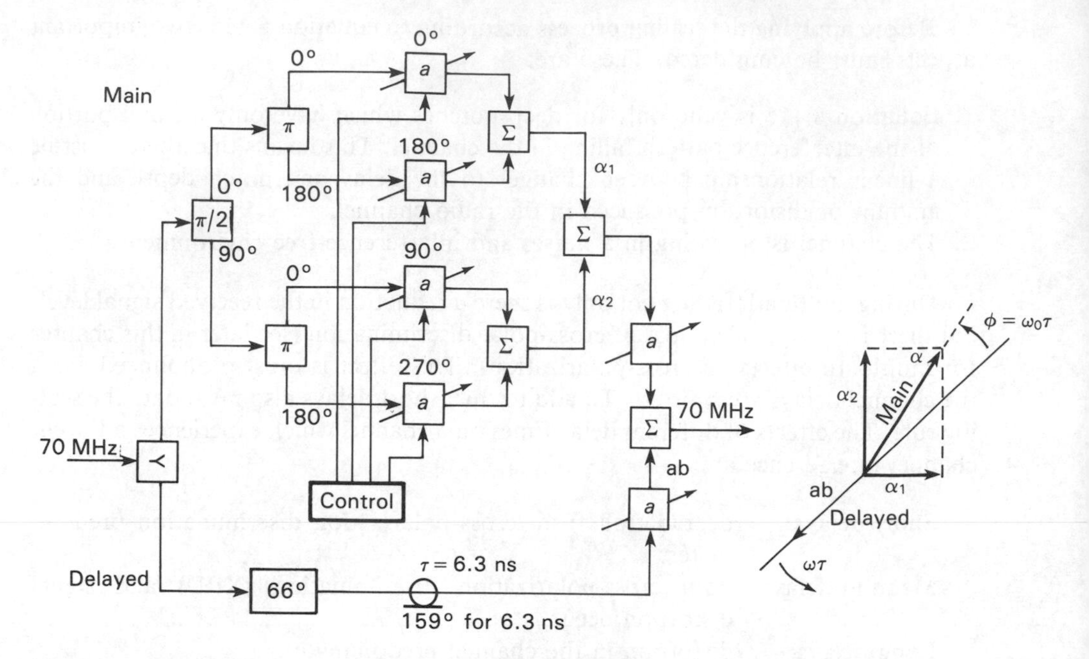

Figure 8.19 IF fading simulator

repeated experiments. The incoming IF signal at a center frequency of 70 MHz is split into an arbitrary phased, adjustable 'main' component and a 'delayed' component which is fixed in delay (τ ns) but adjustable in magnitude. The main component is further resolved into orthogonal components using wideband networks exhibiting flat gain and well-behaved delay. A particular vector is constructed by adjusting the orthogonal components to establish a simulated fade notch frequency: in practice, the phase sense of the 0° and 90° components are independently reversible, by using the control circuit to insert sufficient attenuation in the unwanted path to make it effectively an open-circuit. This permits flexibility in the notch frequency selection. The 6.3 ns fixed delay added to the delayed path imparts a phase shift of 159° to the 70 MHz IF center frequency. This is shown built out to 225°, relative to the 0° transmission path, using a 66° wideband network of the same type. The delayed vector is fixed in the direction opposite to the mid-range position of the adjustable main vector, corresponding to a channel-centered fade ($\phi = 0^\circ$). Since $1/\tau = 1/6.3$ ns $= 158.4$ MHz, a change of 1° in the direction of the main vector corresponds to a frequency displacement of the fade notch location of 0.44 MHz. For $\phi = -45^\circ$, the notch location is displaced 19.8 MHz below the channel center ($f_0 = -19.8$ MHz). The magnitude of the delayed component is then adjusted to achieve the desired notch depth. The main problem which usually occurs with this type of multipath simulator is the error in notch generation, i.e. the difference between the notch frequency and depth produced, when compared with the actual values expected. As discussed above, the notch frequency ω_0, is that frequency at which the signal vectors of the two paths are in antiphase. This frequency depends on the time delay and phase shifter values in use. The accuracy of predicting the notch frequency is determined by the differential phase error between the two simulator paths, together with the calibration accuracy of the phase shifter.

8.3.3.3 Outages due to selective fading

The use of the 'flat' fade margin is *meaningless* when considering the outage behavior of a digital radio system. This is because the frequency-selective behavior of multipath attenuation results in an error ratio greater than that which would be caused by nonselective attenuation of the same mean amplitude. This has been clearly and concisely demonstrated by Giger and Barnett,[26] who obtained field measurements over a 2–2.5 month period from a radio hop carrying a 45 Mbit/s, 8–PSK signal in a 20 MHz band at 4 GHz. The system was designed for zero inter-symbol interference using a roll-off factor $\alpha = 0.5$ for the received raised-cosine Nyquist spectrum. Figure 8.20 shows a plot of the BER against fade depth at the

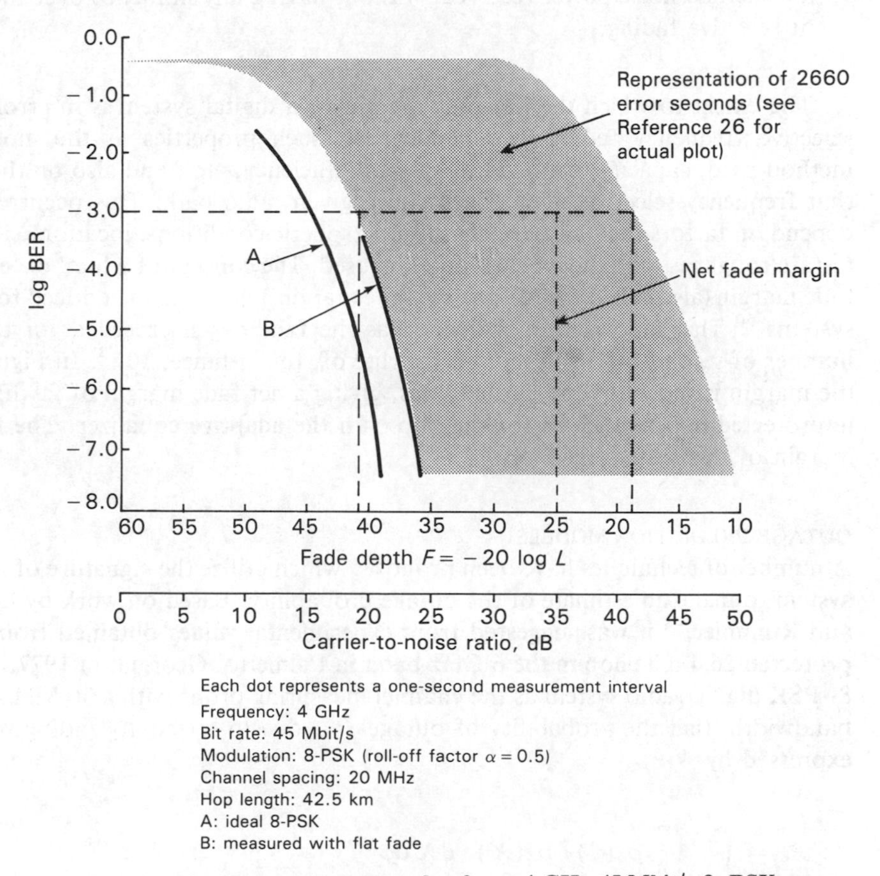

Figure 8.20 Error rate scatter plot for a 4-GHz 45 Mbit/s 8–PSK radio link (Courtesy of IEEE, Reference 26)

channel center frequency. Each dot represents a period of one second during which the 2660 error seconds were observed on the 45 Mbit/s stream. The theoretical 8–PSK curve as shown in Figure 6.5 is also shown in Figure 8.20, together with an experimentally measured flat fade margin, which was obtained by varying artificially the fading of the received signal by using an attenuator. With flat fading an error rate of 10^{-3} should be reached with a fade depth of 41 dB, which is the flat fade margin. However, the results in Figure 8.20 show that this in fact rarely occurred, since the system had an outage or error rate $>10^{-3}$, long *before* the flat fade margin depth of 41 dB was ever reached. In some cases an outage occurred at a fade depth of only 19 dB. From these results the following was concluded:

The main source of degradation is inter-symbol interference.
The concept of flat fade margin is meaningless in a digital radio system.
Increasing the transmitter power to try and reduce the outage time will not produce any improvement, for it will only increase the flat fade margin or reduce the thermal noise power received, without having any influence over the effects of selective fading.

The extent to which the flat fade margin of a digital system is in error during selective frequency fading is dependent on such properties as the modulation method used, capacity, bandwidth, spectral efficiency, etc., and also on the degree that frequency selective effects occur on a given radio path. This occurrence will depend on factors such as path length, atmospheric conditions, location, clearance, terrain roughness and the types of antenna used. The concept of a 'net' or 'effective' fade margin (also called the 'dispersive fade margin') has been introduced for digital systems.[50] This fade margin is defined as the fade depth exceeded for the same number of seconds as a threshold error rate of, for instance, 10^{-3}. In Figure 8.21, the margin is shown by the dashed lines, giving a net fade margin of 29 dB for the unprotected hop and 33 dB for the hop with the adaptive equalizer. The flat fade margin of this system is 46 dB.

OUTAGE PREDICTION MODELS

A number of techniques have been proposed which utilize the signature of the radio system to make an estimate of the outage probability. Based on work by Lundgren and Rummler,[48] it was suggested from experimental values obtained from an unprotected 26.4 mile hop in the 6 GHz band in Palmetto, Georgia, in 1977, using an 8–PSK digital radio system as the channel measuring probe with a 30 MHz channel bandwidth, that the probability of outage P_{os}, due to selectivity fading, could be expressed by

$$P_{os} = \int_{-\pi}^{\pi} \int_{B_c}^{\infty} p_\phi(\phi) \cdot p_B(X) \cdot \mathrm{d}X \, \mathrm{d}\phi \tag{8.116}$$

where $p_\phi(\phi)$ is the probability density function of the angle of the interfering path

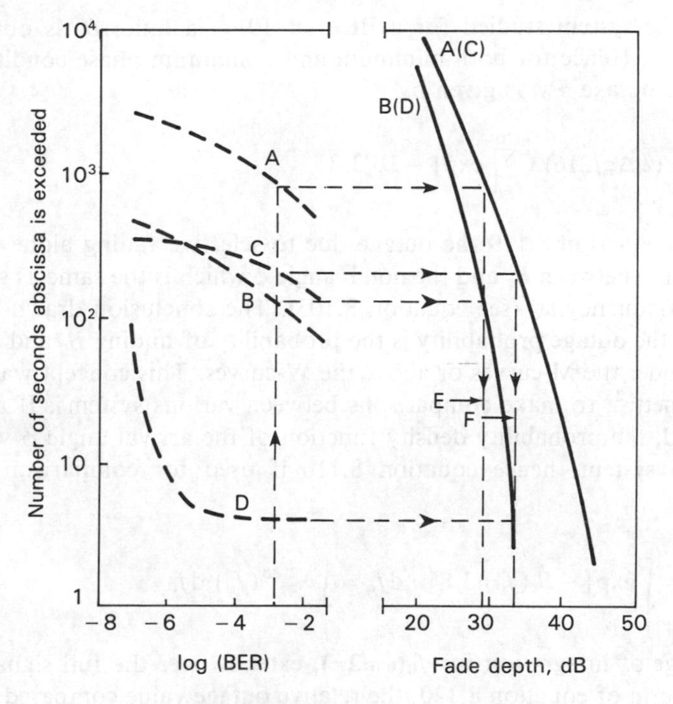

Frequency: 6 GHz
Bit rate: 78 Mbit/s
Modulation: 8-PSK (roll-off factor $\alpha = 0.5$)
Channel spacing: 30 MHz
Hop length: 42.5 km

Curves A: unprotected hop 3 m (10-foot) horn reflector antenna
B: space diversity with IF combiner
C: adaptive IF amplitude slope equalizer
D: with B and C
E: 'net' fade margin for case A
F: 'net' fade margin for case C

Total observation period = 38 days (August–September 1978)

Figure 8.21 Example of cumulative fading and BER distribution for a 6-GHz 78 Mbit/s 8–PSK system (Courtesy of IEEE, Reference 26)

at the center of the channel, which was experimentally found to be

$$p_\phi(\phi) = \begin{cases} 1/(216) & |\phi| < 90^\circ \\ 1/(1080) & 90^\circ \leqslant |\phi| \leqslant 180^\circ \end{cases} \tag{8.117}$$

and $p_B(X)$ is the probability density function of the notch depth for B greater than some notch depth X dB, determined experimentally as

$$p_B(X) = (1/3.8) \cdot \exp(-X/3.8) \tag{8.118}$$

For the system studied for a BER of 10^{-3}, a finite B_c is obtained only for $|\phi_i| < 90°$. Hence for both minimum and nonminum phase conditions, the probability of outage P_{os} is given by

$$P_{os} = (2\,\Delta\phi/216) \cdot \sum_{i=1}^{10} \exp[-B_c/3.8] \tag{8.119}$$

From equation 8.119 the outage due to selective fading alone depends on the relationship between B_c and the notch angle ϕ which is the same as saying the notch relative frequency ω_o (see equation 8.105). The conclusion drawn from this work was that the outage probability is the probability of finding B_c and ω_o values in the region under the M-curves or above the W-curves. This concept was used by Giger and Barnett,[26] to make comparisons between various systems. If two systems are colocated, the probability density function of the arrival angle ϕ will be the same for each system, hence equation 8.116 if used for comparison purposes only becomes:

$$P_{o_{\rm rel}} = \int \exp[-B_c(f_o)/3.8] \cdot \mathrm{d}f_o = \int \lambda_c^{2.28}(f_o)\,\mathrm{d}f_o \tag{8.120}$$

The range of integration for $f_0(\omega_0/2\pi)$, extends over the full signature. From the second form of equation 8.120, the relative outage value compared to say a known system can be determined by squaring the signature $\lambda_c(f_0)$ and then determining the area under this curve. In general, it can be said that a signature which falls entirely below a first signature will result in less outage than that obtained for the first system. If the signature is very 'low' or far removed from the 0 dB (free-space) value, equation 8.120 is not a good measure of the overall relative outage and the flat fade margin and thermal noise usually are the predominant outage mechanisms. The ability to be able to compare different systems partially overcomes the problem of determining the absolute probability of outage, especially as the derivation of an expression for the absolute value, generally applicable to all modulation schemes and environments, is not easy to obtain. This was, however, attempted by Campbell and Coutts,[29] who derived the probability of an outage given selective fading P_{os}, in terms of the mean echo delay τ_0, the system baud period T and a constant K. The constant K is a parameter which is proportional to the area under a normalized signature, where the baud period and echo delay are made equal to unity. Consequently K can be considered as the area under a 'normalized system signature' and is expected to be dependent on the modulation method and equalizer type employed. For the calculations of P_{os}, assumptions about the probability distributions for the relative echo delay and relative echo amplitude and notch frequency offset must be made. Account must also be taken of the differing proportions of minimum and nonminimum phase conditions. If a truncated negative exponential distribution of the relative echo delay (the mean being τ_0) and a uniform distribution for both the relative echo amplitude and notch frequency offset are assumed, then

$$P_{os} = (2K) \cdot [\tau_0/T]^2 \tag{8.121}$$

where for the different modulation techniques K is found to be:[43]

$$K = \begin{cases} 15.4 & \text{for 64–QAM} \\ 5.5 & \text{for 16–QAM} \\ 7.0 & \text{for 8–PSK} \\ 1.0 & \text{for 4–PSK} \end{cases} \tag{8.122}$$

Ruthroff[21] has determined the expression for τ_{max} to be

$$\tau_{max} = 14(d/50)^3 \text{ ns} \tag{8.123}$$

where d is the path length in kilometers.

Jakes[51] has fitted analytical results to system outage measurements on a 40 km path and determined that τ_0 is approximately equal to 0.07 τ_{max}. This leads to the equation:[46]

$$\tau_0 = (d/50)^3 \text{ ns} \tag{8.124}$$

As the cubic dependence of τ_0 on d appears too pessimistic[46] the following expression for τ_0 is taken:

$$\tau_0 = 0.7\ (d/50)^{1.5} \tag{8.125}$$

Thus, substituting equation 8.125 into equation 8.121, expressing the inverse of the baud period as the baud rate, and using baud rate $\times \log_2 M$ = bit rate r_b, the probability of outage P_{os} is given by

$$P_{os} = (2K) \cdot [2d^{1.5} \cdot (r_b/\log_2 M) \times 10^{-6}]^2 \tag{8.126}$$

where r_b is the system bit rate expressed in Mbit/s, d is the path length in km and M is the number of levels in the modulation schemes (4, 8 or 64). The value of K is given by equation 8.122.

If the fraction of time that multipath fading occurs is given by η, the expected outage time can be calculated by multiplying η by equation 8.126, i.e.

Percentage outage time for selective fading

$$= 200\eta \cdot K \cdot [2d^{1.5} \cdot (r_b/\log_2 M) \times 10^{-6}]^2 \text{ percent} \tag{8.127}$$

and η is calculated from equation 8.94 after the flat fading multipath occurrence factor has been determined from equation 8.89.

Example 1

Calculate the expected outage time due to *selective* multipath fading for a 26.4 mile (42.47 km) hop using a 6 GHz 8–PSK digital radio having a bit rate of 78 Mbit/s. The climate is temperate and the terrain roughness factor C is taken as 1.

Solution
The fading occurrence factor determined from equation 8.89 is found to be 0.276. From equation 8.94, η is determined to be 0.074. From equations 8.127 and 8.122, the expected percentage outage time is found to be 0.0215 percent. This represents a figure of 564 seconds errored in the worst month, which is less than the 1000 seconds observed.[48,25]

Example 2

Calculate the expected outage time due to selective fading on a 55 km hop using a QPSK system operating at 11 GHz and with a channel capacity of 140 Mbit/s. Assume a predominantly flat terrain in a temperate climate.

Solution
The fading occurrence factor determined from equation 8.89 is found to be 1.1. Thus from equation 8.94, $\eta = 0.194$, and, from equation 8.127, the expected outage time for the worst month is 0.127 percent. This is about 1.8 times higher than that found from different reports for a 140 Mbit/s digital radio relay system in the 11 GHz band spanning around 50 km.[52,53]

Example 3

Calculate the expected outage time for a 6 GHz 8–PSK 90 Mbit/s radio operating over a hop distance of 47.8 km. Assume that the terrain roughness factor is 18.76 m and the link operates in a temperate climate.

Solution
The value of P_o is determined, from equation 8.89, to be 0.295. From equation 8.94, η is found to be 0.077. Therefore, from equation 8.127, the expected outage time is calculated to be 0.0425 percent during the worst month, giving an outage of 1112 s during this period. From experiments on this link,[54] it is assumed that heavy fading occurred over a two-month period in which 3000 s of BER equal to 10^{-3} were recorded. Thus an average of 1500 s can be taken as the worst month outage, which compares reasonably well with that calculated.

It is expected that equation 8.127 can provide only a rough figure, but coupled with other methods and further experimental results, it may assist in providing a clearer picture of the expected outage of different digital radio systems. The advantage of using this method is that the signature curve of the radio system is not required.

MAJOLI'S METHOD OF PREDICTING PROBABILITY OF OUTAGE DUE TO SELECTIVE FADING[46]

To use Majoli's method a system signature is required, for the method replaces the system signature by a rectangle containing an equivalent area to that of the signature. This rectangle extends from 0 dB on the B_c axis to somewhere near the top of the signature. As the area under the signature is related (but not directly), to the outage probability, Majoli then determines the necessary additional factors required, to relate signature parameters to the probability of outage. One of these parameters T_1, is derived using the assumption that there is a linear relationship between the different path delays τ and the signature height $h(\tau)$. The constant of proportionality in this relationship is T_1, which is related to a normalized signature (usually, $\tau_1 = 6.3$ ns). Another parameter is the relationship of the mean value of the path delay τ_0, which is derived using equation 8.125. As the area under the signature in this context is the channel bandwidth $\times h(\tau_1) \times$ the normalization factor $1/\tau_1$ ns, to obtain a dimensionless quantity (as probability is) this normalized area must be multiplied by the square of the mean delay and a constant of proportionality, which is found to be 4.3. This constant of proportionality is related to the probability density function of the notch depth $(1 - b)$. The expression for the probability of outage is then given by:

$$P_{OS(\text{Majoli})} = 4.3W \cdot \tau_0^2/T_1 \tag{8.128}$$

where W is the bandwidth of the channel in GHz, τ_0 is the mean delay of the reflected wave in ns and is given by equation 8.125, T_1 is a constant of proportionality in ns and is derived from the expression

$$T_1 = \tau_1/h(\tau_1) = 6.3/(\text{signature height from 0 dB}) \text{ ns} \tag{8.129}$$

The signature height may be given in $B_c(\tau_1)$ dB above the 0 dB axis. If this is the case,

$$h(\tau_1) = \text{antilog}[-B_c(\tau_1)/20] \tag{8.130}$$

As mentioned above, the value of τ_1 for normalizing the signature is usually taken as 6.3 ns. To determine the probability of outage, the multipath fading activity factor η is found from equation 8.94 after deriving the flat multipath fade occurrence factor P_o from equation 8.88. Therefore, the expected outage probability is given by:

$$\text{Expected outage probability} = 4.3\eta W\tau_0^2/T_1 \tag{8.131}$$

Example 4

Determine the probability of outage of a 90 Mbit/s digital radio system, operating at 4 GHz and using 64–QAM modulation. The path length is determined to be 50 km. The system signature is given in Figure 8.22.

Solution

From the system signature of the system without equalization, $h(\tau_1)$ is determined to be antilog$[-(7.5/20)] = 0.422$. W is found to be 70 MHz (equivalent rectangular width of signature). The value of τ_0 calculated from equation 8.125 is 0.7 ns. Hence from equation

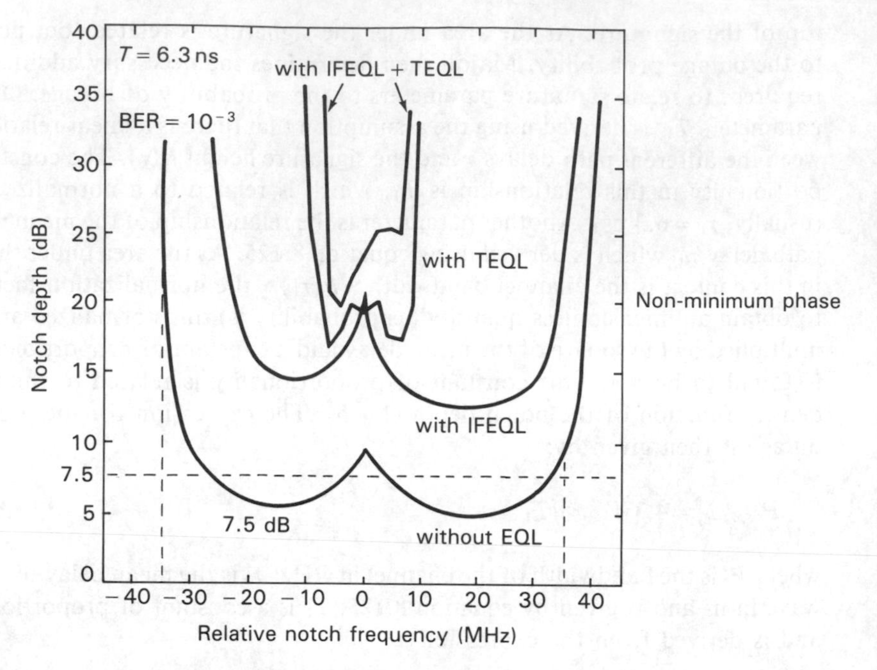

Figure 8.22 Signature used for Example 4

8.129, the value of T_1 is 14.93 ns, and from equation 8.128, $P_{OS(\text{Majoli})} = 9.88 \times 10^{-3}$. (Using equation 8.126, in the method described above, which does not require a signature, the answer is 4.99×10^{-3}.)

Example 5

Calculate the expected outage time due to selective fading on a 55 km hop using a QPSK system operating at 11 GHz and with a channel capacity of 140 Mbit/s. Assume a predominantly flat terrain in a temperate climate.

Solution
From Figure 8.23, the value of $h(\tau_1)$ dB for the 6 ns signature is 9 dB, i.e. $h(\tau_1) = 0.355$. Hence, from equation 8.129, $T_1 = 16.9$ ns, and from equation 8.125, τ_0 is calculated to be 0.807. The signature width W GHz is found to be 0.07 GHz. Hence, from equation 8.128, the probability of outage is calculated to be 0.0116. Using the value of η in Example 2 as 0.194, the value of the expected outage time is determined to be 0.225 percent. This is about three times higher than that determined by experiment.[52]

It should be noted that only the selective fading type of outages can be estimated from any outage calculations based on signatures. Furthermore, the prediction of outages from these methods is meaningful only when the propagation model assumed in the signature is truly representative of the operational conditions. These conditions are heavily dependent on the hop parameters (location, length and roughness, etc.). Experimental results have indicated that *amplitude dispersion* is a good

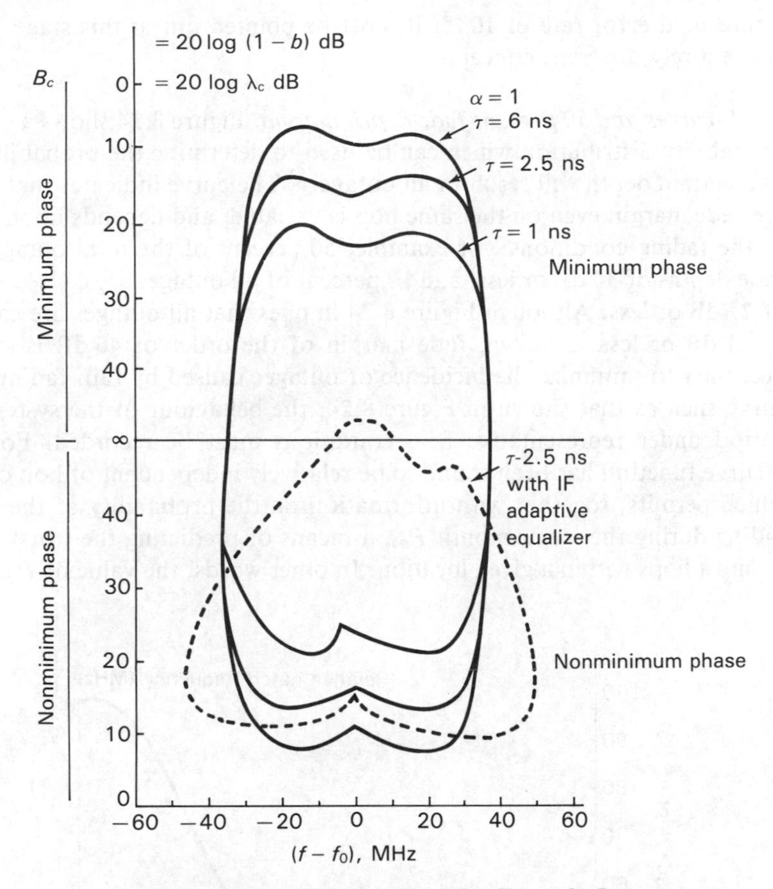

Figure 8.23 Signature used for Example 5
(Courtesy of *Radio and Electronic Engineer*, Reference 52)

indicator of BER performance and may be used in the prediction of outages.[23] Amplitude dispersion ΔA is defined as:

$$\Delta A = |\, 20 \log R_1/R_2 \,| \tag{8.132}$$

where R_1 and R_2 indicate the amplitude of tone signals, usually located at the band edges, separated by $\Delta f = (f_2 - f_1)$ GHz.

S-CURVES, 50 PERCENT FADE DEPTH MARGIN, NET EFFECTIVE MARGIN AND DISPERSIVE FADE MARGIN F_d

An alternative and easier method of predicting system performance than that discussed above, under 'Outage prediction models' in this section, is to use the concepts of fade margins. One type of fade margin is that at which 50 percent of fading occurs above and below a given level. Another type is that of an 'effective' fade margin in which the depth of fade is exceeded for the same number of seconds as a threshold error rate of 10^{-3}. It must be pointed out at this stage, that these are two entirely different concepts.

S-curves and 50 percent fade depth margin. Figure 8.24 shows a curve of a joint probability distribution which can be used to determine the probability that a fade of a certain depth will result in an outage.[53] The curve indicates that the erosion of the fade margin even on the same hop is variable, and depends upon the properties of the fading conditions: for example, 50 percent of the total outage is caused by fade depths of 34 dB or less, and 10 percent of all outages are caused by fade depths of 27 dB or less. Although Figure 8.24 implies that all outages are caused by fades of 41 dB or less, a system fade margin of the order of 46 dB is still considered necessary to minimize the incidence of outages caused by rain fading. To obtain a curve such as that shown in Figure 8.24, the behaviour of the system over a long period under representative field conditions must be recorded. Fortunately, the S-curve function has been found to be relatively independent of hop characteristics, which permits, together with information on the probability of the occurrence of fading during the worst month P_{os}, a means of predicting the worst month outage on most hops within a given location. In other words, the value of P_{os} is determined

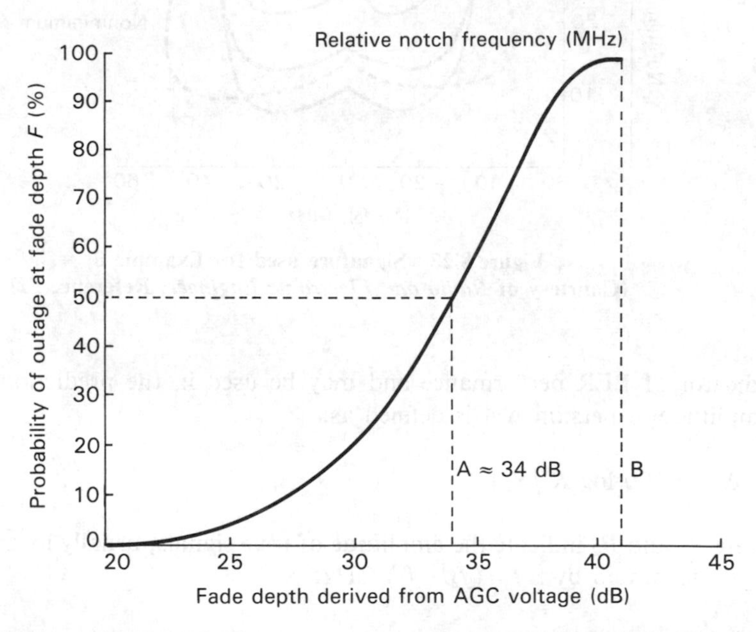

Frequency: 11 GHz
Bit rate: 14 Mbit/s
Modulation: QPSK
Hop length: 54 km
A: a fade of 34 dB has a 50 per cent probability of causing an outage
B: flat fade margin = 41 dB

Figure 8.24 Example of an S-curve or the joint probability distribution showing the effects of multipath propagation on fade margin

directly from the 'S'-curve and η from equation 8.94; the expected outage is found from $\eta \cdot P_{os}$.

Effective or net fade margin F_e and dispersive fade margin. The concept of *effective fade margin*, which is also called the *net fade margin*, has been introduced for digital systems and is defined as the fade depth exceeded for the same number of seconds as a threshold error rate of, for instance, 10^{-3}.[50,26] Figure 8.20 shows the net fade margin defined by a vertical line, to the left of which are as many one-second dots as are above the horizontal line defined by an error ratio of 10^{-3}. In Figure 8.21, this is shown by the dashed lines, giving a net fade margin of 29 dB for the unprotected hop, and 33 dB for the hop with the adaptive equalizer. The flat fade margin of this system is 46 dB (CCIR Report 784–1).[55] The net fade margin may be considered as the composite of the effects of thermal noise, inter-symbol interference due to multipath dispersion, and interference due to other radio systems.[56] At the detector of the radio receiver during fading these three sources will produce three independent power values which due to the their independence, or non-correlation, will add to produce the outage. The dispersion component is determined by the dispersion fade margin, which reflects the impact of multipath distortion upon the design of the system. The dispersive fade margin may be determined from the measured net fade margin by correcting the latter for any thermal noise or interference contributions, if necessary. The minimum value of the dispersive fading margin for a specific system can be determined from the signature. Once the decibel value of any flat attenuation inserted in the path is known, a horizontal line is drawn across the signature at the corresponding value of B_c on the vertical axis, and the difference ΔF dB, between this horizontal line and the horizontal line forming the top of a rectangle with the same area as the signature, is measured. The sum of the value of the flat attenuation α dB and the difference value ΔF provides the value of the dispersive fade margin F_d:

Dispersive fade margin F_d = (Flat attenuation, α dB + difference between flat attenuation and top of equivalent area rectangle ΔF)

$$F_d = \alpha + \Delta F \text{ dB} \tag{8.133}$$

The signature usually taken is that with the delay 6.3 ns, which has a value of B_c corresponding to a BER of 10^{-3}. However, some administrations may have their outage defined at a value of B_c for a BER of 10^{-4}. If this is so, the appropriate signature is taken for a BER of 10^{-4}. The value of the *dispersive fading margin* can be calculated by assuming that the probability of an outage occurring (BER > 10^{-3}) P_{os}, due to selective fading (see equation 8.126), times the factor $10^{(-F_d/10)}$, equals the probability that the dispersive fade depth F_d, is exceeded for the same number of seconds (see equation 8.87), i.e.

$$F_d = 17.15 + 10 \log I_e - 10 \log K - 20 \log [r_b/(140 \log_2 M)] - 30 \log(d/50) \text{ db} \tag{8.134}$$

where F_d is the dispersive fade margin in dB, r_b is the bit rate in Mbit/s, d is the hop length in km, I_e is the improvement factor due to adaptive equalization and M is the number of levels in a modulation scheme.

Relationship between 50 percent fade depth margin and net fade margin. The CCIR Report 338–5,[10] states that for a given link it has been observed that the value of the net fade margin is practically equal to the attenuation on the carrier frequency for which the error ratio is exceeded during about half the time. In a two-ray model this value is given by the attenuation for which the time delay between the two rays is equal to its median value.[57] This means that the dispersive fade margin as given by equation 8.134, plus an amount for the thermal effects, would enable the net fade margin to be determined and equated to the 50 percent fade depth, i.e.

50 percent fade depth margin as shown in Figure 8.24 = Effective fade margin

Calculation of outage time. An approximate value of outage time may be determined by substituting the value of the net fade margin or dispersive fade margin into the expression used to calculate outage for flat fading:

$$\text{Expected outage time} = KQ \cdot f^B \cdot d^C \cdot 10^{F_d/10} \times 100 \text{ percent} \tag{8.135}$$

where KQ, B and C are determined from Table 8.3, f is the carrier frequency (in GHz), d is the path length (in km) and F_d is the net fade margin or, when not available, the dispersion fade margin (in dB).

Example 6

Determine the expected outage time due to selective (dispersive) fading, for a nondiversity 8–PSK system operating at a bit rate of 78 Mbit/s over a 42.47 km hop, using a 6 GHz bearer in a temperate climate in the U.S.A. which has a terrain RMS value of the slopes S equal to 15 m.

Solution
From equation 8.122, $K = 7$, and from equation 8.134, with $M = 3$, the dispersive fade margin F_d equals 25.45 dB. From Table 8.3, column 5, for $S = 15$ m, $KQ = 6.21 \times 10^{-7}$, $B = 1$ and $C = 3$. Thus, from equation 8.135, the outage time is determined to be 0.081 percent. This is equivalent to 2141 s, or twice that observed.

Some of the assumptions which may be made in the formation of an outage model and which help isolate some of the various processes are:[49]

The flat-fading term has a log-normal probability distribution function and is unaffected by diversity.

The selective fading component would have a Rayleigh probability distribution function for nondiversity applications and a third-order pdf for cases where space diversity is used.

The flat and selective fading components are considered to be *statistically independent*.
The activity factor η scales the overall joint probability of outages.
The delays τ in the two-path model have a truncated exponential pdf, which is dependent on path length.
Adjacent channels, which are either co- or cross-polar, are unaffected by the fading in the channel under consideration.
Interference from adjacent channels and from a cross-polar co-frequency channel are treated as added noise which erodes the system fade margin
Cross-polar isolation (XPI), which is discussed in Section 8.3.4.1, is assumed to degrade during fading.
The fading environment could be either in the form of measured distributions of flat and selective fading derived from measurements during a selected period, or assumed forms of distributions which are used with fade depth exceeded for 0.1 percent of the worst month.

Before considering some of the counter-measures to the effects of multipath fading, these effects will be formally discussed, so that some points brought out in the above sections can be reviewed under one heading.

8.3.4 The effects of multipath fading

8.3.4.1 Inter-symbol interference

In quadrature modulation schemes (see Chapter 6), signal distortions may be divided into two classes; 'in-phase distortion', which arises from channel- or propagation-related distortions, but which does not affect the symmetry of the amplitude and group delay characteristics; and 'crosstalk' or 'quadrature distortion', which arises from the asymmetries in the channel. The amplitude slope is the main cause of the BER in systems without adaptive equalization. Interference to the in-phase signal due to amplitude tilt does not produce signal distortion, but with the quadrature signal the distortion is proportional to the first time derivative of the slope. The effects of group delay are similar, except that the distortion on the quadrature channel is proportional to the second time derivative. Parabolic amplitude and group delay imperfections in the channel will, however, affect the in-phase channel but not the quadrature channel. The rapid increase of sensitivity to multipath distortion with increasing symbol rate is related to these channel effects. Two-phase modulation schemes are immune to the distortions that affect quadrature modulation schemes and for this and related reasons are considered more 'robust'. On the other hand, their inefficient use of bandwidth makes their applications limited.

Where coherent carrier demodulation is used in the receiver, multipath effects can cause significant degradations in the carrier phase error. This effect should be considered when comparing alternative carrier recovery techniques. Similarly, for clock recovery from the incoming signal to the receiver. Multipath fading in this case affects the sampling instant and produces a closing of the eye diagram of the recovered information bit stream.

8.3.4.2 Orthogonal polarization discrimination in co-channel digital systems

The spectral efficiency of digital radio systems is, at the present time, only half that attained by FDM–FM analog systems. To double the spectral efficiency of a digital radio system and obtain, by using for example, efficiencies of 4.5 bits/s/Hz using QPRS digital radios,[58] that which is comparable to that achieved by analog systems, cross-polarization is used; one carrier is transmitted on horizontal polarization and the other from a different radio, on vertical polarization. Another reason why cross-polarization is used is to increase the isolation of digital or analog radio systems operating on adjacent RF channels. The problem with the use of a cross-polarized system is that multipath effects can cause the channel RF carrier polarization to change as it is propagated through the atmosphere. Some of the energy propagated in one polarization is transferred to the orthogonal polarization and produces a signal which causes interference in the receiver of the orthogonal polarized channel. This effect is known as cross-polarization interference. As discussed later in the chapter, this effect is most pronounced during periods of rain or hail and may also arise from the characteristics of the antenna systems at each terminal. Usually digital systems, due to their 'ruggedness', permit co-channel carrier-to-interference ratio (C/I) to be as low as 15 dB, without experiencing outage; however, knowledge of the magnitude and occurrence of depolarization is essential for the effective design of such systems and to insure that, if the cross-polarization discrimination degrades sufficiently, the signal on the orthogonal polarization does not cause undue interference and reduce the system performance. The CCIR Report 722–2[10] defines cross-polarization isolation and discrimination.

CROSS-POLARIZATION ISOLATION (XPI)

Consider two orthogonal polarizations V and H, which are transmitted at the same level; the ratio of the co-polarized signal V_c or H_c in a given receiving channel (that is, the intended received signal) to the cross-polarized signal V_x or H_x in that channel (the unintended received signal), is known as the *cross-polarization isolation* (XPI). The two ratios V_c/H_x and H_c/V_x are not necessarily the same. XPI is usually measured in dB, given by 20 log (V_c/H_x) or 20 log (H_c/V_x).

CROSS-POLARIZATION DISCRIMINATION (XPD)

Cross-polarization discrimination (XPD) is the ratio of the co-polarized received signal V_c to the cross-polarized received signal V_x when only one polarization V is transmitted. Thus the co-polar signal V_c and the cross-polar signal V_x are each measured independently and in the absence of any orthogonally polarized transmitted signal H, XPD is usually measured in dB, given by 20 log (V_c/V_x).

When cross-polarization is induced by rain, $V_c/H_x = H_c/V_x = V_c/V_x$. In clear weather conditions, nearly all observations of severe deterioration of XPD have been associated with deep fading of the co-polarized signal caused by multipath fading. The cross-polarized signal has been observed on many links, well correlated with the co-polarized signal during the occurrence of shallow fades (fade depths <10 dB), but not so well correlated for deeper fading. The reduction of XPD may be due to the depolarization of the signal caused by changes in the refractive index along the transmission path, reflection from a tilted atmospheric layer, or scattering or reflection from the ground or water. Apart from causes due to the propagation medium, degradation in the XPD may be caused by the cross-polarization patterns of the antenna, which allows ground reflected or atmospherically reflected indirect signal components to be coupled into it. The direct ray may also be skewed by refractive bending, so that it enters the antenna at an off-axis angle. The requirement for minimizing cross-polarization during multipath fading is that the cross-polarized amplitude and phase patterns are similar to the co-polarized patterns within the angle-of-arrival range of the multipath rays.[59]

8.3.5 Techniques used to reduce the effects of multipath fading on outages

The techniques, used to counter the effects of multipath flat and selective fading, are space and frequency diversity, which act to improve the quality of the received signal, and adaptive channel equalizers, which correct, to a greater or lesser degree, variations in the received signal caused by the propagation medium. Space diversity, together with cross-polarization interference cancelers, also assists in improving the XPD during periods of multipath fading. The effectiveness of any outage reduction technique is usually expressed as an 'improvement factor', which is the ratio of the outage time without the outage reduction technique in service on a single test path, to the outage time when the technique is in service. Normally the outage time must be reduced so that the performance objectives of the link or system can be met.

8.3.5.1 Space diversity

A definition of space diversity is 'the simultaneous transmission of the same signal over a radio channel utilizing two (or more) antennas for reception and/or transmission'. As the name implies, in order to transmit and receive the same information from a source to a sink, two antennas are used, and these are spaced some distance apart. The spacing of the antennas in the receiving or transmitting array is chosen so that the individual signals received are uncorrelated. In practice, it is not possible to achieve a zero correlation coefficient (uncorrelated signals), or even a very low value, but fortunately this does not detract from the substantial benefits obtained by using diversity. For atmospheric fading in a line-of-sight system, a semi-empirical formula has been derived[60] which gives the space correlation coefficient ρ_s in terms of the vertical antenna spacing s meters, the frequency f in GHz, and the path length d in kilometers:

$$\rho_s = \exp[-0.0021\, s \cdot f(0.4d)^{1/2}] \tag{8.136}$$

This equation assumes that the earth-reflected wave is negligible. There is considerable variation in the beneficial effects of space diversity when the space correlation coefficient lies within the range 0.6–1 (largely correlated to completely correlated signals), but the effects remain almost constant when the coefficient is in the range 0–0.6. It is therefore stated in CCIR Report 376–4[55] that it seems reasonable to choose spacings between the antennas so that the space correlation coefficient does not exceed 0.6 – a figure which may therefore be used as a threshold for diversity applications.

In line-of-sight systems during periods of multipath fading the signals received by two vertically separated antennas rarely fade simultaneously when the fades are deep. The available improvement from such a pair of antennas can be defined by the *diversity advantage* I_{os}, which is the ratio of the time percentage for which a given fade level is exceeded for a single channel, or the percentage of time the signal from the main receiving antenna is below the fade margin p_1 percent, to the time percentage of a diversity system, or the percentage of time the signals from the two receiving antennas are below the fade margin simultaneously p_2 percent. These two parameters may be derived from cumulative distribution fading statistics in which the fade depth relative to free space (ordinate) is plotted against the percentage of time the fade depth is exceeded (abscissa) and which contains curves for the single-channel case and for the diversity case. By selecting a particular fade depth corresponding to the fade margin F_m, the values of p_1 and p_2 percent can be directly read and the diversity advantage I_{os} determined. Alternatively, I_{os}, for flat fading, can be determined from the following equation:[61]

$$I_{os} = 100\ (s/9)^2 \cdot (f/4) \cdot a_r^2 \cdot [10^{-4 + F_m/10}]/(d/40) \tag{8.137}$$

where s = antenna spacing center to center, m
f = carrier frequency, GHz
a_r = relative voltage gain of secondary antenna to main antenna $= 10^{[(Ad - Am)/20]}$ where A_d is the power gain of the diversity antenna in dBi and A_m is the power gain of the diversity or secondary antenna in dBi
d = hop length, km
F_m = flat fade margin, dB

If P is the fraction of the month during which the received signal in an unprotected radio channel has a value less than the fade margin set for the system F_m allows (i.e. the fading depth is greater than the fade margin), the fraction of a month of high fading activity during which the signal in a radio channel on a space-diversity protected link is less than that allowed by the fade margin F_m is given by P/I_{os}. This is assuming that a perfect switch is used that will always select the stronger of the two received signals. Then the outage time is reduced by I_{os}:

$$\text{Flat-fade outage time with space diversity} = (\text{Outage time without space diversity})/I_{os} \tag{8.138}$$

The flat fade outage time without diversity can be determined from equation 8.95 or 8.97. Figure 8.25 shows a typical space diversity system, together with the various branching configurations which may be used.

The reduction in outage time when space diversity is used, as shown by equation 8.138, indicates that space diversity is an effective counter-measure against the effects of multipath fading. In digital radio systems, the short-term objectives that are normally the most difficult to meet, owing to the distortions caused by selective fading, are alleviated by this form of diversity. In cases where the performance objectives are not met by an unprotected path, one thinks first of adaptive equalization and, secondly, of space diversity combined with adaptive equalization. The second configuration usually provides sufficient improvement to satisfy the performance objectives, and the improvement factor realized may be larger than the product of the individual improvement factors (the synergistic effect). As mentioned in Section 8.3.4.2, during the time that deep fading occurs the XPD is degraded. By reducing the effective incidence of deep fading using space diversity, the short-term interference effects from cross-polar channels on the same or adjacent frequencies are reduced. This reduction in turn permits a greater spectrum utilization. Likewise, space diversity reduces the interference from other systems, because the digital output power can be reduced. The antenna discrimination may also be relaxed in comparison with nondiversity reception, and this has the effect of permitting a higher nodal capacity or a higher density of radio systems to operate from a radio node, and of permitting an increase in spectrum efficiency. The nodal capacity may be characterized by the total number of megabits that can be transmitted per second from a nodal point. It is thus equal to the product of: the bit rate transmitted on an RF channel, the number of possible channels on a link, and the maximum number of routes possible from a nodal point (which is a function of the angular decoupling).

COMBINERS FOR SPACE–DIVERSITY SYSTEMS

Figure 8.25 shows that the received signals of a space diversity system require to be combined. This is done by using a maximum power *combiner*, which combines the received signals to maximize the combined signal level, or by using a minimum dispersion combiner, which combines the received signals to flatten the amplitude and/or group delay frequency response of the combined signal or, by using a 'hitless' baseband switch, which selects the signal with the lower bit error ratio. If the switchover is initiated by a fast error ratio measurement this form of combining can be very effective. Figure 8.21 shows the advantage of space diversity using an IF combiner over that of an unprotected system. Combiners are in use at RF, IF and baseband frequencies. Maximum power combiners may be RF or IF devices.

Maximum power combiners.[62] The maximum power combiner is a device operating at RF, which shifts the phase of the signal entering through the auxiliary antenna and adds this shifted signal to that entering through the main antenna. The phase shift applied is of such a value that it co-phases, or brings together in phase, the mid-frequency spectral components of the two input signals. The response of the combined channel is similar to that of the two-ray model used to represent selective fading, i.e.

$$H_c(\omega) = \exp(j\theta_c) \cdot a_c[1 - b_c \cdot \exp\{-j(\omega - \omega_0)\tau\}] \tag{8.139}$$

This enables the use of available laboratory simulators to evaluate the effectiveness of maximum power combining.

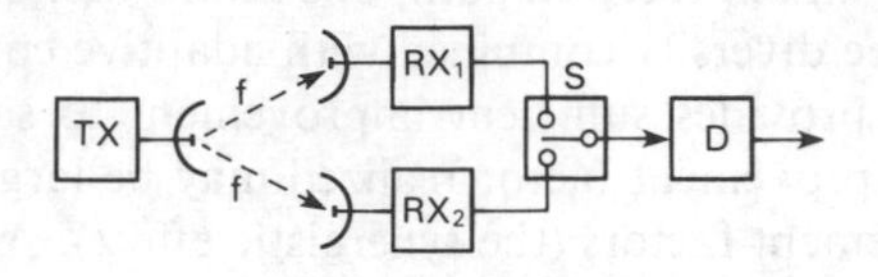
TX
f
f
RX1
RX2
S
D

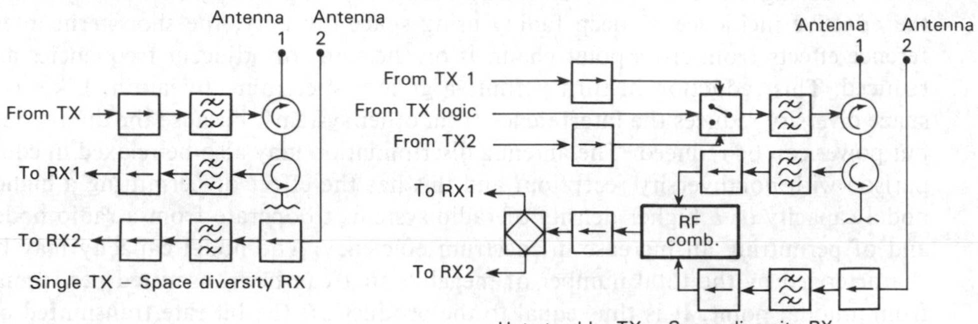
Antenna
1
Antenna
2
From TX
To RX1
To RX2
Single TX – Space diversity RX
Antenna
1
Antenna
2
From TX 1
From TX logic
From TX2
To RX1
RF
comb.
To RX2
Hot stand-by TX – Space diversity RX
RF combining with redundant RX

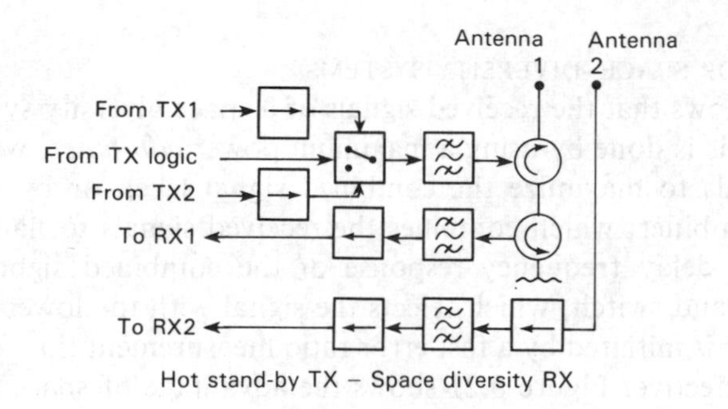
Antenna
1
Antenna
2
From TX1
From TX logic
From TX2
To RX1
To RX2
Hot stand-by TX – Space diversity RX

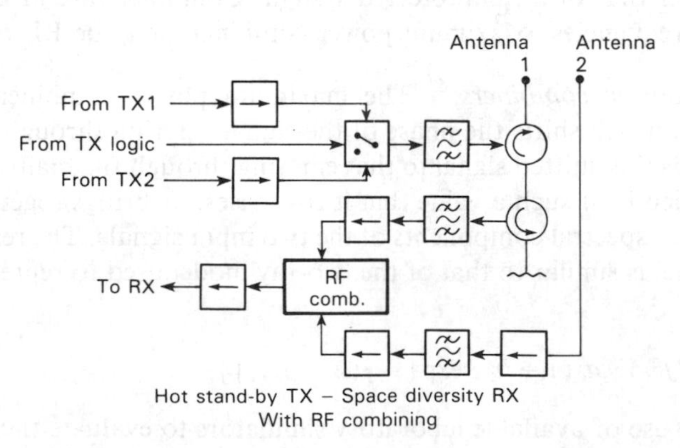
Antenna
1
Antenna
2
From TX1
From TX logic
From TX2
To RX
RF
comb.
Hot stand-by TX – Space diversity RX
With RF combining

(a)

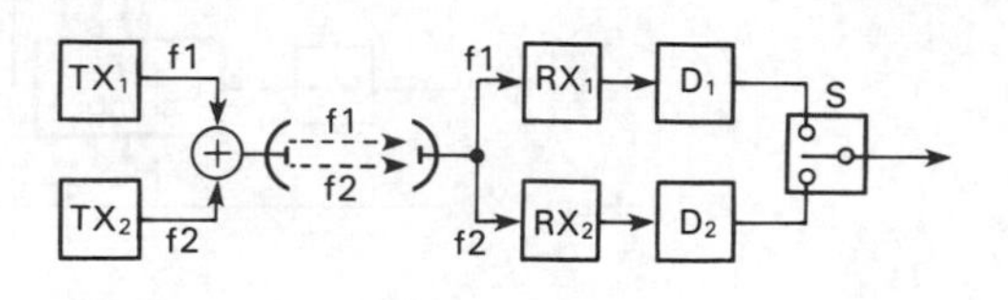

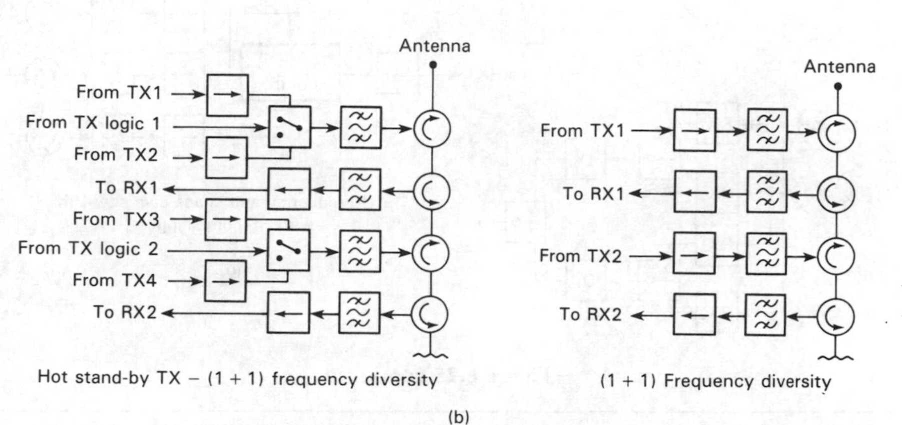

(b)

TX_1
f1
TX_2
f2
f1
f2
RX_1
D_1
S
RX_2
D_2

Antenna 1
Antenna 2
From TX1
To RX1
From TX2
To RX2

(1 + 1) Frequency and space diversity with 2RX

(c)

Figure 8.25 Space and frequency diversity systems and various branching configurations which may be used: (a) space diversity; (b) frequency diversity; (c) space and frequency diversity; (d) space and frequency diversity with four receivers

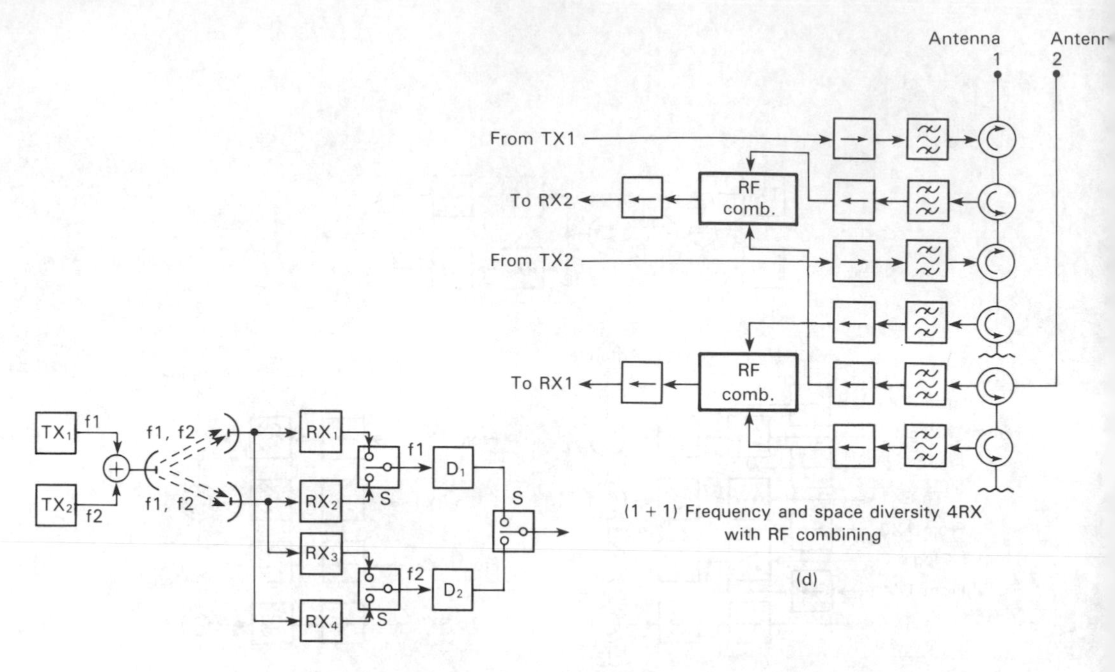

Figure 8.25 continued

Minimum dispersion combiner.[63] This combiner is also an RF combiner. It combines the received signals in such a way that the in-band amplitude dispersion is minimized. This combiner is effective in reducing waveform distortion, but is required to be used in conjunction with a frequency response detector to monitor the spectrum shape. To monitor this spectrum effectively the digital signal is scrambled. When the combined input signal is below a predetermined level, the minimum dispersion combiner is operated as a maximum power combiner.

The minimum dispersion strategy is more attractive in high-capacity radio systems due to its better in-band dispersion equalization. The limitations of this combiner are due to two main factors:

The flat fade margin of the system tends to provide a limit on the effectiveness of the device under conditions where selective fading also occurs. This is because the dynamic range of the device cannot accommodate excessive attenuation due to flat fading and dispersion (variation in frequency response due to selective fading).

A microprocessor control system is required, as the function to be optimized is quite complex. This again tends to limit the dynamic performance of the device to be generally less than the simple co-phase combiner.

Adaptive co-phase IF combining.[64] Space diversity switching, either at IF or RF, is usually unsuitable for a digital radio system because of the excessive number of interruptions that such an arrangement would produce during fading. The use of baseband switches can be made hitless, but is very expensive as two complete receiver-demodulators are required. However, as an alternative to combining at RF, an adaptive in-phase IF combiner can be used. This combiner is hitless and requires duplication of only the mixer preamplifier and channel dropping filter. The combiner consists of a summing junction, an electronic phase-shifter and an electronic control system to maintain the two IF input signals in phase. Because the wanted signals are added coherently, the combiner yields a better signal-to-noise ratio than switched space diversity whenever the difference in signal levels is less than a predetermined level. A consequence of adjusting the phase rather than the delay of the input signal is that the bandwidth of the system is inversely proportional to the absolute time delay between the signals. This can produce significant distortion if the dynamic delay difference between the antennas during multipath fading is large. Phase-adaptive space diversity combining has been shown to be effective in counteracting the effects of multipath fading. The co-phasing combiner maximizes power but does not minimize inband amplitude distortion, which is the prime impairment for 8–PSK systems.

Baseband switch. This unit is usually designed for the specific bit rate of the radio system and performs error-free, or hitless, switching, see Section 11.3.3. This means that the switch operates without loss in bit count integrity and without adding any bit during the switchover. This feature permits the switchover to occur without errors during the transfer from the active to the standby system when a fade or equipment failure occurs, because the equipment is able to compensate automatically a phase shift of up to say, ±4 bits between the failed channel and the standby channel. The hitless mode of operation avoids the loss of synchronization of the multiplexer equipment which may be connected to the radio system, since during the switchover the bit count integrity is maintained. This is done by providing facilities to make a fast assessment of the channel error ratio by means of parity bit insertion. A hitless switch in effect then selects the receiver which has the greater eye opening. The preservation of the actual number of bits, characters or frames, that comprise an original message, or that are transmitted in a unit of time, is known as the bit count integrity. Other features of commercially available equipment are: the baseband interface usually corresponds to the hierachial interface as specified in CCITT Recommendation G703 (see Chapter 5). The baseband processor combines the main signal with other service bits, such as auxiliary signals, digital service channels, parity bits and switching instructions, into an aggregate nonhierarchial signal ready for transmission to the distant terminal. The unit also provides quality evaluation monitor points and operates the automatic switchover on receiving alarms from the transmission equipment. A *multi-line protection* switching system permits up to $N+1$ digital radio bearers to be switched at baseband. This means that there are N active radio systems in use which may be operating not only in the space diversity mode, but in the frequency diversity mode also (this is where different bearer frequencies are used to carry the same information), and one channel is used as the standby. If the channel experiencing fading can switch transparently to a protection channel before the error ratio deteriorates excessively, some improvement in performance can be achieved. The $N+1$ multi-line switch, which may be typically $7+1$

for commercially available equipment, operates in a hitless mode as the 1 + 1 switch described above. The improvements in performance of $N + 1$ multi-line switching will not be as good as that for a 1 + 1 system, since the redundancy is not 100 percent with the standby channel being shared amongst N other systems.

OUTAGES DUE TO MULTIPATH SELECTIVE FADING ON SYSTEMS USING SPACE DIVERSITY

The probability of outage occurring due to selective fading, already discussed in Section 8.3.3.3 ('Outage prediction models'), with particular reference to equations 8.121 and 8.122, can be extended to space diversity systems using baseband bit combining (BBC) or in-phase combining. With the same probability distributions as indicated in Section 8.3.3.3, the probability of outage (BER > 10^{-3}) is given by:[66]

$$P_{os(\text{space})} = 13K^2 \cdot [\tau_0/T]^4 \quad (8.140)$$

where

$$K = \begin{cases} 15.4 & \text{for } 64\text{–QAM} \\ 5.5 & \text{for } 16\text{–QAM} \\ 7.0 & \text{for } 8\text{–PSK} \\ 1.0 & \text{for } 4\text{–PSK} \end{cases} \quad (8.122)$$

Comparisons of predictions with the results of a field experiment have indicated that when both BBC and equalization are employed, more appropriate distributions for the mean echo amplitude are possible, and these provide more accurate performance predictions. Using the same distribution for τ_0 as before (equation 8.125), the expression for the probability of outage becomes:

$$P_{os(\text{space})} = 3.25(P_{os})^2$$

$$= 13K^2[2d^{1.5} \cdot (r_b/\log_2 M) \times 10^{-6}]^4 \quad (8.141)$$

where K is given by equation 8.122, d is the path length (in km), r_b is the bit rate (in Mbit/s), M is the number of levels in the modulation scheme, and P_{os} is the probability of outage with no protection, as given by equation 8.126.

Using the fraction of time that multipath fading occurs or the multipath activity factor η (equations 8.94 and 8.88), the expected outage time can be found:

$$\text{Percentage outage time} = 1300\eta \cdot K^2 \cdot [2d^{1.5} \cdot (r_b/\log_2 M) \times 10^{-6}]^4 \text{ percent} \quad (8.142)$$

Example 7

Calculate the expected outage time for the system given in Example 2, if space-diversity is used. Compare the results and comment.

Solution
Using equation 8.141, the probability of an outage occurring due to selective fading is (1.38×10^{-4}). Hence, with $\eta = 0.194$, the expected outage time is found to be 0.00268 percent or 70 s in the worst fading month. The value given by Hart *et al.*[52] is 80 s, which compares favorably.

Example 8

Calculate the expected outage time if space diversity is used for the case given in Example 3.

Solution
Again using equation 8.141, $P_{os(\text{space})} = 99.01 \times 10^{-6}$, for $\eta = 0.077$; the expected outage time is given by 7.62×10^{-6} or 20 s in the worst month. The value found by Allen *et al.*[54] was an average of 5 for the worst month if a IF combiner was used and 4 if a baseband switch was used. The results compare favorably with the 20 calculated. This means that the improvement factor I_{os} is 55.8

The improvement factor or diversity advantage I_{os} can be calculated from equation 8.138 and 8.141. Thus:

$$I_{os} = 0.308(1/P_{os}) = 38.46 \times 10^9 [K \cdot d^3 \cdot (r_b/\log_2 M)^2]^{-1} \tag{8.143}$$

8.3.5.2 Frequency diversity

Frequency diversity can be defined as 'the simultaneous transmission of the same signal over two or more radio frequency channels which are located in the same frequency band'. Figure 8.25 shows a typical frequency diversity system and the branching configuration which may be used. Also shown in this diagram are hybrid systems in which space and frequency diversity may be combined (quadruple diversity). Another form of frequency diversity is that of *cross-band diversity*, in which the RF bearers are situated in different frequency bands. Although it has been shown that frequency-diversity digital radio systems can give good improvement factors, which are better than those predicted by an analog radio,[67,68] its use is usually restricted due to the inefficient use of the available frequency spectrum. The frequency-diversity improvement factor I_{of}, for a 1 + 1 system, can be estimated from:

$$I_{of} = 0.8(1/fd) \cdot (\Delta f/f) \cdot 10^{F/10} \tag{8.144}$$

where f is the band center frequency (in GHz), d is the path length (in km), $(\Delta f/f)$ is the relative frequency spacing expressed as a percentage[20,61] and F is the fade depth (in dB).

Equation 8.144 is valid for the following parameter ranges:

$$2 \leqslant f \leqslant 11 \text{ GHz} \quad 30 \leqslant d \leqslant 70 \text{ km} \quad \Delta f/f \leqslant 5 \text{ percent} \quad I_{of} \geqslant 5$$

In a multichannel $(N + 1)$ configuration, where a single standby protection channel is provided, additional frequency-diversity improvement can be obtained over a nonprotected channel, without excessive loss in the spectral efficiency, but as the number of active channels increases over 1, the improvement factor decreases from that obtained from equation 8.144. To obtain the improvement factor (equation 8.144), when used in a digital system, the switching system must operate in the hitless mode (see section 11.3.3), so that significant degradation of the traffic channel does not occur. A suitable time considered as adequate for the switching time is 10 ms.

8.3.5.3 Adaptive equalization

The types and basic operation of equalizers were discussed in Section 5.4, to provide some background on the subject. This section continues the discussion, with emphasis on applying equalizers in a digital radio receiver, so that the effects of multipath selective fading can be reduced. For combatting multipath fading there are two general categories:

Frequency domain equalization, which operates at the receiver IF
Time domain equalization, which operates at baseband

Each of these categories may be divided into three groups:

Those which operate best for nonminimum phase-selective multipath fading conditions
Those which operate best for minimum phase fading
Those which operate well in both fading type conditions

Again, these groups can be subdivided into:

Those which operate best for in-band fading conditions
Those which operate best for out-of-band fading conditions

To consider the merits of each type of equalizer as applicable to each of the various modulation schemes is outside the scope of this book. What is intended, however, is to describe some of the more common types of equalizer used with modulation schemes found in commercially available equipment and to remark on their various advantages and disadvantages. The CCIR Report 784–2[97] provides two tables listing the various results of researchers in this field. Tables 8.4 and 8.5 are reproductions of these tables.

For digital systems, time domain signal processing is possibly the better of the two types of equalization technique, since it deals directly with inter-symbol interference. By correlating the interference that appears at the decision instant with the various adjacent symbols producing it, the control information may be derived. This information is then used to adjust the tapped delay line networks and hence provide the appropriate signal for canceling the interference. The advantage of this type of equalizer lies in its ability to compensate minimum or nonminimum phase-induced distortions, which arise from amplitude and group delay deviations in the faded channel. In quadrature modulated systems important destructive effects of fading

Table 8.4 Equalizer characteristics
(Courtesy of CCIR, Report 784–2, Reference 97)

Generic type		Description of equalizer	Complexity of implementation	Fade characteristics, and position of maximum attenuation			
				Minimum phase		Nonminimum	
				Out-	In-band	Out-	In-band
Frequency domain equalizers	F1	Amplitude tilts	S	2	1	2	1
	F2	F1 + parabolic amplitude	S	2	2	2	2
	F3	F2 + group delay tilt	C (M)	3	2	3(1)	2(1)
	F4	F3 + parabolic group delay (For F3 and F4, ratings in brackets apply to 'minimum phase' control assumptions)	C (M)	3	3	3(1)	3(0)
	F5	Single tuned circuit ('Agile notch')	S	3	3	1	0
Time domain equalizers	T1	Two-dimensional linear transversal equalizer	M/C	3	2	3	2
	T2	Cross-coupled decision feedback equalizer	M	3	3	2	1
	T3	T1 + T2 full time domain equalization	C	3	3	3	2

Complexity of implementation
S: simple
M: moderately complex
C: complex

Effectiveness of equalization
3: produces a well-equalized response
2: produces a moderately equalized response
1: produces a partially equalized response
0: not effective

are known to be associated with crosstalk generated by channel asymmetries. Figure 8.26 shows the effects of minimum and nonminimum fading on a pulse entering the in-phase channel I, of a quadrature modulated system, together with the effect on the Q channel.

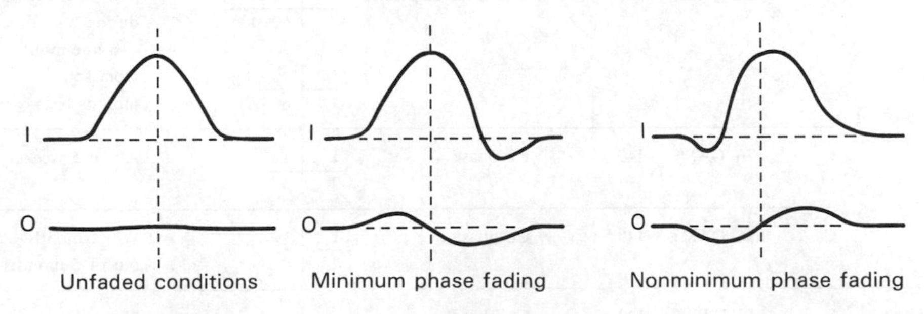

Figure 8.26 Quadrature demodulator minimum and nonminimum phase pulse distortions on I channel

Table 8.5 Outage improvements using various types of equalizer (Courtesy of CCIR, Report 784–2, Reference 97)

Type of equalizer (Note 1) General	Specific	Type of modulation	Information rate, Mbit/s	Type of diversity	Outage improvement factors (Note 2)	Measurement period (Note 8)	Reference
	A	16–QAM	90	Co-phase	1 3.6 6	About 2 months	69
	A	16–QAM	90	Frequency $\Delta f = 59.3$ MHz at 6 GHz	1 100 (Note 10)	1 month	67
					1 48 (Note 11)	1 month	
	A	8–PSK	90	Hitless switch, eye-closure driven	1 6 3–5 175	About 5 weeks	70
	A	QPRS	91	Co-phase	1 40 2 800	3 years total	64
	B	8–PSK	70	Co-phase space-diversity	1 50 (Note 3)	13 months (2 summers)	71
				Cross-band frequency-diversity (7/11 GHz)	1 200 (Note 4)		
				Frequency-diversity $f = 56$ MHz (at 7 GHz)	1 25 (Note 5)	3 months	72 73
Frequency domain	C	16–QAM	200	Co-phase	1 7 1.5		74
	C	16–QAM	200	Minimum dispersion	1 30–40 60–400	Two fade events during 2 months and one month from four months tests	63
					1 20 60		75
	C	16–QAM	140	Co-phase	1 12 2 67	About 5 weeks	76
	C	16–QAM	140	Co-phase	1 3 2 4	Worst month out of 5 months	76
Time domain	E	RB4–PSK	140	Co-phase	1 40–60 (Note 6)	2 months	77

Table 8.5 (*Continued*)

Type of equalizer (Note 1) General	Specific	Type of modulation	Information rate, Mbit/s	Type of diversity	Outage improvement factors (Note 2)	Measurement period (Note 8)	Reference
	B + E	16–QAM	140	Co-phase space-diversity	A_{11}: —, A_{12}: —; A_{21}: 1, A_{22}: 100 (Note 7)	8 months	78
Frequency domain and time domain	C + D	16–QAM	200	Minimum dispersion	A_{11}: 1, A_{12}: 30–40; A_{22}: >180	Two fade events during 2 months	63
					A_{11}: 1, A_{12}: 70; A_{22}: 300	and one month from 4 months tests	75
	C + D	16–QAM	140	Co-phase	A_{11}: 1, A_{12}: 3 (Note 9); A_{22}: 16	Worst month out of 5 months	76
None		16–QAM	140	Hitless switch, parity error driven	A_{11}: 1, A_{12}: 3	Worst month out of 5 months	76
		4–PSK	140	Co-phase	A_{11}: 1, A_{12}: 10–20	2 years	52

Notes

1. Specific types of equalizer are:
   - A: frequency-domain amplitude slope equalizer
   - B: as A, with band-center notch
   - C: 'movable notch'
   - D: linear transversal equalizer
   - E: decision-feedback equalizer
2. Outage improvement factors are presented as a matrix, referred to the following arrangement:

Basic system (A_{11})	(A_{12}) System with space diversity
Basic system (A_{21}) plus equalizer(s)	(A_{22}) System with both space diversity diversity and equalizer(s)

   In each case the matrix element assigned the value unit denotes the system arrangement to which all other arrangements are referred. Note that when $A_{11} = 1$ and $A_{22} > A_{21} \cdot A_{12}$, synergistic effects (combined effect is greater than the sum of the individual effects) are indicated.
3. Flat/net fade margin was 45/31 dB at 7 GHz.
4. Flat/net fade margin was 48/35 dB at 11 GHz (1981) and 43/33 dB at 11 GHz (1982).
5. Flat/net fade margin was 45/31 dB at 7 GHz.
6. Data for RB4–PSK are provisional; firmer data will be published in due course.
7. Flat/net fade margin was 46/36 dB.
8. Administrations have been requested to provide additional details regarding the measurement periods.
9. Flat/net fade margin was 46/20 dB.
10. Flat/net fade margin was 42/35 dB (1980)
11. Flat/net fade margin was 42/30 dB (1982)

As shown in Table 8.5, the improvement gained in a selective fading environment, when equalizers are used in isolation, is not very much. However, there is a dramatic improvement if equalizers are used with space diversity. Measured outage improvement factors have been shown to be as high as 800 : 1, but also as low as 8 : 1, and in addition, the combination usually exceeds the product of the corresponding individual improvements obtained from diversity and equalization separately (the synergistic effect). Explanations put forward for this phenomenon are varied, but do assist in understanding the processes which may be involved. One explanation is that, as it is common during multipath fading for an unfaded signal spectrum from the main antenna to be observed and at the same time for an in-band notch on the diversity antenna to be present (or vice versa), processing the signal from both antennas by a co-phasing combiner produces a resultant signal with a predominantly linear amplitude slope and significantly smaller peak-to-peak amplitude dispersion. Although this remaining distortion can still cause outages, it is almost completely removed by a simple amplitude equalizer. Another explanation is that the co-phasing combiner reduces the proportion of nonminimum phase delay characteristics; consequently, when the combiner is used with an equalizer designed for minimum phase fading, the amount of outage time is reduced. The reduction of nonminimum phase delay is achieved better when an 'out of phase' combiner is used. Possibly the improvement factor for combined space diversity and equalizer arrangements is equal to the product of the space diversity improvement factor I_{of} and the square of the equalizer improvement factor, that is

$$I_{se} = I_{os} \cdot I_e^2 \tag{8.145}$$

where I_{se} is the combined space diversity and equalizer arrangement, I_{os} is the space diversity improvement factor, and I_e is the equalizer improvement factor. From Table 8.5, for the switched diversity case, this appears to be so. Whatever the reason for the synergistic effect – whether it lies in the above explanations, or others, or it is due to the equalization improvement factor[66] – the fact remains that the use of adaptive equalization in combination with space diversity combining is a very powerful method of overcoming the effects of multipath propagation distortion on wideband digital signals.

8.4 RAIN ATTENUATION

Multipath fading and rain attenuation are the most predominant propagation effects on terrestrial line-of-sight radio links operating at frequencies in the gigahertz band, since they determine the variations in transmission loss and thus the repeater spacing, together with the overall cost of a radio relay system. As mentioned in Section 8.3, multipath fading increases as the path length increases, but is not so strongly dependent on frequency; however, rain attenuation tends to increase as the frequency increases, and does not vary much with the path length. Rain attenuation is, in general, negligible below a frequency of 7 GHz, even on hops greater than 50 km.[79]

Above 15 GHz or so, the rain attenuation insures that repeater spacings are kept below 20 km. This reduction in path length, by the same token, reduces the effects of multipath fading, so that it is of minor significance. In the region between 7 and 15 GHz, both multipath and rain fading determine the attenuation statistics. To differentiate between multipath and rain fading occurrences is not difficult, since multipath and rain fading are not often simultaneous occurrences. From this, the overall attenuation distribution can be determined by *adding* the individual time percentages during which either or both effects exceed a particular attenuation level. Hence the outage time expected for a link in which both effects are expected to occur can be determined. It has been shown[79] that, for long paths and low frequencies, multipath fading predominates, whereas, for short paths with higher operating frequencies, rain attenuation has the major influence. In planning digital radio-relay systems, it is important to have attenuation statistics for both multipath (flat and selective fading) and rain fading, which occur during the 'worst month' of the year, so that during this worst month the percentage of time that a preselected threshold is exceeded can be determined. We now consider the CCIR defintion of the worst month, for it is important for multipath, as well as for rain attenuation, calculations.

8.4.1 Worst month

CCIR Report 723–2[10] deals with worst-month statistics. The concept of *worst month* is expressed as: 'for a preselected threshold the worst month is defined as that month with the highest probability of exceeding that threshold; i.e. if X_{ij} is the probability of exceeding a threshold j in the ith month, the worst month for level j is the month with the highest X_{ij} value, X_{hj}. A worst month can therefore be established for each threshold level. The statistic of interest is the highest monthly probability of exceeding the preselected threshold in any one year. This worst month statistic requires that the threshold be specified. Of course this worst month will vary from one threshold to another, e.g. the rain threshold, or the multipath fading threshold, etc. The worst-month distribution for a particular year is given by X_{hj} as a function of j and is the envelope of the highest monthly probability value of all the monthly cumulative distributions from that year. The worst-month statistics may be derived from the observed worst-month values X_{hj} obtained from records over many years and, where possible, are determined from the full set of monthly probability values X_{ij} for each of these years. Finally, the *average annual worst-month distribution* is defined as the average of all of the individual worst month time fractions in excess of j, over a number of years. In practice, most periods of multipath or rain fading are limited to a number of months in a year. If only the limited number of months M are considered, the probability of exceeding a threshold (X_{ij}), in any month during one of the M months is $P(X_j)$. The probability that the worst month value (X_{hj}) exceeds the same threshold X_j, is given by:

$$P(X_{hj} > X_j) = 1 - (1 - P(X_j))^M \tag{8.146}$$

8.4.2 Basic concepts

Heavy rainfall on a radio link absorbs and scatters the power transmitted at frequencies above about 7 GHz and causes the signal to be heavily faded. The fading is worse with horizontal polarization than with vertical polarization. Attenuation caused by sand and dust particles are only of minor importance throughout the world and significant propagation effects are rare. At frequencies around 20 GHz, the attenuation can be 10 dB/km for rain falling at a rate of 100 mm/h, which will severely degrade performance unless the hop lengths are reduced to a few kilometers. Since the cost of a digital radio relay system increases with the number of repeaters, it is important to use the longest possible distance between repeaters and yet maintain the outages within the bounds dictated by the objectives. The path length, between repeater stations, can be accurately determined only if the fading outage due to rain attenuation can be predicted. To make these predictions, a knowledge of the probability distributions of rain attenuation as a function of repeater spacing at various geographical locations is required. Many estimates for the fading statistics on a terrestrial microwave path have been based on the rain attenuation theory of Ryde and Ryde.[81-83] Over the years, as operating frequencies have increased and the path lengths have become shorter, more accurate fading estimates have been developed. Some basic terms used in the literature will now be considered.

8.4.2.1 Definition of a radio path[84]

The *radio path* is defined as the volume of the first Fresnel zone and the rain attenuation is assumed proportional to the number of raindrops in this path volume. The first Fresnel zone is taken, since only the energy contained in its volume contributes significantly to the total energy collected by the receiving antenna. When rain is considered to be falling on a radio path, it is meant that the rain is falling through this volume.

8.4.2.2 Rain rate

Rain rate is a vector which can be written as the product of rain density and the velocity of the raindrops. The rain density is proportional to the number of raindrops per unit volume. The rain attenuation in a radio path depends upon the number and size of the raindrops and not explicitly upon their speed and direction. However, the quantity usually measured is the rain rate, which does depend on the speed and direction of the raindrops. The *point rain rate* is the rain rate measured at a point using a single rain gauge, against being measured over a length. A procedure for calculating a rain attenuation distribution from a point rate distribution is therefore required if reasonably accurate predictions are to be made. The results of rain gauge network measurements have indicated that the measured short-term distributions of point rain rate vary significantly from gauge to gauge. This tends to indicate that the short-term relationship between the path rain attenuation distribution and the rain rate distribution measured by a single rain gauge is *not* unique. The prediction of path rain attenuation distribution from a point rain rate distribution is meaningful only if the time base is sufficiently long to yield stable representative statistics. Accordingly, knowledge of the long-term statistical behaviour of point rain rate and path rain attenuation is essential for radio path design.[86] If there is a uniform distribution of N_D drops of water per cubic centimeter in the space between the two antennas and the drops are spherical with a diameter D and a velocity v_D, then the fraction of volume occupied by the water is defined as the rain density ρ_D, i.e.

$$\rho_D = (\pi/6) \cdot N_D \cdot D^3 \tag{8.147}$$

Rain density is a dimensionless, real, nonnegative quantity. The rain rate R_D for drops with diameter D and velocity v_D is:

$$R_D = \rho_D \cdot v_D \tag{8.148}$$

The direction of the rain rate is the direction of the travel of the drops. In general, a rainstorm has drops of many diameters and the total rain rate R is a summation over the drop diameters present:

$$R = \sum_D \rho_D \cdot v_D \tag{8.149}$$

The lower cut-off threshold in most available rain data is about 0.25 mm/h, since, below this value, rain rates have no significant effects on radiocommunication links at frequencies below 60 GHz.

8.4.2.3 Rain rate sampling volume

The rain density and rain attenuation gradient dB/km are meaningful only if the sampling volume of the rain gauge is large enough to contain sufficient rain drops to yield stable volume average quantities.[86] Measurements indicate that the typical rain density at a 1 mm/h rain rate varies from 50 to 100 raindrops per cubic meter, depending on geographical location. This means that if a volume of 0.1 m^3 was used, it would contain only 2 raindrops at a 0.25 mm/h rain rate. Thus, for a meaningful measurement of rain rate below 1 mm/h, the sampling volume should be at least 1 m^3. If the volume of the gauge is too large, the rain density cannot be considered uniform within the volume of interest; hence the conversion of the rain point rate into rain attenuation gradient will not hold. Tests have indicated that heavy rain has a fine structure of the order of 1 m. This tends to indicate that the sampling volume should not be larger than 1 m^3, i.e.

$$\Delta V = 1 \text{ m}^3 \tag{8.150}$$

8.4.2.4 Rain rate integration time

The optimum integration time, or the time for which the rain is collected, in order to determine the rain rate, is known as the *rain rate integration time*. The optimum integration time is a function of the wavelength, path length and the climate in which the path is located. Lin[86] has determined the appropriate rain gauge integration time t_i, as

$$t_i = \Delta V/(A_g \cdot v_D) \tag{8.151}$$

where A_g is the area of the collecting aperture of a rain gauge, v_D is the average descent velocity of rainfall and t_i is the appropriate integration time. The integration times of most available point rain data are usually in the range $1.5\ \text{s} \leqslant t_i \leqslant 120\ \text{s}$.

8.4.3 Prediction methods[86]

The prediction methods of determining the attenuation gradient A dB/km may be roughly divided up into two groups, both of which use a power law to convert the point rain rate directly to the specific path attenuation, and which both use the same constants k and α, as defined in equations 8.153 and 8.154:

Direct conversion methods. Those which employ reduction coefficients that are derived from the direct conversion of distributed rain rates from point rain rates.

Parametric methods. Those which employ analytical approximations to the point rate distribution and, from this, derive the path average rain rate. These derivations are mostly based on models of the spatial distribution of rain and are characterized by more than one constant.

8.4.3.1 Direct conversion method of predicting attenuation due to precipitation[89]

This prediction method is based on the use of reduction coefficients to determine the effective path length, so that the attenuation exceeded for 0.01 percent of the time can be determined.

STEP 1

Obtain the rain rate exceeded for 0.01 percent of the time (integration time, 1 min), measured at the ground in the location of interest. If this information is not available from local sources of long-term measurements, an estimate can be obtained from the information given in Section 4 of CCIR Report 563–3.[10]

STEP 2

The initial calculation is to determine, from the rain rate exceeded for 0.01 percent of the time with an integration time of 1 min, the specific attenuation γ_R. This is determined from a power law relationship, given by:

$$\gamma_R = k \cdot R^{\alpha}\ \text{dB/km} \tag{8.152}$$

where R is the rain rate in mm/h averaged with an integration time t_i and the parameters k and α are functions of radio frequency and wave polarization, given by:[88]

$$k = [k_H + k_V + (k_H - k_V)\cos^2\theta \cdot \cos 2\tau]/2 \tag{8.153}$$

$$\alpha = [k_H \cdot \alpha_H + k_V \cdot \alpha_V + (k_H \cdot \alpha_H - k_V \cdot \alpha_V)\cos^2\theta \cdot \cos 2\tau]/2k \tag{8.154}$$

where θ is the path elevation angle and τ is the polarization tilt angle relative to the horizontal ($\tau = 45°$ for circular polarization, 0° for horizontal polarization and 90° for vertical polarization).

The values of k and α for a Laws and Parsons raindrop-size distribution and a drop temperature of 20° C have been calculated by assuming oblate spheroidal drops aligned with a vertical rotation axis and with dimensions related to the equivolumetric spherical drops.[87] Table 8.6 provides values of k_H, k_V, α_H and α_V so that k and α can be obtained from equations 8.153 and 8.154. At frequencies other than those listed, interpolation using a logarithmic scale for frequency and k and a linear scale for α, the values of k and α can be obtained. The CCIR Report 721–2,[10] recommends that equations 8.152, 8.153 and 8.154 be used together with the values of k and α in Table 8.6, for evaluating the statistical characteristics of rain attenuation.

Table 8.6 Regression coefficients for estimating specific attenuations in equation 8.152

Frequency, GHz	k_H	k_V	α_H	α_V
7	0.00301	0.00265	1.332	1.312
8	0.00454	0.00395	1.327	1.310
10	0.0101	0.00887	1.276	1.264
12	0.0188	0.0168	1.217	1.200
15	0.0367	0.0335	1.154	1.128
20	0.0751	0.0691	1.099	1.065
25	0.124	0.113	1.061	1.030
30	0.187	0.167	1.021	1.000
35	0.263	0.233	0.979	0.963
40	0.350	0.310	0.939	0.929

Note. For the full range of values from 1 to 400 GHz, the reader is referred to CCIR Report 721–2, Reference 10.

STEP 3

The effective path length d_e of the link is defined as the length of a fictitious path along which a constant specific attenuation would cause the same attenuation as that exceeded for 0.01 percent of the time on the actual path. It depends on the model assumed for the spatial structure of rainfall and is computed by multiplying the actual path length d by a reduction factor r. The value of r is given by:

$$r = (1 + 0.045d)^{-1} \tag{8.155}$$

STEP 4

An estimate of the path attenuation exceeded for 0.01 percent of the time is given by:

$$A_{0.01} = \gamma_R \cdot d_e = \gamma_R \cdot r \cdot d \text{ dB} \tag{8.156}$$

STEP 5

Attenuations exceeded for 1, 0.1, 0.01, 0.001 and 0.0001 percent give the factors 0.12, 0.39, 1, 2.14 and 3.76 respectively, by using the formula of equation 8.157, which is valid in the range 0.001 to 1 percent:

$$A_p = 0.12(A_{0.01}) \cdot p^{[-(0.546 + 0.043 \log p)]} \text{ dB} \tag{8.157}$$

where A_p is the attenuation (dB), exceeded for p percent of the time.

STEP 6

The average annual probability Q of exceeding a given attenuation level X_j may be derived from the worst month probability $P(X_{hj} > X_j)$ redefined as P_w for ease in writing. This is given by:

$$Q = 0.3\ P_w^{1.15} \text{ percent} \tag{8.158}$$

See CCIR Report 723–2, Section 5,[10] for further details.

Example 9

Determine the attenuation due to rain which is exceeded for only 0.001 percent of the time on a 53 km hop, located in England and operating on a frequency of 15 GHz with horizontal polarization. Assume that the rainfall rate exceeded for 0.01 percent of the time is 28 mm/h (1 min integration time), and the antennas are at the same height above mean sea level.

Solution

From Step 1, the values of k and α are determined from Table 8.6 and equations 8.153 and 8.154. Thus, from $k_H = 0.0367$, $k_V = 0.0335$, $\alpha_H = 1.154$ and $\alpha_V = 1.128$, $\theta = 0$, and $\tau = 0$, we find $k = 0.0367$ and $\alpha = 1.154$. From equation 8.152, the specific attenuation $\gamma_R = 1.717$ dB/km. The effective path length is determined from equation 8.155 as $d_e = 0.29542$. Thus, the path attenuation exceeded for 0.01 percent of the time is found from equation 8.156 to be 26.88 dB. To determine the attenuation exceeded for only 0.001 percent of the time, from equation 8.157, the value of the multiplying factor is 2.14; hence the attenuation exceeded for 0.001 percent of the time is 57.53 dB. This value is quite excessive and shows that the flat fade margin has to be at least equal to this value if the link is not to have an outage greater than 0.001 percent of the time.

FREQUENCY SCALING

For single-frequency scaling in regions of low rainfall rate (< 50 mm/h) reasonable results may be obtained over a frequency range of 7–50 GHz. For higher rainfall rates, the method is limited to 30 GHz. Given that

$$(A_1/A_2) = g(f_1)/g(f_2) \tag{8.159}$$

where A_1 and A_2 are the values of the attenuation (in dB) at frequencies f_1 and f_2 (in GHz), and are exceeded with equal probability, and where

$$g(f) = f^{1.72} \cdot [1 + 3 \times 10^{-7} \cdot f^{3.44}]^{-1} \tag{8.160}$$

the value of attenuation at a known different frequency f_1, can then be determined from a known frequency f_2 and attenuation A_2.

Example 10

From the value of the attenuation not exceeded for 0.001 percent of the time, calculated in example 9, find the value of attenuation not exceeded for 0.001 percent of the time, for the same parameters, but using a frequency of 11 GHz.

Solution
From equation 8.160, the value of $g(15)$ is found to be 105.06, and $g(11)$ is 61.76. Hence as $A(15) = 57.53$ dB, from equation 8.159, the value of attenuation not exceeded for 0.001 percent of the time on a 11 GHz 53 km hop is 33.82 dB. This is in alignment with the value of 35–40 dB for the fade margin, usually found on such hops. If this is compared with Exercise 8 at the end of the chapter, it can be seen that the flat fade margin for a similar link is 35 dB. As multipath fading rarely occurs when it is raining, the value of 35 dB for the fade margin in this case should be a design which satisfies both the outage and availability objectives.

POLARIZATION SCALING
Knowing the attenuation on one polarization, the equivalent attenuation on the orthogonal polarization can be determined from:

$$A_V = 300 A_H/(335 + A_H) \text{ dB} \tag{8.161}$$

$$A_H = 335 A_V/(300 - A_V) \text{ dB} \tag{8.162}$$

Propagation measurements have confirmed that horizontally polarized waves are attenuated more strongly by rain than are vertically polarized waves.

8.4.3.2 Parametric method[85,90,91]

As with the ‘direct conversion’ method, the same power law as given in equation 8.152 is used to relate the rain rate R, to the specific attenuation γ_R. The difference in this case is in the way in which the reduction coefficient is derived. Unlike the direct conversion method, it depends on the spatial structure of the rainfall and the integration time. Although there may be a marked dependence of the reduction factor r on frequency, it is assumed that the frequency dependence of the rain attenuation is present only in the coefficients k and α. The values of k and α are derived in the same manner as that for the direct conversion method. The following structure for the reduction coefficient is used:

$$r = [u + \{d(vR + xd + y)/z\}]^{-1} \quad (8.163)$$

with the coefficients u, v, x, y and z depending on integration time. As the spatial structure of rainfall can be different in different geographical areas, i.e. the size of the rain cells (usually less than 15 km) can be greater in different areas for the same rain intensity, the coefficients will depend also on the geographical area. Using the values of the coefficients which are widely used in Europe for integration times of 1 minute or less, the value of r is given as:

$$r = [0.98 + \{d(40.2R - 3d + 200)/28200\}]^{-1} \text{[Europe]} \quad (8.164)$$

From results obtained for different path lengths and 21 different links in Europe, this method[91] claims to be more accurate than the CCIR equation 8.155, for low percentages of time, which are concerned with the availability and performance of radio relay systems. The method gives a very low accuracy for 0.1 percent of the time and for paths less than 9 km in length. For the U.S.A. and Japan, the value of the path reduction factor r is

$$r = [0.85 + \{d(1.14R - 3d + 200)/3929\}]^{-1} \text{[U.S.A./Japan]} \quad (8.165)$$

As with equation 8.164, the method predicts with great accuracy for low percentages of time ($< 10^{-3}$ percent) and with low accuracy for 0.1 percent of the time; it is considered more accurate than the CCIR method given by equation 8.155.

The method of using these results to determine the amount of path attenuation exceeded for a certain percentage of time is the same as that given in Steps 1 to 6 of Section 8.4.3.1, with the exception that equation 8.155 in step 3 is replaced by either equation 8.164 or 8.165.

Example 11

Using the same parameters as in example 9, determine by the parametric method the attenuation due to precipitation in Europe, exceeded for 0.01 percent of the time.

Solution
The value of r determined from equation 8.164 is 0.3152. As $\gamma_R = 1.717$ dB/km, and $d = 53$ km, the value of the attenuation exceeded for only 0.01 percent of the time is 28.68 dB. The CCIR method gave 26.88 dB.

8.4.4 Effect of wet radomes on rain attenuation data

When the aperture of the feed of an unprotected parabolic antenna becomes wet, either by having a layer or drops present, then at frequencies around 11 GHz a problem is caused. This is because water introduces both loss and phase shift around 11 GHz. Its presence at the feed aperture, where the power density of the transmitted signal is high, causes considerable degradation in antenna performance by way of loss, reflection and distortion of the phase of the wavefronts. To overcome this problem together with birds perching on the feed, a radome is used to cover the antenna transmitting face. Under clear weather conditions, this radome produces a small loss. However, when it is wet, it causes considerable loss. Because of the need to have a second radome at the distant terminal, the losses are doubled. Experiments using cone-shaped or hemispherically shaped radomes can produce a loss in the range of 0 to 14 dB, caused by rain running on a pair of them. The variation of loss depends on frequency, rain rate and radome surface aging. At 20 GHz on a section of a 30 m radome, pertaining to an Earth satellite link, a transmission loss of 2–3 dB at a 10 mm/h rain rate was found,[92] when the radome was new. After six months of weathering, the transmission loss had increased to 8 dB at the same 10 mm/h rain rate. For terrestrial links at frequencies of 11–18 GHz, it is expected that the transmission loss from wet radomes will be lower, due to the smaller radome but it should not necessarily be considered negligible, especially since a pair will normally be used.

8.4.5 Effects of rain on orthogonally polarized systems

For dual horizontally and vertically polarized systems, with or without diversity, the main cause of degradation from cross-polarization discrimination (XPD) (see later in this paragraph) for small percentages of the time, is rain on short links (< 20 km). For a given fade margin, a system in which the other channel is working on the same frequency (co-channel), but with the orthogonal polarization, will experience a greater outage time than a system which uses only one channel on each frequency. This effect becomes more pronounced as the fade margin is increased. As mentioned in Section 8.4.4, water or snow on the antenna feed system causes severe degradation in the antenna performance, one of these degradations being the result of cross-polarization. This effect can be reduced by the use of a radome. The ratio of the wanted (co-polar) and unwanted (cross-polar) received signals is referred to as the *cross-polarization isolation (XPI)*, and is the reciprocal of the always positive 'depolarization factor' expressed in dB. The term 'depolarization' is sometimes used to refer to cross-polarization, but is more often taken to mean that there has been a change in the original polarization state. The ratio of the co-polar to cross-polar power levels when only one frequency is used, is the *cross-polarization discrimination (XPD)*. This is equal to the XPI if rain is the cause of the signal degradation. These definitions are given more fully in Section 8.3.4.2. The effects of rain falling as oblate spheroids and producing different attenuation and phase shift for the different polarizations will cause depolarization, thus producing cross-polarization. If the rain, however, falls in the direction which is aligned with one of the polarization axes, which would normally be the vertical axis, then cross-polarization will not occur. In practice, however, there is often a localized mean canting angle as the raindrops fall, caused by wind shear. The canting angles may be widely spread about this mean due to turbulence, so that there is considerable cross-polarization along the path where precipitation occurs. Differential phase shift appears to be the dominant factor in rain-induced cross-polarization at frequencies below 10 GHz, whereas differential attenuation becomes increasingly important at higher frequencies. Cross-polarization discrimination (XPD) decreases with increasing co-polar attenuation (CPA), due to rainfall. For most practical applications, a relationship between XPD and CPA is important in predictions based on attenuation statistics. A general relationship between these two parameters has not yet been fully established, but long-term measurements in the frequency range 3–37 GHz are in agreement with a semi-empirical relation of the general form:[93]

$$\mathrm{XPD} = U - V \log(\mathrm{CPA}) \text{ dB} \tag{8.166}$$

where the coefficients U and V depend primarily on the frequency f (GHz), the path elevation angle θ and the polarization tilt angle τ for linear polarization relative to the horizontal. CCIR Report 722–2[10] provides further details and references on this subject. A rough estimate of the value of XPD can be found for line-of-sight paths with small elevation angles and orthogonal polarization, by using the following values of the coefficients U and V:

$$U = 15 + 30 \log f \text{ dB} \tag{8.167a}$$

$$V = 20 \tag{8.167b}$$

for $8 \leqslant f \leqslant 35$ GHz

Depolarization may be caused by ice. As the ice crystals appear to have their major axis in the horizontal plane, depolarization is minimized for both horizontal and vertical polarizations. For small percentages of the time where this does not occur, as when lightning causes an ice crystal alignment, depolarization may occur. Dry snow near the ground appears to show the same depolarization characteristics as that of ice crystals on slant paths. It causes significant differential phase shift but little differential attenuation. Also, it causes more depolarization than rain for the characteristic low attenuation levels produced by dry snow. Wet snow appears to produce differential attenuation rather than differential phase shift, and to be a less significant polarizing medium than rain. The performance of orthogonally polarized systems can be improved by increasing the output power so as to reduce the effects of thermal noise.

Long-term XPD statistics obtained at one frequency can be scaled to another frequency using the semi-empirical formula:

$$\mathrm{XPD}_2 = \mathrm{XPD}_1 - 20 \log(f_2/f_1) \quad 4 \leqslant f_1, f_2 \leqslant 30 \text{ GHz} \tag{8.168}$$

where XPD_1 and XPD_2 are the XPD values not exceeded for the same percentage of time at frequencies f_1 and f_2.

8.4.6 Scattering by precipitation

In addition to the attenuation produced by rain, hydrometers can cause the radio beam to be scattered. If this scattering is horizontal and perpendicular to the direction of the radio beam, i.e. in the sideways direction, interference between radio paths may be caused. Although the main cause of this problem is rain, scattering from clouds of ice particles and from wet snow or wet hail is also important. Scattering caused by precipitation may establish a coupling between antennas, provided that the radiation lobes of these antennas intercept a common volume in space and that the precipitation occupies some part of the common volume. Coupling between antennas is reduced by signal attenuation along the two paths between the common volume and the antennas. Maximum coupling therefore results from the simultaneous occurrence of strong scattering inside the common volume and low attenuation outside. At frequencies below 8 GHz, where the attenuation is small, Rayleigh scattering applies. The coupling in this case increases approximately as the fourth power of frequency. At the higher frequencies, scatter by the common volume of the antenna side lobes increases less rapidly and the net coupling is rapidly reduced by attenuation outside the common volume. For frequencies above 8 GHz *Mie scatter* theory must be applied in place of Rayleigh scattering,[94,95] by taking into account the average scattering properties of the particles within the scattering volume. The ratio of the received power to transmitted power can be found in CCIR Report 882–1 and 569–3.[10]

8.4.7 Techniques to reduce the effects of rain attenuation

The use of space diversity, frequency diversity, adaptive equalizers and shortening the path length does nothing to reduce the effects of rain fading. The only effective way to reduce the effects of rain is either to drop the frequency or get out of its way. Changing the polarization from horizontal to vertical or reducing the length of the hop may produce some relief. This reduction in attenuation may be determined from equations 8.155 and 8.164 or 8.165.

8.4.7.1 Route diversity

A choice of paths with a separation of several kilometers may reduce the effects of rain attenuation. The diversity gain can be defined as the difference between the attenuation in decibels, exceeded for a specific percentage of time on a single link, and that attenuation exceeded simultaneously on two parallel links, which are separated by several kilometers. This diversity gain tends to decrease as the path length increases beyond 12 km for the frequency range 20–40 GHz, but it is generally greater for a spacing of 8 km than for a spacing of 4 km (for 0.1 percent of the time at 8 km it is 2.8 dB, whereas at 4 km it is 1.8 dB). Increasing the spacing from 8 to 12 km does not provide any improvement. Route diversity gain is not dependent on the frequency in the 20–40 GHz range for a given path diversity geometry.

8.5 GASEOUS ABSORPTION

In this section the attenuation by atmospheric gases will be briefly considered. Figure 8.27 shows the specific attenuation of water and oxygen, together with the total specific attenuation (in dB/km), plotted against frequency. Oxygen does not become a major cause of attenuation until the frequency rises above 50 GHz. At 118.74 GHz an isolated absorption line exists which gives rise to a specific attenuation γ_o of around 1.9 dB/km, while between 50 and 70 GHz a series of very close lines occur with a specific attenuation of around 16 dB/km at ground level (pressure 1013 mb and temperature 15°C). The specific attenuation can be found at ground level, from the following equation:[96]

$$\gamma_0 = [7.19 \times 10^{-3} + 6.09/(f^2 + 0.227) + 4.81/[(f - 57)^2 + 1.50]] \cdot f^2 \cdot 10^{-3} \text{ dB/km for } f < 57 \text{ GHz} \quad (8.169)$$

A similar equation exists for the frequency $f > 63$ GHz, but is not included here, because at this time digital line-of-sight radio links do not operate up to this frequency (see CCIR Report 719–2[10]). Water vapor has three absorption lines, at the frequencies 22.3 GHz, 183.3 GHz and 323.8 GHz. Specific attenuation values for a temperature of 23°C are, for 22.2 GHz, 0.45 dB/km at a water vapor of 19.1 g/m^3 and 0.15 dB/km at a water vapor of 6.4 g/m^3. At 35 GHz the values are 0.24 dB/km for 19.1 g/m^3 and 0.06 dB/km for 6.4 g/m^3.

The sum of the specific attenuations for oxygen and water provide the necessary information to determine a practical value of attenuation for the length of the radio path:

$$A_a = \gamma_a \cdot d = (\gamma_o + \gamma_w) \cdot d \text{ dB} \quad (8.170)$$

where A_a is the total attenuation produced by oxygen and water over the path, d is the path length (in km), γ_o and γ_w are the specific attenuation of oxygen and water respectively (in dB/km). The values of specific attenuation can be found, for practical purposes, from Figure 8.27.

Example 11

Determine the attenuation expected from gaseous absorption for a 53 km hop operating at 15 GHz.

Solution

From Figure 8.27, at 15 GHz the sum of the specific attenuations is found to be approximately 0.029 dB/km; hence additional attenuation due to gaseous absorption is 1.54 dB. Surprisingly, this is not an insignificant amount.

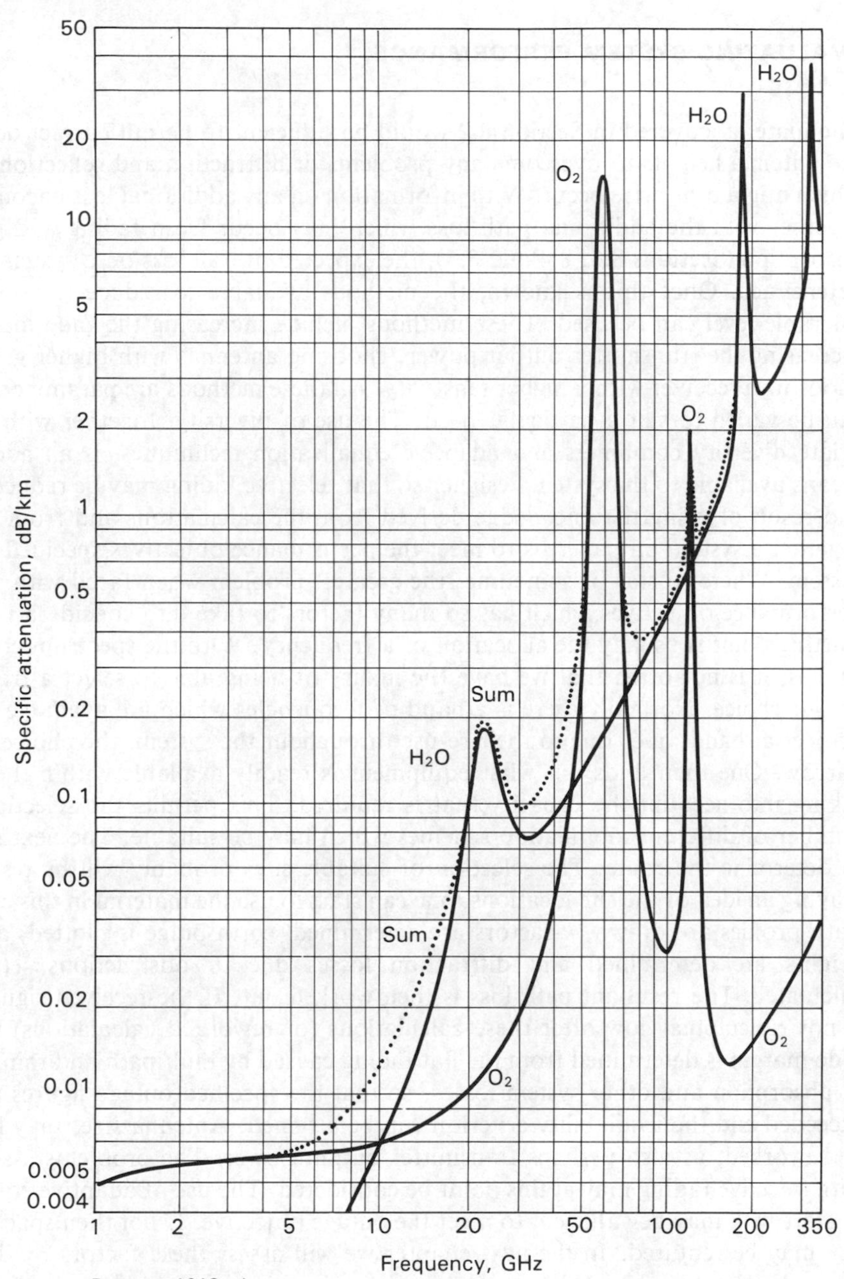

Figure 8.27 Specific attenuation due to atmospheric gases (After CCIR, Report 719–2, Reference 10)

8.6 EVALUATING SYSTEM PERFORMANCE

The material covered in Section 8.2 would be sufficient to permit correct design of the antenna heights to overcome any problems of diffraction and reflection fading which might otherwise occur. With information on any additional loss encountered, together with the additional path loss which may occur from fading and gaseous absorption (Sections 8.3, 8.4 and 8.5), the expected overall loss on a system can be determined. Once this is known, the methods available to reduce this loss to a tolerable level can be used. These methods include increasing the fade margin by increasing the transmitter output power, choosing antennas with higher gain, and choosing a receiver with a higher senstivity. All these methods are parameters which can be varied to suit a particular need. The use of diversity, together with appropriate diversity combiners and adaptive equalization techniques, is an additional means available to the system designer so that selective fading may be reduced. The end result of using the knowledge derived from the calculations and from piecing together a system on paper, is to meet the performance objectives specified for the system. Where to start is sometimes the greatest problem when faced with a set of performance objectives which has so many factors to take into consideration. The starting point is usually the allocation of a frequency. With the spectrum crammed as it is, it is not often that we have the luxury of being able to select a frequency of our choice. Normally there is a band of frequencies which are available for the project at hand, and, due to their re-use throughout the system, the choice is very narrow. One then finds out what equipment is readily available within the band, taking into account the capacity that is required. This permits the selection of a number of different modulation schemes which may be suitable. The next stage is to determine the route. The selection of suitable sites probably will be restricted. Having an idea of the site locations, one can start to use the material in this chapter. Path profiles are drawn, *k*-factors are determined, earth bulge is plotted, antenna heights are determined and diffraction losses due to obstructions, etc., are calculated. The resultant path loss is then worked out. If the received signal level is not ridiculously low after these calculations (or reworked calculations) the flat fade margin is determined from the flat fading caused by multipath and rain as well as absorption and other system losses, so that the specified outage figures are not exceeded and the availability criterion can be attained. Antenna sizes may have to be reworked, as also perhaps transmitter output power. The problems associated with selective fading may at this point be considered. The use of adaptive equalizers to start with may be sufficient to meet the outage objective. If not then space diversity may be required. In the next chapter we will assess these factors by drawing from elsewhere in the book, to come to the point where a system can be designed. Once the system has been designed, however, the problem of interference still needs consideration, for it may be a constraint which necessitates the complete redesign from the very beginning and starting afresh with a new set of frequencies. This aspect will be discussed in Chapter 10.

EXERCISES

1 Given that the atmospheric pressure was found to be 1015 mb at a temperature of 14°C and that the vapour pressure was 11.2 mb, determine the expected refractive index n of the atmosphere at the height that these measurements were taken. (*Answer* 1.000325)

2 Determine the diffraction loss for a 100 km horizontally polarized link over sea water in a temperate climate. The antenna heights at either end of the link are 60 m and 120 m above sea level. Assume a carrier frequency of 4 GHz and a k-factor of 4/3. (*Answer* 46 dB)

3 Which is the better polarization to use for the link described in Question 2? (*Answer* compare with the example given in Section 8.2.2.2, 'Diffraction loss due to grazing over smooth earth')

4 Assume a 50 km radio path with a knife-edge obstruction at a distance 30 km from one end of the link. If the operating frequency is 4 GHz, determine the height the knife edge has to be to obtain a diffraction gain of 1.2 dB. (*Answer* the knife edge should lie 265 meters below the main beam)

5 Assume that a k-factor of 4/3 was used in the example given in Section 8.2.2.3. Calculate the required value of fade margin for the receiver and compare the result with that of Exercise 2 above. (*Answer* 43 dB)

6 Using Majoli's method, determine the amount of multipath fading activity as a percentage of the worst month, expected on a 6 GHz link of 30 km length over a mountain range where the surface roughness is 100 m. (*Answer* 0.206 percent)

7 A 54 km link is to be established in a temperate climate operating at 11 GHz. The terrain is rolling hills with a roughness of 30 m. Determine the required flat fade margin using Majoli's method, to compensate for Rayleigh multipath fading effects only. (*Answer* 38.44 dB)

8 Using the same data as in Question 7, determine from the CCIR method the value of the fade margin. Assume that the link is in N.W. Europe. (*Answer* 34.7 dB)

9 Calculate the outage probability due to selective fading alone (P_{os}), for 200 Mbit/s, 16–QAM, 5 GHz radio operating over a hop length of 63 km. Use equation 8.126. (*Answer* 2.75 percent)

10 Calculate the antenna spacing of a space diversity system operating at a frequency of 6 GHz, over a 50 km hop, if the received signals are to have a correlation coefficient of 0.5. (*Answer* 12.3 m)

11 Determine the diversity advantage for the system given in Example 10, if the fade margin F_m is 35 dB and the main and diversity antennas are of equal size. (*Answer* 70.88)

12 If an unprotected system such as that given in Example 9, which is in a temperate climate with a path roughness of 15 m, experiences only flat fading, determine the expected outage for a flat fade margin of 30 dB. If this system is then upgraded to a space-diversity system in which the antennas are chosen to be the same size and the same flat fade margin F_m, is used, will the system be able to satisfy the CCIR performance objective of: outage <0.006 percent of any month, if the correlation coefficient is chosen to be 0.5? (*Solution:* $P(\text{FAD}_m > F_m) = 0.75 \times 10^{-3}$; flat fade outage time = 0.075 percent, $s = 13.15$ m, $I_{os} = 16.94$, outage = 4.43×10^{-3} percent. CCIR Recommendation for outage = 6.00×10^{-3} percent. *Answer* Yes)

13 If an unprotected system such as that given in Example 9 is upgraded to a space-diversity system with an IF co-phase combiner, determine the expected outage time due to selective fading alone. (*Answer* 0.2458 percent)

14 A 12 GHz horizontally polarized 8–PSK radio link operating at 90 Mbit/s somewhere inland in the region of California, U.S.A., where the terrain is rolling hills of 30 m roughness, is to span 40 km. The antennas are sited such that a 10° angle from the horizontal occurs. Assuming that the rain rate not exceeding 0.01 percent for the worst month is 70 mm/h ($\tau = 1$ m), determine the values of (*a*) (*b*) (*c*) below so that where applicable, the path will work to CCIR outage objectives ($\leqslant$0.006 percent in any month). Also (*d*) give suggestions on how you would go about making this link work to CCIR objectives.

(*a*) the flat fade margin required to overcome flat fade multipath effects (use CCIR method)

(*b*) the fade margin required to overcome the effects of rain fading (use the parametric method)

(*c*) the outage expected due to selective fading

(*Answers* (*a*) 35.06 dB; (*b*) from $\alpha = 1.21677$, $k = 0.01877$, $\gamma_R = 3.3$ dB/km, and $r = 0.4037$, F_m(0.01 percent) = 53.29 dB for 0.001 percent, F_m(0.001 percent) > 114 dB; (*c*) from $\eta = 0.0563$, the expected outage is 0.018 percent. (*d*) To overcome the largest problem of the effects of rain fading, route diversity is possibly the only solution, and to insure that flat fading does not pose a problem, a fade margin of 55 dB is required. This can be less if route diversity is used. As the problem associated with selective fading is not too great, a co-phase combiner used in conjunction with a baseband transversal equalizer (time domain) may be sufficient. The use of space diversity may not be warranted in this case. The answers given are assuming that multipath effects and precipitation do not occur at the same time.

REFERENCES

1. Chatterjee, B., *Propagation of Radio Waves* (Asia Publishing House, 1963).
2. Townsend, A. A. R., *Analogue Line-of-Sight Radio Links: a Test Manual* (Prentice-Hall, 1987).
3. Majoli, L. F., 'A New Approach to Visibility Problems in Line-of-Sight Hops', Proceedings of the National Telecommunications Conference, Washington, November 1979.
4. Majoli, L. F. and Mengali, U., 'Propagation in Line-of-sight Radio links – Part I: Visibility, Reflections, Blackout', Special Supplement to *Telettra Review*, 1985. No. 37.
5. Hall, M. P. M., *Effects of the Troposphere on Radio Communication*, IEE Electromagnetic Waves Series 8 (Peter Peregrinus, 1979).
6. Tartarski, V. I., *Wave Propagation in a Turbulent Medium* (Dover, 1959).
7. Lane, J. A., 'Scintillation and Absorption Fading on Line-of-Sight Links at 35 and 100 GHz', Conference on Tropospheric Radiowave Propagation, IEE, 1968, pp. 166–174.
8. Tartarski, V. I., 'The Effects of the Turbulent Atmosphere on Wave Propagation', NTIS, U.S. Department of Commerce, Springfield, U.S.A., 1971.
9. Clifford, S. F. and Strohbehn, J. W., 'The Theory of Microwave Line-of-Sight Propagation through a Turbulent Atmosphere', *IEEE Transactions on Antennas and Propagation*, 1970, **18**, pp. 264–274.
10. *Propagation in Non-Ionized Media*, Recommendations and Reports of the CCIR, **V**, XVI Plenary Assembly, 1986.
11. Hanle, E., 'The Complex Impedance of the Earth's Surface at Radio Frequencies and its Measurement', *NTZ-Communications Journal*, 1966, **3**, pp. 136–143.
12. Recommendations and Reports of the CCIR, 'Propagation in Non-Ionized Media' **V**, XV Plenary Assembly, 1982.
13. Dougherty, H. T. and Wilkerson, R. E., 'Determination of Antenna Height for Protection against Microwave Diffraction Fading', *Radio Science*, 1967, **2** (New series), pp. 161–165.
14. Vigants, A., 'Microwave Obstruction Fading', *Bell System Technical Journal*, 1981, **60**, No. 6, pp. 785–801.
15. Bullington, K., 'Radio Propagation Fundamentals', *ibid*, 1957, **36**, pp. 593–628.
16. Morita, K., 'Prediction of Equivalent Rayleigh Fading Occurrence Rate on Line-of Sight

Reflected Wave Path', *Review Electronic and Communication Laboratories*, 1972, **20**, No. 7–8, pp. 589–598.
17. Majoli, L. F., 'Propagation in Line-of-Sight Links at 1–10 GHz', Telettra Internal Report No. 940.000.141, 1981.
18. Fabbri, F., 'Anti-Reflecting System for 2 GHz Over-Sea Radio Links', *Alta Frequenza*, August 1973, pp. 393–397.
19. Pearson, K. W., 'Method for the Prediction of the Fading Performance of a Multisection Microwave Link', *Proc. IEE*, 1965, **112**, No. 7, pp. 1291–1300.
20. Barnett, W. T., 'Microwave Line-of-Sight Propagation with and without Frequency Diversity', *Bell System Technical Journal*, 1970, pp. 1827–1871.
21. Ruthroff, C. L., 'Multiple-Path Fading on Line-of-Sight Microwave Radio Systems as a function of Path Length and Frequency', *ibid.*, 1971, pp. 2375–2399.
22. Sasaki, O. and Akiama, T., 'Multipath Delay Characteristics on Line-of-Sight Microwave Radio Systems', *IEEE Transactions on Communications*, 1979, pp. 1876–1886.
23. Serizawa, Y. and Takeshita, S., 'A Simplified Method for Prediction of Multipath Fading Outage of Digital Radio', *ibid.*, 1983, **COM–31**, No. 8, pp. 1017–1021.
24. Majoli, L. F., 'Improving Knowledge of Multipath Fading', National Telecommunications Conference, New Orleans, 1981 (or *Telettra Review* 32).
25. Barnett, W. T., 'Multipath Fading Effects on Digital Radio', *IEEE Transactions on Communications*, 1979, **COM–27**, No. 12, pp. 1842–1848.
26. Giger, A. J. and Barnett, W. T., 'Effects of Multipath Propagation on Digital Radio', *ibid.*, 1981, **COM–29**, No. 9, pp. 1345–1352.
27. Albersheim, W. J. and Schafer, J. P., 'Echo Distortion in the FM Transmission of Frequency-Division Multiplex', *Proc. IRE*, 1952, **40**, p. 316.
28. Turner, D., Easterbrook, B. J. and Golding, J. E., 'Experimental Investigation into Radio Propagation at 11.0–11.5 GHz', *Proc. IEE*, 1966, **113**, No. 9, pp. 1477–1489.
29. Cambell, J. C. and Coutts, R. P., 'Outage Prediction of Digital Radio Systems', *Electronics Letters*, 1982, **18**, No. 25–26, pp. 1071–1072.
30. Rummler, W. D., 'A Multipath Channel Model for Line-of-Sight Digital Radio Systems', International Communications Conference, 1978, **3**, paper 47.5.
31. Rummler, W. D., 'A New Selective Fading model: Application to Fading Data', *Bell System Technical Journal*, 1979, **58**, pp 1037–1071.
32. Sakagami, S. and Hosoya, Y., 'Some Experimental Results on In-band Amplitude Dispersion and a Method for Estimating In-Band Linear Amplitude Dispersion', *IEEE Transactions on Communications*, 1982, **COM–30**, No. 8, 1875–1888.
33. Ramadan, M., 'Availability Prediction of 8 PSK Digital Microwave System during Multipath Propagation', *IEEE Transactions on Communications*, 1979, **COM–27**, No. 12, pp. 1862–1869.
34. Smith, D. R. and Cormack, J. J., 'Measurement and Characterization of a Multipath Fading Channel', *IEEE Conference on Communications* (ICC '82), 1982, Paper 7B.4.
35. Liniger, M., 'One Year Results of Sweep Measurements of a Radio Link', *IEEE International Conference on Communications* (ICC '83), 1983, Paper C.2.3.
36. Greenstein, L. J. and Czekaj, B. A., 'A Polynomial Model for Multipath Fading Channel Responses', *Bell System Technical Journal*, 1980, **49**, No. 7, pp. 1197–1220.
37. Kaylor, R. L., 'A Statistical Study of Selective Fading of Super-High Frequency Radio Signals', *ibid.*, 1953, **32**, p. 1187.
38. Rummler, W. D., 'A Statistical Model of Multipath Fading on a Space Diversity Radio Channel', *IEEE International Conference on Communications* (ICC '82), 1982, **2**, Paper 3B.4.1–6.
39. Rummler, W. D., 'A Comparison of Calculated and Observed Performance of Digital Radio in the presence of Interference', *IEEE Transactions on Communications*, 1982, **COM–30**, No. 7, pp. 1693–1700.
40. Rummler, W. D., 'A Rationalized Model for Space and Frequency Diversity Line-of-

Sight Radio Channels', *IEEE International Conference on Communications* (ICC '83), 1983, paper E2.7.

41. Sylvain, M. and Lavergnat, J., 'Modelling the Transfer Function in Medium Bandwidth Radiochannel During Multipath Propagation', *Annales de Télécommunications (à paraître)*, 1985.
42. Bianconi, G. and Calandrino, L., 'Adaptive Baseband Equalizers in Digital Radio Links, Special Supplement to *Telettra Review*, 1985.
43. Cambell, J. C., 'Outage Prediction for the Route Design of Digital Radio Systems', *Australian Telecommunications Research*, 1984, **18**, No. 2, pp. 37–49.
44. Cambell, J. C., Martin, A. L. and Coutts, R. P., '140 Mbit/s Digital Radio Field Experiment – Further Results', *IEEE International Conference on Communications* (ICC '84), Amsterdam, 1984.
45. Martin, A., 'Amplitude Distortion: a Means of Characterizing Line-of-Sight Microwave Radio Paths during Multipath Fading', *Electronics Letters*, 1984, **20**, No. 19, pp. 798–799.
46. Majoli, L. F. and Mengali, U., 'Propagation in Line-of-Sight Radio Links – Part II', Special Supplement to *Telettra Review*, 1985.
47. Richman, G. D., 'Automated Signature Measurement of Operational Microwave Radio-Relay Equipment using a Novel Multipath Simulator', Proceedings of IEE International Conference on Measurements for Telecommunication Transmission Systems – MTTS 85, Conference Publication No. 256, 1985, pp. 108–111.
48. Lundgren, C. W. and Rummler, W. D., 'Digital Radio Outage Due to Selective Fading – Observation vs Prediction from Laboratory Simulation', *Bell System Technical Journal*, 1979, **58**, No. 5, pp. 1073–1100.
49. Doble, J. E., 'A Flexible Digital Radio System Outage Model based on Measured Propagation Data', The European Conference on Radio Relay Systems, GEC Hirst Research Centre, October, 1986.
50. Dupuis, P., Joindot, M., Leclert, A. and Rooryk, M., 'Fade Margin of High Capacity Digital Radio System', International Conference on Communications' Boston, 1979, Paper 48.6.
51. Jakes, W. C., 'An Approximate Method to Estimate an Upper Bound on the Effect of Multipath Delay Distortion on Digital Transmission', *IEEE Transactions on Communications*, 1979, **COM–27**, No. 1, pp. 76–81.
52. Hart, G., Smith, P. C., McConnell, R. D. and Rolls, A. M., 'Practical Results from the Performance Evaluation of an 11 GHz 140 Mbit/s Digital Radio Relay System', *Radio and Electronic Engineer*, 1983, **33**, No. 5, pp. 181–189.
53. Muir, A. W. and de Belin, M. J., 'The Field Evaluation of a 140 Mbit/s Digital Radio Relay System in the 11 GHz Band', Proceedings of the IEE Conference on Telecommunication Transmission – Into the Digital Era, Conference Publication No. 193, 1981, pp. 106–109.
54. Allen, E. W. and Crossett, J. A., 'Digital Radio Propagation Experiments at 6 GHz – Part II', *Links for the Future*, IEEE International Conference on Communications, 1984, pp. 1455–1459.
55. 'Fixed Service using Radio-Relay Systems', Recommendations and Reports of the CCIR, **IX–1**, XV Plenary Assembly, 1982.
56. Rummler, W. D., 'A Statisical Model of Multipath Fading on a Space Diversity Channel', *Bell System Technical Journal*, 1982, **61**, pp. 2185–2219.
57. Rooryck, M., 'Modelisation de la Propagation avec Trajets Multiples, Applications d'un Modèle à deux Rayons', *Onde Electrique*, August–September 1983, **6**.
58. Barber, S., 'Cofrequency Cross-Polarized Operation of a 91 Mbit/s Digital Radio', *IEEE Transactions on Communications*, 1984, **COM–32**, No. 1, pp. 87–91.
59. Olsen, R. L., 'Cross Polarization during Clear-Air Conditions on Terrestrial Links: A Review', *Radio Science*, 1981, **16**, No. 5, pp. 631–647.
60. Makino, H. and Morita, K., 'Design of Space-Diversity Receiving and Transmitting

Systems for Line-of Sight Microwave Links', *IEEE Transactions on Communications Technology*, 1967, **COM–15**, No. 4, pp. 603–614.
61. Vigants, A., 'Space Diversity Engineering', *Bell System Technical Journal*, 1975, **54**, No. 1, pp. 103–142.
62. Rummler, W. D., 'Modelling the Diversity Performance of Digital Radios with Maximum Power Combiners', *Links for the Future*, IEEE International Conference on Communications, 1984, pp. 657–660.
63. Murase, T., Morita, K. and Komaki, S., '200 Mbit/s 16–QAM Digital Radio System with New Countermeasure Techniques for Multipath Fading', IEEE International Conference on Communications (ICC '81), 1981, **2**, Session 46, pp. 46.1.1–46.1.5.
64. Anderson, C. W., Barber, S. and Patel, R. N., 'The Effect of Selective Fading on Digital Radio', *IEEE Transactions on Communications*, 1979, **COM–27**, No. 12, pp. 1870–1875.
65. Wang, Y. Y., 'Simulation and Measured Performance of a Space Diversity Combiner for 6 GHz Digital Radio', *ibid.*, pp. 1896–1907.
66. Cambell, J. C., 'Digital Radio Outage Prediction with Space Diversity', *Electronics Letters*, 1983, **19**, No. 23, pp. 1003–1004.
67. Dirner, P. L. and Lin, S. H., 'Measured Frequency Diversity Improvement for Digital Radio', *IEEE Transactions on Communications*, 1985, **COM–33**, No. 1, pp. 106–109.
68. Greenfield, P. E., 'Digital Radio Performance on a Long, Highly Dispersive Fading Path', *Links for the Future*, IEEE International Conference on Communications, 1984, pp. 1451–1454.
69. Toy, W. W., 'Effects of Multipath Fading on 16 QAM Digital Radio', IEEE International Conference on Communications (ICC '80), 1980, Paper 52.1.
70. Giuffrida, T. S., 'Measurements of the Effects of Propagation on Digital Radio Systems equipped with Space Diversity and Adaptive Equalization', IEEE International Conference on Communications (ICC '79), 1979, pp. 48.1.1–6.
71. Damosso, E., de Padova, S., Failli, R. and Lingua, B., 'Experimental Results on the Effects of Selective Fading on 70 Mbit/s 8 PSK Digital Radio', IEEE Globecom, 1982, Paper D3.1.
72. Macchi, R., Moreno, L. and Vicini, P., 'Field Evaluation of 70 Mbit/s 7/11 GHz Digital Radio Link', *Alta Frequenza*, 1983, **L11**, No. 6, p. 460.
73. Damosso, E. and Failli, R., 'Experimental Result on Space and Frequency Diversity Improvements in Digital Radio', IEE 3rd International Conference on Telecommunication Transmission, London, 1985.
74. Komaki, S., Horikawa, L., Morita, K. and Okamoto, Y., 'Characteristics of a High Capacity 16 QAM Digital Radio System in Multipath Fading', *IEEE Transactions on Communications*, 1979, **COM–27**, No. 12, pp. 1854–1861.
75. Saito, Y., Morita, K. and Yamamoto, H., '5L–D1 Digital Radio System', IEEE International Conference on Communications (ICC '82), 1982, Paper 2B.1.
76. Martin, A. L., Coutts, R. P. and Campbell, J. C., 'Results of a 16 QAM 140 Mbit/s Digital Radio Field Experiment', IEEE International Conference on Communications (ICC '83), 1983, 1459–1466.
77. Richman, G. D., 'The Behaviour of Spectrum-Conserving RBQPSK Digital Radio during Multipath Propagation', IEEE Globecom, 1982, Paper D3.5.
78. Moreno, L. and Vicini, P., 'Field Trial Results of 140 Mbit/s 16 QAM Digital Radio under Multipath Propagation', IEEE Globecom, 1983.
79. Abel, N., 'Statistics of Multipath Fading and Rain Attenuation on Terrestrial Radio Links operating in the 7 to 15 GHz Range', Proceedings of the IEE International Conference on Antennas and Propagation, Conference Publication No. 195, Part 2, 1978, pp. 64–67.
80. Brussard, G. and Watson, P. A., 'Annual and Annual-Worst-Month Statistics of Fading on Earth-Satellite Paths at 11.5 GHz', *Electronics Letters*, 1979, **14**, No. 9, pp. 278–280.
81. Ryde, J. W. and Ryde, D., 'Echo Intensity and Attenuation due to Clouds, Rain, Hail, Sand and Dust Storms at Centimeter Wavelengths', General Electric Company (G.E.C.) Research Laboratories, Wembley, Report 7831, October 1941.

82. Ryde, J.W. and Ryde, D., 'Attenuation of Centimeter Waves by Rain, Hail and Clouds', G.E.C. Research Laboratories, Wembley, England, Report 8516, August 1944.
83. Ryde, J. W. and Ryde, D., 'Attenuation of Centimeter and Millimeter Waves by Rain, Hail, Fogs and Clouds', G.E.C. Research Laboratories, Wembley, England, Report 8670, May 1945.
84. Ruthroff, C. I., 'Rain Attenuation and Radio Path Design', *Bell System Technical Journal*, 1970. **49**, No. 1, pp. 121–135.
85. Lin, S. H., 'A Method for Calculating Rain Attenuation Distributions on Microwave Paths', *Bell System Technical Journal*, 1975, **54**, No. 6, pp. 1051–1086.
86. Mogensen, G. and Stephansen, E., 'An Estimation of Methods for Prediction of Rain Induced Attenuation on LOS Paths', *IEE International Conference on Antennas and Propagation*, Conference Publication No. 195, Part 2, 1978, pp. 97–101.
87. Olsen, R. L., Rogers, D. V. and Hodge, D. B., 'The aR Relation in the Calculation of Rain Attenuation', *IEEE Transactions on Antennas and Propagation*, 1978, **AP–26**, No. 2, pp. 318–329.
88. Nowland, W. L., Olsen, R. L. and Shkarofsky, I. P., 'Theoretical Relationship between Rain Depolarization and Attenuation', *Electronics Letters*, 1977, **13**, pp. 676–678.
89. Fedi, F., 'A Contribution of Radio Propagation Research to Radiocommunications Development: Prediction of Attenuation due to Rain', *Links for the Future*, IEEE International Conference on Communications, 1984, pp. 163–166.
90. Garcia-Lopez, J. A. and Casares-Giner, V., 'Modified Lin's Empirical Formula for Calculating Rain Attenuation on a Terrestrial Path', *Electronics Letters*, 1981, **17**, pp. 34–36.
91. Garcia-Lopez, J. A. and Piero, J., 'Simple Rain-Attenuation-Prediction Technique for Terrestrial Radio Links', *Electronics Letters*, 1983, **19**, pp. 34–37.
92. Anderson, I., 'Measurements of 20 GHz Transmission through a Wet Radome', International IEEE/G–AP Symposium and URSI meeting, Boulder, 1973, pp. 239–240.
93. Olsen, R. L. and Nowland, W. L., 'Semi-Empirical Relations for the Prediction of Rain Depolarization Statistics: their Theoretical and Experimental Basis', Proceedings of the International Symposium on Antennas and Propagation, Sendai, Japan, 1978.
94. Kerker, M., *The Scattering of Light and other Electromagnetic Radiation* (Academic Press, 1969).
95. Setzer, D., 'Computed Transmission through Rain at Microwave and Visible Frequencies', *Bell System Technical Journal*, 1970, **49**, No. 8, pp. 1873–1892.
96. Liebe, H. J., 'Modeling Attenuation and Phase of Radio Waves in Air at Frequencies below 1000 GHz', *Radio Sciences*, 1981, **16**, No. 6, pp. 1183–1199.
97. 'Fixed Service using Radio-Relay Systems', Recommendations and Reports of the CCIR, **IX-1**, XVI Plenary Assembly, 1986.

9 LINE-OF-SIGHT RADIO LINK ENGINEERING

9.1 TYPES OF CALCULATION

9.1.1 Path calculations

The object of path calculations is to determine all the losses and gains in a system and from these determine the fade margins, probabilities of exceeding a fade margin, probability of outage, diversity requirements and the types and heights of the antennas, etc. The path calculations are interactive and may go through many redesign phases before the final calculations are produced. This is done so that the fade margin is not exceeded for a length of time which is contrary to the relevant performance and availability objectives. The flat fade margin is the difference between the unfaded received signal level and the lowest value of received signal level for a bit error rate of no greater than 10^{-6} and 10^{-3}, measured at the receiver baseband output on a 64 kbit/s channel over periods of a minute and a second, respectively. The existence of two BERs means that there will be two flat fade margins to be taken into account in the calculations. See Section 9.2 for further discussion of path calculations.

9.1.2 Performance calculations

The object of performance calculations is to determine the probability of exceeding the BER objectives, using values of probabilities determined during the path calculations. The BER objectives used are such that the BER should not exceed the following values for a high-grade circuit:

1×10^{-6} during more than $0.4d/2500$ percent of any month for an integration time of one minute, for $280 < d < 2500$ km
1×10^{-6} during more than 0.045 percent of any month for an integration time of one minute, for $d < 280$ km
1×10^{-3} during more than $0.054d/2500$ percent of any month for an integration time of one second, for $280 < d < 2500$ km

1×10^{-3} during more than 0.006 percent of any month for an integration time of one second, for $d < 280$ km

9.1.3 Interference calculations

These calculations are performed in order to verify that the carrier-to-interference ratio at the receiver threshold and the consequent threshold degradation complies with the values used in the path calculations. These calculations are the subject of Chapter 10.

9.2 PATH CALCULATIONS

The path calculations are quite extensive and cover much of the work already dealt with in the previous chapters. To work through these calculations properly, much initial work must be done. This work entails obtaining a pair of frequencies within a given CCIR band, determining the route to be taken, creating from ordnance survey maps the necessary path profiles and performing site surveys. Once this is done, some idea of what the system will look like is established. The work in all may take months and in no event should be treated lightly. An error during this time can cost a considerable amount of resources at a later date. Assuming that a pair or more of frequencies are available and that the two radio terminal points are known, the initial problem is to plan how to get the radio system placed between these two terminals.

9.2.1 Path profiles

On an ordnance survey map, of scale 1 : 10000 or greater the two terminals are marked. If the terminals are on different maps, which they usually are, the points are marked on the respective maps and their latitudes and longitudes are noted. A path profile is what would be created if an enormous knife sliced away the earth along the direction of the radio beam. The first stage of its creation is to draw on the aligned ordnance survey maps a thin pencil line between the two proposed sites. This may mean that the pencil line has to extend over a number of intermediate maps. To assist in overcoming the inherent inaccuracies in such a method, and in reducing the amount of space required, the following method may be used to determine the latitude and longitude of the points where the beam passes over the edges of each map. The method, then, permits the use of one map at a time and allows the line to be drawn across the intermediate maps without any need to align them physically with any other maps.

9.2.1.1 Map crossings

Figure 9.1 shows a diagram of several maps which may typically have to be put alongside each other so that the path between site A and B can be connected by a

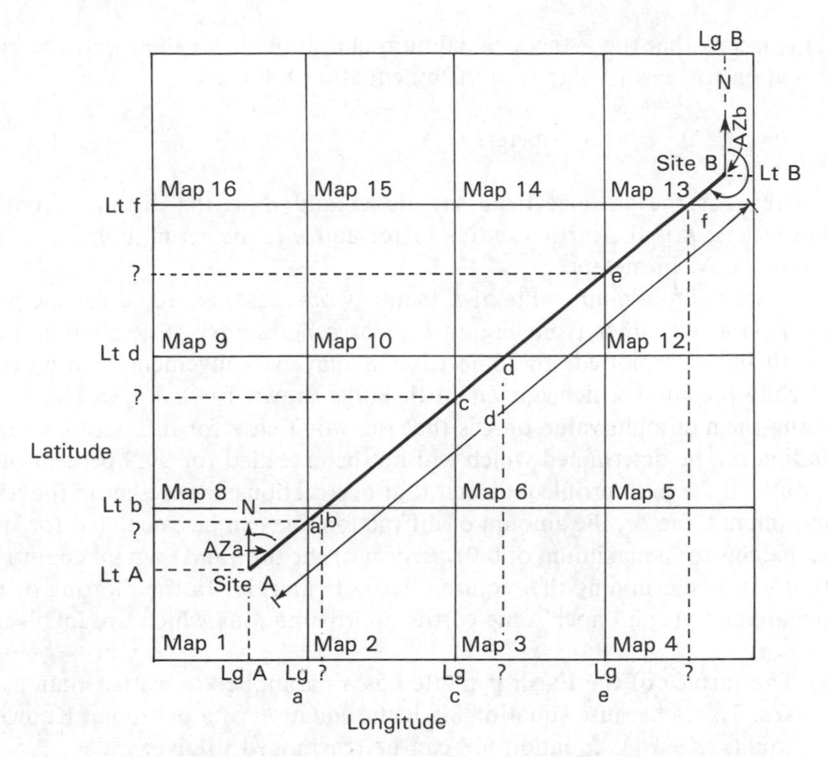

Figure 9.1 Path crossings on various maps for a line-of-sight radio link

straight line. The path of length d is shown to cross each of the maps at points a, b, c, d, e, and f. At the crossings a, c and e, the longitudes are known but not the latitudes. Similarly, at points b, d and f the latitudes are known, but not the longitudes. The Hewlett-Packard HP41 calculator program listing provided in Appendix 3 will provide the unknown latitudes or longitudes of the various map crossings given known longitudes and latitudes. In addition, the program will provide the path length in kilometers and the azimuth at site A and site B. What is required to operate the program are both the latitude and the longitude of both terminal sites. The program is provided for the HP 41 scientific calculator in preference to being listed in BASIC, because it is easier to work with a calculator for this problem than it is with a personal computer.

9.2.1.2 Path profiling approximations

Once the actual path has been drawn on the maps after the various crossings have been computed, starting from say, site A, the height contours are read along the pencil line up to the B site, and a plot of height (ordinate), usually in meters, against distance in kilometers, is progressively drawn up. This represents the basic path profile. On this profile, the heights of the trees are included, as well as any other obstructions which are known to exist, but which are not included in the ordnance survey maps. Once this has been done, the effects of Earth bulge have to be included.

This means that the k-factor maximum and minimum values must be known. The equation for earth bulge is given by equation 8.8 as;

$$h = (4/51) \cdot d_1 \cdot d_2/k \text{ meters} \tag{8.8}$$

where d_1 is the distance from say site A and d_2 is the distance from site B, in kilometers, k is the earth's radius factor and h is the actual bulge at a distance d_1 from site A, in meters.

With the minimum value of k-factor (worst case) expected for the path length, as determined either from Figure 8.11 or other sources (see Section 8.2.2.3), the Earth bulge is plotted. In some cases it may be convenient to use profile paper already prepared which has an earth bulge drawn for $k = 4/3$. The advantage of using the minimum value of k is that the worst case for diffraction or obstruction fading can be determined which will not be exceeded for 99.9 percent of the worst month. If the path profile indicates that obstructions are present in the path for this minimum k-factor, the amount of diffraction loss can be calculated for given antenna heights for a maximum of 0.01 percent of the time and then taken into consideration when determining the required flat fade margin. In the plotting of profiles on rectangular graph paper some of the approximations which are involved are:

1. The surface of the Earth is plotted as a parabolic arc, rather than as a circular arc. This is because equation 8.8 is the equation of a parabola. Knowing that d_2 equals $(d - d_1)$, equation 8.8 can be rearranged to give:

$$h - d^2/51k = -4(d_1 - d/2)^2/51k \tag{8.8}$$

   The focus of the parabola is $1/51k$ and there has been a translation along the x-axis by $d/2$ and along the y-axis by $d^2/51k$. The error involved in this approximation is, however, less then 0.3 percent.

2. The elevations are plotted along vertical lines, and not along the radial line from the Earth's center. In Figure 9.2 the angle $\phi/2$ is appoximately given by:

$$\phi/2 = (d/2) \cdot (1/a_e) = (d/2)/(ka) = 0.5397(d/2) \cdot (1/k) \text{ minutes} \tag{9.1}$$

   For $k = 4/3$ and a 30 km path, the angle $\phi/2$ is found to be 6 minutes. An error greater than this amount would possibly occur during normal drawing.

3. The true length of the path d is not the distance measured along the surface of the earth. The length of path used in calculations is usually measured from a map, or calculated from map co-ordinates, and not corrected for the difference in elevation $(h_2 - h_1)$. The error is small, because the angle θ is seldom larger than 2°. As a result, it is assumed that $D_2 = D_1 = d$.

The next step to be included on the path profile is that of the first Fresnel zone. Assuming that the antenna heights at this stage have not been settled upon, a fair

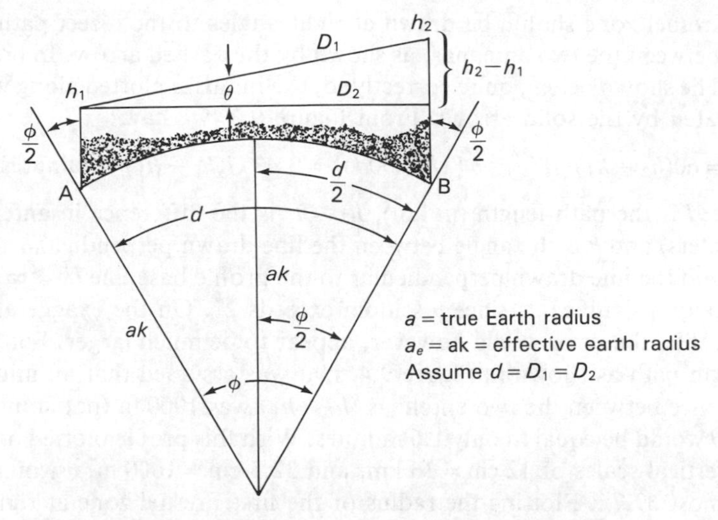

Figure 9.2 Errors in plotting path profiles on linear graph paper

idea of how the path obstructions will affect the first Fresnel zone clearance along the path can be determined by constructing the first Fresnel zone on a piece of tracing paper. This can then be moved up and down to simulate the different antenna heights to determine under the worst case conditions ($k = k_{\min}$) of clearance factors. The clearance factors have been discussed in Section 8.2.2.3, and should be used at this point. The equation to be used to plot the first Fresnel zone is given by:

$$F_1 = 17.32 [d_1 \cdot d_2 / (f \cdot d)]^{1/2} \text{ m} \tag{8.24}$$

where d_1 and d_2 are the distances from each of the terminals (in km), $d = d_1 + d_2$, f is the carrier frequency (in GHz) and F_1 is the first Fresnel zone radius (in meters).

4. Figure 9.3 showns another common error which also arises from the distortion in scales. For a slanting direct path, it could be thought that the radius of the

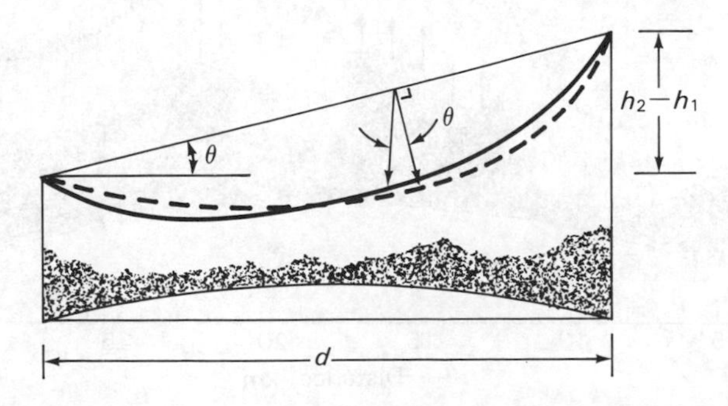

Figure 9.3 Apparent error caused by difference in antenna heights

first Fresnel zone should be drawn at right angles to the direct path or line-of-sight between the two antennas, as shown by the dashed arrow. In practice, and, as will be shown below, quite correctly so, the radius is plotted along vertical lines (indicated by the solid arrow). From Figure 9.3, we have:

$$\theta = 60(h_2 - h_1)(180/\pi) \cdot [1/(1000d)] = 3.4377(h_2 - h_1)/d \text{ minutes} \quad (9.2)$$

where d is the path length (in km), $h_2 - h_1$ is the difference in antenna heights (in meters) and θ is the angle between the line drawn perpendicular to the direct path and the line drawn perpendicular to the profile base line ($k = \infty$). The error is usually insignificant, since θ seldom exceeds 2°. On the exaggerated scale of the profile, the error would however, appear to be much larger. For example on a 36 km path as shown in Figure 9.4, if it was assumed that an unusual height difference between the two antennas ($h_2 - h_1$) was 1000 m (not shown in Figure 9.4), θ would be equal to only 1.6 minutes. With this profile plotted on horizontal and vertical scales of 12 cm = 36 km, and 32.5 cm = 1000 m, θ would appear to be almost 69.7°. Plotting the radius of the first Fresnel zone at right angles to the line of sight would not be of any value whatsoever and would cause large errors to occur. Only when $h_1 = h_2$ would no errors occur by plotting the radius at right angles to the direct beam path.

5. Other errors which may arise when drawing profiles from maps are: the straight line drawn on a map does not represent the true radio path, which follows a great

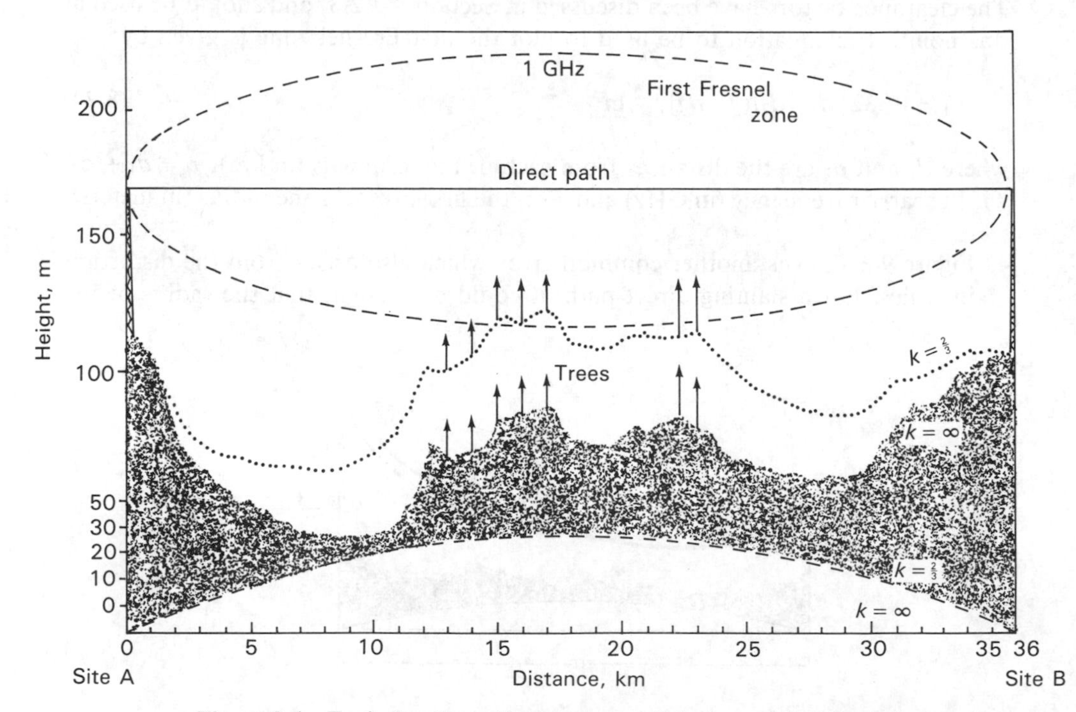

Figure 9.4 Typical path profile showing earth bulge and first Fresnel zone

circle, which is the shortest distance on a sphere which connects two points on that sphere. The plane of the great circle always passes through the center of the sphere.

6. It is very difficult to locate a site on a map exactly. Similarly, it is difficult to determine if a location is in the direct line of the radio beam path.

7. Due to the ever changing environmental conditions, such as tree growth, erection of buildings, etc., maps become obsolete in short periods of time.

8. Paths crossing over the slope of a hill make it extremely difficult to estimate the heights of obstructions or any buildings etc. It is of more benefit if such constructions are emphasized in some way when shown on the profile.

9. The value of the k-factor is assumed to be the same along the entire path. This is not usually the case. Variations do occur, especially if the elevations of the sites differ greatly, or the nature of the terrain changes rapidly along the path. In practice, it has been found that the assumption of constant k-factor does work well.

9.2.1.3 Determining the antenna heights from calculations

Consider the diagram of Figure 9.5. The maximum height B_i, that the radio beam must be at any point d_i, along the path, is given by, the sum of the Earth bulge E_i, the obstruction height O_i, the tree or vegetation height T_i and the first Fresnel zone radius F_1 times its associated clearance factor C. That is:

$$B_i = E_i(k) + (O_i + T_i) + C \cdot F_1$$
$$= (4/51) \cdot (d - d_i) \cdot d_i/k + (O_i + T_i) + 17.32[d_i(d - d_i)/d \cdot f)]^{1/2} \quad (9.3)$$

where d is the path length (in km), d_i is the distance from site A (in km), k is the Earth radius factor, C is the clearance factor, and f is the the frequency (in GHz).

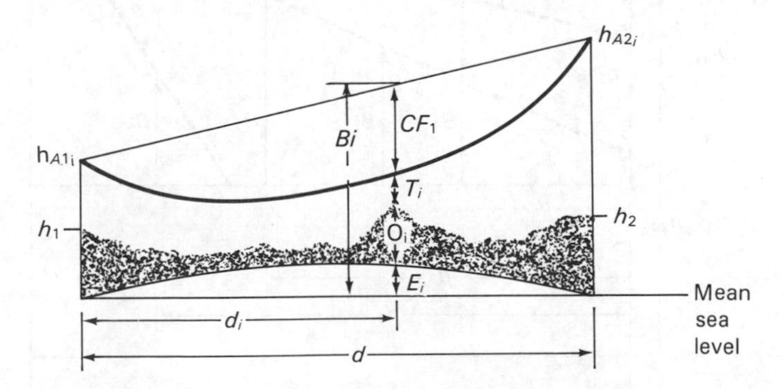

Figure 9.5 Determination of minimum beam height B_i to clear an obstruction

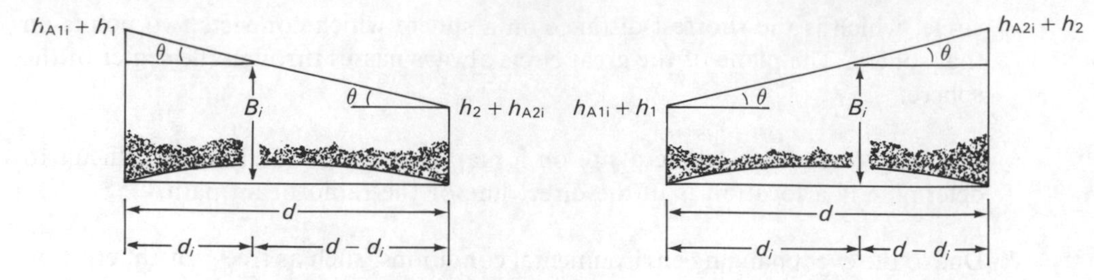

Figure 9.6 Diagrams used to calculate the variable antenna height h_{A1_i} and h_{A2_i} if the far-end antenna is fixed

Figure 9.6 shows two diagrams from which a geometrical relationship between an antenna height and B_i can be determined using simple geometry. That is

$$h_{A2_i} = h_1 + h_{A1} + [B_i - (h_1 + h_{A1})] \cdot (d/d_i) - h_2 \tag{9.4}$$

and

$$h_{A1_i} = h_2 + h_{A2} + [B_i - (h_2 + h_{A2})] \cdot [d/d - d_i)] - h_1 \tag{9.5}$$

where B_i is the beam height as given by equation 9.3, h_1 and h_2 are the site heights above mean sea level (in meters) and h_{A1} and h_{A2} are the fixed antenna heights at each of the two respective sites (in meters). These equations give the required height h_{A2_i} or h_{A1_i} (in meters), of the antenna at each site in order to clear an obstruction, if the other antenna remains fixed.

Figure 9.7 shows the typical effects of how the beam must travel for a fixed

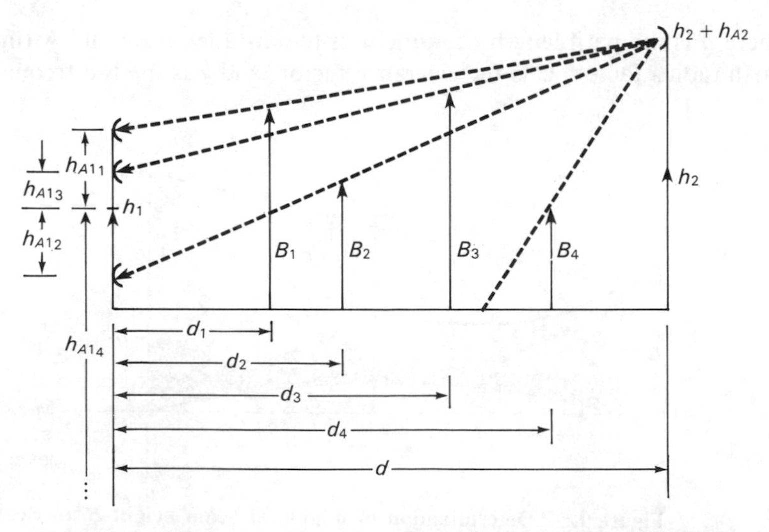

Figure 9.7 Effects of obstructions B_1 to B_4 on antenna height A_{A1_i}

antenna height h_{A2}, if the beam is to clear the obstructions B_1 to B_4 which are located at various points along the path (d_1 to d_4).

For h_{A1_2} and h_{A1_3}, the antenna is required to be placed above the site ground level h_1, if the two obstructions represented by B_1 and B_3 are to be cleared. For the case of the obstructions represented by B_2 and B_4, the antenna may be located at ground level, even though the values of h_{A1_2} and h_{A1_4} indicate that a ground height of h_1 is not necessarily required. In other words, the antenna will permit the correct clearance of the obstacles represented by B_2 and B_4, by being placed further down a hill or at some other suitably placed lower elevation. It is usual to enter equations 9.4 and 9.5 into a computer program, in which all obstruction heights and locations along the path are entered. The program can be devised so that for different clearance factors C, and for different k-factors, the two equations are juggled to produce a minimum antenna height at both terminals which clears all obstructions along the path. The same result can be obtained manually, by fixing an antenna height at one end and selecting from all of the calculations made for each obstruction, the maximum value of antenna height necessary at the other end to clear the obstructions. This set of calculations is then repeated for a different k-factor and for a different value of the fixed antenna height. Once this has been done, the computation is then repeated by fixing the antenna at the other end and varying the height of the previously fixed antenna. Another factor to be taken into account is that of the reflection point. It may happen that, on calculating the value of minimum antenna heights, a ground reflection zone becomes particularly important. If lakes, marshlands, etc., are indicated on the path profile, then, to insure that an undesirable reflection zone is avoided during the process of calculating the antenna heights, the reflection point calculation is usually completed at the same time. The calculation of the reflection point has been discussed in Section 8.2.2.4, 'Calculation of reflection point'.

9.2.1.4 Antenna foreground clearance

For the regions near the antenna at either end of the link, a special first Fresnel zone clearance requirement is necessary. This is due to the near field of the antenna having different characteristics from the far field. The extent of the near field is shown in Figure 9.8. Generally, there is a circular corridor clearance requirement of 10 m radius for the near field requirement and an additional clearance of 10 m to clear properly the antenna near field. The total region extends up to L_1 m in the horizontal plane and L_2 m in the vertical plane from the antenna. The distance L_1 can be determined from

$$L_1 = (d/2) - [(d/4)(d - 4f/3)]^{1/2} \text{ m} \tag{9.6}$$

where d is the path length (in km) and f is the carrier frequency (in GHz).

If L_1 turns out to be negative or imaginary, the value taken should be 10 m. The distance L_2 can be approximately given by the solution of the following equation:

$$[300 \cdot L_2(d - L_2)/fd]^{1/2} + (h_1 - h_2) \cdot L_2/d + L_2(d - L_2)/(12.75k) - 10 = 0 \tag{9.7}$$

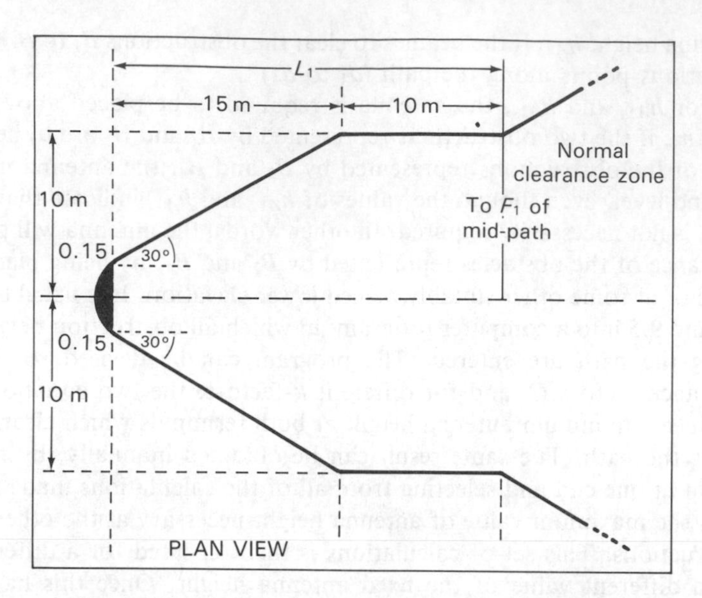

Figure 9.8 Near field antenna clearance requirements

where L_2 is in meters; h_1 and h_2 are the antenna heights above mean sea level (in meters) and k is the k-factor. L_2 will be of the order of $2L_1/3$.

Separate to the above, because of the absence of Fresnel zone considerations, the near field distance, extended in the direction of propagation, for a particular antenna can be calculated from:

$$\text{Near field distance} = D^2/2\lambda = 1.5\,D^2 f \tag{9.8}$$

where D is the antenna dish diameter (in meters) and f is the carrier frequency (in GHz). Clearance should be twice the value given by equation 9.8, to insure that any obstructions are well and truly in the far field. The extent of the region in the vertical plane of the antenna which lies in the near field is given by $\sqrt{2} \cdot D$.

PATH PROFILE DATA

*Radio link route:*____________________________________

Station A

Name of station ________ Address of station ___________________

Ordnance Survey reference_____________ Easting________Northing______

Latitude_____(degrees, minutes, seconds) Longitude_____(degrees, minutes, seconds)

Ground height ______________________ meters above mean sea level (masl)

Antenna height______________________ meters above ground level (magl)

Mast/tower/building height _______________ meters above ground level (magl)

Station B

Name of station ______ Address of station ______________

Ordnance Survey reference ____________ Easting ______ Northing _____

Latitude _____ (degrees, minutes, seconds) Longitude ____ (degrees, minutes, seconds)

Ground height ____________________ meters above mean sea level (masl)

Antenna height ____________________ meters above ground level (magl)

Mast/tower/building height ______________ meters above ground level (magl)

OBSTRUCTION DATA

Notes

Distance from site A = d_i km; height above mean sea level = h_i m

	d_i, km	h_i, m	Notes		d_i, km	h_i, m	Notes
1	_____	_____	_____	10	_____	_____	_____
2	_____	_____	_____	11	_____	_____	_____
3	_____	_____	_____	12	_____	_____	_____
4	_____	_____	_____	13	_____	_____	_____
5	_____	_____	_____	14	_____	_____	_____
6	_____	_____	_____	15	_____	_____	_____
7	_____	_____	_____	16	_____	_____	_____
8	_____	_____	_____	17	_____	_____	_____
9	_____	_____	_____	18	_____	_____	_____

Originated by ____________________ *Date* ________

File ________________________ *Issue* ________

9.2.1.5 Diffraction losses

Section 8.2.2.2 discussed diffraction and the different types of loss which may be caused by it. Section 8.2.2.2 should be used, when the values of the parameters required in the calculation of the losses caused by a knife edge, a smooth earth, or a single rounded obstacle, are obtained from the path profile. The diffraction losses produced by each to these types of obstruction may occur in practice if the antenna heights are not raised high enough, causing the obstructions to penetrate further into the first Fresnel zone than desired. Under these conditions, the amount of additional loss over that of free space, caused by the obstruction or obstructions, is required to be known, so that a correct value of flat fade margin can be calculated. These diffraction loss calculations should be completed using the different value of

k-factor, that is k_{min} and the average k-factor, and the clearance factors, as determined from Section 8.2.2.1, so that the diffraction loss which is expected to occur for 0.01 percent of the time and for at least 50 percent of the time can be determined.

9.2.2 Parameters used in path calculations

This section is intended to clarify the successive steps of the individual path calculations and may be used as part of the design of a radio link. Each step will be numbered and where necessary a brief description of the step provided. At the end of the section a typical path calculation sheet will be provided together with a typical completed sheet for an imaginary design. The parameters used in the calculations are planning values to be expected under normal operating conditions. It is assumed that the following main prerequisites have been specified before calculations begin:

Operating frequencies
Site surveys
Locations of the terminal stations
Antenna heights
Equipment type and parameters (bit rate, modulation, etc. See Section 6.7)

The antenna and feeder types and the branching, connector and adaptor losses should also be specified if possible. If that is not possible at this stage, typical expected values from data sheets may be used.

Link description

The link description is the general section and it should include the link terminal stations, the hop number, the equipment type, the frequency band, the capacity (Mbit/s), the polarization, the date and the orginator's name:

1. *Hop number*. This may be a number designated within the system to identify quickly the hop. It is especially useful where a number of different hops are to be designed, or where hop numbers are already in existence.
2. *Equipment type*. A radio system may consist of different types of equipment. Where a spread sheet of all of the links is to be presented, the identification of the equipment is essential.
3. *Station names*. These may be abbreviated or in full. The station which is designated as the 'A' station or site should be marked clearly. Similarly, the station at the other end of the hop should be designated as the 'B' station. For ease in reading the path calculation sheet, the sheet may be divided into two halves. The left-hand side is the 'A' station and the right-hand side is reserved for 'B' station parameters.
4. *Frequencies*. The transmitting frequency only is given for each site. This is because the receiving frequency at the site A will be the transmitting frequency of the site B and vice versa.
5. *Polarization*. The polarization used for the hop should be designated.

6. *Channel capacity*. The capacity in Mbit/s should be included for the link.
7. *Radio equipment modulation type*. This should be included for equipment located at both sites.
8. *Ordnance Survey map reference or National Grid reference (U.K.)*. Included for each site, for easy reference for future work.
9. *Site elevation*. The elevation of the ground above mean sea level (amsl) for each of the sites.
10. *Latitude and longitude*. This is included for each site so that the distance calculation as well as the azimuths can be completed. The azimuths are used to align the antenna at each site, using only a simple magnetic compass (see Appendix 3).
11. *Path length*. This is the distance between site A and site B along the great circle, in km.
12. *Antenna height*. The antenna heights for each site are included. The antenna heights assist in the calculation of the feeder losses. The length of the feeder can for all practical purposes be considered as 1.5 times the antenna height, in the absence of accurate information.
13. *Diversity antenna height*. If space diversity is to be used, the height of the diversity antenna above ground level (agl) is included for both sites.

Losses

This section considers all of the possible losses incurred over the link. It is used primarily to determine the flat fade margin.

14. *Free space path loss A_0*. This is given from Section 8.2.3.3. as:

$$A_0 = 92.5 + 20 \log f(\text{GHz}) + 20 \log d(\text{km}) \tag{8.79}$$

where f is the frequency of the carrier in GHz, as given in step 4. Usually the band center frequency is used in this calculation, and d is the path length in km as given in step 11.
15. *Feeder type*. This is normally given in the specifications. Values of feeder loss are usually expressed in dB/100 m or dB/100 ft. A typical value for Andrew Antenna Company Elliptical Waveguide EW127A, 10.0–13.25 GHz, is given as 11.32 dB/100 m or 3.45 dB/100 ft.
16. *Feeder length*. Unless this is provided with complete details, it can be estimated by taking the antenna height at each site and multiplying it by 1.5. Where the antenna is mounted on the roof of a multi-storey building and the radio equipment is located in the basement or on some floor other than the roof, the estimate will be based on the difference in length between the roof and the equipment floor plus 20 m. This estimate is for calculation purposes only. The waveguide should not be cut according to these rules of thumb.
17. *Feeder loss*. This is calculated from the information given in steps 15 and 16. For example, if the antenna height up a ground based tower is 100 m, then the cable length is 150 m and, using the figure of 11.32 dB/100 m, the feeder loss will be 16.98 dB. This would be taken as 17 dB. The feeder losses at both sites need to be calculated.

18. *Branching loss*. The branching loss accounts for the losses in the RF filters (transmitter and receiver), circulators and possible extra RF filters which may allow more than one radio system to be connected to the same antenna. A typical range of values of branching loss are 2–8 dB.
19. *Adaptor and connector losses*. These are the losses in the waveguide transitions, adaptors, waveguide pressurizing system and connectors comprising part of the feeder system. A typical range of values is 0.5–1.0 dB.
20. *Attenuator or obstruction loss*. As discussed above, the diffraction or obstruction loss can be calculated from information provided in Section 8.2.2.2, if the path profile indicates that such a calculation is warranted. Sometimes the path length is so short that the ouptut power from a commercially available transmitter is higher than that required to satisfy the fade margin requirement. Rather than leave the transmitter at full power and therefore take the chance of causing problems of interference on future links close by, the transmitter power is padded down. The attenuator used to reduce the transmitter output power is included at this stage of assessment of losses.
21. *Atmospheric absorption loss*. This has been discussed in Section 8.5 and the calculated loss included at this point since it is a constant loss (unlike rain attenuation or multipath fading). Usually the specific attenuation (dB/km) is indicated in the space provided and the loss calculated for the complete link. This loss is not dependent on the site and appears in the center of the path calculation sheet, like the free-space path loss.
22. *Sum of the losses*. This is calculated as the sum of steps 14, 17, 18, 19, 20 and 21.

Gains

These are the power gains found in the system due to the antennas and the transmitter power.

23. *Antenna gain*. This is the sum of the gains of the antennas at each end of the link as well as any passive repeaters or periscope systems. The gain of the antenna is usually expressed as the gain over an isotropic radiator as discussed in Section 8.2.3.2. If the gain of the antenna is expressed over that of a half-wave dipole, then to convert to gain over isotropic, multiply the gain by 1.64 or add 2.15 dB.
24. *Transmitter power* (P_t). As the name implies, this is the power at the output of the transmitter itself and not after any branching circuits or filters. It is usually given in dBm, but sometimes in dBW.
25. *Sum of the gains*. This is calculated as the sum of steps 23 and 24.

Net effects

26. *Total loss* A_l. This is the ratio of the power supplied by the output of the transmitter, before the branching circuits, and the power supplied to the corresponding receiver after the branching circuits, in real installation, propagation and operational conditions. It is the difference in decibels between steps 22 and 23, and

can be expressed as:

$$A_l = P_t - [P_t - (\text{sum of the losses}) + (\text{sum of antenna gains})] \text{ dB}$$
$$= (\text{sum of the losses}) - (\text{sum of the antenna gains}) \text{ dB} \quad (9.9)$$

This means that the total loss is the resulting attenuation between the transmitter (before the transmitter RF filter) and the receiver (after the receiver RF filter). Other names given to the total loss are the net loss or the net path loss.

27. *Receiver input level* P_r (dBm or dBW). This is equal to the transmitter output power P_t (dBm or dBW) as given in step 24, less the total loss A_l as given in step 26:

$$P_r = P_t - A_l \quad \text{dBm or dBW} \quad (9.10)$$

The receiver input level is calculated at the receiver reference point after the receiver branching filter.

28–29. *Receiver threshold levels.* RX_a and RX_b are two practical values of the receiver threshold levels which correspond to the bit error rates (BER) of 10^{-3} and 10^{-6}, respectively. These values take into account the planning of the link performance according to CCIR Recommendation 594 (see Sections 7.3.3.1 and 7.3.3.2). The 10^{-3} threshold level takes into account the degradation of the theoretical carrier-to-noise ratio to produce an intolerable amount of inter-symbol interference on the 64 kbit/s measuring circuit (severely errored seconds) and the 10^{-6} threshold level takes into account the degradation of C/N to produce the degraded minutes objective.

30. *Degradation by radio frequency interference (RFI).* Radio frequency interference and its effects on the 10^{-6} BER objective must be calculated separately. This is discussed in Chapter 7. The calculations of degradation are done after the frequencies of the hop have been set, either by the restriction on available frequencies or by the preliminary interference degradation calculations. A minimum degradation of the receiver input level of 1 dB is allowed in all cases. Usually the minimum requirement which can be tolerated is a value of C/I of 15 dB.

Flat fade margins

31–32. *Gross value of fade margin.* FM_a and FM_b are the differences between the calculated unfaded receiver input level P_r, as determined from step 27 and the receiver threshold levels given in steps 28 and 29. They correspond to the flat fade margins, i.e.

$$FM_a = P_r(\text{step 27}) - RX_a(\text{step 28}) \text{ dB for } 10^{-3} \text{ BER} \quad (9.11)$$

$$FM_b = P_r \text{ (item 27)} - Rx_b(\text{item 29}) \text{ dB for } 10^{-6} \text{ BER} \quad (9.12)$$

Flat fading effects

33. *Multipath flat fading probability* P_o. Another way of describing P_o is to consider it as the multipath fade occurrence factor. This is determined from Section 8.3.2. The value of P_o can be determined from Table 8.3, where

$$P_o = KQ \cdot f^B \cdot d^C \tag{9.13}$$

The Majoli method of determining P_o using equation 8.89 may also be used instead of the CCIR method. Both methods, as previously discussed, take into account the multipath activity factor $\eta \times$ probability of Rayleigh fading P_R.

34–35. *Probability of reaching the thresholds* RX_a and RX_b, *(*P_a and P_b*)*. This is the probability of a flat fade reaching the two input receiver level thresholds RX_a and RX_b by exceeding the fade margins FM_a and FM_b respectively. The two equations using the values given in steps 31 and 32 are:

$$P_a = 10^{-FM_a/10} \tag{9.14}$$

$$P_b = 10^{-FM_b/10} \tag{9.15}$$

36–37. *Fading duration T*. The CCIR (Report 338–5)[1] provides an equation for a given depth of fading, whose duration is distributed according to a log-normal law and whose average value T' seconds is given by:

$$T' = C_2 \cdot 10^{[-\alpha_2 F/10]} f^{\beta_2} \tag{9.16}$$

where F is the fade depth which is equal to or greater than the fade margins given in steps 31 and 32, and f is the center frequency of the band in GHz as determined from item 4. In Table 9.1, which gives values for the constants α_2, β_2, C_2, the parameter d is the path length in km. The constants α_1, β_1 and C_1, relate to the number of fades per hour, N, for fades deeper than 15 dB, where in equation 9.16, T' is replaced by N and C_2 by C_1, α_2 by α_1 and β_2 by β_1. This shown in equation 9.28.

For the two separate fade margins FM_a and FM_b, given in steps 31 and 32, the values of the fading durations T_a and T_b are:

$$T_a = C_2 \cdot 10^{[-\alpha_2 \cdot FM_a/10]} \cdot f^{\beta_2} \qquad \text{BER} > 10^{-3} \tag{9.17}$$

$$T_b = C_2 \cdot 10^{[-\alpha_2 \cdot FM_b/10]} \cdot f^{\beta_2} \qquad \text{BER} > 10^{-6} \tag{9.18}$$

38–39. *Probability of flat fading longer than 10 s*. For the link, the duration of fadings with a value F follows a log-normal distribution of which the standard deviation is given by 5.6 dB.[2] The value of T_a is given by equation 9.17 and T_b given by equation 9.18. The value of C_2 given by the U.S.A. is taken as the median value and not the mean. From equation 2.32, the probability $P(10)$ of a fade depth F having a duration greater than 10 s for the purposes of unavailability rather than

Table 9.1 Parameters for the calculation of fade dynamics

Location	α_1	α_2	β_1	β_2	$\beta_1+\beta_2$	C_1	C_2
France	0.5	0.5	1.4	−1	0.4	–	–
Denmark	0.67	0.33	–	–	–	−0.7(#)	–
U.S.A.	0.5	0.5	1.32	−0.50	0.82	–	56.6($\sqrt{d}$)
Switzerland	0.41	0.59	1.38	−1.38	0	–	–

Note
(#) Based on an assumed value of β_1 of 1.4.
(Taken from CCIR Report 338–5[1])

performance calculations, can be calculated from:

$$P(T_a \geqslant 10) = P(10) = 0.5[1 - \text{erf}(Z_a)] = 0.5\ \text{erfc}(Z_a) \quad (9.19)$$

and for the probability $P(60)$ of a fade depth F having a duration greater than 60 s:

$$P(T_b \geqslant 60) = P(60) = 0.5[1 - \text{erf}(Z_b)] = 0.5\ \text{erfc}(Z_b) \quad (9.20)$$

Z can be calculated from the expression relating the log-normal median value to the log-normal mean and Gaussian variance as given by equation 2.32a and by considering that the standard deviation of the log-normal distribution σ_y is given by the natural logarithm of the Gaussian standard deviation σ_x, i.e. $\sigma_y = \ln \sigma_x$
The expression for Z_a and for Z_b becomes:

$$Z_a = \ln(10/T_a)/[\sqrt{2}\ \ln[10^{(5.6/10)}]] = 0.548\ \ln\ (10/T_a) \quad (9.21)$$

similarly

$$Z_b = 0.548\ \ln\ (10/T_b) \quad (9.22)$$

where T_a and its corresponding value of C_2 for the U.S.A. is obtained from step 36 for BER $> 10^{-3}$ and T_b from step 37 for BER $> 10^{-6}$.
40. *Probability of exceeding BER of 10^{-3}*. This is the probability that an outage will occur, without meaning that this outage will last for 10 s or more. The probability can be given by the product of P_o as determined from item 33 and P_a, as determined from step 34, i.e.:

$$\text{Probability of BER} > 10^{-3} = P_o \cdot P_a = P_o \cdot 10^{(-FM_a/10)} \quad (9.23)$$

41. *Probability P_u of circuit becoming unavailable due to flat fading*. This is the probability that the circuit will have a BER $> 10^{-3}$ for a period of greater than 10 s. It is determined from the product of $P(10)$ as determined in step 38 and the probability of BER $> 10^{-3}$ as given by step 40, that is:

$$\text{Probability of circuit becoming unavailable } P_u = P_o \cdot P_a \cdot P(10) \quad (9.24)$$

42. *Link availability*. This is expressed as a percentage and is given from P_u as determined in step 41, i.e.

$$\text{Availability} = 100(1 - P_u) \text{ percent} \tag{9.25}$$

43. *Probability of circuit BER* $> 10^{-6}$. This is calculated from the product of P_o as determined from step 33 and the probability P_b given by step 35, i.e.:

$$\text{Probability of BER} > 10^{-6} = P_o \cdot P_b = P_o \cdot 10^{(-FM_b/10)} \tag{9.26}$$

44. *Probability of circuit having a BER* $> 10^{-6}$ *for more than 60 s due to flat fading*. This is given by the product of the probability of BER $> 10^{-6}$ as given in step 43 and probability of the fade depth exceeding the fade margin FM_b having a duration lasting more than 60 s, as given in step 39, i.e.

$$\text{Probability of circuit having BER} > 10^{-6} \text{ for } > 1 \text{ minute} = P_o \cdot P_b \cdot P(60) \tag{9.27}$$

45. *Expected fades per hour* For fades greater than 15 dB, the CCIR (Report 338–5)[1] provides an equation using values from Table 9.1 (see steps 36–37), for calculating the number of fades per hour N, i.e.

$$N = C_1 \cdot 10^{-\alpha_1 F/10} \cdot f^{\beta_1} \tag{9.28}$$

As the table is incomplete, for Denmark the value of α_1 can be taken as 0.67, $\beta_1 = 1.4$ and $C_1 = -0.7$. The value of F can be taken as the values of the fade margins FM_a or FM_b. The product of N and T_a or N and T_b provides a value of the total fade duration, which is proportional to the value of P_o obtained in item 33. This step has been included for interest only and is not entered onto the path calculation sheet.

46. *Space diversity improvement factor* I_{os}. This subject has been discussed in Section 8.3.5.1 and is included here as part of the path calculations should diversity be required. Using equations 8.136 and 8.137, the diversity advantage is given by:

$$I_{0s} = 7 \times 10^{2+(FM_a/10)} \cdot [(a_r/d) \cdot \ln(1/\rho)]^2/f \tag{9.29}$$

where FM_a is the flat fade margin for BER $> 10^{-3}$, determined in step 31, f is the band center frequency (in GHz), given in step 4, d is the path length (in km) given in step 1, ρ is the correlation coefficient required to determine the antenna spacing. This is usually in the range 0–0.6, with 0.6 being the greater value permitted for effective diversity, and a_r is the relative voltage gain of the secondary or diversity antenna to the main antenna and is given by $10^{(A_d - A_m)/20}$, where A_d and A_m are the gains (in dBi) of the diversity and main antenna, over isotropic.

DIGITAL MICROWAVE RADIO RELAY SYSTEM: PATH CALCULATIONS

	Station A	*Station B*
Link description		
1. Hop number		
2. Equipment type number		
3. Station name		
4. Frequency, GHz Station A ______ Station B Band center ______		
5. Polarization		
6. Channel capacity, Mbit/s		
7. Transmitter modulation type		
8. Ordnance Survey map reference		
9. Site elevation (amsl), meters:		
10. Latitude/longitude (deg, min, sec)	/	/
11. Path length, km		
12. Antenna height (agl), m		
13. Diversity antenna height (agl), m		
Losses		
14. Free space path loss A_0, dB		
15. Feeder type A: ______ B: ______		
16. Feeder length A: ______ B: ______		
17. Feeder loss, dB		
18. Branching loss, dB		
19. Adaptor and connector loss, dB		
20. Attenuator or obstruction loss, dB		
21. Atmospheric absorption loss, dB		
22. Sum of the losses of all columns	+	+ = dB
Gains		
23. Antenna gains over isotropic, dBi		
24. Transmitter A (assumed = B), dBm		
25. Sum of the gains of all columns	+	+ = dBm
26. Total loss A_l, dB (22) – (25)		
27. Receiver input level, dBm		
28. Received threshold level RX_a, dBm		
29. Received threshold level RX_b, dBm		
30. Degradation of RFI, dB		
31. Flat fade margin FM_a, dB		
32. Flat fade margin FM_b, dB		
Flat fading effects		
33. Multipath fading probability P_o		
34. Probability of reaching RX_a, P_a		
35. Probability of reaching RX_b, P_b		
36. Fading duration T_a, s		
37. Fading duration T_b, s		

38. Probability of fdg duration > than 10 s, $P(10)$ ________
39. Probability of fdg duration > than 60 s, $P(60)$ ________
40. Probability of exceeding BER of 10^{-3} ________
41. Probability of circuit becoming unavailable P_u ________
42. Link availability, percent ________
43. Probability of exceeding BER of 10^{-6} ________
44. Probability of BER > 10^{-6} for more than 60 s ________
46. Space diversity improvement factor I_{os} ________

Example 1

Consider the design of a microwave radio link as described in the link description section of the following path calculation sheet:

DIGITAL MICROWAVE RADIO RELAY SYSTEM: PATH CALCULATIONS

	Station A	Station B
Link description		
1. Hop number	DN1/EG-B/I	
2. Equipment type number	NEC TRP-13GD34MB-500A	NEC TRP-13GD34MB-500A
3. Station name	Epping Green	Barkway
4. Frequency, GHz	Station A, 13.130750	Station B, 12.864750 Band center 13 GHz
5. Polarization	Vertical	
6. Channel capacity, Mbit/s	34	
7. Transmitter modulation type	QPSK	
8. Ordnance Survey map reference:	TL 2917 0670	TL 3720 3645
9. Site elevation (amsl)(meters):	122	155
10. Latitude/longitude (deg, min, sec)	51°44′30″N/0°07′54″E	52°00′07″N/0°00′7″E
11. Path length, km	30.26	
12. Antenna height (agl), m	40	44
13. Diversity antenna height (agl)		
Losses		
14. Free space path loss A_0, dB	144.30	
15. Feeder type A: WC109 B: WC109		
16. Feeder length A: 60 m B: 66 m		
17. Feeder loss, dB	3.0	3.3
18. Branching loss, dB	4.4	4.9
19. Adaptor and connector loss	0.5	0.5
20. Attenuator or obstruction loss	0	
21. Atmospheric absorption loss	0.6	
22. Sum of the losses of all columns	7.9 + 144.91 + 8.7 = 161.51 dB	

Gains

23. Antenna gains over isotropic, dB	44	44
24. Transmitter A (assumed = B), dBm	30	
25. Sum of the gains of all columns	44 + 44 + 30 = 118 dBm	

26. Total loss A_l dB (22) – (25) – P_t	73.5
27. Receiver input level, dBm	43.51
28. Received threshold level RX_a, dBm	-79 dBm (BER = 10^{-4})
29. Received threshold level RX_b, dBm	-76.9 (0–50°C)
30. Degradation of RFI, dB	2 dB degradation for 15 dB C/I
31. Flat fade margin FM_a, dB	35.49
32. Flat fade margin FM_b, dB	33.39

Flat fading effects

33. Multipath fading probability P_o	27.74×10^{-3}
34. Probability of reaching RX_a, P_a	282.49×10^{-6}
35. Probability of reaching RX_b, P_b	458.14×10^{-6}
36. Fading duration T_a, s	1.4514
37. Fading duration T_b, s	1.8483
38. Probability of fdg duration > than 10 s, $P(10)$	0.0635
39. Probability of fdg duration > than 60 s, $P(60)$	0.0045
40. Probability of exceeding BER of 10^{-3}	7.8363×10^{-6}
41. Probability of circuit becoming unavailable P_u	4.9760×10^{-7}
42. Link availability, percent	$100(1 - 4.98 \times 10^{-7})$
43. Probability of exceeding BER of 10^{-6}	12.7×10^{-6}
44. Probability of BER > 10^{-6} for more than 60 s	5.72×10^{-8}
46. Space diversity improvement factor I_{os}	For $\rho = 0.5$, $I_{os} = 100$

Figure 9.9 shows the path profile for the digital radio hop used in the above example. As can be seen from this profile, for a clearance factor of 0.6 and a k-factor (min) of 0.7, the first Fresnel zone is not obstructed in any way. As this is the worst case expected for 99.99 percent of the time, the obstruction loss can be considered to be 0 dB.

The calculations used to derive the values entered in the path calculation sheet are briefly described below with the relevant path calculation step number to which the calculation refers.

Step 14. Using equation 8.79, A_o is calculated as 144.30 dB

Step 17. Using circular waveguide (Andrews-WC 109), which has an attenuation of 4.5 dB/100 m plus the 0.3 dB loss for the circular to elliptical waveguide transition (EW127) for a closed system:

$$\text{Transmitter feeder loss} = (40 \times 1.5 \times 4.5/100) + 0.3 = 3 \text{ dB}$$

$$\text{Receiver feeder loss} = (44 \times 1.5 \times 4.5/100) + 0.3 = 3.3 \text{ dB}$$

Step 18. From the manufacturer's data (NEC), the branching loss for a protected hot standby system is 4.4 dB for the transmit and 4.9 dB for the receive.

Step 19. The adaptor and connector loss is taken as 0.5 dB at each end of the hop,

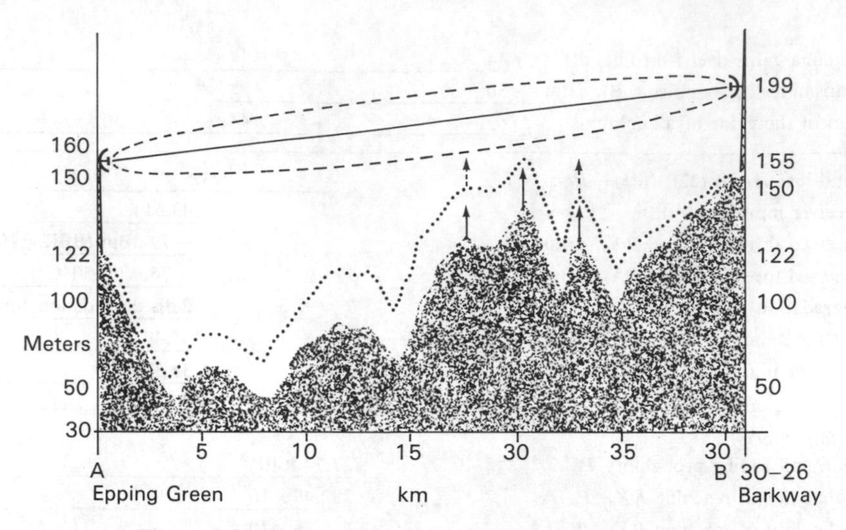

Figure 9.9 Path profile of the link used in Example 1

although it is really not necessary, since the manufacturer states only that given in step 18 as needing to be considered. Even so this loss may be considered as a safety factor loss.

Step 20. From the path profile shown in Figure 9.9, the obstruction loss is zero dB, for a clearance factor of 0.6 times the first Fresnel radius and at a k-factor (min) of 0.7.

Step 21. From Figure 8.27, the atmospheric absorption loss is found to be 0.02 dB/km at 13 GHz. Thus for 30 km, the loss is 0.6 dB.

Step 31. The flat fade margin $FM_a = P_r - RX_a = -43.51 - (-79) = 35.49$ dB.

Step 32. The flat fade margin $FM_b = P_r - RX_b = -43.51 - (-76.9) = 33.39$ dB.

Step 33. Using equation 9.13 and $KQ = 1.4 \times 10^{-8}$, $B = 1$ and $C = 3.5$, $P_o = 27.74 \times 10^{-3}$.

Step 36. Using $C_2 = 56.6\sqrt{d}$, $\alpha_2 = 0.5$ and $\beta_2 = -0.5$, $T_a = 1.4514$ s.

Step 37. Similarly to step 36, using the same constants, $T_b = 1.8483$ s.

Step 38. Using equations 9.21 and 9.22, the values of Z_a and Z_b are calculated as 1.0577 and 0.9252 respectively. This produces $P(T_a \geqslant 10)$ or $P(10)$ and $P(T_b \geqslant 60)$ or $P(60)$ as 0.00635 and 0.0045 respectively.

Step 46. Using a correlation factor $\rho = 0.5$, and a diversity antenna which is the same size as the main antenna, the diversity advantage I_{0s} for FM_a is found from equation 9.29 as 100.

9.3 PERFORMANCE CALCULATIONS

The performance calculation sheet provides details of the calculated performance expected per radio hop as well as for the system. A margin relative to the specified objective is also calculated. Both the performance and path calculations are based on equipment planning parameters which are to be expected under normal operating conditions. A description of each of the performance calculation steps is given below as for the path profile calculations.

1. *Section length*. This is the sum of all of the hop lengths in km. If only one hop is considered it is that hop length.

2. *Number of hops – in the section*

3. *Number of repeaters*. The number of repeaters is equal to the number of hops when digital regeneration takes place at each repeater station.

4–5. *The probability of exceeding the BER*. This is the sum of all of the individual probabilities for the different hops for each of the BER probabilities. This means that there will be two summations. One for obtaining the sum of the probabilities of BER > 10^{-3}, each one individually determined from step 40. The other for calculating the sum of all of the probabilities of BER > 10^{-6}, each one being given by step 43, i.e.

$$\text{Total probability of BER} > 10^{-3} = \sum_{i=1}^{H} P_{oi} \cdot 10^{-(FM_{ai}/10)} \quad i = 1, 2, \ldots, H \tag{9.31}$$

$$\text{Total probability of BER} > 10^{-6} = \sum_{i=1}^{H} P_{oi} \cdot 10^{-(FM_{bi}/10)} \quad i = 1, 2, \ldots, H \tag{9.32}$$

where H is the number of hops in the link.

6. *Selective fading outage in worst month*. This is calculated using either the Majoli method described in Section 8.3.3.3 or by using equation 8.127, which is:

$$\text{Probability of BER} > 10^{-3} \text{ for selective fading}$$
$$= 2 \sum_{i=1}^{H} \eta_i \cdot K_i [2d_i^{1.5} \cdot (r_{bi}/\log_2 M) \times 10^{-6}]^2 \tag{8.127}$$

where η is the fraction of time which multipath propagation occurs and is given by equation 8.94, i.e.

$$\eta = \begin{cases} 1 & \text{for } P_o > 10 \\ 0.182\ P_o^{0.7} & \text{for } 0.1 < P_o < 2 \\ 1.44\ P_o & \text{for } P_o < 10^{-2} \end{cases} \tag{8.94}$$

The value of P_o is determined in step 33 of the path calculations.

7. *Total outage.* This is the sum of the percentages of step 4 and step 6 of the performance calculations.

8. *Objective.* The CCIR objective for a high-grade circuit for severely errored seconds is:

(a) The BER should not exceed a value of 1×10^{-3} for more than $0.054L/2500$ percent of any month for an integration time of one second. The circuit or link length L, being less than 2500 km, but greater than 280 km.
(b) The BER should not exceed a value of 1×10^{-3} for more than 0.006 percent of any month for an integration time of one second, the circuit or link length being less than 280 km.

In the real world where there may be only one hop to consider which is far less than 280 km, although not strictly to the CCIR recommendations, direct apportionment to length may be considered. Thus

Objective for $d < 280$ km $= 0.006 \times d/280$ percent

The CCIR proposal for a local-grade performance objective applicable to each direction and each 64 kbit/s channel is:

The BER should not exceed a value of 1×10^{-3} for more than a fraction A of any month with an integration time of one second. Values of A in the range 0.005–0.015 have been proposed.

This objective can be compared against that calculated in steps 4 and 6 of the performance calculations (equation 9.31).

9. *Objective.* The CCIR objective for a high-grade circuit for degraded minutes is:

(*a*) The BER should not exceed a value of 1×10^{-6} for more than $0.4L/2500$ percent of any month for an integration time of one minute, where the link length L is less than 2500 km, but greater than 280 km.
(*b*) The BER should not exceed a value of 1×10^{-6} for more than 0.045 percent of any month for a link length less than 280 km.

Similar to that proposed (non-CCIR) above for severely errored seconds, for a hop length d, the degraded minutes objective may be considered as:

Degraded minutes objective for $d < 280$ km $= 0.045 \times d/280$ percent

The CCIR proposal for a local-grade performance objective applicable to each direction and to each 64 kbit/s channel is:

The BER should not exceed a value of 1×10^{-6} for more than B percent of any month with an integration time of one minute. Values of B in the range 0.5 to 1.5 have been proposed.

This objective can be compared against that calculated in step 5 of the performance calculations (equation 9.32).

10. The margin with respect to the objective can be determined from;

$$\text{Performance margin} = 10 \log (\text{objective percentage})/(\text{calculated percentage}) \text{ dB} \quad (9.33)$$

11. For each hop, the performance with an equalizer improvement factor I_e, can be calculated by dividing the probability of BER $> 10^{-3}$ as given in step 40 of the path calculations, by the improvement factor.

12. For each hop, the performance with an equalizer and space diversity can be calculated by dividing step 9 of the performance calculations by the diversity improvement factor I_{os} as given in step 46. Knowing that the synergistic effect will produce a better result, we can use Table 8.5 to obtain a combined figure of the improvement factor in which equalizer, diversity combiner and diversity are all used.

13. For each hop, the performance with an equalizer improvement factor I_e can be calculated by dividing the probability of BER $> 10^{-6}$ as given in step 43 of the path calculations, by the improvement factor.

14. As for step 10 of the performance calculations the improved probability of exceeding the BER $> 10^{-6}$ can be determined for the case where adaptive equalizer, diversity combiner and space diversity are used.

9.4 AVAILABILITY CALCULATIONS

As discussed in Section 7.5, the causes of unavailability which are statistically predictable, unintentional and resulting from the radio equipment, power supplies, propagation, interference and from auxiliary equipment and human activity must be used in the estimate of unavailability. Unavailability is defined as the BER $> 10^{-3}$ for more than 10 s, and/or the digital signal is interrupted (i.e. alignment or timing is lost).

15. *Equipment unavailability*. From Section 7.8 (equation 7.33), the system unavailability is determined as (100 – system availability) percent.

16. *Rainfall unavailability*. From equation 8.157 determine the expected outage time for a fade margin of $FM_a = A_p$, i.e. the equation:

$$[FM_a/0.12(A_{0.01})] = p^{-(0.546 + 0.043 \log p)} \tag{9.34}$$

will require to be solved for p, the percentage outage time. As rain usually lasts longer than 10 s, the value of p obtained will represent the unavailability for the worst month.

17. *Multipath unavailability (flat).* Although this unavailable time does not usually occur when it is raining, it will contribute to unavailability at another time. It may be calculated for a section by calculating for each individual hop the value of P_u as determined in step 41 (equation 9.24) and summing the values. This value, if multiplied by 100, will give the percentage of unavailable time, i.e.

Percentage probability of a link becoming unavailable

$$= 100 \sum_{i=1}^{H} P_{ui} = 100 \sum_{i=1}^{H} P_{oi} \cdot P_{ai} \cdot P(10)_i \tag{9.35}$$

where $i = 1, 2, \ldots, H$ and H is the number of hops in the link.

18. *Multipath unavailability (selective).* This can be determined from the product of the selective fading outage, as determined from step 6 of the performance calculations and $P(10)$, as determined from step 38 of the path calculations.

$$\text{Unavailability} = 100\ P(10) \cdot (\text{Probability of BER} > 10^{-3}\ \text{selective}) \tag{9.36}$$

19. *Human error.* This is usually given a value of 0.3 percent of the time, but is usually discounted if the station is unmanned.

20. *Total unavailable time.* This is the sum of steps 15, 16, 17 and 18 of the performance calculations.

21. *Objective.* The CCIR unavailability objective is given as 0.2 percent for a 2500 km end-to-end radio circuit, which includes the local ends as well, but not the human error amount. Considering the discussion in Section 7.5.3, the unavailability figure that a trunk network may be expected to contribute is 0.06 percent of the total unavailability figure of 0.2 percent; the back-haul system at each end contributes 0.0225 percent, and the local distribution network contributes 0.0325 percent. These values may be considered as those required for a 2500 km network. The link length of 2500 km may be an unrealistic value for a small country, which has for example, an end-to-end length of 1200 km. A more realistic approach would be to allocate the 0.06 percent dedicated to digital radio relay trunk circuits, to a proportion of this maximum length which is applicable only to the radio system. This may be, for example, only 600 km. The other 600 km may be the largest distance expected using optical fiber. Thus the radio relay trunk allocation for 600 km may be taken as 0.06 percent and for circuits or hops shorter than 600 km, the percentage unavailability

is scaled accordingly. The allocation of the 0.2 percent unavailability figure to the various circuit elements (back-haul, local, etc.) depends on the authority to whom the system belongs. For this discussion, we adopt the following values;

Trunk unavailability objective for $L < 600$ km	$= 0.0600 \cdot L/600$ %/km
Back-haul unavailability objective (each end)	= 0.0225 percent
Local network unavailability (each end)	= 0.0325 percent
Other systems (optical fiber +) $L < 600$ km	$= 0.0300 \cdot L/600$ %/km
Total ($L = 600$ km)	= 0.200 percent

Proportioning the unavailability percentage with distance is possibly the most practical way of overcoming the system design problems which are associated with a trunk network that has alternative routing. A call set up somewhere in a country can reach its destination via many different routes. Each route will have a different path length and hence a different unavailability allocation. The longest route, however, which is considered the worst case is then allocated a total unavailability value of 0.06 percent. This method although practicable can be expensive, especially for short hops, where the unavailability percentage is extremely small. To meet these small objectives means that a more expensive design has to be considered. For the case where there is no alternative routing expected, and a call set up has to traverse a fixed number of hops comprising the radio link, it would appear that a more cost-efficient method would be to allocate 0.06 percent for the total circuit length and divide the 0.06 percent by the number of hops to give an unavailability percentage allocation to each hop. The problem which may be encountered in this case is when the occasion arises to expand the link or modify it in some way, causing the availability allocation for the trunk route to be contravened. Whatever way is chosen depends to some degree on the application and configuration and also on the policy of the authority concerned. In the example used where the longest radio relay link length is considered to be 600 km, and 0.06 percent unavailability has been allocated, a 30 km link would be expected to have an unavailability of $0.06 \times 30/600 = 0.003$ percent. Example 1 shows on the completed path calculation sheet that the probability P_u of becoming unavailable (step 41) is calculated to be 0.000678 percent. This hop, therefore, will satisfy the unavailability objective, but at the cost of increasing the antenna sizes from 1.2 m (41 dBi) to 1.8 m (44 dBi) and changing the feeder from an elliptical type (EW127A) with 11.32 dB/100 m to the circular type (WC 109) with an attenuation of 4.5 dB/100 m. Note that dBi means the gain of a practical antenna over isotropic, as given by equation 8.66. If the less expensive, small antenna and the elliptical waveguide had been used, the probability P_u of the circuit becoming unavailable, can be calculated to be equal to 0.002 percent, which is quite large. The reason why this is considered to be quite large is that the unavailability due to rain fading, which is usually much larger than that caused by multipath flat fading, has not yet been considered and there is only 0.001 percent remaining (0.003 – 0.002 percent), before the unavailability percentage allocation has been exceeded.

DIGITAL MICROWAVE RADIO RELAY SYSTEM: PERFORMANCE AND AVAILABILITY CALCULATIONS

Performance calculations

1. Section length, km	______
2. Number of hops in the section	______
3. Number of repeaters	______
4. Probability of BER $> 10^{-3}$	______
6. Probability of BER $> 10^{-3}$ (selective fading)	______
7. Total outage (BER $> 10^{-3}$)	______
8. Objective (severely errored seconds)	______
10. Margin over objective, dB (BER $> 10^{-3}$)	______
11. Equalizer improvement BER $> 10^{-3}$	______
12. Equalizer and diversity improvement (BER $> 10^{-3}$)	______
5. Probability of BER $> 10^{-6}$	______
6. Probability of BER $> 10^{-6}$ (selective fading)	______
Total of BER $> 10^{-6}$	______
9. Objective (degraded minutes)	______
Margin over objective, dB (BER $> 10^{-6}$)	______
13. Equalizer improvement BER $> 10^{-6}$	______
14. Equalizer and diversity improvement BER $> 10^{-6}$	______

Availability calculations

15. Equipment unavailability, percent	______
16. Rainfall unavailability, percent	______
17. Multipath unavailability, percent (flat)	______
18. Multipath unavailability, percent (selective)	______
20. Total unavailability, percent	______
21. Objective, percent	______

Example 1 (continued)

The performance and availability calculations are listed in the calculation sheet below for Example 1 on page 552, which has already been considered for path calculations, together with an explanation of the derivation of the values in accordance with Steps 1–14 and 15–21 above.

DIGITAL MICROWAVE RADIO RELAY SYSTEM: PERFORMANCE AND AVAILABILITY CALCULATIONS

Performance calculations	
1. Section length, km	30.26
2. Number of hops in the section	1
3. Number of repeaters	0
4. Probability of BER $> 10^{-3}$	7.8363×10^{-6}
6. Probability of BER $> 10^{-3}$ (selective fading)	2.5590×10^{-6}
7. Total outage (BER $> 10^{-3}$)	10.395×10^{-6}
8. Objective (serverly errored seconds)	6.4286×10^{-6}
10. Margin over objective, dB (BER $> 10^{-3}$)	−2.0 dB without equalization, 4.9 dB with.
11. Equalizer improvement BER $> 10^{-3}$	2.0791×10^{-6}
12. Equalizer and diversity improvement BER $> 10^{-3}$	—
5. Probability of BER $> 10^{-6}$	12.7×10^{-6}
6. Probability of BER $> 10^{-6}$ (selective fading)	25.144×10^{-6}
Total BER $> 10^{-6}$	37.844×10^{-6}
9. Objective (degraded minutes)	48.632×10^{-6}
Margin over objective, dB (BER $< 10^{-6}$)	1.09 dB
13. Equalizer improvement BER $> 10^{-6}$	5.82×10^{-6}
14. Equalizer and diversity improvement BER $> 10^{-6}$	
Availability calculations	
15. Equipment unavailability, percent	0.1473×10^{-3}
16. Rainfall unavailability, percent	5.5436×10^{-3}
17. Multipath unavailability, percent (flat)	0.0054×10^{-3}
18. Multipath unavailability, percent (selective)	0.0163×10^{-3}
20. Total unavailability, percent	5.71×10^{-3}
21. Objective, percent	3×10^{-3}

Comments

1. The availability figure is almost twice that permitted by the availability objective for this hop. The largest contribution is from the effects of rain on the hop.
2. There exists a need for a duplicated system operated in the standby mode if the equipment unavailability is to be reduced to an acceptable level.
3. An adaptive baseband equalizer is required to insure that the severely errored seconds objective is met (BER $> 10^{-3}$)

The calculations used to derive the values entered in the performance and availability sheet are briefly described below with the relevant performance and availability step number to which the calculation refers.

Step 1. This has already been calculated in step 11 of the path calculations.

Step 4. This has already been calculated for the one-hop case in step 40 of the path calculations.

Step 5. This has already been calculated for the one-hop case in step 43 of the path calculations.

Step 6. For the one-hop case, $i = 1$; and, from equation 8.127 with $K = 1$ for QPSK, $r_b = 34$ Mbit/s, $M = 2$ and η is calculated to be 0.039 946. The value of the selective fading outage probability is thus determined to be 2.5590×10^{-6}.

As the selective fading and flat fading outages are not less than the objective value, there is a need for an adaptive baseband equalizer, but no need for diversity. The probability of the BER $> 10^{-6}$ due to selective fading is to be determined using Majoli's method since there is no equation to assist in this case. This means that the BER $= 10^{-6}$ signature is required. The manufacturer's figures in lieu of the signature curve are: width = 16.5 MHz with a notch depth of 19.5 dB. These figures are with an equalizer. For the notch depth, a height of 18.5 dB is assumed. From equation 8.128, using $T_1 = 6.3/(\text{antilog}\ (-18.5/20)) = 53$ ns, $\tau_o = 0.7(d/50)^{1.5} =$ 0.33 ns, $W = 16.5$ MHz and $\eta = 0.0399$, the value of the probability (BER $> 10^{-6}$) $= 4.3 \times 0.0399 \times 16.5 \times (0.33) \times (0.33) \times 10^{-3}/53 =$ 5.8234×10^{-6}.

Without the equalizer the manufacturer's figures are 30 MHz for the width of the signature and 12 dB for the notch depth, with an assumed height of 11 dB. This gives $P_{os}(\text{BER} > 10^{-6}) = 25.14 \times 10^{-6}$.

Step 11. Assuming an equalizer improvement of 5, the equalizer improvement on BER $> 10^{-3}$ will be the unequalized value of the probability of BER $< 10^{-3}$ divided by 5, i.e. a 4.9 dB margin over the objective.

Step 18. From step 38 of the path calculations $P(10)$ is found to be 0.0635, and from step 6 of the performance calculations, multipath unavailability is found to be 1.63×10^{-5} percent.

Step 15. To determine the equipment unavailability, the starting point is to determine the mean time between failures for the associated equipment to derive the 64 kbit/s test circuit. The MTBF for typical equipment is given in Table 7.10. From the 34 Mbit/s radio (MTBF = 50 years), the associated mutiplex equipment is one third-order multiplex (8 to 34 Mbit/s), (MTBF = 8.2 years), one second-order multiplex (2 to 8 Mbit/s), (MTBF = 9.4 years) and one primary multiplex (64 bit/s to 2 Mbit/s), (MTBF) = 4.5 years).

Thus the system MTBF for both ends of the hop is given by;

$$1/(\text{MTBF}) = 2(1/50 + 1/8.2 + 1/9.4 + 1/4.5) = 0.941$$

Therefore the MTBF is 1.0626 years = 9314.5 $h = T$. If the mean time to repair t is given as 8 hours, the system unavailability for an unprotected system is given by equation 7.28 as;

$$\text{Unavailability} = 100\ t/(t+T) = 85.814 \times 10^{-3} \text{ percent}$$

This figure of unavailability is completely unacceptable. To improve it, a simple redundant system is required using a hot-standby radio and multiplex equipment. Using equation 7.33, the new unavailability figure (100 − availability) = 0.1473×10^{-3}. This is acceptable. Usually the multiplex equipment is not considered in these calculations unless they are specifically a part of a special package.

Step 16. From the path profile the angle of the direct ray θ, due to the difference in antenna heights, is 0.108°. As Table 8.6 does not provide values of k_H, k_V, α_H and α_V, for 13 GHz, linear interpolation is necessary to determine between the 12 and 15 GHz values, and logarithmic interpolation between 12 and 15 GHz for values of k_H and k_V. Thus; $k_H = 0.022$, $k_V = 0.019$, $\alpha_H = 1.20$, $\alpha_V = 1.18$. As the polarization is vertical, $\tau = 90°$. Therefore, $k = 0.019$, and $\alpha = 1.18$. The link is in England with rain rate of 28 mm/h exceeded for only 0.01 percent of the time (1 min integration time), and from equation 8.152, the specific attenuation is found to be $\gamma_R = 0.97$ dB/km. From equation 8.164, the value of r is found to be 0.977, and from equation 8.156, the attenuation not exceeded for 0.01 percent of the time is: $A_{0.01} = 0.97 \times 30.26 \times 0.977 = 28.7$ dB. From the path calculations, step 31, the fade margin $FM_a = 35.49$ dB; hence the equation to be solved, after substituting these values into equation 9.34, is the quadratic equation $1.013 + 0.546 \log p + 0.043(\log p)^2 = 0$. This produces the worst case solution of $p = 5.5436 \times 10^{-3}$ percent.

EXERCISE

1 Consider Example 1. If the value of the feeder loss is changed to 11.32 dB/100 m and the antennas at each end of the link are changed to have a gain of 41 dBi instead of the present 44 dBi, calculate the link availability after the effects of only multipath fading have been considered. (*Answer* 99.998 percent)

REFERENCES

1. 'Propagation in Non-Ionized Media', Recommendations and Reports of the CCIR, **V**, XVI Plenary Assembly, 1986.
2. *Ibid*, XIV Plenary Assembly, 1978, Report 338–3, Section 5.3.

10 INTERFERENCE AND FREQUENCY PLANNING

10.1 INTRODUCTION

For some time to come digital radio systems will have to exist with analog radio systems. This situation generates interference problems which would not exist if all the systems in a given area were either analog or digital. Three of these problems which have to be considered are:

Digital-to-digital interference
Analog-to-digital interference
Digital-to-analog interference

A digital system is more tolerant to interference from other sources than its analog counterpart, so that the major problem lies with interference into the analog portion of the system from the digital system. To operate correctly, a digital system usually requires a carrier-to-interference ratio (C/I) of 15–20 dB, according to the modulation scheme. However, in a complex network, as there are many different interference configurations, the digital bearer must maintain the 15–20 dB C/I even under fading conditions, which means that the interfering channel level is to be 15–20 dB below the receiver threshold of the channel being interfered with. This, in fact, is what sets the stringent antenna, coupling, filtering, etc., requirements of a digital system to be like those expected in an analog system. For example, if the receiver threshold level is -79 dBm for a BER $> 10^{-3}$, and the fade margin is 36 dB, then, to make the fade margin of 36 dB fully usable for flat fading, a C/I of 15 dB must still be available at this receiver threshold. Consequently, the interference level must be 15 dB below the receiver threshold, i.e. -94 dBm, and this sets the requirement for antenna decoupling to be the sum of the fade margin plus the C/I, which will be $36 + 15 = 51$ dB. The effects of cross-polarization now become apparent, when considering that any interfering cross-polarized signal has to be at least 51 dB below the normal received signal. In an overall network design, the carrier-to-interference ratios are reduced by keeping the power of other transmitters to the bare minimum, by proper choice of antenna systems, etc., so that the minimum fade margin required to meet the CCIR objectives is attained. As mentioned above, the tolerance of digital channels to interference depends on the modulation system. In particular, a modulation system which requires a low C/N for a certain

BER > 10^{-3} threshold is also more tolerant to interference. The most tolerant types (or robust types) of modulation are 2–PSK or 4–PSK. The offset QPSK is slightly more sensitive to interference and the FM, 8–PSK, 16–QAM types, etc., require a much larger C/I. In addition to the type of modulation, the tolerance to interference is also dependent on other equipment parameters, so that the calculated value of C/I to produce a given system degradation is only approximate. For digital radios, the manufacturer usually provides a set of curves which show the BER dependence on co-channel and adjacent-channel interference. Figure 10.1 shows the effect of C/I on the BER of a 4–PSK 'GTE–CTR 148', 34 Mbit/s radio receiver and the effect of C/I on the BER of an offset 4–PSK receiver (GTE–CTR 148 (offset)). It is not uncommon to have a situation in which more than one interfering channel exists at the same time, especially at congested microwave radio nodes. Where all interfering frequencies are the same (co-channel operation), the interfering signals are added power-wise to obtain the total interfering power. To use a specific C/I curve, such as one of those shown in Figure 10.1, the total interfering power I must be considered, in computing C/I for the digital radio receiver suffering interference. If, for example, there was in a system only one co-channel interferer, which produced a value of C/I of 15 dB at the receiver input, and then after some time this was increased to two co-channel interferers, the second being of the same power at the receive terminal as the first, the interfering power will have been increased by 3 dB, which brings the value of the C/I measured at the receiver terminal to 12 dB. To obtain the same 15 dB at the receiver terminal would mean that both individual values of C/I of the interfering co-channel sources would have to be 18 dB. Where there are different types of interference present, such as a co-channel and an adjacent-channel interferer, or various combinations for more than two interferers, the situation becomes quite complex. Usually, BER curves are provided by the manufacturer for the case of 1 co-channel and 1 adjacent-channel interferer, so that applying the power addition law to each type of interferer of the same type, the most complex situation can be reduced to the basic 1 co-channel and 1 adjacent-channel case. In the general case, at a station where there are many radio receivers, the interference situation is improved if all the received signals entering that site are equal in level. This is, the received level of each of the various hops, at the site is designed to be the same level as any other received signal coming into that site. The system engineering must take this factor into account and try to equalize the receive levels by selecting the proper transmitter power options, transmitter antenna size, feeder type or length and by the insertion at the transmitter, of suitable RF attenuators. As C/I does not depend on antenna gain (both C and I increase by the same amount after passing through the antenna), nor on the feeder loss, the receiving level for interference calculations must be considered as the receiving level measured at the antenna input. The following simple calculations demonstrate why it is important to have, at any one site, the received levels from different sources at the same power level. If the flat fade margin FM_a, is not to be degraded due to a C/I received at the antenna input terminal, the antenna decoupling or the reduction of power of a signal from one antenna into another V dB, is given by

$$FM_a + C/I = V \text{ dB} \tag{10.1}$$

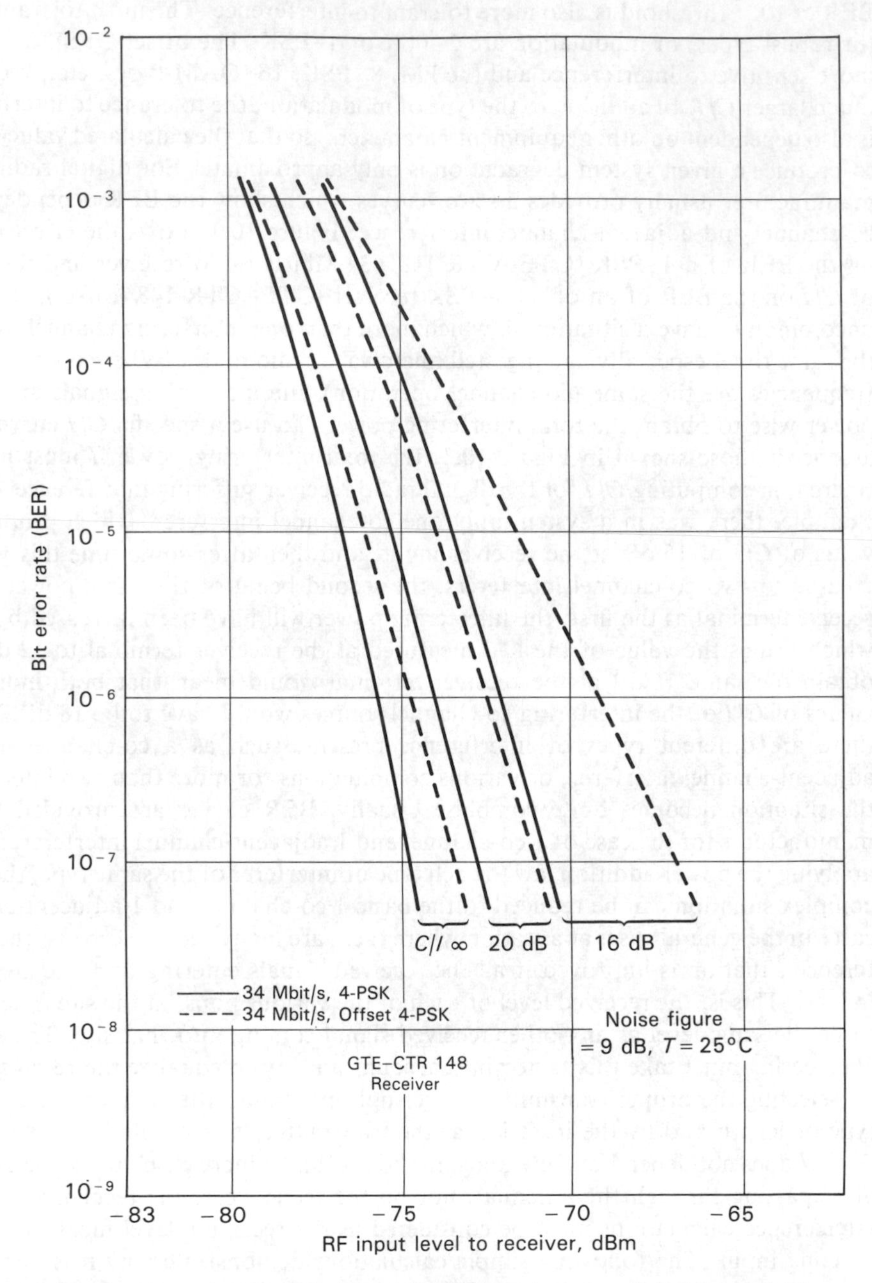

Figure 10.1 4−PSK and offset 4−PSK BER plotted againt RF input levels for different C/I
(Source: GTE Telecommunications CTR 148 digital radios)

For the simple case of two antennas on a tower, where C_1 is the received signal level at antenna 1, and C_2 is the received signal level at antenna 2, if it is assumed that the antenna decoupling is the same for each antenna, the C/I at the receiver terminals of each of the receivers 1 and 2 is given by;

$$(C/I)_1 = C_1 - (C_2 - V) \text{ dB} \quad (10.2)$$

$$(C/I)_2 = C_2 - (C_1 - V) \text{ dB} \quad (10.3)$$

Where the signal levels at each of the antennas are the same, $C_1 = C_2$, and from equations 10.2 and 10.3, C/I is the same at each of the receiver inputs. However, if the received signal levels at each antenna are different, i.e., $C_1 > C_2$, then C/I at receiver 1 is greater than C/I at receiver 2, meaning that receiver 2 may suffer, due to bad planning, a reduction in the level of its fade margin.

Example 1

For an extreme case of bad planning, consider that at a receiving site there are two antennas mounted on the same tower, and that both antennas have the same decoupling. If the minimum value of C/I which can be tolerated by each system is 15 dB and a flat fade margin of 35 dB has been chosen for each receiver, determine the expected value of C/I at each receiver, if the received signal level at antenna 1 is -38 dBm and at antenna 2 is -75 dBm.

Solution
From equation 10.1, the coupling V of each antenna is calculated to be $35 + 15 = 50$ dB. From equation 10.2, the value of $(C/I)_1$ at receiver 1 is calculated to be 87 dB, and from equation 10.3 at receiver 2, the $(C/I)_2$ is found to be 13 dB. This shows that due to the stronger receiving level at antenna 1, the C/I at receiver 2 has been degraded to the point where it has become intolerable. A simple solution to overcome this problem is to pad down the transmitter belonging to the first receiver, to a value which permits the received signal at antenna 1 to be -75 dBm.

If this transmitter is padded down to give -75 dBm, and if equal (although they do not have to be) antennas of 46 dBi were used and feeder losses in each case amounted to 10 dB, the received signal level at each receiver would be -39 dBm, although the received signal level at the antenna was -75 dBm. The value of C/I at the receiver for an *equal* received signal level at both antennas is found to be 50 dB, which equals the antenna decoupling V.

10.1.1 Sources and effects of RF interference

10.1.1.1 Sources of interference

There are three types of main interfering signal sources. These are: from a cross-polarized channel, from an adjacent co-polarized channel, and from the reception of a frequency on a hop facing the opposite direction of the hop being interfered with. A cross-polarized channel, whose cross-polarization discrimination has been reduced either by multipath fading or by rain fading can cause considerable problems if the frequency on the orthogonal polarization is the same as that frequency of the hop being interfered with (the victim carrier). The cross-polarized channel using the same frequency is used to enhance the utilization of the frequency spectrum and is known as 'frequency re-use'. Under normal propagation conditions, the decoupling between the two polarized signals is more than adequate as shown above, but, when degraded, the level of co-channel interference is correspondingly increased. It is generally agreed, however, that even under such conditions, a system can be designed to meet availability requirements by careful choice of repeater spacings, transmitter power, receiver noise figure, antenna size, etc. Adjacent-channel interference may arise between two systems which share the same general area but which do not share the same paths or sites (*inter-system interference*) or between two systems which again share the same general area and which do share the same paths or sites (*intra-system interference*). The term 'adjacent' implies that the carrier frequencies are not the same, whereas, the term co-channel implies that they are the same.

Table 10.1
(Source: CCIR Report 779, Reference 1)

Location on Figure 10.2	Examples of various sources	Power	Classification
	Sources fading with the wanted signal		
1	Cross-polarized channel(s) (Note 2)	S_{1_1}	Main interferer
2	Front-to-back radiation	S_{1_2}	Weak interferer
4	Overreach (3 hops)	S_{1_4}	Weak interferer
8	Front-to-back re-radiation and over-reach	S_{1_8}	Weak interferer
7	Adjacent co-polarized channel(s) (may be subject to frequency-selective fading)	S_{1_7}	Main or weak interferer
	Transmitter noise	N_t	
	Total power (Note 1)	$S_1 + N_t$	
	Sources not fading with the wanted signal		
	Receiver noise	N_r	
x	Transmitter into receiver	S_{2_x}	Weak interferer
6	Extraneous reflections	S_{2_6}	Weak interferer
5	Opposite hop overreach (3 hops)	S_{2_5}	Weak inteferer
3	Opposite hop front-to-back reception	S_{2_3}	Main interferer
D	Other digital systems	S_{2_D}	Weak interferer
A	Other analog systems	S_{2_A}	Weak interferer
S	Satellites	S_{2_S}	Weak interferer
	Total power (Note 1)	$S_2 + N_r$	

Notes

1. S_1 and S_2 represent the ensemble of each of the two groups of interference, evaluated under faded conditions. It should not be construed that a summation of interference powers is equivalent to Gaussian or white noise for the purpose of calculating error ratio.
2. Cross-polarization discrimination is known to be a function of rainfall attenuation and of multipath propagation.

Main and weak interfering sources, are listed in Table 10.1. Reference should also be made to Figure 10.2, which indicates the possible types of interfering source. These can be divided into two groups (see Table 10.1) as follows:

$S1_j$ Sources which share, either partially or completely, a common path with the wanted signal (intra-system) and to a large extent fade simultaneously with the wanted signal, for example during rain attenuation

$S2_j$ Sources which do not have a common path with the wanted signal (inter-system) and in general do not fade simultaneously with the wanted signal, or which fade less than it does.

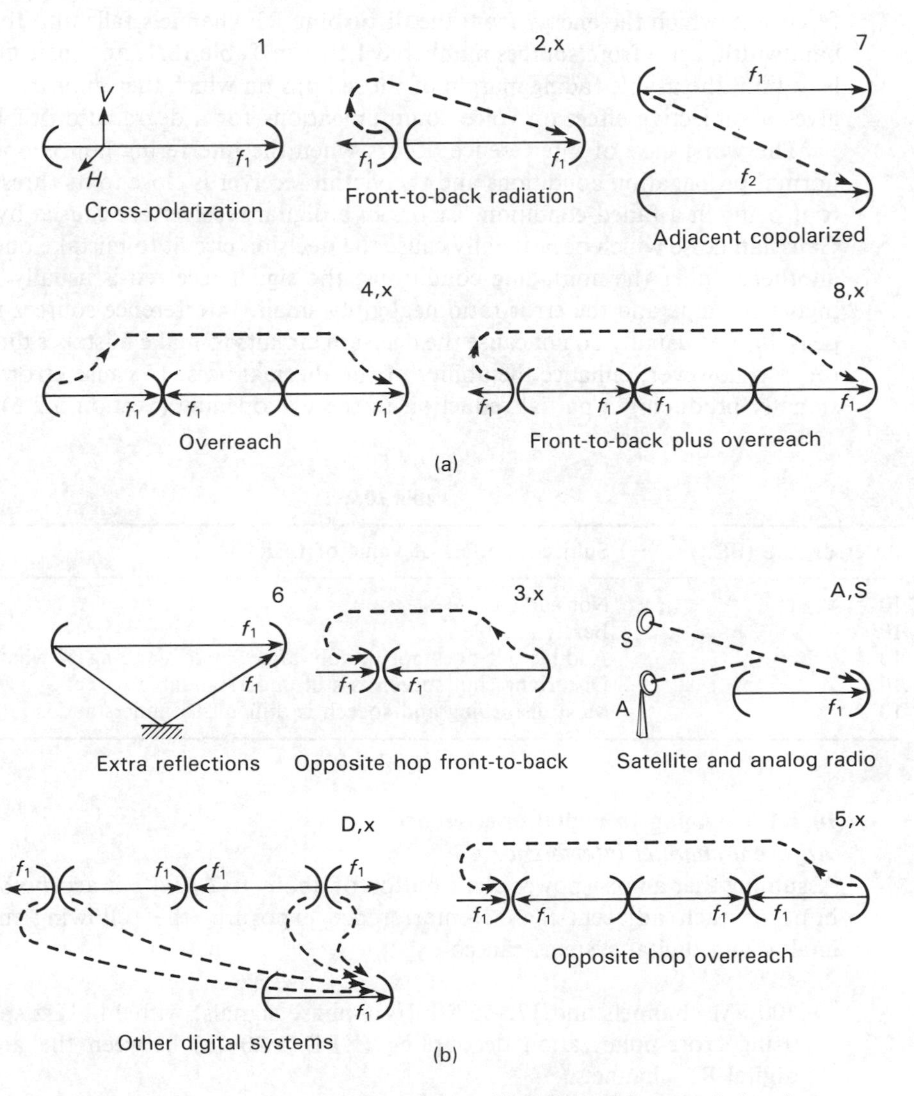

Figure 10.2 RF interference sources: (a) fading with wanted signal; (b) not fading with wanted signal

10.1.1.2 Effects of RF interference

Each of the interfering sources listed in Table 10.1 produces an increase in the error ratio by degrading the flat fade margin. To overcome this degradation means that the transmitter power, or the level of the received power, must be increased to produce a high BER so that the BER objective can still be maintained. The difference between the carrier power levels which correspond to the worst BER in the presence and absence of interference determines the receiver threshold deterioration. Adjacent-channel interference or that interference from adjacent co-polar RF channels (inter-channel interference), is of particular concern in the choice of the spacing of channels. The main contributors of co-channel interference or that type of interference in which the energy from the disturbing RF channels falls into the receiver bandwidth, arise from sources numbered 1 to 3 in Table 10.1, and their main effect is to limit the usable fading margin of those hops on which they appear. Table 10.2 gives a subjective effect on voice communications for a degradation of BER.

The worst case of interference occurs when the interfering hop is operating in normal propagation conditions and the victim receiver is close to its threshold, due to it being in a faded condition. Errors in a digital receiver are caused by peaks in Gaussian noise which occasionally cause the decision circuit to mistake one level for another. Under the nonfading conditions, the signal received is usually 30–50 dB higher in value and the error ratio negligibly small. Interference sources which are peak-limited usually do not cause the decision circuits to make mistakes directly, but they do however, enhance the ability of the thermal noise to cause errors by occasionally producing a partial reduction in the eye opening (Section 5.2.5).

Table 10.2

Bit error rate (BER)	Subjective effect of value of BER
10^{-6}	Not audible
10^{-5}	Barely audible
10^{-4}	Audible but does not unduly affect understanding of what is spoken
10^{-3}	Disturbing, but speech is still understandable
10^{-2}	Most disturbing and speech is difficult to understand

10.1.1.3 Analog to digital interference

Adjacent-channel interference

Assuming that a noise power contribution of about 1–2 $pW0p$ is acceptable in each hop for each adjacent-channel-interference exposure, the following mixture of analog and digital systems can co-exist:

300 FM channels and 17.152 Mbit/s 4-phase signals, with 14 MHz spacing, by using cross-polarization decoupling (XPD $\geqslant$ 20 dB) between the analog and digital RF channels.

960 FM–FDM channels and 34 Mbit/s 4-phase signals, using 28 MHz spacing.

At the lower radio frequencies, where sharp cut-off RF filters are feasible, there is no need for cross-polarization. Also 70 Mbit/s 8–PSK and QPR systems can co-exist with cross-polarization decoupling (XPD = 20 dB) if the digital radio has a 35 MHz IF and a 24.5 MHz RF transmit filter bandwidth.

1800 FDM–FM channels and 34 Mbit/s signals with 29 MHz spacing co-exist with difficulty, since the digital-to-analog interference is difficult to suppress; however, using 8–PSK and QPR systems at 70 Mbit/s and 40 MHz spacing with XPD = 20 dB, co-existence is possible.

Using a modulation index of about 0.2 to 0.3, 8 level FM digital radio at 17 Mbit/s with a peak-to-peak deviation of 7 to 10 MHz can co-exist with a 300 FDM–FM channel analog radio if 14 MHz spacing is used; alternatively with 20 MHz spacing a 960 FDM–FM radio can co-exist. Sufficient cross-polarization decoupling is required.

Similarly, using a peak-to-peak deviation of 13–16 MHz and 28 MHz or 29.65 MHz spacing, a 960- or 1800-channel FDM–FM analog radio can co-exist. Cross-polarization decoupling is required to the 1800-channel case.

10.1.2 Attenuation tuning[3]

At a frequency-congested node, there is often a requirement to add yet another radio hop. In this situation, it is more than likely that asymmetric interference will exist. It is possible to reposition into the receiver the attenuation normally placed into the transmitter to pad down the power, and thus reduce the effects of interference from other co-frequency channels working into the node. When raising the output power from the transmitter of the new link, one must not cause the C/I value of any existing links to be exceeded. This must be considered in calculating whether or not the link may be added to the existing system. If it is assumed at the outset that it is possible to add the new link, but the C/I value produced by the existing system indicates that it will not function properly, then the following method may be used. Consider Figure 10.3(a) which shows two links that have been symmetrically attenuated, using the pads R_1 to R_4. The dotted paths shows the interfering signal paths. Consider only the receiver at site A; then the received signal power C_2 at site A is given by

$$C_2 = P_1 + G_1 + G_2 - A_0 - R_1 - R_2 \quad \text{dB} \tag{10.4}$$

and the interference power I_2 is given by

$$I_2 = P_4 + G_4 + G_2 - A_0 - R_4 - R_2 \quad \text{dB} \tag{10.5}$$

Thus the carrier-to-interference ratio $(C/I)_2$ at receiver 2 is

$$(C/I)_2 = (P_1 - P_4) + (G_1 - G_4) - (R_1 - R_4) \tag{10.6}$$

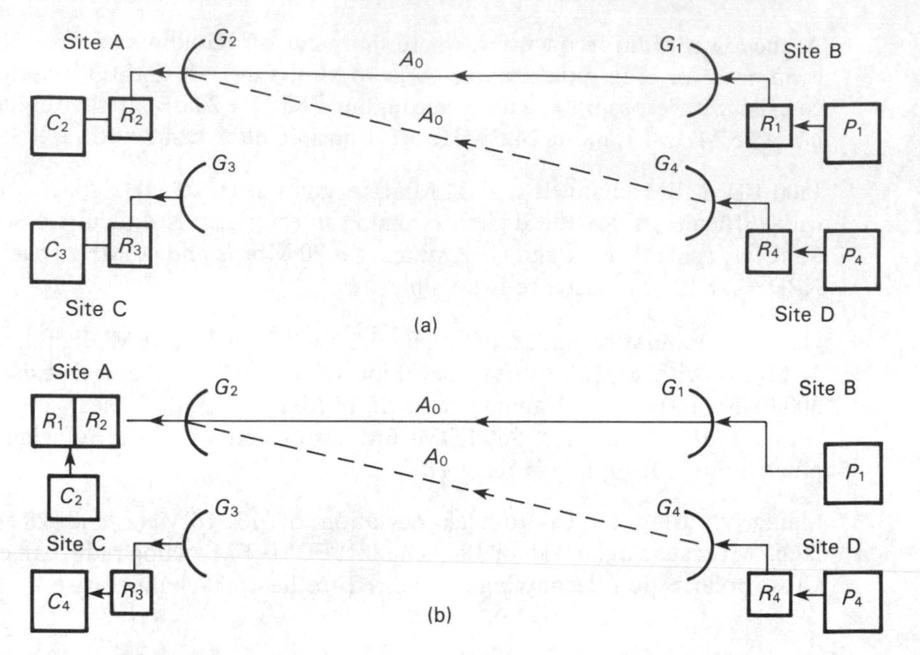

Figure 10.3 Interference paths to demonstrate the principle of attenuation tuning

where G_1 to G_4 are the antenna gains, and A_0 is the free-space path loss, which is assumed for convenience to be the same for both links. Consider Figure 10.3(*b*), where the transmitter power at site B is no longer padded down, but the pad is transferred to the receiver circuit of site *A*. The received power C_2 is now computed to be

$$C_2 = P_1 + G_1 + G_2 - A_0 - R_1 - R_2 \text{ dB}$$

This is the same as that given by equation 10.4, as expected. However, the interference power is now reduced by an additional amount R_1 to give;

$$I_2 = P_4 + G_4 + G_2 - A_0 - R_1 - R_2 - R_4 \text{ dB} \tag{10.7}$$

and to give a $(C/I)_2$ at site A as

$$(C/I)_2 = (P_1 - P_4) + (G_1 - G_4) + R_4 \text{ dB} \tag{10.8}$$

This means, on comparing equation 10.6 with 10.8, that C/I has been raised by the amount R_1, without any loss in received signal power at site A. The effective isotropic radiated power (EIRP) at site B has been raised, which will give rise to a higher level of interference at site C. Providing that this higher level is below the degradation value permitted, the link between sites C and D will still be able to operate.

10.1.3 Interference performance of digital radio systems

As discussed in Section 6.3.5, the value of C/N in the theoretical carrier-to-noise-ratio/probability-of-error curves for the various modulation schemes is the value for a system noise bandwidth which is the *minimum* required for the transmission of the signal. This minimum symbol bandwidth is known as the *double-sided Nyquist bandwidth* and is equal to the system symbol rate expressed in hertz. An alternative name for the double-sided Nyquist bandwidth is the *symbol rate bandwidth.* For example, if a 140 Mbit/s signal is to be transmitted using a 64–QAM radio transmitter, the symbol rate is $140/\log_2 64 = 23.333$ Mbit/s and the double-sided noise bandwidth is equal to 23.333 MHz. Because of problems in the design of filters, for transmitting in this bandwidth, this minimum bandwidth is never achieved on practical radios, except for those using partial response signaling. Strictly defined, C/N is the ratio of the unmodulated IF carrier power to the IF noise power. As discussed in Section 6.3.5, since some modulation systems such as PSK, APK, etc., have no identifiable carrier component and the transmitted spectrum conforms to a truncated $\sin x/x$ spectrum, C/N may be referenced sometimes to the *unmodulated* IF carrier power and sometimes to the *modulated* IF carrier power. Whichever is used must be specified, if results are to be compared, because the received IF signal power with modulation will be different from that without modulation. The amount depends on the modulation scheme. Again if C/N were determined at the demodulated output, thc value would be different from that measured at the output of the narrowest IF filter in the receiver chain, because the radio filtering after the measurement point alters the effective noise bandwidth and hence the total noise power. Again, to overcome this problem, the results of measurements of C/N should also be reported together with the associated measurement noise bandwidth, unless the carrier-power/noise-spectral density is used. The noise spectral density η, is independent of bandwidth, as discussed in section 6.3.5, and is given in watts/Hz. The effects on C/N and the probability of error (symbol error in M-ary schemes, see section 7.7.1) due to interference are similar to the effects of flat fading. Any interference signal present at the receiver antenna which has a signal spectrum that is contained within the passband of the RF and IF filters, will pass directly through the receiver into the demodulator. If the spectrum is outside either of these filters, the interfering signal will suffer attenuation distortion, which is proportional to the amount of out-of-band spectrum present. Assuming that no attenuation distortion to the interfering signal occurs, C/I will remain unchanged as it passes through the receiver circuits and both wanted and unwanted signal components are processed equally. The process of symbol regeneration with this interference on it, or increased noise due to low C/N, which results from flat fading, will be affected by the additional unwanted components being superimposed on the symbols and a continuum of levels will occur rather than the two or more allowed values. Due to the interfering signal having the same effect as white noise, its distribution can be assumed Gaussian, and the probability of error may be computed for a given level of interference. The theoretical curves for each modulation scheme presented in Chapter 6 actually show the probability of symbol error for a given carrier-to-noise ratio. Figure 10.4 shows typical curves (BER against C/N) for practical receivers. The

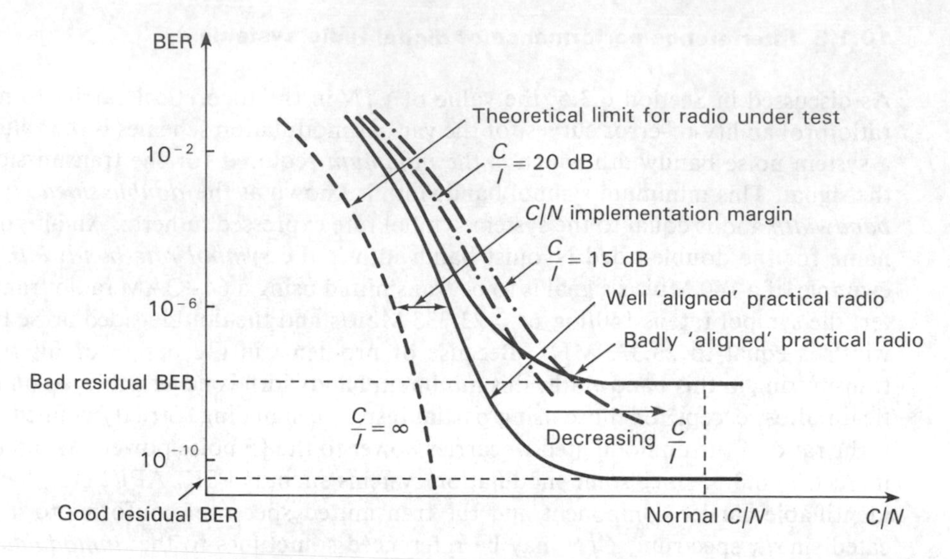

Figure 10.4 Effect of interference on BER: residual values for good and bad receivers

theoretical curve is shown by the dashed lined for comparison with that of the well aligned and adjusted receiver and similarly for a badly aligned or adjusted one.

For the well aligned receiver the performance curve is shifted to the right from the theoretical curve due to the imperfections and receiver intrinsic noise. This shift is called the 'C/N implementation margin'. There is also a tailing-off of the practical curve at the low BER end, due to the inherent ISI, noise and other practical imperfections. The residual BER of the well-aligned practical radio is, however, very low, typically $< 10^{-10}$. For the badly aligned system, the overall performance is degraded from that of the aligned system and is $> 10^{-10}$. The effect of interference is to shift the practical implementation curves further to the right. The amount of shifting is proportional to C/I. This is what one would expect, for the increase in interference (as C/I falls) is similar to the increase in thermal noise for a decrease in C/N. To maintain the same BER implies an increase in C/N. If the noise N is considered to be that at the output of the receiver, it can be represented by:

$$\begin{aligned} N &= 10 \log (kTWF) \text{ dBm} \\ &= -174 \text{ dBm} + 10 \log W + F \text{ dB} \end{aligned} \tag{10.9}$$

where k = Boltzmann's constant (1.38×10^{-23} J/K).

T = the absolute temperature in kelvins.

$kT = -174$ dBm at 15.34°C.

W = the IF bandwidth in hertz.

F = receiver noise figure in decibels.

Using equation 10.9, the carrier-to-noise ratio can be replaced by the RF input level to the receiver, as shown in Figure 10.1. For example, if the noise figure $F = 9$ dB, and the IF bandwidth is 28 MHz, then $N = -90.53$ dBm. Comparison of Figure 6.5 with Figure 10.1 shows that the implementation margin at a BER of 10^{-7} is 0.8 dB.

Some comparisons of various modulation schemes in the presence of carrier wave interference and also on a Rayleigh fading channel are given in Table 10.3.[4]

Table 10.3 illustrates the effect of an in-band carrier-wave interferer (10 dB or 15 dB down in power from the desired signal), on the value of C/N required for a BER $= 10^{-4}$. This situation can be modeled either for interfering side lobes from adjacent channels or for the main lobe from a co-channel interferer. Of the schemes for which data are available,[4] noncoherent FSK and BPSK are the most robust and show the least degradation from ideal performance, while the higher-level modulation schemes exhibit the most degradation.

Table 10.3 Performance of representative modulation schemes in the presence of carrier wave interference and on a Rayleigh fading channel

Type	Modulation scheme	C/N for BER $= 10^{-4}$ $C/I = 10$ dB	$C/I = 15$ dB	Average C/N for BER $= 10^{-2}$
	OOK: coherent detection	20	14.5	17 (#)
AM	OOK: envelope detection	–	–	19 (#)
	QAM		–	14
	FSK: noncoherent detection ($d = 1$)	14.7	13.3	20
	CP–FSK: coherent detection ($d = 0.7$)	–	–	13(*)
FM	CP–FSK: noncoherent detection ($d = 0.7$)	–	–	18(*)
	MSK ($d = 0.5$)	–	–	14
($d = 0.5$)	MSK: differential encoding	–	–	17
	BPSK: coherent detection	10.5	9.2	14
	DE–PSK	11	9.7	17
	DPSK	12	10.3	17
PM	QPSK	12.2	9.8	13.5
	DQPSK	> 20	14	20
	8–PSK: coherent detection	20	15.8	16.5
	16–PSK: coherent detection	–	>24	21
AM/PM	16–APK	–	–	18

Notes
*Assumes three-bit observation interval.
#Assumes optimum variable threshold.
d is the modulation index.

10.2 CHOICE OF INTERMEDIATE FREQUENCIES FOR HIGH-CAPACITY DIGITAL RADIO RELAY SYSTEMS[1]

CCIR Report 788[1] considers the choice of intermediate frequencies (IF) on the basis that it influences the feasibility, size, cost and reliability of such subsystems as modulators, amplifiers, mixers and demodulators. In addition, for digital radio systems operating above 100 Mbit/s, the IF is likely to affect the unwanted radiation and interference characteristics of the system and hence influence the channel filter characteristics. In the transmit modulator, the IF output of the modulator contains the phase-modulated IF signal whose frequency spectrum extends down to baseband frequencies, with components folding about zero frequency to produce unwanted signals in the IF passband. Filters can adequately suppress the baseband components, but the in-band fold-over components, when a switched modulator is used, cannot be filtered out and may be of sufficient level to contribute to the overall channel distortion. In addition, higher harmonics may leak into the IF passband and be transmitted. The suppression of these sources of interference is more easily done at high values of IF. The value of the transmitter IF should preferably be chosen to satisfy:

$$f_{IF} > \text{about } 3\, r_s \qquad (10.10)$$

where r_s is the digital symbol rate.

The IF used in FDM–FM systems is chosen to permit the interleaving of carrier frequencies and local oscillators in order to reduce potential interference. This arrangement is not that beneficial to a high-capacity digital system since the ratio of signal bandwidth to channel spacing is much higher and in-band oscillators therefore pose a possible interference problem. However the IF may be chosen as:

$$f_{IF} = (n - 0.5)\Delta F \qquad (10.11)$$

where ΔF is the channel spacing, and n is an integer.

A consideration in the choice of IF is therefore one of interference to and from adjacent channels and local transmitters, a factor which depends on the rejection provided by the channel filter and whose primary function is the determination of the digital channel performance. The frequencies at which interference may enter a system from outside the allocated RF band are the receiver local oscillator frequency; the receiver image channel frequency; and the IF.

The image channel is generally the most sensitive of these three and in the absence of adequate protection by means of channel filters, it may be necessary to protect the image frequency by suitable choice of IF. To totally protect both the image channel and the local oscillator from out-of-band interference, it would be necessary to choose an IF so that the image-channel and local-oscillator frequencies fell within the occupied band. However, such an arrangement exposes the image channel to increased interference from local in-band transmitters. The situation can be eased by appropriate additional filters which make the choice of IF less subject to interference criteria.

The possible causes of out-of-band radiation are: the transmitter local oscillator; and a transmitter unwanted sideband.

These radiations can be controlled by rejection in the transmit channel filter, which is aided by a high IF and by additional filters. Some intermediate frequencies in use are 140 MHz, 360 MHz, 1.2 GHz, 1.25 GHz and 1.7 GHz.

10.3 INTRODUCTION TO MICROWAVE FREQUENCY COORDINATION AND INTERFERENCE CALCULATIONS[5]

In the past the design of a radio link was concerned mainly with insuring that the path clearance criteria were met, and that the path and performance calculations met the objectives set for the system. Nowadays in most countries the engineering of a microwave route takes into consideration additional factors, such as:

- Other microwave frequencies which may be in use in the area concerned
- Radar stations which may cause interference to the system
- The establishment of an additional route which does not cause interference problems to existing systems, nor have problems due to interference from those systems
- The knowledge of existing satellite earth stations within the required coordination distance of the proposed microwave station
- The frequencies selected for the proposed route to permit coordination with existing locations at connection points or at points where the new route crosses other routes
- The possibility of the proposed site interfering with synchronous stationary satellites or future satellites above the equator

Usually, the authority concerned with the registration of all frequencies for a country or area has a complete record of existing microwave emitters, with such pertinent information as the latitude and longitude of each site location, power emitted in dBm or dBW, antenna types and gain, and the channel frequencies.

10.3.1 Calculations of interference for coordination

The procedures for coordinating a proposed path or making a study of an existing path to determine if harmful interferences may exist are started by considering the azimuths and path distances of all operating links in the immediate area of the proposed hop. Figure 10.5 shows the interference possibilities between stations of two paths.

The simple case for two hops, as shown in Figure 10.5, will be discussed, although in a real situation, the number of hops will usually be more than two. The same principles apply, however, to any number of hops in a given coordination area. The azimuths of sites 1 to 2, sites 1 to 3, sites 1 to 4, sites 2 to 3 and sites 2 to 4,

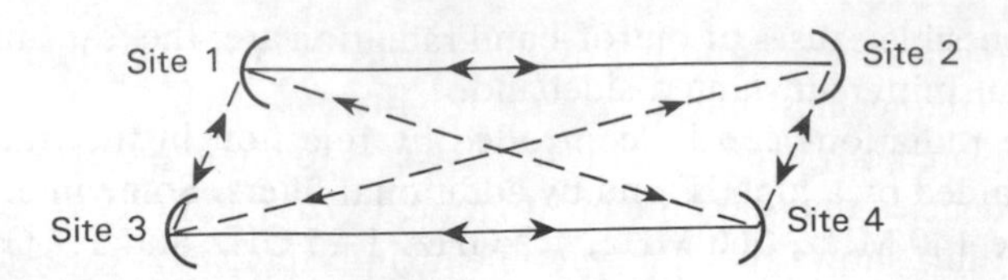

Figure 10.5 Interference possibilities between two hops

are all calculated together with the free-space loss for each path. Knowing the transmitted power from each of the sites, together with each of the transmitting and receiving antenna gains, we can determine the received level of each signal into a particular receiving site. At this point, the antenna decoupling factor or its discrimination against the interfering points have not been included. To do this part of the calculation, the discrimination angles between the normal transmission path and the interference paths must first be determined and each angle compared with the polar pattern of each antenna which permits the gain of each antenna that is a source of possible interference to the receiving site to be determined. With this information the values of C/I are calculated for the following paths: site 1 into site 4 from site 3; site 1 into site 3 from site 4; site 2 into site 3 from site 4; and site 2 into site 4 from site 3. The reciprocals are also considered, to determine C/I for site 4 into site 1 from site 2, etc. As can be judged from the number of calculations for only two hops, the number of calculations involved for larger numbers of hops can be handled only by a computer. Each value of C/I is then compared against that which is acceptable for the particular site. This means that if each site contains a different type of receiver, the permitted C/I for each site may be different. In a computer, a look-up table for specific frequencies[6] is usually loaded into the data base.

After the comparison of the permitted C/I, those interference paths which exceed the objective for proper operation of the existing or proposed hops are noted. It is not unusual for a computer interference analyzer program to consider a coordination distance of up to 200 km radius and check up to ten proposed paths against all existing paths at one time. Once the interfering paths have been identified, each path is considered individually for line-of-sight conditions and any diffraction attenuation which may exist for the normal and minimum value of k-factor. For an interference path which is under 120 km from the proposed hop, if line-of-sight conditions exist any time for $k = 4/3$ to $k = \infty$, the case is considered to be line-of-sight for the purposes of interference studies. If the path has earth blocking at $k = \infty$, the tropospheric scatter and diffraction loss to be expected at various percentages of the time may be calculated. After the search for a nonacceptable C/I due to other microwave radios operating in the area, the next task is to consider the proposed path in terms of any interference which may be caused by radar or other emitters; where such interferers exist, the same calculation process as described above is used to determine C/I for each site and check whether the value exceeds that permitted. Similarly, after this exercise has been completed, a search for interference from satellites must be made. The total inteference study, when done for a number of hops, is no mean feat, and is usually handled by a Government body. Private data banks on frequencies may be set up for the cases where the frequencies allocated to that private body are few. However, access to all the information existing on relevant frequencies may be difficult to come by, for military or other reasons. If the proposed link shows that the amount of interference is excessive, arising from or to existing links, a redesign is warranted. This may only be a matter of changing the output power of the proposed hop transmitters, or of rearranging the hop attenuators (if they exist), as described in Section 10.1.2. However, it may mean that either the frequencies must be changed or different radio equipment, antennas, etc., are used.

Below is a step-by-step method which may be used to calculate the interference into a particular digital radio receiver at a nodal point. These calculations are included in this chapter, for a case where the reader may wish to calculate manually the interference produced by a suspect hop (or hops), in the near vicinity, which may cause the objective C/I at threshold and the consequent threshold degradation to be exceeded over those values used in the path calculations given in Chapter 9. For each node considered, the interference signals from the outlying transmitters into the receivers at the nodal points are calculated. Interference from the transmitters of the nodal stations into the outlying receivers for which wanted and interference signals propagate along the same path, are fading-correlated to a large extent and are disregarded.

Step 1. Hop number of the wanted carrier as used in the path calculations for the same proposed hop

Step 2. Hop number of the interfering carrier used on the path calculations for that hop

Step 3. Threshold RF receiving level at the specified BER in the absence of RF interference for the hop specified in step 1

Step 4. Transmitter output power of the interfering station for the hop designated in step 2

Step 5. Free-space path loss along the interfering path

Step 6. Feeder loss at the interfering station

Step 7. Feeder loss at the disturbed station

Step 8. Nominal antenna gain at the interfering station

Step 9. Nominal antenna gain at the disturbed station

Step 10. Branching loss

Step 11. Adaptor and connector loss

Step 12. Great circle distance between the interfering path and receiving path at the receiving antenna of the disturbed station

Step 13. Relative polarizations of the interfering and receiving signals (X = cross-polar, C = co-polar)

Step 14. Antenna discrimination. For co-polar signals, the figure is derived from the co-polar radiation pattern of the receiver antenna of the disturbed station. For cross-polar signals the figure results from the combination of (*a*) the cross-polar discrimination at the receiving station at the angular distance given in step 12; (*b*) co-polar discrimination at the receiving station at the angular distance given in step 12, plus the on-beam attenuation of spurious polarization at the interfering station.

Step 15. Selectivity. This is the influence of the overall selectivity on the ratio:

$$\text{(Energy received from wanted spectrum)}/\text{(Energy received from interfering spectrum)} \quad (10.12)$$

It is assumed that the wanted and interfering carriers are modulated with random bit sequences.

Step 16. Individual carrier-to-interference ratio. This is equal to $(C - I)$ dB, where C is the receiver threshold, and I is given by

$$\begin{aligned} I &= \text{Transmitter output} - \text{free space loss} - \text{feeder loss} \\ &\quad + \text{antenna gain} - \text{branching loss} - \text{adaptor/connector loss} \\ &\quad - \text{antenna discrimination} - \text{selectivity} \\ &= \text{step } 4 - \text{step } 5 - \text{step } 6 - \text{step } 7 + \text{step } 8 + \text{step } 9 - \text{step } 10 \\ &\quad - \text{step } 11 - \text{step } 14 - \text{step } 15 \end{aligned} \quad (10.13)$$

Step 17. RF interference peak factor. The peak factor of the overall interference from various similar interferers at the same receiver is derived from:

$$PF = 10 \log \left\{ \frac{[1 + (I_2/I_1)^{1/2} + (I_2/I_1 \times I_3/I_2)^{1/2} + \cdots +]^2}{[1 + I_2/I_1 + (I_2/I_1 \times I_3/I_2 + \cdots +]} \right\} \quad (10.14)$$

Step 18. The cumulative carrier-to-interference ratio is given by:

$$(C/I)_{\text{total}} = C/(I_1 + I_2 + \cdots + I_n) \quad (10.15)$$

for n different interferers.

Step 19. The final degradation of the BER threshold is derived from the overall C/I (step 18) and its peak factor (step 17). See also Step 19 in Section 10.3.2.1.

10.3.2 Explanation of the calculations

10.3.2.1 Peak factor[7]

The peak factor is the peak value of the interference envelope R divided by the RMS value of the modulation envelope. This parameter is used to determine the C/I for a given C/N at a specific BER, for coherent PSK (CPSK) modulated systems.[5] An unfiltered angle-modulated signal has a peak factor of zero. CCIR Report 388–5[2] shows curves of C/I versus C/N for bit error probabilities of 10^{-3}, 10^{-5}, 10^{-7} and 10^{-9} respectively. The curves apply to single- or multiple-interference cases. The advantage of using these curves is that an exact statistical distribution of the interference does not have to be determined. This method,[5] requires knowledge only of the value of C/N at the demodulator input, the peak-to-RMS value or peak factor and the ratio of the powers of the wanted signal and interference C/I. For M-ary CPSK systems:

$$C/I(M\text{-ary}) = C/I(\text{binary}) - 10 \log \left[\sin^2(\pi/M)\right] \tag{10.16}$$

that is, for a QPSK system, where $M = 4$, the value of C/I for a particular C/N and peak factor would be determined first for the binary system and then converted to QPSK. The value of the probability of error can also be scaled for a specific C/I to an equivalent P_e for an M-ary system, by the following:

$$P_e(M\text{-ary}) \text{ at } (C/N)_1 = 2\ P_e(\text{binary}) \text{ at } (C/N)_1 + 10 \log \left[\sin^2(\pi/M)\right] \tag{10.17}$$

Figure 10.6 shows one of the above-mentioned sets of curves for determining the value of C/I when the peak factor and C/N at the demodulator input are known. For a single angle-modulated signal (FM, PM, CPSK or DPSK) interfering with binary CPSK, the results can be directly obtained from Figure 10.6, or for different BER, those in CCIR Report 388–5, when the curves for zero peak factor are used. From these the following general conclusions can be drawn:

When the interfering signal power is equal to, or larger than, the thermal noise power, the effect of the angle-modulated interference is considerably less than that of an equal amount of white Gaussian noise power.
When the interfering signal power is small compared to the thermal noise power, the effect on error rate can be safely estimated by equating the interfering signal to that of white Gaussian noise of equal power.
At a given C/I, the higher the M-ary level, the more vulnerable is the system to interference.

Step 19. In addition to the above method (CPSK BER versus C/I, using the peak factor), a method which can be used with more than one interferer, was developed earlier by Prabhu.[8] This method determines the value of C/I for M-ary CPSK and also families of curves are presented with one interferer for 8– and 16–CPSK. The prime purpose of the interference calculations is to obtain a value of the degradation in the received signal level due to interference. This degradation will directly affect the value of the effective fade margin, and consequently the overall performance of the hop. As mentioned previously, when considering the path calculations, this degradation factor for a specific value of C/I (usually 15 dB – worst case) is taken. In conditions where it is probable that the value of C/I is less than 15 dB, the effects of C/I on the C/N threshold to determine the degradation of BER must be calculated. The measurement of interference effects is not that easy in the real-life situation, and results mainly rely on calculations. In addition to the effects of interference, the BER versus C/N curve, for a practical off-the-shelf type of equipment, may already suffer a degradation due to the imperfections and misalignment of the equipment. This has been discussed in Section 10.1.3. The degradation of the receive

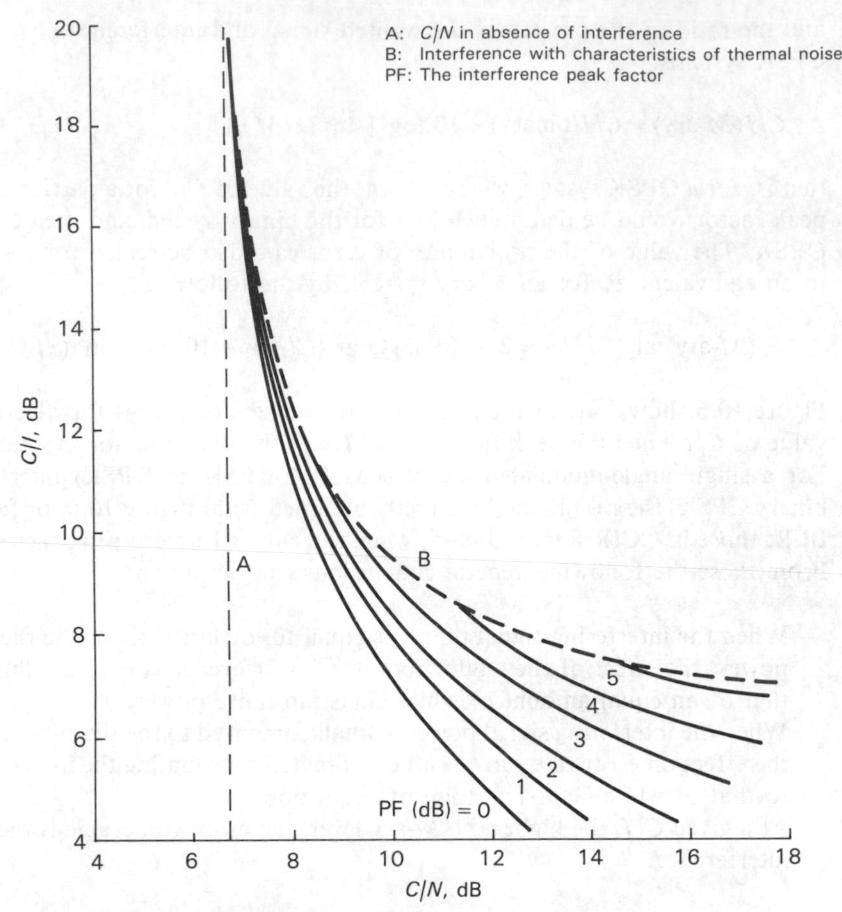

Note: The curves are theoretical and take no account of practical system restraints.

Figure 10.6 *C/I* versus *C/N* for 10^{-3} BER and different interfering peak factors (Courtesy of CCIR, Report 388–5, Reference 2)

threshold (or C/N), should be taken from this practical level, or from the level which is guaranteed by the manufacturer. In one of the tables below, the guaranteed value of BER for a given C/N is given, so that any calculations made to determine the real receiver threshold degradation will be valid in the field situation.

10.3.2.2 Probability of error for different *C/I* for different modulation schemes

QASK

For QASK signals[9] a method for determining the upper bound on P_e has been developed. However, for $C/I > 25$ dB the bound becomes too pessimistic to be of any use. Some typical values are given in Table 10.4.

Table 10.4 Approximate values of C/N for different P_e in 16–QASK systems

P_e	C/I	C/N	P_e	C/I	C/N
10^{-3}	∞ (exact)	17.8	10^{-3}	30	19.5
10^{-3}	∞ (practical)	18.4	10^{-3}	25	21.9
10^{-4}	∞ (exact)	19.0	10^{-4}	30	20.6
10^{-4}	∞ (practical)	19.9	10^{-4}	25	24.0
10^{-6}	∞ (exact)	20.8	10^{-6}	30	23.0
10^{-6}	∞ (practical)	21.7	10^{-6}	25	29.8

DPSK[10,11]

For M-ary DPSK where $M = 4$, Rosenbaum has produced graphs of the expected P_e versus C/N for different C/I. Table 10.5 tabulates some typical results from the graphs presented in his paper[10].

Table 10.5 Approximate values of C/N for different P_e and C/I in 4–DPSK Systems

P_e	C/I	C/N	P_e	C/I	C/N
10^{-3}	∞	12.6	10^{-3}	12	18.2
10^{-3}	15	15.4	10^{-3}	18	14.1
10^{-3}	20	13.6			
10^{-4}	∞	14.2	10^{-4}	12	20.5
10^{-4}	15	17.4	10^{-4}	18	16.2
10^{-4}	20	15.3			
10^{-6}	∞	16.2	10^{-6}	12	23.2
10^{-6}	15	20	10^{-6}	18	19.9 (2 interferers)
10^{-6}	20	17.5			19.8 (6 interferers)

QAM

For M-ary QAM where $M = 16$, some typical figures applicable to NEC 140 Mbit/s, 16–QAM radio are given in Table 10.6.

Example 1

Determine the C/N degradation at a P_e of 10^{-3}, for a NEC 16–QAM, 140 Mbits/s, 15 GHz radio in which the C/I has been determined to be 20 dB from the interfering effects of a co-channel.

Solution

From Table 10.6, the guaranteed value of C/N at a P_e of 10^{-3} for no interference ($C/I = \infty$) is given as 20.5 dB. At a C/I of 20 dB, the value of C/N for the same P_e of 10^{-3} is found to be 21.7 dB. Thus the C/N degradation is (21.7 – 20.5) = 1.2 dB more than the guaranteed C/N for $C/I = \infty$.

Example 2

Determine the C/N degradation at a BER of 10^{-3}, for a NEC 16–QAM 140 Mbit/s, 15 GHz radio in which the C/I has been determined to be 6 dB from the interfering effects of an adjacent-channel interferer, whose carrier is 40 MHz away from the victim carrier.

Table 10.6 16–QAM P_e versus approximate values of C/N for different C/I

P_e	C/I(co-channel)	C/N	P_e	C/I (adjacent)	C/N(40 MHz separation)
10^{-3}	∞ (exact)	17.6	10^{-3}	∞ (exact)	17.6
10^{-3}	∞ (practical)	18.5	10^{-3}	∞ (practical)	18.5
10^{-3}	∞ (guaranteed)	20.5	10^{-3}	∞ (guaranteed)	20.5
10^{-3}	30	18.8	10^{-3}	10	18.8
10^{-3}	25	19.5	10^{-3}	6	20.0
10^{-3}	22	21.1	10^{-3}	4	21.2
10^{-3}	20	21.7			
10^{-4}	∞ (exact)	19.0	10^{-4}	∞ (exact)	19.0
10^{-4}	∞ (practical)	20.2	10^{-4}	∞ (practical)	20.2
10^{-4}	∞ (guaranteed)	22.5	10^{-4}	∞ (guaranteed)	22.5
10^{-4}	30	20.8	10^{-4}	10	20.8
10^{-4}	25	21.7	10^{-4}	6	22.6
10^{-4}	22	23.6	10^{-4}	4	25.7
10^{-4}	20	26.3			
10^{-6}	∞ (exact)	20.9	10^{-6}	∞ (exact)	20.9
10^{-6}	∞ (practical)	23.0	10^{-6}	∞ (practical)	23.0
10^{-6}	∞ (guaranteed)	25.4	10^{-6}	∞ (guaranteed)	25.4
10^{-6}	30	24.0	10^{-6}	10	24.5
10^{-6}	25	26.4	10^{-6}	6	31.2
10^{-6}	22	34.0	10^{-6}	4	∞
10^{-6}	20	∞			

Notes
The value of input receiver level (R_x) for the 4 GHz band is given by $C/N - 95 = R_x$ dBm.
The value of input receiver level (R_x) for the 15 GHz band is given by $C/N - 92.4 = R_x$ dBm.
The guaranteed value of BER is measured at the baseband and is for a one-hop system (no interference).

Solution
From Table 10.6, the guaranteed value of C/N at a BER of 10^{-3} for no interference ($C/I = \infty$) is given as 20.5 dB. At a C/I of 6 dB, the value of C/N for the same BER of 10^{-3} is found to be 20 dB. Thus the C/N degradation is $(20.5 - 20.0) = 0.5$ dB less than the guaranteed C/N for $C/I = \infty$.

Example 3

Determine the total degradation in the fade margin for the same radio specified in Examples 1 and 2 if both co-channel and adjacent-channel interferers are present at the levels given in both of those examples.

Solution
The total degradation will be the sum of the degradations of Example 1 and Example 2, i.e. $-0.5 + 1.7 = 1.7$ dB, as -0.5 dB cannot be considered as a degradation. If this degradation means that the receiver signal level must be raised by 1.7 dB from the case where no interference occurs, the flat fade margin FM_a must be decreased by this amount unless provision has already been made for this degradation (see item 30 of the path calculations, Section 9.2.2).

10.3.2.3 Antennas

ANTENNA SIZE

The antenna size refers to the diameter of the radiating aperture for a parabolic antenna. The size of the radiating aperture is *less* than the actual physical diameter of the dish. A normal parabolic antenna will incorporate a rolled rim or turnback-type rim to provide strength or to provide an attachment surface for an absorber lined shroud. The thickness of the rolled rim can be typically up to 50 mm, and the thickness of the turnback rim can be typically 150–200 mm.

FREQUENCY BAND

This is defined to be that continuous band of frequencies over which an antenna will operate. In general the antenna will operate over one band of frequencies only. Higher demands upon existing tower installations have meant that an important consideration in antenna design is to have an antenna which can support more than one frequency band. Often these bands are widely separated, and the antenna in this case is known as a 'dual-band' antenna. In addition to the two-band facility, the same antenna may be designed to provide cross-polarized operation. In this case the antenna will be equipped with four output/input ports. The factors which influence the final frequency band of an antenna are based fundamentally on the requirement of the antenna performance characteristics. A good VSWR, or a low VSWR, is difficult to achieve over a wide frequency band. Physical characteristics, such as the change of impedance of the waveguide with the frequency, or the waveguide fundamental mode operation will prevent a good VSWR from being obtained over a wide frequency band. Even if it were possible to design an antenna with a wide frequency band, the practical aspects of separating the transmit and receive frequency channels in the waveguide branching networks would lead to poor isolation between the channels. For these reasons and others, the operating band of a microwave antenna fed by a waveguide, is not more than ± 6 to 7 percent of the center frequency.

ANTENNA GAIN

The transmit or receive gain of an antenna is the ability of the antenna to direct the RF energy in a specific direction, or to receive energy from a specified direction, respectively. For line-of-sight radio links as opposed to multipoint radio systems, such as mobile radios, etc., the requirement is for the transmitting antenna to emit energy in single direction only, that direction being towards the next station down the line in the radio relay system. The definition of gain is the difference in power density at the far field point, between the antenna under consideration and the power density at that same point, if the antenna was replaced by a fictitious antenna which radiated energy equally in all directions (isotropic). Such an antenna cannot be realized in practice, but is normally used as a reference. The gain of an antenna is influenced primarily by its frequency of operation and its diameter. The theoretical gain of a parabolic antenna is given as

$$\text{Parabolic antenna gain } G_{\max} = 20 \log D - 20 \log \lambda + 10 \log n + 9.943 \text{ dB} \tag{10.18}$$

where D is the dish diameter, λ is the wavelength of the center frequency and n is the aperture efficiency given by,

$$n = S_e/S \tag{10.19}$$

where S_e is the effective area of the receiving antenna in the direction of the transmitter and S is the physical area of the antenna. The units of D and λ are the same, and the unit of S_e and S is the square of the unit of D and λ.

The aperture efficiency indicates that the gain is less than ideal in practice, for the ideal case would be where the aperture area is the same as the physical dish area S. A typical antenna efficiency is 50–60 percent. The factors which limit the efficiency are:

Illumination efficiency. This is dependent on the radiation pattern of the feed and is a measure of how uniform the aperture field is.

Spillover efficiency. This is also dependent on the radiation pattern of the feed and is a measure of how much energy radiated from the feed does not strike the reflector and therefore will not contribute to the forward radiation.

Ohmic loss. This is due to the ohmic loss between the antenna waveguide flange and the radiating feed horn aperture.

Cross-polar loss. This loss is due to the property of any antenna to convert some of its co-polarized energy into a small amount of cross-polarized energy. Thus, a small portion of the co-polar energy is lost.

Mismatch loss. This is a loss that arises from the antenna possessing a VSWR greater than unity. The antenna will thus reflect some of the energy back into the waveguide.

Ohmic, cross-polar and mismatch losses represent those losses that are present, to a greater or lesser degree, in any antenna. In addition to these losses are those which are configuration-dependent and may not require to be taken into account. They are:

Shield loss. If an absorbant lined shroud is attached to the periphery of the antenna, a significant improvement in the far-out side-lobe and back-lobe attenuation is produced. If the gain of the antenna under test is assessed by the method of directivity integration (which will be discussed later), the integration of the pattern of a shielded antenna can lead to an erroneous value for directivity. This value, normally higher than the correct value, is due to the integration taking into account a pattern area with an artificially low level.

Radome loss. If a radome is used, the loss is difficult to obtain, because the radome possesses an ohmic or dissipative loss and also will in general degrade the directivity of the antenna. The directivity, however, can be assessed by integration with and without a radome fitted. The integration with a radome fitted is usually calculated from available loss tangent figures. In addition to these considerations, the radome loss is not constant with time, and is heavily dependent on the ageing of the radome.

Strut loss. The feed of a parabolic antenna is supported, so that the radiating aperture of the feed is located near to the parabola's focus. In certain circumstances, these supports can be projected in a plane, so that the effect on the radiation pattern is not observed (e.g. if the supports are located in the 45° plane). In this case, the blockage effect of the struts has to be considered by reducing the antenna gain by the proportional area of the struts projected onto the antenna aperture.

Other losses are implicitly contained in the radiation pattern. These include losses associated with surface inaccuracies such as deviations from a true parabolic surface, focal length setting errors, feed misalignments, feed/waveguide band scattering effects and vortex plate effects.

ANTENNA GAIN MEASUREMENT

The measurement of gain is complex. Of the different methods which may be available, four will be briefly discussed.

Gain by comparison. The received signal obtained from the antenna under test is compared with the signal received from an antenna with a known gain, which has been obtained using the same test set-up. From the difference in signal power levels the gain of the unknown antenna can be deduced. The antenna of known gain can be a standard pyramidal horn for which accurately known gain values can be determined by rigorous calculation. In principle, this method is accurate. The problems which can occur are due mainly to the effects of supporting steelwork on the gain and radiation pattern of the reference horn and the fact that in general the difference between the beam width of the reference horn and that of the antenna is very large. If the gain is measured in a far-field outdoor test range, ground reflections can distort the results. Small variations in the received level can also be obtained from VSWR effects, due to the differences in the antennas' impedance.

Gain by directivity integration. This method relies on the radiation pattern of the antenna to be considered. The directivity g can be expressed as:

$$g = 4\pi P / P_t \tag{10.20}$$

where P is the power density in the beam region, per unit solid angle, and P_t is the total power radiated. P is obtained from the measurement of the beam level on the antenna test range equipment. P_t can be expressed as the integral of the function of the radiation pattern and is derived from the integration of the radiation pattern of the antenna over a whole sphere. The impracticality arises on a test facility, where movement for testing the antenna is made in one plane only. Examinations of the radiation pattern of a nearly circular symmetric antenna suggests that the integration over the whole sphere can be reduced to one over a single angle only, providing the average directivity value is taken between two orthogonal planes, where the electric field vector is either horizontal or vertical. The integration required for the directivity value can be found either by hand digitizing the radiation pattern (as discussed later) of the antenna, using the pattern recorded on rectangular graph paper, or by performing a real-time integration on the test range, so that the antenna pattern in decibels as a function of angle is fed into analog-to-digital convertors. The outputs at various angles from the A/D convertors are then fed into a microprocessor where up to several hundred data points are stored. After taking the radiation pattern, the computer can be requested to perform the integration upon the stored data points. Once the directivity value has been determined, the gain of the antenna is found by subtraction of the appropriate loss terms, which have been determined by calculation, from measured information. The factors which can influence this method are: pattern distortion caused by ground reflections; pattern compression caused by overloading the receiver; inaccuracy of the received signal level output and of the angular information obtained from the synchro-motors on the antenna turntable; digitizer errors; inaccuracy of assumptions made in deriving the theoretical results; inaccuracies associated with no uniform amplitude and phase distributions over the antenna under test; and inaccuracies in the derivation of the loss terms. The total error that could occur is found by taking the RMS sum of all of the error terms. This will be stated as the boundary of accuracy of the results of gain measurement by directivity integration.

Gain by absolute power measurement. This method makes use of the Friis transmission formula to equate received power to a combination of transmit power, receiver antenna gain, transmitter antenna gain, distance of separation and frequency. If the transmit antenna is taken as a gain reference, the transmitter power is known and the receiver power is measured. From this the gain of the antenna under test can be determined. The problems associated with this method are: the significant difference between the transmitted and received powers when the distance of separation is large can lead to difficulties in obtaining accurate measurements of receiver signal levels. Also the proximity effect of the support steelwork and the VSWR interaction effects at both transmitter and receiver antennas add to the inaccuracies. Other inaccuracies are those due to ground reflections and to problems associated with the near field. If the separation distance is not greater than the far field distance criteria (far field distance $= 2\,D^2/\lambda$), a correction factor for the near field effects of radiation from apertures has to be taken into account. These near field effects are that all points of the transmitter antenna aperture contribute to all points of the receiver antenna aperture. The correction factors are known accurately for fictitious devices such as uniform amplitude taper aperture distributions, but only approximately for real-life aperture distributions. Even if the far field criterion is satisfied, there is a small error from ignoring the separation distance. Due to these factors this method is infrequently used.

Durcell's method. This is another infrequently used method. In it the antenna under test is oriented to radiate onto a flat metallic sheet. The sheet is perpendicular to the antenna's bore-sight axis (main beam axis), so that it will give rise to a significant and measurable VSWR contribution in the antenna under test. The change in observed VSWR can be equated to the change made in separation distance and an equation for gain can be derived. This method has been found suitable for low-gain horns. The problems of applying this method arise from the nonperpendicularity of the sheet to the antenna axis, from the lack of flatness of the sheet causing phase errors, which reduce the quantity of signal power coupled back into the antenna under test, and from the interaction effect which is significant for close distances of separation. This last effect is that energy reflected by the antenna under test is reradiated and received for a second time, etc., by the antenna under test.

ANTENNA RADIATION PATTERN

As the antenna is a reciprocal device, it operates identically, whether transmitting or receiving. For present purposes, the antenna will be treated in this section as being only receiving. The radiation pattern is defined to be the response of the antenna to a signal of constant power level being transmitted towards the antenna from different directions. The antenna response power is measured at the output flange of the antenna. In measuring, it is normal practice to rotate the antenna under test, in a horizontal plane, through a full 360°, the response power being measured at the output port flange by a receiving set. The transmitting antenna during these tests is fixed at a convenient location some distance away. Measurements in other planes can be accomplished by simply rotating the antenna under test and changing the latitude of the transmitting antenna after each series of horizontal plane rotational measurements are made on the receiving antenna. In this way, the radiation pattern over the complete sphere can be obtained. This procedure, is however, extremely lengthy and fortunately can be simplified in several ways. First, it is only the feeds of the antennas, if they are assumed circularly symmetric, that need to be rotated and secondly, due to the use of either a vertical or a horizontal electric field in most terrestrial microwave radio applications, the antenna pattern needs to be measured only in these two planes. The E- and H-plane patterns give the response of the antenna, when the electric field E, of both transmitting and receiving antenna (antenna under test), is horizontal (horizontal polarization) and similarly, when the electric field of both transmitting and receiving antenna is vertical (vertical polarization). If it is required that the antenna should be dually polarized, or to be singly polarized and operate in a mixed polarization environment, then individual patterns for horizontal and vertical polarization are required where both transmitting and receiving antennas have the same polarization, and individual patterns for horizontal and vertical polarization are required, where both transmitting and receiving antennas are orthogonal. In this way, there are two co-polarized plane and two cross-polarized plane patterns required for each frequency of interest. The radiation patterns are processed to form a radiation pattern envelope (RPE). Basically, the RPE is prepared from pattern information as follows: the E and H co-polarized patterns are taken, at not less than three frequencies, these being the bottom, mid and top frequency of the specified antenna operating band. The six patterns are overlaid on a light table facility and the worst-case peaks are drawn on an intermediate sheet of paper. Since the antenna will rarely be exactly symmetrical about the bore-sight, owing to scattering and blockage effects and random surface errors, etc., due account is taken of the left- and right-hand sides of the pattern. A symmetrical group of straight line segments is devised (apart from the main beam region) and arranged to fit over the worst-case peaks overlay, leaving a specified margin. The amount of the margin depends on the quality of the dish employed and also on the angular region under consideration. The series of straight line segments is drawn on a specially printed graticule and is the co-polarized RPE. A cross-polarized RPE is devised using the same procedure as above and is superimposed upon the same form as the co-polarized RPE.

The RPE produced as described provides a system planner with a very convenient graphical representation of the radiation pattern envelope. However, a variety of factors can affect the accuracy of the recorded radiation pattern. Similar errors to those discussed for the measurement of the antenna gain also occur. These errors arise systematically due to the amplitude and phase front of the received signal striking the aperture of the antenna under test. This may be due to the far field criterion not being entirely satisfied or because the transmitting antenna is too large compared with the receiving antenna. Random errors occur in the amplitude and phase of the received signal due to the effects of ground reflections. For accurate reading, the antenna must be rotated in the plane which intersects the antenna under test and the transmitting antenna. This is in practice difficult to achieve. The effects of low signal-to-noise ratio and compression and nonlinearity effects in the receiver also add to errors. The efficient RF sealing of the cables, waveguide, receiver apparatus, etc., must be insured, especially when measuring the pattern of low-back-lobe antennas if errors from these sources are not to become significant. The systematic amplitude variation should be less than ± 0.25 dB and the random amplitude variation expected should be within ± 0.5 dB, if a reasonably accurate pattern is to result. The systematic variation of the phase front should be within 22.5° (far field criterion). The systematic errors mentioned above can be minimized by the appropriate choice of transmitting antenna size and of the distance between the two antennas. The random errors also can be reduced by aligning the transmit antenna such that the first null of the antenna is directed to that point on the ground that would be the specular reflection point for the ground reflected ray.

REGIONS OF THE RADIATION PATTERN

Figure 10.7 shows antenna radiation patterns for 13 GHz systems and for 13 GHz low-capacity systems.[12] Figure 10.8 shows the radiation pattern of a typical antenna with a directivity of 41.3 dB and an efficiency of 62.57 percent. This antenna has no shields or radomes attached. In Figure 10.8, the main beam and the first ten near-in side lobes are influenced primarily by the size and frequency of the antennas, and secondarily by the form of the illumination from the primary feed. Surface errors also cause significant distortion of the pattern in this region. The intermediate region is influenced mainly by the illumination of the edges of the dish, scattering from bends and, with increasing angle, the shape of the radiation pattern from the primary feed. Scattering from other obstacles, such as vortex plates, attachment bolts, guy wires, etc., also affects the amplitude levels in this region. The spillover region is caused by the increasing level of feed radiation as the angle is increased, just before the pattern is cut off by the interception of the reflector. As a rule of thumb, the level will usually be at 0 dBi, to within 1 or 2 dB, on an unshielded antenna. When the angle is increased further, the rear lobe region is entered and the behaviour depends on the edge illumination. As the angle approaches 180° degrees, the level begins to rise to a peak, called the back lobe, exactly at 180°. Again, as a rule of thumb, in a standard antenna, the level of the back lobe is -7 dBi to within 1 or 2 dB. Inclusion of shields or edge geometry attached to the periphery of the reflector can reduce rear-lobe and back-lobe levels drastically. This is described in more detail later in the chapter.

Many antennas have a dynamic range in excess of 70 dB, i.e. the ratio of the main beam level to a side lobe in the rear-lobe region. Due to the impracticality of designing a receiver with this dynamic range, which will maintain an acceptable

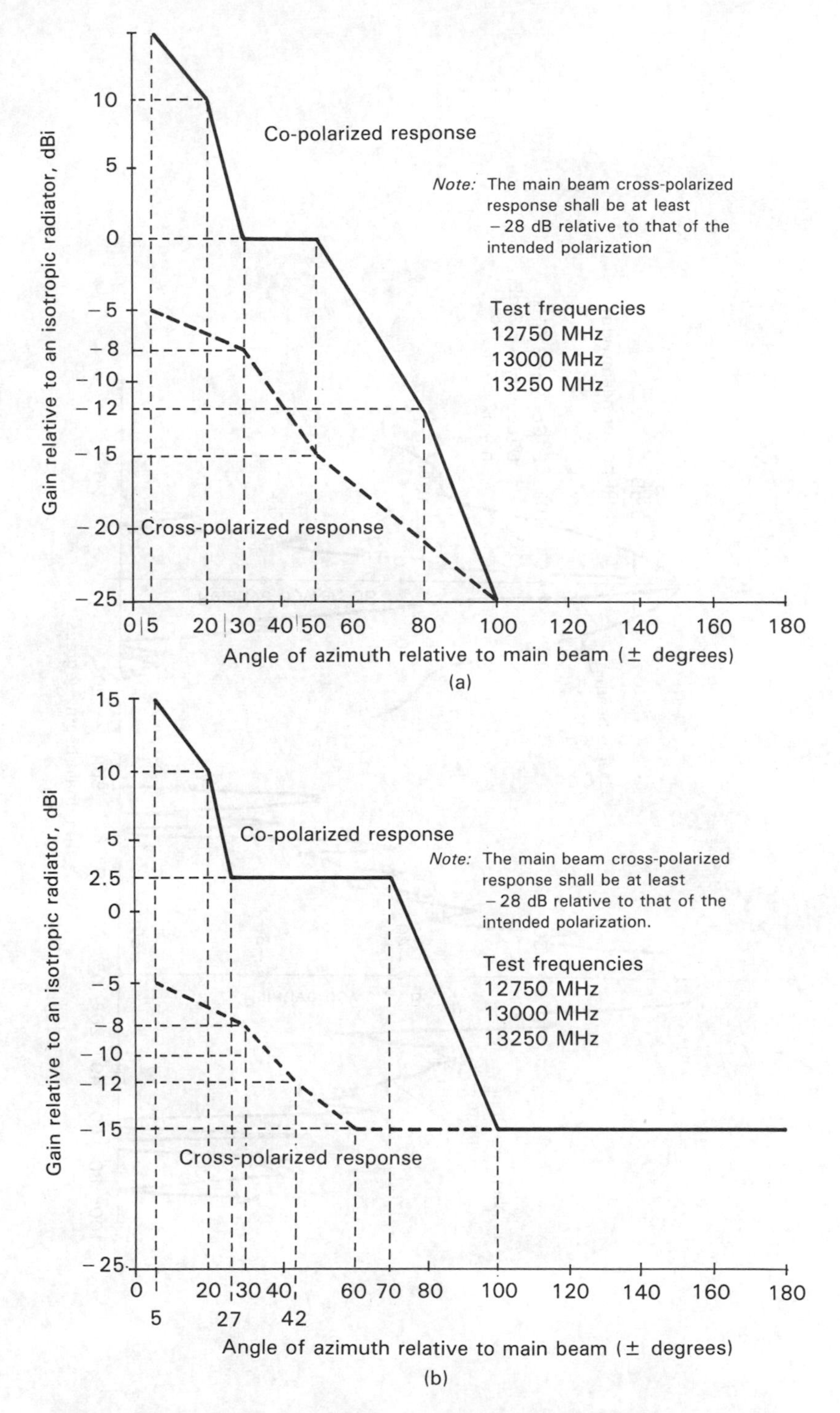

Figure 10.7 Antenna radiation patterns for: (a) 13 GHz systems, and (b) 13 GHz low-capacity systems (Courtesy of Department of Trade and Industry, Reference 12)

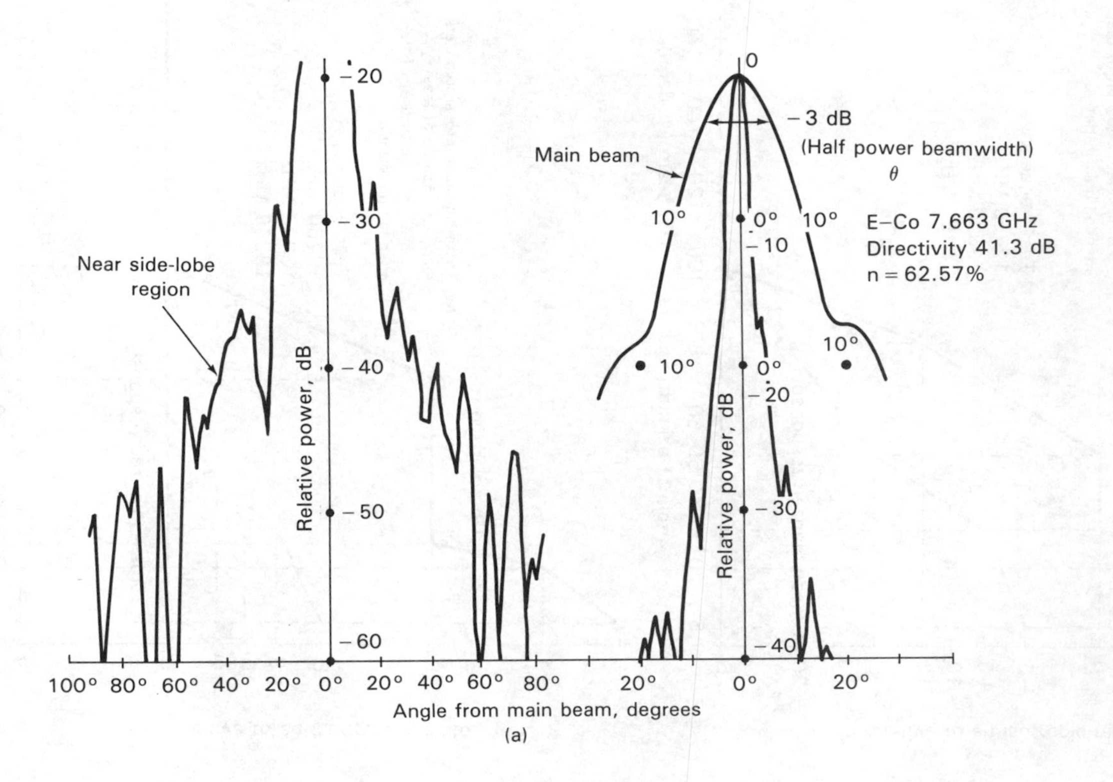

Near side-lobe region
Relative power, dB
−20
−30
−40
−50
−60
100° 80° 60° 40° 20° 0° 20° 40° 60° 80°
Main beam
0
−3 dB
(Half power beamwidth)
θ
E–Co 7.663 GHz
Directivity 41.3 dB
n = 62.57%
10°
0°
−10
−20
−30
−40
20° 0° 20°
Angle from main beam, degrees

(a)

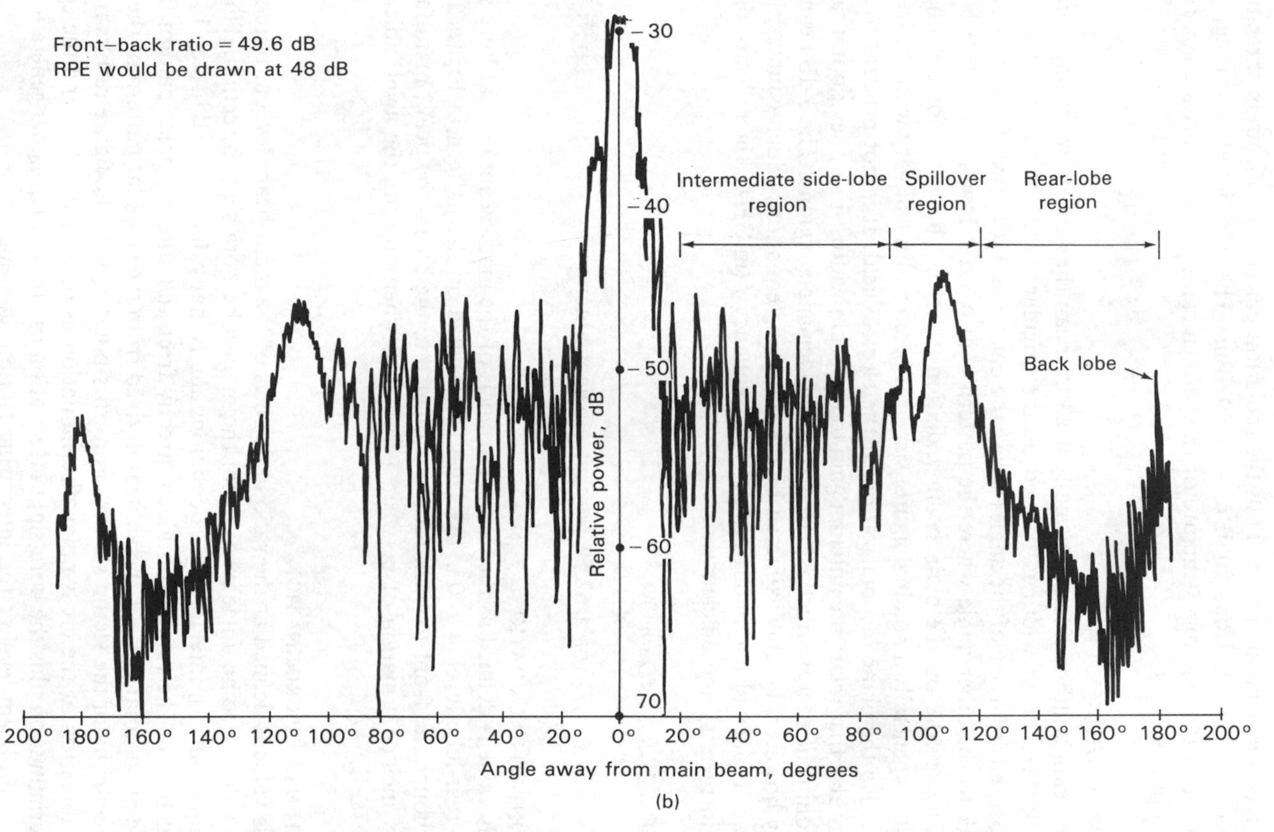

Figure 10.8 Typical radiation pattern of an Andrew HP6–71 antenna; (a) angle from main beam; (b) angle away from main beam

linearity performance, receivers which have a dynamic range of only 40 dB are usually chosen. To extend the dynamic range to that of the antenna, the transmitter power is increased when the antenna pattern starts to approach the bottom limit of the receiver dynamic range. Thus the radiation pattern is built up in three transmit power settings, 0 dBm, 20 dBm and 40 dBm. The additional advantage of this method is that the fine structure of the pattern can be recorded if a chart recorder is employed.

BEAM WIDTH

The beam width is a term normally used to mean the *half-power beam width*. The half-power beam width is defined as the angular width of the main lobe when measured from one of the angles where the pattern is 3 dB down from the bore-sight peak to the other angle where the pattern is also 3 dB down. This angle is shown in Figure 10.8 on the main beam radiation pattern. The value for antenna beam width is generally a nominal figure, since the beam width will vary with the frequency and also the plane of polarization of the electric field. It is not prudent to use the formulae that relate the antenna gain to the beam width, since they do not take into account the energy which is radiated into wider-angle regions. The 3 dB beam width may however, be approximated by the following equation, which equates the half-power beam width θ (in degrees), to the frequency f (in GHz), the reflector diameter D (in meters) and efficiency n.

$$\theta = 21.885/(fD \cdot n^{1/2}) \tag{10.21}$$

$$= 27.67/fD \text{ for a 62.57 percent efficient antenna}$$

FRONT-TO-BACK RATIO

This value is defined as the ratio in decibels of the main-beam level at bore-sight to the rear-lobe level at 180°. The level is indicated on the pattern in Figure 10.8(*b*). Although the front-to-back ratio is indicated to the back-lobe peak, manufacturers may include a margin of between 1 and 2 dB when quoting the level indicated on the RPE.

ANTENNA DISCRIMINATION

This value is defined as the ratio in decibels of the main-beam level at bore-sight to the value at some angle away from the main-lobe bore-sight direction. This value is important for interference calculations and is that value required in step 14 in Section 10.3.1. The value is obtained directly from the antenna polar pattern. For the case of co-polar signals, the value is derived from the co-polar pattern of the receiving antenna of the victim station. For cross-polar signals the figure results from the combination of the cross-polar discrimination as discussed below and the co-polar discrimination plus the bore-sight attenuation of cross-polar signal generation from the main-beam level at the interfering station antenna.

CROSS-POLAR DISCRIMINATION

This is defined as the ratio in decibels of a desired polarization and an undesired polarization. In the past, the cross-polar discrimination was taken implicitly to be the ratio of the two polarizations at bore-sight. Thus, if a dually and linearly polarized antenna is considered, the cross-polar discrimination is the ratio of the response of the antenna (aligned to view a pure polarized signal), when it has a polarization which is the same as the transmit source (co-polarized) and when it has a polarization which is opposite to that of the transmit source (cross-polarized). In digital applications however, there has been consideration of the cross-polarization discrimination due to the effects of multipath and rain. Suggestions have been made that an antenna should be designed to meet a certain cross-polar discrimination with an angular contour extending in both azimuth and elevation, since findings have suggested that the effects of cross-polar discrimination (XPD) can be reduced with antennas having high cross-polar discrimination within a contour. In addition to this, it has also been suggested that the antenna should not possess an on-axis null in the cross-polar pattern since the rapid phase change resulting in the bore-sight region can make the cross-polar attenuation between two antennas less sensitive than the co-polar attenuation between them, causing a corresponding deterioration of the XPD. The definition of the XPD has also been the subject of discussion (see CCIR definition in Section 8.3.4.2). In theory, an aperture antenna, when fed by an open-ended circular waveguide or a choked horn, can possess exceptional values of XPD along the principal planes. It has also been suggested that if a circularly symmetric reflector is fed from a circular-type feed, then cross-polarized radiation will be generated, which is dependent upon the difference between the co-polarized patterns in the two principal planes and the curvature of the field lines of the feed, but which will lie in the 45° planes, or the so-called 'cardinal planes'. This means that there will be four cross-polarized lobes arranged on the diagonal planes symmetrically about the main co-polarized lobe. In a typical antenna the ratio of co-polarized to cross-polarized lobe (in the 45° plane) can be as low as 23 dB for a standard-performance antenna, and possibly approaching 30 dB or better for a high-performance antenna. It is difficult to produce an antenna that has better 45° ratios than this without seriously affecting the gain performance, since the feed pattern would have to be severely nonoptimal in this respect. It is this ratio of co-polarized to cross-polarized lobe levels in the 45° plane, which is used in the loss term taken for cross-polar radiation. Generally, the performance of the antenna is of most importance in the two principal planes. Initially, it may be thought that the two principal planes would track through the nulls that exist between the four cross-polar lobes. In practice, principal plane cross-polarization does exist and is dependent on the level of the 45° plane cross-polarization generated by the feed. The feed is the major factor in determining the 45° cross-polarization of the antenna. Systematic distortion in the reflector may generate cross-polarization, together with random distortions in the reflector which tend to redistribute cross-polarized radiation nearer the principal planes, and to contribute to principal plane cross-polarization. Another contribution is that of internal distortions within the feed that may give rise to higher-order modes, such as TM_{01} or cross-polarized versions of the main mode. Scattering from vortex planes, attachment bolts, struts, waveguide bends, guy wires, etc., also make their contribution. However, the main contributor is the feed itself, and principal plane cross-polarized radiation will depend on the level of the radiation existing in the 45° planes. With the above mentioned 45° cross-polar levels of 23 and 30 dB for standard- and high-performance cross-polar antennas, typical principal-plane cross-polar levels can generally be specified at 30 dB and 35 dB respectively, although variations are possible.

ANTENNA VSWR

The voltage standing wave ratio (VSWR) of an antenna is a measure of the energy that is reflected back from the antenna and which does not contribute to the usefully radiated energy. Thus the reflected energy is lost energy, although it may again be reflected back from the source to end up being partly transmitted. The double reflected signal, although reduced in magnitude, does have a time delay in comparison to the main signal or the primary incident signal into the antenna. The effect of this doubly reflected signal is to cause distortion in the primary signal. This form of distortion is known as echo distortion, and its effect depends on the distance between the antenna and source end of the feeder, as well as the magnitude of the reflected signal. This form of distortion, although detrimental to FM systems, does not affect digital transmission to the same extent. The VSWR of the antenna is rarely directly measured. A more usual measure is that of 'return loss' RL, which is the dB ratio of the reflected and incident signals on the antenna under test. That is

$$RL = 20 \log (\text{VSWR} + 1)/(\text{VSWR} - 1) = 10 \log (P_f/P_r) \text{ dB} \tag{10.22}$$

where P_f is the forward power and P_r is the reflected power.

The return loss can be measured over a swept frequency so that the behaviour of the antenna return loss over the whole operating band can be considered. In general, manufacturers specify the maximum VSWR specification guaranteed over the frequency band. The causes of VSWR being greater than the ideal value of unity are any complex impedance discontinuity which may arise from the input flanges, dents along the waveguide or feeds, waveguide bend distortion, transitions between rectangular and circular waveguides, tuning pins within a feed, the feed aperture, the feed window, the reflector/vortex plate combination and scatter contributions from waveguide bends and guy wires. Many of the above sources of impedance discontinuities can be eliminated by proper design and manufacturing procedures.

INTER-PORT ISOLATION

This measurement can be applied only to a dual polarized feed or antenna. The inter-port isolation is the ratio of the power transmitted into one of the antenna ports and the small portion of power received in the other port. A finite amount of inter-port isolation, which should ideally be infinite, is an indication that distortions of some kind are partially depolarizing the signal. The smaller the inter-port isolation the larger the problem. Alternatively, the polarization separating mechanism, which may be not operating properly, could be the cause of a reduced value of inter-port isolation from that expected. Some types of distortion that can occur are those caused by the circular waveguide not being linear, the circular transitions not being concentric, the pins used being oval or not exactly in the plane of polarization, the threads of the tuning screws, the machined orthogonal parts not being quite orthogonal, imperfect soldering of the waveguide feed, and the interior of the circular waveguides not having been machined properly. The inter-port isolation can be measured with the feed radiating into free space, when the term 'open-circuit isolation' is used or with a shorting plate placed on the feed aperture, when the term 'short-circuit isolation' is used. The latter test is particularly sensitive and is a useful diagnostic aid during feed development and manufacture. Open-circuit inter-port isolation is of concern to antenna users because it gives an indication, (but not a measurement) of a depolarization mechanism within the feed, other than that produced by any physically realizable aperture, which would radiate energy into the four 45° plane cross-polar lobes, described above, under cross-polar discrimination. The distortion causing inter-port isolation arises from a small amount of one polarization being converted into the orthogonal polarization. A typical standard antenna inter-port isolation value is 35 dB, although a high cross-polar discrimination antenna may require an open-circuit inter-port isolation value of 40 dB or even 45 dB. In a dual-band antenna, the inter-port isolation has to be measured at both frequency bands of operation. Inter-port isolation is reciprocal, in the sense that if the signal is entered into the input of port A and measured on port B, the same result is obtained as that from measuring the signal on port A if the signal were entered into port B. Any differences which may arise are due to slightly different phasing or VSWR effects.

10.3.2.4 Selectivity

This section is concerned with the determination of the ratio of the energy received from a wanted spectrum to that received from an interfering spectrum. The selectivity is mainly concerned with adjacent-channel interference. The calculation of co-channel interference will also be included as it is the amount of in-band power received by the victim receiver and is dependent on the relative bandwidths of the transmitting interferer and the victim receiver. The main concern is that interference problems can be created by the power from an emission in one band overlapping into an adjacent band and interacting with a receiver tuned close to the band edge. In order to eliminate or restrict this adjacent-band interference problem, sharing criteria are required which specify power flux density limits and minimum frequency and distance separation limits. Adjacent-band problems may occur at any interface of bands allocated to radioastronomy, passive space research, aeronautical and marine communication, etc. The selectivity used as one of the subtractive terms in equation 10.3 of step 16 in Section 10.3.1, for the determination of the interference power, is alternatively known as the *frequency dependent rejection* (FDR): this is also a measure of the rejection produced by a receiver selectivity curve on unwanted transmitter emission spectra. CCIR Report 654–1[5] describes the method of calculating the FDR (discussed below in this section), in addition to the frequency distance (FD), which is a measure of the minimum distance separation that is required between a victim receiver and an interferer as a function of the difference between their tuned frequencies. The FDR can be expressed as:

$$\mathrm{FDR}(\Delta f) = 10 \log\left[\left(\int_0^\infty P(f)\,\mathrm{d}f\right)\Big/\left(\int_0^\infty P(f)\cdot H(f+\Delta f)\,\mathrm{d}f\right)\right] \qquad (10.23)$$

where $P(f)$ = emission spectra density generally normalized to unity maximum power spectral density (in W/Hz).

$H(f)$ = receiver selectivity.

and $\Delta f = f_t - f_r$ Hz (10.24)

where f_t = interferer tuned frequency in Hz

f_r = receiver tuned frequency in Hz

The FDR can be divided into two terms, the on-tune rejection (OTR) and the off-frequency rejection (OFR), which is the additional rejection that results from off-tuning the interferer and receiver; i.e.

$$\mathrm{FDR}(\Delta f) = \mathrm{OTR} + \mathrm{OFR}(\Delta f) \tag{10.25}$$

where OTR is given by equation 10.23 with $\Delta f = 0$, and OFR by equation 10.23 as it stands, except that the numerator value is multiplied by $H(f)$, the receiver selectivity. Thus

$$\mathrm{OTR} = 10 \log\left[\int_0^\infty P(f)\mathrm{d}f \Big/ \int_0^\infty P(f)\cdot H(f)\mathrm{d}f\right] \tag{10.26}$$

$$\mathrm{OFR}(\Delta f) = 10 \log\left[\left(\int_0^\infty P(f)\cdot H(f)\,\mathrm{d}f\right)\Big/\left(\int_0^\infty P(f)\cdot H(f+\Delta f)\,\mathrm{d}f\right)\right] \tag{10.27}$$

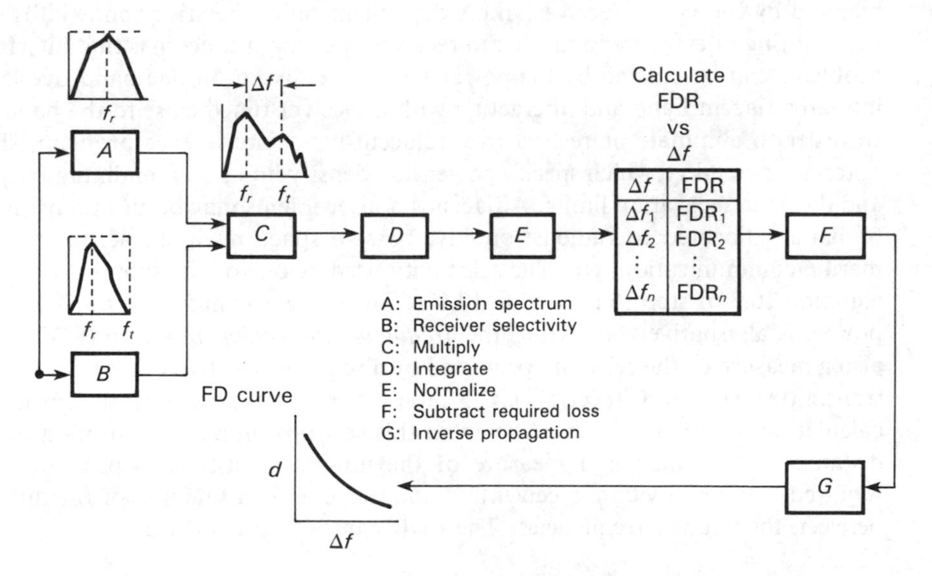

Figure 10.9 Block diagram of calculation process of FDR and FD (Courtesy of CCIR, Report 654–1, Reference 5)

The OTR is also called the correction factor and can be approximated by:

$$\text{OTR} \simeq K \log(B_t/B_r), \quad B_r \leqslant B_t \tag{10.28}$$

where B_r = receiver 3-dB bandwidth (in Hz).
B_t = transmitter 3-dB bandwidth (in Hz).
$K = 10$ for noncoherent signals.
$K = 20$ for pulse signals.

The calculation of OTR can aid in the determination of the co-channel interference power, since the OTR is the rejection obtained when the receiver and unwanted transmitter are tuned to the same frequency. To determine the FDR, the emission spectrum and receiver selectivity are specified at a discrete number of frequencies (emission spectrums and receiver responses are discussed below). Extrapolation slopes define these functions beyond the discrete number of frequencies and interpolation is used to define functional values between discrete data points. The integration, as specified in equation 10.23, is evaluated numerically by the trapezoidal method using a variable integration step size. Figure 10.9 shows a block diagram of the calculation process.

10.4 METHODS OF PREVENTING INTERFERENCE

As mentioned in Section 10.3.2.3, the antenna not only radiates the wanted signal in the main lobe, but also radiates part of the energy in all other directions. The side-lobe radiation is a source of interference to other radio receivers. To reduce the mutual interference between antennas of different links operating in the same frequency bands, the interfering signal may be suppressed by using electromagnetic radiation screens installed at a given distance from the antenna to be protected, or mounted directly on the antenna.

10.4.1 Use of special screens to improve interference immunity in radio links (CCIR Report 831)[5]

In the design of small high-performance screens, the shape of the screen edge is based on the field being zero at the calculation point. The field of source S at point P is calculated by summation of the fields from the unblanked portions of the concentric rings shown in Figure 10.10. For the ring radius r and its aperture φ', the parametric equation which assures that the field is zero at the calculation point is given by:

$$r = [\lambda_0 \cdot r_0 \cdot R_0 \cdot \arctan Z)/(\pi(r_0 + R_0))]^{1/2} \tag{10.29}$$

where $Z = [\rho' \sin \Psi + \rho \cos \Psi)/(\rho' \cos \Psi - \rho \sin \Psi]$

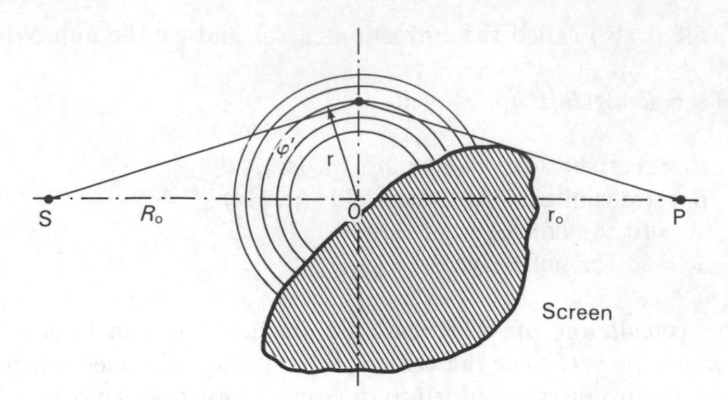

Figure 10.10 Parameters used in equation 10.18 (Courtesy of CCIR, Report 831, Reference 5)

and

$$\varphi' = (2\pi/\rho_{max}) \cdot (\rho^2 + \rho'^2)^{3/2}/(\rho^2 + 2\rho'^2 - \rho\rho') \tag{10.30}$$

The parameters r, R_0, r_0, φ' in the parametric equations 10.29 and 10.30, are given in Figure 10.9, and λ_0 is the wavelength of the incident wave. By inserting the functional relationship $\rho = f(\Psi)$ into equation 10.29, the corresponding shape of the screen contour is found. Further details on calculations and experiments for different functional relationships between ρ and Ψ are provided in CCIR Report 831. However, for screens which are to protect high-gain antennas (40 dB and up), the double-ring screen may be used. This is shown in Figure 10.11. If the 'double-ring' screen is suspended directly from the antenna mast, the mast and the supporting elements reduce the efficiency of the screen in the region of greatest suppression. The placement of the screen is in the direction of the interfering side lobe, and effectively acts as a blinker, or curtain screen. This screen is placed vertical to the ground and at a calculated distance from the antenna. The 'double-square' screen in which the

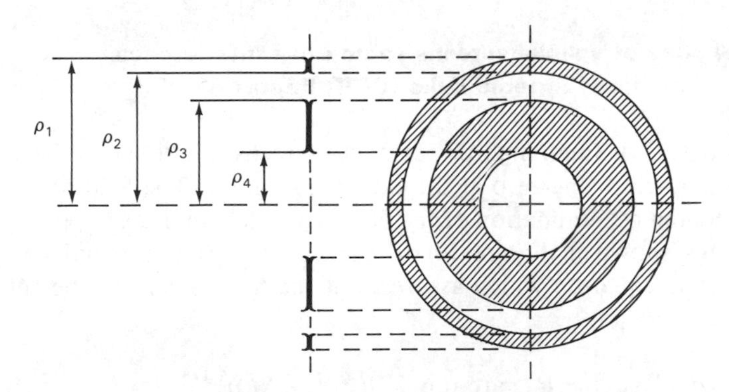

Figure 10.11 'Double-ring' screen (Courtesy of CCIR, Report 831, Reference 5)

antenna supports are blanked by the screen curtain produces only a slightly reduced efficiency from that of the annular screen and has the advantage of possessing a nearly axisymmetrical suppression pattern. Both the double-ring and double-square screens may be used to protect antennas from interference arriving from any direction, including the front half-space, outside the angles $\theta \simeq 0.7\,\theta_0$, where $2\,\theta_0$ is the null beam width of the antenna to be protected. Screens mounted on the antenna reduce the side-lobe radiation in particular sectors. One of the simplest types is the disc screen mounted on the rear side of the reflector at a certain distance from the edge of the aperture, and in the same plane as the antenna face. At a distance of $2\lambda_0$ from the rim of the antenna and of such a size that it overlaps the rim of the antenna by $2\lambda_0$, the screen increases the immunity of the antenna to interference by 8–10 dB in the section $180 \pm 35^\circ$. The greatest suppression (15–18 dB) around the narrower sector of 180° is produced by a slotted screen which is installed at a distance of $2\lambda_0$ from the antenna rim, and which overlaps the rim at this recessed distance by $1.07\lambda_0$. This means that the slotted screen is smaller than the unslotted screen. The slotted screen may be successfully used to suppress the overflow lobe field of antennas used in trans-horizon links and designed for use with an offset reflector having a square or rectangular aperture.

10.4.2 Use of interference cancelers (CCIR Report 830, Reference 5)

In cases where the traditional methods of insuring the trouble-free operation of radio systems by geographical or frequency separation are inapplicable, interference cancelers are another means of enabling interactive radio links to operate simultaneously. Another application of interference cancelers occurs in systems in which the main communication channel is combined with additional information transmitted in the same band width as the main information. In this case, interference cancelers allow the main and additional channels to be separated with a negligible level of crosstalk and a proportionate increase in channel capacity. Two types of interference canceler which are applicable to interference problems are:

Single-channel devices, which process the additive mixture of the wanted and unwanted signal reaching the receiver input.
Multichannel devices, in which, at the channel inputs, either one or several unwanted signals from other channels arrive in addition to the wanted signal.

The main use at the present time is to solve the problem of protecting earth stations from radio relay interference due to analog FM or AM radio systems.

10.4.3 Frequency separation as a means to counteract adjacent-channel interference

CCIR Report 651–1[5] also provides a method of determining the frequency distance (FD), which is a measure of the minimum distance separation that is required between a victim receiver and an interferer as a function of the difference between their tuned frequencies. When a maximum value I_m of acceptable receiver interference power from an interferer is specified, the required loss can be given by:

$$L_b(d) + \text{FDR}(\Delta f) = P_t + G_t + G_r - I_m = \text{specified required loss} \qquad (10.31)$$

where P_t = interferer transmitter power (in dBm).
G_t = gain of interferer antenna in direction of the victim receiver (in dBi)
G_r = gain of victim receiver antenna in the direction of the interferer (in dBi)
$L_b(d)$ = basic transmission loss for a separation distance d between interferer and victim receiver (in dB).

The receiver performance is acceptable only if the left-hand side of equation 10.31 is greater than or equal to the right-hand side. The required loss combines a distance-dependent term $L_b(d)$ and a frequency-dependent term FDR(Δf). A curve of the various combinations of the two separations in frequency Δf, and distance d, which achieves a specified required loss, is an FD plot. Usually the ordinate is the distance separation and the abscissa is the separation in tuned frequency of the interferer and victim receiver. The FD curve separates the acceptable receiver performance region from the nonacceptable receiver performance region. An FD plot shows the trade-offs in either frequency or distance separation which allow interferer and victim receiver to co-exist in a space frequency environment. Figure 10.12 shows a typical FD plot.

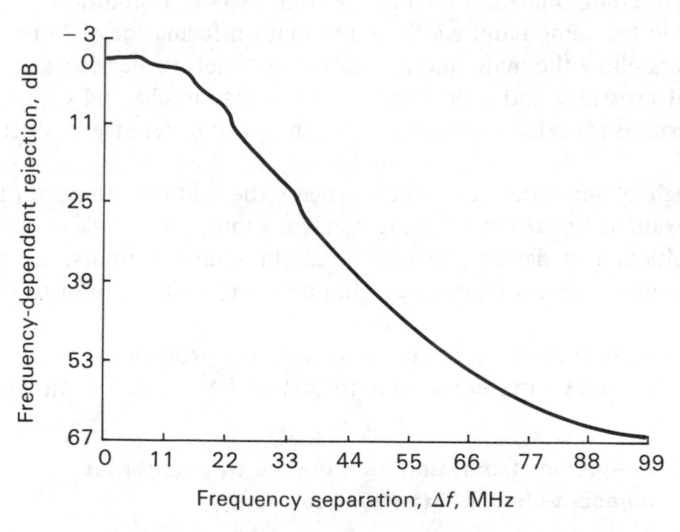

Figure 10.12 Typical frequency distance plot (FD)

10.5 EMISSION SPECTRA AND RECEIVER SELECTIVITY CURVES[13-15]

The Radio Regulations, Geneva 1982, define the *necessary bandwidth* as follows: 'For a given class of emission the width of the frequency band which is just sufficient to insure the transmission of information at the rate and with the quality required under specified conditions'. The maximum frequency error of the unmodulated carrier frequency relative to its nominal value shall not exceed 200 kHz for analog systems and 100 kHz for digital systems. The CFR[13] states in Section 94.71, for operational fixed-radio facilities operated in the microwave spectrum of 928–929 MHz and above 952 MHz, that besides providing a schedule of maximum bandwidth for different frequency bands, the mean power of emissions shall be attenuated below the mean output power of the transmitter in accordance with the following schedule (applicable to transmissions using digital modulation):

1. For operating frequencies below 15 GHz, in any 4 kHz band, the center frequency of which is removed from the assigned frequency by more than 50 percent up to and including 250 percent of the authorized bandwidth, as specified by the following equation, but in no event less than 50 dB: $A = 35 + 0.8(P - 50) + 10 \log W$. (Attenuation greater than 80 dB is not required.) In this equation A = attenuation (dB) below the mean output power level (in dB), P = percentage removal of center frequency from the assigned frequency, and W = authorized bandwidth (in MHz).
2. For operating frequencies above 15 GHz in any 1 MHz band, the center frequency of which is removed from the assigned frequency by more than 50 percent up to and including 250 percent of the authorized bandwidth, as specified by the following equation but in no event less than 11 dB: $A = 11 + 0.4(P - 50) + 10 \log W$ (Attenuation greater than 56 dB is not required.)
3. In any 4 kHz band, the center frequency of which is removed from the assigned frequency by more than 250 percent of the authorized bandwidth, at least [43 + 10 log(mean output power in watts)] dB, or 80 dB, whichever is the lesser attenuation.
4. On any frequency above 40 GHz the carrier harmonics of a system operating under the provisions of Section 94.91 in Reference 13 (special provisions for low power, limited coverage systems in the band segments 21.8–22.0 GHz and 23.0–23.2 GHz), shall be attenuated at least [33 + 10 log 10(mean output power in watts)] dB.

10.5.1 Spurious emissions[14]

A spurious emission, as defined in the Radio Regulations, Geneva 1982, is an emission on a frequency or frequencies which are outside the necessary bandwidth and the level of which may be reduced without affecting the corresponding transmission of information. Spurious emissions include harmonic emissions, parasitic emissions, intermodulation products and frequency conversion products but exclude out-of-band emissions. An out-of-band emission is defined as an emission on a frequency or frequencies immediately outside the necessary bandwidth which results from the modulation process by excluding spurious emissions.

10.5.1.1 Spurious emissions from the transmitter

Spurious emissions from the transmitter are any individual unwanted emissions present at its output port, which is considered to be at the output of the transmitter or diplexer if fitted. Typically, spurious emissions from a transmitter shall not exceed −79 dBW within the frequency group in which the transmitter lies, e.g. lower or upper. Spurious emissions at any other frequency shall not exceed −96 dBW. (*Note.* At certain frequencies the CEPT value of −120 dBW may be desirable when shared antenna working is planned.)

10.5.1.2 Spurious emissions from the receiver

Spurious emissions from the receiver are any individual emission present at its input port which is considered to be at the input of the receiver or diplexer if fitted. Typically, spurious emissions from the receiver including those at the receiver local oscillator frequency shall not exceed −96 dBW.

10.5.2 Equivalent isotropically radiated power (EIRP)

This is defined as the product of the power supplied to the antenna and the antenna gain in a given direction relative to an isotropic antenna. For the purpose of the Radio Regulatory Specification,[14] the EIRP shall not normally exceed 45 dBW under any conditions of modulation for the frequency range 12750–13250 MHz. In Reference 13, Section 94.73, this value is given as 50 dBW for the frequency range 10680–13250 MHz.

10.5.3 Emission spectra

The power spectral density of the radiated signal can be controlled by suitable baseband conditioning and by RF filtering. Figure 10.13 shows the CFR[13] mask described in Section 10.5 (Schedule, 1) imposed on a PSK carrier. Also shown are adjacent carriers from identical systems. The energy falling into adjacent channels must be 80 dB below the authorized carrier level. It is assumed that the carrier frequency in this case is in the 13 GHz band.

Example 4

Determine the FCC emission spectrum mask, for a 16–QAM transmitter operating in the 15 GHz band (14.500 GHz to 15.350 GHz). The transmitter RF bandwidth W is 40 MHz.

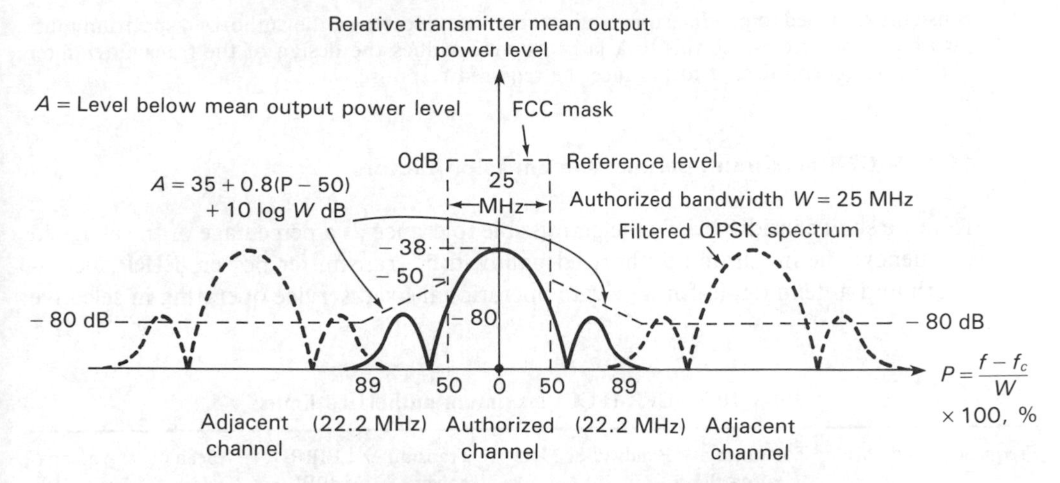

Figure 10.13 CFR mask (35 + 0.8(P − 50) + log W) dB and QPSK and QPSK adjacent-channel spectra

Solution

From Section 10.5 (Schedule, 2), at frequencies of 50 percent separation from the carrier frequency (20 MHz), the attenuation is (11 + 10 log 40)dB down on the carrier, i.e. 27 dB. At 122.5 percent separation from the carrier (49 MHz), the attenuation is found to be: $A = 11 + 0.4(P - 50) + 10 \log W = 27 + 0.4(72.5) + 16 = 56 + 16$ dB. The value, 56 dB, is the maximum value of attenuation required. Hence the spectrum mask stays at 56 dB for frequencies further removed from the carrier. All levels are referred to a 0 dB reference unmodulated carrier level. The power spectral density (C/W), for the signal, measured at the center frequency in an arbitrary, comparatively narrow 1 MHz band (b MHz), is given as 10 log (b/W), referenced to the carrier C, i.e., for this case, 10 log (1/40) = − 16 dB. Thus the unfiltered 16–QAM power spectral density maximum, lies 16 dB below the unmodulated carrier power (0 dB reference). Figure 10.14 shows a diagram of this mask and spectrum, which shows the stringent filtering of the 16–QAM spectrum required, before the signal emission is permitted to leave the transmitter. The combination of the transmitter and receiver filters

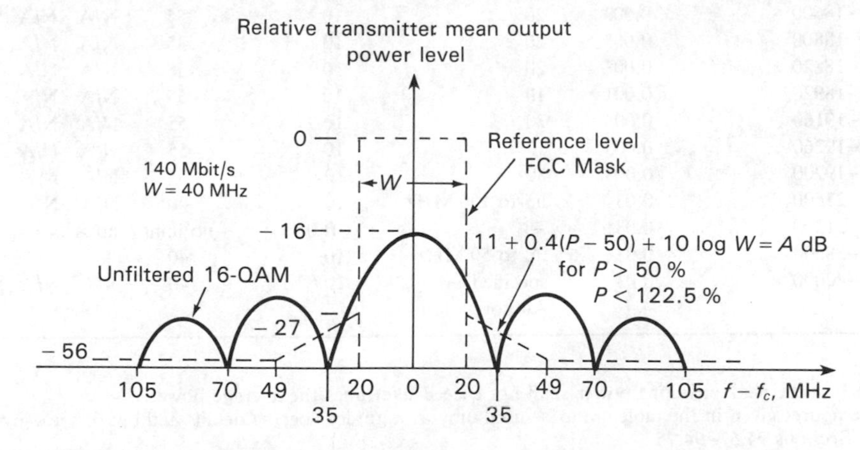

Figure 10.14 Solution to Example 4 for 16–QAM 15 GHz FCC emission spectrum mask. Message bit rate = 140 Mbit/s; bandwidth = 40 MHz

is usually designed to produce the required response. However, the authorized spectrum must also be considered, since this to a large extent specifies the design of the transmitter filter within the general design to produce the required response.

10.5.4 CFR maximum permissible emission factors

Here we shall consider briefly the permissible tolerance as a percentage of the assigned frequency, the maximum authorized bandwidth, transmitter power, EIRP, beamwidth and antenna gain for a private operational fixed service operating in selective

Table 10.7 CFR (FCC) maximum authorized limits

Frequency band, MHz	Tolerance as % of carrier	Bandwidth, MHz	Transmitter power, watts	EIRP, dBW (Note 1)	Beam width, degrees		Antenna gain, dBi	
					A	*B*	*A*	*B*
928–929	0.0005	0.025	5	17	14	20	N/A	
952–960	0.0025	25, 50, 100, 200 kHz	20	40	14	20	N/A	
1850–1990	0.002	5 or 10	20	45	5	8	N/A	
2130–2150	0.001	0.8 or 1.6	20	45	5	8	N/A	
2150–2160	0.001	10	20	45	5	8	N/A	
2180–2220	0.001	0.8 or 1.6	20	45	5	8	N/A	
2450–2460	0.001	0.8	20	45	5	8	N/A	
2500–2690	0.0025	to 2680 MHz = 6	20	45	5	8	N/A	
6525–6875	0.005	5 or 10	20	50	1.5	2	N/A	
10500–10680	0.0003	5	see Note 4	40	N/A	N/A	38	38
12200–13150	0.005	10 or 20	10	50	1	2	N/A	
13200–13250	0.03	25	10	50	1	2	N/A	
17700–18140	0.003	80	10	55	N/A	N/A	38	38
18140–18142	0.003	2	10	55	N/A	N/A	38	38
18142–18580	0.003	6	10	55	N/A	N/A	38	38
18580–18600	0.003	20	10	55	N/A	N/A	38	38
18600–18800	0.003	20	10	35	N/A	N/A	38	38
18800–18820	0.003	20	10	55	N/A	N/A	38	38
18820–18920	0.001	10	10	55	N/A	N/A	38	38
18920–19160	0.003	20	10	55	N/A	N/A	38	38
19160–19260	0.003	10	10	55	N/A	N/A	38	38
19260–19700	0.003	80	10	55	N/A	N/A	38	38
21200–23600	0.03	up to 100 MHz	10	40	N/A	N/A	38	38
31000–31300	0.03	–	0.05	no limit	no *A*/*B* = 4		N/A	
31300–38600	0.03	up to 50 MHz	10	40	–		–	
38600–40000	0.03	specified in Authorization	10	40	N/A	N/A	38	38

Notes
1. EIRP – the peak envelope power shall not exceed five times the average power.
2. The figures given in the table are to be used only as a guide. Specific details and options are given in CFR 47, Sections 94.67–94.75.
3. N/A = not applicable.
4. See CFR 47, Section 94.73

bands between 928 MHz and 40 GHz. The reader is referred to CFR (FCC) Section 94.67 to 95.75 for detailed explanations of each of these variables within the specified bands.

10.6 CHOICE OF RADIO FREQUENCY AND SPECTRUM UTILIZATION

A digitally coded analog signal has a greater bandwidth than the signal itself. As discussed in Chapter 2, a 4 kHz voice-band signal is PCM-encoded at a rate of 64 kbit/s. This corresponds to approximately 32 kHz of bandwidth of a Nyquist filter, transmitting the binary equivalent of this signal. The increase in bandwidth tends to reflect the efficiency of the radio spectrum utilization. Although an FM–FDM wideband modulated transmitter requires more bandwidth than its base-band signal, the same information, when transmitted using digital radio systems, requires even more bandwidth. This is so when only the bandwidth of one radio channel is considered, as generally done in using the 'bits/Hz' efficiency measure, i.e. the information capacity in bits/s divided by the radio channel bandwidth in hertz as discussed in Section 6.7.2. The total efficiency, however, depends also on the geographical area 'occupied' by one radio relay system. Because digital systems are able to tolerate a much higher interference level than their analog counterparts, and because orthogonal polarizations are easily used without the adverse effects which apply to analog systems, smaller protection ratios, and thus shorter distances between systems with the same carrier frequency, can be employed. As a consequence of these considerations, digital radio relay systems can approach the overall efficiency of those using analog techniques. In practice, small- to medium-capacity digital systems, which have a radio channel bandwidth about equal to a corresponding FM system, can achieve better overall efficiency. The CCIR recommends that the spectrum efficiency should be at least 2 bit/s/Hz. The FCC Regulations (CFR 47, Section 21.222) specifies that below 15 GHz and in the band 17.7–19.7 GHz, the bit rate must be equal to or greater than the RF bandwidth and independent of the polarization used, of frequency re-use, or of how the system is configured. The choice of frequency band has a great effect on the general characteristics of the system.

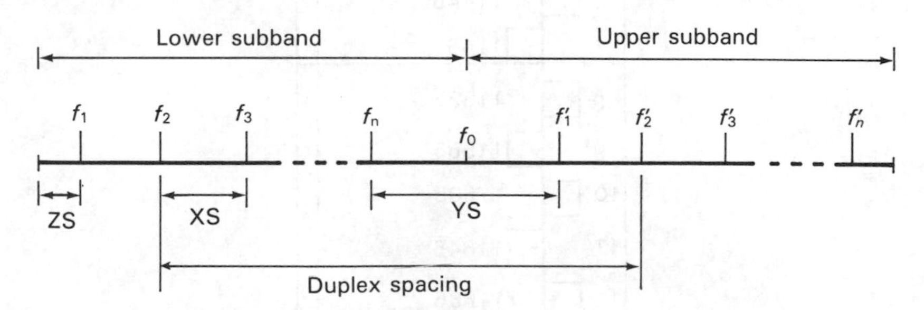

Figure 10.15 Radio channel arrangements

8 GHz Band
OIRT.2 Standard
1800 channels

7900–8400 MHz

H (V)*	V (H)*	(MHz)
1		7926
2		7954
3		7982
4		8010
5		8038
6		8066
7		8094
8		8122
f_0		8150
1′		8192
2′		8220
3′		8248
4′		8276
5′		8304
6′		8332
7′		8360
8′		8388

56 MHz
266 MHz
70 MHz
500 MHz

Note: * alternative polarization arrangement

11 GHz Band
CCIR Recommendation 387.3
CCIR Recommendation 389.2

10700–11700 MHz

H (V)*	V (H)*	(MHz)
1		10715
2		10755
3		10795
4		10835
5		10875
6		10915
7		10955
8		10995
9		11035
10		11075
11		11115
12		11155
f_0		11200
1′		11245
2′		11285
3′		11325
4′		11365
5′		11405
6′		11445
7′		11485
8′		11525
9′		11565
10′		11605
11′		11645
12′		11685

80 MHz
530 MHz
90 MHz
1000 MHz

12 GHz Band
Japan Standard

1220–1244 MHz

H (V)*	V (H)*	(MHz)
1		12210
2		12230
3		12250
4		12270
5		12290
6		12310
f_0		
1′		12330
2′		12350
3′		12370
4′		12390
5′		12410
6′		12430
7′		12440

20 MHz
120 MHz
240 MHz

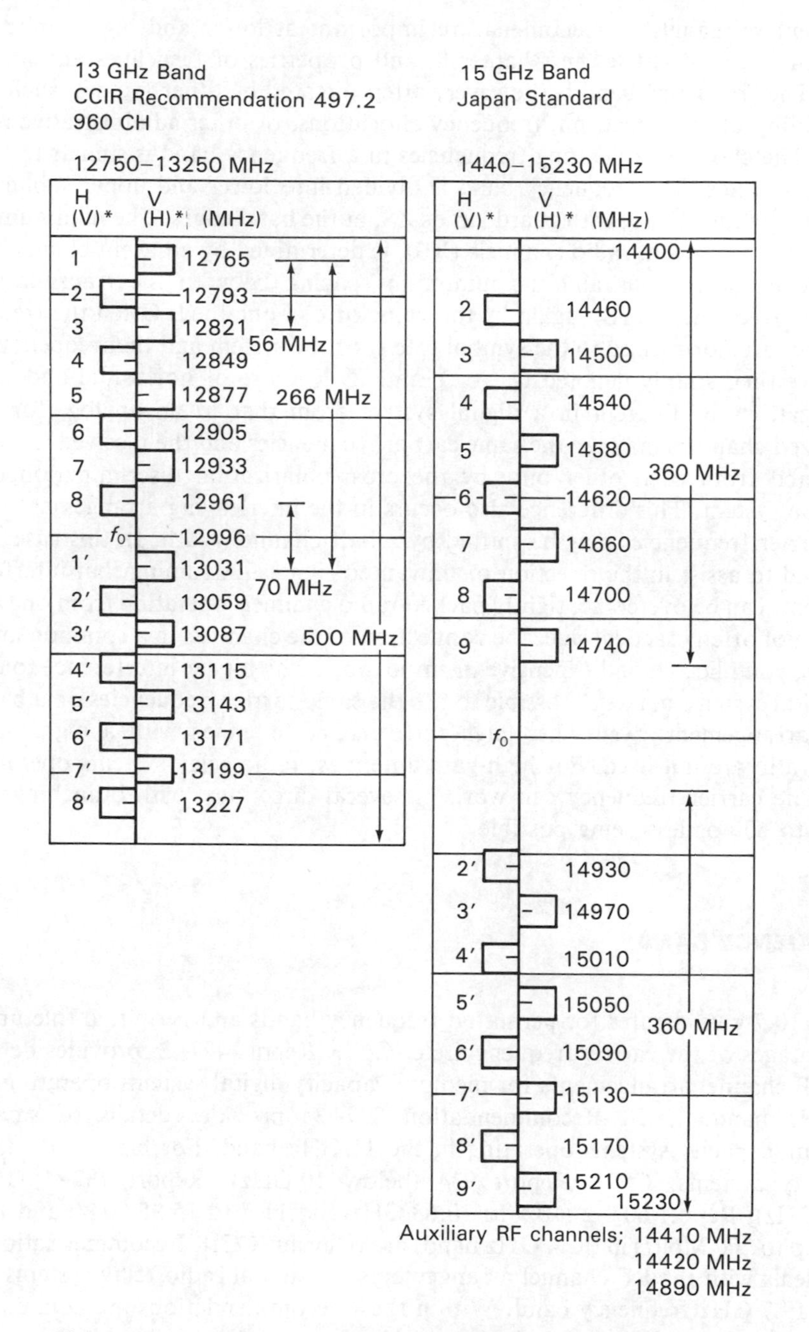

Figure 10.16 Frequency band radio channel arrangements for various bands

Different propagation phenomena are important at lower and higher microwave frequencies, as discussed in Chapter 8, and properties of feed lines and antennas vary. The frequency band is, however, often dictated by other factors, such as the availability of free spectrum, frequency allocations, or other administrative regulations. The choice of operating frequencies in a frequency band is similar to that of analog systems. The frequency band is divided into lower and upper subbands as shown in Figure 10.15, with guard bands ZS, at the band edges. The minimum spacing between adjacent radio channels (XS), is determined by adjacent-channel interference considerations and the minimum spacing between the transmitter and receiver frequencies (YS), again by the value of C/I obtained. Often the frequency spacings are normalized to the symbol rate r_s of the system and the frequency spacings are then simply denoted by X, Y and Z. The use of horizontal and vertical polarizations is different in a digital system from that of an analog, for cross-polarized channels may use the same carrier frequencies and the received signals are separated from each other only by the cross-polarization discrimination of the antenna system. This difference also occurs in the interleaving arrangement, where the carrier frequencies may be shifted by a half-channel-width. In this case, filters are used to assist in the rejection of unwanted signals. Assuming that interference problems can be overcome, tightly packed radio channels operating from one antenna are not often practical since the connection of the channels to a common antenna may be complicated and expensive or impossible. The higher interference tolerance of digital systems makes it possible to use the same carrier frequencies on a back-to-back arrangement, even when high-performance antennas with a high back-to-front ratio are not used. For high-gain antennas, radio relay systems operating at the same carrier frequency can work in several directions, with branching angles down to 60° or less being possible.

10.7 FREQUENCY BANDS

Table 10.7 gives figures for permitted frequency bands and permitted tolerances as percentages of the carrier frequency, etc. CCIR Report 497–2, provides details of the RF channel arrangements for medium-capacity digital systems operating in the 13 GHz band. CCIR Recommendation 387–3[1] provides details of small- or medium-capacity systems operating in the 11 GHz band. For high- and medium-capacity systems, CCIR Report 934 (below 10 GHz), Report 782–1 (10.7 to 11.7 GHz), Report 607–2 (10.5 to 10.68 GHz and 11.7 to 15.35 GHz) and Report 935 (up to 200 Mbit/s in the 4 GHz band) are relevant. CCIR Recommendation 595[1] also deals with the RF channel arrangements for digital radio relay systems in the 17.7–19.7 GHz frequency band. Within these recommendations are contained the necessary information to plan the frequencies for digital radio relay systems, taking into account the recommended pairing of frequencies, i.e., the transmitting and receiving frequency pair, and the required guard bands between adjacent channels and between adjacent bands. For example, consider that the frequency band of 13 GHz is chosen for a digital application, in which the capacity may be up to 960 telephone channels. Using the information contained in Annex I of CCIR Recommendation 497–2,[1] a radio frequency arrangement having the following characteristics may be used:

$$\text{Lower part of the band: } f_n = (f_o - 259 + 35n) \quad \text{MHz} \tag{10.32}$$

$$\text{Upper part of the band: } f'_n = (f_o + 21 + 35n) \quad \text{MHz} \tag{10.33}$$

where $n = 1, 2, 3, 4, 5$ or 6.

f_0 = a reference frequency (MHz) near the center of the 12.75–13.25 GHz band.

f_n = the center frequency (MHz) of an RF channel in the lower half of the band.

f'_n = the center frequency (MHz) of an RF channel in the upper half of the band.

If the digital radio system is expected only to operate at 34 Mbit/s (480 telephone channels), the CCIR Recommendation 497–2 provides the following center frequencies of the radio channels:

$$\text{Lower part of the band: } f_n = (f_o - 259 + 28n) \quad \text{MHz} \tag{10.34}$$

$$\text{Upper part of the band: } f'_n = (f_o + 7 + 28n) \quad \text{MHz} \tag{10.35}$$

where $n = 1, 2, 3, 4, 5, 6, 7$ or 8.

If the center frequency f_o, is chosen to be 12996 MHz, then, for example, the transmitting and receiving pair (or receiving and transmitting pair), for channel 6 ($n = 6$), is found from equations 10.34 and 10.35, to be:

$$f_6 = (12996 - 259 + 28 \times 6) = 12905 \text{ MHz}$$
$$f'_6 = (12996 + 7 + 28 \times 6) \quad = 13171 \text{ MHz}$$

Figure 10.16 shows the frequency band allocations for 8, 11, 12, 13 and 15 GHz using CCIR or Japanese standards.

REFERENCES

1. 'Fixed Service using Radio-Relay Systems', Recommendations and Reports of the CCIR, **IX–1**, XV Plenary Assembly, 1982.
2. 'Frequency Sharing and Co-ordination between Systems in the Fixed-Satellite Service and Radio-Relay Systems', Recommendations and Reports of the CCIR, **IV** and **IX–2**, XVI Plenary Assembly, 1986.
3. Morris, P., Private communications to the author, Mercury Communications Ltd, 1987.
4. Oetting, J. D., 'A Comparison of Modulation Techniques for Digital Radio', *IEEE Transactions on Communications,* 1979, **COM–27**, No. 12, pp. 1752–1762.

5. 'Spectrum Utilization and Monitoring', Recommendations and Reports of the CCIR, **I**, XV Plenary Assembly, 1982.
6. Turner, H. E., 'A Computer System for Microwave Frequency Co-ordination and Interference Calculations', *IEEE Transactions on Communications,* 1972, **COM–20**, No. 2, pp. 179–189.
7. Rosenbaum, A. S. and Glave, F. E., 'An Error-Probability Upper Bound for Coherent Phase-Shift Keying with Peak-Limited Interference', *ibid.*, 1974, **COM–22,** No. 1, pp. 6–16.
8. Prabhu, V. K., 'Error Rate Considerations for Coherent Phase-Shift Keyed Systems with Co-Channel Interference', *Bell System Technical Journal,* 1969, pp. 743–767.
9. Javed, A. and Tetarenko, R. P., 'Error Probability Upper Bound for Coherently Detected QASK Signals with Co-Channel Interference', *IEEE Transactions on Communications, 1979,* **COM–27,** No. 12. pp. 1782–1785.
10. Rosenbaum, A. S., 'Error Performance of Multiphase DPSK with Noise and Interference', *IEEE Transactions on Communications Technology,* 1970, **COM–18**, No. 12, pp. 821–824.
11. Jones, J. J., 'FSK and DPSK Performance in a Mixture of CW Tone and Random Noise Interference', *ibid.,* 1970, **COM–18,** No.10, pp. 693–695.
12. 'Antennas for Private Fixed Radio Services in the Band 12750 MHz to 13250 MHz', Performance Specification MPT 1406, Department of Trade and Industry, Radio Regulatory Division, London, 1984.
13. 'Telecommunication', Code of Federal Regulations (CFR), 47, October 1985.
14. 'Radio Regulatory Specification for Analogue and Digital Private Fixed Radio Relay Equipment Operating in the Band 12750 MHz to 13250 MHz', MPT 1403, Department of Trade and Industry, Radio Regulatory Division, London, 1983.

11 DIGITAL MICROWAVE RADIO EQUIPMENT AND SYSTEM TESTS

11.1 INTRODUCTION

In this final chapter, the basic building blocks of a digital transmitter and receiver will be considered, together with a brief outline of the tests required to prove that the radio system is performing to expectations.

The differences between digital and analog radio equipment are most significant at the baseband where the modem design and the baseband filtering and equalization circuit design are completely different for both digital and analog radio systems. The difference in design of these baseband units arises from the difference in nature of digital and analog signals. At IF and RF, the differences are not so pronounced and are mainly due to the different modulation methods used. As discussed in Chapter 5, the baseband signal which is received from, or transmitted to, a coaxial cable, must first be processed so that the signal is compatible with the system which will process or carry the information further. In the transmitter, the received digital signal from the cable must again be processed to prepare it for further digital processing, before modulation can take place. The transmitted bit stream is almost always scrambled in the transmitter baseband circuits, to prevent repetition of similar bit patterns, which would otherwise cause discrete spectral lines to appear in the emitted spectrum. After the scrambling, the signal is coded for modulation. As the absolute phase of the transmitted signal cannot normally be detected at the receiver, differential encoding is often used to permit phase changes in the carrier to provide the required phase information. Such is the case for phase-modulation schemes. Bit streams of the coder output are then used to control the modulator. The modulator may operate at a constant intermediate frequency or at the RF output of the transmitter. In analog systems, the IF modulator is more usual, whereas for digital systems using quadrature modulation techniques, the IF is mandatory. This is the reason why the double-sided bandwidth is always used in these systems. For generation of 4–PSK signals, the four phase states may be generated at the RF by using a series connection of two phase shifters, in which the phase of the reflected signal can be changed by 90 and 180° by changing the point of reflection. Simplified block diagrams of various types of transmitter are shown in Figure 11.1. The emitted spectrum can be shaped only by the antenna branching RF filter where an RF

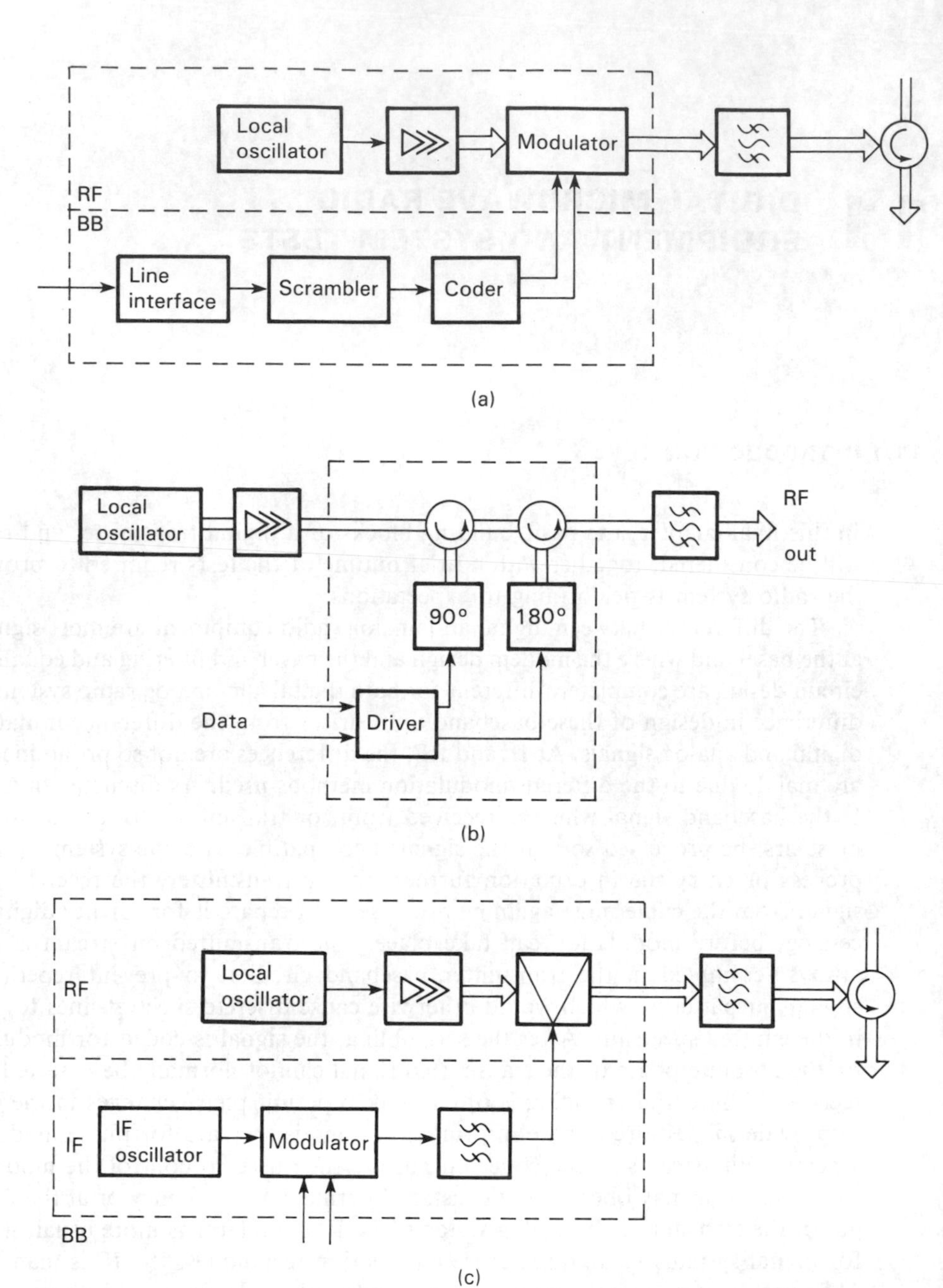

Figure 11.1 Transmitter with: (a) RF modulator, (b) RF modulator for 4–PSK, and (c) IF modulator. BB = baseband

modulator is used. This filter is usually more difficult to design than its IF counterpart, because of the need to restrict the spectrum to be within the limits discussed in Chapter 10, and yet not cause undue distortion of the transmitted signal by being too small in bandwidth. Another factor to be taken into account in an RF filter is the loss of signal power caused. To overcome this problem, an RF amplifier may be used after the modulator. This amplifier, however, causes nonlinear distortion and the spectrum to spread. In turn this causes the construction of the filter to be more expensive to maintain the required specifications. If an IF modulator is used, the spectrum shaping can be done by an IF filter, which is less expensive. The use of the IF filter for spectrum shaping in turn relaxes the requirements of the RF filter. However, any nonlinearities produced by the up-convertor or the following RF amplifier may cause problems due to the relaxed specifications of the RF filter. If the receiver demodulation is performed at IF, the demodulator is preceded by an IF filter which determines the selectivity, and by an IF amplifier with automatic gain control, which produces a constant-level signal that is independent of variations in the received carrier level. After demodulation, baseband processing is performed and the decoding and descrambling of the digital signal performed, to recover the original bit stream. The signal is then processed to provide the required line code, so that the signal can be properly matched to the transmission medium. Figure 11.2 shows the basic receiver block diagram.

At repeater stations, part of the baseband units may be omitted although the digital regeneration circuits remain. These modified baseband circuits are then coupled to the corresponding transmitter baseband circuits. In an analog radio, the direct connection of receiver IF to transmitter IF is usual, in order to reduce the amount of noise which would otherwise accumulate due to successive demodulation and modulation to and from the baseband. Digital signals, however, do not have

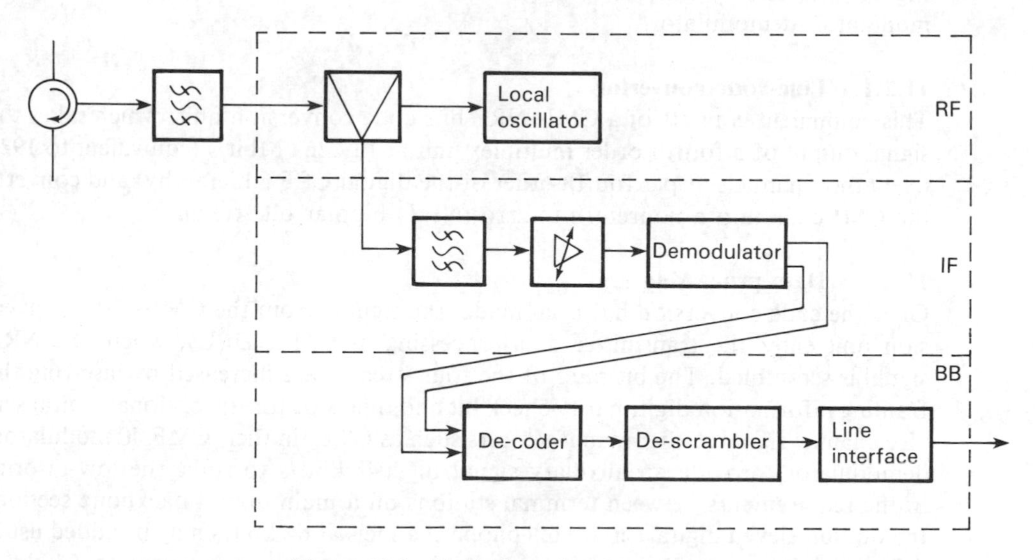

Figure 11.2 Typical receiver block diagram

the accumulated noise problem due to the digital regenerator circuits present in the baseband equipment.

11.2 TRANSMITTER CIRCUITS

In this section and in Section 11.3 which deals with the receiver circuits, the 16–QAM transceiver will be discussed. The commercially available 16–QAM radio can transmit 140 Mbit/s and may come in different system configurations, such as: hot-standby terminal station, as discussed in Section 7.8.1 with hitless switching by a baseband bit combiner, a hot-standby regenerative repeater station with hot-standby switching of the RF branching circuit, or a baseband back-to-back repeater station with a drop-and-insert facility at 34 Mbit/s. In the 10.7–11.7 GHz band the bandwidth of the RF channel is 40 MHz. The basic transmitter circuit usually consists of the following modules:

Transmitter baseband circuit
Baseband data processing unit
Modulator
Transmitter IF filter and amplifier
Up-convertor
RF amplifier and branching filter

11.2.1 Transmitter baseband circuits

Figure 11.3 shows the block diagram of an NEC, MDAP–140MB, 16 QAM modulator–demodulator.

11.2.1.1 Line-code convertor

This equipment consists of a CMI–NRZ line code conversion unit, which takes the signal output of a fourth-order multiplex unit at 139.264 Mbit/s (equivalent to 1920 telephony channels as per fourth order of the digital CEPT hierarchy) and converts the CMI code into a nonreturn-to-zero (NRZ) bipolar bit stream.

11.2.1.2 Data processing unit

Once the code conversion has been made, the signals from the CMI–NRZ conversion unit enter the transmitter data-processing unit (TX DPU), where the NRZ signal is scrambled. The bit rates of the four streams are increased by inserting the framing information digit, parity check bit and time slot for the optional digital service channel signal. Other manufacturers such as GTE, in their CMF 40 modulator-demodulator, provide an auxiliary signal of 2048 kbit/s to solve the low-priority traffic requirements between terminal stations on a main digital backbone section. In addition eleven digital service telephone channels at 64 kbit/s may be added using delta modulation or PCM. Digital switching instructions equivalent to 64 kbit/s

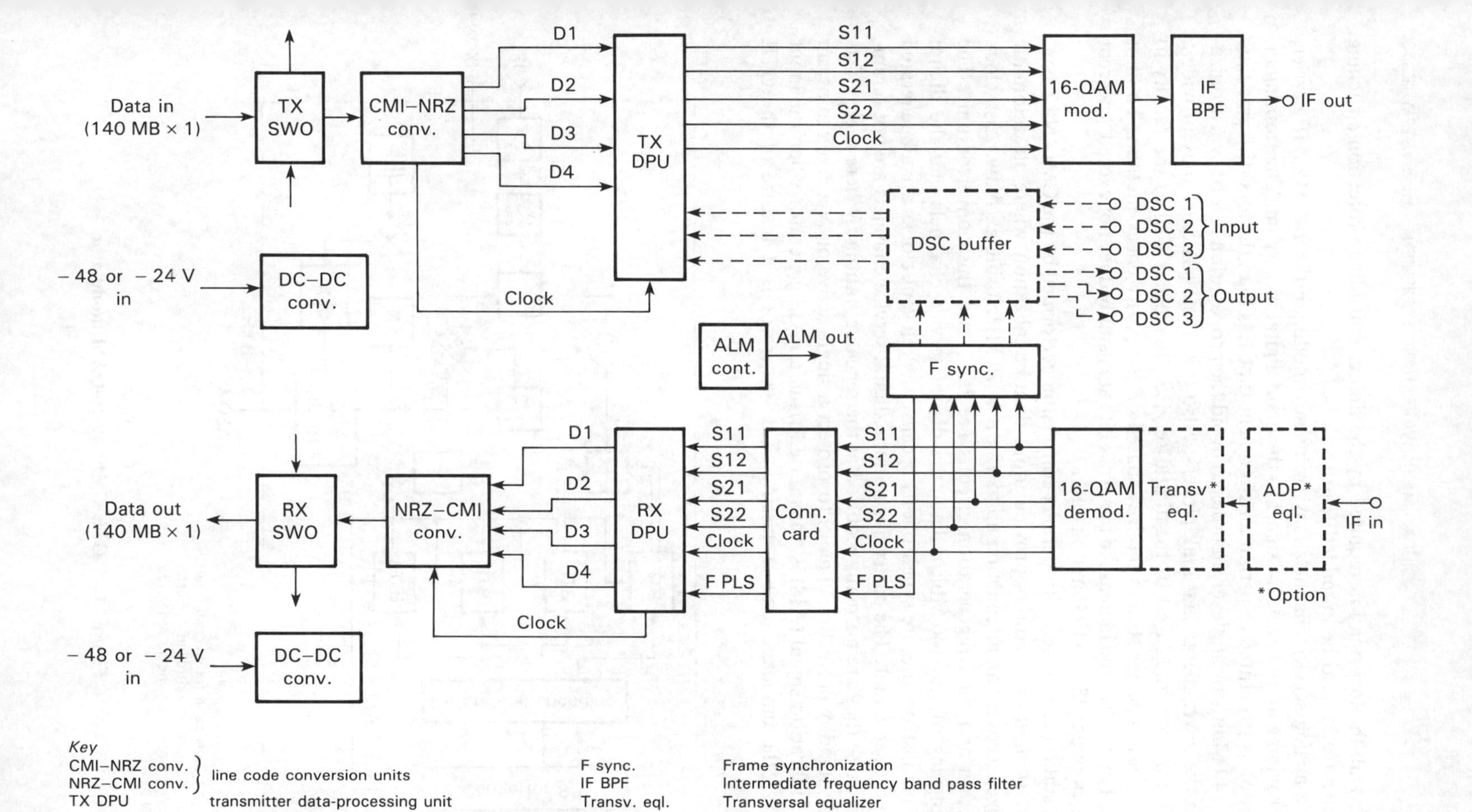

Figure 11.3 NEC MDAP-140 MB digital baseband modulator–demodulator unit

carrying all the information required by the digital multi-line protection switching equipment may also be an added feature.

The analog service channels may frequency modulate the 140 MHz IF carrier, thus providing an independent signal to the main digital bit stream. The framing information is structured to permit an alignment time less than the loss of frame time of the 140 Mbit/s multiplexer, in order to avoid the propagation loss of synchronization in the lower hierarchial multiplexers. Also the frame structure may provide a facility for drop and insert of the digital service channels as well as the facility to monitor at a repeater station the BER based on the parity check method. To limit the RF bandwidth, the increase in the gross bit rate usually does not exceed 4 percent of the nominal 139.264 Mbit/s bit rate.

In addition to the serial-to-parallel conversion performed by the CMI–NRZ convertor, some manufacturers provide input data equalization where an automatic equalizer recovers an attenuation up to 12 dB at 70 MHz using a $\sqrt{f}$ law. Also clock extraction and data regeneration circuits are built into this convertor unit. The multiplexing of the 140 Mbit/s data, the digital service channels and the digital switching instructions are synchronous because the relevant clocks are all generated within the same unit. The parity bits and the digital service channels are not scrambled because they are normally required at the repeater stations. Where there exists a single standby protection channel to protect a single active or N active radios, an alarm indication signal (AIS) is transmitted from an AIS generator when the input signal to the baseband unit is missing. This generator is a feature of the GTE

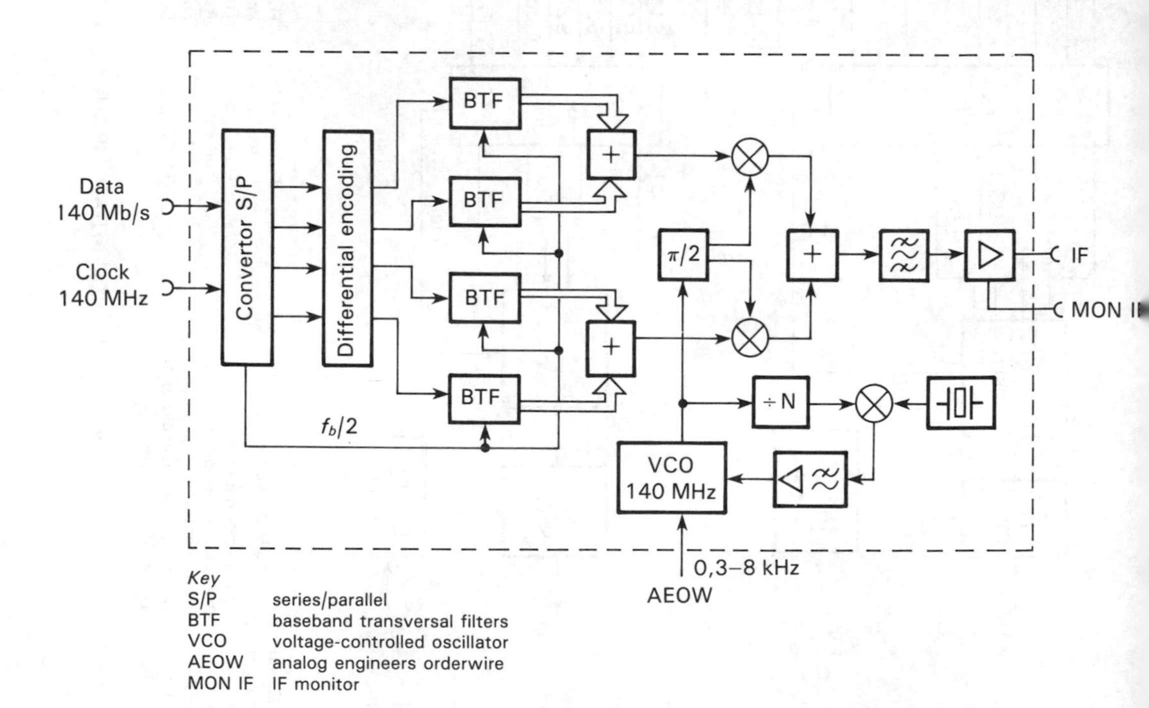

Figure 11.4 GTE CMF–40 16-QAM modulator

CMF–40 modulator-demodulator unit. Figure 11.4 shows the GTE 16–QAM modulator. This differs from that of the NEC radio, in that the series-to-parallel conversion is not performed in the CMI–NRZ convertor, but in the 16–QAM modulator.

11.2.2 16–QAM modulator

The 16–QAM modulator, besides the above mentioned series-to-parallel conversion, performs the following functions: differential encoding, baseband filtering of the four binary streams through digital transversal filters, conversion of the four binary streams into two multilevel signals (4-level PAM), local generation of 140 MHz carrier and modulation of the two quadrature components of the carrier and further combination to obtain the 16–QAM signal. To maintain the spectrum shaping obtained in the baseband while operating with multilevel signals, the amplitude modulation (AM) over the two quadrature axes is kept linear. The transversal filters used to filter each of the four binary streams (35 Mbit/s) produces a raised-cosine spectrum with a roll-off factor of 0.5. This type of filtering gives also the advantage of equalizing the residual group delay of the channel. The IF filter at the output of the modulator eliminates the unwanted signal spectrum. The generation of the 140 MHz (or 70 MHz) IF local oscillator frequency is obtained through a phase-locked loop oscillator circuit. The IF carrier can be frequency modulated by an analog signal to provide analog service channels.

11.2.3 Transmitter up-convertor, amplifier and branching filter

The output IF signal from the IF filter enters the up-convertor circuits to produce an output signal at carrier frequency. Figure 11.5 shows a block diagram of the GTE CTR 188 11 GHz transmitter.

11.2.3.1 Up-convertor

The input signal delivered by the 16–QAM modulator enters the up-convertor circuit. The IF signal has already been filtered at this stage for the best compromise between inter-symbol interference and out-of-band attenuation, in order to minimize adjacent-channel interference into other systems and to comply with the spectrum restrictions placed by the regulatory body, as dictated by the output power spectrum mask. The up-convertor includes a linearizer circuit after the IF signal has been amplified. This linearizer has a distortion characteristic which is complementary to that of the RF power amplifier. The result is the linearization of the complete transmit chain, which permits the power back-off to be kept to a small value. This predistorted signal is again amplified and fed into the up-convertor or mixer. The mixers in some designs may use a pair of varactor diodes in a balanced configuration. The heterodyne frequency is generated by a voltage-controlled oscillator from a phase-locked loop circuit.

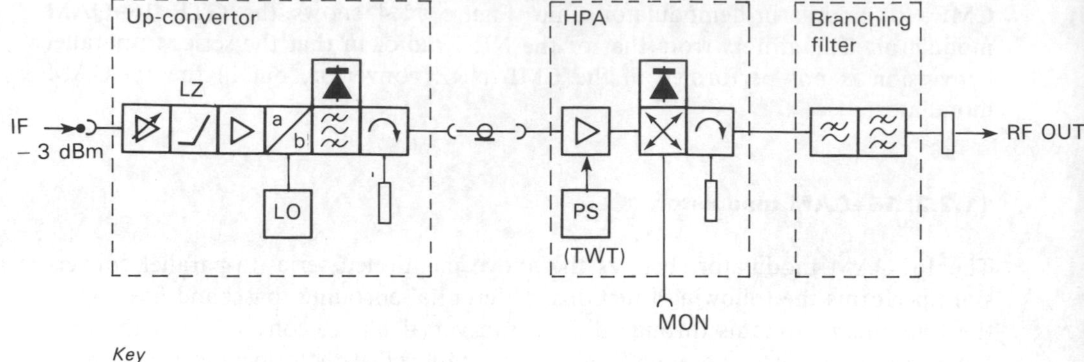

Key
HPA High power amplifier
LO Local oscillator
PS High voltage power supply for TWT
TWT Traveling wave tube
LZ Linearizer with distortion complementary to high power amplilfier.

Figure 11.5 GTE CTR 188 11 GHz transmitter

11.2.3.2 Power amplifier

At the output of the up-convertor a filter selects the wanted sideband. The diode detector permits the output signal to be monitored for expected power level via a directional coupler. The RF signal at the output of the directional coupler is then fed into a power amplifier. Two types of power amplifier are common: that which makes use of gallium arsenide field-effect transistors (GaAsFET) which may deliver an average power of 25 dBm at the branching chain output, and a traveling wave tube (TWT) amplifier which can deliver a 33 dBm power level to the branching chain output. The output of the amplifier enters into a directional coupler and a diode detector which again allows in-service power measurement, alarms and monitoring.

11.2.3.3 Branching filter

A terminated circulator isolates the power amplifier from the following low-pass filter and branching filter, which are cavity filters and which couple into the waveguide to the antenna. The number of cavities used depends on the design of the IF filter. A five-cavity filter is used in the GTE transmitter.

11.3 RECEIVER CIRCUITS

Figure 11.2 shows the diagram of a typical receiver. The baseband circuits in both transmitter and receiver are digital logic circuits, which perform the signal processing required between the line interface and the modem. A line interface unit regenerates the received line signal and performs code conversion between the line code and the unipolar binary pulses used in processing; it also provides overvoltage protection and attenuation equalization if required, due to longer cable sections. The converse operations are true for the leaving bit stream, or that bit stream coming out of the receiver baseband unit or baseband demodulator. If the transmitted bit rate is not a bit rate which is produced by an accepted hierarchial multiplex structure, a multiplex unit is required. In this case the received nonsynchronous bit streams (usually two) are multiplexed to produce one bit stream with a bit rate which is slightly higher than the sum of the two bit rates. Additional information bits are added, so that on the receiver side correct demultiplexing is possible. A codec is required to code the transmitted bit stream into a form which is suitable for the modulator and to decode the bit stream coming from the demodulator. These functions are performed by comparing values of successive pulses (differential encoding/decoding) and, based on the resulting difference, choosing the proper output value. Scrambling and descrambling functions may also be included in the codec by the use of shift registers and feedback connections. The codec in the transceiver under discussion is the data-processing unit (DPU). The RF units, as mentioned previously, are similar to those found in an analog radio system, except for the modulator if it is operating at RF as shown in Figure 11.1(*b*). However, where there are similarities between the two types of radio system, some differences do exist and must be taken into account. Differences exist in the design of the filters. Filters have to be designed with the properties of the digital RF signal kept in mind. In the transmitter section, the spectrum shaping is often an essential requirement, whereas, in the receiver side, the design of the RF filter is more lax due to the usually more stringent design considerations placed on the IF filter. The IF filter dictates the receiver selectivity. Both transmitter and receiver filters will suffer from small variations in their amplitude and group delay characteristics over the bandwidth which contains the most significant part of the spectrum. When amplification of the modulated RF signal is required, it must be done by an amplifier which has well-controlled nonlinearities, or as mentioned above, by providing a pre-distortion which cancels the effects of these nonlinearities. Most digital RF signals have small envelope variations which cause a small amount of spectral spreading and which in turn reduce, on detection, the amount of eye opening (Section 5.2.5). Baseband units in a digital transmitter or receiver are built up from digital circuits which are either working or not. They cannot slowly degrade as in analog circuits. In addition, digital circuits have fewer adjustable components. This means that repairs to digital circuits are usually done by replacing the whole transmitter or receiver, replacing a module within the transmitter or receiver circuits, or by replacing a printed circuit card or submodule within the modules. Maintenance is thus easier and outage time is reduced. Sophisticated supervisory and remote control systems are easier to implement in digital systems and this makes possible the quick location of failing equipment.

11.3.1 Receiver RF circuits

Figure 11.6 shows the block diagram of the CTR 188 11 GHz receiver with space diversity combiner block diagram. The receiver assembly comprises four functional

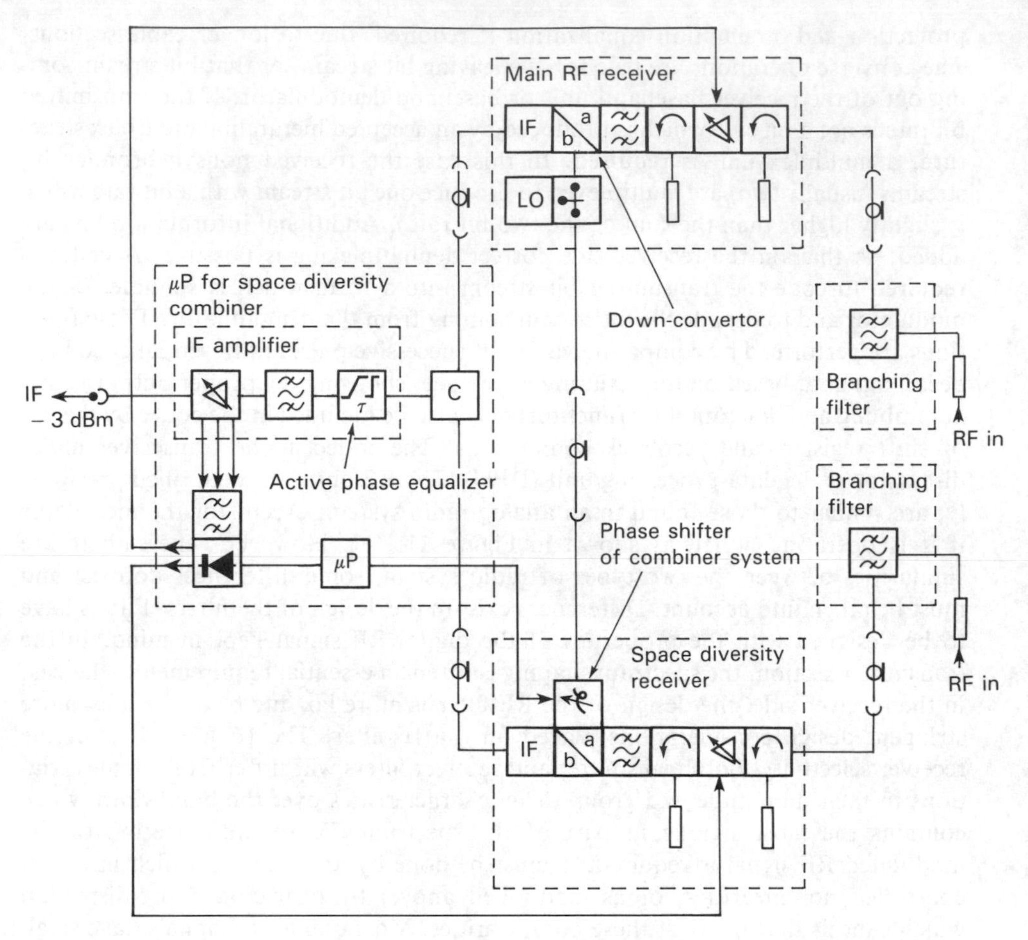

Figure 11.6 CTR 188 11 GHz 16–QAM receiver: RF block diagram MP = microprocesor

blocks, which are used in accordance with the system requirements:

Main RF receiver
Space diversity receiver
IF module
Microprocessor for space diversity combiner

The space diversity option using the IF combiner is of the minimum dispersion type, which is discussed in Section 8.3.5.1, 'Minimum dispersion combiner'.

When space diversity is used, the RF signals coming from the main and diversity antennas feed the receive branching chains of circulators and RF filters before entering into the two RF receivers. The RF receivers usually contain low-noise front-end pre-amplifiers which have automatic gain control (AGC) circuits built into them.

These amplifiers are usually two-stage amplifiers built with GaAsFET devices. Circulators are provided before and after the AGC amplifier to permit monitoring of the signal level at these points. A band rejection filter follows each RF amplifier to eliminate the image noise band produced by the same pre-amplifier. The signals enter the down-convertors with high amplitude linearity and mix with the voltage-controlled phase-locked local oscillator. The wideband balanced receiver mixer usually utilizes microwave integrated circuits these days. The local oscillator often is built from a low-frequency crystal oscillator, operating at 10 MHz or so, and the necessary phase-locked multiplier to achieve the required local oscillator frequency. For the space diversity system shown, the local oscillator feeds the down-convertor of the main signal and the phase shifter of the combiner system, which is connected to the down-convertor of the space diversity signal. The two IF signals at 140 MHz obtained from the mixer circuits enter the IF pre-amplifiers before entering the IF combiner. The amplitude of the two diversity signals and their relative phase is controlled by a microprocessor, to minimize the channel dispersion at the common IF output, counteracting the effect of selective fading.

The output signal from the IF combiner feeds the IF filter, the active phase equalizer and the main IF amplifier which has the AGC control circuit built into it. The main IF amplifier has an auxiliary output that is connected to the IF analyzer and microprocessor unit, when the space diversity system is required. The IF analyzer reads out the IF amplitudes through selective detectors, which monitor the IF levels at the center of the IF band and at the band edges. The output of the detectors is computed by the microprocessor that provides real-time control of the IF combiner circuits to obtain the maximum flatness possible for the IF output. The IF output then passes to the demodulator. An IF monitoring output is available at this point.

11.3.2 Receiver baseband circuits

The main functions of the receiver baseband circuits are to provide:

- IF adaptive equalization
- Demodulation of the IF signal
- Adaptive baseband equalization and data regeneration
- Baseband processing including differential decoding and parallel-to-series conversion
- Demultiplexing of the aggregate signal into the main original signals and service signals
- Monitoring the BER

Usually the IF and baseband adaptive equalizers are optional features on commercially available radios. The main objectives to be realized in the demodulator section are to: produce linear demodulation over the quadrature axis; filter the IF to eliminate the out-of-band emissions; perform clock extraction through IF envelope detection; effect carrier recovery through baseband processing using the modified Costas loop technique (see Section 6.5.2.3); and use, if required, a baseband adaptive equalizer of the decision feedback type, to enhance the demodulation process and produce an effective equalizer improvement factor of around 8. Figure 11.7 shows a block diagram of the GTE CTR 188 16–QAM demodulator. The basic structure of this demodulator is classical and is similar to that of the NEC 16–QAM radio.

The demodulation is obtained by mapping the multilevel PAM signals on the two quadrature axes. After baseband filtering and regeneration, four binary streams at 35 Mbit/s are obtained. The differential decoding and the parallel-to-series conversion produces the 140 Mbit/s data stream. The carrier recovery required for coherent demodulation is obtained through the modified Costas loop technique. The adaptive baseband decision feedback type of equalizer is used to reduce the effect of inter-symbol interference due to amplitude and phase distortions incurred during transmission of the signal carrier through the transmission channel. The clock extraction at the symbol rate is obtained through the envelope detection of the IF

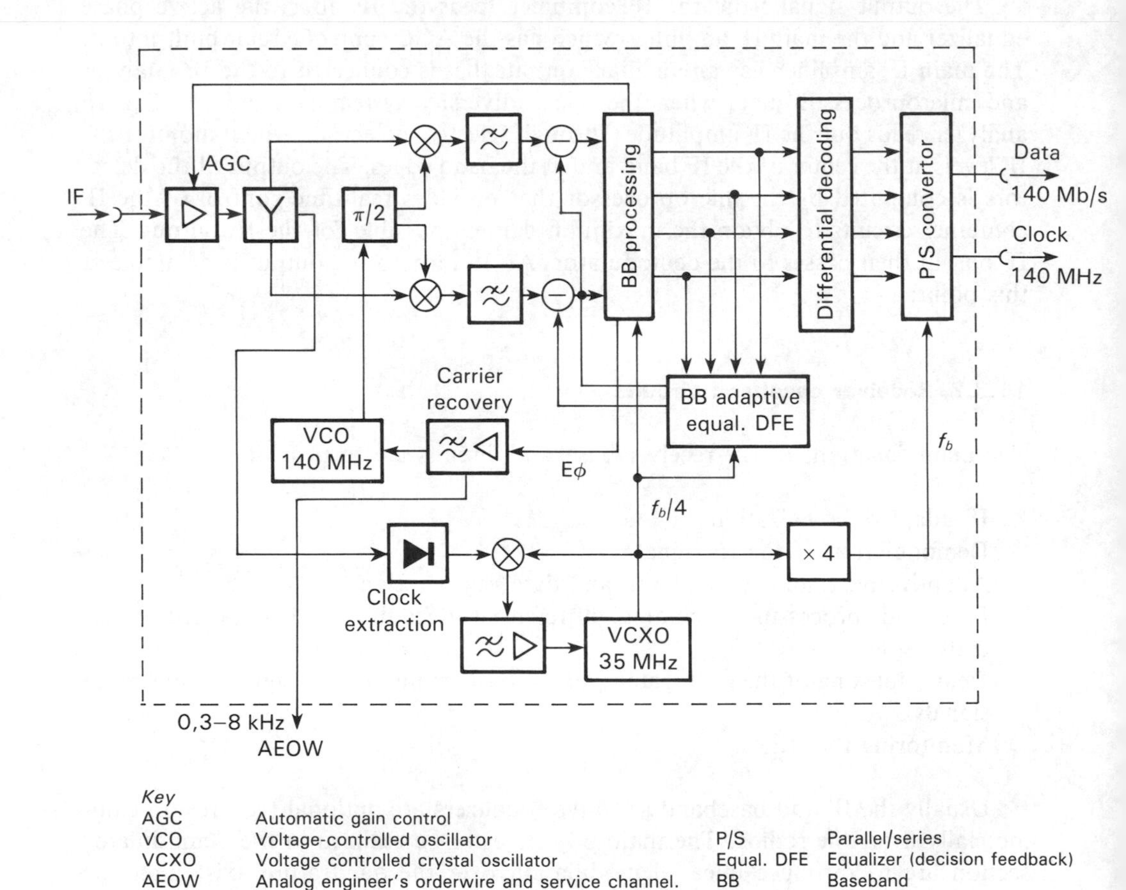

Key

AGC	Automatic gain control		
VCO	Voltage-controlled oscillator	P/S	Parallel/series
VCXO	Voltage controlled crystal oscillator	Equal. DFE	Equalizer (decision feedback)
AEOW	Analog engineer's orderwire and service channel.	BB	Baseband

Figure 11.7 GTE CMF 40 16–QAM demodulator

received signal. Figure 11.7 also shows the demodulation from the IF carrier of the analog service channels. Figure 11.3 shows that the parallel-to-series conversion is performed by the NRZ-to-CMI line code conversion unit. Figure 11.7 shows this process occurring before the code convertor circuit is reached. The function of the NRZ-CMI code conversion unit, besides performing if required the parallel-to-series conversion, is described below. In Figure 11.3, the block marked F SYNC extracts the framing information and monitors the parity bits as well as integrating the parity error count for performance monitoring. When the error count exceeds a predetermined value an alarm is raised. This unit also extracts the service channel information if used. The digital service channel can also be extracted by this unit. The NRZ-CMI convertor combines the four bit streams into one by parallel-to-series conversion and also converts the bipolar binary NRZ digital signal into the CMI coded digital signal, ready for transmission to the outside world. Different manufacturers produce different layouts or arrangements of the baseband processing circuits, as can be seen from the structure of the block diagram shown in Figure 11.7. To complete the processing following the demodulator block diagram shown in Figure 11.7, the block diagrams of Figures 11.8 and 11.9 show the bit extraction

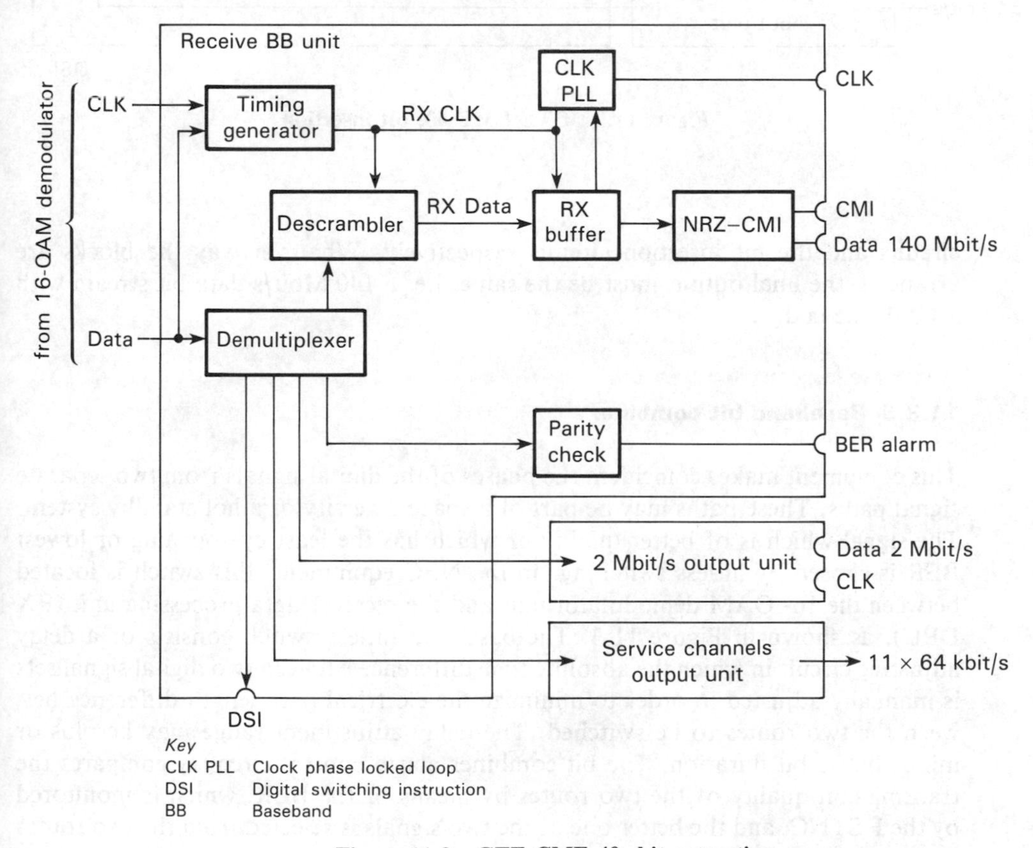

Figure 11.8 GTE CMF 40, bit extraction

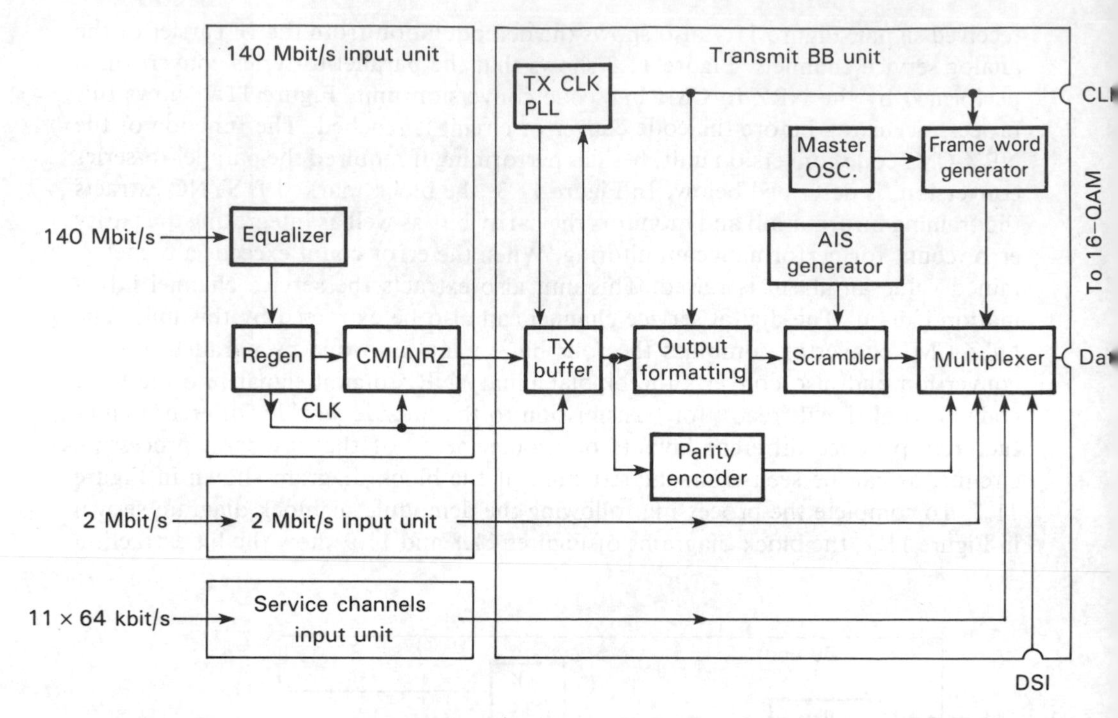

Figure 11.9 GTE CMF 40, bit insertion

circuits and the bit insertion circuits respectively. Whatever way the blocks are arranged, the final output must be the same, i.e. a 140 Mbit/s data bit stream with a CMI line code.

11.3.3 Baseband bit combiner

This equipment makes coincident the phases of the digital signals from two separate signal paths. These paths may be part of a space diversity or a hot standby system. The signal which is of better quality or which has the least eye opening or lowest BER is chosen by hitless switching. In the NEC equipment, this switch is located between the 16–QAM demodulator unit and the receiver data processing unit (RX DPU), as shown in Figure 11.3. The baseband hitless switch consists of a delay adjusting circuit in which the absolute time difference between two digital signal sets is manually adjusted in order to minimize the electrical path length difference between the two routes to be switched. The delay adjustment range may be plus or minus half-a-bit duration. The bit combiner alarm control circuitry compares the transmission quality of the two routes by means of the BER, which is monitored by the F SYNC, and the better one of the two signals is selected from the two routes

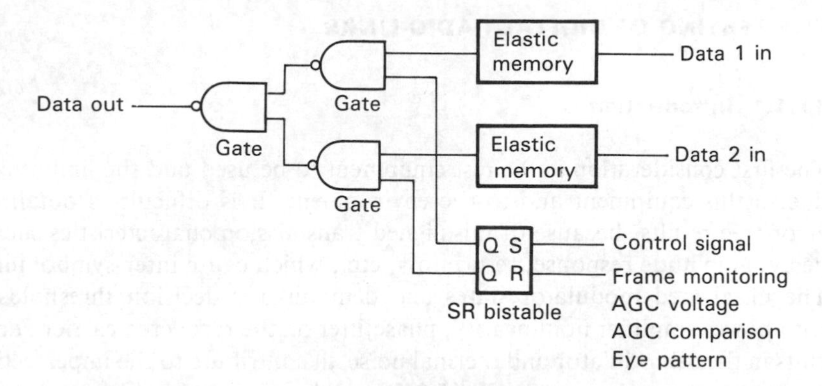

Figure 11.10 Principle of hitless switching

by using a gate switch as shown in Figure 11.10. This gate switch is controlled by the alarm control circuitry (see Figure 11.11) which, as mentioned performs the comparison. The timing of the two signal sets is kept within ±2.5 bits and no disruption to the data is made during a switchover. Figure 11.11 shows the block diagram of the baseband switch for a 1 + 1 protected system, i.e. an active system and a standby unit which carries the same traffic but is not on-line. In the remainder of this chapter we shall briefly consider the tests to be performed on the digital radio and the digital radio link.

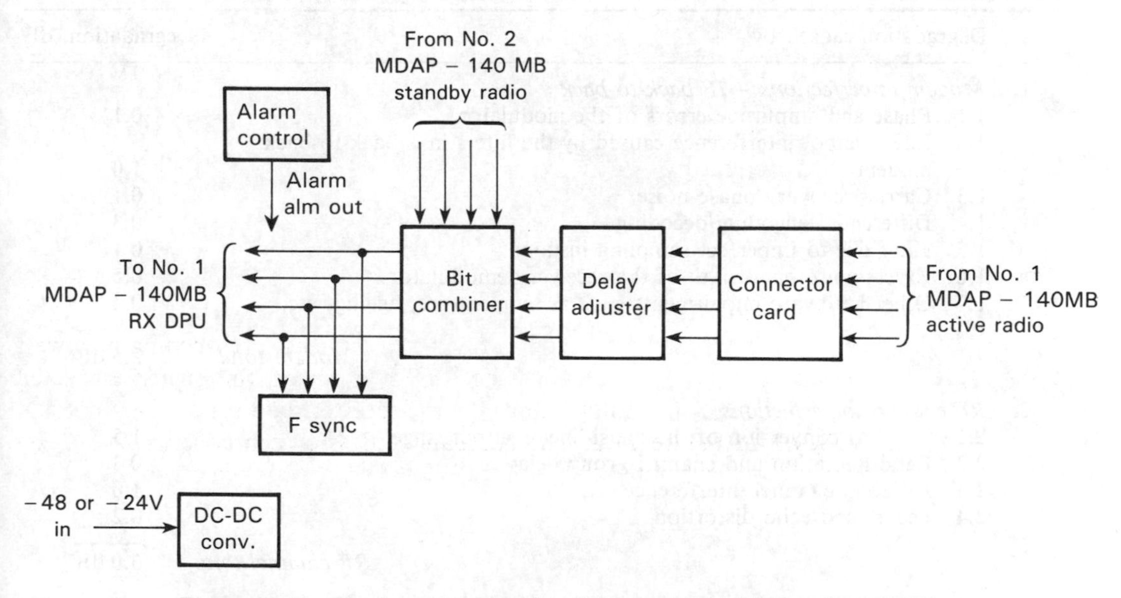

Figure 11.11 Baseband bit combiner (hitless switch) NEC DAP-140 MB for 1 + 1 protection switching

11.4 THE TESTING OF DIGITAL RADIO LINKS

11.4.1 Introduction

The first consideration is the test equipment to be used and the limitations of tests due to this equipment and to the environment. It is difficult to obtain long-term error-free results, because of misaligned transmission characteristics such as group delay, amplitude response, return loss, etc., which cause inter-symbol interference. The misaligned modulator states and demodulator decision thresholds, together with power amplifier nonlinearity, phasejitter on the recovered carrier and clock circuits in the demodulator and thermal noise all contribute to the imperfections which cause partial eye closure (Section 5.2.5) and therefore reduce the receiver threshold. This in turn reduces the margin available to low received-signal levels and interference. Unlike propagation effects, these mechanisms give rise to long-term degradation and are responsible for the low-level continuous-background BER which is present even with high received-signal levels. All these degradations can be quantified in terms of the C/N penalty or 'implementation margin' resulting from eye closure. A typical radio system degradation budget is given in Table 11.1 below. From Table 11.1 the C/N penalty due to individual impairments is 5.5 dB. This means that the value of C/N will have to be 5.5 dB higher than the theoretical value of C/N for a given BER. This in turn means that a higher transmitter power, or lower receiver noise figure will be required for a given fade margin or system

Table 11.1 System degradation budget in a typical digital radio system

	Degradation caused by	Degradation, dB
1.	*Modem imperfections – IF back-to-back*	
1.1	Phase and amplitude errors of the modulator	0.1
1.2	Inter-symbol interference caused by the filters in a back-to-back modem	1.0
1.3	Carrier recovery phase noise	0.1
1.4	Differential encoding/decoding	0.3
1.5	Jitter due to imperfect sampling instants	0.1
1.6	Excess noise bandwidth of the receiver demodulator	0.5
1.7	Other hardware impairments such as temperature, ageing, etc.	0.4
	Modem total	2.5 dB
2.	*RF channel imperfections*	
2.1	AM/PM conversion of the quasi-linear output stage	1.5
2.2	Band limitation and channel group delay	0.3
2.3	Adjacent-channel interference	1.0
2.4	Feeder and echo distortion	0.2
	RF channel total	3.0 dB
	Total modem and RF channel degradation	5.5 dB

gain. Every extra decibel of transmitted power is very expensive particularly when linearity is important, as in QAM systems.

11.4.2. Methods of making *C/N* measurements

There are two methods which may be used to measure *C/N*. These are the traditional method of fade simulation, and the method of additive noise. Whatever method is used, the carrier power must be defined as being either the *modulated* or *unmodulated* signal, particularly when the results are to be compared. The modulated carrier power will normally be less than the unmodulated power due to the spectrum truncation caused by the filtering in the radio transceiver and also due to amplitude modulation in the modulation scheme, for example QAM. Differences of 1–2 dB or more are normal.

11.4.2.1 Fade simulation method

This method relies on the reduction of the carrier level *C* entering the receiver and the subsequent recording of the BER using a BER test set at the output of the receiver baseband on a specific 64 kbit/s test channel. An attenuator is connected at the RF input of the receiver so that the incoming signal can be attenuated into the receiver noise. A power meter is used to check the effective *C/N* in the IF section of the receiver, as this is the only place where *C/N* is valid. The power measurement made at the IF is in fact $(C+N)$. To determine the noise power of the receiver *N*, the far end transmitter is powered down and a power reading is then taken at the same point in the IF section. This will record the value of the noise power level *N*, which permits *C/N* to be determined. Alternatively, the received signal level entering the front end of the receiver can be measured and the received signal level then plotted against the BER. In this case, the levels are very low, so it is normal to use a calibrated attenuator placed between the receiver front end and the waveguide to the antenna and to rely on the attenuator settings for the equivalent value of the received signal level. It is assumed that the unattenuated received carrier level does not change while the test is being performed. The noise measurement at IF will most likely be higher than the noise level at the regenerator following the demodulator, because of the additional filtering which may exist. This can be accounted for by making the noise level measurement in an equivalent noise band W_e, after inserting the appropriate filter into the power meter when taking measurements at IF. The improvement in the *C/N* will be given by 10 log W/W_e, where *W* is the IF bandwidth. Figure 11.12 shows the test set-up for taking measurements to plot *C/N* against the BER.

As mentioned above, it is assumed that, during the measurement of *C/N* and the associated BER, the carrier signal received into the antenna is constant during the tests. The noise level recorded is a function of the receiver itself and hence remains relatively constant; however the received carrier level even on a good day will vary by 1–2 dB. This will of course lead to measurement errors and unstable BER readings. This is especially so at low BER values where long gating times are

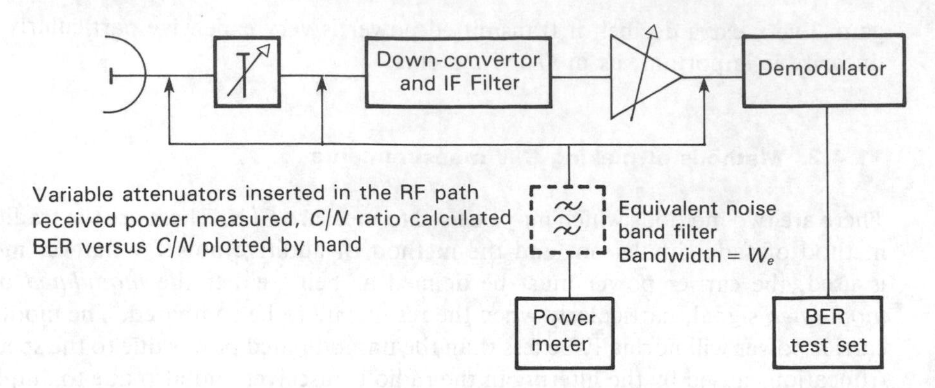

Figure 11.12 Fade simulation method of measuring *C/N* against the BER

required for statistically relevant measurements. By referring to Figure 10.1, it can be seen that for a 1 dB change in the value of C/N or the received input level, there is a difference of an order in BER. In addition to this, as the input received signal is reduced close to the threshold level, any variation in the received carrier can produce a loss of synchronization which would invalidate any BER readings taken. This method does have the advantage of checking the maximum amount of flat fade depth, or minimum received signal level that the working system has, for a given BER, and would always have to be checked at some time during the testing stage.

The advantages of using the fade simulation method are: it tests the overall system implementation margin (see Figure 10.4) due to impairments in the modulator/demodulator stages and due to transmission, and it tests the receiver noise figure, the transmitter power, the transmitting and receiving antenna gain and the receiver AGC action. An additional advantage lies in the use of a simple low-cost attenuator rather than more complex test equipment, but this must be weighed against the accuracy of the microwave attenuators, the repeatability of the measurement due to the many uncontrolled variables (unknown noise parameters and bandwidths), and the time required to perform the test.

11.4.2.2 Additive noise testing methods[1]

In additive noise testing, the digital radio receiver operates at normal unattenuated received signal levels, so that the effect of the receiver noise is negligible and the possible loss of synchronization is minimized. The IF signal in the receiver path is connected through the HP 3708A carrier/noise test set, which is set to the appropriate system bandwidth so that the test set can add noise with a high crest factor (>15 dB) to the IF signal. A high noise crest factor (>20 dB) permits accurate measurements of BER down to 10^{-12}. By the test set adding the noise, C/N is accurately known and the BER is checked using a pattern generator at the transmit terminal and an error detector at the output of the receiver demodulator. Figure 11.13 shows the test set-up using the HP3708S, which performs automatic measurements with the HPIB controllable BER testers. The measurements can be made manually with a wide range of nonprogrammable BER testers. The advantage

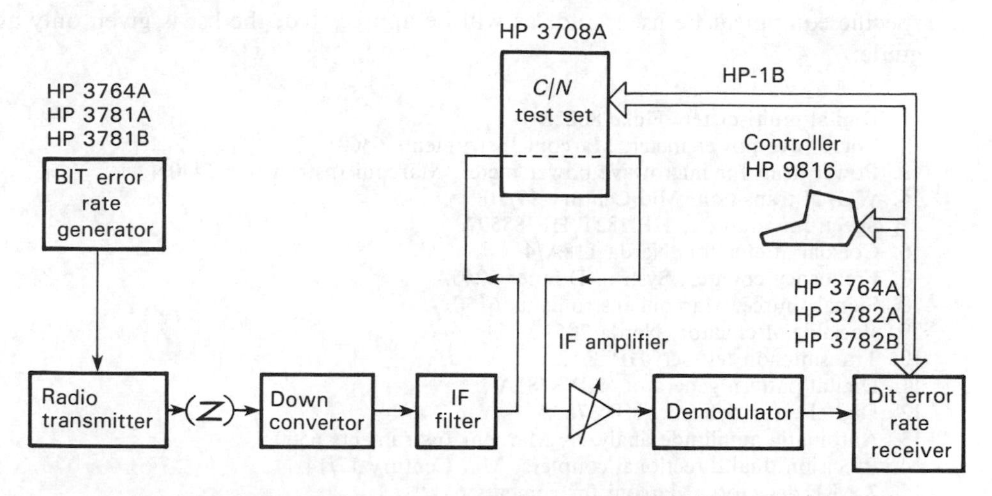

Figure 11.13 Additive noise testing using the HP3708A carrier/noise test set
(Source: Reference 1)

of using this method and this test equipment is that as the carrier varies due to propagation conditions, the test equipment injects into the IF a noise power which tracks the carrier to keep a constant C/N over the measurement period. The value of C/N does, however, require to be related to a specified bandwidth, for example, the bandwidth of the practical receiver or to thc thcoretical minimum Nyquist bandwidth (or symbol rate bandwidth) so that the relevant noise bandwidth filters can be inserted by the test equipment. For determining the very low background BER at high C/N, a long gating time is usually required. This time may be several days. The background BER is required to be known since there may be many radios in the radio relay section, that may make the error performance unacceptable. The measurement time can be reduced by using a C/I test, in which an interfering signal is added to produce a controlled amount of eye closure. The increased BER is then easily measured and the true background BER can be estimated. This has been discussed briefly in Section 6.3.4. and is discussed further in Hewlett-Packard Product Note 3708–3 (Publication No. 5953–5490).

11.4.3 In-station tests

In this section we consider briefly the test equipment to be used and the tests to be performed, to complete the in-station tests on a radio transmitter and receiver terminal. As manufacturers specify the details of particular tests to be carried out on their equipment in the field, only the principles will be discussed here.

11.4.3.1 Test equipment used in in-station testing

Below is a list of test equipment which may be used. It is not mandatory that this specific equipment be used, and, as will be appreciated, the list is given only as a guide:

1. Digital multi-meter, Fluke 8022B
2. Microwave power meter, Marconi Instruments 6460
3. Power head for microwave power meter, Marconi Instruments 6440N
4. WG–N transition, Mid Century 17/10
5. Spectrum analyzer, HP 182T/HP 8559A
6. Coaxial attenuator, Narda 118A/4
7. Frequency counter, Systron Donner 6245A
8. Signal source, Marconi Instruments 6150
9. Variable attenuator, Narda 705–3–69
10. Transmission test set, HP 3552A
11. Digital pattern generator, HP 3781A
12. Digital error detector, HP 3782A
13. Automatic amplitude analyzer, Marconi Instruments 6500
14. Precision dual directional couplers, Mid Century 17/14
15. 2 × RF detectors, Marconi Instruments 6511
16. Oscilloscope, Tektronix 2215
17. Portable radios (walkie-talkies)

The portable radios provide an independent means of communicating between the indoor environment and the external environment.

11.4.3.2 Power supply voltages

1. Connect the digital multi-meter between earth and the test points or between the test points designated by the manufacturer and record all measurements taken.
2. Check all readings against those as specified by the manufacturer or factory test results and take remedial action where necessary.
3. All records should be kept for maintenance personnel or for future reference.
4. Connect the oscilloscope to the power supply at the radio terminal end and record the value of the AC ripple observed. This value should be less than 0.2 percent of the supply voltage.

11.4.3.3 Transmitter output power

1. Switch off the site transmitter, or transmitters feeding into a single antenna, and disconnect the waveguide feeder from the output of the transmitter antenna coupling unit.
2. If a hot-standby unit is being used (see Section 7.8.1), connect the output of the antenna coupling unit with a waveguide to *N*-type transition and connect a 20 dB coaxial attenuator to the transition.
3. Connect the power meter head to the *N*-type connector on the coaxial attenuator and connect the power meter head to the power meter.
4. Switch on the transmitters and select one particular transmitter manually.
5. Measure the output power and compare against that expected. Record reading.
6. Manually switch over to the other transmitter, confirming that the appropriate lamps illuminate.
7. Measure the output power and compare against that expected. Record reading.
8. Switch off the transmitters but do not at this stage restore to the original connections.

11.4.3.4 Transmitter output frequency

1. Disconnect the power meter and power meter head from the 20 dB coaxial connector.
2. Connect the frequency counter to the coaxial connector.
3. Switch on the transmitter and select one of the transmitters if a hot-standby unit is used.
4. Measure the transmitter unmodulated output frequency after disconnecting any baseband data.
5. Confirm that the output frequency is the frequency recorded on the transmitter identification plate and is within the limits specified by the manufacturer.
6. Manually change to the other transmitter and disconnect any baseband data entering into it.
7. Measure the unmodulated frequency of this second transmitter.
8. Confirm that the output frequency is as recorded on the transmitter identification plate and is within the tolerance provided by the manufacturer.
9. Record the output transmitter frequencies on test result sheets.

11.4.3.5 Calibration of AGC level against input receive signal level

In most equipment configurations, the receiver has an associated meter that indicates the received signal strength. These AGC meter readings are relative indications of the received carrier level and can be used to determine the received signal level, under operating or traffic-carrying conditions. An accurate graphic record of AGC meter readings versus RF input level must, however, be made first. The results of this curve can also be used to establish when the antenna systems are optimally aligned. The procedure for developing an AGC curve requires that an RF signal generator should be connected to the input of the receiver at either the directional coupler or at another calibrated RF input test point, to simulate a received signal. With no RF input, the AGC meter should indicate near zero. The frequency of the output level produced by the generator is arranged to be the center frequency of the receiver under test, and its level is precisely adjusted to the nominal input signal level, with compensation being made for losses in the directional coupler, hybrids and circulators (the total items comprise the antenna coupling unit). Rather than rely on the attenuator setting shown on the front panel of the RF signal generator, for obtaining the power level entering the receiver, a calibrated switchable attenuator may be connected between the RF signal generator and the receiver input: the output power of the RF signal generator is fixed and the switchable attenuator used to vary the power level. Only after recording the power at the output of the lead connected to the output of the attenuator, with a power meter for various attenuation settings, can the input power to the receiver be accurately known. This method only works satisfactorily in a field environment for frequencies at or below 1 GHz.

1. Set up the test equipment as described above with the stepped attenuator set to 10 dB and the output power of the RF generator set -10 to -20 dBm at the center frequency of the receiver.
2. Adjust the frequency of the RF signal generator to be exactly that of the receiver under test. This may be done by watching the AGC meter peak.

3. Record the frequency by disconnecting the RF signal generator from the attenuator and connecting a frequency counter.
4. Remove the frequency counter and connect the RF signal generator back to the switchable attenuator. The output lead from the attenuator (where the power calibration was made) should be connected onto the 20 dB fixed attenuator already fitted to the hot-standby unit output flange.
5. Insure that the receiver front panel meter switch is in the AGC position.
6. If two receivers are used, turn the front panel switch to record the AGC readings of both receivers for a known input RF level.
7. Record the AGC readings for a RF input level of – 30 dBm and thereafter to an input level of – 90 dBm in 5 dB steps.
8. Plot the AGC calibration curve as readings are taken. This will ensure that mistakes in readings are minimized. The RF signal level into the receiver should be on the horizontal axis of the plot and extend from – 100 dBm at the extreme left to – 20 dBm at the extreme right. The panel meter reading should be plotted on the vertical axis, with the maximum number of divisions equally divided into, say, ten sections.

In some radios the AGC meter reading may increase with a decrease in the applied signal strength.

11.4.3.6 Transmitter output spectrum

Section 10.5 dealt with the permitted emission spectrums for different digital radio transmitters. In this section the test procedure used to verify that the transmitter emission spectrum does not exceed the mask established by regulation, is set out.

1. Disconnect the frequency counter from the 20 dB attenuator, as for the test procedure given in Section 11.4.3.4.
2. Connect the input of the spectrum analyzer to the 20 dB attenuator. Note that it is most important that this attenuator be *in situ*, so as not to overload the spectrum analyzer.
3. Manually switch to one of the hot-standby transmitters.
4. Adjust the spectrum analyzer tuning and attenuation to display the transmitter carrier.
5. Reconnect the data into the transmitter baseband using the pattern generator. The necessary line code is provided in Table 5.5 for the different bit rates, and the pseudo-random test sequence pattern length is given in Table 4.10. For 34 Mbit/s, this is the HDB3 code and the pattern length is $2^{23} - 1$.
6. Adjust the spectrum analyzer sweep width to display the transmitter modulated output spectrum.
7. Compare the displayed spectrum against the regulation mask.
8. Confirm the symmetrical shape of the spectrum.
9. Measure the first side-lobe peak level relative to the carrier peak level. This should be at least – 25 dBc (dB referred to carrier level).
10. Record the results on a test record sheet.
11. Manually switch to the second transmitter.
12. Repeat steps 4 to 10.

11.4.3.7 Spurious emissions

1. Insure that the transmit carriers are unmodulated by disconnecting the baseband pattern generator.
2. With the test equipment set up as in procedure 11.4.3.6, select the operating carrier frequency band range on the spectrum analyzer, e.g. for a 13 GHz band the range should be 9–15 GHz. Adjust the spectrum analyzer sweep to 1 s/division, with a sweep range of

20 MHz/division, resolution of 300 kHz, reference attenuator to 0 dB and reference level to −10 dBm.
3. Switch to one of the hot-standby transmitters and inspect the radio spectrum output of this transmitter for spurious emissions in the band, (e.g. 12.750 GHz to 13.250 GHz). This may be done in three sweeps centered at one-third-point frequencies throughout the band (e.g. 12.800, 12.980 and 13.160 GHz). This will cover the band with some overlap and allow measurement down to −60 dBc.
4. Confirm that any spurious emissions are at a signal strength of less than −60 dBcm (see Section 10.5.1).
5. Manually switch to the other transmitter and repeat the test.
6. Record the results on the appropriate test result sheets.
7. Restore the baseband connections and disconnect the spectrum analyzer from the 20 dB attenuator.

11.4.3.8 Analog service channel sensitivity

This procedure is designed to set the deviation of the carrier so that analog service channels can be transmitted properly. The method used to set the deviation sensitivity is the 'Bessel zero' method. The Bessel zero is defined as the point at which all of the carrier energy has been distributed amongst the sidebands. At this point the modulation index equals 2.045 and the carrier amplitude is zero. Since frequency modulation varies in proportion to the level of the modulating frequency, and the required level-versus-frequency characteristic of the modulator clearly defines the point at which the Bessel zero will occur, a criterion may be established to which the deviator sensitivity may be adjusted. The sensitivity of the deviator determines the degree of the frequency shift for a given input amplitude and frequency. To set the desired system deviation requires the use of a modulating source and a spectrum analyzer to monitor the carrier frequency component. This method is similar to that used in setting the deviation sensitivity of analog radio systems.

1. Set up the test equipment as in Section 11.4.3.6 and insure that there is no data signal entering the transmitter baseband.
2. Manually switch to one of the transmitters and tune the spectrum analyzer to display the transmit carrier.
3. From the transmission test set, apply an audio signal at the manufacturer's recommended impedance, audio frequency and level, into the appropriate input port of the transmitter.
4. Slowly increase the audio output signal level from the transmission test set and observe the first point at which the carrier disappears, as displayed on the spectrum analyzer. Be sure that it is the first null and not a second or higher null. The output level from the transmission test set should be noted and compared against that recommended by the manufacturer. If it is not within limits, set up the audio oscillator level to that recommended and adjust the transmitter deviation for the first spectral carrier null.
5. Switch to the second transmitter and repeat the measurement.
6. Disconnect the transmission test set and record the final results on the test record sheets.

11.4.3.9 Loop test or back-to-back test

Depending on the radio equipment this test may be performed by connecting the transmitter IF to the receiver IF. If this is not possible, the transmitter RF signal at frequency f_1, say, must be converted into the receiver frequency f_1' by a suitable convertor device connected between the transmitter output and the receiver input. Such pieces of equipment are often supplied by the manufacturer as standard items of test equipment required. For this procedure, this item of equipment will be called a 'turn-around oscillator'.

1. Switch off the transmitter.
2. Terminate the waveguide to *N*-type transition in a coaxial load or into a load which is appropriate to the working frequency band.
3. Connect the turn-around oscillator to the low-level monitor output of the transmitter. Where there is no monitor point (which would be unusual), the output of the transmitter must be connected to a feedthrough attenuator of at least 70 dB.
4. Restore power to the terminal.
5. Connect to the baseband data input a pattern generator and to the baseband data output an error detector.
6. Set the pattern generator and the error detector to the pseudo-random test sequence pattern length as given in Table 4.10 and to the appropriate line code as provided in Table 5.5. Again for a 34 Mbit/s radio, this would be $2^{23} - 1$ and HDB3 respectively.
7. Select one of the transceivers and confirm synchronization is achieved and that an error-free data signal is received for a period of no less than 30 minutes.
8. Select 1×10^{-5} error rate on the pattern generator and confirm that 1×10^{-5} error rate is recorded on the error detector.
9. Disconnect the pattern generator and confirm that the AIS is received.
10. Repeat the tests for the other transceiver.
11. Enter all of the results in the test record sheets.

11.4.3.10 Alarms

The test set-up provided in Section 11.4.3.9 is ideal for testing some of the alarms on the system. The names of the alarms may change, but overall their function remains the same in most commercially available equipment. In this procedure some of the more usual alarms will be tested for proper operation:

1. With the equipment looped through the turn-around oscillator and the pseudo-random bit stream data passing through the system as described in Section 11.4.3.9, disconnect the output of the pattern generator to the first channel under test.
2. Verify that the following alarms illuminate:
   No input.
   Alarm indication signal (AIS).
   Disconnect the data connection from the input to the first transmitter and check that the appropriate transmitter alarm illuminates.
   Disconnect the data connection from the input to the other transmitter and check that the appropriate alarm on this transmitter illuminates.
   Confirm that the 'no input to the receiver' lights, for both channels, are on.
   With dual receivers confirm that the receiver autochange lamp is illuminated.
3. Enter the results on the test record sheets.
4. Disconnect all test equipment and restore all connections to normal.

11.4.3.11 Hot-standby changeover switching

1. Manually select transmitter 1 on the front panel of the digital radio terminal.
2. Confirm that the channel 1 test lamp illuminates.
3. Manually select transmitter 2.
4. Confirm that channel 2 test lamp illuminates.
5. Select 'force test' to transmitter 1.

6. Confirm that 'force test' switch overrides manual switching.
7. Select 'force test' switch to transmitter 2.
8. Confirm that 'force test' switch overrides manual switching.
9. Record results on the test record sheets.

11.4.3.12 Antenna and feeder VSWR

1. Turn off the transmitter.
2. Disconnect the feeder from the transceiver and connect a dual directional coupler to the feeder.
3. Connect a waveguide to *N*-type transition to the directional coupler.
4. Terminate both the forward and reverse outputs of the directional coupler with RF detectors.
5. Connect a microwave signal source (Marconi MI 6150) to the waveguide to N-type transition.
6. Connect the forward power RF detector to the R input of the automatic amplitude analyzer (MI 6500).
7. Connect the reverse power RF detector to the B input of the automatic amplitude analyzer. Set the start frequency to the beginning of the frequency band under consideration (e.g. 12.7 GHz) and the stop frequency of the swept band to the other frequency extreme of the band (e.g. 13.5 GHz).
8. Measure the VSWR over the specified band and locate the maximum VSWR recorded.
9. Photograph the VSWR display on the automatic amplitude analyzer and attach the photograph to the test record sheets.
10. Record the maximum VSWR value and the frequency at which this occurs. The VSWR is expected to be less than 1.3.

11.4.3.13 Waveguide pressure sealing

1. Disconnect the RF detectors, transition, automatic amplitude analyzer and directional coupler from the waveguide feeder.
2. Connect the waveguide feeder to the radio, and pressurize the waveguide system.
3. Isolate from the pressure source and leave the waveguide for at least 12 hours and then measure the waveguide pressure after this time. No pressure change should then be detected unless the temperature has radically altered.

11.4.4 End-to-end tests

This section surveys the test equipment to be used in the end-to-end tests and the procedures in principle which may be followed in carrying out these tests. The same situation applies here, as for the in-station tests, namely that the equipment manufacturers will provide in detail the specific tests to be performed on their equipment.

Before these tests can be started, the in-station tests must be completed for both ends of the hop. The BER tests may be carried out either from end-to-end (which is more preferable) or as loop-back tests. Path 1 is through the first transmitter and receiver pair at each ends of the link, and path 2 is through the second transmitter and receiver pair of a hot-standby unit located at the same terminals of path 1 at each end of the link. Once these tests have been performed, the same total end-to-end tests are then conducted in the opposite direction.

11.4.4.1 Test equipment used in end-to-end or hop tests

Below is a list of suggested test equipment which may be used. The specific brand or type of equipment should not be treated as mandatory, but as a guide to the type of equipment which can be used.

1. Digital pattern generator, HP 3781A
2. Digital error detector, HP 3782A
3. Variable waveguide attenuator, Mid Century MC 17/3
4. 2 × Transmission test set, HP 3552A
5. Jitter test set, HP 3785A
6. HP 85 computer with input/output interface unit cables
7. Polaroid camera to photograph oscilloscope displays
8. Oscilloscope (up to 100 MHz for 34 Mbit/s radios) and G.703 pulse mask
9. Portable radios (walkie-talkies)
10. Digital pattern-generator/error-detector W & G PF–4 (used as an alternative to items 1, 2 and 6 when conducting BER test in loopback)

11.4.4.2 Antenna alignment

1. Check the antenna element to determine that the polarization is the same as that required. If the element is horizontal, the polarization is horizontal; similarly if the element is vertical.
2. At station A, monitor the AGC voltage and pan the antenna for maximum signal strength. Note that for some types of radio, this may mean that the AGC voltage must be at a minimum.
3. Tilt the antenna for maximum received signal strength.
4. Repeat antenna panning and tilting until no further improvement in the signal strength is obtained.
5. Lock the panning frame in both planes of movement.
6. Record that the antenna alignment has been done at this site.
7. Repeat steps 1 to 6 at station B.

11.4.4.3 Path signal strength measurement

1. With the link connected and operating under no-fade conditions, monitor the AGC voltage on the digital radio terminal meter.
2. Using the results of the in-station AGC curve tests, convert the AGC voltage to the received signal strength under no-fade conditions. (Under no-fade conditions, the meter reading will remain constant over a 5 minute interval).
3. Repeat steps 1 and 2 for path 2.
4. Enter the results on the test record sheets.

11.4.4.4 Link jitter tests

1. Set up the test equipment so that a pattern generator is connected to the baseband of the transmitter at station A. The pseudo-random test sequence line code is provided in Table 5.5 for the different bit rates and the test sequence length is given in Table 4.10. For 34 Mbit/s, this is HDB3 and the pattern length is $2^{23}-1$.
2. At station B connect the jitter test set to the corresponding receiver baseband output and make two measurements of the output peak-to-peak jitter as defined by Table 4.11, with the appropriate filters inserted in the measuring set. For example, the maximum peak-to-

peak jitter in unit intervals for a 34 Mbit/s radio is 1.2 UI with the low pass (LP) plus high pass 1 (HPI), filter inserted and 0.12 UI with the LP plus HP2 filter inserted.
3. Record the jitter received at station B on the test record sheets.
4. Repeat the test for path 2.

11.4.4.5 Pulse output shape

1. At station A connect a pattern generator to the baseband data input of the terminal. Set the pattern generator for "all '1's output".
2. At station B connect an oscilloscope to the baseband data output. Use a matched termination to couple the oscilloscope to the baseband output, by using a 'T' BNC section. Set the oscilloscope to display one data pulse.
3. Confirm that the pulse is within the permissible limits according to the CCITT pulse form (G.703)(2); see Figure 11.14 and 11.15 for 8 Mbit/s and 34 Mbit/s recommended pulse shapes.
4. Photograph the oscilloscope display (if required) and attach the photograph to the test result sheets.
5. Repeat for path 2.
6. Enter results up in the test result sheets.
7. Restore all equipment connections to normal.

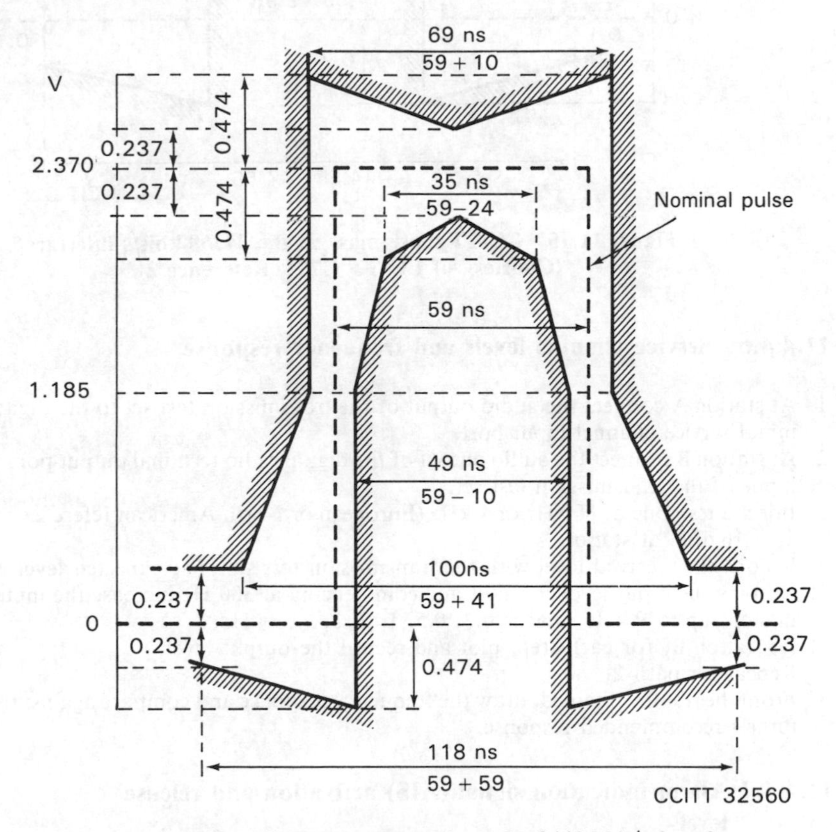

Figure 11.14 CCITT pulse mask at the 8448 kbit/s interface (Courtesy of CCITT G703, Reference 2)

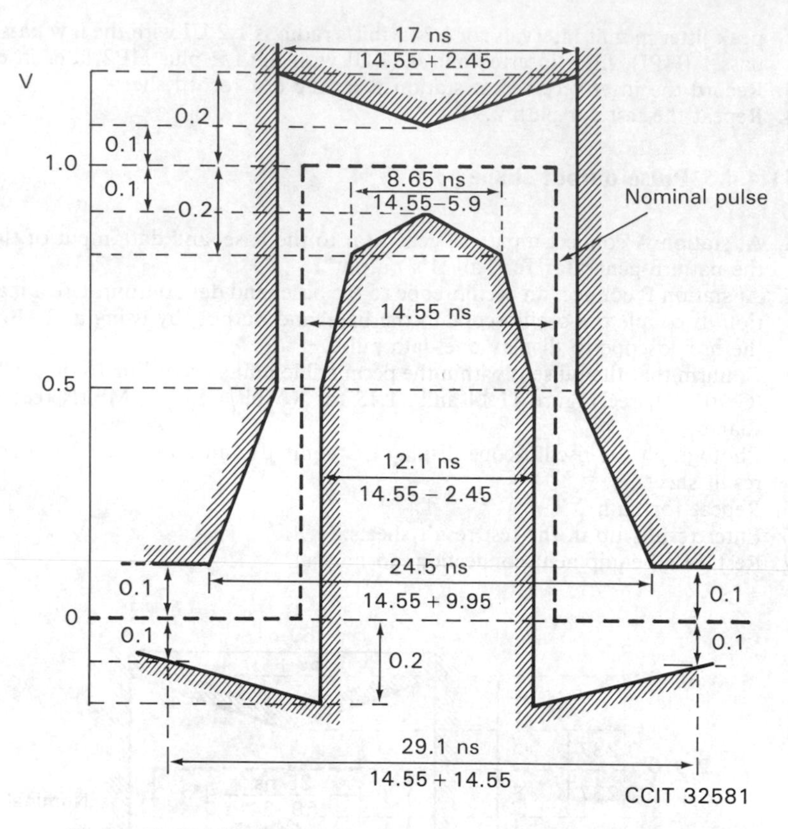

Figure 11.15 CCITT pulse mask at the 34368 kbit/s interface (Courtesy of CCITT G703, Reference 2)

11.4.4.6 Service channel levels and frequency response

1. At station A connect the audio output of the transmission test set to the digital radio terminal service channel input port.
2. At station B connect the audio output of the digital radio terminal output port to the audio input of the transmission test set.
3. Inject a test tone of 800 Hz or 1 kHz (European or North American reference respectively), at – 16 dBm at station A.
4. Record the received level with the transmission test set. The expected level is + 7 dBm.
5. Keeping the input level constant and commencing at 300 Hz increase the input frequency in steps up to 9 kHz, that is 0.3, 0.5, 1, 2, 3, ..., 9 kHz.
6. At station B, for each step, plot and record the output level.
7. Repeat for path 2.
8. From the results obtained, draw the frequency response and compare against the manufacturer's recommended response.

11.4.4.7 Alarm indication signal (AIS) activation and release levels

1. At station B, after turning off the transmitter, disconnect the waveguide from the hot-

standby unit output flange and insert a variable waveguide attenuator. Restore the transmitter power.

2. At station A connect a pattern generator to the digital radio terminal baseband data input port. Set the pattern generator to the appropriate bit rate, line code (as per Table 5.5) and pattern length (as per Table 4.10), no errors.
3. At station B connect an error detector to the baseband data out port and set the error detector to synchronize with the incoming data.
4. Using the variable waveguide attenuator, increase the attenuation until AIS is just introduced.
5. Record the BER just prior to AIS activation.
6. Reduce the level of attenuation until AIS is released and record the BER of the received signal immediately after BER release.
7. Repeat the test three more times and calculate the average of all four readings for each condition.
8. Repeat for path 2.
9. Enter the results in the test record sheets.

Note. This test is sometimes difficult to perform due to the signal strength fluctuations over the transmission path causing difficulty in determining the actual switching points.

11.4.4.8 Link BER test

Tests on short links may initially be made with equipment configured at the distant end, such that the baseband of the receiver is looped back into the transmitter. This reduces the equipment and time to test the link. However, if errors are occurring, the full end-to-end BER tests must be carried out:

1. At station A connect a pattern generator data output to the digital distribution frame at the point on which the baseband of the transmitter terminal is terminated.
2. Select the appropriate bit rate, line code and pseudo-random pattern length, no errors, on the digital pattern generator (as per step 2 in Section 11.4.4.7).
3. At station B (or A if looped back), connect to the data input of a digital error detector, the baseband of the radio baseband data output port termination on the digital distribution frame.
4. Set the error detector to the same bit rate, line code and pattern length as that of the pattern generator.
5. Select receiver 1 at station B and transmitter 1 at station A.
6. If the error detector is the PF–4, set the 'evaluation' switch to 10^9.
7. Ensure 'REP' light is on, using $1 \times$ REP switch.
8. Set date and time using switch 21 with PARAM and PRINT buttons.
9. Set error threshold (ETH) to Er 1–9 using switch 21 with PARAM and PRINT buttons.
10. Set error interval (EFI) to 1.0 using switch 21 with PARAM and PRINT buttons.
11. Leave switch 21 on ETH/Depend.
12. Press START/CLEAR to begin test.
13. After test has been running for 24 hours press STOP to end test.
14. *Note.* If unfamiliar with the equipment, insert a few errors with the buttons marked 1x, SET, ON/OFF to check the operation.

11.4.4.9 Automatic changeover

1. Set up the test equipment as in 11.4.4.7 (AIS tests).
2. Manually select path 1 at both station A and station B by the manual front panel switch on the digital radio terminals.
3. Confirm that error-free data are received.

4. At station A remove the data input to transceiver 1.
5. Confirm that station A changes to path 2 and station B is unchanged.
6. Restore data input to transceiver 1 at station A.
7. At station B, remove data output connection from transceiver 1 and confirm that the digital radio terminal switches to path 2.
8. Confirm that station A is unchanged.
9. Enter the results on the test record sheets.

11.4.4.10 Other tests

The above completes the basic tests required for end-to-end testing. Additional tests may be required depending on the equipment used, and the facilities which this equipment offers. To validate any tests, whether they be in-station or end-to-end, a prerequisite to proper testing is that the test equipment is still within the period not requiring calibration. If the equipments have passed the date at which they are due for calibration, then they must not be used in any performance or alignment tests.

As can be noticed from the above tests, with a digital signal there are no critical level adjustments and the proper operation of the system is seen by bit-error-ratio (BER) measurements, without the need for complicated measurements of baseband characteristics, or deviations, etc.

Usually when setting up the equipment initially, and performing the above tests for the purpose of establishing the proper operation of the equipment, strapping alterations required in conjunction with nonstandard equipment configurations or conditions of usage are also carried out. These strappings may include the change of line interface into balanced 75-ohm interface, choice of line code, change of word form for scrambling and unscrambling in a codec, elimination of alarms in an incompletely equipped sub-rack, strapping in the demultiplexer, etc. The options are usually quite large in number and details of what is necessary for a particular terminal are worked out in advance between the user and the manufacturer, so that the installation personnel have the proper strapping diagrams before attempting work in this area. In the above tests, only a few alarms have been mentioned. Again there may be other alarms which require to be tested other than those given. Noise measurements on the service channel may also be required.

REFERENCES

1. *How Good is your Digital Radio? – An Introduction to C/N and C/I Performance Testing of Digital Radio using the HP 3708A* (Hewlett Packard).
2. *'Digital Networks – Transmission Systems and Multiplexing Equipment'*, Recommendations G700–G956, CCITT Red Book, **III.3**, VIII Plenary Assembly, 1984.

APPENDIX A

Table A.1 Values of complementary error function $\text{erfc}(t) = \frac{2}{\sqrt{4\pi}} \int_t^\infty e^{-t^2}\,dt$

t	erfc(t)	t	erfc(t)	t	erfc(t)	t	erfc(t)
0.000	1.000 000	0.005	0.994 358	0.010	0.988 717	0.015	0.983 076
0.020	0.977 435	0.025	0.971 796	0.030	0.966 159	0.035	0.960 523
0.040	0.954 889	0.045	0.949 257	0.050	0.943 628	0.055	0.938 002
0.060	0.932 378	0.065	0.926 759	0.070	0.921 142	0.075	0.915 530
0.080	0.909 922	0.085	0.904 318	0.090	0.898 719	0.095	0.893 126
0.100	0.887 537	0.105	0.881 954	0.110	0.876 377	0.115	0.870 806
0.120	0.865 242	0.125	0.859 684	0.130	0.854 133	0.135	0.848 589
0.140	0.843 053	0.145	0.837 524	0.150	0.832 004	0.155	0.826 492
0.160	0.820 988	0.165	0.815 493	0.170	0.810 008	0.175	0.804 531
0.180	0.799 064	0.185	0.793 607	0.190	0.788 160	0.195	0.782 723
0.200	0.777 297	0.205	0.771 882	0.210	0.766 478	0.215	0.761 085
0.220	0.755 704	0.225	0.750 335	0.230	0.744 977	0.235	0.739 632
0.240	0.734 300	0.245	0.728 980	0.250	0.723 674	0.255	0.718 380
0.260	0.713 100	0.265	0.707 834	0.270	0.702 582	0.275	0.697 344
0.280	0.692 120	0.285	0.686 911	0.290	0.681 717	0.295	0.676 537
0.300	0.671 373	0.305	0.666 225	0.310	0.661 092	0.315	0.655 975
0.320	0.650 874	0.325	0.645 789	0.330	0.640 721	0.335	0.635 670
0.340	0.630 635	0.345	0.625 618	0.350	0.620 618	0.355	0.615 635
0.360	0.610 670	0.365	0.605 723	0.370	0.600 794	0.375	0.595 883
0.380	0.590 990	0.385	0.586 117	0.390	0.581 261	0.395	0.576 425
0.400	0.571 608	0.405	0.566 810	0.410	0.562 031	0.415	0.557 272
0.420	0.552 532	0.425	0.547 813	0.430	0.543 113	0.435	0.538 434
0.440	0.533 775	0.445	0.529 136	0.450	0.524 518	0.455	0.519 921
0.460	0.515 345	0.465	0.510 789	0.470	0.506 255	0.475	0.501 742
0.480	0.497 250	0.485	0.492 780	0.490	0.488 332	0.495	0.483 905
0.500	0.479 500	0.505	0.475 117	0.510	0.470 756	0.515	0.466 418
0.520	0.462 101	0.525	0.457 807	0.530	0.453 536	0.535	0.449 287
0.540	0.445 061	0.545	0.440 857	0.550	0.436 677	0.555	0.432 519
0.560	0.428 384	0.565	0.424 273	0.570	0.420 184	0.575	0.416 119
0.580	0.412 077	0.585	0.408 059	0.590	0.404 063	0.595	0.400 092

Table A.1 (*Continued*)

t	erfc(t)	t	erfc(t)	t	erfc(t)	t	erfc(t)
0.600	0.396 144	0.605	0.392 220	0.610	0.388 319	0.615	0.384 442
0.620	0.380 589	0.625	0.376 759	0.630	0.372 954	0.635	0.369 172
0.640	0.365 414	0.645	0.361 680	0.650	0.357 971	0.655	0.354 285
0.660	0.350 623	0.665	0.346 986	0.670	0.343 372	0.675	0.339 783
0.680	0.336 218	0.685	0.332 677	0.690	0.329 160	0.695	0.325 667
0.700	0.322 199	0.705	0.318 755	0.710	0.315 334	0.715	0.311 939
0.720	0.308 567	0.725	0.305 219	0.730	0.301 896	0.735	0.298 597
0.740	0.295 322	0.745	0.292 071	0.750	0.288 844	0.755	0.285 642
0.760	0.282 463	0.765	0.279 309	0.770	0.276 178	0.775	0.273 072
0.780	0.269 990	0.785	0.266 931	0.790	0.263 897	0.795	0.260 886
0.800	0.257 899	0.805	0.254 936	0.810	0.251 996	0.815	0.249 081
0.820	0.246 189	0.825	0.243 321	0.830	0.240 476	0.835	0.237 655
0.840	0.234 857	0.845	0.232 083	0.850	0.229 332	0.855	0.226 604
0.860	0.223 900	0.865	0.221 218	0.870	0.218 560	0.875	0.215 925
0.880	0.213 313	0.885	0.210 723	0.890	0.208 157	0.895	0.205 613
0.900	0.203 092	0.905	0.200 593	0.910	0.198 117	0.915	0.195 664
0.920	0.193 232	0.925	0.190 823	0.930	0.188 436	0.935	0.186 072
0.940	0.183 729	0.945	0.181 408	0.950	0.179 109	0.955	0.176 832
0.960	0.174 576	0.965	0.172 342	0.970	0.170 130	0.975	0.167 938
0.980	0.165 768	0.985	0.163 620	0.990	0.161 492	0.995	0.159 385
1.000	0.157 299	1.050	0.137 564	1.100	0.119 795	*1.150*	*0.103 876*
1.200	0.089 686	1.250	0.077 100	1.300	0.065 992	1.350	0.056 238
1.400	0.047 715	1.450	0.040 305	1.500	0.033 895	1.550	0.028 377
1.600	0.023 652	1.650	0.019 624	1.700	0.016 210	1.750	0.013 328
1.800	*0.010 910*	1.850	8.8889−3	1.900	7.2096−3	1.950	5.8207−3
2.000	4.6778−3	2.050	3.7419−3	2.100	2.9795−3	2.150	2.3614−3
2.200	1.8629−3	2.250	1.4627−3	*2.300*	*1.1432−3*	2.350	8.8929−4
2.400	6.8853−4	2.450	5.3060−4	2.500	4.0697−4	2.550	3.1068−4
2.600	2.3605−4	2.650	1.7850−4	2.700	1.3434−4	*2.750*	*1.0062−4*
2.800	7.5017−5	2.850	5.5658−5	2.900	4.1100−5	2.950	3.0205−5
3.000	2.2092−5	3.050	1.6081−5	*3.100*	*1.1649−5*	3.150	8.3987−6
3.200	6.0247−6	3.250	4.3040−6	3.300	3.0593−6	3.350	2.1637−6
3.400	1.5220−6	*3.450*	*1.0673−6*	3.500	7.437 −7	3.550	5.163 −7
3.600	3.563 −7	3.650	2.433 −7	3.700	1.683 −7	*3.750*	*1.137 −7*
3.800	7.60 −8	3.850	5.27 −8	3.900	3.53 −8	3.950	2.27 −8
4.000	1.60 −8	*4.050*	*1.03 −8*	4.100	9.3 −9	4.150	4.0 −9
4.200	3.3 −9	4.250	2.0 −9	*4.300*	*1.0 −9*		
1.163	1.0003−1						
1.820	1.0057−2						
2.325	1.0089−3						
2.750	1.0062−4						
3.123	1.0028−5						
3.458	1.0053−6						

Table A.1 (*Continued*)

t	erfc(t)	t	erfc(t)	t	erfc(t)	t	erfc(t)
3.766	1.000 −7						
4.050	1.03 −8						
4.300	1.0 −9						

Notes

1. Values in italics are the closest values in the table to one part in 10, 10^2, 10^3 ... 10^9
2. The minus sign at the end of a number, followed by a digit x, denotes the number $\times 10^{-x}$ e.g. $5.27{-}8 = 5.27 \times 10^{-8}$

Definitions of certain functions

Gaussian pdf

$$= f(x) = \frac{1}{\sigma\sqrt{2\pi}} \exp(-x^2/2\sigma^2) \qquad -\infty < x < \infty$$

Gaussian CPDF

$$= F(x) = P(X \leqslant x) = \frac{1}{\sigma\sqrt{2\pi}} \int_{-\infty}^{x_1} \exp(-x^2/2\sigma^2)\, \mathrm{d}x$$

Complementary CPDF

$$= P(x > X) = 1 - P(X \leqslant x) = 1 - \frac{1}{\sigma\sqrt{2\pi}} \int_{-\infty}^{x_1} \exp(-x^2/2\sigma^2)\, \mathrm{d}x$$

$$= \frac{1}{\sigma\sqrt{2\pi}} \int_{x_1}^{\infty} \exp(x^2/2\sigma^2)\, \mathrm{d}x$$

Error function

$$= \mathrm{erf}(t) = \frac{2}{\sqrt{\pi}} \int_0^t \exp(-t^2)\, \mathrm{d}t$$

Complementary error function

$$= \mathrm{erfc}(t) = 1 - \mathrm{erf}(t) = \frac{2}{\sqrt{\pi}} \int_t^{\infty} \exp(-t^2)\, \mathrm{d}t$$

As

$$\int_{x_1}^{\infty} \exp[-x^2/(mn^2)]\ dx = \int_{x_1}^{\infty} \exp[-[x/n\sqrt{m}]^2]\ dx$$

$$= n\sqrt{m} \int_{x_1/n\sqrt{m}}^{\infty} \exp(-x^2)\ dx$$

Complementary CPDF

$$= \frac{1}{\sigma\sqrt{2\pi}} \int_{x_1}^{\infty} \exp(-x^2/2\sigma^2) = \frac{1}{\sqrt{\pi}} \int_{x/\sigma\sqrt{2}}^{\infty} \exp(-x^2)\ dx$$

Hence complementary CPDF $= P(x > X) = (1/2) \cdot \mathrm{erfc}[x/\sigma\sqrt{2}]$

APPENDIX B

PATH CROSSING PROGRAM: FOR HP41C CALCULATOR PROGRAM NAME 'MAP'

See Section 9.2.1 for introduction and purpose of this program. This program is designed for both an HP41 calculator and HP82143 peripheral printer. If a printer is not available then the program can be changed to operate without the printer by deleting lines 03, 05, 07, 09, 11, 12, and 17.

Operation

The latitudes of site A and site B are entered with the following signs:

Northern latitudes are *positive*
Southern latitudes are *negative*

The longitudes of site A and site B are entered with the following signs:

Western longitudes are *positive*
Eastern longitudes are *negative*

After the inputs have been entered, the program will automatically calculate the distance in kilometers, between the two sites and the azimuths at sites A and B. After this, the program will request an input for the known longitude of the map crossing. If this is what is required, after entering the information, the relevant map crossing latitude will be calculated. This program is contained in 'C' which can be automatically called by entering 'XEQ (ALPHA) C (ALPHA)'. If the longitude is what is required to be calculated, the program contained in 'D' is called up by entering 'XEQ (ALPHA) D (ALPHA)'. There will be two answers for the longitude for a given latitude. The distance calculation is contained in 'B' and may be called up at any time after the initial information has been entered.

Example

Find the distance between, and the azimuths at, site A (5°22′05″N, 10°21′11″E) and site B(10°02′03″S, 30°15′05″E). If a map crossing is required to be determined knowing the longitude is 15°E, determine the latitude at which the path crosses on the map. If the path crosses another map at a latitude of 5°S, determine the longitude of the crossing on this map.

1. XEQ (ALPHA) MAP (ALPHA)
Response: printer prints out DISTANCE,AZIMUTHS,MAP CROSSINGS;
calculator asks for longitude of site B in the format; degrees, minutes and seconds D,MS

2. – 30.1505 R/S
Reponse: printer prints out LNGb = – 30.1505;
calculator asks for latitude of site B (LATb D,MS?)

3. – 10.0203 R/S
Response: printer prints out LATb = – 10.0203;
calculator asks for longitude of site A (LNGa D,MS?)

4. – 10.2111 R/S
Response: printer prints out LNGa = – 10.2111;
calculator asks for latitude of site A (LATa D,MS?)

5. 5.2205 R/S
Response: printer prints out distance $D = 2789$ km, azimuth at site A, AZa = 127.4849° or 127°48′49″, azimuth at site B, AZb = 306.5917 or 306°59′17″.
calculator asks for longitude of map crossing (LNG X = ?)

6. – 15.0000 R/S
Response: printer prints out longitude of map crossing LNG X = – 15.0000 and then after calculating prints out the required latitude as LAT X = 1.4450. As the answer is positive, the latitude is North, i.e. 1°44′50″N

7. XEQ (ALPHA) D (ALPHA)
Response: calculator asks for latitude of known map crossing (LAT Y?)

8. – 5.0000 R/S
Response: printer prints out LAT Y = – 5.0000 then calculates the required longitudes, before printing:

LNG1 Y = 169.1016 or 169° 10′16″W, which is over the other side of the world, and
LNG1 Y = – 23.3729 or 23° 37′29″E.

New map crossings can be calculated for other values by calling up the programs ‘C’ or ‘D’ as described above, without entering the intial A and B site, longitude and latitude.

```
01◆ LBL "MAP"
02 SF 12
03 ADV
04 "DISTANCE,"
05 ACA
06 "AZIMUTHS,"
07 ACA
08 "MAP"
09 ACA
10 "CROSSINGS"
11 ACA
12 ADV
13 CF 12
14◆ LBL 67
15◆ LBL 10
16◆ LBL A
17 ADV
18 CF 02
19 "LNGb D,MS?"
20 PROMPT
21 "LNGb="
22 ARCL X
23 AVIEW
24 HR
25 STO 23
26 CHS
27 STO 00
28 "LATb D,MS?"
29 PROMPT
30 "LATb="
31 ARCL X
32 AVIEW
33 HR
34 STO 22
35 STO 05
36 "LNGa D,MS?"
37 PROMPT
38 "LNGa="
39 ARCL X
40 AVIEW
41 HR
42 STO 21
43 ST+ 00
44 "LATa D,MS?"
45 PROMPT
46 "LATa="
47 ARCL X
48 AVIEW
49 HR
50 STO 20
51 STO 04
52 SIN
53 RCL 22
54 SIN
55 *
56 RCL 22
57 COS
58 RCL 20
59 COS
60 *
61 RCL 00
62 COS
63 *
64 +
65 ACOS
66 E4
67 *
68 90
69 STO 07
70 STO 08
71 STO 09
72 /
73 STO 24
74 2
75 ST* 08
76 ST* 09
77 ST* 09
78 ST/ 00
79 ST/ 04
80 ST/ 05
81 XEQ 00
82 STO 02
83 XEQ 00
84 ST+ 02
85 -
86 XEQ 01
87 STO 03
88 RCL 02
89 XEQ 01
90 STO 01
91◆ LBL 11
92◆ LBL B
93 RCL 24
94 "D="
95 ARCL X
96 AVIEW
97 RCL 01
98 HMS
99 "AZa="
100 ARCL X
101 AVIEW
102 RCL 03
103 HMS
104 "AZb="
105 ARCL X
106 AVIEW
107 GTO C
108 RTN
109 RTN
110◆ LBL 12
111◆ LBL C
112 "LNG X="
113 PROMPT
114 "LNG X="
115 ARCL X
116 AVIEW
117 HR
118 RCL 21
119 -
120 STO 06
121 ABS
122 CHS
123 STO 04
124 RCL 06
125 RCL 02
126 *
127 X<0?
128 GTO 15
129 RCL 02
130 ABS
131 RCL 08
132 -
133 GTO 16
134◆ LBL 15
135◆ LBL a
136 RCL 02
137 ABS
138 CHS
139◆ LBL 16
140◆ LBL b
141 STO 05
142 RCL 20
143 STO 00
144 RCL 07
145 ST+ 00
146 ST+ 04
147 ST+ 05
148 XEQ 02
149 XEQ 00
150 XEQ 00
151 +
152 CHS
153 X>0?
154 GTO 15
155 RCL 07
156 +
157◆ LBL 19
158◆ LBL e
159 HMS
160 "LAT X="
161 ARCL X
162 AVIEW
163 RTN
164◆ LBL 15
165◆ LBL a
166 RCL 07
167 -
168 GTO 19
169◆ LBL 13
170◆ LBL D
171 2
172 STO 11
173 CF 22
174 "LAT Y?"
175 PROMPT
176 "LAT Y="
177 ARCL X
178 AVIEW
179 HR
180 STO 03
181◆ LBL 18
182◆ LBL d
183 STO 05
184 COS
185 1/X
186 RCL 20
187 STO 04
188 COS
189 *
190 RCL 02
191 SIN
192 *
193 ASIN
194 STO 06
195 CHS
196 RCL 08
197 +
198 STO 01
199 RCL 20
200 RCL 22
201 +
202 RCL 02
203 STO 00
204 *
205 FS?C 22
206 GTO 15
207 SF 22
208 CHS
209◆ LBL 15
210◆ LBL a
211 CHS
212 X>0?
213 GTO 15
214 RCL 06
215 ST- 00
216 GTO 16
217◆ LBL 15
218◆ LBL a
219 RCL 01
220 ST- 00
221◆ LBL 16
222◆ LBL b
223 XEQ 02
224 XEQ 00
225 CF 02
226 2
227 *
228 CHS
229 RCL 21
230 +
231 HMS
232 "LNG1 Y="
233 ARCL X
234 AVIEW
235 DSE 11
236 GTO 30
237 STOP
238◆ LBL 30
239 RCL 03
240 GTO 18
241 RTN
242◆ LBL 00
243 RCL 04
244 RCL 05
245 +
246 RCL 04
247 RCL 05
248 -
249 FS?C 02
250 GTO 15
251 SF 02
252 SIN
253 X<>Y
254 COS
255 GTO 16
256◆ LBL 15
257◆ LBL a
258 COS
259 X<>Y
260 SIN
261 X=0?
262 GTO 17
263◆ LBL 16
264◆ LBL b
265 /
266 RCL 00
267 TAN
268 /
269 ATAN
270 RTN
271◆ LBL 17
272◆ LBL c
273 RCL 07
274 RTN
275◆ LBL 01
276 RCL 20
277 RCL 22
278 +
279 X>0?
280 GTO 15
281 RDN
282 RCL 08
283 +
284 RTN
285◆ LBL 15
286◆ LBL a
287 RDN
288 X>0?
289 RTN
290 RCL 09
291 +
292 RTN
293◆ LBL 02
294 2
295 ST/ 04
296 ST/ 00
297 ST/ 05
298 RTN
299 STOP
300 END
```

INDEX